Contents

PART 2
Contracts

71

Chapter 5　Introduction to Contracts

73

Chapter 9   Consideration   141

Chapter 10   Lawful Subject Matter   153

Chapter 17   Specifications   305

Chapter 18   Agency   327

PART 4
Property   353

Chapter 19   Tangible Property   355

Chapter 20   Intellectual Property   383

About the Author

Cynthia M. Gayton, Esq. holds a Bachelor of Arts degree in International Affairs from The George Washington University and a Juris Doctor degree from George Mason University in Arlington. She resides in Arlington, Virginia and Harper's Ferry, West Virginia. Cynthia is a member of both the State Bar of Virginia and the District of Columbia Bar. Ms. Gayton operates a sole practice that specializes in intellectual property and corporate law. In addition, Ms. Gayton is an adjunct professor of engineering law at The George Washington University School of Engineering and Applied Sciences where she has taught since 2003. She worked as an associate at Morgan Lewis & Bockius, concentrating in complex antitrust litigation. At the American Institute of Architects, she was as-

sociate counsel. Ms. Gayton is the author, "Knowledge Management in the Large Law Firm" available at www.knowledgeboard.com. Finally, she is the author of several articles published by *VINE: The Journal of Information Knowledge Management Systems,* including: Beyond Terrorism: Data Collection and Responsibility for Privacy," Vol. 36, no. 4, 2006; "Alexandria Burned—Securing Knowledge Access in the Age of Google," Vol. 36, no. 2, 2006; and "Legal issues for the knowledge economy in the 21st Century," Vol. 36, no. 1, 2006.

Preface

UPDATES TO THE EIGHTH EDITION

This Eighth Edition of *Legal Aspects of Engineering* reflects many legal and professional changes in the industry. In addition to an overall consolidation and content re-arrangement, this edition has:

- Twenty one new cases in the main volume, in addition to four new NSPE cases in Appendix B.
- Five new sections covering:
- The new additions to the Federal Rules of Civil Procedure, including electronically stored information and document retention (Chapter 4),
- Statutes of repose (Chapter 13),
- The economic loss rule (Chapter 14),
- Project delivery systems (Chapter 16),
- Intellectual property and criminal law (Chapter 20).
- An expanded and re-organized Performance, Excuse of Performance, and Breach; Remedies; and Intellectual Property chapters in response to student interest as well as changes in the law.

Acknowledgments

I would like to thank the following persons and institutions:

- Dr. Michael Stankosky, Associate Professor of Systems Engineering at The George Washington University
- Thomas Jason Edwards
- The George Washington University, School of Engineering and Applied Sciences, Department of Engineering Management and Systems Engineering
- The Gelman Library at The George Washington University
- Bolivar-Harper's Ferry Public Library
- And especially Richard C. Vaughn for his past work on this title.

Overview of Engineering and the Law

The first jobs most engineers hold after they receive their bachelor's degrees are in the employment of others. They become members of management teams. Many of them rise to higher positions of management, where they continue to use their engineering backgrounds even though their titles may imply only management responsibilities. Other engineers find that solving engineering problems is so exhilarating that they go on to solve such problems for others as consultants. Still other engineers pursue careers in the military, academia, and in federal, state, or local governments.

In any of these endeavors, the engineer's relationships with others are prescribed by rules of law and ethics. Such rules provide the respective rights and responsibilities of the parties. A knowledge of these rules, then, is valuable to the engineer. In this first part we study these rules, as well as relationships and controversies that spring from them.

The Engineer in Management

HIGHLIGHTS

- Engineering Management depends on four factors:
  1. Communication skills
  2. People management
  3. Ability to focus on the bottom line
  4. Knowledge of the law

- As a manager, it is necessary to develop what may be called "executive qualities" which are comprised of the following attributes:
  - Strong leadership
  - Effective delegation
  - Accurate decision making
  - Discipline

Legal knowledge assists engineers in performing their duties efficiently and effectively in the face of a legal dilemma.

Engineers are problem solvers. Expertise in stating and solving problems results from specialized training in problem-solving techniques. It is this ability to state and solve problems that employers hope to find when they hire engineers, and it is because of the many uses for this knowledge that there is presently a substantial demand for engineering graduates.[1]

Most training equips engineers to solve problems of a mathematical nature—problems that may readily be reduced to symbolic form. However, not

---

[1]According to the Bureau of Labor Statistics, the demand for engineers will increase between 9 and 17% from 2004 - 2014. (http://stats.bls.gov/oco/ocos027.htm, accessed 6/19/07)

all problems lend themselves to such an attack. The stress resulting from a given force applied to a particular design of beam is easily stated in mathematical terms. It is a little more difficult, but usually still possible, to assume probabilities and solve for the number of parts to be run or for warehouse space required for next year's production. It is exceedingly difficult, however, to formulate or state laws governing the relationships between people in terms of x's and y's with proper coefficients and thereby solve legal and ethical problems. Most such problems involve the interpretation of human laws and the use of discretion and judgment in determining rights and responsibilities.

Often, the solutions of legal problems and problems involving human relations are no less important to a successful engineer than the solutions of mathematical problems. In most engineering jobs, the engineer is part of a management team. Before turning to the aspects of law with which engineers should be familiar, consider certain management skills they should strive to acquire.

ENGINEERING MANAGEMENT

For the vast majority of engineering graduates, the first job secured is merely a stepping stone to higher positions. Most people, including engineers, are ambitious.

Normally, entry-level engineering jobs require a large amount of technical skill. As engineers move up their career ladders, the percentage of time in which seasoned engineers use their technical skills usually decreases. Regardless of the ladder the engineer has chosen to climb—research, manufacturing engineering, consulting, sales, or any other job—progression to higher levels depends on at least four factors in addition to engineering ability:

- Communication skills
- People-management skills
- Ability to focus on the bottom line
- Knowledge of the law

Communication Skills

An idea initiated by an engineer may have very great latent value, but until it is used or communicated in some way, the idea is worthless to the engineer and to others. In addition, the mental work necessary to develop the idea in a detailed description or a diagram is in itself of value. Nearly everyone has had the experience of gaining new insight, or of discovering added features of an idea, when faced with the task of trying to explain it to someone else.

People-Management Skills

A promotion from a strictly technical position to one of greater responsibilities almost always leads to handling people. Being "boss" isn't easy. The manager who takes time to explain to subordinates "why" and to keep them informed is likely to be more successful than one who does not. Some people-management issues are discussed in Part 3, Chapter 18 (Agency) as well as Part 6, Chapters 24 and 25, (Labor and Worker's Compensation, respectively).

Ability to Focus on the Bottom Line

Most operations are undertaken with a profit motive. Even in situations when operations that are not expected to make a profit (such as a service department), cost is usually important. If the selling price of a company's product is unchanged, money saved in manufacturing or raw material cost represents added profit; conversely, added cost decreases profit. Many engineers have won promotions and many consultants earn their livelihoods on their ability to analyze operations and reduce costs. Part 3 (Engineering Contracts) provides a good overview of how costs and contract specifications are used to develop payment methods, including lump sum, cost-plus percentage, cost-plus fixed fee, and unit-price methodologies.

Knowledge of Law

Engineers are not expected to become attorneys from an exposure to one survey course in law, any more than attorneys could become engineers by taking one survey course in engineering. However, engineers should be aware of the effects of carelessness in dealing with others. Engineers should know when they need the advice of an attorney. A legal background is a "preventive" asset; that is, with a basic knowledge of law, engineers should be equipped to help prevent costly lawsuits against their company. Meticulous reading of contracts before signing is an important preventive measure. It is surprising how little attention is paid to contracts and supporting documents, particularly in view of the fact that these documents outline the rights and responsibilities of the parties.

EXECUTIVE QUALITIES

In recent years there has been an increasing trend toward filling top management or executive positions with engineers. Companies have recognized the value of the engineer's analytical approach to executive problems.

Although there is a good deal of truth to the often quoted comment that "there is always room at the top," those who get there usually possess special abilities. Engineering training is beneficial to the executive aspirant, but so is a knowledge of many other fields.

What makes an executive? Why does one person achieve this goal while many others strive and fail? At first glance, the behavior of one successful executive appears to have little in common with that of another who is equally successful. However, closer examination reveals certain similar behavior patterns. Each usually possesses the four qualities just mentioned for successful managers, often to a high degree. Other qualities, too, are seemingly common to most top executives, and deserve consideration.

Leadership

The quality known as leadership is difficult to define. It is clearly evident in one person and strangely lacking in another. Psychologically, leadership indicates an identification of the group with the one who leads—it is necessary that the leader be considered by the group as one of them. It is also required that the leader have one or more qualities esteemed by the group.

Consider This Quotation

The good leader takes his or her place at the center of a circle, not at the top of a pyramid. The good leader is a visionary, able to project out into the future a goal and then serve as enabler, facilitator, encourager on the way to achieving that goal.[2]

Indeed, it is doubtful that anyone is truly a "born leader." It is more probable that leadership qualities result from training acquired through study and observation. Leadership qualities are enhanced by practice.

Opportunities to practice the poise and purposefulness of leadership occur in limitless ways. One of the main reasons job applications often contain space for listing organizational activities is to assist the employer in determining the amount of practice in leadership the candidate has had.

Two outstanding leadership characteristics are the ability to keep the ultimate goal uppermost in mind, and the ability to pursue it enthusiastically. Enthusiasm is infectious. A speech delivered in a monotone makes for dull listening; however, the same speech using the same words but delivered with enthusiasm can move people to action.

Leadership stems partly from the ability to stand firm on principles. There is a popular misconception that a leader should not admit mistakes, although few people make perfect decisions all the time. Not only must leaders admit their own mistakes; they must take responsibility for the mistakes of their subordinates, because those actions result from the leaders' direction or lack thereof. Placing blame on others is not respected by leaders in top management.

Delegation

A characteristic most top executives share is the ability to delegate authority and responsibility to others. It is difficult for anyone to rise to the top of any organization without the ability to delegate. There is not enough time for one person to effectively and thoroughly perform the requirements of a top management job. Executives who delegate few tasks rob themselves of time needed for adequate thought before making decisions. Also, the failure to delegate routine tasks to others can be a barrier to executives' promotions; if no one can be found who has performed a portion of the executives' tasks with the authority necessary for that performance, it is natural to leave the executives where they are.

Specialization is an inherent advantage of effective delegation. No one is a specialist in everything. By assigning some of their tasks to others, executives can obtain the advantage of specialized treatment.

Delegation, as the term is used here (and in most businesses), means more than merely assigning tasks to be performed. Delegation includes clothing the delegee with the necessary authority to carry out the assigned function. It is this parting with a portion of authority that causes shortsighted executives—consciously or unconsciously—to oppose delegation to others. It is this very aspect of delegation, however, that contributes to the growth of assistants. The able executive realizes this and takes full advantage of it in helping others develop. Chapter 18 is an important chapter to review in order to understand the concept of agency as an engineer moves from completing tasks herself to delegating those tasks to others.

---

[2]William J. Byron, Research Professor, Sellinger School of Business; from address given at Center for Professional Development graduation ceremony at Georgetown University, Washington, DC, upon receipt of Dorothy M. Brown Leadership Award (June 6, 2003). William J. Byron, *Vital Speeches of the Day,* vol. LXIX no. 20, 8/1/03.

Decision Making

All of us make decisions involving choices from among alternatives. Our choices are not always correct. One attribute that seems characteristic of those people who reach top management is their ability to be right a higher percentage of the time than the average person. Of course, top management decisions are decisions about particularly difficult problems. Decisions run all the way from a single-variable problem to multivariable problems where little, if anything, is fixed or known. Generally, routine decisions are delegated to others; the top manager is the one who makes the decision when major uncertainties exist. The decision to expand a plant or install new production facilities based on an apparently expanding market is such a decision.

Many of the assumptions can be reduced to probabilities. If enough of this can be done, the problem can be programmed for a computer, which will then give the executive some answers. However, the answers are based on assumptions and probabilities, and the executive must decide whether to go ahead. The risk is still the executive's, not the computer's.

A few top executives possess such vast knowledge and the ability to analyze and synthesize that they can make rapid-fire policy decisions that are nearly always right. However, such people are rare. Generally, people in top management do not make hurried policy decisions. There is often grumbling from below because of apparently undue procrastination. Despite the grumbling, such delay is usually the course of wisdom, because the risks are frequently sizable. A decision based on inadequate facts or erroneous assumptions is hazardous, and delay in waiting for more facts is often inescapable. Even the rare management genius who makes correct decisions rapidly usually has had many years of experience in more methodical decision-making that has equipped him for this present role.

> Top management decisions generally consist of five elements, dealt with in sequence:
> 1. A gathering of facts
> 2. A recognition of limiting conditions
> 3. Assumption of facts and conditions as they are expected to be, and recognition that these are assumptions
> 4. Analysis of the facts, limits, and assumptions
> 5. Decision

Discipline

Discipline is a necessary component of any well-run organization. People must be taught; old patterns must be changed. Most top executives are masters of the use of reward and reprimand in changing the behavior of subordinates. To be effective, executive orders must imply some form of negative or positive consequence: reprimand for disobedience, rewards for outstanding performance, if the effort required for the performance is to be continued.

The extent to which reward and reprimand are necessary depends to a great degree on the personal stature of the executives. If they are held in high regard by their subordinates, a word or so is all that is necessary.

In addition to people's drive for food, water, and the means of satisfying other basic needs, they have a whole host of derived needs, not the least of which is the need of recognition. Every person needs recognition or respect from others—lack of it causes loss of self-respect and, eventually, diminished effort. Recognition can be either tangible or intangible, and both forms are required. Verbal praise sounds hollow after awhile if it is not accompanied by some material reward. Similarly, material rewards without praise for accomplishments are incomplete.

It has often been stated that rewards should be public, with criticism private. The truth in the statement is inescapable. Most top managers observe this principle in the interest of preserving their organizations.

These are only a few of the principles that guide top executives in their management of discipline. Most of these guidelines are understood and observed without conscious thought when disciplinary occasions rise.

Many qualities can make a person successful in top management. Only a few have been mentioned here. Nevertheless, these basic qualities must be mastered by executive aspirants. The purpose of a business and its management is to produce something. Converting time and raw material into goods and services requires production facilities. Assembly of the machines and equipment required to produce a product is normally undertaken as an engineering project. Not only must the original facilities be planned and built but every design change or functional change of the product also requires changing machines and equipment. The job of setting up production facilities becomes, then, not a "one-shot" enterprise, but an almost continuous replanning and re-arrangement.

The burden of deciding when and how much to change—and what to change—falls on top management; the job of planning and carrying out the details of rearrangement is usually assigned to the engineering department.

ENGINEERING PROJECTS

A large proportion of the capital wealth of the United States has resulted from engineering projects, as well as the intellectual property inherent in them (see Chapter 20). Civil engineering projects—roads, bridges, buildings, and the like—are most familiar to the public. As a result, whenever the term **engineering project** is used, visions of a dam or expressway cloverleaf are likely to come to mind. The value of civil projects cannot be denied, but contributions by other engineering fields are also significant, even though the public is not as aware of these activities or results.

Since the development and adoption of mass-production methods in the United States, an interdisciplinary approach has developed. People are needed who can apply knowledge of civil, mechanical, electrical, chemical, industrial, and other engineering fields to manufacturing problems. This combining of engineering talents to solve manufacturing problems has come to be known by many names, but the term **manufacturing engineering** seems more appropriate than most. Typically, manufacturing engineering is concerned with the process required to mass-produce some product. It starts with an analysis of an idea and continues as long as there are engineering problems to be solved. The following discussion covers the various stages of an engineering project in a manufacturing engineering context. However, the same basic concepts also apply to other types of engineering projects.

Project Phases

Any engineering project involves three phases or stages of development:

1. Conception of the idea
2. Reduction of the idea to practice
3. Refinement of the idea and ensuring that the project works.

The stages are fairly distinct, and a particular engineering group may have responsibilities in one or more of the stages.

Conception of the Idea.   Just about every product and convenience we enjoy started as someone's idea. Neither products nor the processes by which they are manufactured can be developed without someone's original idea. Not all ideas are practical, however. A large number of those that are adopted require substantial alterations and financial investment before they are acceptable. Many ideas appear attractive in the beginning, only to be demonstrated as impractical by objective examination. This objective examination of a possible engineering project is known as a **feasibility study.**

A feasibility study is a preliminary examination of a proposed idea. It is meant to answer such questions as: What will it cost to produce various quantities per year? Can we market enough to make a reasonable return on the required investment? How many can be sold at a given price? What processes will be better in the long run? The answers given determine whether it is desirable to proceed to the next stage—actually setting up to produce.

Reduction to Practice.   Turning someone's idea into a reality can be quite complex in a manufacturing situation. Planning is necessary, and requires vision, and it continues until all the pieces are firmly in place. Even then, maintenance should be planned. Changes are made easily in the planning stage—it costs little to erase a machine location on a layout and place the machine in another location. Even rearrangements of the entire process are inexpensive at this point. It is here that questions pertaining to equipment sizes, locations, and added features must be answered, and the answers justified, if the process is to be successful. Layout changes after the process equipment has been placed are very expensive. For this reason, questions that should have been raised in the planning stage but were never brought up reflect on the process engineer's ability.

The process engineer designs a layout of the process, complete with machines and equipment, and writes specifications for the machines to function as desired. The specifications are then sent out, proposals received, and contracts awarded to the successful bidders. Engineers are the owner's agents; as such they often must supervise the building of machines or other structures to fit the layout, and then supervise the installation. Engineers also must control the times of completion of the elements of layout. Rarely is a process completed and functioning properly within the original timeframe. There is nearly always at least one contractor who is late. A wise engineer will allow some time for this delay in the schedule.

Refinement and Oversight.   It is probably safe to state that in every manufacturing process ever installed there were special problems to be solved before full-scale production could begin. The presence of "bugs" in a newly installed process is normal but they must be removed before the process can be considered complete. The engineer who set up the process is the logical person to remove these bugs before the operation is turned over to the production people. These stages will be elaborated upon in Chapter 16.

Law and Engineers

In any engineering project, engineers act as professionals; they are the representatives, or agents, of the owner. Their function is to act in the best interests of the owner—to get the best possible results with a minimum of delay and problems. Engineers must deal with the rights of others.

There are two fundamental reasons why engineers should have some knowledge of the law. Specifically, to protect and ensure:
1. The rights of others, and
2. His or her own rights.

An engineer is a guardian of the owner's rights and, in a manner of speaking, of the rights of others with whom the owner deals. Because court proceedings are costly in both time and money, engineers generally should avoid entanglements that would lead to litigation. And, because violation of the rights of others is likely to lead to court proceedings, engineers must know the characteristics of these rights, if their preventive job is to be well done.

For construction projects, the relationships between the owner and contractor are set forth in a series of documents drawn up by the engineer. Documents such as instructions for bidders, requests for proposals, proposals, general terms and conditions, specifications, drawings, and sometimes purchase orders, order acknowledgments, invoices, and the like, compose parts of the contract. Careless errors in the preparation of these documents can cause legal controversies, or place the owner and engineers in indefensible positions when controversies arise. Engineers must formulate the documents in such a way that the owner's position is protected. Imposing an undue hardship on the contractor may lead to unnecessary litigation. Similarly, ambiguities in the terminology or leaving too much to future agreement can lead to unnecessary litigation. Hence, specificity and realistic goals are what the contract usually requires.

In some respects the engineers' position is between the owner and the contractor. When disputes arise, engineers are likely to be called on to mediate or participate in the controversies. To do a reasonable job in this position, engineers must be acquainted with the legal rights and responsibilities of both parties. Engineers don't have to be attorneys, but some knowledge of the law is essential, and engineers should be able to recognize situations in which it is necessary to consult an attorney. Remember, engineers can't very easily recognize legal troubles unless they have some knowledge of the rights involved.

As with other human endeavors, obtaining expert advice as early as possible can avoid problems or serve to provide damage control. There is a second reason for engineers to acquire a knowledge of the law. Besides their professional activities, engineers are citizens as well. The law controls many of our private day-to-day dealings with others. When we buy insurance or apply for a credit card, our rights and responsibilities should be clear.

Engineers are members of society as well as professionals who possess technical skills. As educated members of society and professional people, their knowledge and abilities should extend well beyond their technical skills. One popularly accepted criterion of the cultured person is the ability to analyze and discuss news events with some perception. Much of the news presented to us online or via radio, television, newspapers, and magazines has legal significance. An awareness of current events and especially their legal ramifications is essential.

LEGAL ANALYSIS

Engineers have backgrounds in analysis of scientific or technical things. In the design of a bridge or automotive component, for example, an engineer analyzes forces and reactions to them, statically and dynamically. Using the results of analyses, engineers adjust the designs to serve both their employers and the public appropriately. Increasing reliance on computer simulation may ease the engineer's duties from a drafting standpoint, but do not relieve her from the responsibility for the reliability of the final product.

In the legal setting, engineers are often still concerned with engineering analysis in tort, particularly product liability cases, which are discussed more thoroughly in Chapter 22. They may conduct investigations known as **failure analyses** to find why some malfunction occurred. Scientific analysis may be required in a variety of criminal-case settings. Engineers are accustomed to the kind of analysis where most things are reasonably precise and even the provision for error is usually predictable. Legal analysis is less precise.

Although legal analysis is similar to technical analysis in that both methods require the analyst to consider all potential outcomes when presented with a problem, the difference lies in the unpredictable nature of human decision-making. A software engineer can predict the effect of bad code in a program and debug it. A civil engineer can consider the wind effects on a bridge and design to accommodate it. Human behavior is harder to predict with certainty, rendering an often confusing and unintelligible legal construct.

Yet there are guidelines. If there is a case before a civil or criminal court, lawyers look to legal **precedents** or prior cases that were resolved under similar circumstances and facts, which may help predict the result of a case in which he is involved. Statutes, passed by elected officials, serve as legal guides to federal and state governments. The Federal Trade Commission and Department of Justice use formulas to predict the effect of mergers.

The thing to remember about law is that as a member of society you deal with it every day. If you look hard enough at the tax code, you will see a series of "if/or"; "if/but", and "only if" statements (not that you would want to do such a thing, however). When you fill out your 1040 form, you are progressing through parts of the United States tax code that have been summarized on a few sheets of paper. From this perspective, the transition from technical analysis to legal analysis is not that far of a leap. Your engineering background in problem-solving will aid you significantly.

Review Questions

1. Why should an engineer have some knowledge of the law?

2. Name at least three additional qualifications an engineer should possess for success in management. Name at least three additional qualities of successful executives.

3. What are the stages of an engineering project? What would each stage be composed of in the proposed manufacture of, say, table lamps?

4. What is manufacturing engineering?

Ethics  CHAPTER 2

- Individual states regulate professional registration requirements.

- Professional engineering societies can deny membership to those who violate ethical codes.

- An engineer can be prosecuted *by the state* for representing that she is a registered engineer.

The role of an engineer today is quite different from that of a few generations ago. More than ever before, practicing engineers have an increased responsibility to the public. As a result, membership in professional societies, and the professional codes on which they are based, are increasingly important. Professional codes are not enforceable by law, but can be enforced by a professional society when an engineer has committed ethical violations, and may cause the engineer to lose his license as a result. In addition, engineers can be subject to lawsuits, often with criminal penalties, for failure to comply with state licensing requirements. The *State of Arizona v. Wilkinson* case at the end of Chapter 3 addresses this dilemma.

The focus in this chapter is how professional engineering standards are defined by a number of different sources, including federal laws, state registration requirements, and professional societies.

FEDERAL LAWS

A separate, recognized field of learning and the presence of societies of its members do not make a "profession." People do not have "professional" status merely because they have graduated from a school and joined a society. Perhaps it is best to consider for a moment the meaning of the term **profession.** It is not difficult to find definitions of the term. *Webster's Dictionary* says it is "a calling requiring specialized knowledge and often long and intensive academic preparation." *Black's Law Dictionary* calls it "a vocation or occupation requiring special, usually advanced, education, and the labor and skill involved in a profession is predominantly mental or intellectual rather than physical or manual."

Although each definition serves a purpose, both are brief at the expense of completeness. A more complete definition was given by the U.S. Congress in the Labor Management Relations Act. In this act, Congress defined the term **professional employee** as follows:

The term **professional employee** means:

a. any employee engaged in work
   1. predominantly intellectual and varied in character as opposed to routine mental, manual, mechanical, or physical work;
   2. involving consistent exercise of discretion and judgment in its performance;
   3. of such a character that the output produced or the result accomplished cannot be standardized in relation to a given period of time;
   4. requiring knowledge of an advanced type in a field of science or learning customarily acquired by a prolonged course of specialized intellectual instruction and study in an institution of higher learning or a hospital, as distinguished from a general academic education or from an apprenticeship or from training in the performance of routine mental, manual, or physical processes; or
b. any employee who
   1. has completed the courses of specialized intellectual instruction and study described in clause (4) of paragraph (a); and
   2. is performing related work under the supervision of a professional person to qualify himself to become a professional employee as defined in paragraph (a).

Besides the four requirements just stated, various other criteria are frequently added to the list:

1. Registration requirements for practicing the professions;
2. Representation of members and control of activities by a professional society;
3. The public service nature of the occupation; and
4. Adherence to a code of ethics.

STATE REGISTRATION LAWS

The state laws of each of the United States require engineers to be registered before they are allowed to practice professional engineering in the state. Registration in any one state does not give engineers the right to act as professional engineers in another state; however, many states have reciprocal agreements whereby registration is much simplified if engineers are already registered in another state.

The primary purpose of the state engineering registration laws is to protect the public from shoddy engineering practices. To this end, it is necessary that prospective licensees convince a board of examiners that they are qualified to practice professional engineering. The usual method for this is a scrutiny of the candidates' past engineering work and training, and a qualifying examination. This examination commonly lasts two days and is either oral and written or entirely written. It usually covers the basic sciences and specialization in a particular field of engineering.

Full registration as professional engineers allows licensees to act as professional engineers within the state and to resort to the courts to collect fees for their services. Penalties in the form of fines and/or confinement are usually specified for practicing without a license.

PROFESSIONAL SOCIETIES

Each of the recognized branches of engineering has formed at least one society of its members. In addition to these bodies, there are three organizations that represent and serve all engineers.

Accreditation Board of Engineering and Technology

In 1932, the Engineers Council for Professional Development (ECPD) was formed. For 47 years it was concerned with accrediting engineering curricula and with other non-accrediting activities such as guidance, ethics, and the development of young engineers. In 1979, ECPD restructured itself into an accreditation board and joined with the American Association of Engineering Societies (AAES) for its other functions.

Accredited engineering and technology curricula are examined at least once every six years by an examiner from one of two commissions of the Accreditation Board of Engineering and Technology (ABET): the Engineering Accreditation Commission examines engineering curricula, while the Technology Accreditation Commission examines technology programs. As of October 2006, there were 1,787 accredited engineering programs and 670 accredited technology programs.

American Association of Engineering Societies

The American Association of Engineering Societies (AAES) was founded in 1979, superseding the Engineers Joint Council (founded in 1945). Its objectives are to advance the science and practice of engineering in the public interest; and to act as an advisory, communication, and information exchange agency for member activities. As noted earlier, AAES also acquired some of the functions of the Engineers Council for Professional Development when it restructured.

National Society of Professional Engineers

The National Society of Professional Engineers (NSPE), formed in 1934, is concerned with the social, economic, political, and professional interests of all engineers. NSPE activities were largely responsible for the passage of engineer registration laws in the various states, public recognition of engineering as a profession, and recognition of the value of engineering activities.

PUBLIC SERVICE

Professions do not magically spring into existence on the day a law is passed or upon the creation of a professional society. The acceptance of a field as a profession requires continuing efforts to maintain high standards of public service.

CODE OF ETHICS

Definition

According to *Webster's Dictionary,* ethics is "the discipline dealing with what is good and bad and with moral duty and obligation."

Ethics are the ground rules of our moral conduct. They consist of our attitudes toward honesty, integrity, trust, and loyalty. They are exhibited in our day-to-day contacts with others. No laws compel engineers to take an interest in community affairs or to give a completely unbiased report of the results of an investigation. The manner in which engineers act depends on their own moral code or ethics.

Establishment of Moral Patterns

A person does not acquire a code of ethics or sense of moral duty by reading a passage in a textbook and then deciding to abide by what was stated. Rather, a personal code grows out of the experiences and observations of one's life.

Examples set by parents, friends, classmates, teachers, and professors all contribute to establishing acceptable moral behavior. These factors contribute to an engineer's moral structure; to these will be added experiences and observations on the job.

In July of 2007, the NSPE revised its well-known and respected code of ethics. This code is not meant as a body of inflexible rules. It is meant as a guideline, to be worked into and be the basis for the engineer's professional and moral standards. See Appendix A for complete listing of NSPE Code of Ethics for Engineers.

Law and Ethics

Not all unethical actions are illegal; indeed, many unethical acts do not involve any fine or imprisonment, as do criminal acts. Neither the examining boards nor the courts have any rights or responsibilities in the ethical practice of engineers. That is not to suggest, however, that the courts are not concerned with the negligence of engineers. As will be discussed in Chapter 3, the courts indeed hear such cases. Whatever formal reproof there is for moral misbehavior must come from the engineering societies.

Although the engineering profession does not have societies the equivalent to the American Medical Association or the state bar associations for handling ethical infractions, it has professional engineering societies. These societies have an important role in monitoring unethical behavior. Two cases are provided in Appendix A from the NPSE to give you an idea about how this organization resolves hypothetical and real ethical dilemmas.

Codes of professional ethics are not meant to be of practical value to an individual, yet in most cases they are. People who are honest and loyal in their adherence to a code of ethics in dealing with others often find that, as a result, others are honest and loyal in dealing with them.

Gifts and Favors

A strict interpretation of the NSPE Code of Ethics indicates that anything offered to an engineer by a present or prospective contractor should be rejected. This would seem to include all manner of gifts, favors, and evidences of hospitality. If a ball point pen or a cigar is acceptable, why wouldn't also a set of golf clubs or a silver tea service? If there is no stigma attached to a free lunch, then why not also an evening of nightclubbing at the vendor's expense? If an inexpensive favor is to be condoned, where should the line be drawn?

Whenever engineers accept a gift, favor, or hospitality from a contractor or potential contractor, their freedom of action is inhibited. The obligation to deal with the particular contractor may not be very evident—many times it acts only as a subtle reminder. Nevertheless, engineers are human beings capable of being persuaded even against their best engineering judgment. Influence of this nature is against the interests of the engineers' employer. In recognition of this fact, many large companies have adopted policies restricting or eliminating receipt of gifts from vendors.

Recruiting Practices

In recent years engineers have been in short supply. It is occasionally contended that anyone with a pulse and a diploma certifying him to be a graduate of an engineering curriculum is eligible to be hired as an engineer. The demand for engineers is not likely to decline substantially in the foreseeable future.

As might be expected, this unprecedented demand for engineers has led to some peculiar and even reprehensible recruiting practices. (See Case No. 99-5 in Appendix B, which discusses a firm's attempt to increase its staff by soliciting specific engineers by direct mail.) Proper use of the talent and problem-solving ability of qualified engineers can make or save money for a company. Therefore, pressure is brought to bear on those whose duty it is to acquire such staff. The results are not always in keeping with the highest ethical practices. For instance, many consider it quite unethical for a company to contact engineers working for another company in an effort to lure them away. If the engineers make the first move, however, the resulting job change usually is considered to be above reproach.

The blame for a company's loss of engineers may sometimes be laid at its own doorstep. A distressingly large number of companies have a tendency to overlook contributions made by their engineers. If engineers' method or design changes result in a sizable saving to the company, the engineers may feel that there should be some recognition of them as a result—after all, they could have accepted paychecks and performed only as their managers required. The company, on the other hand, may feel that engineers are paid both for periods in which their contributions are outstanding and for many other periods in which seemingly little is accomplished.

Engineers, however, owe a duty of full service to their employer not to mention a contractual duty in the form of a non-competition or non-disclosure agreement, which engineers ignore at their peril. (See Chapter 14, Remedies and the case *Advanced Marine Enterprises, Inc., et al. v. PRC Inc.*). During the first few months after new engineers are hired, their contributions are not likely to be great, yet their employer has invested time and money in them. If engineers quit before they have repaid this investment, the company loses money. Engineers who have frequently "job-hopped" may find that even though a demand for engineers exists, they have a rough time getting another job. The NSPE's Code of Ethics addresses the professional activities of engineers, not the activities of students or others in seeking employment.

An engineer's ethics will be tested many times during their careers. For example, consider an engineer who knows of a design problem with the steering system for a new car. What does the engineer do if management decides the $15 per unit required for the alternative design is too much? Have a discussion with the supervisor about the design? Send an email questioning the decision and emphasizing that lives are at stake? Create a blog with information about the chosen design and the costs of alternatives? At what stage has the engineer done enough to discharge his or her ethical obligations?

Professional Lifestyle of an Engineer

A degree in engineering is a foundation; a life's career in the engineering profession can be built on it. Engineers can reasonably expect to be treated as professionals by their clients and superiors. However, members of the public are often unaware of contributions made by members of the engineering profession. The service rendered by a doctor, an attorney, or a member of the clergy is obvious to the public. A person experiencing pain seeks out a doctor for diagnosis and treatment. One who has been accused of a crime requires personal contact with an attorney. A family experiencing domestic difficulties may turn to a member of the clergy for aid.

The public has little knowledge, however, of the engineer whose work makes crossing a bridge or riding an airplane safe and convenient. An engineer's contributions are often unseen and it is for this reason, among others, that engineers must act professionally if they are to establish and maintain the respect accorded members of other professions!

An engineering education is never finished. Engineers have a duty to their clients and profession to learn new developments in the engineering field. For dedicated professionals, the day does not end when the office doors close behind them. There is always more to learn.

Other situations may test an engineer's ethics. Suppose the engineer knows of a discharge of a carcinogenic chemical into a nearby river that is used for drinking water. Does the engineer have a duty to report the discharge? If so, to whom—the engineer's supervisor, the Environmental Protection Agency, the local authorities using the river for drinking water, or the press? No clear answers exist. Hence, the engineer must base such decisions on personal ethics and values.

The Ethics Tool

Professional engineers are taught to be ethical. Their code of ethics is as much a tool as is their knowledge of the grain structure of steel or the deflection of a beam. Proper use of the tools that engineers possess will give them a rewarding career; improper use may lead to frustration and disaster.

Review Questions

1. A process engineer was about to recommend the purchase of equipment for a new manufacturing process for his employer. A vendor calling on him told him of a new and apparently cheaper means of accomplishing the same result. The engineer was familiar with the type of work in question but had never heard of the new process. The vendor invited him to go (at the vendor's expense) to several sites where the new process was being used. Should the engineer go to see the new process? At the vendor's expense? Which canons or rules have a bearing on the situation?

2. If the engineer took the trip mentioned above and the vendor suggested an evening at a local night club to avoid the boredom of a hotel room (with the vendor picking up the night club tab), should the engineer accept or decline? Why or why not?

3. In going through the files on a process in which her employer finds himself in trouble, an engineer finds several instances of poor judgment and miscalculation by her predecessor who set up the process. Most of the present problems in the process are caused by the previous engineer's errors. The previous engineer left the company and is now working for another firm. The present engineer's assignment is to improve the process. How should the improvements be justified to her employer? What, if anything, should the engineer say or do about her predecessor's mistakes? Which canons or rules apply?

4. About a month ago an engineer made an outstanding improvement in a process. His company produces approximately 300,000 parts per year through the process, and the direct labor savings alone amounts to approximately $0.40 per part. The supervisor of the department in which the process is located has complimented the engineer on his achievement, but no one else in the plant has done anything more than mention it to him. What, if anything, should the engineer do?

5. What are some of the characteristics of a "professional person"?

6. Just before Christmas an engineer receives a package from a vendor with whom she has dealt in the past. The engineer is now concerned with work entirely outside the field of the vendor's interest. The package contains eight place settings of sterling silver. Should the engineer keep the gift or return it? Why?

7. An engineer is approached by a friend who argues stoutly for joining an engineer's union. The friend points out that promotions will be based largely on seniority, that wages will be paid according to the class of work undertaken (which is likely to improve the engineer's economic situation), and that overtime will be paid for all work in excess of 40 hours per week or 8 hours per day. Should the engineer join the union? What are the arguments for doing so and for not doing so?

8. Refer to Case No. 01-5 in Appendix B. The Board identified a situation that might compromise an engineer's judgment in this case. What was it? Can you think of a situation where this approach may adversely affect the client? The public?

Development of Law

CHAPTER

3

- The U.S. legal system evolved from the English Common Law and Roman Civil Code.

- Sources of law include Constitutional Law, Statutory Law, and Case Law.

- *Stare Decisis* is the method by which courts abide by previous decisions.

Our activities are regulated by laws. As we live and work we become familiar with many laws, particularly those concerning the physical world—the laws of nature. We know that if we are near the earth when we drop an object, it will fall to the earth in accordance with the law of gravity. We can even predict accurately how fast the object will fall and where it will strike the earth, if we consider the laws of motion and the retarding forces. Such natural laws form a particular kind of universe of laws. They are not human-made laws, only human-discovered. These laws would exist even if we passed a legislative act against them.

In this text we are concerned with human-made laws, the laws governing relationships among people. As we will talk about it here, **law** refers to a set of rules and principles set up by society to restrict the conduct and protect the rights of its members.

A person living alone who had no contact with others would have no need for human-made laws. Add another person and the need for law would become apparent. Each of the two would have rights that might be infringed on by the other. As a result, each must control his or her behavior in such a manner that the other's rights are protected. In such a simple society the relationships would not be complex; simple rules would be sufficient.

BEGINNINGS OF LAW

Laws began as social customs. It was considered proper to behave in a certain manner in particular circumstances. Initially, the tribal chief, and then later the religious leader or priest, was charged with the preservation of these customs, including punishment for infractions. The fallibility of humans made it difficult for leaders to enforce human-made laws, so the incorporation of a divine legal origin attempted to remove the imperfect human element to enable enforcement. Thus, many of our early laws were said to have resulted from divine manifestation.

Divine Laws

Two codes of divine laws have made major contributions to the laws of Western civilization. The first of these is the Code of Hammurabi, originating around 1750 B.C., which sets forth the most basic principles of justice. Hammurabi is based on the idea of "an eye for an eye."

> The concept of divine manifestation is evidenced in our own Declaration of Independence: "We hold these truths to be self-evident, that all Men are created equal, that they are endowed by their Creator with certain inalienable Rights, that among these are Life, Liberty, and the Pursuit of Happiness."

The second set of divine laws that influenced our law is the Ten Commandments. Present influences of these laws in our legislation and court decisions are easily found. As rules, these Mosaic laws have an outstanding feature—they are short and simple. In oral cultures, each person was presumed to know the laws under which he or she lived. It was necessary that this presumption be made, because if it were not, anyone could plead ignorance of the law and thereby avoid it.

The presumption persists today that knowledge of the law is presupposed, even though they are not easily memorized. Ignorance is not a defense to violating the law.

Civil and Common Law Foundation

Two great systems of law are used in more than 40 percent of the world's countries: the **common law** and the **civil law.** The common law is used in most English-speaking countries and former English colonies. Civil law (sometimes called the Continental system), originated about 450 B.C. as the Law of the Twelve Tables.[2] This was the law of the Roman Empire as it expanded and contracted during the following ten centuries. During that time, statutes were passed and meanings clarified in court decisions. Under Emperor Justinian in the sixth century, these laws were boiled down to their essentials and published as the *Corpus Juris Civilis,* the body of the civil law. This civil law spread to other countries throughout the empire and became the foundation of the legal systems of continental Europe and those countries' former colonies. (See Appendix C, from CIA Factbook.)

> According to the Université d'Ottawa's research on world legal systems, almost 40 percent of the world's legal systems are either civil law or common law systems, representing almost 30 percent of the world's population.[1]

In operation, Roman-based civil law and English-based common law are quite different, though in practice the results are frequently the same. Civil law is based on the idea of comprehensive and complete written codes or statutes, the court's task being to

---

[1] Université d'Ottawa, www.droitcivil.uottawa.ca/world-legal-systems/eng-tab2.php, last viewed July 6, 2007.

[2] See www.fordham.edu/halsall/ancient/12tables.html last viewed August 6, 2007.

apply the correct statute to the particular set of facts in the case. Common law is built on the use of cases—prior decisions in similar situations.

Most of the law brought to the United States by its early settlers was common law. However, in the states settled by French and Spanish settlers, remnants of civil law may be found. Thus, in Louisiana, Texas, and California, principles of civil law continue to have some influence.

Today, the term **civil law** has a dual meaning. As discussed, it started as a legal code based on the Roman example. It is used today to describe our system of private law as opposed to criminal law, discussed later in this chapter. When the term **civil law** is used in the remainder of the text, it will refer to the system of private law.

Enforcement

In governing the behavior of individuals in a society, a human-made rule has little practical value unless it can be enforced. Enforcement takes one of three forms:

- *Punishment.* Fines or imprisonment are the usual means of punishing someone who has committed a crime.
- *Relief.* In actions involving private rights, the relief sought is usually money damages for the person who has been harmed at the hand of another, or a court order preventing future harm.
- *Social censure.* Frequently, the strongest enforcement factor is the fear of social ostracism—the fear of public opinion. Although public opinion is not a recognized legal means of enforcement, it is an effective method of monitoring the acts of others, even if the acts are not illegal.

Few of us possess a formal knowledge of the laws that govern our actions, yet we obey them. Even the worst criminals obey nearly all the laws most of the time. We comply unconsciously. The laws have become a part of each of us. If you recall from the beginning of the chapter, laws started as social customs, and an element of that continues to this day. Laws that cannot be agreed upon by most of society are difficult, if not impossible, to enforce.

UNITED STATES LEGAL STRUCTURE

In the United States, our legal system is composed of three broad bodies of law:

1. Laws embodied in the U.S. Constitution and the individual state constitutions
2. Federal and State Statutes
3. Court decisions or case law

Constitutional Law

Constitutional law sets up the operation of a government, including its powers and limitations. It states fundamental principles in the relationships between citizen and state, including rights that may not be infringed. Rights may arise under both state and federal constitutions. The extent of the importance of the U.S. Constitution and the U.S. Supreme Court will be discussed further in Chapter 4.

Statutory Law

A statute is a law stating the express declaration of the will of a legislature in the subject of the law enacted. A state law prohibiting gambling and setting forth a maximum penalty of $500 or six months in jail for violations would be an example of statutory law. Similarly, federal laws, such as the Occupational Health and Safety Act, and municipal laws (or ordinances) regulating traffic are statutes. Administrative regulations, such as those of the Federal Trade Commission, have the same force and effect as statutes. Treaties also have generally the same force as statutes passed by Congress.

One of the many reasons for passage of a statute is dissatisfaction with the common law in a particular field. Frequently, when decisions have for some reason become muddled in dealing with a problem, an appeal is made to the legislature for a law to clarify the issues involved. Passage of a statute voids the common law covering the same point within the legislature's jurisdiction. The statute must not, of course, conflict with the U.S. Constitution or with the state constitution involved. If, when a case arises, the courts find that a statute conflicts with the Constitution, the Constitution is protected, and the statute is held invalid.

Another reason for passing a statute is the legislature's ability to develop a comprehensive set of rules to take care of a specific problem. Suppose that the problem of low-level radioactive waste disposal is being considered in a given state. That state's legislature can enact a statute dealing with all the relevant questions of who makes the decisions, who handles the waste, where waste storage is to be located, when it should be done, how the waste should be handled, who is liable to whom, what amounts of liability are to be imposed, and what remedies are available. Needless to say, it would be quite likely that many years and several complex decisions at the least would be needed before the judiciary could develop similar rules.

Court Decisions or Case Law

When a potential litigant brings an action against someone in the U.S. court system, she (or as is most likely the case, her attorney) has to decide whether the case should be brought in a court of law or equity based on the type of relief sought. If the litigant brings the case in a court of law, or brings an action at law, the litigant anticipates recovering money, or its equivalent. If the litigant brings the case in a chancery court, or brings an action in equity, the litigant's recovery will not be satisfied by money damages. In the United States, the court decisions made in either of these forums is the basis of our case law. In either action, the litigant's means of recovery is derived from our common law heritage.

History.   U.S. law originated as the common law brought over from England by the colonists. In the first century or so following the Norman Conquest, what we know as common law began. It arose from the practice of judges to write their opinions, giving the general principles and reasoning they followed in deciding cases. When the facts were similar, judges tended to follow earlier opinions of other judges. These decisions were followed by the courts throughout England. Hence, the term **common law** was used to describe these decisions.

Today, common law is used by a court when no statute or Constitutional law covers the particular legal problem involved.

Application of Prior Cases—Stare Decisis.   Abiding by previous decisions is known as ***stare decisis.*** The main feature of common law is that the law itself is built on case decisions. When a case is decided, that decision becomes the law for that court and other courts within its jurisdiction in deciding similar future cases. A judge uses earlier cases as the foundation for decision.

About 200 years after the Norman conquest, a jurist named Henry Bracton compiled the decisions that had been rendered under the king's court systems.[3] This was the beginning of case reporting. These and later decisions formed the common law of England and, eventually, of nearly all of the United States.

Shortly before the American Revolution, Sir William Blackstone, an English jurist, completed his *Commentaries on the Laws of England.* In this law book—the first major contribution since the time of Bracton—Blackstone clarified and made intelligible English common law. An American jurist, James Kent, made a similar contribution in this country about 60 years later. Common law is defined in Kent's *Commentaries* as "those principles, usages, and rules of action applicable to the government and security of persons and property, which do not rest for their authority upon any express or positive declaration of the will of the legislature."[4] In both books the principles of common law were extracted from recorded decisions. These two books provided the basis for further development of common law in both countries.

> A court's jurisdiction, in the manner used above, indicates the area in which the court may operate. For example, the jurisdiction of the Supreme Court of Ohio is the State of Ohio, and a common law decision by the Ohio Supreme Court would be binding on all other courts in Ohio.

> In the cases in this text, you will see many references to previously decided cases. Many other references have been omitted to allow easier reading of each case.

It is a common misperception that courts do not make the laws, they merely enforce them. The statement is largely true of Constitutional law and statutory law, but not for the common law. As decisions are made for new types of cases, new legal interpretations are made by the courts.

Business Custom.   In deciding cases the courts often will make use of a relevant business custom. For instance, as will be discussed later, silence on the part of one to whom an offer is made usually cannot constitute acceptance. However, if there has been a history of dealings between persons or if there is a practice in a particular business such that silence constitutes acceptance, the court will consider this and decide accordingly. Terminology peculiar to a trade is given its trade usage interpretation in court.

Business customs played an extremely important role in legal evolution, because these customs developed into what is often called the **law merchant.** During the Middle Ages and then the Renaissance, commerce and trade flourished in certain parts of Europe. Merchants often needed some certainty about the rules that would be applied to their disputes by courts in various locations. Eventually, a body of law developed out of the business customs, and this set of laws was used by various courts in deciding commercial cases between merchants. Some of these customs have been codified in the Uniform Commercial Code (discussed in Chapter 15, Sales and Warranties).

Changes in the Law.   Frequently one hears the complaint that the law is behind the times, that it is slow to change in a rapidly changing world. The complaint is fairly well founded; law is slow to change. However, it is usually far better to have a law or fixed principle on which one can depend than to have a law or principle that may this time be decided one way and the next time, another.

---

[3]For more information about Henry Bracton, see a short description in New Advent Catholic Encyclopedia, Vol. II, 1907, online edition, K. Knight, 2003 www.newadvent.org/cathen/02726c.htm last viewed August 6, 2007.

[4]For more on Kent's *Commentaries,* see www.constitution.org/jk/jk_000.htm last viewed August 6, 2007.

The law does change, slowly, to reflect changes in society and changes in technology. By way of illustration we can look back a few years to the changes that became necessary when the automobile took over personal transportation from the horse and buggy. Similarly, we can look ahead to changes that are likely to be needed when space travel becomes a commercial reality.

Laws usually are changed in two ways: by overruled decisions and by the enactment of statutes. As statutory law was discussed above, overruled decisions are discussed below.

Overruled Decisions.   When a case goes to court, the attorneys for both parties in the case have usually conducted relevant legal research. It is likely that both will be armed with decisions in previous cases on which the judge in the present case is expected to base the case's decision. Let us assume that one of the cases used is based on facts quite similar to those in the case at hand. Let us further assume that the decision in the earlier case was handed down by the state supreme court. If the present case is in a lower court and the facts of the case are in all ways the same, the judge should follow the prior decision. However, because there are nearly always differences in the facts of two cases, let us assume that the judge, on the basis of slightly different facts, did not follow the earlier case, and that an appeal resulted. The case at hand is finally taken to the state supreme court. If the supreme court justices see the facts in the two cases as being essentially the same but render a decision different from the decision in the precedent case, the earlier case has been either distinguished on the factual differences or overruled. The law has been changed. There is no effect as to the parties in the earlier case; that case was decided by the law of that time. The law was not changed until the state supreme court changed it in the process of overruling its earlier precedent.

In the interest of retaining stability in the law, courts are quite reluctant to overrule prior decisions. At the same time, they recognize that change is inevitable.

Differences between Common Law and Equity.   In the beginning, common law was administered by royal writs. These writs were orders, in written form, to a sheriff or other officer to administer justice in a particular way or to summon a defendant before the royal justices. We still use writs—written and sealed court commands or mandates ordering some specified action to be done. Following a court judgment, a writ of execution, for instance, may be issued to an officer, telling the officer to take possession of and sell some of the loser's property to satisfy the judgment.

Equity is a legal system based on principles of "fairness" permitting remedies other than money damages to redress perceived wrongs.

The difference between common law and equity lies in the remedies afforded under each legal proceeding. For example, common law could not:

- Prevent a wrong from taking place,
- Order persons to perform their obligations, or
- Correct mistakes.

It became apparent that these gaps in the law had to be filled. To accomplish this, the new legal system had to have two components: (1) a separate court system, and (2) distinct equitable remedies.

Courts of Chancery.   The Courts of Chancery are so named because the King would delegate the hearing of unusual cases for which no remedy was available at law to his chancellor, and then to vice-chancellors. The new courts took the name of courts of chancery. In the United States, there used to be separate court rooms and judges for equity cases. In a few states, the court rooms are the same and the judges are the same ones who deal with other types of law, but the equity procedure is different. In most states, and in the federal court system, law and equity courts have been completely combined.

Equitable Remedies.   To obtain **equitable relief,** the plaintiff is required to meet the applicable standard. Different courts have different standards for various types of equitable relief. For example, to obtain a preliminary injunction, most courts require that the person seeking relief show the following:

1. A substantial likelihood of success on the merits
2. Irreparable injury
3. That the harm if the relief requested is denied outweighs the harm if the relief is granted
4. That the public interest is not disserved by granting the relief.

Irreparable injury usually refers to the court's inability to completely remedy the situation through an award of money damages, such as when such damages cannot be calculated with accuracy or could not provide a complete remedy. A nuisance—such as smoke or noise that makes a home uninhabitable—would be a basis for getting into equity jurisdiction with this reason.

Because equitable remedies are often based on fairness, an oft-quoted maxim, an enforceable rule in equity is that the party seeking equitable relief must come to the court with "clean hands." This principle is more fully explored in the case *United States of America ex rel. James Zissler v. Regents of the University of Minnesota,* at the end of the chapter. Equitable remedies are also discussed in Chapter 14, Remedies.

To reiterate, the remedies available in a court of equity are quite different from those offered in common law. They are ***in personam*** remedies, that is, they are directed to a person, whereas common law acts ***in rem,*** or on an object. Probably the most common and well-known equity remedies are the injunction and specific performance. The list of equity remedies, however, is quite long and includes divorces, mortgage foreclosures, accountings, reformation of contracts, and many others. In fact, an equity court can act in any way necessary to secure a right or remedy a wrong. The very word "equity" implies that justice will be done, and if a remedy must be invented to serve the purpose, that will be done.

The injunction exists in two general forms: temporary and permanent. A **temporary injunction** (a restraining order or injunction *pendente lite*) is readily obtainable for cause. An attorney can request a temporary injunction from the court. As soon as the judge signs the temporary injunction, it is an act in contempt of court for anyone who has knowledge of the order and who is subject to the order to fail to obey it. The object of a temporary injunction is to hold the status quo until a hearing can be held on the merits of the case. The complaint may be dismissed at the hearing, some other type of remedy may be given, or a permanent injunction may result when the facts are heard. If a permanent injunction is issued, the order is effective as long as the cause of the injunction exists.

Specific performance usually arises in connection with contracts involving land. Land is considered to be unique; no one piece of land is exactly like another. When a contract is made for the sale of a piece of land and the seller refuses to deed the land to the buyer, the seller may be forced to do so by an order for specific performance from a court of equity. Property other than real estate is treated the same way only when it is recognized as being unique, for example, an antique or a rare painting.

Equity is reluctant to give a remedy that will require continuous supervision over an extended period. In some cases such remedies have been given, but where another remedy will suffice, the other remedy is preferred.

The remedies afforded in equity courts are not limited to equity remedies. Once a cause of action is legitimately in an equity court, the court will settle all the issues involved, including the remedies afforded at common law and the equity remedy. For instance, an equity decree might include money damages for injuries already suffered as well as an injunction against further injury. Equity, however, will not give a remedy that is directly contrary to common law; neither will equity act when an adequate remedy exists under a statute or under common law.

Benefits. There are two readily apparent benefits to courts of equity: (1) speed and (2) privacy.

Speed. A court of equity generally acts more rapidly than does a court of law. A temporary injunction may be obtained with little delay. In addition, there is usually no jury involved in equity cases, which eliminates the time-consuming selection of jurors and deliberation over the evidence.

Privacy. Because there is no jury involved, a case in equity may be decided more privately. This feature is particularly significant when a case involving a trade secret is tried. (See Chapter 20, Intellectual Property and the case at the end of the chapter, *Coca-Cola Bottling Company of Shreveport, Inc. v. The Coca-Cola Company,* 107 F.R.D. 233, 227, U.S.P.Q 18 [1985]). If a jury were to hear the facts, there would be 12 more people to hear the secret. The secret would become virtually public information.

Criminal Law

In the United States we have many rights and freedoms. We can do as we please up to the point where the things we do infringe on the rights or freedoms of others. As you may recall from the beginning of the chapter, many laws were originally ethical or moral codes agreed to by society. A crime is an antisocial act, but, of course, not all antisocial acts are crimes. Telling true but unflattering stories about another to secure a promotion would probably be considered antisocial by most, but it would not be a crime. Society is injured when a crime is committed. The main purpose of trying a person accused of a crime and punishing him or her when guilt is determined is to prevent recurrences of the criminal act.

One main distinction between a criminal action and a civil action is that for the criminal action the state undertakes prosecution, whereas a citizen undertakes civil action. A criminal action is undertaken to punish the wrongdoer; a civil action is pursued to get compensation for a loss suffered or to prevent a loss from being suffered.

The same act may constitute both a tort and a crime—as such it gives rise to both a civil and a criminal action against the person who committed the act. In fact, it takes some thought to conceive of a **tort** (an injury to another's person or property) that may not be a crime and a crime that would in no way be a tort. Torts are discussed further in Chapter 21.

A crime is an act prohibited by statute. As you will see in *State of Arizona v. Wilkinson,* failure to register as a residential contractor or engineer is a crime. Although this case is about a residential contractor, the laws under which the defendant was sued are the similar to those under which a person lacking a professional engineering license would be subject. As you read *State of Arizona v. Wilkinson,* notice that the Supreme Court of Arizona takes great pains to distinguish between what recovery, or in this case, restitution, the victims can receive under criminal law, and that which can be recovered under civil law. As time passes and society becomes more complex, we tend to increase the number of laws defining certain acts as crimes. When a criminal statute is made, it is generally necessary to answer at least three questions in the statute:

1. What is the act (or omission) that is to be prohibited?
2. Who can commit the crime (or, conversely, who cannot)?
3. What is the punishment for commission of the act?

Often there is both a physical and a mental component of a crime. The mental element is known as criminal intent. Many state statutes eliminate a requirement for this proof on the basis that the act speaks for itself and that a person will intend the natural results of such acts.

Degrees of Crime

Treason

The highest crime is **treason.** The United States Constitution states: "Treason against the United States shall consist only in levying War against them, or in adhering to their Enemies, giving them Aid and Comfort. No Person shall be convicted of Treason unless on the Testimony of two Witnesses to the same overt Act, or on Confession in open Court."

Felony

Felonies compose the second level of crime. Early common law punished felonies with a sentence to death. Today a felony is generally defined as an act that is punishable by death or imprisonment in a penitentiary for a term of longer than one year.

Misdemeanor

The lowest level of crime is known as a **misdemeanor.** It consists of all prohibited acts less than felonies. Traffic violations are misdemeanors, as are zoning law violations and breaches of peace. Punishment for a misdemeanor usually consists of a fine or jail sentence or, generally, anything less than death or imprisonment in a penitentiary for a term of longer than one year.

Punishment for a criminal offense ordinarily takes one of the following forms: death; imprisonment; fine; removal from office; or disqualification to hold and enjoy any office of honor, trust, or profit under the Constitution or laws of the state.

UNITED STATES OF AMERICA ex rel. JAMES ZISSLER v. REGENTS OF THE UNIVERSITY OF MINNESOTA, 992 F. Supp. 1097 (D. Minn. 1998)

Study terms: Equitable relief, unjust enrichment, breach of fiduciary duty, equitable defense, statute of limitations, False Claims Act

Introduction

In February 1995, Plaintiff James Zissler ("Zissler") filed suit against the Defendant the Regents of the University of Minnesota ("the University"), on behalf of the United States as a qui tam relator under the False Claims Act ("FCA"), 31 U.S.C. § 3731. In December 1996, Plaintiff the United States of America ("the government") intervened in the action. The government then filed suit against the University, alleging claims for violation of the FCA; unjust enrichment; payment by mistake; disgorgement of profits; and breach of fiduciary duties. The government's claims were on allegations that the University fraudulently submitted grant applications to the National Institute of Health ("NIH"); improperly sold unlicenced (sic) biological drugs resulting in illegal profits to the University; made fraudulent submissions for unlicenced (sic) and unreimbursable drugs to Medicare; and received illegal kickbacks related to home infusion services. On July 23, 1997, this Court granted the University's Motion to Dismiss the claims brought against it under the False Claims Act. Currently before the Court are the University's Motion for Partial Summary Judgment, based largely upon the statute of limitations, and its Motion to Dismiss the government's equitable claims. For the reasons set forth below, the Court will grant the Motion in part and deny it in part.

Facts

A. Nature of the Government's Allegations Against the University

1. Misuse of Federal Grant Money

Between 1969 and 1993, the University received approximately $19 million in NIH grant funds for a research project entitled "Studies of Organ Transplantation in Animals and Man" ("Transplant Grant"). A central component of the Transplant Grant was the study, development, and production of Antilymphocyte Globulin ("ALG"), a drug used to reduce organ transplant rejection reactions after surgery. The University made false statements to the NIH in connection with the Transplant Grant by representing that it earned no grant-related income, when, in fact, it earned over $80 million in program income from the sale of ALG. The University also falsely inflated salary, equipment, and other grant-related costs charged to the NIH and other federal agencies in connection with the "Program for Surgical Control of Hyperlipidemias" ("POSCH") Grant and twenty-eight other federal grants.

2. Illegal Profits

In January of 1971, the Food and Drug Administration ("FDA") designated ALG as an investigational new drug ("IND"). Even though it is illegal to sell investigational drugs for a profit, the University sold twenty different ALG products to approximately 280 different purchasers throughout the world and made $80 million on the sales. The University concealed the illegal profits it earned from ALG sales from the federal government.

3. Medicare Reimbursement

The University submitted false claims to the Medicare Program for payment relating to ALG, Procuren, and Perfusate. . . .

B. The Government's Knowledge of the University's Misconduct

1. Knowledge of ALG Sales

a. The FDA

After a 1984 inspection, the FDA learned that the University was selling ALG, and it asked University officials at the ALG program to explain why such sales should not be considered the improper commercialization of an IND. In 1987, after the University repeatedly failed to respond to this demand, the FDA ordered an investigation into the University's interstate shipment and sales of ALG. Wayne Schafer ("Schafer"), an investigator for the FDA at its Minnesota branch office, conducted this investigation.

In his October 1987 report, Schafer indicated that the University had been selling ALG and that production of ALG had increased, "as well as the charges for the product, to the point where the 'program' is a large manufacturing operation with about 40 employees." In 1986, the University shipped 29,001.87 grams of ALG at a cost of $215/gram, and in 1987, it shipped 23,532 grams, at a cost of $215/gram, for a combined total revenue of approximately $11,500,000. In addition, Schafer reported that the University was about to solicit bids for a new $8.5 million facility for the production of ALG.

Officials at the ALG program told Schafer that "they do not make a profit on the products and at times have had considerable deficits. The products are reportedly being sold on a cost recovery basis." . . .

Schafer passed his report on to Michael Dubinsky ("Dubinsky"), the Director of Regulations and Bioresearch Monitoring, Office of Compliance, Center for Biologics Evaluation and Research of the FDA. Dubinsky reviewed this report in late 1987 or January 1988. . . .

The University repeatedly told the government that it was not earning any profits from the sale of ALG, when in fact it was earning large profits. . . .

In October and November 1992, the University disclosed documents to the FDA which showed that the University was profiting from the sale of ALG. FDA inspectors had not seen these documents previously, even though many of them were dated before previous inspections had been conducted.

b. The NIH

In 1977, the NIH issued a report on the ALG Transplant Grant, indicating that the University was engaged in large-scale production of ALG and that several hundred patients were involved in studies of ALG at several hospitals. The report noted that "the material will be provided free to participating centers and the University of Minnesota will attempt to work out some means of recovering costs."

■ ■ ■

In its annual grant applications and renewals to the NIH, the University never reported that it had earned any money from its sales of ALG. . . . In October 1993, the University disclosed to NIH that it had received substantial program income from the Transplant Grant.

2. Knowledge That the Transplant Grant Had Supported the ALG Program

The FDA was aware that NIH funding was being used to operate the ALG program. In a May 31, 1979 letter, Condie told the FDA that either additional NIH funding would be obtained to operate the ALG program, or it would be shut down. . . .

Approval of ALG Sales

On April 25, 1989, the University requested that the FDA allow it to sell ALG on a cost recovery basis for $134 per gram. . . .

■ ■ ■

The FDA approved the University's request to sell ALG at $134 per gram on May 1, 1989.

C. Applicable Regulations

Based upon ALG's status as an IND, it is subject to the Federal Food, Drug, and Cosmetic Act ("FDCA"), 21 U.S.C. § 301 et seq., and the Public Health Service Act ("PHSA"), 42 U.S.C. § 262.

■ ■ ■

In addition, as a recipient of NIH grant money, the University was subject to applicable NIH regulations. The NIH requires that officials at the applicant institution certify as true each grant application, progress report, and grant continuation application. . . .

Analysis

I. Standard of Review—Motion for Summary Judgment

Federal Rule of Civil Procedure 56 states:

[Summary] judgment shall be rendered forthwith if the pleadings, depositions, answers to interrogatories, and admissions on file, together with the affidavits, if any, show that there is no genuine issue as to any material fact and that the moving party is entitled to judgment as a matter of law.

Fed. R. Civ. P. 56(c). . . .

II. Statute of Limitations

The University argues that the six-year statute of limitations found in 28 U.S.C. § 2415(a) bars the government's claims for unjust enrichment, payment by mistake, equitable relief ancillary to the FCA, disgorgement, and breach of fiduciary duty. The government disputes that any of these claims are barred by the applicable statute of limitations. . . .

A. NIH Claims—Unjust Enrichment, Payment by Mistake, and Breach of Fiduciary Duty

In its claims for unjust enrichment, payment by mistake, and breach of fiduciary duty, the government seeks to recover NIH grant money that it paid to the University because the University failed to inform the NIH that it had received program income from its sale of ALG. The University argues, and the government does not dispute, that 28 U.S.C. § 2415(a) applies to these claims. This statute states, in relevant part, that:

Subject to the provision of section 2416 of this title, and except as otherwise provided by Congress, every action for money damages brought by the United States . . . which is founded upon any contract express or implied in law or fact, shall be barred unless the complaint is filed within six years after the right of action accrues. . . .

. . . Neither party addresses the issue of when any of the government's claims against the University accrued. The Court believes that these causes of actions accrued on the date the University engaged in the improper conduct which gives rise to the claims—when it sold ALG for a profit, received program income from its sale of ALG, failed to report program income, and failed to report profits from the sale of ALG.

. . . Thus, the issue becomes when the proper government official knew, or reasonably should have known, the facts material to these rights of action [pursuant to 28 U.S.C. §2416(c)].

1. Proper Government Official

In order to determine if and for what period the statute of limitations was tolled, the Court must first determine who is the "government official charged with the responsibility to act in the circumstances." 28 U.S.C. § 2416 (c). There is no Eighth Circuit case law interpreting this phrase. The legislative history for § 2416(c) indicates:

This provision is required because of the difficulties of Government operations due to the size and complexity of the Government. It is not intended that the application of this exclusion will require the knowledge of the highest level of the Government. Responsibility in such matters may extend down into lower managerial levels within an agency. As a general proposition, the responsible official would be the official who is also responsible for the activity out of which the action arose. Such an official is the one likely to know whether the material fact does involve the possibility of a cause of action which may be asserted against the Government.

S. Rep. No. 1328 (1966), reprinted in 1966 U.S.C.A.A.N. 2502, 2507.

The government argues that the NIH official who needed to know about the University's program income from the sale of ALG is John Garthune ("Garthune"), who was the Assistant Director for Grants and Contracts of the Division of Extramural Activities and the Acting Chief Grants Management Officer at the National Institute of Diabetes and Digestive and Kidney Diseases (NIDDK) of the NIH during the applicable times. As Acting Chief Grants Management Officer for NIDDK, Garthune had the authority to take action on issues involving grantee program income. . . .

The University's contention that Schafer and Dubinsky are the officials whose knowledge could stop the tolling of the statute of limitations is flawed . . .—it imputes the knowledge of the FDA to the NIH, and vice versa. The University contends that "the knowledge of the FDA officials is attributable to NIH officials because they are merely two branches of the same federal agency." The language of § 2416(c), however, does not support this proposition. . . .

Based upon the evidence that was presented in this Motion, the Court concludes that, with respect to the government's claims for unjust enrichment, payment by mistake, and breach of fiduciary duty, the statute of limitations was tolled until John Garthune knew, or reasonably should have known, the facts material to these causes of actions.

2. Garthune's Knowledge of the Facts Material to These Causes of Action

The six-year statute of limitations is tolled during the period during which Garthune either: (1) did not know the facts material to the right of action; or (2) reasonably could not have known of the facts material to the right of action.

■ ■ ■

The Court finds that genuine issues of material fact exist regarding whether Garthune knew, or reasonably should have known, that the University had received unreported program income from the Transplant Grant by January 26, 1988. Garthune asserts that he did not know that the University had received program income from the Transplant Grant until 1993 and that he has no knowledge of any alleged discussions between John Wu, a case clerk at the NIH, and FDA officials regarding ALG sales that were supported by the Transplant Grant. . . .

Based upon the evidence presented in this Motion, the Court concludes that a reasonable fact finder could conclude that Garthune did not know, and reasonably could not have known, about the program income the University had earned from the sale of ALG before January 26, 1988. The Court, therefore, will deny the Defendant's Motion with respect to the portion of the government's claims for unjust enrichment, payment by mistake, and breach of fiduciary duty that deal with recovering NIH grant money paid to the University.

B. Disgorgement

The government alleges that the University used federal grant money to illegally sell ALG for profit, in violation of the PHSA and the FDCA, and it seeks to divest the profits that the University earned in direct contravention of these laws. The government has brought two claims requesting the divestment of profits from the University—a claim for equitable relief ancillary to the FCA (Count VI) and a claim for disgorgement of illegal profits (Count VII).

1. Applicable Statute of Limitations

The University argues that the government's claim for disgorgement (Count VII) is not an independent cause of action, but merely a remedy that the government could possibly assert under its claims for unjust enrichment and breach of fiduciary duty. As such, the University contends that it "is based on the existence of a contract implied in law," and is barred by the six-year statute of limitations in 28 U.S.C. § 2415(a).

The Court finds that the government's (sic) reliance upon the statute of limitations in 28 U.S.C. § 2415(a) is misplaced. Section 2415(a) states, in part, that:

Subject to the provision of section 2416 of this title, and except as otherwise provided by Congress, every action for money damages brought by the United States . . . which is founded upon any contract express or implied in law or fact, shall be barred unless the complaint is filed within six years after the right of action accrues.

28 U.S.C. § 2415(a).

Disgorgement of illegal profits, however, is an equitable remedy, not a request for money damages. See *United States v. Incorporated Village of Island Park,* 791 F. Supp. 354, 370 (E.D.N.Y. 1992). . . . Even if this Court were to construe the government's request for disgorgement as a remedy for its claims for unjust enrichment and breach of fiduciary duty, those claims would not be claims for money damages, but for equitable relief. By the statute's plain meaning, § 2415(a) does not apply to the government's claim for disgorgement of illegal profits.

It is a well established rule that "an action on behalf of the United States in its governmental capacity . . . is subject to no time limitation, in the absence of congressional enactment clearly. In the instant case, the University has not identified a statute of limitations that applies to the government's request for disgorgement of illegal profits, whether it be a remedy for its other claims or a separate cause of action, and therefore, the Court concludes that no statute of limitations applies to it. . . ."

The Court, therefore, will deny the University's Motion with respect to its argument that the statute of limitations bars the government 's disgorgement claim (Count VII).

2. Laches

The University argues that laches should bar the government's claim for disgorgement of ALG profits.

The doctrine of laches "is an equitable defense to be applied when one party is 'guilty of unreasonable and inexcusable delay that has resulted in prejudice' to the other party." *Bostwick Irrigation Dist. v. United States*, 900 F.2d 1285, 1291 (8th Cir. 1990) (quoting *Goodman v. McDonnell Douglas Corp.*, 606 F.2d 800, 804 (8th Cir. 1979)). The Eighth Circuit has clearly held that, "whatever the application of [laches] to private parties, we have recognized the long-standing rule that laches does not apply in an action brought by the United States."

In an attempt to circumvent this rule, the University argues that the government's claim for disgorgement is brought on behalf of private individuals, and it asks this Court to apply an exception to the rule that courts have created for cases in which the EEOC brings a claim that asserts the rights of individual claimants.

■ ■ ■

The University also contends that this Court, as part of its inherent power of a court sitting in equity, can dismiss stale claims. The disgorgement claim, the University argues, is "beyond stale" because the FDA knew of the sales of ALG for over twenty years before the disgorgement claim was commenced.

Courts sitting in equity have the power to dismiss stale claims, even when no statute of limitations would apply. . . . As the Court has previously noted, there is no direct evidence that the FDA knew of the profits the University was earning from its sale of ALG, while there is ample evidence that the University actively concealed such profits from the government for years.

Moreover, the Court questions the prejudice that the University has allegedly suffered from this delay. . . . This Court does not accept the University's assertion that it was solely the government's delay in filing suit that allowed the sale of ALG to continue. The University was responsible for the administration and oversight of the Transplant Grant; its action or inaction in adequately monitoring this program, therefore, caused the prejudice of which it complains. Accordingly, the Court will deny the University's Motion with respect to its claim that laches or the delay of the government in filing suit should bar its disgorgement claim.

V. Availability of Equitable Remedies

The University contends that this Court cannot award the equitable remedy of disgorgement of ALG profits because the government has an adequate remedy at law. The University fails to identify the government's adequate remedy at law, although it implies that the government's claims for unjust enrichment, payment by mistake, and breach of fiduciary duty are the adequate remedies at law.

It is well established that a party is entitled to equitable relief only if there is no adequate remedy at law. *See Lewis v. Cocks*, 90 U.S. (23 Wall.) 466, 470, 23 L. Ed. 70 (1874);). The mere existence of a possible remedy at law, however, is not sufficient to warrant denial of equitable relief. *Interstate Cigar Co. v. United States*, 928 F.2d 221, 223 (7th Cir. 1991). . . .

The Court finds that the University has failed to show that the legal remedies available to the government are adequate, thus precluding equitable relief. In the instant case, the government is seeking monetary damages to recover the NIH grant payments it made to the University based upon the University's fraudulent statements that it was not earning profits or program income from the sale of ALG. The Court does not believe, however, that the recovery of these damages would "do full justice to the litigant parties." *Mitsubishi,* 14 F.3d at 1519. Courts frequently allow equitable remedies, such as constructive trusts or disgorgement of profits, where one party has wrongfully or fraudulently obtained money and then used it to obtain a greater profit. . . . Because the University has failed to offer any explanation as to why the legal remedies available to the government are adequate, the Court will deny its Motion to dismiss equitable claims based upon the adequacy of the government's legal remedies. . . .

VII. The Government's Unclean Hands

The law of equity requires that

he who comes into equity must come with clean hands. This maxim is far more than a mere banality. It is a self-imposed ordinance that closes the doors of a court of equity to one tainted with inequitableness or bad faith relative to the matter in which he seeks relief . . . .

Precision Instrument Mfg. Co. v. Automotive Maintenance Mach. Co., 324 U.S. 806, 814, 65 S. Ct. 993, 997, 89 L. Ed. 1381 (1945).

Courts in equity have a wide range to use their discretion to refuse aid to an unclean litigant. *Precision Instrument,* 324 U.S. at 814, 65 S. Ct. at 997. In determining whether a party's misconduct prevents it from seeking equitable relief from the court, that party's misconduct must not necessarily have been of such a nature as to be punishable as a crime or as to justify legal proceedings of any character. Any wilful act concerning the cause of action which rightfully can be said to transgress equitable standards of conduct is sufficient cause for the invocation of the maximum by the chancellor. . . .

The University contends that the government has unclean hands and should be barred from seeking equitable relief. The government has unclean hands, the University argues, because it knew for over twenty years that the University was selling ALG and it approved the sale of ALG in 1989 for $134 per gram.

The Court does not believe that the government has unclean hands which would bar it from seeking disgorgement of the profits the University earned from the sale of ALG. The government is seeking disgorgement of profits, not the proceeds the University earned from its sale of ALG. Thus, the Court does not believe that the government's knowledge of sales, and its failure to file suit against the University, despite this knowledge, constitutes a "wilful act concerning the cause of action which rightfully can be said to transgress equitable standards." *Precision Instruments,* 324 U.S. at 815, 65 S. Ct. at 997. In its broad discretion, the Court will not preclude the government from seeking equitable relief because of its unclean hands, and it will deny the University's Motion on this ground.

Conclusion

Based on the foregoing, and upon all the files, records, and proceedings herein, IT IS ORDERED that the University's Motion to Dismiss and for Partial Summary Judgment is GRANTED with respect to the government's claim for Equitable Relief Ancillary to the False Claims Act (Count VI), and is DENIED in all other respects.

> **STATE OF ARIZONA v. THE HONORABLE MICHAEL O. WILKINSON, Judge of the SUPERIOR COURT OF THE STATE OF ARIZONA, in and for the COUNTY of MARICOPA, Respondent Judge, JOHN R. PORTER, Real Party in Interest, 202 Ariz. 27; 39 P.3d 1131 (2002)**

Study terms: Contracting without a license, misdemeanor, restitution

We granted review to consider whether and to what extent the courts can order restitution for victims of an unlicensed contractor who performs incomplete and faulty work. We conclude that a trial court may award restitution when and to the extent that the criminal act of contracting without a license directly causes a victim's economic loss.

I.

John R. Porter, representing himself to be a licensed contractor, separately contracted with T.S. and N.L. (the victims) to perform remodeling work on their homes. T.S. purchased needed materials and paid Porter $2,854.77. N.L. paid Porter at least $9,040.27. In both instances, Porter failed to complete the work and did some of the work improperly. The victims each filed a complaint with the Registrar of Contractors, alleging both poor workmanship and non-performance. The Registrar of Contractors' investigation revealed that Porter did not hold a valid contractor's license.

Porter was charged and convicted in Phoenix Municipal Court of two counts of acting in the capacity of a contractor without holding a contractor's license, a class one misdemeanor.[5] Arizona Revised Statutes (A.R.S.) §§32-1151, 32-1164. The municipal judge conducted a restitution hearing pursuant to A.R.S. section 13-603, and ordered Porter to pay $22,429.11 to T.S. and $22,365.67 to N.L. The judge calculated these awards by adding the amounts each victim had paid to Porter to the estimated cost of repairing Porter's faulty work and finishing work he left incomplete.

Porter appealed the restitution order to the Maricopa County Superior Court. Judge Wilkinson vacated the restitution awards, stating that the victims' economic losses were caused not by Porter's failure to procure a contractor's license, but by "shoddy and incomplete work." State v. Porter, No. LC 1999-000438, Minute Entry Order at 2 (Dec. 4, 1999). Having no further right to appeal, the State filed a special action in the court of appeals. The court of appeals accepted jurisdiction but denied relief, holding that the economic loss the victims suffered was a "remote, indirect, or consequential result" of Porter's crime, and therefore beyond the scope of criminal restitution. State v. Wilkinson, 198 Ariz. 376, 381 P22, 10 P.3d 634, 639 P22 (App. 2000).

We granted review to determine whether a victim can recover restitution from a person convicted of contracting without a license in violation of A.R.S. section 32-1151. At the Court's request, the parties separately argued the question whether Arizona's restitution statutes are consistent with the Arizona Constitution's guarantee of the right to a civil jury trial.

II.

A. To implement the important constitutional right of crime victims to recover prompt restitution, the legislature enacted several statutes that define the circumstances under which and the extent to which a court may award

---

[5]The court also convicted Porter of one count of advertising to provide contracting services without first obtaining a contractor's license. Ariz. Rev. Stat. (A.R.S.) § 32-1165. The trial judge did not base any restitution award on this conviction.

restitution. Section 13-603 directs the court to "require the convicted person to make restitution" to the victim, "in the full amount of the economic loss as determined by the court. . . ." A.R.S. § 13-603.

> Economic loss includes any loss incurred by a person as a result of the commission of an offense. Economic loss includes lost interest, lost earnings and other losses which would not have been incurred but for the offense. Economic loss does not include losses incurred by the convicted person, damages for pain and suffering, punitive damages or consequential damages.

A.R.S. § 13-105.14. Section 13-804.B further defines the scope of restitution by directing the court to consider "all losses caused by the criminal offense or offenses for which the defendant has been convicted." A.R.S. § 13-804.B.

These statutes, considered together, define those losses for which restitution should be ordered. First, the loss must be economic. Second, the loss must be one that the victim would not have incurred but for the defendant's criminal offense. As the court of appeals noted, however, "'but for' causation does not suffice to support restitution, for if it did, restitution would extend to consequential damages. Yet our criminal code expressly provides the contrary." *Wilkinson*, 198 Ariz. at 380 P19, 10 P.3d at 638 P19. By eliminating consequential damages, the statutory scheme imposes a third requirement: the criminal conduct must directly cause the economic loss. . . . We hold, therefore, that the statutes direct a court to award restitution for those damages that flow directly from the defendant's criminal conduct, without the intervention of additional causative factors.

B. Applying the above standards, the court of appeals concluded that Porter's victims could not recover any restitution. We disagree in part.

When Porter, presenting himself as a licensed contractor, entered agreements with T.S. and N.L. to provide contracting services, he violated A.R.S. section 32-1151. . . . Porter's criminal actions directly caused those losses. Indeed, the original conception of restitution, and the form with the most direct link to criminal conduct, is that

> of forcing the criminal to yield up to his victim the fruits of the crime. The crime is thereby made worthless to the criminal. This form of criminal restitution is sanctioned not only by history but also by its close relationship to the retributive and deterrent purposes of criminal punishment.

United States v. Fountain, 768 F.2d 790, 800 (7th Cir. 1985). Under Arizona's statutes, these victims are entitled to recover their payments to Porter as restitution.

A different result obtains, however, as to the expenses the victims incurred because Porter failed to complete the work he contracted to do or did so in a faulty manner. We agree with the court of appeals that Porter's criminal conduct of contracting without a license did not cause these losses. . . . Therefore, the losses incurred as a result of Porter's poor and unfinished work constitute indirect damages and cannot qualify for restitution.

C. Our conclusion that the restitution statutes encompass only damages directly caused by the criminal conduct involved not only remains faithful to the statutory language, but also prevents the restitution statutes from conflicting with the right to a civil jury trial preserved by Arizona Constitution Article II, Section 23. Potential problems arise if we too broadly combine civil liability with criminal sentencing. As the court of appeals has noted:

> If reparations as a condition of probation are to include elements beyond mere "special damages" we believe a trial court must use great caution. The sentencing phase of a criminal case is not the ideal forum for the disposition of a [civil] case. Both parties are deprived of a jury; the defendant may be limited in showing causation or developing a defense of contributory negligence or assumption of risk.

State v. Garner, 115 Ariz. 579, 581, 566 P.2d 1055, 1057 (App. 1977).

By limiting restitution to those damages that flow directly from a defendant's criminal conduct, the legislature focused upon the primary purposes of restitution: reparation to the victim and rehabilitation of the offender. . . . The penalty thus fits squarely within the goals of criminal punishment and does not deprive him of a civil trial to which he might otherwise be entitled.

D. The majority opinion of the court of appeals also suggested that the reason the victims cannot recover damages for the harm caused by Porter's unworkmanlike performance is that such damage is not an element of the crime of which he was convicted. . . . The test is whether particular criminal conduct directly causes the victim's loss. In this case, damage caused by Porter's unworkmanlike performance does not meet the statutory requirements for restitution because the criminal conduct did not directly cause the damage, not because the statute omits unworkmanlike conduct as an element of the crime of which Porter was convicted.

■ ■ ■

V.

For the foregoing reasons, we vacate the opinion of the Court of Appeals, vacate the judgment of the Superior Court, and remand to the Phoenix Municipal Court for a new restitution hearing, the restitution award to be made in a manner consistent with this opinion.

Review Questions

1. Why do we need human-made laws? What purpose do they serve?

2. What are the four basic types of laws in our legal system? What does each consist of?

3. What is the meaning and significance of the term *stare decisis?*

4. How do changes in common law take place?

5. What factors led to the establishment of equity as a separate system of law?

6. What types of remedies are offered by common law? By equity?

7. Party A owns a factory that emits large quantities of foul-smelling smoke. People in a nearby housing development are annoyed by the odors whenever the wind shifts to a certain direction. Party B, a resident in the housing development, has lodged a complaint against A for public nuisance. In what kind of court would the case be likely to be tried in your state? Why? What would be the probable result of the legal action?

8. What are the general requirements for equity jurisdiction? How might these requirements be met?

9. Refer to *In the case United States of America ex rel. James Zissler v. Regents of the University of Minnesota* to answer the following questions:
   a. What equitable defense did the University use to bar the Government's claim for disgorgement?
   b. The University claimed that the Government was not entitled to equitable relief. What were some of the University's reasons?

10. Refer to *State of Arizona v. Wilkinson.* If it was discovered that Porter was a member of the NPSE, which canons and rules of practice did he violate?

11. Describe and give an example of each of the three degrees of crime under United States law.

Courts, Trial Procedure, and Evidence

HIGHLIGHTS

■ The establishment of the United States judicial system is outlined in Article 3 of the United States Constitution.

■ Whether a case can be brought before the United States Supreme Court, federal court, or state court is determined by that court's jurisdiction.

■ When an action is brought in federal court, the litigation is governed by federal rules of civil procedure or criminal procedure, and federal rules of evidence.

■ Because of his or her expertise in a particular engineering field, an engineer may be requested to act as an expert witness, or provide advice and consultation for a trial.

The fact that one person accuses another of a wrong does not mean that the accusation is true. Court trials with a judge, a jury, and attorneys for both plaintiff and defendant have not always been used. Disputes have historically been settled on the battlefield. In many of the early civilizations trial by battle was replaced either by trial by ordeal or trial by jury.

Our present jury trial system developed with the common law. Its continuance in criminal and civil cases is guaranteed in the Sixth and Seventh Amendments of the U.S. Constitution.

COURTS

In the United States we have a system of courts for each state, as well as a system of federal courts. The systems are somewhat similar. Both the federal and state systems provide an ultimate tribunal: a supreme court. At the next lower level are the appellate courts (in the federal system and some state systems) that handle appeals from the lower courts. The lower courts are the district courts in the federal system, and usually county courts in the state systems.

Federal Courts

Supreme Court.   The United States Supreme Court is our highest tribunal and final court of appeal. In addition, it is the sole judicial body permitted to interpret the United States Constitution. Article 3 of the U.S. Constitution provides for our Supreme Court and whatever inferior federal courts Congress may from time to time require. The jurisdiction (for types of cases) is limited in the Constitution to nine categories. The only cases that may originate in the Supreme Court are ones involving ambassadors, public ministers, and consuls, or those in which a state is a party. In these the U.S. Supreme Court has original jurisdiction; all other cases go to the Supreme Court by appeal.

Cases are appealed to the Supreme Court from either U.S. courts of appeals or from state supreme courts in the normal course of events. However, cases may go directly from any court to the U.S. Supreme Court if the question to be settled involves the U.S. Constitution or is of very great public interest. Appeal is usually made by a petition to the court for a *writ of certiorari*. If the petition by the appellant is successful, a *writ of certiorari* will be issued to the lower court demanding that the case be sent up for review. Only a very small portion of such petitions are successful.

Courts of Appeals.   Thirteen United States courts of appeals exist. Appeals on federal questions are normally settled by three justices, who then form a panel of each court. These courts were first established by Congress in 1891 because of the burden of appeals on the U.S. Supreme Court. The courts of appeals function as appellate courts only, and do not conduct trials. Issues between parties are appealed on the basis of a conflict in the law, the facts of the issue having been decided previously in a lower court.

District Courts.   The trial courts of the federal court system are the United States district courts. The districts presided over by the U.S. district courts are formed in such a way that no state is without a federal district court. The number of justices in a U.S. district court is determined by statute and is based on the number of federal cases arising in the district.

For a case to be tried in a federal court, it must meet one of thirty-eight enumerated criteria, among which include cases that:

- Are an action against a foreign state
- Involve a federal question
- Are between citizens of two different states or between a citizen of a state and a foreign country *and* the amount in controversy exceeds $75,000
- Involve admiralty or maritime controversies
- Is a bankruptcy case
- Involves any act of Congress regulating commerce, and the amount in controversy exceeds $10,000

Most of the controversies handled by the district courts involve federal statutes or the Constitution (that is, "federal question" cases) or "diversity" cases (that is, those between citizens of different states).

Common examples of cases tried in United States district courts are admiralty, patent, copyright, trademark, restraint of trade, tax cases, and infringement of personal rights. Many other cases arise from decisions of administrative boards (such as the National Labor Relations Board) that operate as quasi-judicial entities and look to the federal courts for enforcement of their orders. Claims against the federal government may be filed either in a U.S. district court or in the United States Claims Court.

> The U.S. Court of Claims heard cases against the United States brought directly by claimants from 1835 to 1982. It was abolished in 1982. The successor courts that were established are the U.S. Claims Court (trial jurisdiction) and the U.S. Court of Appeals for the Federal Circuit (appellate jurisdiction).

State Courts

State court systems are far from uniform throughout the United States. Not only do the systems differ, but the names of the courts differ as well. A few generalities can be stated.

Each state has a final court of appeal, usually called the supreme court of the state. Nearly always the highest court confines its work to appeals of cases tried in lower courts. Courts of intermediate appellate jurisdiction are interposed between the supreme court and lower courts in many states.

The next lower tier of courts consists of the trial courts of general jurisdiction, known variously as circuit courts, courts of common pleas, county courts, superior courts, or, in New York, the supreme court. Probate courts or surrogate courts handle cases of wills, trusts, and the like. Such courts are limited in geographical jurisdiction to a particular county, district, or other major political subdivision of the state.

At the lower end of the judicial hierarchy are various municipal courts. Police courts, justices of the peace, small-claims courts, juvenile courts, and recorders' courts are common examples of these courts of very limited jurisdiction.

Jurisdiction

The **jurisdiction** of a court means its right or authority, given either by a legislature or a constitution, to hear and determine causes of action presented to it. A court's jurisdiction relates to geographical regions (for the location of persons and subject matter), to types of cases, and possibly to the amount of money or types of relief concerned.

Courts have jurisdiction over property, both real and personal, (otherwise known as *in rem* jurisdiction) located within their assigned geographical limits. Even though the owner may not be available, his or her property may be taken in satisfaction of a judgment. A court has no authority over property lying outside its territorial limits. Let us assume that a Tennessee court has awarded Black $5,000 as a result of a damage action against White. If White does not pay and cannot be made to pay in satisfaction of the judgment, justice for Black is rather hollow. The Tennessee court could take and sell that part of White's property that could be found in Tennessee until the judgment was satisfied. However, it could not touch any of White's property in, say, Georgia.

A court has jurisdiction over all persons (otherwise known as *in personam* jurisdiction) who have had sufficient contacts with the state such that the exercise of jurisdiction does not offend traditional notions of fair play and substantial justice, whether the persons are residents or not. Jurisdiction over a person is exercised by serving that person with a summons or other legal process—a defendant would be served with a summons, a witness with a subpoena. How service may be made is a matter clarified by each state's statutes. Generally, a sheriff or other officer is directed to serve the process on the person; however, if the person cannot be found there is usually an alternate means of service. In certain types of cases, such as quieting title to real property, divorce, probate of a will, and others in which the thing involved is within the court's jurisdiction, a process may be served even though the person on whom it is to be served is outside the court's jurisdiction.

The means by which this is done is known as **constructive service.** This service on an absentee is made by publication of the process in a local paper. In most damage actions, however, absence of both defendant and property acts as a serious obstacle—action against the defendant would be pointless unless some recovery could be anticipated.

Jurisdiction of courts is also limited as to types of cases that they have authority to handle. A probate court, for instance, ordinarily has no authority to handle criminal cases; criminal courts usually do not handle civil suits.

A monetary limitation is placed upon many of the courts. Justice-of-the-peace courts and small-claims courts are limited to cases involving no more than the statutory limit—up to $5,000 in some states.

PRETRIAL PROCEDURE

As with many areas of law, the rules relating to procedural issues vary. In the federal civil courts, the Federal Rules of Civil Procedure guide the process under which lawsuits progress.

Pleadings

A lawsuit begins when a complaint is filed with the appropriate court. A complaint should set forth the facts giving rise to the cause of action asserted and should set forth the relief requested by the plaintiff.

Following its receipt of a complaint, the court issues a summons to the defendant. The summons is simply a notice to the defendant that he or she has been sued. A copy of the complaint may or may not accompany the summons, depending on the jurisdiction. If the complaint does not accompany the summons, it is made available to the defendant by the clerk of the court. If the defendant is served a summons but does not answer it, the result may be a judgment against the defendant by default. (See *Bage v. Southeastern Roofing* at the end of the chapter).

Once a complaint has been filed and the defendant has been served according to the statutes or rules relating to service of process, the defendant has a set period of time in which to answer the plaintiff's allegations. Usually, the defendant files an answer that responds to the allegations in the complaint. Generally, most answers deny at least some of the complaint's allegations (or there would be no dispute) and include affirmative defenses, which amount to reasons why the defendant should not be held liable. Often, a defendant also files one or more **counterclaims,** by which the defendant seeks a recovery from the plaintiff. (In federal court, the plaintiff then needs to file the plaintiff's "answer" to the defendant's counterclaim.)

A defendant also often includes various **motions** with the answer. For example, a defendant may contest the court's jurisdiction over him or her with a motion to dismiss for lack of personal jurisdiction. A defendant may test the legal sufficiency of the plaintiff's complaint by filing what is called a **motion to dismiss** for failure to state a claim on which relief may be granted. Such a motion, as one might guess, argues that the law does not recognize the claim asserted in the plaintiff's complaint.

Generally, the rules covering pleadings allow the allegations to be fairly general and do not require a great deal of specificity. Sometimes, however, a pleading will be so vague that it is difficult to formulate a response. In such situations, a motion for a more definite statement is appropriate.

Discovery

After the lawsuit begins and before trial, both sides usually engage in **discovery,** in which each side learns about the other's evidence, versions of the facts, and legal theories relating to the case. On December 1, 2006, new amendments to the Federal Rules of Civil Procedure went into effect. These amendments concern the discovery of "electronically stored information" or "ESI." More about ESI is discussed under the subheading "Request for the Production of Documents," below.

Depositions.   The Federal Rules of Civil Procedure provide a number of mechanisms for each side to learn about the other's case. One of the most commonly used discovery procedures is that of depositions on oral questions. A **deposition** is simply a procedure in which the witness, whether a party to the case or not, orally answers questions that are put to the witness by an attorney for one of the parties in the case. The questions, answers, and any objections by other attorneys, are all taken down and transcribed by a court reporter. Before the deposition begins, the witness takes an oath to tell the truth. The transcription thus provides a complete written record of the witness' testimony. More recently, litigants have begun to use videotape depositions. Videotape depositions have the advantage of providing a record of not only the questions and answers but also the facial expressions of the witness and any expressions of nervousness, such as sweating, excessive blinking, and fidgeting, and are often used at trial.

Request for the Production of Documents.   Another method of discovery is the use of a request for the **production of documents.** To use this method, one party prepares a request that lists the categories of documents that it wants the opponent to produce for inspection and copying. The opponent then reviews the list of categories and responds whether such documents will be produced. The use of such requests for the production of documents allows each side to see the other's documents, which are often quite revealing. Moreover, documents do not forget, as real people sometimes do. Thus, documents are often critical in discovery. The rules that govern the discovery of physical documents also apply to electronic documents. Due to the nature of electronic documents, however, a special committee devoted to the review of the Federal Rules, called the Advisory Committee on Civil Rules, worked extensively on modifying the Rules to accommodate electronic discovery. The Committee published a list of proposed amendments which were reviewed and approved by the United States Supreme Court. These new Rules are worth mentioning because they affect how companies comply with requests for production of documents as well as how companies retain electronically stored information for purposes of litigation.

Before December 2006, there was no formal acknowledgement that electronically stored information was discoverable. However, most discovery requests included the production of electronic files in all formats including email. Now, the Rules formally acknowledge electronically stored information as discoverable and cover all current formats of computer-generated information as well as information stored in any formats known in the future.

The Rules also require that the parties get together early in the discovery process to identify the scope and format of ESI that will be produced. The parties also are required to identify document retention issues as well as attorney-client privilege protection.

In the intervening years between the time when the Advisory Committee first identified ESI as a problem for discovery requests and December 2006, parties incurred significant expenses producing electronic documents which was often difficult to access or costly. For example, Gray Company, which has been incorporated since 1986, has, over the course of several decades, migrated its production data from one format to another, but has maintained its accounting system in a custom designed program format. Blue has sued Gray for breach of contract and fraud, requesting accounting information from Gray going back 10 years starting from when the companies first entered into their business agreement. Blue is requesting accounting data to be produced in the most current version of a popular spreadsheet program. Gray can respond to the request, but only if Blue will accept physical print-outs of the information. If Gray can show that the information Blue requests is not reasonably accessible because of undue burden or cost, the court may order discovery for such information in the requested format only for good cause.

Another frequent occurrence in the pre-December 2006 era was inadvertently produced attorney-client protected or trial preparation related information. Under the new Rules, if a party notifies the receiving party of the grounds for privilege, the receiving party must return, destroy or sequester that information.

Unfortunately, many companies do not have formal document retention policies. These policies set forth generally how and when the business destroys, archives, or retains documents for business purposes. When a company is sued, the plaintiff often asks for information relating to the defendant's document retention policy. Most companies have informal retention policies or only comply to retention guidelines mandated by government agencies such as the IRS or Department of Defense. Responsible companies draft document retention policies that include the retention of electronic information. In many instances, companies have incurred sanctions by a court when a party has destroyed or lost relevant documents. Under the new Rules, if, under the normal business operations of a company, information is overwritten or deleted, the court will not impose these sanctions. This "safe harbor" provision only applies if these normal business operations are conducted in "good faith." In addition, the Rules require the parties to institute a "litigation hold" procedure once a suit has been filed. This procedure requires that the parties inform employees that they should not delete, destroy or modify data that may be discoverable.

Interrogatories. Yet another type of discovery is the use of interrogatories. Interrogatories are essentially written questions that are submitted to the other side. After receiving a set of interrogatories, the other side must respond with any appropriate objections and with answers that are sworn to by the party or an officer of the party. Generally, interrogatories are much cheaper to use than taking the deposition of the other side, but the amount and quality of information received through interrogatories are generally much less and not as good as that received through depositions. Besides the types of discovery procedures just outlined, other procedures are available.

Pretrial Motions

Many cases are resolved short of a full trial. Perhaps the most common method of resolving a case before trial is through a motion for summary judgment. Under Federal Rules of Civil Procedure (FRCP) 56, a **summary judgment** is to be rendered by the court if the pleadings, depositions, answers to interrogatories, and any affidavits on file with the court show that "there is no genuine issue as to any material fact" and that the moving party is entitled to a judgment as a matter of law. To determine whether there are any genuine issues of material fact, the court must look to the substantive laws that relate to the claims asserted in the pleadings. For example, in a case involving a complaint that asserts breach of contract and a counterclaim that asserts fraud, the court will look to contract law and tort law regarding fraud to determine what the issues are. Once the court determines the issues, it looks to the evidence on file (much of which has been developed through discovery) and then determines whether there is evidence sufficient to raise issues of fact for each and every issue of the claim or counterclaim that is the subject of the motion. If there is no genuine issue of material fact for each of the elements of the breach of contract and fraud claims the court grants summary judgment for both.

Another motion that is often used to dispose of some or all of a case is the motion to dismiss for failure to state a claim on which relief may be granted. Essentially, this motion is used to take the position that, as a matter of law, there can be no recovery on the claim asserted, even if the facts alleged are true. This motion is often used when new legal theories are being asserted, such as when a court in one jurisdiction allows a certain type of claim that has not yet been recognized in the jurisdiction in which the case is pending.

Joining Additional Parties and Claims

The rules of procedure that apply to cases in the federal courts allow for joining additional claims (FRCP 18) and additional parties (FRCP 20) besides those involved in the plaintiff's complaint and the defendant's answer and counterclaims. The rules enable either the plaintiff or the defendant to assert as many claims as each party may have against the other, whether or not these claims are related to the main case filed by the plaintiff. The reason for allowing the parties to add multiple claims is the idea that no inconvenience can result from joining such matters in the pleadings but only from the trial of two or more matters that have little or nothing in common. Accordingly, the rules also provide for separate trials of the different claims or issues to avoid prejudice or to expedite the case, or merely for convenience.

The rules also allow either party to bring in additional parties. For example, suppose you burn your hand on a toaster. If you decide to sue the store that sold you the toaster, the store may then decide to bring the manufacturer of the toaster into the case as an additional party. The store will want to do this to hold the manufacturer responsible and to gain reimbursement for any liability to you. Likewise, the manufacturer may then proceed to bring in still more parties, such as other manufacturers who manufactured the heating control element or the wiring of the toaster. The manufacturer originally named may try to point the finger at other manufacturers of such component parts, arguing that those manufacturers were instead at fault. To pursue the toaster example even further, suppose that a manufacturer of the heating control component has a breach of contract claim against the manufacturer of the toaster. The component manufacturer can then pursue a breach of contract claim against the manufacturer of the completed toaster even though this claim has practically nothing to do with the original claim brought by you for personal injuries due to the toaster.

TRIAL PROCEDURE

The Jury

Parties to a case can choose whether they will have a jury decide the facts of the case. Both parties may agree to submit all the issues, including issues of fact, to the judge. If there are questions of fact (for example, "Did the plaintiff act reasonably?"), both parties must be in agreement if they are to dispense with the jury.

A trial jury is known as a **petit jury.** It usually consists of 12 persons but may consist of fewer people for civil cases or for crimes less serious than capital offenses.

Selection of the individual members of the jury involves both attorneys and the judge. Grounds for challenging prospective jurors are established by statute. The two opposing attorneys may use these grounds to disqualify prospective jurors being examined. For example, grounds for disqualification include a financial or blood connection between the prospective juror and one of the litigants. In addition to disqualifying a juror on the basis of some bias, each attorney may usually disqualify a limited number of prospective jurors arbitrarily by exercising the right of **peremptory challenge.** The judge supervises the qualification proceedings. When the jury has been impaneled, the case is ready for trial.

Courtroom Procedure

When the pleadings are complete, the case has come up on the court's docket, and the jury has been chosen and sworn in, the trial begins. Following opening statements by the attorneys, witnesses are sworn in and the evidence is examined. Each attorney sums up the case to the jury, the judge charges the jury, and the jury retires to reach a verdict. In the judge's charge to the jury, he or she sums up the case and instructs the jury about the issues to be decided. After reaching a verdict, the jury returns to the courtroom and the foreman announces the decision. The judge then enters the judgment, that is, the official decision of the court in the case.

New Trial or Appeal

Within a certain time after the judgment has been entered, a **new trial** or an appeal may be requested. Generally, a new trial is concerned with error in the facts of the case, whereas an **appeal** is concerned with a misapplication of the law.

A successful motion for a new trial may be made on the basis of an almost unlimited number of circumstances. It is argued that if a new trial is not granted, there will be a miscarriage of justice. The following are examples of reasons given for requests for new trials:

- Unfairness or bias in selection of members of the jury
- Prejudice stemming from financial or blood relationship between a jury member and a party to the trial
- Misconduct by a jury member, for example, talking with a witness or an attorney in the case
- Error by the judge in failing to allow evidence that should have been admitted or in admitting evidence that should have been excluded
- False testimony (perjury) of a witness
- Unforeseen accident preventing the appearance of a witness

If the motion to the trial court for a new trial is unsuccessful, appeal may be made to a higher court to order a new trial.

Either party may appeal a decision in a civil case; only the defendant may appeal from an adverse criminal judgment. Appeals are based on questions of law—the trial court has decided issues of fact.

Reasons for appeal are presented to the appellate court based on objections or exceptions to the ruling of the trial court. These objections are usually concerned with objections to and rulings on the admissibility of evidence, errors in the conduct of the trial, and instructions by the judge. Evidence presented and testimony taken are usually included along with the judge's instruction to the jury.

The **appellant** (the appealing party) is usually required to post an appeal bond when the case goes to a higher court. The purpose of the appeal bond is to insure that the appellant will pay court costs and the judgment to the appellee if the trial court decision is upheld.

The appellant prepares and files with the appellate court a brief of the case. The brief contains a statement of the case from the appellant's position and a list of errors forming the basis of appeal. The **appellee** (the party not appealing) then prepares and files a reply brief.

Equity Suits

An equity procedure may be different from that of a civil procedure. In an equity suit the judge decides both questions of fact and questions of law. There is no jury unless the judge specifically requires a jury recommendation on a question. Even then, the verdict of the jury is only a recommendation, and the final conclusion as to fact rests with the judge. If factual questions are long and involved, the judge may appoint a "master" to take testimony and make recommendations.

Equity acts *in personam*. The decree directs a person to do or not to do a certain thing. The decrees are either interlocutory or final. An **interlocutory decree** reserves the right of the court to act again in the case at some later time. A temporary injunction is an example of such a decree.

Cases

In both common law and equity the doctrine of *stare decisis* is followed. Similar prior cases are followed in deciding present issues. Because cases must be known to be followed, it may be useful to know how cases are recorded and reported.

The West Publishing Company publishes reports of all state cases that reach the appeal courts, and all federal cases. Consider the two cases at the end of this chapter. What you see at the top of the body of the case is called a **citation,** where you can see the names of the parties, where the case is reported, and the date the case was decided. The case of *Scordill v. Louisville Ladder Group,* 2003 U.S. Dist. LEXIS 19052 (ED La. 2003), is a case in which there are two plaintiffs, John and Cynthia Scordill. The defendant is Louisville Ladder Group. The case is reported in an online case reporting service, LEXIS, which is part of the ReedElsevier publishing company, and the legal half of the LEXIS/NEXIS service. The citation indicates that the case was decided in 2003 in a United States District Court. In this case, the United States District Court of the Eastern District of Louisiana. The number 19052 is a unique case indicator which, along with 2003 U.S. Dist. LEXIS, will bring you directly to this specific case if you conduct a cite search in LEXIS. *Bage, LLLC, Respondent, v. Southeastern Roofing Co. of Spartanburg, Inc.,* 646 S.E.2d 153. "The case of *Bage, LLLC, Respondent, v. Southeastern Roofing Co. of Spartanburg, Inc.* 373 S.C. 457; 646 S.E.2d 153; 2007 S.C. App. LEXIS 83 (Ct. App. 2007), is a case in which Bage and Southeastern Roofing were plaintiff and defendant, respectively. The case was reported in the 373rd volume of South Carolina Reports on page 457 as well as the 646th volume of the Southeastern Reporter, second series on page 153. In addition, the case was reported in LEXIS' database of South Carolina appellate decisions for 2007.

The case reporters used by the West Publishing Company for reporting state cases divide the United States into seven districts. Each district covers several states. In addition to the North Eastern Reporter, there is the Pacific Reporter, the South Western Reporter, the Southern Reporter, the South Eastern Reporter, the North Western Reporter, and the Atlantic Reporter. The cases in these reporters are state cases that have been appealed from lower court decisions. Such cases are important because appellate decisions become controlling law for that type of case in that state's courts. A particular decision in an appealed case may even be the basis for decisions on similar cases in other states or in the federal courts.

Another set of reporters covers federal case decisions. The Federal Reporter reports U.S. circuit court of appeals cases; U.S. district court opinions are found in the Federal Supplement; Supreme Court decisions are reported in the Supreme Court Reporter. Federal special court cases and decisions of administrative boards are found in other series of West volumes. In some instances, a case is reported exclusively by a reporting service such as LEXIS, or may only be available by the court.

TYPES OF EVIDENCE

Evidence is used to prove questions of fact. Each party must be willing to prove that what she said is true. The judge or jury then has the task of determining the true situation. Evidence is the means of establishing proof and, in federal cases, its admissibility is governed by the Federal Rules of Evidence.

In a criminal case the evidence must prove guilt "beyond a reasonable doubt" for the defendant to be found guilty. A civil case, by contrast, is won or lost on the comparative weight of the proof.

The burden of proof in a criminal case is always assumed by the state. In a civil case the burden usually rests on the plaintiff to prove his charge but, under certain circumstances, the burden may shift to the defendant. One such circumstance occurs when a counterclaim is made.

Real Evidence and Testimony

Evidence may be classified in many ways. It is classed as **real evidence** if it is evidence the judge or jury can see for themselves. For example, a fire extinguisher shown in court to be defective would be real evidence; so would a defective cable or a ladder that broke because of a defective rung. **Testimony** consists of statements by witnesses of things that have come to their knowledge through their senses. Testimony might be used to prove that a driver was operating a car unsafely and thus contributed to the cause of an accident.

Judicial Notice

Certain facts are so well known and accepted that the court will accept them without requiring proof. The court takes such **judicial notice** of logarithm tables, provisions of the federal or a state constitution, Newton's laws of motion, and the like. Evidence would not be required to prove that gasoline is combustible, but the presence of gasoline in a particular situation might require proof.

Conclusive and Prima Facie Evidence

Conclusive evidence is evidence that is incontestable. It is evidence that can in no way be successfully challenged. The existence of a written contract is conclusive evidence that someone wrote it.

Prima facie evidence is something less than conclusive. It is rebuttable. It is capable of being countered by evidence from the opposing side but, if allowed to stand, is sufficient to establish some fact. A signature on a written document would be evidence of this nature.

Direct and Circumstantial Evidence

Direct evidence goes to the heart of the matter in question. It is evidence that, if uncontested, would tend to establish the fact of the issue. A witness to a signature could establish the fact of signing by a particular party.

By contrast with direct evidence, **circumstantial evidence** attempts to prove a fact by inference. Circumstantial evidence might be used to show that the person now denying the contract would not have acted as he or she did at some past time unless he or she had entered into the contract. A net of circumstances is often woven to show a high probability that a particular version of an issue is true.

Best or Primary Evidence

The court will require the **best evidence** possible in a particular case; that is, the highest and most original evidence available. **Secondary evidence** will not be used unless, for some reason, the primary evidence is not obtainable. In a case involving a written document, for instance, the document is the best evidence of its existence and provisions. If the original document were destroyed or lost, a copy of the document or, if no copy exists, testimony as to its existence and contents probably would be admissible. The document, itself, would be best or primary evidence; testimony would be secondary.

Hearsay Evidence

Hearsay evidence is second hand. It is testimony about something that the witness has heard another person say. Most hearsay evidence is objectionable to the court for three main reasons:

- The person whose observation is quoted is not present in the court to be seen by the jury.
- The original testimony was not under oath.
- There is no opportunity for the original testimony to stand the test of cross-examination.

There are a few exceptions to the exclusion of hearsay evidence by the court. The rules vary somewhat in this respect from state to state. Hearsay is sometimes permitted where other evidence is completely lacking. Hearsay testimony about dying declarations, as well as excited utterances, are usually admissible. Business records are often admitted.

Parol Evidence Rule

A contract or statement that has been reduced to writing is the best evidence of the meanings involved. Testimony (**parol evidence**) that would tend to alter these meanings is objectionable. Only when the terminology is ambiguous or when unfamiliar trade terms are used will testimony be allowed to clarify the meaning, and then an expert witness may be called on for an interpretation.

The parol evidence rule is confined to interpretations of wording of a document; it does not apply to a question of the validity of the instrument. In a question about the reason why a person entered into a contract—for instance, a claim of duress—oral evidence would be allowed. Similarly, an attack on the legality of consideration offered would permit testimony.

Opinion Evidence

During the taking of testimony in a trial, the objection is occasionally heard that the "counsel is asking for a conclusion of the witness." An ordinary person appearing on the witness stand generally is not allowed to express opinions or conclusions in evidence. Such is the rule, but, as with many other rules of law, there are exceptions. In some instances the nature of the testimony requires opinions to be given—otherwise, the evidence will not be clear. In fact, just about any perception of anything is a conclusion based on sensory responses and experience.

The main objection to **opinion evidence** is that logical deduction or reasoning is being required of a witness. If such reasoning and conclusions are required in the progress of a trial, an expert from the field of knowledge involved should be called on to express an opinion about the facts or the meaning of a series of facts. If an opinion is necessary, it is desirable to have the best possible opinion.

Engineers are qualified by training and experience to act as expert witnesses in certain types of cases. Occasionally, it is by giving expert testimony that members of engineering faculties obtain fees that enable them to afford to remain teachers.

Testimony

The determination that a witness is competent to testify is part of the court's (judge's) function. Competence is determined by the witness's mental capacity or mental ability. The witness is required to testify only as to information having some bearing on the case at hand. If the testimony given under examination or cross examination gets too far afield, either the opposing attorney or the judge may object.

Irrelevancy refers to the lack of relationship between the issues of the case and testimony requested or given. In a case involving machinery specifications, a question pertaining to an engineer's home life is irrelevant.

The right of a witness not to testify on certain matters is known as **privilege.** According to the Fifth Amendment to the U.S. Constitution, no person may be compelled to testify against himself or herself. Communications between certain people need not be revealed in court. Examples of such privileged communications are those between husband and wife, doctor and patient, and attorney and client. Neither party may be made to testify unless the privilege is waived by the party affected by the trial.

THE ENGINEER AS AN EXPERT WITNESS

Most opinion evidence is excluded from a trial; the opinion of an average person acting as a witness is inadmissible. Opinion implies conjecture, and the law looks with disfavor on indeterminate factual situations. Nevertheless, such factual situations do arise and the truth in them must be determined as closely as possible. Such questions as the adequacy of design of a structure or the capabilities of a specially designed machine often must be answered.

Expert

Attorneys and judges are usually quite learned people. Knowledge of the law requires a broad general knowledge of many specialized fields to understand the factual situations presented in cases. Most lawyers, however, would admit that their knowledge of a particular technical field is quite general and that, therefore, they would be incapable of drawing intelligent conclusions on complicated technical questions. For such purposes an **expert** is needed.

An expert was once facetiously defined as "any person of average knowledge from more than 50 miles away." Such a criterion would hardly stand up in a court of law. An expert is a person who, because of technical training and experience, possesses special knowledge or skill in a particular field that would not be possessed by an average person.

Qualification before the court as an expert witness is part of the expert testimony. A seminal case regarding expert testimony is *Daubert v. Merrell Dow Pharmaceauticals, Inc.,* 509 U.S. 579, 113 S. Ct. 2786 (1993) where the Supreme Court considered what criteria must be met by expert testimony as well as expert reports in order to be admissable. During a *Daubert* hearing as well as at trial, the prospective expert witness can expect to be asked questions about his or her background, projects in which he or she has been involved, training, whether he or she is registered, and other similar questions. The court then determines whether that prospective witness is qualified to give expert testimony. Further discussion regarding expert testimony is provided in the case *Scordill v. Louisville Ladder Group, LLC* at the end of the chapter.

Expert Assistance

There are three important ways in which an expert may assist an attorney in a case involving a technical matter:

1. Advice and consultation regarding technical matters in the preparation of the case,
2. Assistance in examining and cross-examining technical witnesses, and
3. Expert testimony on the issues involved.

Advice and Consultation.   In engineering, as in other technical fields, it is necessary for a person who is thoroughly familiar with technical concepts to explain them to laypeople. A jury is made up of a cross-section of a community (or some approximation to that). Part of engineers' function as an expert witness is to present to their attorney and to the judge and jury the facts of the case in such a manner that laypeople can understand them. The minimum requirements for this communication are technical proficiency and the ability to effectively express the knowledge.

The basis of any effective presentation is investigation. Engineers, as expert witnesses, should be so thoroughly familiar with the facts of the case that nothing the opposition can propose will come as a surprise. Drawings and specifications may have to be carefully read, materials tested, and building codes or other laws examined. All should be analyzed for the presence of flaws, if engineers are to do an effective job. Assistance in the preparation of the attorney's brief requires the best of engineers' investigative powers. (See *Harborview Office Center, LLC v. Camosy Incorporated* at the end of this chapter and *D.A. Elia Construction Corporation v. New York State Thruway Authority* in Chapter 17).

An engineer's obligations to client and attorney requires an objective approach to the case. Not only should the facts to substantiate a client's claim be present, but opposing facts should also be shown. There are two sides to any controversy. In the interest of winning the case, opposing arguments must be considered and rebuttals prepared.

Trial Assistance.   In a case involving technical fields, both parties normally obtain experts to aid them and to testify in their behalf. In addition to assistance in preparation of a case and testimony in court, the expert may also be valuable in suggesting questions the attorney should ask. Direct-examination questions are prepared in advance, but most cross-examination questions and redirect and recross questions must be planned during the course of the trial.

A flaw in the technical argument posed by the opposition might escape an attorney's notice, but it should not escape the notice of technical experts in that field. The questions suggested may be aimed at the opposition by way of cross-examination, or they may be in the form of direct questions to be asked of the experts when they are put on the stand to counter the opposition's evidence.

Testimony and Depositions.   The judge, attorneys, and jurors are laypeople as far as the technical expert's field is concerned. Experts, therefore, must present their information in a simplified manner so that it will be readily understood by laypeople. Such presentation is quite akin to teaching. A simple foundation must first be laid and then the complexities built on it. The preparation of such a presentation is not always easy.

Such testimony necessitates comparison of technical principles with everyday occurrences and the use of pictures, slides, models, and drawings to make a meaning clear. Many hours of preparation are required for an hour of effective presentation. Each part of the presentation must be as nearly perfect as possible—incapable of being successfully questioned by opposing counsel.

It is almost essential for expert witnesses to be present at the entire trial if they are to be effective on the witness stand. Prior evidence established by opposing counsel may require alterations in the expert's presentation and changes in the attack on the opposing witnesses.

Honesty in answers is one of the prime requisites of any witness under examination. The opposing attorney will look (with expert assistance) for any point on which he or she can attack the expert's testimony. Once found, and properly worked on, a small loophole in the presentation can destroy the effect of laboriously developed testimony. If engineers do not know the answer to a particular question, the least damaging answer is a simple "I don't know." If the question involves prior testimony, witnesses may request the reading of that prior testimony before answering the question. If the question requires calculations or a consultation between witnesses and the attorney or other experts, time for such calculation or consultation may be requested from the court.

The demeanor of expert witnesses on the stand is important. Their appearance and answers to questions posed should inspire confidence in their ability. A professional appearance, professional conduct, and a professional attitude toward the entire proceedings come through clearly to the others in the courtroom. During cross-examination the opposing counsel usually tries to belittle or pick apart testimony damaging to the opponent's case; usually the more damaging the testimony, the greater the effort to reduce it to a shambles. Failing this, the attorney may attempt merely to enrage witnesses, in hopes that an opening in the testimony may occur.

Calm and considered answers by witnesses are the best defense against the opposing counsel's attack. Courtesy and self-control must be exercised.

Reference to writings by acknowledged authorities in the technical field is advisable. Often it seems that a quotation excerpted from a textbook has more weight than an oral statement on the witness stand by the author.

If a witness is unable to attend a trial to testify, then testimony may be taken in some place other than the courtroom prior to the trial. The testimony is taken under oath and recorded for use in the trial. Both opposing attorneys must be present, and the rights to direct examination, cross-examination, redirect, and recross are the same as they would be in the courtroom. At the time of the trial the deposition is read into the court record and becomes part of the proceedings.

A deposition is generally considered to be less effective than testimony given in open court where jurors can see and hear the presentation even though it is videotaped. However, there are times when it is the only means available, and is far preferable to the alternative of omission.

Deciding Whether to Take the Case

Engineers who are asked to be expert witnesses should decide whether they really believe in the prospective client's position. The quality of support rendered by engineers in the case often depends on how firmly they believe in the case of their client. On the witness stand, engineers will be required to tell the truth of the case as they see it. Conviction that their client is right is often apparent in the manner in which the testimony is given.

Fees

The fee engineers charge for acting as expert witnesses in a case should correspond to the fees they would normally charge for other consulting work. Charges should be based on the amount of time required and the relative importance of the engineers' role in the proceedings. In no case should the fee be contingent on the outcome of the case. Contingency fees are frowned on by the court as tending to cause inaccurate testimony; opposing counsel will not hesitate to take full advantage of this arrangement.

The fees for professional engineering services vary depending on where the engineer's business is located, the engineer's expertise, and the services involved. A typical minimum charge in 2003 ranges from $175 to $300 per hour or so, in addition to such expenses as required travel and hotel accommodations.

BAGE, LLC RESPONDENT, V. SOUTHEASTERN ROOFING CO. OF SPARTANBURG, INC., 373 S.C. 457; 646 S.E.2d 153; 2007 S.C. App. LEXIS 83 (Ct. App. 2007)

Study terms: Service of process, jurisdiction, default judgment, agent express and implied warranties, negligence

Southeastern Roofing appeals the circuit court's order finding the company was properly served by service of process on its employee, Debbie Green, and determining that it failed to show good cause to allow relief from an entry of default under Rule 55(c), SCRCP. We affirm.

Factual/Procedural Background

In August 2003, BAGE, L.L.C. ("BAGE") entered into a written contract with Southeastern Roofing Company of Spartanburg, Inc. ("Southeastern Roofing"). Under the agreement, Southeastern Roofing was to perform re-roofing work on a commercial office building BAGE owned. More specifically, the company was to remove the outer layers of the existing roof system and install of a new, modified bitumen roof on the structure. Southeastern Roofing was to immediately commence the project after the contract was signed and to complete the job within approximately six weeks. Work on the roof, however, did not begin until the end of October 2003 and continued only sporadically through the winter and into the spring of 2004.

From almost the moment Southeastern Roofing started operations on BAGE's building, significant leaks in the roof began to occur. BAGE repeatedly contacted Southeastern Roofing, demanding the leaks be stopped. Despite BAGE's requests, the necessary repairs were never made and water continued to infiltrate and further damage the building. BAGE ultimately filed suit, claiming breach of contract, breach of express and implied warranties, and negligence.

At the fledgling stages of the litigation, BAGE's counsel spoke with Southeastern Roofing's general manager, Jamie Cubitt, who initially agreed to accept service of process by mail. When Cubitt failed to return the summons and complaint, BAGE sought to serve the company through its registered agent. After discovering that the agent listed with the Secretary of State was no longer affiliated with Southeastern Roofing, BAGE pursued service via a private process server.

On July 9, 2004, the process server arrived at Southeastern Roofing's office with the intendment of serving Cubitt with BAGE's summons and complaint. Cubitt was not in the office at that time, and the server was instead met by Debbie Green, another Southeastern Roofing employee. Green signed for the service of process.

Upon returning to the office, Cubitt instructed another employee, Cheri Barnette, to send a copy of the summons and complaint to Southeastern Roofing's insurance agency. These documents were faxed to the insurance company on July 13, 2004. No cover letter was included in this facsimile nor was any follow-up with its insurance carrier ever taken by Southeastern Roofing.

Southeastern Roofing never responded to the complaint. BAGE filed an affidavit of default and motion for an entry of default on September 7, 2004. That same day, an entry of default was dated and filed with the court. By a motion filed on September 20, 2004, Southeastern Roofing moved to set aside the order granting the entry of default.

A hearing on the motion to set aside the entry of default was held before the circuit court on December 8, 2004. Southeastern Roofing argued (1) the service of process had been improper and thus deprived the court of personal jurisdiction and (2) "good cause" existed to set aside the entry of default under SCRCP Rule 55(c).

By an order dated April 28, 2005, the judge denied Southeastern Roofing's motion to set aside the entry of default. In regard to service of process, the order specifically found (1) Green was an office manager at Southeastern Roofing for the purposes relevant to service of process and (2) Green had Cubitt's specific authorization to accept service of process. With respect to the Rule 55(c) motion, the judge found Southeastern Roofing had failed to show good cause as to allow relief from the entry of default. Southeastern Roofing timely moved for reconsideration of the order denying its motion to set aside default. This motion was denied.

On April 27, 2006, following a damages hearing before the Richland County master-in-equity, BAGE obtained a default judgment against Southeastern Roofing in the amount of $1,151,888.84. This judgment was properly filed with the court on May 3, 2006.

Standard Of Review

"Questions of fact arising on a motion to quash service of process for lack of jurisdiction over the defendant are to be determined by the court." *Brown v. Carolina Emergency Physicians, P.A.*, 348 S.C. 569, 583, 560 S.E.2d 624, 631 (Ct. App. 2001). . . .

In reviewing a trial judge's exercise of discretion, the issue before an appellate court is not whether it believes good cause existed to set aside the entry of default, but whether the trial judge's determination is supported by the evidence and not controlled by an error of law. *Pilgrim v. Miller*, 350 S.C. 637, 640-41, 567 S.E.2d 527, 528 (Ct. App. 2002).

LAW/ANALYSIS

I. Service of Process

Southeastern Roofing argues the delivery of process to Green was insufficient as a matter of law. We disagree. . . .

Our supreme court has enunciated:

> Service on a corporation may be made by hand delivering a copy of the summons and complaint to an officer of the corporation or to an authorized agent of the corporation. Rule 4(d)(3), SCRCP.

■ ■ ■

Roche v. Young Bros., Inc. of Florence, 318 S.C. 207, 210, 456 S.E.2d 897, 899-900 (1995).

■ ■ ■

Not every employee of a corporation is an "agent" of the corporation for the purposes of service of process. *Brown v. Carolina Emergency Physicians,* P.A., 348 S.C. 569, 583-84, 560 S.E.2d 624, 631-32 (Ct. App. 2001). If the employee in question is not a managing or general agent, the question is whether the individual possessed "specific authorization to receive process." *Id.*

■ ■ ■

This court addressed a factually similar issue in *Schenk v. National Health Care,* Inc., 322 S.C. 316, 471 S.E.2d 736 (Ct. App. 1996). In Schenk, the plaintiff attempted to serve process on the defendant's registered agent by means of a private process server. When the server took the summons and complaint to the address recorded with the secretary of state, an individual different from the one on file, Carol Grant, informed him the listed agent was no longer there and had retired from the company. Grant signed for the papers, assuring the server that she was duly authorized to accept service as the office manager. She assured him that she would pass the documents along to the company's corporate headquarters. The defendant never answered the complaint and an entry of default ensued. A motion to set aside default for improper service was denied by the circuit court. On appeal, this court found service of process had been effective and affirmed the circuit court. Noting Grant was the defendant's office manager and that she assured the process server of her authorization to accept service of process, we concluded she was a managing agent for the defendant under Rule 4(d)(3), SCRCP. *Id.* 322 S.C. at 319-20, 471 S.E.2d 738.

Although, "[w]ithout specific authorization to receive process, service is not effective when made upon an employee of the defendant, such as a secretary," *Brown,* 348 S.C. at 584, 560 S.E.2d at 632 (citing *Moore,* 322 S.C. at 523-24, 473 S.E.2d at 67), Green indubitably served in a much greater capacity at Southeastern Roofing than simply that of a secretary. The testimony given by Green during her deposition clearly refutes the contents of her prior affidavit submitted by Southeastern Roofing and its assertion that she was only a "secretary/receptionist." During Green's deposition, the following colloquy took place... [which] demonstrates Green's specific authorization to accept the service of process:

Q. . . . Did you tell Mr. Cubitt that the person that was there had a summons and complaint for Mr. Cubitt?

A. Yes, Sir. I just told him that he had some—the server had some papers from him for some—for him to sign for.

■ ■ ■

Q. Did Mr. Cubitt indicate to you it was ok if you—

A. Yeah.

Q. —Accepted the papers?

A. Mm-Hmm (Affirmative Response).

In his order denying the motion to set aside the entry of default, the circuit judge found:

■ ■ ■

3. After examination by counsel for Plaintiff in her deposition, and on further examination by counsel for Defendant, Ms. Green reconfirmed her conversation with Mr. Jamie Cubitt about being "okayed" to "receive whatever it was that the server was there to give you."

■ ■ ■

Furthermore, we would note the efficacy and rationale behind the requirement for service of process—hat the defendant has notice of the proceedings—has luculently* been met in this case. . . . The history and dealings between the parties through July 9, 2004 reveal that BAGE's counsel spoke with Cubitt in regard to this suit and that BAGE attempted to serve process through the postal service. Accordingly, it seems disingenuous for Southeastern Roofing to claim Cubitt, in his conversation with Green, was unaware the papers being served by the process server were the summons and complaint at issue. Moreover, Southeastern Roofing's manager readily admits to having knowledge of the summons and complaint long before the deadline to answer passed. In his affidavit, Cubitt states: "[O]n or about July 10, 2004, I was notified that a Summons and Complaint concerning this case was left at SE Roofing's place of business the prior day."

II. Entry of Default

Southeastern Roofing argues the court erred in not setting aside the entry of default as provided under Rule 55(c), SCRCP. We disagree.

South Carolina Rules of Civil Procedure provide: "For good cause shown the court may set aside an entry of default . . ." Rule 55(c), SCRCP. Thus, under the rule, the standard for granting relief from an entry of default is "good cause," *Wham v. Shearson Lehman Bros., Inc.*, 298 S.C. 462, 465, 381 S.E.2d 499, 501 (Ct. App. 1989), and is more lenient than the standard for granting relief from a default judgment under Rule 60(b) SCRCP.

■ ■ ■

The decision of whether to grant relief from an entry of default is solely within the sound discretion of the trial court. *Wham*, 298 S.C. at 465, 381 S.E.2d at 501. An abuse of discretion in setting aside an entry of default arises when the judge issuing the order was controlled by some error of law or when the order, based upon factual, as distinguished from legal conclusions, is without evidentiary support.

■ ■ ■

In this case, over two months elapsed between the time Southeastern Roofing was served with the summons and complaint and when it moved for relief. Although the summons and complaint were allegedly sent to Southeastern Roofing's insurance agent, any negligence by the insurance agent in handling these documents is imputable

---

* [Author's note: clearly or lucidly]

to Southeastern Roofing. Pursuant to the findings of the circuit court, Southeastern Roofing mishandled the service of process of this lawsuit:

> It is apparent from the Affidavits filed on behalf of the Defendant that Defendant simply failed to give this matter the proper attention which needed to be given when an entity is served with a Summons and Complaint. . . .

Based on the alleged defective workmanship of Southeastern Roofing, the circuit court did not believe Southeastern Roofing had a meritorious defense.

The circuit court found the evidence did not show the existence of good cause as to allow Southeastern Roofing relief from the entry of default. In making this determination, the circuit judge relied upon *Stark Truss Co. v. Superior Construction Corp.*, 360 S.C. 503, 602 S.E.2d 99 (Ct. App. 2004). In Stark, the circuit court refused to relieve the defendants from an entry of default, finding they had failed to present sufficient proof of good cause under Rule 55(c), SCRCP. The Court of Appeals affirmed. Noting the defendant's only explanation for not filing an answer within thirty days was that the company's president was "struggling with some depression and had a lot things slip through his fingers" and gave no reason why its attorney failed to file an answer, this court found "there was evidence to support the circuit court's refusal to set aside the entry of default." *Id.*, 360 S.C. at 510, 602 S.E.2d at 103.

As determined by the circuit judge, Southeastern Roofing's failure to properly respond to the summons and complaint resulted from its failure to take proper actions in order to assure a response was timely filed. The company offers no good or valid explanation as to why it failed to respond. There is no indication that there was any follow-up by the employees of Southeastern Roofing with its insurance agent after the summons and complaint were allegedly faxed to the agent. It is axiomatic that Southeastern Roofing simply did not give this matter the proper attention it required.

Conclusion

We come to the ineluctable* conclusion Green qualifies as both a managing and authorized agent of the appellant. Concomitantly, the appellant was properly served when the summons and complaint were delivered to Green as an agent of the appellant. There is copious evidence in the record to support the circuit judge's finding that appellant did not prove the existence of good cause to allow the court to set aside the entry of default. We hold there was no abuse of discretion by the trial judge in refusing to grant the appellant relief under SCRCP, Rule 55(c).

Accordingly, the decision of the circuit court is **AFFIRMED.**

JOHN SCORDILL, III, and his wife CYNTHIA SCORDILL v. LOUISVILLE LADDER GROUP, LLC et al 2003 U.S. Dist. LEXIS 19052 (ED La. 2003)

Study terms: Daubert factors, expert testimony, generally accepted engineering principles

Before the Court are the motions of defendant, Louisville Ladder Group, LLC, to exclude the expert testimony of Greg Garic and for summary judgment. Defendant Louisville Ladder filed two motions for summary judgment. . . . For the following reasons, the Court denies defendant's motion to exclude the expert testimony of Greg Garic, grants defendant's motion for summary judgment on plaintiffs' inadequate warning and design defect claims, and denies defendant's motion for summary judgment on plaintiffs' manufacturing defect claim.

---

* [Author's note: inescapable or unavoidable.]

I. Background

John Scordill is a welder. He purchased two stepladders at a Home Depot in 1997 or 1998. Both ladders were Davidson Model 592-61 stepladders, ladders that stand six feet tall and that are made of fiberglass rails and aluminum steps. The ladders were manufactured in 1996 by Louisville Ladder at a manufacturing facility in Monterrey, Mexico. While working on a job in Orleans Parish on February 16, 2002, Scordill placed one of the stepladders—the incident ladder—alongside a wall. He climbed up to the second rung of the ladder and turned around so that his back was to the ladder. He then reached up with his right hand to weld an I-beam to metal plates that had been installed the day before. He leaned his left elbow against the wall to steady himself, his left hand grabbing his right wrist to support the welding gun in his right hand. Scordill avers that the ladder then buckled beneath him. Scordill fell and sustained numerous injuries.

Plaintiffs, John and Cynthia Scordill, sued defendant Louisville Ladder in state court, alleging claims of unreasonably dangerous manufacturing, unreasonably dangerous design, and failure to adequately warn. Plaintiffs assert that the ladder failed along its left front rail, just below the first rung of the ladder. Defendant removed the case to this Court. Defendant moves the Court to exclude the report and testimony of plaintiffs' expert witness, Greg Garic. Defendant also moves the Court for summary judgment on each of plaintiffs' claims.

II. Motion To Exclude Expert Testimony

The Federal Rules of Evidence govern defendant's motion to exclude the report and testimony of plaintiffs' expert, Greg Garic. Rule 702 provides:

If scientific, technical or other specialized knowledge will assist the trier of fact to understand the evidence or to determine a fact in issue, a witness qualified as an expert by knowledge, skill, experience, training, or education, may testify thereto in the form of an opinion or otherwise, if (1) the testimony is based upon sufficient facts or data, (2) the testimony is the product of reliable principles and methods, and (3) the witness has applied the principles and methods reliably to the facts of the case.

FED. R.EVID. 702. This rule applies not only to testimony based on scientific knowledge, but also to testimony of engineers and other experts that is based on technical or specialized knowledge. The rule requires the trial court to act as a "gate-keeper," ensuring that any scientific or technical expert testimony is not only relevant, but also reliable. *See Daubert v. Merrell Dow Pharmaceuticals, Inc.*, 509 U.S. 579, 589, 125 L. Ed. 2d 469, 113 S. Ct. 2786 (1993).

Defendants assert, first, that Greg Garic is not qualified to be an expert and, second, that his proffered testimony is not reliable. As to his qualifications as an expert, Garic concedes that he has neither designed nor manufactured a ladder. Garic is only vaguely familiar with the standards that American National Standards Institute ("ANSI") has set for ladders, and he has never tested a ladder for compliance with these standards. . . .

On the other hand, Garic does have knowledge and experience that bears on why the ladder collapsed. Garic has a master's degree in mechanical engineering. He worked for NASA for 15 years, where he was honored for his contributions to the space shuttle program. He is currently employed by Stress Engineering Services, Inc., and he has significant experience in stress analysis. Garic's primary areas of practice are stress analysis, fracture mechanics, finite element modeling, and fitness for service assessment. . . . Defendants do not dispute Garic's expertise in stress analysis. Accordingly, the Court finds that Garic is qualified to testify as an expert in mechanical engineering and stress analysis.

Next, the Court turns to defendant's argument that Garic's expert report and testimony are not reliable. In *Daubert*, the Supreme Court identified factors that bear on the issue of reliability, including: "(1) whether the expert's theory can be or has been tested; (2) whether the theory has been subject to peer review and publication; (3) the known or potential rate of error of a technique or theory when applied; (4) the existence and maintenance of standards and controls; and (5) the degree to which the technique or theory has been generally accepted in the scientific community." A Rule 702 inquiry into the reliability of expert testimony is a flexible and necessarily fact-specific inquiry. . . . Plaintiffs, as the party offering the expert, bear the burden of proving by a preponderance of the evidence that the proffered testimony is reliable.

The Court notes that its role as a gatekeeper does not replace the traditional adversary system and the place of the jury within the system. As the *Daubert* Court noted, "vigorous cross-examination, presentation of contrary evidence, and careful instruction on the burden of proof are the traditional and appropriate means of attacking shaky but admissible evidence." . . . As a general rule, questions relating to the bases and sources of an expert's opinion affect the weight to be assigned that opinion rather than its admissibility and should be left for the jury's consideration.'" *United States v. 14.38 Acres of Land, More or Less Sit. in Leflore County, Miss.*, 80 F.3d 1074, 1077 (5th Cir. 1996) (*quoting Viterbo v. Dow Chemical Co.*, 826 F.2d 420, 422 (5th Cir.1987)).

A. Testimony of Greg Garic

To prepare his expert report in this case, Garic examined the ladder involved in the incident—the incident ladder. Garic reviewed Scordill's deposition testimony and also met with Scordill to further clarify his description of the accident and the events leading up to it. Garic also examined the second ladder—the exemplar ladder—that Scordill purchased in 1997 or 1998 along with the incident ladder. . . .

In his expert report and deposition testimony, Garic identifies three manufacturing defects. First, Garic determined that the step-rivets for the bottom two steps of the incident ladder were closer to the edge of the fiberglass flange than allowed by the manufacturer's specification. Garic observed that the exemplar ladder contains the same alleged manufacturing defect. Second, Garic observed a split in the incident ladder rail's flange-to-web junction. It is undisputed that the incident ladder did not exhibit such a split on the morning of the incident. The exemplar ladder, however, exhibits cracking in the fiberglass in the same area where the fiberglass of the incident ladder split. As a result, Garic concluded that the cracking is the result of poor manufacturing practices and that even if the cracking was not visible in the incident ladder before the incident, its flange-to-web junction did not comply with manufacturing specifications. Third, Garic observed that at several points along the left rail, the height of the flange fell slightly below the lower tolerance limit of 1.109 inches. . . .

In his supplemental report, Garic ultimately concludes that "the likely root cause of the failure is the cracking of the fiberglass in the inside corner of the rail, as was observed in the exemplar ladder." He further opines that "had the rivet been solidly positioned, additional support would have been provided to the step, which could have prevented or slowed its deflection and ultimate failure."

B. Reliability

To determine whether an expert's testimony is sufficiently reliable, the Court should first consider whether the *Daubert* factors noted above are appropriate, and then it can consider whether other factors are relevant to the case at hand. Defendant argues that Garic failed to apply the scientific method by not properly testing his hypotheses regarding the cause of Scordill's accident. As Garic noted in his deposition, however, he developed and then analyzed various hypotheses regarding the initial point of failure of Scordill's accident. He applied generally accepted engineering principles to project the logical sequence of events from each potential initial point of failure and then determined whether the observed damage was consistent with the mechanics of each potential accident scenario. He indicated that through this process, he eliminated certain hypotheses and refined his conclusions as much as possible given the informational constraints. Also, as discussed above, Garic concluded that additional testing, such as finite element analysis, would not be feasible or meaningful in this case. Plaintiffs further contend that they cannot reasonably conduct meaningful testing of the incident ladder because it was destroyed in the accident, and any subject ladder utilized in testing would materially differ from the incident ladder, thus producing irrelevant results. Addressing the second *Daubert* factor, the defendant contends that nothing Garic has done could withstand peer review. Because Garic's opinion is based on the very specific facts of this case, it does not lend itself to peer review. Garic has not generated a study that is subject to repetition but instead has applied generally accepted engineering principles and concepts utilized in stress analysis to the facts of this accident. As a result, the Court concludes that the second *Daubert* factor does not apply. The Court also finds that the third and fourth *Daubert* factors are inapplicable, because they involve the known or potential rate of error of the technique utilized by the expert and whether there exist standards controlling the technique's operations. In this case, Garic did not develop a particular scientific procedure to reach his conclusions. The fifth standard is applicable to the extent that Garic's analysis applied generally accepted engineering principles, in addition to his own training and experience in the field of stress analysis, to the factual situation specific to this case. . . .

In his supplemental expert report, Garic describes how he reached his conclusions by applying his education, skill and experience to Scordill's description of the sequence of events and his observations of the incident ladder. . . . The Court finds that Garic 's testimony can assist the jury in understanding the evidence and in making its finding of facts in this case.

Louisville Ladder quarrels with the bases and sources of Garic's opinions and offers conflicting expert testimony. As noted above, however, the questions raised by the defendant should affect the weight that the jury gives to the testimony rather than its admissibility. . . . The Court finds that Garic's testimony is sufficiently relevant and reliable to reach the jury. As a result, the Court denies defendant 's motion to exclude Garic's testimony. . . .

IV. Conclusion

For the foregoing reasons, the Court denies defendants' motion to exclude the expert testimony of Greg Garic. Further, the Court grants defendants' motion for partial summary judgment and dismisses with prejudice plaintiffs' inadequate warning and design defect claims, and the Court denies defendant's motion for summary judgment on plaintiffs' manufacturing defect claim.

HARBORVIEW OFFICE CENTER, LLC, Plaintiff-Appellant, v. CAMOSY INCORPORATED, PARTNERS IN DESIGN ARCHITECTS, INC., ET AL. 2006 WI App 56; 290 Wis. 2d 511; 712 N.W.2d 87; 2006 Wisc. App. LEXIS 149

Study terms: Discovery, spoliation of evidence, imputation, egregiousness, prejudices, interrogatories, request for production of documents, document retention

OPINION

Harborview Office Center, LLC appeals from trial court judgments dismissing its causes of action against Camosy Incorporated, Partners in Design Architects, Inc. (PID), Bollig Lath & Plaster Company, Inc. and Klein-Dickert Milwaukee, Inc (collectively, the "respondents"). The trial court dismissed Harborview's claims as a sanction for its spoliation of evidence. Because the trial court properly applied the correct legal standard for egregiousness to the essential facts of record and reasonably concluded that dismissal was an appropriate sanction, we affirm the judgments against Harborview.

BACKGROUND

As the trial court noted, the record in this case is voluminous. . . .

In early 1997, Harborview entered into an agreement with Camosy whereby Camosy agreed to serve as general contractor on a project to construct a three-story office building in Kenosha, known as the Harborview Office Center. Each of the remaining respondents provided services that contributed to the construction of the Center. Camosy constructed the building based upon the plans and specifications PID, the architectural firm, prepared. Klein-Dickert installed the aluminum windows used in the exterior skin of the building. Bollig installed the Exterior Insulation and Finishing System (EIFS). Dryvit Systems, Inc., which is not a party to this action, supplied the EIFS used in the construction of the building.

The EIFS consists of several layers of materials designed to enclose, insulate and waterproof a building. The first EIFS component that is installed is a water resistant gypsum sheathing called Dens-Glass Gold. The Dens-Glass is screwed directly to the metal studs of the building. After installation, the joints of the Dens-Glass are sealed with a fiberglass tape. The next component is a water resistant coating called Backstop, which is applied to the surface of the Dens-Glass. If this water resistant coating is applied properly, it is virtually impossible for water to penetrate the Dens-Glass. The third component is the waterproof adhesive, which is troweled on to the Backstop membrane. The fourth component is the expanded polystyrene layer (EPS), which is plastic foam board insulation approximately one and one-half inches thick. The EPS is pressed into the waterproof adhesive. Next, of particular interest in this case, three-quarter inch deep horizontal and vertical V-shaped grooves are cut into the EPS to add architectural interest. Then, a waterproof base coat and embedded fiberglass mesh are simultaneously applied to the entire face of the EPS boards and over the V-grooves. Finally, for appearance purposes, a stucco-like finish coat is applied over the base coat.

Construction of the Center was completed in late 1997. The first signs of water infiltration were discovered in late 1997 or early 1998, prior to tenant occupancy. Harborview discovered additional water leaks with each new rainstorm. Each of the respondents participated in varying degrees in the early efforts to identify the causes of the leaks and to remedy the problems. While these efforts reduced the water infiltration, they did not eliminate it completely.

On December 14, 2001, Harborview filed suit against Camosy, PID, Bollig and Klein-Dickert Co., Inc., arguing negligence and breach of contract. The initial complaint and the two amended complaints that follow alleged:

> [Harborview] has experienced significant and recurring water infiltration problems at the office building since construction was completed in early 1998. With nearly every rain, windows leak, and over time ceiling tiles and carpet have become stained, window sills have separated and drywall has cracked and softened, and other inconveniences and problems have occurred.

Harborview claimed that it would be necessary to remove and replace all of the office windows to resolve the water infiltration problems. The respondents filed several cross-claims against each other seeking contribution and/or indemnification.

Over the next two years the parties retained experts to comment on the validity of the allegations pertaining to water infiltration, to identify the design and construction deficiencies that were the causes of the water infiltration and to evaluate the process required to correct those deficiencies. In August 2002, Harborview retained an architect and engineer, Brian Fischer. Fischer would later oversee the remediation project. In his March 2003 report, Fischer listed several deficiencies he thought contributed to the water infiltration.[2] Fischer concluded that in order to correct these deficiencies it would be necessary to reinstall and/or replace parts of the window system. Fischer later testified that at this point he did not consider that cracks in the V-grooves were the cause of the water infiltration.

■ ■ ■

---

[2]Fischer listed the following six deficiencies: (1) the installed sill flashings which were not the specified sill flashings, (2) the sill flashings had no "end dams," (3) the upturned legs on some of the sill flashings were bent or crushed by window installation, (4) the head flashings did not have "vertical legs," (5) the sealant around the windows was improperly applied on the EIFS finish coat instead of the recommended base coat, and (6) the sealant joints around the windows were improperly constructed.

On May 22, Harborview, by way of letter, invited the respondents to be present at an inspection and demonstration of the work to be performed during the remediation project. The letter states that the primary purpose of the demonstration was

> to inspect the condition of the [EIFS] at the window opening and attempt to identify the most appropriate and efficient means to remove the existing sealant and finish coat of the [EIFS.] We also anticipate that we will remove a portion of the [EIFS] at its interface with the pre-cast concrete element at the base of the building in order to determine how the [EIFS] was terminated. This will permit us to identify appropriate and effective repair procedures. In the process of removing the window, we anticipate that we will inspect the head jamb and sill flashing.

Harborview conducted the demonstration on May 28, 2003, with the respondents, several of the experts and counsel in attendance.

According to the respondents' trial briefs and arguments, the remediation project would proceed as follows: After the windows were replaced, the EIFS would be ground down inside the window jambs so that the area could be primed and sealant applied between the smoothed surface of the EIFS and the aluminum frame of the window. A five-inch high section of the EIFS around the entire perimeter of the Center would be removed for the purpose of correcting the metal flashing that lay between the lower edge of the EIFS and the top of the concrete apron surrounding the building. Fischer later explained in his deposition that the original remediation project contemplated some work on the V-grooves, but it would be primarily for cosmetic purposes, "in some of those areas . . . they would grind out some of this finish coat material to try to improve some of the workmanship, some places where it was too thick, some places when it was troweled that was pulled over. They were primarily going to do that for aesthetic purposes."

Thomas Bychowski, the project manager for the company involved in the reglazing of the Center during the remediation project, testified that when he initially went out to the site to measure the windows, he noticed that there were cracks in the V-grooves . . . He apparently brought the concerns up to Harborview's ownership and suggested water testing as a remedy. Harborview's ownership informed him that they had done enough water testing and wanted to move forward with remediation. He was told that the cracks were "superficial" and an EIFS installer and expert had analyzed the cracks.

The remediation project commenced with job site preparations in early July. On July 17, 18 and 21, Fischer and other contractors involved in the remediation process surveyed the EIFS on the exterior of the building. At the conclusion of the survey, Fischer had not concluded that the cracks noted in the EIFS were a source of water infiltration. He testified that

> when we did that survey that all of the places where there were cracks, we had not concluded that those were necessarily sources of water entry, that they might simply be hairline cracks in that finish coat, so that if they ground those down and cleaned them up, that would take care of that situation.

The first windows were removed on July 24. The first new windows were installed on August 4. Following the installation of the windows, Fischer conducted water tests to ensure that there was no further water infiltration into the Center.

The water tests were conducted in areas of twelve windows each. The areas were designated as Area I through Area XIV. The first water test was conducted on Area I on August 19 ... On August 25, Area III was water tested and failed. Further investigation revealed water behind the underlying foam insulation. Some of the V-grooves in that Area were reworked and on August 26 another water test was conducted. For the second day in a row, Area III failed the water test.

By this point, given the results of the water testing, Fischer concluded that the defects in the V-grooves were a possible source of water infiltration in the building. He reported this to Harborview ownership. Fischer then gave the orders to rework all of the horizontal and vertical V-grooves, in part, because of the possibility that water was intruding through the cracks in the V-grooves. The rework of the V-grooves included grinding off the finish coat, placing new mesh and base coat into the area and then refinishing. On August 28, after its V-grooves had been reworked, Area III passed the water testing.

From that point forward, all V-grooves in a given area were reworked and new windows put in place prior to water testing that area. By September 12, sixty to seventy percent of the grinding work on the V-grooves was complete. By September 18, the reworking of the V-grooves on the building was "substantially completed," with ninety-eight percent of the V-grooves ground down.

On September 17 and 18, the parties conducted the depositions of Debra Hertzberg, Neil Guttormsen and Charles Gierl, each of whom is a Harborview principal. Hertzberg, Guttormsen and Gierl testified that somewhere between two and three weeks prior to their depositions, Fischer recommended that the exterior of the EIFS be repaired. Both Guttormsen and Gierl testified that the contractors had indicated that there were cracks in the EIFS and that these defects potentially were a source of water infiltration. Hertzberg, Guttormsen and Gierl authorized Fischer to go ahead with the additional work.

On September 19, Bollig sent Harborview a letter stating that during the depositions on September 17 and 18, it was disclosed that more extensive work was being performed upon the EIFS than originally proposed, demonstrated and disclosed. On September 30, Harborview wrote to Bollig asserting that Harborview's "recent letter contained several misunderstandings." Harborview alleged that the parties had long known that cracks in the EIFS were a source of water infiltration and that repair work was going to be done on those cracks.

On November 5, Bollig forwarded to Harborview a set of interrogatories and a request for the production of documents. Bollig specifically asked Harborview for information pertaining to defects in the EIFS. Harborview failed to answer the discovery requests or seek an extension; therefore, on January 7, 2004, Bollig filed a motion to compel. That same day Bollig received Harborview's responses to the discovery requests. Bollig alleged that these responses were "purposefully vague." The court heard the motion to compel on January 16, 2004. The court rejected as untrue Harborview's claim that no new defects were discovered during the remediation project. The court granted Bollig's motion to compel and, among other things, ordered Harborview to respond to Bollig's discovery requests and pushed back the date of trial.

In September 2004, after months of further discovery, the respondents filed motions to dismiss for spoliation of evidence. Following oral arguments by the parties, the court granted the respondents' motions. The court acknowledged that pursuant to *Garfoot v. Fireman's Fund Ins. Co., 228 Wis. 2d 707, 599 N.W.2d 411 (Ct. App. 1999)*, the sanction of dismissal is warranted only upon a finding of egregiousness. The court found that Harborview's conduct was egregious. The court determined the second amended complaint, filed in November 2002, focused on the windows as the source of the water infiltration problem. The court explained that as of the May 28, 2003 demonstration, no work on the V-grooves was planned and any concerns about the V-grooves were purely cosmetic. The court found that Harborview knew, at least as of August 21, 2003, that the V-grooves were a source of a water problem and did not notify the respondents. The court held that Harborview had an obligation to stop work and notify the respondents once it realized that leaking in V-grooves was occurring. The court explained that the case was in litigation and that Fischer, an expert, was aware of the risks of destroying evidence. The court concluded that the first time the respondents had notice that the V-grooves were a source of the water problems was at the September 17 and 18 depositions. The court found that by this point, ninety-eight percent of the V-grooves had been reworked and therefore the respondents' ability to gather evidence concerning the sources of water infiltration and to allocate responsibility amongst themselves was compromised.

■ ■ ■

Discussion

Harborview challenges the circuit court's dismissal of its claims against the respondents on several grounds. Harborview alleges that the record does not support the court's finding of egregiousness, the court's imputation to Harborview of the conduct of its expert, and the court's determination that the destruction of the evidence impaired the respondents' ability to defend against its claims. . . . We will analyze each of Harborview's claims after setting forth the well-settled law governing sanctions for spoliation of evidence.

Spoliation Law

"There is a duty on a party to preserve evidence essential to the claim being litigated." *Sentry Ins. v. Royal Ins. Co. of Am.*, 196 Wis. 2d 907, 918, 539 N.W.2d 911 (Ct. App. 1995). As both parties recognize, the decision whether to impose the sanction of dismissal for a violation of that duty is committed to the trial court's discretion. The question is not whether this court as an original matter would have dismissed the action; rather, it is whether the trial court erroneously exercised its discretion in doing so. *See id.* We affirm discretionary rulings if the trial court examined the relevant facts, applied a proper standard of law, and utilizing a demonstrably rational process, reached a reasonable conclusion. On a motion for sanctions, if there are factual disputes or conflicting reasonable inferences from undisputed facts, an evidentiary hearing, rather than simply oral argument based on briefs, affidavits and depositions, is necessary to resolve those disputes.

The sanction of dismissal should rarely be granted. In *Garfoot,* we reaffirmed the proposition that dismissal as a sanction for destruction of evidence requires a finding of egregious conduct, which means a conscious attempt to affect the outcome of litigation or a flagrant, knowing disregard of the judicial process. . . .

In this case, the trial court observed these legal standards, examined the relevant facts and reached a reasonable conclusion. The court explicitly acknowledged the *Garfoot* standard for dismissal as a sanction. The court carefully applied that standard to the circumstances of the case. As the court stated, "I have a box full of files in my office. I didn't read things just once. I read them several times. I have read every brief, every deposition excerpt that was provided, every affidavit, and this is truly a case that I have agonized over."

Finding of Egregiousness

The record supports the conclusion that Harborview's conduct was egregious because it was a flagrant, knowing disregard of the judicial process. Simply stated, Harborview filed its complaint against the respondents in December 2001; from at least that point forward, Harborview had a duty to preserve evidence essential to the claims being litigated and it knowingly failed to do so.

As of December 2001, the parties had yet to determine the exact cause of the water infiltration. The source of the problems was therefore a key issue in the litigation. The respondents concede that early in the litigation the parties discovered the cracks in the V-grooves and that as a general matter they knew those cracks can cause water leaks. Indeed, the respondents must make such a concession. Camosy's expert, Robert Kudder, testified that when he inspected the building in August of 2002, he observed cracks in the V-grooves. Kudder also indicated that such cracks, depending on the circumstances, can lead to water infiltration. However, the record demonstrates that prior to the start of the remediation project, the parties focused on the windows and window perimeters as the sources of the water infiltration at the Center.

The complaints fail to mention the cracks in the V-grooves as defects contributing to the water problem. Harborview's consultant, John Lampe, determined that water was passing into the building as a result of the defects in window flashing and window perimeter caulking. . . . In his March 2003 report, Fischer, Harborview's expert, reported six deficiencies relating to the windows and window perimeters and recommended removing and possibly replacing all of the windows. He acknowledged that going into the remediation project he did not suspect that the cracks in the V-grooves were the cause of the water infiltration. Bychowski, the project manager for the remediation project's reglazing company, testified that he was informed that the cracks in the EIFS had been analyzed and were only "superficial." Robert Nikolai, Camosy's senior project manager, testified in April 2003 that while cracks in the V-grooves can cause leaks, he believed that the windows were the most likely cause of the leaks at the Center.

Fischer's testimony shows that the remediation project, as originally planned, did not contemplate the extensive work that was eventually done to the V-grooves. Even after Fischer had surveyed the building in late July 2003, he thought that the cracks were not necessarily sources of water entry and "that if they ground those down and cleaned them up, that would take care of that situation."

Fischer admitted that the water tests he chose to conduct in August 2003 had a diagnostic purpose. They gave him the opportunity to ascertain that there were sources of water infiltration other than the windows themselves. It was based upon the results of those tests that he was able to conclude that water was infiltrating the building through the cracks in the V-grooves and that the V-grooves needed more extensive repair work. Fischer testified that he knew when he gave the orders he was permanently altering the physical condition of the V-grooves and was precluding others from water testing them in their original state.

The record shows Harborview was aware of the need to carefully preserve evidence during the remediation project that concerned the sources of the water infiltration. In May 2003, Harborview filed a motion to establish a protocol for evidence discovery and retention. Harborview also invited the respondents to a demonstration of the work to be performed during remediation.

The record further reveals that Fischer informed Harborview ownership of his discovery that cracks in the V-grooves were potentially a source of water infiltration and recommended to Harborview ownership that the V-grooves be substantially reworked. With Harborview's approval, Fischer gave the orders to rework all of the V-grooves.

The record shows that Fischer also was well aware of the contentious litigation surrounding the source of the water infiltration when he discovered that the V-grooves were leaking. At the July 24 job progress meeting, the contractors were advised that the "building is the subject of litigation; therefore, any comments that must be made regarding existing conditions should be discussed in the job trailer with no bystanders present." On July 31 they were advised that "no one is to discuss the project with anyone that is not part of the construction crew/personnel." Furthermore, for the purposes of preserving evidence for litigation, Fischer had documented the remediation project by photograph and had put some physical evidence removed from the site into storage.

However, the record demonstrates that neither Harborview ownership nor its experts or counsel notified the respondents before September 12 at the earliest[4] that water testing had identified the cracks in the V-grooves as a source of water infiltration and the cracks were being reworked for that reason. Furthermore, the trial court had to issue an order in January 2004 requiring Harborview to turn over the information pertaining to the cracks in the V-grooves, which it considered to be newly discovered defects. Given all of these circumstances, we comfortably conclude that Harborview's conduct was, at a minimum, in flagrant, knowing disregard of the judicial process.

Harborview maintains that its conduct was not egregious. Harborview concedes that "Fischer's decision to repair the cracks in the V-grooves was volitional and that the cracks in the V-grooves could not be tested in their original condition after they had been repaired."

■ ■ ■

---

[4] . . . For purposes of this appeal, we will assume without deciding that Fischer notified the respondents on September 12. Fischer had a duty to stop work and notify the respondents once he discovered that the V-grooves were a new source of water infiltration in late August. He did not and by September 12 a majority of the V-grooves already had been reworked.

Harborview mistakenly directs us to several facts of record that it claims show that it did not act egregiously. Harborview points out, among other things, that it performed the work in public, it was open to questions from respondents, the respondents had Fischer's cell phone number and several of the respondents and their experts and attorneys witnessed the work being done. While all of this may have been true, there is no evidence in the record that either Harborview or its experts notified the respondents before September 12 that water testing had identified the cracks in the V-grooves as a source of water infiltration and were being more extensively repaired for that reason. The respondents, therefore, would not have had any reason to ask questions or suspect anything was amiss. Further, it was Harborview's duty to stop the work on the V-grooves and notify the respondents of the discovery that the V-grooves may have contributed to the water problems.

Finding of Prejudice

Harborview argues that the trial court erred in finding that the destruction of evidence impaired the respondents' ability to present a defense and as a result dismissal was an inappropriate sanction. Harborview suggests that an action may be dismissed only where the plaintiffs knowingly took steps to destroy material evidence before the defendants had the opportunity to physically inspect it. Harborview alleges that the respondents can still determine the amount of water that passed through the cracks in the V-grooves, the respondents "passed on opportunities to test and inspect the cracks in the V-grooves," and the respondents' own initial repair work impaired their ability to conduct tests.

We are not persuaded. First, we remind Harborview that a court may impose a sanction of dismissal even if the destruction of evidence has not impaired the opposing party's ability to present a defense. Second, Harborview misses the point. As Fischer admitted, water testing the building permitted him to discover a new source of water infiltration other than the windows: the cracks in the V-grooves. Harborview's reworking or repairing of the V-grooves permanently altered their physical condition. Because of Harborview's actions, the respondents can no longer test or analyze the V-grooves in their original state and we will never be able to know the extent to which this damaged the respondents' ability to present a defense.

Imputation of Fischer's actions

Harborview maintains that the trial court erred in imputing to it Fischer's actions. Harborview contends that Fischer's conduct should not be imputed to it because it reasonably relied upon him and it was unaware that the respondents had not been notified of the destruction of evidence. We are not persuaded.

Harborview hired Fischer as its expert in the remediation project. As noted, Fischer was well aware of the litigation between the parties. Fischer notified Harborview ownership that he had discovered that the cracks in the V-grooves were a potential source of water infiltration and recommended a different course of action than originally planned. Harborview authorized Fischer to go ahead with the proposed course of action on the V-grooves. No one chose to notify the respondents. Further, we have imputed the conduct of an expert to a party. *See Garfoot, 228 Wis. 2d at 728* (holding that the distinction between independent contractor and master/servant for liability purposes does not determine the outcome on a motion for sanctions and imputing the conduct of an independent contractor and a technician who were retained to further the offending party's interest and who were acting in furtherance of those interests).

■ ■ ■

Conclusion

We are sensitive to the fact that Harborview has spent over 1,700,000 dollars in remediation and now may not be able to recover its expenses. However, when the trial court dismissed this case as a sanction for spoliation of evidence, the court articulated the proper legal standard for dismissal announced in *Garfoot,* applied the standard to the essential facts of record and reached a reasonable conclusion. The record supports the findings of egregiousness and prejudice and the imputation to Harborview of the experts' conduct. The record also demonstrates that an evidentiary hearing is unnecessary. The judgments of dismissal are affirmed.

By the Court.–Judgments affirmed.

Review Questions

1. Describe the federal court system. Describe the court system in your state.

2. Assume that you have been the unfortunate victim of someone's negligence. What steps will you or your attorney take in attempting to get compensation for the loss you have suffered?

3. Distinguish between the proof required in a criminal case and the proof required in a civil case.

4. How would you prove the existence and terms of an oral agreement if there were no third parties present to overhear the agreement?

5. Why does a court hold in disfavor:
   a. Oral testimony as to the meaning of a written document?
   b. Hearsay testimony?
   c. Expression of an opinion by a witness?

6. Why is circumstantial evidence so often used in criminal trials?

7. In the case of *Scordill v. Louisville Ladder Group*, what are the *Daubert* factors? Which factors did the court apply to this case?

8. According to *Bage, LLC v. Southeastern Roofing Co.*, when can an employee accept service of process on behalf of his or her employer? Can a secretary accept service of process?

9. In *Harborview Office Center, LLC v. Camosy Incorporated, et al.*, what does the court mean by the phrase "we have imputed the conduct of an expert to a party?" Why does Harborview think the trail court erred in imputing expert Fischer's actions to it? Of what is Fischer accused?

Contracts

The world of the engineer is an environment of serious communications, nearly all of which have contractual implications. A contract to build a bridge, road, or building may require numerous subcontracts—contracts with material suppliers, labor contracts, leasing contracts, easements, utility contracts, and many others. A contract to manufacture and ship parts to an appliance manufacturer on a continuing basis may similarly precipitate additional contracts. The day-to-day engineering management of such contracts requires a very large number of activities and communications, almost all of which involve contracts. For example, the quality of a final product is often spelled out or implied in the contract for that product. The product's quality is determined by the production processes used, the manner in which they are used by the production work force, and the materials on which the processes operate. All of the design and operating decisions involved with such production have quality overtones and for that reason, if for no other, are important contract considerations. The next few chapters briefly examine the law of contracts, then, the following section looks at engineering contracts.

Introduction to Contracts

■ There are five elements of a valid and enforceable contract:
1. Competent Parties
2. Agreement
3. Consideration
4. Lawful Purpose
5. Form

Modern civilization is a world of contracts. Each one of us depends on them. Every purchase is a contract, whether it is a pair of socks, a restaurant meal, or a battleship. When you turn on your television set to watch your favorite program, it is done in execution of a contract. The utility company has agreed to furnish electric power and you have agreed to pay for it. You go to work as your part of a contract with your employer. On the job, you do what your employer wants you to do, and at the end of a week or a month your employer pays you for it. It is all part of the same contract. If your employer pays you by check or electronic deposit it is because your employer has a contract with a bank to safeguard its money and give it out on order. If your employer pays you in cash, the currency represents a contract between the bearer and a Federal Reserve Bank. When you die and are buried, the security of your last resting place may depend on the terms of the contract under which the land was obtained. You can't avoid contracts, even by dying.

Even a simple action such as leaving your watch at the jeweler's for repair involves you and the jeweler in a complicated legal situation. Everything is resolved painlessly when you pick up the watch and pay the jeweler a week later. In the meantime, the relationship between you and the jeweler involved the following:

- Personal property (the watch)
- Agency (quite likely a clerk represented the owner)
- The law of bailment (personal property was left with another for repair)
- Insurance law (had the watch been lost)
- A contract (the jeweler's agreement to repair and your agreement to pay for the service)

All contracts are agreements and, morally at least, all agreements are contracts, but moral duties are not always enforceable under the law. Suppose that Black accepted Dr. White's invitation to dinner on Tuesday, then forgot the engagement and did not appear. Black broke a social contract and in doing so committed a serious breach of social ethics. But Black's contract was not an enforceable one in the sense that a court of law would award damages or order Black to appear and, no matter how much pain and suffering Black caused Dr. White, the doctor could not collect damages.

Now, suppose White is a practicing engineer and Black is an attorney looking for help, that is, a prospective client. Black calls and makes an appointment with White for 3:00 p.m. on Thursday. In breaching this appointment, Black may be breaching a contract. The commodity in which White trades is time and his ability as an engineer. Many jurisdictions would hold that, unless White could otherwise gainfully use the time set aside for Black's appointment, White could collect for that time. In deciding the case, the court would examine the understanding of the parties and also business practices in the area, and these factors would largely determine the case's outcome.

If these two parties entered into an agreement whereby Black agreed to purchase a new wireless network from White for $15,000, such an agreement would be a contract, and enforceable in a court of law. Each party to the contract would have an action at law available if the other failed to perform as agreed.

In modern America, we take individual freedom for granted. Many of our opportunities and freedoms can be traced to the law of contracts and a strong public policy of freedom of contract. In effect, the courts allow the parties a great deal of freedom to agree to the terms and conditions by which goods are made and sold or by which services are rendered. Such freedom, however, is not without restraint; an agreement to perform a criminal act, for example, will not be enforced.

In the following chapters, consider whether the policy of allowing the parties to freely contract is or should be subject to other public policies. Also consider whether you agree with the cases in which a court found the parties' agreement to be void as against a public policy (other than freedom of contract).

DEFINITION OF CONTRACT

The only kind of an agreement the law recognizes as a contract is defined in *Black's Law Dictionary* as "an agreement between two or more persons which creates an obligation to do or not to do a particular thing." The Uniform Commercial Code (UCC), which has been adopted by every state and the District of Columbia, defines a **contract** as "the total legal obligation that results from the parties' agreement as affected by [applicable] law." The term **agreement,** in turn, is defined as "the bargain of the parties in fact as found in their language or by implication from other circumstances including course of dealing or usage of trade or course of performance. . . . Whether an agreement has legal consequences is determined by the applicable law of contracts."

Stated more simply, a contract could be defined as "an agreement enforceable by law."

SOURCES OF CONTRACT LAW

The area of law dealing with contracts developed under common law through the decisions of English and, later on, American judges. Many "modern" rules of contract law have their roots in cases that are centuries old. The common law rules were distilled into a multi-volume treatise called the *Restatement of the Law of Contracts*. Later, a second *Restatement* was published. The *Restatement* (Second) remains an important authority for the common law on contracts. Certain types of contracts, however, are governed by distinct statutes and treaties. For example, the UCC controls contracts for the sale of goods. Another more recent example of a source of contract law is the United Nations Convention on Contracts for the International Sale of Goods ("CISG"). This Convention was signed by the United States in 1980, and took effect in 1989. Certain contracts involving the international sale of goods with buyers and sellers located in different countries will now be governed by the Convention's rules (instead of the common law or the UCC). Most of the following discussion focuses on the general common law rules and the UCC.

ELEMENTS OF A VALID CONTRACT

Analysis of valid contracts reveals that they are made up of the following five elements, each of which must exist for the contract to be enforced:

1. Competent Parties
2. Agreement
   a. Offer
   b. Acceptance
3. Consideration
4. Lawful Purpose
5. Form

Each element is only briefly introduced here but is treated in more detail in the next several chapters.

Parties

For a contract to be thoroughly binding it must be made by at least two parties, neither of whom will be able to avoid the contractual duties by pleading as a defense an incompetence to contract. The "garden variety" contract involves two parties as promisor and promisee. However, it is more difficult to determine the liability of the individuals when they concern several parties and the individual interests of some have been merged. When there are merged interests, the relationships of the parties are treated as joint, several, or in some cases as joint and several. Any legal entity can be a party to a contract. Thus, an individual, corporation, partnership, or joint venture can enter into a contract. For example, an employment agreement (written or not) exists between an engineer (an individual) and the engineer's employer, such as a corporation. The federal, state, and local governments also enter into contracts for a vast array of different goods and services.

Individuals who merge their interests into a joint relationship to form a party to a contract may be thought of as partners in the promises made. If the individuals together agree to "bind themselves" or "covenant" to do a certain thing in the terms of an agreement, it is treated as a joint contract. The liability is much the same as it is in a partnership. That is, should the parties breach the contract, each is liable for complete performance of the contract until the full value has been satisfied.

Say that Black and White agree jointly to hire Gray, a contractor, to make alterations in an existing building and to pay him $4,000 for the alterations. If, after completion of the alterations, payment is refused by Black and White, Gray must sue them jointly. If Black cannot pay part of the debt, White may have to pay the entire amount. If either White or Black dies, the survivor will be liable for the full amount of the debt.

Restriction of the liability of individuals merging their interests is a feature of the **several contract.** If, in the contract just described, Black and White had agreed to "bind themselves severally," or "covenant severally" to pay $4,000 for the alterations that Gray was to make (restricting themselves to, say, $2,000 each), the liability of Black and White each would be limited to the amount stated. In case of default of payment, Gray could take only separate actions against Black and White.

If merging individuals agree to "bind themselves and each of them" or to "covenant for themselves and each of them," the contract is treated as a joint and several contract. If the Black and White contract just described were joint and several, and Gray found court action necessary, such action could be taken either jointly or severally, but not both ways.

In a third-party beneficiary contract, the purpose of the contract is usually to benefit a third party. If the intent to benefit the third party is clear from the agreement, the third party can enforce those benefits in a court of law. If, however, the consideration under the contract benefits one of the parties to the contract and only incidentally benefits a third party, no right of court action is available to the third party. Probably the most common example of a third-party beneficiary contract is life insurance.

Agreement

The agreement consists of an offer and acceptance. The **offeror** states the terms of the proposed contract. The **offeree** must accept the terms as they are proposed to complete a binding contract. Contracts can be accepted in at least two ways: *unilaterally* or *bilaterally*. A **unilateral contract** is a promise of a consideration made without receiving a promise of consideration from another party. A common example of the unilateral contract is the reward notice. White publishes a promise of a reward of $500 for the return of a diamond-studded wrist watch. Black, on finding the watch, accepts the offer of a reward by returning the watch and then claims the reward. To obtain the reward in a legal action, Black must have been aware of the reward promised when Black returned the watch. A **bilateral** (or **reciprocal**) **contract** is a mutual exchange of promises, present consideration, or both, by two or more parties.

> An **offeror** is the person who makes an offer to someone else.
> An **offeree** is the person to whom an offer is made.

Consideration

Consideration in a contract consists of the money, promises, and/or rights given by each party in exchange for the money, promises, and/or rights that each party receives from the other.

Lawful Purpose

The purpose and consideration in the contract must be lawful if the contract is to be enforceable in a court of law. A contract to perform an unlawful act will not be upheld in a state that imposes a penalty for committing the same act.

Form and the Statute of Frauds

Difficulties in proving the existence, validity, and terms of contracts have prompted the requirement for certain types of contracts to be in written form pursuant to the Statute of Frauds, covered in Chapter 11. Failure to comply with this requirement renders such contracts unenforceable by the court.

As soon as the offeree expresses acceptance of the terms offered, a contract is created. Usually, the contract at this stage is executory, and as long as something remains to be done by either or both of the parties under the terms of the agreement, it remains so. Even when one of the parties has performed that party's obligations completely and the other party has not yet completed performance, the contract is partially executed, but in a legal sense it is still executory. After all parties to the contract have performed all the actions required according to the agreement, the contract is fully executed.

A contract to purchase a refrigerator is executory when the sales agreement is made between the buyer and the seller. It is partially executed (but legally executory) when the refrigerator is delivered to the buyer and fully executed when the buyer has finished paying for it.

CLASSIFICATIONS OF CONTRACTS

Contracts can be classified in numerous ways. Unilateral and bilateral contracts have already been discussed. Below are several other classifications. Broadly, contracts can be classified by how they are created (express, implied, or quasi-contract as well as unilaterally or bilaterally); whether or not the contracts can be enforced by law (valid, unenforceable, voidable, and void) and how the contract was drafted (formal or informal).

Contract Creation

How a contract is created not only has an affect on the parties involved, but sometimes determines whether it can be enforced. **Express contracts** involve overt statements and communications to reach an agreement by the parties, the terms being expressed between them orally, in writing, or both.

Implied contracts are based on implications of fact. If the parties, from their acts or conduct under the circumstances of the transaction, make it a reasonable or necessary assumption that a contract exists between them, it will be held that such a contract does in fact exist.

Quasi-contracts are contracts implied in law. They are generally based on the theory of unjust enrichment, which in effect says that it would, in all justice and fairness, be wrong to allow one person to be enriched by another without having to pay for it. The theory amounts to a legal fiction created by the courts to permit recovery in cases where, in fact, there would be no recovery otherwise. Black contracts orally to pay $50,000 for White's services for a period of 18 months. The Statute of Frauds requires such a contract to be in writing if it is to be enforceable. In ignorance of this requirement, White performs the services for Black and now sues for payment. White cannot recover on the original contract but may recover the reasonable value of the services under a quasi-contract on the theory that it would be unjust to allow Black to be enriched by White's services without having to pay for them. (See *Board of County Commissioners v. Village of Smithfield* at the end of the chapter.

In construction projects, it is not unusual to see a contractor or subcontractor recover for goods and services previously provided despite a breach of contract. Such a recovery is for the value of the goods and services, not for the contractually agreed-on amount.

Legal Status

Contracts also can be classified as to legal standing into valid, unenforceable, voidable, and void.

A **valid contract** is an agreement voluntarily made between competent parties, involving lawful consideration, and in whatever form may be prescribed by law for that particular kind of subject matter. It contains all the essential elements previously described. Courts will enforce such a contract.

An **unenforceable contract** is one that creates a duty of performance that may be recognized in a court of law but that, because of some defect in the contract, may not be enforced by the court.

A detailed contract with a number of different provisions may have one or two of them found to be unenforceable. For example, suppose the contract requires the payment of money on the delivery of goods and, if the payment is late, the principal amount plus interest at a rate of 20 percent per year is due. Many states have laws restricting the rate-of-interest charges; charge of 20 percent in some states would constitute "usury" and would not be enforced. However, depending on the applicable state statute on interest, the contract's provisions regarding the required payment of the principal amount would still be enforceable.

Sometimes, however, the whole contract may be unenforceable. A common example of such a contract is one that is not made in accordance with the Statute of Frauds. An oral contract that, because of the subject matter, should have been in writing is an unenforceable contract.

A **voidable contract** is one in which one or both of the parties may avoid the contract if either so desires. Black, a minor, purchases a bicycle from White, an adult. Black may return the bicycle and demand the money back or keep the bicycle and pay the price. White has no similar right of avoidance.

A **void contract** is, strictly speaking, not a contract. A contract to commit a crime is an example of a void contract. The courts ordinarily treat contracts for an unlawful purpose as nullities.

Formality

Contracts also may be classified as **formal** and **informal.** Most contracts are informal in nature. Contracts to perform services or to sell goods to another are usually informal contracts. One particular type of formal contract (so called because the contract derives its validity or effect from its form) is the **negotiable instrument.**

Negotiable Instrument

A **negotiable instrument,** sometimes called **bearer paper,** is a contract for the payment of money. It derives its negotiable characteristics from its form. The requirements for a contract to be a negotiable instrument are set forth in the Uniform Commercial Code. Briefly, the agreement must include the following features:

- Be in writing and signed
- Offer an unconditional promise or order
- Agree to pay, in money, an ascertainable amount
- Be surrendered on demand or at a fixed or determinable future time
- Be payable to order or bearer
- Identify a drawee, if there is one

KUENZI v. RADLOFF
34 N.W. 2d 798 (Wis. 1948)

Study terms: Valid contract, estimated value

The plaintiff is licensed under provisions of sec. 101.31 as a professional engineer. He has 40 years experience in designing buildings and was licensed in 1932.

In December 1945 the defendants were considering the erection of a building in the city of Waupun to be occupied by bowling alley and a tavern. The defendant Radloff consulted the plaintiff and as a result of this consultation the plaintiff wrote the following letter:

Dec. 29, 1945

Mr. Harold Radloff
Waupun, Wisconsin

Dear Sir:

I wish to confirm our conversation of some time ago wherein I named you a fee of 3% of the estimated value of the project for services in making up plans for the construction of a proposed bowling alley to be built at Waupun, Wis. This also includes the services of securing a full approval of the Industrial Commission.

Respectfully submitted
Yours truly,
Arthur Kuenzi

Accepted
O. A. Krebsbach
H. Radloff

Upon receipt of the signed proposal from the defendants the plaintiff proceeded with the design and preparation of the plans for the proposed building. The defendants from time to time during the preparation of the plans consulted with the plaintiff and his associates and changes were made in accordance with the suggestions made by the defendants. The plans were completed early in March 1946 and were presented to the Industrial Commission and duly approved by it and then promptly delivered to the defendant, Radloff.

On April 11, 1946, an application for the allocation of construction materials, to which application was attached a copy of the plans, was made on behalf of Radloff and Krebsbach to the Civilian Production Administration. A second application to the Civilian Production Administration, signed by both defendants, was submitted to the Civilian Production Administration on April 25, 1946. In each of these applications the cost of the structure, including fixtures and building service, is stated to be $80,000. The following statement was made in the application.

"A site has been obtained and an architect engaged for the construction of such building and the plan submitted to a contractor who has in turn ordered various materials for the construction thereof. All of said obligations and commitments made previous to March 26, 1946." Both applications were denied on May 1, 1946.

On April 28, 1946, the plaintiff sent to the defendant Radloff an invoice for $1,350 based on the estimated value of the building of $45,000. The plaintiff also demanded payment from the defendant Krebsbach before the commencement of this action. The plaintiff received the following letter from the defendant Radloff:

Monday morning

Dear Sir:

   Sorry to keep you waiting but we are still working through Washington to get started building.

   We will make payment to you just as quickly as possible. Milan Nickerson, one of our partners, dropped out, didn't want his money laying idle so he went into the cement block business.

   We are picking out another good partner and will get in touch with you or write when we have the partners lined up.

   This Nickerson was undecided for some time and that's why we didn't send you any money, until we have the other party lined up.

   How is the steel coming? We will have to pay you and have it on hand when it comes and wait for the permit to start to build. You said in your last letter of quite a while ago that the steel would be here within a couple of weeks.

Yours very truly,
H. Radloff

P.S. Keep this under your hat about a third party and if you should happen to know of someone who has 15 or 20 thousand and wants to put it in a good business of a bowling alley and tavern let us know.

The letter was undated; neither the plaintiff nor the defendant, Radloff can fix the date on which it was sent. Evidently it was sent after the receipt of the invoice from the plaintiff because payment is promised.

Upon notice of the denial of their application by the Civilian Production Administration on May 1, 1946, the defendants abandoned the project, and the building for which the plans were prepared has never been erected.

Upon the facts it appears as a matter of law that the plaintiff had a contract with the defendants for the making of plans for the construction of the proposed building; that he proceeded to carry out his part of the contract by preparing the plans, procuring their approval by the Industrial Commission, and delivering them to the defendants; that they were accepted by the defendants and used by them in their efforts to procure a priority order from the Civilian Production Administration.

The defendants seek to defeat the plaintiffs' claim on a number of grounds. . . .

■ ■ ■

It will be observed that the contract provided that the fee should be 3% of the estimated value of the project for services in making up plans for the construction of the proposed bowling alley. The defendants place a great deal of emphasis on the term "estimated value." The defendants made the claim, and the court sustained it, that this meant the value of the building after its erection. In so holding it is considered that the trial court was in error. The basis on which plaintiffs' fee was to be computed was the estimated value of the project, not of a building that might never be erected. In its opinion the trial court said: "The court refused to allow this testimony as to cost due to the fact that the letter that comprised the basis for the case said value of the project as the basis rather than cost of construction," and excluded all testimony as to cost except that of plaintiff, and restricted the evidence to what the building would be worth on the site after it was constructed. Just how the value of a building that has never been constructed can be determined does not appear. "Value," in the sense in which the trial court used the term, means market value, what the property could be sold for after the building was erected. Even if that were the test, evidence of the cost would be relevant and material, but it is not the test in this case.

It is considered that this case is ruled by *Burroughs v. Joint School District,* 155 Wis. 426, 144 N.W. 977. It was there held that if when the term "value" is applied to a particular contract, or conditions growing out of it, it leads to results clearly not contemplated by the contract read as a whole, and it is susceptible of another meaning that harmonizes with all the provisions of the contract, such other meaning should be given to it. In that case a building contract provided for payment in each month of a sum equal to 90% of the value of the work done and material furnished during the preceding month, as assessed by the architects. In that case the word "value" was construed to mean not market but contract value. It is considered that in this case the term "estimated value" of the project referred to the estimated cost of the material and services necessary to complete the building according to the plans. There is not a scintilla of evidence in this case that plaintiff was to wait for his compensation until the completion of the building and an appraisal thereof. The idea that the base on which the fee was to be computed can be established by the sale of a mythical building owned by a mythical owner and sold to a mythical buyer, is too elusive and indefinite a standard to apply to practical affairs. So in this case we hold that the term "estimated value of the project" means the estimated cost of completing it.

Upon this point the plaintiff testified as follows: "I made an estimate of $45,000 being the value of this building, which is the same amount as the estimated cost. My estimate was the sum of $45,000 as I recall it. I made a charge against the defendants based on $45,000 and charged 3% of that amount or $1,350."

Two witnesses were called on behalf of the defendant, who testified that the value of the completed building would be fifteen to twenty thousand dollars, around twenty thousand dollars. This was on the theory that the building when completed would have to be rebuilt if used for any other purpose than a bowling alley. One witness testified: "You would have quite a job getting $20,000 for it, because it would have to be torn apart and fixed over for something else." This sort of evidence comes far short of establishing the estimated value of a project.

Considerable evidence was received in regard to the income tax returns of the plaintiff. So far as we are able to ascertain this evidence was immaterial. Whether the proper income tax returns were made or not is a concern of the Department of Taxation, and has no relevancy on the question raised in this case. It was introduced in an effort to establish that the real plaintiff was a co-partnership, a matter that has already been considered.

It is considered that, having fully performed the contract between the parties, including the procurement of the endorsement of the Industrial Commission, and defendants having accepted and acted on the plans as delivered to them, the plaintiff is entitled to compensation on the basis of the lowest estimated cost of the project appearing in evidence, to-wit the sum of $45,000 with interest and costs.

The judgment appealed from is reversed and the cause is remanded with directions to the trial court to enter judgment for the plaintiff as indicated in the opinion.

BOARD OF COUNTY COMMISSIONERS OF JEFFERSON COUNTY, ET AL., v. VILLAGE OF SMITHFIELD, ET AL., 2006 Ohio 6242, 2006 Ohio App. LEXIS 6206

Study terms: Quasi-contracts, unjust enrichment, *quantum meruit,* meeting of the minds

Defendants-appellants, the Village of Smithfield, et al. (hereinafter collectively referred to as Smithfield), appeal the decision of Jefferson County Common Pleas Court finding that appellants owe plaintiffs-appellees, the Board of County Commissioners of Jefferson County, et al. (hereinafter collectively referred to as Jefferson County), $267,354.24 plus interest for water purchased by Smithfield from Jefferson.

In September 1999, the water wells for Smithfield went dry. Smithfield contacted Jefferson County and asked if they could provide them with potable water. Utilizing an existing line, Jefferson County began supplying Smithfield with water. Foreseeing that Smithfield was going to be using Jefferson County as a permanent supplier of water, Jefferson County expended $195,000.00 in constructing a more permanent water line to Smithfield.

The agreement between Jefferson County and Smithfield to provide water was never reduced to writing or formalized by either party. Jefferson County continued to supply water to Smithfield from November 1, 1999 to present. However, Smithfield did not pay Jefferson County the entire amount billed.

Due to the growing arrears, representatives from both parties met in June 2002 to discuss the matter. Smithfield's position was that it was not paying the full amount due to billing discrepancies.

When Smithfield continued to refuse to pay the full amount, Jefferson County filed a complaint against Smithfield on April 23, 2004. Jefferson County set forth three claims—complaint on account, breach of contract, and unjust enrichment—and sought $256.364.50 based on a rate of $3.00 per thousand gallons.

On July 19, 2004, Smithfield filed a motion for summary judgment based on R.C. 735.05. R.C. 735.05 requires that expenditures above a certain amount be "authorized and directed by ordinance of the city legislative authority." Since Smithfield never passed an ordinance pursuant to R.C. 735.05 authorizing the purchase of water from Jefferson County, Smithfield argued that the alleged agreement was invalid and unenforceable. . . . Jefferson County countered that R.C. 735.05 was not applicable when both of the involved parties are political subdivisions. On September 24, 2004, the trial court denied Smithfield's motion based on *Bd. of Cty. Commrs.*

The matter then proceeded to a bench trial on April 13, 2005. The parties stipulated to the amount of water sold to Smithfield and the amount of money Smithfield had paid. The principal dispute was what price the parties had agreed to for the water. Jefferson County contended that it was $3.00 per thousand gallons and Smithfield argued that it was $2.00 per thousand gallons.

On July 21, 2005, the trial court sided with Jefferson County and found the agreed upon rate to be $3.00 per thousand gallons. The trial court awarded Jefferson County the balance due of $67,354.24. This appeal followed.

Smithfield raises two assignments of error. Smithfield's first assignment of error states:

"THE TRIAL COURT ABUSED ITS DISCRETION IN FAILING TO GRANT SUMMARY JUDGMENT IN FAVOR OF THE VILLAGE OF SMITHFIELD."

An appellate court reviews a trial court's decision on a motion for summary judgment de novo. . .

Since Smithfield never passed an ordinance pursuant to R.C. 735.05 authorizing the purchase of water from Jefferson County, Smithfield argues that the alleged agreement to provide water was invalid and unenforceable. Jefferson County argues that R.C. 735.05 is not applicable when both of the involved parties are political subdivisions.

R.C. 735.05 states, in pertinent part:

> "The director of public service may make any contract, purchase supplies or material, or provide labor for any work under the supervision of the department of public service involving not more than twenty-five thousand dollars. When an expenditure within the department, other than the compensation of persons employed in the department, exceeds twenty-five thousand dollars, the expenditure shall first be authorized and directed by ordinance of the city legislative authority."

Smithfield cites *Millersburg v. Wurdack* (1919), 22 Ohio N.P. (n.s.) 49, 30 Ohio Dec. 218, in support of its position. *Millersburg* interpreted G.C. 4328, the predecessor to R.C. 735.05. In *Millersburg,* the Village of Millersburg contracted with Wurdack, a non-public entity, to supply water to the Village. The contract exceeded the statutory amount which required authorization by a municipal ordinance. Because the Village had not passed an ordinance authorizing the contract, the court applying and interpreting G.C. 4328, found the contract void and unenforceable.

In support of its position, Jefferson County cites this Court's decision in *Bd. of Cty. Commrs. v. Bd. of Twp. Trustees* (1981), 3 Ohio App.3d 336, 3 OBR 391, 445 N.E.2d 664. Jefferson County was also a party in that case. Jefferson County had a contract with the Township of Island Creek to provide water to its fire hydrants for a period of five years. After five years, Jefferson County continued to provide the water for an additional seven years without a written contract. Island Creek refused to pay for the additional seven years of water supplied based on the lack of a written contract and R.C. 5705.41, the statute for townships that is the corollary of R.C. 735.05. Jefferson County sought payment on theories of quasi-contract. The trial court awarded judgment in favor of Jefferson County despite noncompliance with R.C. 5705.41 and the lack of a written contract.

On appeal, this Court noted that Island Creek's arguments were generally correct when dealing with a dispute between a government entity and a private or non-public entity. However, this Court held that the statute did not apply when dealing with a dispute between two political subdivisions. Specifically, this Court explained:

"[Island Creek's] citations of authorities as set forth generally are a correct proposition of law applicable to a political subdivision. However, the fact situation in the instant case does not fit the usual pattern where the political subdivision is dealing with individuals and companies, where the public at large must be protected. These statutes cited by the appellant are designed for the protection of the public and it is essential to a valid contract that such procedure be substantially followed.

"The Ohio rule exempting municipalities from liability by quasi-contract is based on a policy which attempts to protect the taxpayer from the fiscal irresponsibility of government officials.

"The uniqueness of the instant case is that there are two public subdivisions involved. A fair question arises—should the taxpayers of one subdivision suffer by giving free water service to the taxpayers of the other subdivision?

"We cannot escape the conclusion that the authorities and doctrines of law cited by the appellant should not apply to the facts of the instant case. If the township escapes liability, then it is county taxpayers who will suffer. It is they who will pay for the fiscal irresponsibility of the township trustees, who knowingly accepted hydrant service from [Jefferson County] and now refuse to compensate it by payment of a reasonable rate for its services." *Bd. of Cty. Commrs.,* 3 Ohio App.3d at 338, 3 OBR 391, 445 N.E.2d 664.

The case cited by Smithfield is an eighty-seven year old common pleas court case from a county in another appellate district. The case cited by Jefferson County is an appellate case from this district and, therefore, is binding precedent in this matter. Consequently, R.C. 735.05 does not apply to this case since both parties are political subdivisions and Smithfield may be subject to liability on the theory of quasi-contract or quantum meruit even in the absence of a written contract.

Accordingly, Smithfield's first assignment of error is without merit.

Smithfield's second assignment of error states:

"ASSUMING A QUASI-CONTRACT, THE TRIAL COURT ERRED IN DETERMINING THE PRICE OF $3.00 PER THOUSAND GALLONS OF WATER."

■ ■ ■

On appeal, Smithfield concedes that Jefferson County met its burden in proving that there was an agreement between the two to supply and purchase potable water. However, Smithfield argues that there was never a meeting of the minds on a key term of the agreement-price. Jefferson County contends that Smithfield agreed to pay $3.00 per thousand gallons of water, the same rate it charged to its other bulk rate customers. Smithfield, on the other hand, contends that the agreement was for $2.00 per thousand gallons. Smithfield bases the $2.00 price on its past dealings with Jefferson County concerning water. Before its own wells went dry, Smithfield argues that Jefferson County used to purchase water from Smithfield and resell it to Piney Fork for $2.00 utilizing Smithfield's water lines. Although Jefferson County can no longer use Smithfield as a water source to supply Piney Fork, Smithfield contends that Jefferson County still utilizes Smithfield's waterlines to supply Piney Fork, albeit from a different source.

At trial, both parties stipulated that Jefferson County had provided Smithfield with 185,022,000 gallons of water from November 1, 1999 up through and until December 31, 2004. . . .

[On behalf of both parties, several witnesses testified about the agreed upon rate.]

As the trial court aptly observed, Smithfield offered no credible evidence that the rate was $2.00 per thousand gallons, although there was ample evidence that a $2.00 rate was desired. At best, this is the type of case where the evidence was susceptible to more than one interpretation. Therefore, in that type of situation, this Court must construe it consistently with the lower court's judgment. In addition, the weight to be given the evidence and the credibility of the witnesses are primarily for the trier of the facts. In sum, the trial court's judgment was not against the weight of the evidence.

Accordingly, appellant's second assignment of error is without merit.

■ ■ ■

Having found both Smithfield's assignments of error without merit, the judgment of the trial court is hereby affirmed.

Review Questions

1. Name the essential elements of a valid contract. Identify these elements in *Kuenzi v. Radloff.*

2. Distinguish between a joint contract and a several contract. Give an example of each.

3. What is a quasi-contract? Who creates it? For what purpose?

4. Draw up a valid negotiable instrument with John Doe as payee and Richard Roe as maker or drawer.

5. Distinguish between unenforceable contracts and voidable contracts. Give an example of each.

6. Consider the word "value". How many ways are there to estimate the value of a project that exists only in the planning stage? How many ways are there to estimate the value of something that has been in existence for a period of time—for example, a two-year-old computer?

7. How many contracts did you enter into today? For example, did you withdraw money from an automatic teller machine? What contracts were involved? Did you purchase a movie ticket? Write down at least three contracts into which you entered today and classify them into the categories set out in the section "Different Classifications of Contracts."

8. In *Board of County Commissioners v. Village of Smithfield, et al.,* why was the contract between the parties determined to be a "quasi-contract"? What was missing?

Parties CHAPTER 6

HIGHLIGHTS

■ There are classes of individuals protected by law against predatory contracting practices including:
1. Infants or minors
2. Incompetent and insane persons

These persons can avoid contracts under certain circumstances.

■ Business entities can enter into binding contracts although individual persons have liability for contract breaches. In some instances, liability is limited.

There must be at least two parties to a contract. There may be more than two; in fact, the only ceiling for parties is the number it is practicable to identify.

No person may make a contract with himself or herself. Consider Black, who is executor of White's estate and an ordinary citizen. If Black, as executor of White's estate, agrees with Black, as an individual, to do something, no valid contract will result.

The law offers its protective shield to those who, due to their immaturity or for some other reason, are held to be incompetent to contract. The law does not usually offer its protective shield unless it is asked to do so, however. A person must plead incapacity to contract to receive the protection.

Among those whose ability to contract is in some way limited are infants, insane persons, intoxicated persons, corporations, governments, and professional people. To illustrate the rights and defenses of the parties to a contract, infants' contracts will be treated at a greater length than the contracts of other incompetent parties.

INFANTS

Age of Infancy

According to common law, **infancy** ends when a person reaches his or her 21st birthday. There is no legal adolescence. On one day a person is an infant according to law; on the next day that person is an adult. The only status that might come close to a legal adolescence is that of emancipated minor. Emancipation of a minor (the legal meaning of infant is the same as for minor) takes place when a minor's parents or guardians surrender their rights to the minor's care, custody, and earnings. Such an emancipated minor is usually treated somewhat more sternly by the courts than a non-emancipated minor. The minor's ability and knowledge are considered to be less than those of an adult. A day later (at common law) the minor is vested with the full powers, rights, and responsibilities of an adult.

Statutes in many states have altered the age that ends infancy or have provided for removal of the incapacity under other circumstances. Several states provide that legal majority is reached on a person's 18th birthday. Other states make the age 19. Many statutes provide that marriage removes the infant's incapacity. Statutes of various states provide means by which an infant may remove the incapacity by request to a court of law. This is often done when an infant undertakes a business venture, such as is the case with young film or recording stars.

Partnership

When an adult becomes a partner, the adult stands to lose some or all of his or her personal fortune as well as any investment in the business if the partnership becomes bankrupt. This is because all of the partners remain personally liable for the debts of the entire partnership. However, when a minor becomes a partner in a business venture, the minor can lose in bankruptcy only whatever value the minor has contributed.

Agency

Generally, an agency contract in which a minor is the principal is void (not voidable) under common law. A recent trend, however, is to consider such contracts as voidable, which seems more logical. Infancy is no bar to acting as an agent for another, however, because it is the principal who is bound rather than the agent. An adult, therefore, may enter into a binding contract with a third person through an infant agent.

Disaffirmance

In general it is true that if one party to a contract is not bound, neither is the other. However, the law recognizes certain exceptions to this generality. Infants' contracts form such an exception.

An infant may avoid the obligations under almost any contract. The minor needs merely to notify the other party of a disaffirmance of the contract. The infant's right to disaffirm is a personal right; no one else may disaffirm for the minor.

If a minor elects to disaffirm a contract that is still completely executory, no major problem is involved. Because no consideration has changed hands, none needs to be returned. The problems arise when consideration has been given. If a minor disaffirms a contract with an adult after having given the adult consideration, return of the consideration to the minor is mandatory. The minor must also return whatever consideration the minor has received if it is possible to do so. It has been stated, however, that a minor's right to avoid a contract is a higher right than the adult's right to get back the consideration.

If the consideration received by the minor has been demolished or depleted in value in the minor's hands, the minor may still return it and demand the return of his or her consideration. Destruction of the subject matter in the minor's hands is only further evidence of incapacity. Even if the minor cannot return any of the consideration received, the minor may still be successful in getting back what was given. White, a minor, buys a used car from Black Auto Sales. White pays $1,500 down and agrees to make a series of monthly payments for the car. Two weeks later White loses control of the car while driving it and, when it hits a tree, the car becomes a total loss. White could return the wreck and get the $1,500 back. If the car was not insured for theft and White lost it to a thief, White could still disaffirm and get the downpayment back.

Although the preceding example is representative of the rulings in a number of courts, other courts require the minor to return the consideration received or its equivalent in value to disaffirm. The minor retains the right to avoid the contract, but the courts will not allow the minor to harm the other party by disaffirming. In one such case, a minor had taken a car to a garage for repairs. When the repairs were finished, the minor refused to pay for them and demanded the car.[1] In requiring the minor to pay for the repairs, the court compared the legal protection offered to minors to a shield. The judge stated, in effect, that although a defensive shield is afforded by law, it is not intended to be used as a sword against another. There was nothing defensive in the acts of the minor in this case. In fact, it represents a bald swindle more than it does a contract. If the repairs had amounted to replacement of rod bearings, main bearings, and piston rings, for instance, the inherent difficulty of returning the consideration to the adult would be quite apparent.

If the infant poses as an adult to get the adult to deal with him or her, the court may demand that the adult be left unharmed or, at least, that the harm be minimized. It would seem unreasonable to allow an infant to harm another by deceit and then to protect the infant.

If the subject matter of the infant's contract is something other than real property (real estate) the infant may disaffirm the contract at any time before the infant becomes an adult. In addition, the infant has a reasonable period of time after reaching adulthood to disaffirm the contract. The infant may disaffirm either by an express statement or by implication. If the contract is executory and the infant does nothing about it within a reasonable time, it will be implied that the infant has disaffirmed it.

If the subject matter is real property, the infant must wait until the infant becomes an adult to disaffirm the contract. There seems to be good reason to distinguish between real and personal property in infants' contracts. Land sold by the infant will always be there. This is not necessarily true of personal property with which the infant has parted.

Ratification

An infant's contract can be ratified as well as disaffirmed. However, to agree to be bound by the terms of the agreement, the infant must wait to become an adult. This seems reasonable: If an infant, in the eyes of the law, is incapable of making a binding contract, the infant could hardly be expected to ratify one previously made. Contracts involving both real and personal property must

---

[1] *Egnaczyk v. Rowland,* 267 N.Y.S. 14 (1933).

await ratification until the infant reaches adulthood. Ratification may be either express or implied. The law merely requires the infant to indicate the intent to be bound if the infant so elects.

Binding Contracts

An infant cannot avoid certain contracts he or she has made. Most prominent among such contracts are agreements by which the infant obtains the necessities of life. If a minor is not provided with such things as food, clothing, or lodging by a parent or guardian, the infant may make contracts for such things. If the minor is already supplied with the necessities, the minor cannot be made to pay for an additional supply of them. When payment is enforced by law, it is payment for the reasonable value of the goods or services, thus putting the liability on a quasi-contractual basis.

If the infant has reached the minimum age to enter the armed forces, an enlistment agreement is not avoidable. Marriage cannot be avoided by a minor who is old enough to marry according to the state law.

Parents' Liability

Ordinarily a parent (or guardian) is not liable for an infant's contract unless the parent has been made a party to it. For this reason, many who deal with infants insist that the adult responsible for the infant's welfare also agree to be bound. Only in cases where the parent (or guardian) has failed to provide the infant with necessities will the parent be required to pay reasonable value to third parties for needs supplied.

INSANE AND INCOMPETENT PERSONS

Early court decisions involving contracts in which one party was insane declared such contracts to be void. The courts agreed that there could not be a meeting of the minds if one of the minds did not exist legally *(non compos mentis)*. Court decisions today usually hold that such contracts are voidable at the option of the insane person when that person becomes sane. An increasing number of cases involving those suffering from Alzheimer's disease test the boundaries of incompetence when the afflicted also has prolonged periods of lucidity. Consider the case *Turja v. Turja* at the end of the chapter where a testator, likely suffering from Alzheimer's disease, revised her will to the detriment of two of her three children.

The law today recognizes that there are various forms and degrees of insanity. An insane person may have sane intervals in which the person is capable of contracting. A person may be sane in one or more areas of activity and insane in others. If a contract was made during the person's rational moments, providing the person had not been adjudged insane, it is binding despite previous or subsequent irrational behavior. Quite obviously, there can be some difficulty in establishing proof of sane behavior of an insane person or the limits of an area of activity in which the person is rational. These are questions of fact, however, and usually are left for a jury to decide on the basis of expert testimony of psychiatrists and psychologists.

Although it is generally true that courts will not examine the value of the consideration exchanged in a contract, they appear to make an exception where an insane or incompetent person's contract is involved. If it appears that the other party to the contract took advantage of the person's mental condition, the courts will allow that person to avoid the contract. If it can be shown that the other party had no knowledge of the condition of the other bargainer's sanity and did not take advantage of the incompetent party, the courts will usually let the contract stand. This is especially true where the parties cannot be returned to their previous status (such as with contracts for services performed where the work has been completed).

INTOXICATED PERSONS

If a person tries to avoid a contract by pleading intoxication, that person must prove he was so inebriated that he could not be expected to understand the consequences of entering into a contract. A minority of the courts take a very dim view of a suit pleading intoxication. They hold that intoxication is a voluntary state and that the person should have had foresight enough to avoid getting drunk.

In most jurisdictions, however, intoxication of a party at the time the party entered into a contract is a sufficient ground to avoid the contract. If, however, an innocent third person would be harmed by avoidance of the contract, the contract usually will be allowed to stand. Even in jurisdictions where intoxication is frowned on by the courts as a means of avoidance of a contract, it can be used to show susceptibility to fraud.

CORPORATIONS

State laws provide means whereby an organization may become incorporated. The purpose and scope of its activities are set forth in the organization's articles of incorporation or charter. The corporation is an artificial person and possesses full power to act within the limits of its charter. If a contract of the corporation goes beyond its charter limits, it is an *ultra vires* contract. Before the corporation is given the right to do business in a state, it has no enforceable capacity to contract. Contracts entered into by the not-yet-existent corporation cannot form the basis of a suit by the corporation. Moreover, the individuals supposedly acting on the corporation's behalf (the **incorporators**) remain liable, not the corporation. Until a corporation is formed (or registered as a foreign corporation if it was formed in another state), someone must make contracts for it. The corporation may take over these contracts after it is permitted to operate lawfully. The capacity to contract is also lost if the corporation's charter or its right to operate as a foreign corporation is suspended for some reason.

Ultra Vires Contracts

There appears to be considerable variation in court decisions in cases involving *ultra vires* contracts. They can, however, be classified in a general way and some general statements can be made about them.

It is probably best to consider *ultra vires* contracts from the standpoint of their stage of completion. If an *ultra vires* contract is entirely executory, the courts probably will not enforce it. If it has been completely executed, the courts will tend to leave it alone. When the contract performance has been completed by one of the parties but not by the other (that is, the contract is partially executed), the courts generally give a remedy. Some courts allow recovery on the contract, and others place the remedy on a quasi-contractual basis, holding that the contract itself is void but allowing recovery on the basis of unjust enrichment.

The Corporate Veil

The corporation's status as an artificial, yet legally recognized, entity (often called the **corporate fiction** or the **corporate veil**) affords limited liability to stockholders and the corporate officers. A stockholder generally risks losing only the amount invested in the stock. The same is true of corporate officers (president, vice-presidents, secretary, treasurer, and—perhaps—some managers) if the corporation has been fair and honest in its dealings with others.

The corporate veil protects these officers. But there is a limit to this protection and a reason to pierce the corporate veil when the people who run the corporation have purposely deceived, defrauded, and injured others. When such a situation occurs, these corporate officers (who are charged with the knowledge of the corporation's affairs) may be held personally liable for the liabilities of the corporation. The title of engineering vice-president may sound sweet, but with that title comes the joint and several responsibilities for the acts of the corporation. See the case of *Mobridge Community Industries v. Toure* at the end of this chapter.

OTHER BUSINESS ENTITIES

Besides the corporate form, a number of other legal forms are used to conduct business. The sole proprietorship is probably the most common. In a sole proprietorship, the owner and the business are treated as one and the same. Other forms include partnerships, which can be general or limited, and joint ventures. In a general partnership, all partners are individually liable for the debts of the partnership. (See *District Board of Trustees v. Don R. Morgan* at the end of the chapter).

A limited partnership, however, includes both "general" partners and "limited" partners. The **general partners** run the business of the partnership and remain individually liable for the partnership's debts. A **limited partner,** however, is not personally liable for the partnership's debts; at the same time, however, the limited partner cannot get very involved in the business operations, because to do so can result in personal liability. Hence, limited partnerships are somewhat like corporations in terms of allowing individuals to invest in the business and share in the profits while not subjecting themselves to individual liability (so long as there is no active involvement in the day-to-day operations of the business).

A limited liability company, or LLC, has features of both a corporation and partnership. As the name implies, liability is limited to the company itself. From a tax perspective, profits and losses from the business "pass through" to the individual members, which is a feature similar to that of a partnership. Members of an LLC can be living persons as well as corporations and other business entities.

GOVERNMENTS

The 11th Amendment to the Constitution prohibits suits against the United States or its individual states directly. Because of this prohibition, people who do business with federal, state, or municipal governments have a practical need to know the contracting restrictions imposed, especially because it may difficult to have the contracts enforced. For example, when a city charter indicates the manner in which contracts are to be made, the city will not be bound to contracts made in some other fashion. If the Greenville city charter requires that certain contracts must be accepted by a public works board, acceptance by Mayor Brown will not bind the city. Someone performing a contract on the basis of the mayor's acceptance might find himself or herself an unwilling donor to the municipality. Federal and state statutes also may require that a given contract be submitted for competitive bids, and that specified performance and payment bonds be given. If the bidding procedures are not followed, a court may declare the contract void or voidable. It should be noted that parties aggrieved by the actions of federal agencies, and in some instances the United States itself, may be able to bring a suit if the party has "standing" to do so. Suits of this kind follow the guidelines of the Administrative Procedure Act (discussed further in Chapter 23, Administrative Law).

> "The Judicial power of the United States shall not be construed to extend to any suit in law or equity, commenced or prosecuted against one of the United States by Citizens of another State, or by Citizens or Subjects of any Foreign State."
> United States Constitution, 11th Amendment, ratified February 7, 1765.

PROFESSIONAL PERSONS

Nearly all recognized professions have restrictions placed on them by state statutes. The statutes customarily require professional persons to be registered or licensed as such before they are allowed to contract for their professional services. One who performs professional services without being registered or licensed may not resort to the courts to collect fees for such services, and such a nonprofessional may even be penalized for making such a contract. (See *State of Arizona v. Wilkinson*, in Chapter 2.) White, an engineer working for Black Manufacturing Company, designs a product that is built by Black and later involved in an injury to Brown. Brown brings a negligence action, alleging that faulty design of the product caused his injury. If the design is proved to be faulty, against whom does Brown have a right to recover? The general answer is that Black Manufacturing Co. would likely bear the loss. The reason for the liability is the concept of *respondeat superior*—the employer is responsible for the acts of employees during their employment. However, if White made the design while working as a consultant for Black, White would be liable for the design faults. In a similar situation, if White were an employee of Green Consultants in making the design for Black, Green is then liable, again under *respondeat superior.*

JOHN A. TURJA; RICHARD H. TURJA v. STEPHEN F. TURJA 118 F.3d 1006 (4th Cir. 1997)

Study terms: Probate exception, diversity jurisdiction, lack of testamentary capacity

The appeal involves the venerable, but infrequently discussed, probate exception to a federal court's diversity jurisdiction. Two brothers brought this action against a third brother alleging that he exercised undue influence over their mother, which led her to execute a will, a trust agreement, and related documents in his favor. After a bench trial, the district court concluded that the mother lacked the requisite mental capacity and was unduly influenced in executing the trust documents, and so voided them. The district court, finding that it lacked jurisdiction because of the probate exception, refused to exercise jurisdiction over the will or rule on the mother's capacity to execute it. The contesting brothers appeal, asserting that the district court erred in refusing to exercise jurisdiction over the will, and in not awarding them attorneys' fees under the trust agreement. We affirm.

I.

John, Richard, and Stephen Turja are the only children of Dick and Marion Turja. John, a resident of Hawaii, and Richard, a resident of Utah, brought this suit, based on diversity of citizenship, against Stephen, a Virginia resident.

The family originally understood that the Turja estate (primarily the family home with a stipulated present value of $301,000) would go to the surviving parent, and after that parent's death, be split among the three sons. Since 1978, Stephen has lived with his parents, and in more recent years spent a substantial portion of his time caring for their needs. Indeed, in 1983, Stephen quit his full time job to assist his parents, who were becoming disabled. By 1986 Marion Turja, who had been close to all of her children and grandchildren, had become increasingly disoriented, confused, and drew away from them. At the time her husband died in 1991, Marion was quite disoriented and likely suffering from Alzheimer's disease.

On August 5, 1992, Marion executed a will bequeathing her china to John, forgiving a $10,000 loan to Richard, and leaving her house and residuary estate to Stephen. Later in the same month, Stephen took his mother to see a trusts and estates lawyer, Warren Grossman; Grossman was told that Marion wished to establish a living trust in which she would place most of her property, and at her death leave her china to John, forgive Richard's debt, and leave the house and her residuary estate to Stephen. Concerned that Marion was incompetent, Grossman contacted Marion's doctor and asked him to assess her testamentary capacity. After several visits, that doctor determined that Marion suffered from dementia and lacked the capacity to execute legal documents. Grossman then refused to create the trust. However, Stephen managed nonetheless to have a trust created that would mirror the distribution set out in the August 5 will; Marion executed the trust in November 1992 and additional related documents in January 1993, apparently again at Stephen's urging.

Marion died in August 1993. Stephen did not contact his brothers or his mother's sisters to inform them of his mother's death, and was the only person who attended her burial. A month later, he conveyed the family residence to himself pursuant to the trust agreement. Richard first learned of his mother's death through one of his wife's friends, who worked for the insurance company that handled Marion's death benefits.

After discovering what had happened, on November 1, 1993, John and Richard brought suit in the Circuit Court of Arlington County, Virginia. On December 30, 1993, Stephen offered Marion's will for probate in that court. Subsequently, John and Richard nonsuited their state action and, on August 24, 1995, filed this action in federal court against Stephen, individually and as executor of their mother's estate.

The complaint alleged six counts: Count I, lack of testamentary capacity; Count II, undue influence; Count III, fraud; Count IV, action to set aside deed and other transfers; Count V, constructive trust; and Count VI, unjust enrichment. The district court dismissed the first two counts without prejudice insofar as they involved Marion's will; the court reasoned that it had no subject matter jurisdiction over those claims because "federal courts may not hear probate matters as part of their diversity jurisdiction."

The court tried the remainder of the case. After a two-day bench trial, the court found: that Marion lacked the mental capacity to execute the trust documents, that Stephen exercised undue influence over his mother in order to get her to execute them, and "that the amendment to the Turja trust and other related legal documents . . . should be set aside as null and void." The court did not reach the other causes of action, finding resolution of them unnecessary to its holding.

John and Richard moved for attorneys' fees, under a provision in the trust agreement that provided for attorneys' fees to a prevailing party in the case of "any dispute arising out of this trust." The district court denied fees, holding that because it had invalidated the trust agreement, the provision in it regarding attorneys' fees could not be enforced.

John and Richard appeal; Stephen filed no cross appeal.

II.

John and Richard assert that the district court erred in refusing to exercise jurisdiction over their claims that Marion lacked testamentary capacity and was unduly influenced in executing her will.

They concede that this court has recognized the "probate exception" as a jurisprudential limit on diversity jurisdiction. However, they contend that the probate exception does not apply here because the relief they seek in connection with the will, which at the time the suit was filed only controlled personal property worth $100, was "incidental" to the other relief sought here. . . .

Moreover, a federal court does not gain jurisdiction to determine a will's validity merely because the issue is "incidental" to other claims. Instead, we must look at the contours of the "probate exception," for if John and Richard's claims regarding their mother's will fall within this exception to federal diversity jurisdiction, the district court may not address them no matter how close their connection to claims (like those involving creation of the trust) over which a federal court does have jurisdiction.

The leading Supreme Court precedent is *Markham v. Allen,* 326 U.S. 490, 90 L. Ed. 256, 66 S. Ct. 296 (1946). In *Markham,* the Court determined that although "a federal court has no jurisdiction to probate a will or administer an estate . . . federal courts of equity have jurisdiction to entertain suits 'in favor of creditors, legatees and heirs' and other claimants against a decedent's estate 'to establish their claims' so long as the federal court does not interfere with the probate or assume general jurisdiction of the probate or control of the property in the custody of the state court."

In *Foster,* 200 F.2d at 947, we further explained that:

The law is well settled that the federal courts have no jurisdiction over matters within the exclusive jurisdiction of state probate courts. However, as to matters which do not involve administration of an estate or the probate of a will, but which may be determined in a separate action *inter partes* in the courts of general jurisdiction of the state, the federal courts do have jurisdiction if the requisite diversity of citizenship exists.

In Virginia, the probate court has jurisdiction to determine "whether the writing, or any part of it, is the true will of the deceased and whether the writing is testamentary in character." This jurisdiction is exclusive. We have previously recognized this feature of Virginia law, holding that federal courts have no jurisdiction over a suit brought to set aside a will based on claims of undue influence and mental incompetence. . . .

There are good reasons for the probate exception's limit on federal jurisdiction. As Judge Posner explained in one of the cases upon which John and Richard principally rely, reserving probate matters to state courts generally promotes legal certainty, judicial economy, and resolution by a court expert in those matters. Although in this case, honoring the probate exception will result in a somewhat inefficient use of judicial resources, in most lawsuits that will not be the case. Furthermore, we note that John and Richard could have avoided the duplication here by pursuing their case where they originally initiated it—in state court. Finally, we are confident that the Virginia probate court will take into account the district court's factual findings regarding the trust documents when evaluating Marion's testamentary capacity to execute her will. . . .

Thus the judgment of the district court is

AFFIRMED.

THE DISTRICT BOARD OF TRUSTEES, ETC. v. DON R. MORGAN, ETC., 890 So. 2d 1155; 2004 Fla. App. LEXIS 19293

Study terms: Ratification, certificate of authorization, void *ab initio*, licensing statutes, partnership, fraudulent inducement, voidable, misrepresentation, non-compliance

The District Board of Trustees of St. Johns River Community College appeals the final judgment rendered by the trial court after a jury trial. The appellee is an architect who had entered into a contract for the provision of professional services to the College, and the underlying dispute involved whether the architect was entitled to payment for certain work performed. While we affirm the judgment, we write to address an issue that in some respects brings us into conflict with our sister court in the First District. The issue concerns whether a contract entered into between the College and a partnership comprised of two fully licensed Florida architects who never obtained the certificate of authorization described by section 481.219, *Florida Statutes* (1999), is void *ab initio,* and is therefore unenforceable by the architects. We conclude that the contract was only voidable, not void, and was accordingly enforceable while it was still in existence.

We recite only the facts necessary to adjudicate this particular issue on appeal. Essentially, in April of 1999, the College entered into an AIA "Standard Form of Agreement Between Owner and Architect" with Morgan-Stresing Associates ("MSA"). The agreement, which was signed both by the appellee, Don Morgan, and by his purported partner, Paul Stresing, concerned the design and preparation of bidding documents associated with the construction of the St. Johns River Community College's visual and performing arts complex. Mr. Morgan and Mr. Stresing are both licensed Florida architects, having met the requirements of section 481.213, *Florida Statutes* (1999). The problem is that their partnership, MSA, never obtained a certificate of authorization, as is required by section 481.219, *Florida Statutes* (1999).

Section 481.219 allows architecture to be practiced through a corporation or a partnership, subject to certain conditions having been met. *Subsection (2)* of the statute reads in pertinent part as follows:

> For the purposes of this section, a certificate of authorization shall be required for a corporation, partnership, or person practicing under a fictitious name, offering architectural services to the public jointly or separately. However, when an individual is practicing architecture in her or his own name, she or he shall not be required to be certified under this section.

After the architects had completed about 70 percent of their contract work, they were terminated by the College. When this suit was commenced, the College counterclaimed and asserted a number of affirmative defenses in which, among other things, it pointed out that MSA had failed to obtain a certificate of authorization. Thus, they posited, the contract was void *ab initio,* and was obtained by fraud in the inducement. MSA conceded that it had actively represented to the College that both Mr. Morgan and Mr. Stresing would devote their time as principals to the project. They claimed, however, that they were in the process of obtaining the certificate of authorization at the time of the termination. They argued that the College was simply looking for a way to use the plans prepared by Mr. Morgan and Mr. Stresing without paying full price.

It appears that the Board of Trustees had knowledge as early as December of 1999, that Mr. Morgan and Mr. Stresing were not "partners," and that a certificate had not been obtained. There is also testimony, however, that the two men were told by the College to continue to work together on the project, and to try to develop a resolution of the partnership issue. In June of 2000, the College wrote to the attorney representing the architects telling them to cure the certificate of authorization problem by June 30. MSA failed to meet that deadline. . . .

The evidence also indicated that five weeks later Mr. Morgan and his attorney attended a meeting of the Board of Trustees, and Mr. Morgan and Mr. Stresing provided proposed modifications to the original contract in which they offered . . . different solutions to the problem caused by the lack of a certificate of authorization. . . . The Board rejected the proposals, terminated the contract, and hired a new architect. Significantly, however, there was evidence that the College consciously elected to use the MSA plans, and turned those plans over to its new architect. In addition, the College paid consultants money that was owed to MSA.

The jury found on the counterclaim of the College that the architects had in fact obtained the contract by fraud, and that the College had suffered damages of $61,476.58, as a result. The jury also found, however, that the College breached the contract it had with MSA, and that MSA suffered damages of $413,049.68. After the damages were adjusted accordingly, and a final judgment in favor of the architect was rendered, this appeal ensued.

The College argues that the final judgment should be reversed because the contract between the College and MSA was in violation of statute, and was, therefore, void and not enforceable by MSA. The College relies on the failure of MSA to obtain a certificate of authorization in support of its theory that the contract was void. The foundation for this proposition is *O'Kon and Company, Inc. v. Riedel,* 588 So. 2d 1025 (Fla. 1st DCA 1991).

O'Kon was a Georgia corporation engaged in the business of providing architectural and engineering services. It decided to venture into Florida for work. In addition to hiring a licensed Florida architect, O'Kon hired an unlicensed architect living in Florida to work on the plans. O'Kon, however, had not registered in Florida, and had not received a certificate of authorization, contrary to section 481.219. After the developer decided to abandon the project and not pay its architects, O'Kon filed a claim of lien against the real property that was the subject of its contract, and also sought to collect on its contract claim, even though it was unlicensed and unregistered in this state. O'Kon argued that all the architectural work in the case was either done by the Florida licensed architect or supervised by him (O'Kon being the principal), and that registration was a mere ministerial and technical act. There was, however, testimony that one unlicensed architect had, indeed, worked on the plans.

■ ■ ■

[T]he court held that the employment of an unlicensed architect to work on the plans was a violation of the state's "licensing statute," and justified the conclusion that the contract was unenforceable. *See O'Kon and Co., Inc. v. Riedel,* 588 So. 2d 1025 (Fla. 1st DCA 1991)("*O'Kon II*").

The problem is that both Mr. Morgan and Mr. Stresing, the sum total of persons offering architecture services through MSA, are both fully licensed by the State of Florida as architects. The licensing statute for architects is section 481.213, not section 481.219. There was no evidence submitted below that any other person was employed by MSA for the purpose of being involved with the preparation of the plans and specifications for the Community College project. Thus, Mr. Morgan and Mr. Stresing did not violate a "licensing statute" when they failed to obtain a certificate of authorization. They both had licenses. While we have no difficulty in concluding that unlicensed persons have no right to enforce a contract for services that require a license, the present case is simply not one involving unlicensed architects. Because there was substantial competent evidence that all persons working on the project under the MSA banner were fully licensed; and that the College was fully aware of the lack of a certificate of authorization for months before they terminated the contract; and that the College accepted the work produced by MSA well after it became aware of the lack of a certificate, there was no risk that professional architectural services would be foisted on the public, in general, and on the College, in particular, by unlicensed persons. To blindly adopt the *O'Kon II* rationale under these circumstances would, in our judgment, be unjust, and would elevate form over substance.

We find ourselves in closer agreement with . . . Judge Allen's well-reasoned special concurrence in *O'Kon II*. As Judge Allen pointed out, the language of section 481.219 does not compel a conclusion that a failure to obtain the certificate of authorization invalidates a contract by the architectural organization *ab initio.* In fact, section

/ 481.205, *Florida Statutes,* appears to indicate that enforcement of the licensing and other statutes contained in Chapter 481 is placed in the hands of the Board of Architecture and Interior Design. Thus, if there is a disciplinary action to be taken against Mr. Morgan, Mr. Stresing or MSA for failure to obtain a certificate of authorization, then the Board of Architecture and Interior Design is the body designated by the legislature to take that action. Voiding the contract after hundreds of thousands of dollars of work was performed and accepted just does not pass the smell test. As the trial judge cogently noted in denying the motion of the College for a directed verdict:

> But it seems patently unfair to contend they (MSA) didn't have the legal authority to produce the product, and then you (the College) take the product, and then claim they can't sue you to get paid for the property that you've taken.

The fairer way to conceptualize an agreement tainted by non-compliance with section 481.219 is to consider the contract to be voidable, in much the same way that fraudulent inducement renders a contract voidable, but not void. . . .

Here, there was ample evidence from which the jury could find that the College was aware of the non-compliance with the statute, yet chose to continue to operate under the contract and to benefit by it. While the College may well have been able to void the contract when it learned of the statutory non-compliance, it did not have the luxury of waiting month after month to pull the trigger, all the while accepting the work of the architects in the interim.

■ ■ ■

[W]e note that the College sought and received damages for fraudulent inducement. The fraud alleged was the misrepresentation of the architects about their business relationship. Consistent with the proposition that fraudulent inducement renders a contract voidable, not void, Florida law provides an election of remedies where fraudulent inducement is claimed. The wronged party may either repudiate the contract and seek rescission, or ratify the contract and seek damages. As the Supreme Court has indicated, "This principle ensures that a party who 'accepts the proceeds and benefits of a contract' remains subject to 'the burdens the contract places upon him.'" *See Mazzoni Farms,* 761 So. 2d at 313 *(quoting Fineberg v. Kline,* 542 So. 2d 1002, 1004 (Fla. 3d DCA 1988)). Here, the College ratified the contract. Having done so, it cannot be heard to complain because the jury agreed and found that the contract was enforceable by both parties.

In substance, therefore, what we hold is that there is no automatic avoidance of a contract because of a violation of *section 481.219;* in as much as such contracts are only voidable, at the election of the non-offending party. The contract might very well be voided, depending on the circumstances, but that conclusion is frequently fact dependent. Such is the case here. The jury heard and considered the evidence, and found that while the architects may have committed a fraud in inducing the contract, the College had breached the contract while it was still viable. As there is substantial competent evidence in the record to support that finding, we have no reason to second-guess the outcome.

To the extent that the opinion in *O'Kon II* renders void and unenforceable all architectural contracts that are entered into by an architectural business organization that fails to obtain a section 481.219 certificate of authorization, we certify conflict.

AFFIRMED.

Review Questions

1. Black, a minor, sells a piece of real estate that Black inherited. Black spends the money rather foolishly and, by the time of adulthood, Black is deeply in debt. Black's creditors force Black into bankruptcy. The creditors find that the land Black sold as a minor is quite valuable; in fact, the value is sufficient to pay Black's debts and leave a substantial remainder. Black refuses to disaffirm the sale. May the creditors disaffirm for Black? Why or why not?

2. What is an *ultra vires* contract? Give an example of such a contract.

3. Gray, a salesperson, contracted with Green to sell Green a piece of equipment at a price about 10 percent below the usual market price for such equipment. The contract occurred in a local restaurant during an extended lunch hour. Gray had two or three martinis before eating lunch. Gray now claims he was drunk at the time the contract was made and, on this basis, seeks to avoid the contract. Green has established that all of Gray's other acts (including paying both restaurant tabs) gave no indication that Gray was under the influence of alcohol. Is it likely that Gray will be held to the contract? Why or why not?

4. Are contracts for professional services voidable by the recipient of the services if the professional is not licensed?

5. In the case of *Turja v. Turja,* why did the district court refuse to rule on whether the mother had the requisite mental capacity to execute her will? Recall federal jurisdiction from Chapter 4: Why is this case before the U.S. Court of Appeals for the 4th Circuit?

6. What recommendations would you have made to the board to ensure that your arrangement with it was not in violation of the statute in *District Board of Trustees v. Don R. Morgan?*

7. The court proposes in *District Board of Trustees v. Don R. Morgan* that the Board of Architecture and Interior Design was the more appropriate venue to try and enforce licensing issues. Assume for purposes of this question that Mr. Morgan and Mr. Stresing were engineers in as well as architects. Did they violate any NSPE canons or rules?

Agreement

HIGHLIGHTS

■ An agreement is comprised of an offer and acceptance.

■ The Uniform Commercial Code supplements the common law in order to assist states in standardizing contract-related agreement terms.

■ New electronic contracting statutes help fill gaps left open in the UCC and common law regarding electronic transactions.

A contract is an agreement, although not all agreements are contracts. According to *Black's Law Dictionary,* an **agreement** is "the coming together in accord of two minds on a given proposition." An agreement consists of an offer and an acceptance of that offer (refer back to the definition of a contract in Chapter 5). The Uniform Commercial Code (UCC) speaks in terms of the parties' "bargain." The UCC does not define what constitutes an offer or an acceptance. Instead, the common law remains effective as to such issues (see Figure 7.1). Hence, this chapter's discussion addresses the common law rules on the formation of contracts.

CONTRACT FORMATION

To form a contract, there needs to be an offer and an acceptance of that offer. What amounts to an offer or an acceptance is considered here, as are issues regarding just when and how an offer may be accepted.

§ 1-101. Short Title
This Act shall be known and may be cited as the Uniform Commercial Code
§ 1–102. Purposes; Rules of Construction; Variation by Agreement.
  (1) This Act shall be liberally construed and applied to promote its underlying purposes and policies.
  (2) Underlying purposes and policies of this Act are
      (a) to simplify, clarify and modernize the law governing commercial transactions;
      (b) to permit the continued expansion of commercial practices through custom, usage and agreement of the parties;
      (c) to make uniform the law among the various jurisdictions.
  (3) The effect of provisions of this Act may be varied by agreement, except as otherwise provided in this Act and except that the obligations of good faith, diligence, reasonableness and care prescribed by this Act may not be disclaimed by agreement but the parties may by agreement determine the standards by which the performance of such obligations is to be measured if such standards are not manifestly unreasonable.

Figure 7.1

Offer

An **offer** contemplates a future action or restraint of action. It is a proposal to make a contract. In general, an offer states two things: what is desired by the offeror and what the offeror is willing to do in return.

The *Restatement (Second) of Contracts* defines an offer as follows:

> [An offer is] the manifestation of willingness to enter into a bargain, so made as to justify another person in understanding that his assent to that bargain is invited and will conclude [the bargain].

The offer does not have to be a formal statement to be binding on the offeror. When the intent to offer is conveyed from the **offeror** (the person making offer) to the **offeree** (the recipient of the offer), an offer occurs. Under certain circumstances the law will consider a series of acts by the parties to be an offer and acceptance of the offer resulting in a binding contract. A man entering a hardware store where he is well known, grasping a shovel marked $14.95, motioning to the owner who is busy with another customer, and then leaving the store with the shovel, has made an offer to purchase. Although no word was spoken, an offer to buy the shovel at the suggested price of $14.95 was made. By allowing the man to leave with the shovel, the hardware store owner accepted the offer, and a contract resulted.

Intent.   The usual test of both offer and acceptance is the **reasonable man standard.** Would a reasonable man consider the statements and acts of the party to constitute a binding element of the agreement necessary in a contract? The circumstances surrounding the statements and acts of the parties are examined to determine the existence of an intent to contract. In considering the circumstances, the courts gauge the parties' intentions objectively, rather than each party's "subjective" view of what they meant and understood the other to mean. In one case, a blacksmith shop owner, enraged at the loss of a harness, stated in his irate ravings about the alleged thief that he would pay anyone $100 for information leading to the capture of the thief and an additional $100 for a lawyer to prosecute him.[1] The court held that the language used, under the circumstances, would not show an intention to contract to pay a reward. In another case, Justice Oliver Wendell Holmes, Jr. stated, "If, without the plaintiff's knowledge, Hodgdon did understand the transaction to be different from that which his words plainly expressed, it is immaterial, as his obligations must be measured by his overt acts".[2]

It is often quite difficult to determine by the language used by the parties whether a valid offer and acceptance have occurred. In the bargaining that often occurs in attempting to reach an agreement, words are used that, removed from context, sound as though the parties had formed a contract. One person may say to another, "I would like to sell my car for $1,500," or "Would you give me $1,500 for my car?" These are not offers. The courts, in recognition of business practices, view such statements as solicitations to offer and not as binding offers.

Advertisements.   When making an offer, the offeror must be prepared for acceptance of the offer. Acceptance would complete the contract, and, if the offeror could not perform the contract, the offeree would have a legal action for damages available against the offeror. For this reason, advertising—prices marked on merchandise in stores and similar publicity—is considered by the courts to consist of mere invitations or inducements to offer to purchase at a suggested price. The courts generally do not view such conduct as amounting to an offer. An expression by a customer of willingness to buy the merchandise at the price stated does not form a contract. The customer, not the store, has made the offer. The store has the right to accept or reject the offer. It is possible for the store to have sold the last of its merchandise in that line and thus to find performance to be impossible. If an offer of the price by a customer were to be construed as acceptance and the store could not perform, it would be liable for breach of contract. To hold that the store has made a contract in all cases in which a customer tenders the price of certain merchandise would therefore be unreasonable.

Of course, some advertisements are worded (whether purposely or accidentally) so as to constitute an offer. Acceptance by anyone will then form a binding contract. A common example of an advertisement that constitutes an offer is an ad offering a reward. To accept, anyone with knowledge of the advertisement may perform the act required in the ad, thus completing the requirements for a valid contract.

An advertisement (or solicitation or request) for bids on construction work generally does not constitute an offer. Unless the advertisement specifically states that the lowest bid will be accepted, the advertiser has the right to reject any or all bids submitted. Similarly, a quotation of a specific price usually does not amount to an offer. Such requests for bids and price quotations are usually viewed as preliminary negotiations only.

Continuing Offer.   An offer generally provides the offeree with a continuing power to accept the offer until the offer is revoked. When the offeror agrees not to "revoke" the offer (that is, to keep the offer open for some set period of time) and such an offer is based on consideration, it is an option, a contract in itself. The offeror basically agrees to perform or not to perform certain acts on acceptance by the offeree, providing only that the offer be accepted within the specified time limit.

---

[1] *Higgins v. Lessig*, 49 Ill. App. 459 (1893).

[2] Mansfield v. Hodgdon, 147 Mass. 304, 17 N.E. 544 (1888).

Termination of Offer.   An offer may end in various ways: by acceptance, by revocation or withdrawal, by rejection, by death of a party, or through the passage of time. The expiration of an option supported by consideration, however, ends only on expiration of the time period for which it was to be held open. An option cannot be revoked, and rejection of the offer or death of one of the parties usually does not end its life.

> The phrasing of a modification of an acceptance resembles this: "I accept your offer providing you will do [an additional condition] in addition." Such a statement may not constitute acceptance, but rather a counteroffer. The Uniform Commercial Code is more likely than the common law of contracts to lead to a finding of acceptance in situations such as this.

The offeror has the right to revoke the offer at any time before acceptance. The withdrawal or revocation must be communicated to the offeree before the offeree's acceptance. If the offeree has knowledge of the sale to another of the goods offered on a continuing offer, the offeree has effective notice of withdrawal of the offer.

Rejection by the offeree ends the offer. The rejection of the offer often takes the form of a qualified acceptance of the offer or a request for modification of the offer. Either act usually amounts to rejection. Mere inquiry as to whether the offeror will change the terms, though, does not constitute a rejection of the offer.

Generally, the death of either party before acceptance of an offer terminates the offer. This is the rule even when the offeree has not received notice of the death. The *Restatement (Second) of Contracts* describes this rule as a "relic" of the view that it is necessary for a meeting of the minds to occur for a contract to be formed. If one of the minds is no longer among the living, so the reasoning goes, no contract can occur.

If a time limit is specified in a continuing offer, the offer terminates on the expiration of the period of time in which the offer was to remain open. If no time was specified in the offer, the offer will be open for acceptance for a reasonable time, the reasonableness of the time depending on the circumstances and subject matter of the offer. One would expect that a reasonable time to decide whether to purchase a truckload of ripe bananas (about to lose their value) would be much shorter than a reasonable time to decide whether to buy furniture.

Acceptance

Generally, for a contract between two parties to exist there must have been an acceptance of an offer. However, an offer may be accepted and a contract thus formed only during the life of the offer. With the exception of an option contract, as mentioned earlier, once an offer has been rejected or revoked, or its time has run out, it cannot be accepted.

Similarly, once an offer has been accepted, it is impossible for the offeror to withdraw the offer. The rules sound simple, but in many situations it is often difficult to determine whether an acceptance has taken place. Such uncertainty often arises because of various ambiguities in the parties' statements, letters, or conduct.

An offer may be accepted only by the person to whom it was made. A person other than the offeree who obtains the offer cannot make a valid contract by accepting. The best that such a person can do is to make a similar offer to the original offeror for acceptance. An offer must be communicated to the offeree by the offeror or the offeror's agent to constitute an acceptable offer. Black tells Gray, who is a friend of both Black and White, of Black's intent to offer White a television set for $500. Gray reveals this to White. White cannot at this point create a valid contract with Black by stating White's acceptance. White must either wait for Black to communicate the offer to her or offer Black $500 for the TV set. At this point there is no offer to be accepted. However, Black could have used Gray as his agent, requesting Gray to give the offer to White. Black's comments then would have constituted a valid offer.

Acceptance of an offer is held to consist of a state of mind evidenced by certain acts or statements. The mere determination to accept is not sufficient; it must be accompanied by some overt act that reveals that the offeree accepts the offer. Without such an overt act, it is impossible for the offeror or the court to determine whether the offer has been accepted, despite the offeree's secret intent to accept, unless previous dealings have established the offeree's silence or inactivity as acceptance. Generally, the offeror may specify the manner in which the offer may be accepted, such as by letter, email, signature of the contract document, or the like. If the offeror does not specify the mode of acceptance, the offeree may accept in any manner and by any medium reasonable under the circumstances. In *Wachter Management Company v. Dexter & Chaney, Inc.*, at the end of the chapter, Wachter entered into a service contract with Dexter & Chaney for the installation of software as well as software maintenance, training and consulting services. The software packaging had language asserting that upon opening the sealed package, the purchaser agreed to the company's licensing agreement. The majority and dissenting opinions for the case disagreed about whether Wachter expressly agreed to the terms of the shrinkwrap license. The Kansas City Supreme Court looked to traditional contracting analysis and concluded that the shrinkwrap was an offer to modify the terms of the contract set forth in the original service agreement.

The communication of offer and acceptance creates a contract, regardless of how it is done. When the offer and acceptance are oral, the contract is formed as soon as the offeree speaks the words of acceptance.

One sticky problem often arises in regard to the offeree's communication of acceptance. Suppose the parties are not in voice contact with each other. For example, suppose they are communicating by mail. Then let us suppose the acceptance is made in timely good order but gets lost in transit. Is there a contract or not? This is rather important information for the offeror, because the same offer probably will be made to someone else if the original offer is rejected.

After some years of confusion on this point the law is now pretty well settled—there is a contract. It sprang into being when the letter of acceptance was mailed. The offeror can protect himself or herself from a possible breach of contract action by including a requirement that acceptance be made in a writing received in the offeror's office by some fixed time to be effective.

Time of Revocation

An offer may be withdrawn by the offeror up until the time it is accepted. The offeror's revocation, however, is never effective until it has been communicated to the offeree. When the negotiations are carried on by mail or email, a question may arise about whether a contract exists. To point up this difficulty, consider the following example. Black and White have been negotiating by mail for the sale of White's land. White finally makes an offer that Black finds acceptable, and Black sends a letter at 12:00 noon to White, accepting the offer. Previous to this, White has sent a letter to Black revoking the offer, but the withdrawal is not received by Black until some time later than noon on the day Black mailed his acceptance. The courts would hold that there is a contract between the parties, and even the loss of the letter of acceptance would not change the holding, providing only that Black could prove sending such a letter and that the letter constituted acceptance of the contract. As you might conclude, these rules encourage offerors to convey a revocation of the offer quickly, lest the offeree accept in the meantime.

Silence as Acceptance

Unless there is a customary practice or agreement between the parties to the contrary, silence generally cannot be construed as acceptance. Wording an offer to the effect that "If I do not hear from you, I will consider my offer accepted by you" has no legal effect upon the offeree. Mere silence generally does not form a binding contract.

The "Battle of the Forms"

In many situations, the parties informally negotiate only a very few terms of a contract, such as price, quantity, and time of delivery. The parties then exchange preprinted forms—the buyer sending a purchase order and the seller sending an order acknowledgment or invoice. The preprinted forms usually contain the buyer's terms of purchase and the seller's terms of sale. Almost always, these forms are written with a very one-sided view in favor of the party sending the form. Thus, the buyer's terms almost never match the seller's terms. Each party wants its form to apply rather than the other's. The question arises as to whether a contract was ever formed and, if so, what its terms were. Because the parties often at least partially perform the contract by delivering goods or by accepting or paying for all or some of the goods, it is usually clear that there was some contract, whatever its terms.

Because the common law applied a "mirror image" rule (requiring the acceptance to "mirror" all of the terms and conditions of the offer), the party sending the last form usually retained the advantage. Why? The buyer's purchase order usually amounts to an offer. The seller's invoice or order acknowledgment, however, usually acts simultaneously as a counteroffer and a rejection of the original offer. (Remember, the seller's terms almost always vary substantially from the buyer's terms.) If the buyer sends no further forms and makes no further offers, and the seller then delivers goods that are accepted by the buyer, the buyer will be deemed to have agreed to the seller's terms by way of performance.

The UCC approaches the problem involving the battle of the forms differently from the common law approach. (See Figure 7.2.) Notice that section 2-207(2) creates a division between the

§ 2-207. Additional Terms in Acceptance or Confirmation
(1) A definite and seasonable expression of acceptance or a written confirmation which is sent within a reasonable time operates as an acceptance even though it states terms additional to or different from those offered or agreed upon, unless acceptance is expressly made conditional on assent to the additional or different terms.
(2) The additional terms are to be construed as proposals for addition to the contract. Between merchants such terms become part of the contract unless:
    (a) the offer expressly limits acceptance to the terms of the offer;
    (b) they materially alter it; or
    (c) notification of objection to them has already been given or is given within a reasonable time after notice of them is received.
(3) Conduct by both parties which recognizes the existence of a contract is sufficient to establish a contract for sale although the writings of the parties do not otherwise establish a contract. In such case the terms of the particular contract consist of those terms on which the writings of the parties agree, together with any supplemental terms incorporated under any other provisions of [the UCC].

Figure 7.2

rules applicable to contracts "between merchants" and contracts not between merchants. Generally speaking, a merchant is one who regularly deals in goods of the kind that are the subject of the contract. If the parties are not merchants, either party's added terms on the paperwork (e.g., order or order acknowledgment) are treated as "proposals" for addition to the contract. If, however, the purchase order states that the seller must accept all the buyer's terms to complete the contract, then an order acknowledgement cannot alter those terms without defeating the contract. On the other hand, the parties' subsequent conduct in performing the contract may establish the existence of some contract, probably governed largely not by either of the printed forms but by the UCC. If the parties are merchants, the new terms presented by the seller's order acknowledgment apply unless any of the following occurs:

- The offer limits acceptance.
- The new terms materially alter the buyer's terms.
- The buyer objects.

A careful buyer using a printed form usually limits the terms of the acceptance and also objects to the addition of any new terms. Generally, such new terms will not be added and the buyer's terms often will control. Thus, the UCC adopts a rule favoring buyers in many situations (at least in those situations involving merchants).

ELECTRONIC CONTRACTING

The general contracting principles discussed above govern most traditional day-to-day transactions. What they do not cover entirely are those contracts entered into when the parties are conducting business via electronic means, and not at arms-length, where the ability to negotiate is prohibitive and often undesirable. The UCC attempted to address this problem under proposed UCC Section 2B, which was later codified into state acts governing electronic transactions and a federal law regarding electronic signatures. (See Uniform Computer Information Transactions Act (UCITA) and the Uniform Electronic Transactions Act (UETA), and Electronic Signatures in Global and National Commerce Act (E-SIGN Pub. L.No. 106-229, 114 Stat. 464 (2000)).

UETA and UCITA

These uniform state acts were created to provide some unity in the formation, execution, and enforcement of electronic contracts. UETA was drafted in 1999 by the National Conference of Commissioners on Uniform State Laws (NCCUSL). Several jurisdictions have adopted UETA, often without any significant changes. UETA allows electronic transactions to be conducted in ways similar to those conducted under traditional contracting practices. Electronic signatures are validated under it.

The NCCUSL also drafted UCITA in 1999. The only states that have adopted it are Maryland and Virginia, with several others still considering it. UCITA involves computer information transactions only, specifically computer software, digitized music, diskettes, and compact disks. In addition, it governs those contracts entered into on a vendor's website or internal computer system, as well as service contracts.

E-SIGN

E-SIGN went into effect on October 1, 2000. It governs the legal effect, validity, and enforceability of electronic contracts and electronic signatures. There are several exemptions where E-SIGN has no effect. These exemptions include transactions involving wills and trusts, family law issues such as adoption and divorce, and judicial orders and filings. If a state has adopted UETA in its entirety as well as a law similar to E-SIGN, the state law governs.

Concerns about Electronic Contracting

For the most part, traditional contract law applies to electronic transactions. There are a few unique situations that occur in the online world that are not faced in the offline world. These situations are discussed below.

Clickwrap Agreements.   In the offline world, when a person makes a software purchase, they are often unaware that simply by opening the box or breaking a seal, they are agreeing to the terms of a licensing agreement. Most licensing agreements are either printed on the box or diskette/CD packaging. If a potential purchaser does not want to be bound by the terms of the license, he or she does not open the box or tear the seal. Upon opening and removing a seal, the purchaser has in effect agreed to the terms of the agreement. In the online world, this dilemma has been manifested in "clickwrap" agreements where a purchaser has to accept in some fashion the terms of the licensing agreement, often before he or she can download the software. UCITA governs these agreements. It requires that a potential purchaser have an opportunity to review the terms of a license before being obligated to pay.

Signatures.   The Statute of Frauds, discussed in Chapter 11, requires that certain contracts be signed in order to be enforceable. This requirement is not relieved online. What is relieved is the requirement for a handwritten signature for some types of contracts. Under the UCC, a signature can also be a symbol,[3] a digitized handwritten signature, or a digital signature using key-based technology.

Acceptance.   Under traditional contracting principles, an offerer can revoke his or her offer any time before the offeree accepts it. An acceptance goes into effect upon its mailing. UCITA says that an acceptance goes into effect when it is received by the other party.

   The complexity of modern business often leads to situations in which it is unclear exactly what terms govern the parties' relationship. For example, assume that the buyer and seller agree over the phone on price, quantity, and date of delivery, then exchange conflicting documents in both paper and electronic forms. Whose terms govern? As you consider the cases at the end of the chapter, determine whether the results were fair and how the outcomes could have been avoided. At least one scholar has noted that it is lucky that most business persons do not read the other's forms and that most transactions do not develop into disputes.

---

[3]UCC § 1-201(39).

MELVIN v. WEST 107 So. 2d 156 (Fla. Dist. Ct. App. 1958)

Study terms: Oral agreement, real property, authorization, consideration, nonexclusive contract, "ready, able and willing"

West, a real estate broker, sued Melvin, a property owner, to recover a real estate commission allegedly earned by West for effecting the sale of certain of Melvin's lands to one Buchanan and his associates. By stipulation, the case was tried before the Circuit Judge without a jury and judgment was for the plaintiff. The defendant has appealed from the judgment, assigning as grounds for reversal that the evidence does not support the judgment and that the evidence shows that a nonlicensed employee of the plaintiff actively participated in procuring the listing of the land and in procuring the purchaser, and hence the plaintiff is barred by law from recovering a commission.

According to the evidence most favorable to the plaintiff, the defendant owned a piece of land which plaintiff understood contained approximately 3,000 acres and was for sale at $1,100 an acre. The defendant orally listed this property for sale with the plaintiff, and several other brokers, with the specific understanding that "the first registered real estate broker who brought in a check and a signed sales contract would be the successful negotiator of the deal."

About a month after West had been given the listing, he informed Melvin that he had a prospect who was interested in buying the 3,000-acre tract at $1,100 an acre. In reply, Melvin told West that the price was not $1,100 an acre for 3,000 acres, as understood by West, but was $3,500,000 whatever the acreage. West conveyed this information to his prospect, Buchanan, and thereafter and with full knowledge of the fact that the price of the property had been fixed at $3,500,000, West and Buchanan inspected the property.

Shortly after viewing the property, West and Buchanan conferred with Melvin about the purchase of the property by Buchanan and two associates, but came to no definite agreement in regard to a sale. A few days after this conference, Buchanan informed West that he had ascertained that the tract did not contain 3,000 acres, as had been originally represented by West to him, but a lesser acreage, and hence that he and his associates were not interested in buying the property at the quoted price of $3,500,000 but might be willing to pay $2,750,000, and that if the counter offer was not acceptable they "would have no further interest in the matter."

After receiving this counter offer from Buchanan, West phoned Melvin and told him that if he was interested "in helping save the deal" he and West should "get (their) heads together and see if (they) could work out some way to work the situation out . . . (that) Mr. Buchanan was still interested in the property as such . . . and if the price and terms could be worked out to (Buchanan's) satisfaction . . . (he, West) felt the deal could still be made."

The evidence shows that a few hours after this conversation, West, Melvin, and L. V. Hart, one of West's employees, met in West's office at Ft. Lauderdale to discuss what might be done to "save the deal." The meeting ended when Melvin decided, with West's approval, that since he and Buchanan "were the two principals in the situation" he, Melvin, would negotiate directly with Buchanan, if someone would drive him to Miami Beach for the purpose. Thereupon Hart, West's employee, agreed to drive Melvin to Miami Beach where Buchanan was found and a conference was held, which began at approximately 5 o'clock in the afternoon and ended about three hours later, when Melvin finally agreed, after much discussion, to reduce the property from $3,500,000 to $2,900,00, provided $600,000 was paid in cash and the remainder was paid in specifically agreed annual installments, with interest, over a 10-year period; and also agreed that since Buchanan had no authority to commit his associates to the purchase of the property at the new price and terms agreed on at the conference he, Buchanan, might have "until two o'clock the next day (Friday, March 16, 1958) to obtain the permission of (his) associates to continue with the business on the basis of this present agreement and to authorize the contract to be drawn up on this basis"; that in the event permission was obtained "to continue with the business," Buchanan would inform West and would "deposit" with him a check in the sum of $50,000 as a "binder payment"; and that Mr. and Mrs. Melvin would meet with West and Buchanan, at Buchanan's office, on Saturday morning, March 17, 1956, to execute a contract of purchase, to be prepared by West's attorney, which would bind the parties to the sale and purchase of the property.

The evidence is to the effect that on Friday morning, March 16, Buchanan got in touch with his two associates and obtained their permission to purchase the property at the price and terms agreed to by Melvin at the Thursday afternoon conference; that Buchanan called West, at about 11:30 Friday morning, to advise him that he had had "the matter confirmed and . . . had a check available"; and that West immediately phoned the Melvin residence to inform Melvin of what Buchanan had said, but was told by the person answering the phone that Melvin was away from the premises. West continued calling the Melvin residence at frequent intervals throughout the remainder of the day but was never able to contact Melvin. That night, at a few minutes past midnight, Melvin called West to inform him that on that day he had sold the property to two prospects produced by a real estate salesman, one Harris, who also had an open listing on the property; that a downpayment check had been offered and accepted, and that a sales contract had been executed by seller and buyers and hence that he could not sell the property to Buchanan.

The testimony of plaintiff's witness, Harris, the real estate salesman who produced the purchasers to whom the property was sold, was to the effect that on Thursday, March 15, 1956, he had interested two prospects in buying the property at the price of $3,500,000 that was being asked by Melvin; that around 6 o'clock Thursday evening his prospects had offered him a "binder" check that he had refused because "they had not stepped on the property"; that he thereupon attempted to call Melvin between 8:45 and 9 o'clock on Friday morning to inform him that he had the property sold and to request Melvin to meet with the purchasers so that the transaction might be closed without delay; that he met with Melvin one hour and ten minutes later and introduced him to the purchasers; that Melvin then went with Harris and the purchasers to the property, where a purchase contract was executed and a check was delivered between 2:45 and 3 o'clock in the afternoon.

The evidence shows, that after receiving this information late Friday night, West went to Miami Beach early Saturday morning to inform Buchanan that Melvin had sold the property to another prospect. While there West presented a contract of purchase for Buchanan's signature, had Buchanan execute it, and received from Buchanan his personal check in the sum of $50,000 that was dated March 16, 1956, and was made payable to the order of "Wm. H. West, Realtor." According to West, he accepted the check because "Buchanan felt that there might still be a possibility that Mr. Melvin would change his mind and that the other transaction would not go through . . . (and) he wanted to be on record that he had carried out the part of the deal that he had made with Mr. Melvin."

About thirty days later, after it appeared certain that Melvin had no intention of honoring his oral promise to Buchanan, West returned the check to Buchanan and instituted the present suit to recover a commission for selling the property.

The first question on the appeal is whether the foregoing evidence, when viewed in the light most favorable to the plaintiff, is sufficient to support the judgment rendered in his favor.

Two types of brokerage contracts are generally used in the business of selling real estate. Under the first type, the seller employs a real estate broker to procure a purchaser for the property, and the broker becomes entitled to his commission when he produces a purchaser who is ready, able, and willing to purchase the property upon the terms and conditions fixed by the seller, leaving to the seller the actual closing of the sale. Under the second type, the seller employs a broker to effect a sale of the property, and the broker, to become entitled to his commission, must not only produce a purchaser who is ready, able, and willing to purchase the property upon the terms and conditions fixed by the seller but must actually effect the sale, or procure from the purchaser a binding contract of purchase upon the terms and conditions fixed by the seller. . . .

Where a broker employed to effect a sale procures a purchaser who is ready, able, and willing to purchase the involved property upon the terms and conditions fixed by the seller and before he can effect the sale or procure a binding contract of purchase the seller defeats the transaction, not because of any fault of the broker or purchaser but solely because he will not or cannot convey title, the broker will be entitled to his commission even though the sale has not been fully completed, if the buyer remains ready, able, and willing to purchase the property upon the terms and conditions fixed; the strict terms of the contract between the seller and the broker requiring him to actually complete the sale or procure a binding contract of purchase being deemed, in such case, to have been waived by the seller. . . .

As pointed out in *Hanover Realty Corp. v. Codomo,* Fla., 95 So. 2d 420, the rule excusing the broker from complete performance where performance has been made impossible by arbitrary acts of the seller is simply an application to brokerage contracts of the rules relating to contracts generally to the effect that "where a party contracts for another to do a certain thing, he thereby impliedly promises that he will himself do nothing to hinder or obstruct that other in doing the agreed thing," and "one who prevents or makes impossible the performance as happening of a condition precedent upon which his liability by the terms of a contract is made to depend cannot avail himself of its nonperformance."

It is vigorously contended by West that when Melvin went to Miami Beach, with West's acquiescence, and made the oral promise to Buchanan to reduce the price of the property from $3,500,000 to $2,900,000 and to allow Buchanan until 2 o'clock the following day to ascertain from his associates whether or not they would be interested in buying the property at the new price, West thereupon became entitled to his commission, because, in contemplation of law, he became "the first broker to sell, because he fully performed his contract to sell when (Melvin) took over and made the oral agreement with Buchanan." To support his contention, he relies upon what is said in *Knowles v. Henderson,* supra, to the effect that where a broker employed to sell property finds a purchaser who is ready, able, and willing to buy at terms fixed by the seller, but before he can effect the sale or procure a binding contract of purchase the seller arbitrarily refuses to go through with the transaction, the broker will be entitled to his commission, if the purchaser remains ready, able, and willing to buy at the terms fixed.

We fail to see how the facts of the case at bar bring the case within the exception stated in *Knowles v. Henderson,* supra.

It is perfectly plain that when Melvin went to Miami Beach to confer directly with Buchanan and thereby attempt "to save the deal," West had not become entitled to a commission, because he had not found a purchaser who was ready, able, and willing to purchase the property at the price demanded by Melvin. Since Buchanan and his associates had flatly refused to buy the property at $3,500,000 and Melvin had rejected Buchanan's counter offer of $2,750,000, the most that can be said for West's legal position at the time was that he had three possible prospects, one of whom was willing to confer further about the property but had no power to bind his associates beyond the counter offer that had already been made and rejected.

It is equally plain that since West had utterly failed to perform his nonexclusive oral contract within the terms of his listing, Melvin was not legally obligated to West to negotiate with Buchanan, at the conference, in any particular manner, except to refrain from entering into an arrangement with West's prospects for the purpose of fraudulently preventing West from earning a commission.

So far as can be ascertained from the record, nothing was said or done by Melvin throughout his conference with Buchanan that could be deemed a waiver of the terms of West's original brokerage contract, or that would have, or could have, prevented West from becoming entitled to a commission if a sale had been made to his prospects at the new purchase price, prior to the time some other broker had effected a sale of the property within the terms of his listing. For, as has already been stated, when the conference ended between Melvin and Buchanan no sale had been made of the property. Melvin had orally promised to reduce the purchase price of the property from $3,500,000 to $2,900,000—which Buchanan had been unable to accept because he had no authority from his associates to do so—and had given Buchanan until 2 o'clock the following afternoon to ascertain from his associates whether or not they were interested in purchasing the property at the reduced purchase price.

Even though the oral promise made by Melvin to extend the time for Buchanan to talk with his associates was not enforceable as an option, since no consideration was given for the promise . . . it is impossible to understand how, if West had been negotiating with Buchanan in place of Melvin, he could have accomplished as much as, or more than, was accomplished by Melvin in an effort to effectuate a sale. Manifestly, West could not have lawfully made the concession as to price that was made by Melvin, since he had no authority, in respect to price, except to offer the property for sale at $3,500,000—a figure that had already been presented to Buchanan and rejected by him. And, assuming for the sake of argument that West had been given authority, prior to the conference, to make concessions in regard to price and terms and thereby bind Melvin, it is plain that Buchanan could not have accepted a single concession on behalf of his associates, since at the time of the conference he had no authority to bind his associates beyond the amount of the counter offer of $2,750,000 that had already been made to, and turned down by, Melvin. Consequently, when the conference between Melvin and Buchanan ended, West occupied no weaker position, because of anything said or done by Melvin during the conference, than he had occupied prior to that time; indeed, it appears that he was in a stronger position. Before the conference, all that West had were three prospects whom he had been unable to interest in the property at the price and terms fixed by the seller; hence he had failed to fulfill the terms of his nonexclusive contract. After the conference, West not only still had his three prospects but also had, in effect, an oral nonexclusive contract to sell that authorized him to sell the property to Buchanan and his associates for $2,900,000 (while other brokers had to sell, under the terms of their contracts, for $3,500,000) provided, of course, he actually effected a sale of the property or procured a binding contract of purchase, prior to the time some other broker did, and provided the sale was effected prior to the time the gratuitous offer to sell the property at the reduced price terminated or was lawfully withdrawn by the seller.

As we have already stated, Harris, who also had a nonexclusive contract to sell the property on a "first come, first served" basis, called Melvin not later than 9 o'clock on Friday morning March 16, 1956, to inform him that he had sold the property at the fixed price of $3,500,000; and at 10:10 on the same morning Harris introduced his purchasers to Melvin. All of this occurred more than an hour before Buchanan advised West that he and his associates were willing to buy the property and hence occurred prior to the time that West could have notified Melvin of the fact, even if the latter had chosen to remain at home throughout the whole of Friday morning. The contract of sale was executed by the seller and the purchasers produced by Harris at 2:45 or 3:00 Friday afternoon. As a matter of law, the execution of this contract terminated all nonexclusive brokerage contracts that were then outstanding. . . . We find nothing in the oral concessions made by Melvin at the conference that could be construed as an agreement on his part to pay West a commission regardless of whether a sale was made in the meantime by some other broker; nor was there anything said or done by Melvin to indicate that Melvin intended to give West an exclusive agency to sell the property at the new price stated at the conference.

We conclude, therefore, that the trial court erred in giving judgment to the plaintiff; and consequently, that the judgment should be reversed and the cause remanded for further proceedings according to law.

The conclusions we have reached make it unnecessary for us to consider the second point urged for reversal by the appellant.

Reversed.

WACHTER MANAGEMENT COMPANY, Appellee, v. DEXTER & CHANEY, INC., Appellant, *282 Kan. 365; 144 P.3d 747; 2006 Kan. LEXIS 656; 61 U.C.C. Rep. Serv. 2d (Callaghan) 150*

Study terms: Software license, breach of warranty, shrinkwrap, Uniform Commercial Code, fraudulent inducement, integration clause/incorporation by reference

Wachter Management Company (Wachter) filed an action for breach of contract, breach of warranty, and fraudulent inducement against Dexter & Chaney, Inc. (DCI). DCI filed a motion to dismiss the action based on improper venue. The district court denied DCI's motion, holding that a choice of venue provision contained in a "shrinkwrap" software licensing agreement was not enforceable. DCI brings this interlocutory appeal pursuant to *K.S.A. 60-2102(b)*.

Facts

Wachter is a construction management company incorporated in Missouri with its principal place of business in Lenexa, Kansas. DCI is a software services company that develops, markets, and supports construction software, project management software, service management software, and document imaging software for construction companies like Wachter. DCI is incorporated in Washington with its principal place of business in Seattle.

Beginning in April 2002, DCI approached Wachter for the purpose of marketing its software to Wachter. Wachter expressed some interest in DCI's software but delayed negotiations to purchase the software until August 2003. After detailed negotiations, DCI issued a written proposal to Wachter on October 15, 2003, for the purchase of an accounting and project management software system. The proposal included installation of the software, a full year of maintenance, and a training and consulting package. The proposal did not contain an integration clause or any provision indicating that it was the final and complete agreement of the parties, nor did the proposal contain any provision indicating that additional terms might be required. An agent for Wachter signed DCI's proposal at Wachter's Lenexa office on October 17, 2003.

Thereafter, DCI shipped the software and assisted Wachter in installing it on Wachter's computer system. Enclosed with the software, DCI included a software licensing agreement, also known as a "shrinkwrap" agreement, which provided:

> "This is a legal agreement between you (the 'CUSTOMER') and Dexter & Chaney, Inc. ('DCI'). By opening this sealed disk package, you agree to be bound by this agreement with respect to the enclosed software as well as any updates and/or applicable custom programming related thereto which you may have purchased or to which you may be entitled. If you do not accept the terms of this agreement, promptly return the unopened disk package and all accompanying documentation to DCI.

■ ■ ■

> "CUSTOMER ACKNOWLEDGES HAVING READ THIS AGREEMENT, UNDERSTANDS IT, AND AGREES TO BE BOUND BY ITS TERMS AND CONDITIONS. CUSTOMER ALSO AGREES THAT THIS AGREEMENT AND THE DCI INVOICE ENUMERATING THE NUMBER OF CONCURRENT LICENSED USERS TOGETHER COMPRISE THE COMPLETE AND EXCLUSIVE AGREEMENT BETWEEN THE PARTIES AND SUPERSEDE ALL PROPOSALS OR PRIOR AGREEMENTS, VERBAL OR WRITTEN, AND ANY OTHER COMMUNICATIONS BETWEEN THE PARTIES RELATING TO THE SUBJECT MATTER OF THIS AGREEMENT."

The software license agreement also contained a choice of law/venue provision providing that the agreement would be governed by the laws of the State of Washington and that any disputes would be resolved by the state courts in King County, Washington.

In February 2005, after encountering problems with the software, Wachter sued DCI in Johnson County, Kansas, for breach of contract, breach of warranty, and fraudulent inducement, seeking damages in excess of $350,000. DCI moved to dismiss Wachter's petition, alleging improper venue based on the provision of the software licensing agreement which provided that King County, Washington, was the proper venue. In response, Wachter argued that the software licensing agreement was an unenforceable addition to the parties' original contract.

The district court denied DCI's motion, finding that the parties entered into a contract when Wachter signed DCI's proposal and concluding that the software license agreement contained additional terms that Wachter had not bargained for or accepted. The district court certified its ruling for an interlocutory appeal, and the Court of Appeals granted DCI's request for interlocutory appeal. We transferred the matter to this court on our own motion pursuant to *K.S.A. 20-3018(c)*.

Analysis

We review the district court's decision on a motion to dismiss using a de novo standard of review.

■ ■ ■

DCI argues that the district court erred when it refused to recognize the applicability of "shrink-wrap" license agreements. DCI raises four arguments. First, DCI contends that Wachter accepted the terms of the license agreement by opening and using the software. Second, DCI argues that other courts have upheld the terms of shrinkwrap agreements. Third, DCI claims that the district court's refusal to apply the licensing agreement provides Wachter with an undeserved windfall because its use of the software was unfettered by any license terms. Fourth, DCI asserts that the venue clause in its license agreement is applicable to noncontract claims that are related to or limited by the contract.

Wachter counters by arguing that DCI's license agreement is a unilateral alteration of the contract created when Wachter accepted DCI's proposal. Disputing DCI's contention that it assented to the additional terms by opening, installing, and using the software, Wachter claims that it received no independent consideration for the amendments to the contract. Wachter further argues that the Uniform Commercial Code (UCC), *K.S.A. 84-1-101 et seq.*, precludes unilateral alterations to contract terms.

We must begin our analysis with a determination of whether the UCC applies. The UCC applies to transactions involving goods. *K.S.A. 84-2-102.* Goods are defined as "all things (including specially manufactured goods) which are movable at the time of identification to the contract for sale other than the money in which the price is to be paid, investment securities (article 8) and things in action." *K.S.A. 84-2-105.*

Computer software is considered to be goods subject to the UCC even though incidental services are provided along with the sale of the software. *Systems Design v. Kansas City P.O. Employees Credit Union*, 14 Kan. App. 2d 266, 272, 788 P.2d 878 (1990). The Systems Design court noted that modifications and corrections to the computer programs for improving the system operation were incidental to the sale of the software because, without the purchase of the software, the services would have been unnecessary.

The *Systems Design* analysis applies to the facts in this case. Although DCI's proposal included maintenance, training, and consulting services, these services would not have been necessary if Wachter had not purchased DCI's software. Because the services were incidental to Wachter's purchase of computer software, we conclude that the software at issue in this case qualifies under the definition of goods, and the UCC applies.

Pursuant to *K.S.A. 84-2-204*, a contract for the sale of goods is formed "in any manner sufficient to show agreement, including conduct by both parties which recognizes the existence of such a contract." Unless otherwise unambiguously indicated by the language or circumstances, an offer to make a contract shall be construed as inviting acceptance in any manner and by any medium reasonable in the circumstances. The Kansas Comment to *84-2-206* provides that the offeror is the master of the offer and may require a specific manner of acceptance if it is unambiguously conveyed in the language of the offer or other circumstances.

In this case, DCI issued a written proposal to Wachter containing an itemized list of the software to be purchased, the quantity to be purchased, the price of the software, the time period for execution, and the cost for the incidental maintenance, training, and consulting services. DCI's proposal requested Wachter to accept its offer to sell Wachter software by signing the proposal above the words "[p]lease ship the software listed above." Accordingly, Wachter accepted DCI's offer to sell the software to it by signing the proposal at Wachter's office in Lenexa. Thus, a contract was formed when Wachter accepted DCI's offer to sell it the software, indicating agreement between the parties. See *K.S.A. 84-2-204.*

K.S.A. 84-2-201 requires contracts for the sale of goods over $500 to be in writing and signed by the parties. The signed proposal constitutes a written contract in this case. The cover letter with DCI's proposal stated that it included "modules and licenses." However, DCI did not attach a copy of its Software Licensing Agreement to the proposal or incorporate it by reference in the proposal to indicate that there would be additional contract language regarding the licenses. Consequently, the parties' contract did not contain the terms of the Software Licensing Agreement. Wachter was advised of the terms of the Software Licensing Agreement after DCI shipped the software in partial performance of its duties under the contract. Because the Software Licensing Agreement was attached to the software rather than the original contract, it must be considered as an attempt to amend the contract.

The UCC addresses the modification of a contract. *K.S.A. 84-2-207* provides in pertinent part:

"(1) A definite and seasonable expression of acceptance or a written confirmation which is sent within a reasonable time operates as an acceptance even though it states terms additional to or different from those offered or agreed upon, unless acceptance is expressly made conditional on assent to the additional or different terms.

"(2) The additional terms are to be construed as proposals for addition to the contract. . . .

"(3) Conduct by both parties which recognizes the existence of a contract is sufficient to establish a contract for sale although the writings of the parties do not otherwise establish a contract. In such case the terms of the particular contract consist of those terms on which the writings of the parties agree, together with any supplementary terms incorporated under any other provisions of this act."

K.S.A. 84-2-209 provides:

"(1) An agreement modifying a contract within this article needs no consideration to be binding.

"(2) A signed agreement which excludes modification or rescission except by a signed writing cannot be otherwise modified or rescinded, but except as between merchants such a requirement on a form supplied by the merchant must be separately signed by the other party.

"(3) The requirements of the statute of frauds section of this article *(section 84-2-201)* must be satisfied if the contract as modified is within its provisions.

"(4) Although an attempt at modification or rescission does not satisfy the requirements of subsection (2) or (3) it can operate as a waiver.

"(5) A party who has made a waiver affecting an executory portion of the contract may retract the waiver by reasonable notification received by the other party that strict performance will be required of any term waived, unless the retraction would be unjust in view of a material change of position in reliance on the waiver."

Proposed amendments that materially alter the original agreement are not considered part of the contract unless both parties agree to the amendments. *Southwest Engineering Co, Inc. v. Martin Tractor Co., Inc.,* 205 Kan. 684, 694, 473 P.2d 18 (1970). UCC 2-209 requires express assent to the proposed modifications.

DCI argues that Wachter expressly consented to the shrinkwrap agreement when it installed and used the software rather than returning it. However, continuing with the contract after receiving a writing with additional or different terms is not sufficient to establish express consent to the additional or different terms. *Step-Saver, 939 F.2d at 98; Arizona Retail Systems,* 831 F. Supp. at 764; *Klocek,* 104 F. Supp. 2d at 1341; *Orris,* 5 F. Supp. 2d at 1206.

In *Step-Saver*, the plaintiff ordered software from the defendant by telephone and followed up with a written purchase order. The defendant shipped the software along with an invoice. A license with additional terms was printed on the box containing the software. The additional terms stated that opening the box indicated an acceptance of the additional terms. Without determining exactly when the contract was formed, the *Step-Saver* court concluded that the parties' performance in ordering, shipping, and paying for the software demonstrated the existence of a contract and limited its analysis to the terms of the purchase order.

The *Step-Saver* court noted that *UCC 2-209* allows the parties to amend an agreement without additional consideration and then applied the principles of *UCC 2-207* to conclude that the parties must expressly intend to modify a previous agreement. The applicable sections of *UCC 2-207* provide:

"(1) A definite and seasonable expression of acceptance or a written confirmation which is sent within a reasonable time operates as an acceptance even though it states terms additional to or different from those offered or agreed upon, unless acceptance is expressly made conditional on assent to the additional or different terms.

"(2) The additional terms are to be construed as proposals for addition to the contract." *K.S.A. 84-2-207.*

The *Step-Saver* court stated that *UCC 2-207* provides a default rule in the absence of a party's express assent to proposed changes in an agreement. According to the default rule, the terms of the agreement include the terms both parties have agreed upon and any terms implied by the provisions of the UCC. 939 F.2d at 99. Treating the shrinkwrap license as a "written confirmation containing additional terms," the Step-Saver court concluded that continuing with the contract after receiving the shrinkwrap license was not sufficient to establish an express assent to the new or additional terms contained in the shrinkwrap license. 939 F.2d at 98, 105-06. Because the plaintiff never expressly agreed to the additional terms contained in the shrinkwrap license, the terms did not become part of the parties' agreement. 939 F.2d at 106. The court also found that the integration clause in the license agreement and the "consent by opening" language did not render the software provider's acceptance conditional because it gave no real indication that it was willing to forgo the transaction if Step-Saver rejected the additional terms. 939 F.2d at 102.

■ ■ ■

DCI does not address the application of the UCC or the analysis of *Step-Saver, Arizona Retail Systems, Klocek,* or *Orris.* Instead, DCI relies on *Hill v. Gateway 2000, Inc.,* 105 F.3d 1147 (7th Cir. 1997); *ProCD v. Zeidenberg,* 86 F.3d 1447 (7th Cir. 1996); and *Mortenson Co. v. Timberline Soft-ware,* 140 Wn. 2d 568, 998 P.2d 305 (2000), for the proposition that shrinkwrap agreements are valid, and the terms contained within them are enforceable, because the purchaser accepts the terms when it uses the product.

In *ProCD,* a consumer purchased a software database program at a retail store. A license enclosed in the package with the software limited its use to non-commercial applications. The software also required a user to accept the license agreement by clicking an on-screen button before activating the software. Contrary to the license, the consumer made the database available on the internet at a reduced price. ProCD sued, seeking an injunction against further dissemination of its database.

The *ProCD* court determined that the *vendor* is the master of the offer under the UCC and may invite acceptance by conduct or limit the kind of conduct that constitutes acceptance. The court found that ProCD proposed a contract that invited acceptance by using the software after having an opportunity to review the license. If the buyer disagreed with the terms of the contract, he or she could return the software. Holding that the consumer was bound by the terms of the license agreement, the *ProCD* court stated that "[n]otice on the outside, terms on the inside, and a right to return the soft-ware for a refund if the terms are unacceptable (a right the license expressly extends), may be a means of doing business valuable to buyers and sellers alike." 86 F.3d at 1451.

In *Hill,* a consumer ordered a Gateway computer over the telephone. When the computer arrived, the box contained Gateway's standard terms governing the sale. According to Gateway's standard terms, the consumer accepted the terms by retaining the computer for 30 days. When the consumer was not satisfied with the operation of the computer, he sued Gateway on behalf of a class of similarly situated consumers. Relying on the *ProCD* court's analysis that the *vendor* is the master of the offer, the *Hill* court enforced the arbitration clause found in Gateway's standard terms even though the consumer was not aware of the terms until he received the computer. *105 F.3d at 1150.* The *Hill* court noted that there are many commercial transactions in which money is exchanged for products with disclosure of certain terms of the sale following the execution of the sale. 105 F.3d at 1148 (citing *ProCD,* 86 F.3d 1447 [7th Cir. 1996], and *Carnival Cruise Lines, Inc. v. Shute,* 499 U.S. 585, 111 S. Ct. 1522, 113 L. Ed. 2d 622 [1991]). The Hill court announced its policy that "[p]ractical considerations support allowing vendors to enclose the full legal terms with their products. . . . Customers as a group are better off when vendors skip costly and ineffectual steps such as telephonic recitation [of standard terms], and use a simple approve-or-return device." 105 F.3d at 1149.

Both *ProCD* and *Hill* can be distinguished from this case. The buyers in *ProCD* and *Hill* were both consumers who did not enter into negotiations with the vendors prior to their purchases. Wachter, on the other hand, participated in detailed negotiations with DCI before accepting DCI's proposal to sell Wachter the software. The *ProCD* and *Hill* courts concluded that the offer to sell software to the consumers was not accepted until the consumers opened the packaging with the terms of the sale enclosed and retained the product after having an opportunity to read the terms of the sale. Accordingly, the contract was not formed until the last act indicating acceptance occurred. Here, however, the last act indicating acceptance of DCI's offer to sell software to Wachter occurred when Wachter signed DCI's proposal. Thus, the contract was formed before DCI shipped the software and Wachter had an opportunity to consider the licensing agreement.

Although *ProCD* and *Hill* are distinguishable, the third case cited by DCI is factually similar to this case. In *Mortenson,* a construction contractor purchased software to assist with its bid preparation. The contractor issued a purchase order and the developer shipped the software accompanied by a shrinkwrap license, which included a limitation of remedies clause. Applying *UCC 2-204,* the *Mortenson* court held that the initial purchase order and the shrinkwrap license were part of a "layered contract" where the initial purchase order was not an integrated contract because it did not contain an integration clause and required additional terms to be determined later. *140 Wn. 2d at 580.* The *Mortenson* court adopted the contract formation analysis from *ProCD* and *Hill* and concluded that Mortenson's use of the software constituted its assent to the terms of the shrinkwrap license. *140 Wn. 2d at 584.*

Two of the Washington Supreme Court justices dissented from the *Mortenson* opinion, stating that "the majority abandons traditional contract principles governing offer and acceptance and relies on distinguishable cases with blind deference to software manufacturers' preferred method of conducting business." 140 Wn. 2d at 589, 599. The dissent distinguished its analysis by noting that under traditional contract law principles the *offeror,* not just the *vendor,* is the master of the offer. 140 Wn. 2d at 590. Observing that the construction contractor made an offer to purchase when it sent a purchase order to the software developer, the dissenting justices concluded that the parties formed a contract when the software developer accepted the terms of the purchase order by signing it. Because the contract was formed before the software developer delivered the product with the shrink-wrap license agreement, the dissenting justices treated the shrinkwrap agreement as a proposal to modify the terms of the contract pursuant to *UCC 2-209.* The dissent relied on *Step-Saver* and *Arizona Retail Systems* for its conclusion that the parties must expressly consent to any modifications. Without evidence that the construction contractor expressly consented, the dissenting justices stated that they would have remanded the matter to determine whether the parties' conduct constituted assent. 140 Wn. 2d at 598.

Although the facts in *Mortenson* are similar to the facts in this case, we disagree with the *Mortenson* court's analysis of when the contract was formed. We adhere to the traditional contract principles outlined by the dissenting justices in Mortenson and the decisions in *Step-Saver, Arizona Retail Systems, Klocek,* and *Orris.* The offeror, whether the seller or the buyer, is the master of the offer.

In this case, DCI and Wachter negotiated prior to entering into a contract for the sale of software. DCI's written proposal following the parties' negotiations constituted an offer to sell. Wachter accepted that offer when it signed the proposal, requesting shipment of the software. The contract was formed when Wachter accepted DCI's proposal. Because the contract was formed before DCI shipped the software with the enclosed license agreement, the Software Licensing Agreement must be treated as a proposal to modify the terms of the contract. There is no evidence that Wachter expressly agreed to the modified terms, and Wachter's actions in continuing the preexisting contract do not constitute express assent to the terms in the Software Licensing Agreement. Thus, the forum selection clause in the Software Licensing Agreement is not enforceable against Wachter. We affirm the district court's denial of DCI's motion to dismiss and remand the matter for further proceedings.

DISSENT BY: LUCKERT

Dissent

LUCKERT, J., dissenting: I would reverse the district court's conclusion that the choice of venue provision in the software licensing agreement was not enforceable. I disagree with the majority's analysis that the license agreement was a modification of the contract. Rather, the original offer included the license or, at least, expressed the intent of the parties that a license was a part of the offer. Wachter assented to and accepted these terms by its conduct.

DCI's letter transmitting the proposal notified Wachter that "[t]he proposal includes modules and licenses." Wachter did not question, object to, or offer an alternative to the proposal. Instead Wachter signed the proposal, thus accepting the offer which included licenses. See *K.S.A. 84-2-204(1)* ("A contract for sale of goods may be made in any manner sufficient to show agreement, including conduct by both parties which recognizes the existence of such a contract."). In turn, DCI sent the software. Wachter opened the software and installed it. In doing so, it accepted the goods. If Wachter felt that the goods were nonconforming because of the license agreement, Wachter was required by *K.S.A. 84-2-204* to reject the goods. Wachter's failure to object after a reasonable time for inspection constituted an acceptance of the goods. See *ProCD, Inc. v. Zeidenberg*, 86 F.3d 1447, 1452-53 (7th Cir. 1996); *Salco Distributors, LLC v. iCode, Inc.*, 2006 U.S. Dist. LEXIS 9483, 2006 WL 449156 (M.D. Fla. 2006) (unpublished opinion).

Alternatively, by DCI's referring to "licenses" in the proposal, there was an expression of intent either that there was to be a layered contract or that a contract was not formed until Wachter accepted the license agreement. When sending the software, DCI gave Wachter the option of accepting the terms of the license or "promptly return[ing] the unopened disk package." DCI, as the offeror, could invite acceptance by conduct and could propose limitations on the kind of conduct that constitutes acceptance. *K.S.A. 84-2-204*. The notice on the shrinkwrap advised Wachter of the specifics of the license agreement and notified Wachter that by opening the software it was accepting the terms. Again, Wachter did not question or object to the terms and did not offer alternative terms. Rather, it accepted the terms of the license with notice through DCI's communication that the proposal included licenses that were a part of the contract. Wachter could have prevented formation of the contract by returning the software. Instead, Wachter opened the software. Wachter's conduct was sufficient to evidence its agreement with the terms of the contract.

The majority's reliance on *Step-Saver Data Systems, Inc. v. Wyse Technology*, 939 F.2d 91 (3d Cir. 1991), is misplaced, as the facts of that case are clearly distinguishable. *Step-Saver* involved a value-added retailer who included the software in an integrated system which was sold to an end user. The party contesting applicability of the licensing agreement had been assured the license did not apply to it at all. Additionally, the seller of the program twice asked the buyer to sign an agreement comparable to their disputed license agreement. Both times the buyer refused, but the seller continued to make the software available. Thus, the facts regarding acceptance varied significantly from those in this case, where the communication between Wachter and DCI made clear that the proposal included a license agreement and Wachter never attempted to reject the goods or the license or to negotiate alternative terms.

The other cases cited by the majority rely on *Step-Saver*. Furthermore, those cases are based upon the theory of modification of a contract. Under the facts of this case, where the negotiations included a clear communication that the proposal included licenses and the parties' contract did not include an integration provision, the rationale of these cases does not apply. This case does not involve contract modification but contract formation. Furthermore, even if the issue were one of modification, Wachter's acceptance, made with notice that the license was a part of the proposal, was an assent to the modification. The communication made Wachter aware that the continuation of the contract depended upon its acceptance of the terms of the license. Thus, under *K.S.A. 84-2-209* dealing with continuance of a contract after a proposed modification, Wachter accepted the modification (if the license were so considered) and became bound by the terms of the license agreement. In this regard, the communication of the proposal from DCI to Wachter, which placed Wachter on notice that a continuation of the contract required acceptance of the license, distinguishes this case from *Klocek v. Gateway, Inc.*, 104 F. Supp. 2d 1332 (D. Kan. 2000); *United States Surgical Corp. v. Orris, Inc.*, 5 F. Supp. 2d 1201 D. Kan. 1998); *and Arizona Retail Systems v. Software Link*, 831 F. Supp. 759 (D. Ariz. 1993).

Under the facts of this case on this issue of contract formation rather than contract modification, I find persuasive the analysis of *Hill v. Gateway 2000, Inc.*, 105 F.3d 1147, 1149-50 (7th Cir.), cert. denied 522 U.S. 808 (1997); *ProCD, Inc. v. Zeidenberg*, 86 F.3d at 1452-53; *Brower v. Gateway 2000*, 246 A.D.2d 246, 250-51, 676 N.Y.S.2d 569 (1998); *and Mortenson Co. v. Timberline Software*, 140 Wn. 2d 568, 583-84, 998 P.2d 305 (2000).

NUSS, J., and BEIER, J., join in the foregoing dissent.

Review Questions

1. Black is in need of a turret lathe similar to a used one White is trying to sell. White offers Black the lathe for $2,500. Black: "I accept, providing you will deliver the lathe and have it in my shop tomorrow noon." White: "Sorry, but my truck will not be back in town before tomorrow night." Black: "Then I will pick it up myself, because I need it immediately." White: "I believe I will wait a while longer before selling it." Black: "You can't; we made a contract." Was there a contract? Why or why not?

2. On January 3, Gray sends Brown the following offer by first-class mail: "I offer you my vertical milling machine in the same condition it was when you last saw it for $4,200. This offer will remain open for your certified mail acceptance until January 15." On January 9, at 10:00 a.m., Brown sends the following facsimile to Gray: "I accept the offer of your vertical milling machine for $4,200. Am sending truck for it immediately." Is there a contract? Why or why not?

3. Expanding on question 2, assume that Brown received a withdrawal from Gray at 9:00 a.m., January 9, but, because the original letter stated that the offer was open until January 15, Brown sent acceptance by email an hour later. Was there a contract? Why or why not?

4. Referring again to question 2, assume that Brown's facsimile acceptance on January 9 was lost within Gray's office and not delivered to him until January 16. Was there a contract? Why or why not?

5. Green required parts for an appliance he intended to manufacture. He sent out blueprints and specifications to several manufacturers, including the White Manufacturing Company, requesting bids on 100,000 of such parts. White Manufacturing Company submitted what turned out to be the lowest bid. The bid was refused, however. Can White Manufacturing Company demand and get the job on the basis of its bid? Why or why not?

6. Brown lost an expensive watch that had his name engraved on it. He placed an advertisement in a local paper offering a $100 reward for its return. Black found the watch, noticed Brown's name on it, found Brown's address in a telephone book, and returned the watch to him. Later, he read the reward notice and demanded the reward. Is he entitled to the $100? Why or why not?

7. The price of White's product is $50. Newspaper advertising inadvertently lists the price as $30, precipitating a deluge of orders. White honors the advertised price even though his cost is $35. He has begun an action against the newspaper for $20 for each unit sold. Who will win? Why?

8. Referring to the case of *Melvin v. West*, summarize the two types of brokerage contracts commonly used in selling real estate. Which type of contract existed between Melvin and West?

9. The Supreme Court of Kansas in *Wachter Management Company v. Dexter & Chaney, Inc.* concluded that the contract between the parties was formed when? How should the Software Licensing Agreement be treated as it relates to the contract between the parties? According to the dissenting opinion, did Wachter expressly agree to the license agreement?

Reality of Agreement

HIGHLIGHTS

■ Parties enter into agreements voluntarily. If not, a contract may be held void or unenforceable.

■ There are five categories in which reality of agreement is questionable:
1. Mistake
2. Fraud
3. Misprepresentation
4. Duress
5. Undue influence

Chapter 7 stated that there must be an agreement or meeting of the minds between contracting parties. The meeting of minds must be voluntary and intentional. It is implied that neither of the parties has been prevented from learning the facts of the proposed contract. In short, the assumption is made that all contracting parties have entered into the agreement freely and "with their eyes wide open."

This assumption especially applies when the contract is reflected by a written and signed instrument. Generally, an agreement in writing binds the parties to the agreement according to the wording on the paper. A person is held to have read and understood the terms set forth. If the person did not understand the document, he or she should not have signed it. If one wishes to enter into a written agreement with another but disagrees with some of the terms, that person has the right to eliminate these terms, get the other party to the contract to agree to the change, and obtain the other's signature to the change before signing. Most written contracts are either typed or hand-printed. If, in reaching an agreement, one of the parties changes the contract in the party's handwriting with the other's knowledge of the handwritten change, this writing will stand in lieu of the preexisting printed or typed part with which it is in conflict.

If the assumption that the contracting parties freely and voluntarily entered into the contract is not true, the contract may be held void or unenforceable. Although the law will uphold a person's right to contract as part of that person's freedom and will hold that person, as well as the party with whom that person bargains, to the agreement they have made, it would be unjust to do so in the absence of free and voluntary action. Where one person knowingly attempts to take advantage of another, it would not seem right to enforce the attempt.

The rules and principles of contracts are very practical and logical as, of course, they should be when they govern practical events. Probably there is no better demonstration of the straightforwardness of contracts than in cases where assent by one of the parties is not voluntary and intentional.

Contracts in which the reality of agreement is questionable fall into five general categories:

1. Mistake
2. Fraud
3. Misrepresentation
4. Duress
5. Undue influence

These are treated separately in this chapter. After considering these situations, the available remedies are briefly discussed.

SITUATIONS IN WHICH ASSENT IS NOT VOLUNTARY

Mistake

A mistake may be either unilateral or mutual. A **unilateral mistake** is a mistake that one party to a contract makes about the material circumstances surrounding the transaction. Usually the law does not allow relief for unilateral mistakes. A **mutual** or **bilateral mistake** occurs when both of the parties were mistaken about some material fact. When both are mistaken, the law will not bind either of them.

> A material fact is one that is substantial and if known, would influence whether a party would enter into an agreement.

Unilateral Mistake. When a mistake results from poor judgment of values or negligence by the injured party, the law usually will hold the injured party to the bad bargain. Black sells White a used automatic screw machine for $15,000 and White finds later that a similar machine could have been purchased for $10,000. White cannot avoid the contract because of the difference in price.

In certain instances of unilateral mistake, a party may rescind the contract. If one party to the contract, with full knowledge of the other's mistake, took advantage of the situation, the courts usually do not hold the injured party to the bargain. Such a situation might easily fit into the picture of fraud given later in this chapter. Certain types of clerical or calculation errors will also allow rescission. If a contractor, in preparing a bid on a job, makes a clerical error or a mathematical miscalculation, a court may hold that there is no binding contract. This is particularly likely if the error is an obvious one and the other party must have known or reasonably suspected an error in the quotation.

Mutual Mistake.   If both parties to a contract are mistaken about a material fact regarding a matter, the courts generally hold the contract to be voidable. Two types of situations commonly occur: mistakes about identity and mistakes about existence. Suppose Black agreed to purchase White's computer network for $15,000, and the two parties had two different types of computer networks in mind. The courts would say that the minds of the two parties had not met and that, therefore, no enforceable contract resulted. In this example, the two parties may have contracted for the sale of the computer network at some location away from the place where the computer network was kept. If the computer network had in fact been destroyed by a fire, the contract would not be enforceable. The subject matter would have ceased to exist (at least in the manner assumed by the parties). The assumed existence of the subject matter would have been a mutual mistake.

When the value is something different from that assumed by both of the parties, however, a different view is taken by the law. For instance, Black purchases farmland from White at a reasonable value for farmland. Later, a valuable mineral deposit is discovered in the land. If Black had no prior knowledge of such deposits, the sale stands, and it is Black's gain. There are, of course, many instances of purchases of valuable paintings and antiques in which the parties involved knew nothing of the true value of the subject matter. Such sales are generally quite valid.

When the mistake concerns the legal rights and duties involved, the mistake generally is not a ground for rescission of the contract. Every person is supposed to be familiar with the law and is charged with that knowledge.

Fraud

Five basic elements must be established to prove fraud. There must have been:

1. a false representation
2. of a material fact
3. made with the intent that it be relied on and
4. reliance on the misrepresentation by the injured party
5. to his or her detriment.

A false representation may, but need not, be an outright misstatement. It may, instead, take the form of an omission of information that should have been passed along to the injured party. A false representation also might be an artful concealment, such as painting a car engine black in order to conceal erosion to a potential buyer. In certain instances, particularly where a previous relationship of trust has existed between the parties, there may be a duty to speak and reveal all information concerning the subject matter. The courts have been quite general in describing how fraud may arise, preferring not to set up a fixed blockade for the sharp operator to attempt to avoid. The false representation must be of a "material" fact that is not obviously untrue. For instance, if you are told that a car you are thinking of buying is in perfect condition when, in truth, it has a broken rear window that anyone could see, the statement would not support an action for fraud.

The false representation must be material. If the seller represented the car as having brand-X spark plugs when the spark plugs were really another brand of equivalent quality, the statement would hardly be considered material. Also, the representation must falsely represent a fact. When you stop to think of it, it may be a little difficult to define what is fact. Certainly, your actions yesterday and today may be stated as facts. But, aside from a few accepted certainties, it is often impossible to state something about the future as fact. In attempting to get Black to buy land, White tells Black, "I intend to start a housing development in a section very near this land." If Black buys in reliance on the statement and no housing development is begun, there might be some question about whether White ever intended to build. Was the statement a statement of fact? It could very well be, because it is a statement of a present intention.

If a statement is made in such a way as to express an opinion, it is not a statement of fact and cannot support a claim for fraud. Sales talk, puffing, statements that a certain thing is the "best," or is "unsurpassed," or a "good buy" are not fraudulent statements at common law. An exception occurs when a person maintains that he or she is qualified to give an expert opinion on the subject. If a party is injured in reasonable reliance on such an opinion, the injured person may be able to recover for fraud.

Generally, then, the representation of material fact must be a false representation, other than opinion, having to do with something past or present. It is not necessary for the statement to be the sole inducement to action by the injured party. It is only necessary that, if the injured party had known the truth, that party would not have taken the action actually taken.

Knowledge of the truth or falsity of a statement made does not necessarily determine the existence of fraudulent representation. If the person making the representation knows it to be false, it is fraudulent. If the person does not know whether it is true but makes the representation in reckless disregard of the truth and to persuade the other party, the party making the statement may be charged with having acted fraudulently.

A misrepresentation for the purpose of fraud is normally made with the intent that the party to whom it is made will act on it. If the other party does not act on it, there is no fraud. If a third party (for whom the representation was not intended) overhears it and acts on it to his or her detriment, it is nevertheless not fraud. Action must be taken by the person for whom the false representation was intended.

The injured party must have acted in reliance on the representations made. People are assumed by the courts to be reasonable, and the courts assume that they will take ordinary precautions when dealing with others. One is charged with the knowledge or experience that anyone in the same trade or profession would normally possess. For instance, an auto mechanic should be far less gullible than a lingerie salesperson when purchasing a used car. If White has conducted her own investigation before entering into an agreement rather than relying on another's representations, it is far more difficult for her to sustain a charge of fraud.

The false representations of material fact for the purpose of fraud may be proved; they may have been made with full knowledge of their falsity and with intent to deceive; action may have been taken in reliance on the misrepresentations, but if no injury resulted there can be no recovery for fraud.

Black is induced by White, as a result of various fraudulent misstatements, to purchase 1,000 shares of "Ye Olde Wilde Catte" oil stock at $1 per share. On discovery that Ye Olde Wilde Catte stock isn't worth the paper the shares are printed on, Black sues, but before the case comes up on the court docket the oil company strikes oil and the stock price increases to $2 per share. Black quite obviously would not continue the action, but even if Black wanted to, Black couldn't recover damages because there was no injury.

Misrepresentation

To distinguish between fraud and misrepresentation, consider the following example. Black contracts to buy a hunting dog from White for $200. During the negotiations, White tells Black that the dog is three years old. A few months later the dog becomes ill and Black takes him to a veterinarian. After treatment the vet tells Black that Black has a fine 10-year-old dog. Black has been a victim. Black has either been defrauded or has been victimized by an innocent misrepresentation.

The distinction depends on the presence or absence of White's knowledge and intent. If White was present 10 years ago when the dog was born, White can certainly be charged with knowledge of the dog's age. If White knew the dog's age, the statement to Black indicating that the dog's age was three years could have been made only with an intent to deceive Black. If, however, the dog was acquired by White six months ago from Gray, who stated the dog's age to be two years, the statement made by White was probably an innocent misrepresentation (or mistake, in some courts). (See *Performance Abatement Services, Inc. v. Lansing Board of Water and Light, et al.*, in Chapter 17 for additional discussion about innocent misrepresentation.)

Although no fraud results from an innocent misrepresentation, the injured party has the right to rescind the contract, giving up the consideration that was received. However, in situations involving an innocent misrepresentation, there is usually no independent right to sue for damages, as there is in cases of fraud. Please note, however, that in some cases misrepresentations about the productivity of a business may constitute fraud, regardless of intent. (See *Tri-State Trucking of Rhode Island, LLC v. Tri-State Trucking, Inc.* at the end of this chapter.)

Negligent Misrepresentation.   But what happens with the misrepresentation was not innocent, but does not rise to the level of fraud? In those instances, an action for **negligent misrepresentation** may be brought. Like fraud, this cause of action is intended to protect against damages due to reliance on another's false statements. However, only the negligence of the speaker is required. If the speaker negligently made false statements, then there may be liability. (See *Deann Thomas v. Lewis Engineering, Inc.* at the end of this chapter.) A further discussion of this tort can be found in Chapter 21, Common Torts.

Duress

Duress is forcing a person to consent to an agreement through active or threatened violence or injury. The threat must be accompanied by the apparent means of carrying it out. To be duress, the threat must place the victim in a state of mind such that the person no longer has the ability to exercise free will.

In the early court decisions, a holding of duress was limited to cases in which the victim's life or freedom was endangered or in which the victim was threatened with bodily harm. Modern courts take the view that duress is the deprivation of a person's free will; the means of accomplishing this is immaterial as long as unlawful injury is involved. Threat of injury to other persons, even to property, could conceivably be the means employed to gain the desired end. White is an art lover and has a collection of paintings. Black, to obtain White's signature on a promissory note, threatens to cut up one of White's paintings with a pocket knife. This threat could constitute duress.

Some courage is inferred by the courts. It must be shown not only that threats of unlawful violence occurred, but that they were sufficient to overcome the person's free will. When duress has been used, the contract is voidable and the victim can disaffirm the contract or affirm the contract.

Undue Influence

Duress and undue influence are alike in at least one respect. When either has been successfully exercised, the free will of the victim has been overcome. The means of accomplishing the objectives are somewhat different. Duress requires violence or a threat of violence or harm. **Undue influence** usually results from the use of moral or social coercion, most often arising where a relationship of trust or confidence has been established between two people. Common relationships of this nature are those between husband and wife, parent and child, guardian and ward, doctor and patient, and attorney and client. It also may arise where a person is in dire need, or in some physical or mental distress that would put the potential victim at the mercy of the other party.

To establish undue influence, all of the following must be shown:

- There was a person who, because of a special relationship, was in position to be influenced.
- Improper pressure was exerted by the person in the dominant position.
- The pressure influenced the victim—that the victim acted on it and was injured as a result.

REMEDIES

In situations involving a mutual mistake or an innocent misrepresentation, the most common remedies used by the courts are rescission and reformation. **Rescission** involves placing the parties in the respective positions they would have enjoyed if there had been no contract. This usually means a return of the amounts paid and/or the goods delivered. Where the contract involves services already rendered, rescission is sometimes not practical. **Reformation** involves the modification of the contract to reflect the intentions of the parties. For example, a contract for a road might be understood to involve an asphalt surface. If the written agreement specified a concrete surface, however, that portion of the contract would be subject to reformation to clearly reflect the parties' agreement.

Several remedies are available to a person who has been defrauded:

- That person may wish to continue the contract with full knowledge of the fraud.
- If the contract has been completed, that person may wish to retain what was received and institute a case for what damages have been suffered.
- The defrauded person may rescind the contract, return the consideration received, and get back what was given.
- If the contract is still executory, the defrauded party may merely ignore the obligations under it, relying on the proof of fraud if the other party brings an action to enforce the contract.

In any event, if the defrauded party wishes to bring a court action after discovering the fraud, the injured party must do so in a reasonable time. If the injured party does not take action within a reasonable length of time, the right to take action will be lost. If a party waits too long, the state's statute of limitations may prevent bringing the claim. As in cases involving fraud, a party subjected to duress may disaffirm the contract. Moreover, fraud and duress often amount to torts (discussed in a later chapter); thus, a victim of fraud or duress also may sue for damages.

MCMULLEN v. JOLDERSMA 435 NW 2d 428 (Mich. 1988) Per Curiam

Study terms: Fraudulent representation, innocent misrepresentation, fraudulent concealment, priority of contract, "silent fraud", reliance, rescission

Plaintiffs appeal as of right a December 20, 1986, circuit court judgment of no cause of action against defendants Paul and Mary Joldersma and separate circuit court orders granting defendants Buehler Realty, Inc., and James

M. Parrette's motion for summary dispositions. Plaintiffs present a myriad of issues for our review; However, we find that none require reversal.

This case arose out of dispute over alleged fraudulent representations and omissions surrounding the sale of a party store located in Newaygo County, Michigan. On July 23, 1981, defendants Paul and Mary Joldersma sold their party store to plaintiffs on a land contract. In September, 1984, plaintiffs filed suit against the Joldersmas, Buehler Realty, Inc., and realtor James Parrette, alleging that defendants fraudulently concealed the material fact that the State of Michigan had plans to construct a highway bypass as part of M-37 running north from Newaygo which would substantially divert all traffic from both M-82 and M-37 away from the party store. Plaintiffs further alleged that the construction of the bypass, which was completed in 1984, destroyed the value of their business. Plaintiffs sought rescission of the land contract, restitution, and exemplary damages for mental distress.

On April 18,1985, plaintiffs filed an amended complaint alleging the following: fraud against the Joldersmas, Parrette, and Buehler (Counts I and IV); innocent misrepresentation against the Joldersmas, Buehler, and Parrette (Counts II and V); and breach of contractual duty of care against Buehler and Parrette (Count III).

On May 8, 1985, defendants Buehler Realty and Parrette moved for summary disposition pursuant to MCR 2.116(C) (7), (8), and (10). On June 14, 1985, defendants Paul and Mary Joldersma followed suit and filed their motion for summary disposition on the basis of MCR 2.116(C) (7) and (10).

On June 18, 1985, an initial hearing on all defendants' motions was held before Judge Terrence Thomas. Judge Thomas took the various motions under advisement and, while a decision was pending, removed himself from the case because of a potential conflict of interest. On February 24, 1986, a hearing was held before Judge Charles Wickens on the various summary disposition motions. On March 13, 1986, Judge Wickens issued his opinion granting summary disposition to defendants Buehler Realty, Inc., and its agent, James Parrette. In the same opinion, defendants Joldersmas' motion was denied because the court concluded that a factual issue was presented.

Plaintiffs then proceeded to a three-day bench trial against the only remaining defendants, Paul and Mary Joldersma, in which plaintiffs primarily sought rescission of the land contract and damages for fraud. About six months after trial, Judge Wickens issued his opinion of no cause of action and awarded costs to the Joldersmas.

Plaintiffs argue first that the court erred in granting summary disposition in favor of Buehler Realty, Inc, and James M. Parrette. As there is no mention whatsoever in the court's opinion regarding the statute of limitations as a basis for summary disposition, we decline to address that aspect of plaintiffs' argument.

In granting summary disposition in favor of these defendants, the court reasoned: "There is no privity of contract between the Plaintiffs and the real estate agency that would justify this action. There is no activity by either of these defendants growing out of the scope of their agency relationship with the sellers which would warrant an action against the real estate agency separate from the sellers."

Although not expressly stated, the tenor of the Court's ruling indicates that it was based on MCR 2.116 (C) (8).

In Count IV of the first amended complaint, plaintiffs alleged "silent fraud" or fraudulent concealment of a material fact against Parrette and Buehler, In Count V, the theory of innocent misrepresentation was alleged.

We first address the former theory, i.e., "silent fraud" or fraudulent concealment of a material fact. As correctly noted by defendants, Michigan courts have held a seller liable to a buyer for failing to disclose material defects in the property or title. . . . However, to date there are no cases holding that a real estate agent is similarly liable for such a fraud. We do not believe that, by virtue of their agency relationship as real estate agents for the sellers, defendants were duty bound to disclose the pending bypass plans to the buyers. Neither Parrette nor Buehler was a party to the underlying business transaction nor was this a situation where the challenged information was subsequently acquired rendering previous misstatements untrue or misleading. . . . Under the circumstances, the imposition of such a duty would necessarily conflict with the duty defendants owed to the seller. Moreover, we find It noteworthy that plaintiffs were not without representation. Indeed, they were represented by an attorney and, also, a certified public accountant, Roy Heppe. By her own testimony, Virginia McMullen testified that prior to concluding the transaction she asked Heppe "to check out the area and about the flow of traffic and so forth." Thus, we find defendants owed no duty, either legal or equitable, to disclose the fact that there existed plans for the re-construction project. Consequently, plaintiffs would not be able to prevail on this claim.

Although plaintiffs focus on Count IV, we also find that Count V, innocent misrepresentation, was properly dismissed. In order to properly state a claim under this theory, privity of contract must be established. . . . We find that such a requirement was lacking as between plaintiffs and Parrette and Buehler and, thus, summary disposition was properly granted as to this claim as well, Similarly, plaintiffs also claim that the trial court erred in determining that defendants Paul and Mary Joldersmas' actions in failing to disclose the reconstruction bypass plan constituted actionable fraud. We disagree.

In *Taffa v. Shacket*, . . . the elements for establishing fraud or silent fraud were set forth: "(1) a material representation which is false; (2) known by defendant to be false, or made recklessly without knowledge of its truth or falsity; (3) that defendant intended plaintiff to rely upon the representation; (4) that, in fact, plaintiff acted in reliance upon it; and (5) thereby suffered injury. . . . The false material representation needed to establish fraud may be satisfied by the failure to divulge a fact or facts the defendant has a duty to disclose. Such an action is one of fraudulent concealment."

After hearing all the evidence, the trial court concluded there was "no fraud or innocent or intentional concealment or misrepresentation on the part of Defendant sellers" The trial judge found no existing facts that created for the sellers a duty to disclose the possibility that the relocation of M-37 might take place in the future, pending federal funding. Additionally, the pertinent facts with respect to the proposed project were of public record and plaintiffs employed Roy Heppe, a CPA, to investigate these facts for them. Finally, the court concluded that even if the relocation of M-37 would have been material to plaintiffs' store operation, they did not rely on the sellers to disclose this information but, rather, on Heppe, in whom they had placed their reliance.

We are not persuaded, after thoroughly examining the entire record . . . and giving due consideration to the trial court's opportunity to judge the credibility of the witnesses . . . that the court erred in this instance. Testimony clearly established that approval for the project was still ongoing several months after plaintiffs purchased the store. Indeed, final approval from the federal government did not obtain until September 10, 1982, well over a year after the purchase (July 23, 1981) Moreover, such information was a matter of public record and had been so for years.

Not insignificantly, plaintiffs wrote to the state Department of Licensing and Regulation on June 20, 1983, requesting a full inquiry regarding Heppe's alleged malpractice with respect to the purchase. Excerpts from this letter further evidence plaintiffs' reliance on Heppe regarding the decision to purchase the property: "In summary: *Our purchase was contingent upon the findings and recommendations of Mr. Heppe, whom we paid, for his professional expertise.* Because of this, we now find ourselves with a party store that was over priced—which we cannot sell because of the land contract balances and the fact it is to be by-passed by the Michigan State Highway Department.

"When construction of the by-pass is started, I can think of no conceivable way a payment of $1167 per month plus expenses can be made. Mr. Joldersma knew this—Mr. Parrette knew this—we paid Mr. Heppe—*now* we know it. *We trusted Mr. Heppe completely*—placed our lifesavings, assets and future in his hands. In addition, I terminated my employment at Ford Motor Company after 24 years (30 year retirement and benefits) to purchase this store.

"In conclusion, may I ask—is not this C.P.A. responsible? Do we have recourse? In our personal opinion, we feel he may be guilty of breach of ethics." (Emphasis added.)

We are unpersuaded that the court erred in finding that plaintiffs did not in fact rely on the statements or omissions of the Joldersmas, James Parrette, or Bonnie Robbins. Rather, plaintiffs relied upon the recommendations of their agent, Roy Heppe. As reliance upon a false representation or omission of fact is a requisite element, . . . plaintiffs simply cannot assert that a fraud was perpetrated upon them by parties upon whom they did not rely. For this reason we similarly find no merit to plaintiffs' claim that the Joldersmas are responsible for any fraudulent concealment by their agent James Parrette and listing agent, Bonnie Robbins.

Plaintiffs next claim that they should not be charged with knowledge of the public record regarding the bypass plans. . . . Significantly, plaintiffs admit that prior to entering into the business transaction, they took note of the "rickety old bridge." A reasonable inquiry would have revealed the existence of the bypass project.

Finally, with respect to this issue, plaintiffs rely upon the following rule quoted in their brief: "The law of constructive notice can never be so applied as to relieve a party from responsibility for actual misstatements and frauds, and to prevent a representee from having a right to rely upon representations under such conditions. *Thus, one to whom a misrepresentation is made is not held to constructive notice of a public record which would reveal the true facts.*" (Emphasis added by plaintiffs.)

However, in this case defendants did not make "actual misstatements" regarding the bypass plans. Indeed, the very essence of this case is their failure to say anything at all.

We have carefully reviewed plaintiffs' next argument—that a seller can be liable for a misrepresentation of a future event and find it has no merit. At the time of the purchase, the bypass project was a future possibility—contingent upon federal approval and funding. Because of this fact, the Joldersmas' failure to inform plaintiffs did not constitute a fraudulent omission.

We have similarly reviewed plaintiffs next allegation—that defendants failed in their burden of proving that plaintiffs did not rely upon defendants' misrepresentations or omissions of fact—and find no reason to set aside the verdict on this basis.

Plaintiffs also claim that a defrauded party to a contract can rely on more than one representation and, thus, reliance on Heppe's recommendation does not relieve the Joldersmas of liability because plaintiffs also relied on a continuation of traffic as represented by the Joldersmas. However, the court expressly found plaintiffs did not rely upon statements or omissions of the Joldersmas. Having previously concluded that the court did not err in its findings, we find no basis for this claim.

Plaintiffs also contend that the Michigan Department of Transportation's adoption of Proposal B (Bypass Plans) was a material fact. Be that as it may, materiality is but one element of fraud. The court concluded, and we do not disagree, that plaintiffs failed to establish all of the requisite elements necessary to prevail.

As to plaintiffs' next issue, we agree that in the context of a rescission action, as here, it is irrelevant to what extent plaintiffs attempted to mitigate their damages.

Plaintiffs also assert that they have demonstrated the requisite legal damage suffered as a result of their reliance upon the alleged misrepresentation of defendants. However, defendants' expert, Eric Adamy, testified that in his opinion the store was not nearly as profitable after plaintiffs assumed the business for reasons other than the relocation of M-37. According to Adamy, plaintiffs did not adequately promote the store, they did not maintain an adequate inventory, there was decreased emphasis on electronic sales, there was an economic recession, and finally, the number of party stores in Newaygo County increased. The court apparently gave considerable weight to this testimony and concluded that plaintiffs had not been damaged by the highway bypass. We do not find error.

We cannot lose sight of the fact that during the last full year the Joldersmas operated the store, the gross sales were $365,727.02. After the first full year plaintiffs operated the business, gross sales dropped to approximately $200,234 (year ending June 30, 1982), followed by another decline to $163,730 for the next year (ending June 30, 1983). By the close of the next fiscal year (ending June 30, 1984), sales had increased slightly to $174,104. By the end of the fiscal year, sales had declined to $152,000. Significantly, however, the bypass project was not constructed until the first six months of 1984 which lends credence to the court's conclusion that the relocation of M-37 had no significant impact on the decline in sales. Rather, it was owing to factors other than the bypass.

Plaintiffs further contend that the trial court erred in excluding a witness, Calvin Deitz, from testifying in plaintiffs' case in chief. However, by failing to state any authority for their contention that the court's ruling was error, this issue is not properly before us. As we have consistently stated, a party may not leave it to this Court to search for authority to sustain or reject a position. . . .

We have carefully reviewed plaintiffs' final allegation of error and are unpersuaded that the court abused its discretion by admitting testimony concerning financial operations for the party store after the date of closing. As the parties agree, plaintiffs sought rescission, an equitable remedy. Not only is this information relevant as to plaintiffs' fraud claim, the remedy of rescission returns the parties to the status quo, i.e., it places the parties in the position they occupied before the transaction in question. . . .

Rescission of a land contract should not be granted where the result would be inequitable. The decision rests within the sound discretion of the court and each case must be decided on its own particular facts. . . . Here, it was necessary for the court to examine all the facts to determine the appropriateness of rescission. After doing so, the court concluded, inter alia, that the plaintiffs could not restore defendants to the status quo and were not entitled to rescission. We cannot say, given all the circumstances of this case, that the court erred so as to require reversal by allowing the challenged testimony. Moreover, and contrary to plaintiffs' claim, we find no evidence that the court was biased in defendants' favor.

We are not unmindful of the hardship that has befallen plaintiffs with respect to acquisition of the party store and we express no opinion, beyond the scope of our appellate review, as to the origin. However, this case was heard on the merits and we find no basis in law or equity to upset the final verdict.

AFFIRMED.

DEANN THOMAS, Appellant-Plaintiff, v. LEWIS ENGINEERING, INC., Appellee-Defendant., 848 N.E.2d 758; 2006 Ind. App. LEXIS 1069

Study terms: Negligent misrepresentation, reliance, knowledge, privity

Statement of the Case

Deann Thomas appeals from the trial court's grant of summary judgment in favor of Lewis Engineering, Inc. ("Lewis") and from the denial of Thomas' cross-motion for summary judgment on her complaint alleging negligent misrepresentation. Thomas presents a single issue for review, namely, whether the trial court erred when it entered summary judgment in favor of Lewis.

We affirm.

Facts and Procedural History

In May 2002, Eric Owens marked with a string what he believed to be the western boundary of his property in Hendricks County, where he intended to build a fence. Thomas, the adjacent property owner west of Owens' property, informed Owens that the marked line was on Thomas' property. According to the findings of fact made by the trial court in the underlying quiet title suit, "Owens hired Lewis for the purpose of locating Owens' west boundary line when Owens was considering the construction of [his] fence." Despite Thomas' protest, Owens built the fence along the boundary he had marked. In October 2002, Lewis prepared a retracement survey of Owens' parcel. Lewis then provided a copy to Owens. The survey indicated that the location of the proposed fence was not on Thomas' property.

In December 2002, Owens filed suit against Thomas "related to the placement of the fence and the ownership and location of the property line separating their [respective] parcels." Thomas counterclaimed to quiet title and alleged trespass. After a bench trial in that case, the trial court entered judgment against Owens and in favor of Thomas on Owens' second amended complaint and on Thomas' counterclaim.

In August 2004, Thomas filed her complaint against Lewis, alleging negligent misrepresentation with regard to the retracement survey performed for Owens and seeking to recover the fees and costs Thomas spent to defend against Owens' suit and to prosecute her counterclaim. Lewis filed a motion for summary judgment, and Thomas filed a response and counter motion for partial summary judgment. After a hearing on both motions, the trial court granted Lewis' motion and denied Thomas' counter motion. Thomas appealed.

Discussion and Decision

Standard of Review

When reviewing summary judgment, this court views the same matters and issues that were before the trial court and follows the same process. Specifically, we determine whether there is a genuine issue of material fact and whether the moving party is entitled to judgment as a matter of law. We construe all facts and reasonable inferences to be drawn from those facts in favor of the non-moving party.

Negligent Misrepresentation

Thomas contends that the trial court erred when it granted Lewis' motion for summary judgment on her negligent misrepresentation claim. Specifically, Thomas alleges that "Lewis owed Thomas a duty, as a matter of law, to perform its work for Owens in a professional, non-negligent manner, and thereby not cause her to expend time, energy, and money to defend her property from a lawsuit based on Lewis's negligent and deficient work-product." We cannot agree.

The tort of negligent misrepresentation, as embodied in the Restatement (Second) of Torts, provides in part:

> (1) One who, in the course of his business, profession, or employment, or in any other transaction in which he has a pecuniary interest, supplies false information for the guidance of others in their business transactions, is subject to liability for pecuniary loss caused to them by their justifiable reliance upon the information, if he fails to exercise reasonable care or competence in obtaining or communicating the information.

> (2) Except as stated in Subsection (3), the liability stated in Subsection (1) is limited to loss suffered (a) by the person or one of a limited group of persons for whose benefit and guidance he intends to supply the information or knows that the recipient intends to supply it; and (b) through reliance upon it in a transaction that he intends the information to influence or knows that the recipient so intends or in a substantially similar transaction.

> (3) The liability of one who is under a public duty to give the information extends to loss suffered by any of the class of persons for whose benefit the duty is created, in any of the transactions in which it is intended to protect them.

Restatement (Second) of Torts § 552 (1977).

Indiana has not adopted Restatement Section 552 without limitation. Indeed, the condition of Indiana law regarding the tort of negligent misrepresentation has been aptly described as one of "relative chaos." *Tri-Professional Realty, Inc. v. Hillenburg*, 669 N.E.2d 1064, 1068 (Ind. Ct. App. 1996). But it is clear that to date Indiana has not recognized that a duty exists to support the tort outside the limited context of an employment relationship. Instead, we have held that a professional owes no duty to one with whom he has no contractual relationship unless the professional has actual knowledge that such third person will rely on his professional opinion.

In considering the privity exception, the court in *Essex* [v. *Ryan*, 446 N.E.2d 368 (Ind. Ct. App. 1983)], recognized the difference between "*knowledge* that a third party will rely on the opinion given and an *expectation* that unidentified others might rely on it." *Essex* 446 N.E.2d at 372 (emphasis in original). Mere foreseeability is not enough. Instead, the actual knowledge exception to the privity rule has been applied only where there has been contact between the professional and the third party.

The court in *Essex* acknowledged that its holding was limited to the particular facts, namely, a surveyor's liability to unknown third parties. But the reasoning in that case applies equally to the facts in this case. Here, Thomas had no contractual relationship with Lewis. Thus, in the context of Thomas' negligent misrepresentation claim, Lewis did not owe a duty to Thomas unless Lewis actually knew that Thomas would rely on Lewis' survey.

Thomas noted that the marking of Owens' western property line necessarily marked the eastern boundary of Thomas' property. And the adjacent property owner's identity was easily ascertainable by searching public records. But absent privity, the rule applied in *Essex* and *Tri-Professional* requires the professional not merely know or be able to readily ascertain the identity of the third party but that the professional have actual knowledge that the third party will rely on its opinion or services. There is no evidence that Lewis knew that Thomas would actually rely on Lewis' survey.

Thomas attempts to create a duty based on the recipe set out in *Webb v. Jarvis*, 575 N.E.2d 992 (Ind. 1991). There, the supreme court required the balancing of three factors before a court may impose a duty. The three factors are: (1) the relationship between the parties, (2) the reasonable foreseeability of harm to the person injured, and (3) public policy concerns. *Id*. But the court in *Webb* used that formula to determine duty in the context of an ordinary negligence claim. Here, Thomas alleged negligent misrepresentation, not negligence.

Even if we were to apply *Webb* in determining a duty, Thomas' claim would still fall short. Regarding the first factor, there is no evidence that Thomas had a relationship or even any contact with Lewis. And as to the third factor, public policy concerns weigh against creating a duty here. This court stated in *Essex* that it "[was] not convinced that the economic benefits accruing to consumer plaintiffs would outweigh the hazards of potential liability which abolition of the privity requirement would impose upon providers of professional opinions." *Essex*, 446 N.E.2d at 373. The court quoted Justice Cardozo's opinion in *Ultramares Corp. v. Touche*, 255 N.Y. 170, 174 N.E. 441 (N.Y. 1931), when he stated:

> If liability for negligence exists, a thoughtless slip or blunder . . . . may expose [professionals] to a liability in an indeterminate amount for an indeterminate time to an indeterminate class. The hazards of a business conducted on these terms are so extreme as to enkindle doubt whether a flaw may exist in the implication of a duty that exposes to these consequences.

Essex, 446 N.E.2d at 373. The *Essex* court then noted further that

> Justice Cardozo's caution is particularly relevant today given the increasing litigiousness of our society and the rising cost of malpractice insurance. Section 552 does not, in our view, sufficiently guard against imposing unwieldy duties upon providers of professional opinions . . . We believe the privity requirement, subject to an actual knowledge exception, properly balances the competing interests of consumer and professional in a cause such as the one before us.

Id.

We agree that the privity requirement, with the actual knowledge exception, adequately balances the interests of consumers and professionals. Thomas asks us to find that a professional owes a duty to every unknown third party impacted by that professional's work. But as we recognized in *Essex*, such a holding would open the floodgates of litigation. Expanding the scope of a professional's liability may aid some consumers but would likely result in an increase in costs for professional services for all consumers. And consumers of other professional services would undoubtedly attempt to apply any expansion of the rule on these facts to claims against professionals other than surveyors as well.

The second factor, the foreseeability of the harm to an adjacent property owner, weighs in favor of Thomas on these facts. As noted above, the location of Owens' boundary necessarily located the boundary of the adjacent property as well, and Thomas' identity as the adjacent property owner was easily ascertainable from public records. But the lack of a relationship between the consumer and a professional and the public policy interests weigh strongly in favor of not finding a duty on these facts. After balancing the interests of consumers and professionals, we conclude that the rule applied in Essex is a sound rule that should be maintained, namely, that a professional owes a duty to a third party outside of a contractual relationship only if the professional has actual knowledge that the third party will rely on the professional's opinion or service.

Conclusion

We conclude that the duty rule applied in Essex is the correct one to apply in negligent misrepresentation cases. Specifically, a professional owes no duty to one with whom it has not contracted unless the professional has actual knowledge that the third party would rely on the professional's opinion or service. Because Thomas had no relationship with Lewis, she had no right to rely on its survey. And, in fact, Thomas did not rely on Lewis' survey. Instead, she argued that the survey was inaccurate when she defended against Owens' suit and filed her own counter suit. In sum, Thomas has shown neither a duty arising from a relationship with Lewis nor a duty arising from Lewis' actual knowledge that Thomas would rely on its survey. Thus, Thomas has failed to state a valid claim under Indiana law. As such, the trial court did not err when it granted summary judgment in favor of Lewis on Thomas' negligent misrepresentation claim.

Affirmed.

TRI-STATE TRUCKING OF RHODE ISLAND, LLC AND WE DISPOSE, LLC v. TRI-STATE TRUCKING, INC., EDWARD SCANLON, JR., 2007 R.I. Super. LEXIS 74

Study terms: Fraudulent inducement, misrepresentation, lack of capacity to sue, parole evidence rule, corporate charter, revocation

Before this Court after a bench trial are various claims for damages and other relief arising out of the sale of a waste hauling business. Also, before the Court is a motion for the release of funds from an escrow account. These claims arise from the sale of substantially all of the operating assets of Tri-State Trucking, Inc. (TST) to Tri-State Trucking of Rhode Island, LLC (TSTRI, LLC) on or about February 2005.

I. Facts and Travel

TSTRI, LLC and We Dispose, LLC are two entities in the waste removal business, and are controlled by Peter Calcagni. The Court will refer to these two entities and Mr. Peter Calcagni collectively as Calcagni in this decision, unless the context requires otherwise. TST formerly was in the waste disposal business until it sold substantially all of its operating assets to Calcagni.

There are two issues raised by Calcagni's complaint which are before the Court at this time. The first is the allegation that TST and its principals fraudulently induced Calcagni into entering into the various agreements by misrepresenting the average monthly revenue of the waste hauling business. Calcagni contends that he is entitled to a revision of the initial purchase price based upon the actual revenue numbers. TST denies that any misrepresentation occurred.

■ ■ ■

[An additional dispute involved the application of an "adjustment clause" in the Agreement.]

The paperwork memorializing the asset sale is hardly a model of clarity or precision, but resolution of this dispute requires the Court to interpret it. The closing of the asset sale occurred in early February 2005. The transaction provided that TST would sell substantially all of its equipment, customer lists, and goodwill to TSTRI, LLC, an entity which Peter Calcagni caused to be created for the purpose of purchasing those assets. In exchange, the Asset Sale Agreement (Agreement) provides that TSTRI, LLC would pay the sale price of $3,403,843 for the assets. This was to be paid via a $720,000 down payment at closing, and a promissory note (Note) for the balance. The Agreement also provides that Peter Calcagni would personally guarantee the obligation represented by the Note.

Following the closing, relations between the parties broke down due to various disputes arising from the asset sale and loan transaction. Calcagni invoked the adjustment clause and contends that he is entitled to approximately $600,000 in reductions to the purchase price. Because of that contention, he also argues that he has overpaid on his obligation under the Note. TST disagrees, however, believing that Calcagni is in default under the Note. TST has exercised its claimed right to accelerate the due date on all payments under the Note. TST brought its two count complaint in C.A. No. 06-4986 seeking damages arising from the Note and the appointment of a receiver for TSTRI, LLC.

Therefore, the Court must first consider the allegations of fraud and misrepresentation in order to determine the initial purchase price. Then, it must ascertain whether any reductions are required to that price under the adjustment clause, and if so, the amount of any reductions. The Court will also address Calcagni's contention that TST lacks capacity to sue because its certificate of incorporation has been revoked. Finally, the Court will consider whether Calcagni is in default under the Note and whether any remedies should follow from the Note.

II. The Initial Purchase Price

A. Findings of Fact

The Note provides that Calcagni shall pay the principal amount of $2,683,843 plus interest at 3.5 percent annually. This principal amount is consistent with the Agreement, which provides for a down payment of $720,000 against the stated purchase price of $3,403,843. The Note was executed and delivered at the closing, and provides for 48 equal payments of $60,000 per month until the Note is paid in full . . . The total of principal and interest, therefore, would be exactly $3.6 million.

Calcagni disputes that the "actual" initial purchase price was $3,403,803 as stated in the Agreement. Rather, Calcagni contends that the price was conceived as a function of the monthly revenues of TST's waste hauling business. Peter Calcagni testified that he values a business by multiplying the average monthly revenue of that business by some number—in this case, fifteen—in order to arrive at the purchase price. He claims that he relied upon the average monthly revenue amount provided in "Schedule 1.1(b)," which was provided by TST, was present at the closing, and is referenced by the Agreement in many places. That document gives an average monthly billing amount of $197,725, which when multiplied by 15, equals $2,965,875. Therefore, Calcagni contends that this figure is the initial purchase price represented by the agreements.

Calcagni then claims that Richard Nicholson, a principal of TST, misrepresented the amount of the monthly billings in Schedule 1.1(b). Therefore, Calcagni contends that he was induced to pay a higher price than he otherwise would have paid. Based on his review of the "Quickbooks" records, which were maintained by TST on a computer that was eventually transferred to Calcagni, the average monthly billing was actually $194,259.

Mr. Fred Guarino, the accountant who prepared that report for Calcagni, concluded in his report that Schedule 1.1(b) overstated the actual average monthly revenues by $3466. He also testified about the methodology used to reach those numbers. He had analyzed the three-month period before the closing to reach his $194,259 figure. Mr. Nicholson, conversely, could not remember during his testimony which time period he used to reach the $197,725 figure contained in Schedule 1.1(b).

Using the multiple of 15, the discrepancy revealed by the Sales Analysis Report would amount to a change of $51,992 in the sale price. However, this does not explain the difference between the approximately $2.9 million initial purchase price that Calcagni asserts, and the approximately $3.4 million price contained in the Agreement that Calcagni signed.

Calcagni contends that the difference is explained by the concept of "imputed interest." He contends that interest for the life of the Note was aggregated and added to the purchase price to reach the sum of $3,403,843. That imputed interest would be the difference between the stated amount in the Agreement, and the multiple of the monthly billings, which equals $437,968. Therefore, Calcagni's position is that the "actual" purchase price for the waste hauling business was approximately $3 million and that any adjustments should begin at that number. Once the initial price is corrected for any misrepresentation, and the adjustment clause applied, the "imputed interest" would apparently be added back to the resulting price under Calcagni's methodology.

The evidence does not reveal how the $437,968 interest figure was reached—neither the amount used as principal, the interest rate, or the number of monthly payments, are evident to this Court. It suffices to note that Calcagni's imputed interest figure is wholly inconsistent with the Note's terms of $2,683,843 principal, at 3.5% interest annually over 48 months, which would result in $ 196,157 of interest over the life of the loan.

B. Interpreting Contracts and the Parole Evidence Rule

Because the Agreement and the Note were executed as part of the same transaction, the Court will consider them as one instrument for purposes of interpretation.

As a general rule, where there exists an integrated, written agreement, parole evidence may not be used to vary, alter, or contradict that written agreement. By excluding evidence of prior or contemporaneous negotiations, the rule reinforces the importance and primacy of written instruments.

Calcagni's interpretation of the initial purchase price is clearly an attempt to modify the terms of the writings. The writings contain terms for a purchase price, loan principal, an interest rate, and periodic payments ... In sum, everything contained in the writings provides for a $3,403,843 purchase price, or $3.6 million including interest during the life of the loan. Therefore, unless the parole evidence rule is inapplicable, the Court must accept the terms of the writing as conclusive.

Two exceptions to the rule are arguably relevant. First, parole evidence may be used to interpret the meaning of ambiguous terms of an instrument. However, the Court finds that neither the price term in the Agreement nor the principal term in the Note are reasonably susceptible to multiple interpretations.

Second, such evidence may be used to show "illegality, fraud, duress, mistake, lack of consideration, or other invalidating cause." Restatement (Second) of Contracts § 214(d). Therefore, in order to be entitled to relief, Calcagni must demonstrate that he was fraudulently induced to enter the contract.

C. Fraud or Misrepresentation

Courts have recognized that misrepresentations about the productivity of a business may constitute fraud. A plaintiff seeking to establish common law fraud must show that "the defendant made a false representation intending thereby to induce plaintiff to rely thereon, and that the plaintiff justifiably relied thereon to his or her damage." *Zaino v. Zaino*, 818 A.2d 630, 638 (R.I. 2003).

As described above, Calcagni alleges fraud based upon a $3,466 discrepancy between the average monthly revenue stated in Schedule 1.1(b), and the Sales Analysis Report of TST's financial records. If the parties intended that the purchase price would solely be a multiple of the monthly revenues, as Calcagni contends, then the alleged misrepresentation would require a $51,992 adjustment to the sale price after applying the multiple of fifteen.

If the Agreement had explicitly provided for a sale price of fifteen times the figure in Schedule 1.1(b)—$2,965,875, then perhaps Calcagni's position would be tenable, and he could plausibly seek a $ 51,992 adjustment to the initial price. In this case, however, Calcagni cannot demonstrate that he detrimentally relied upon the Schedule 1.1(b). If he truly relied solely on Schedule 1.1(b), he could not possibly have agreed to a figure as high as the $3.4 million figure contained in the Agreement.

Calcagni's claim that the stated purchase price includes $ 437,968 of hidden, imputed interest is not credible because it is contradicted by the terms of the writings he signed, and which he is presumed to have read. . . .

Rather, there are two possible explanations for any harm befalling Calcagni. Peter Calcagni might have misunderstood the effect of the agreements that he was signing—specifically the concept of imputed interest—at the time of the closing. If so, that misunderstanding caused him to agree to a price much higher than the fifteen times average monthly revenue that he claims he intended. While unfortunate, such circumstances would not entitle him to relief. By manifesting his assent to the contract, Calcagni caused TST to proceed with the transaction as detailed in the writings. Therefore, even if Calcagni misunderstood that transaction, he must be bound by its terms.

Alternatively, Peter Calcagni might have completely understood the effect of the agreements he was signing, but after the fact, became dissatisfied with the arrangement and sought alternation of the purchase price. In that case, not only would Calcagni lack entitlement to relief, but would also be guilty of committing a fraud upon this Court. In either case, however, he is not entitled to a revision of the initial purchase price stated in the Agreement.

Finally, the Court notes that the "average monthly revenue" is not an absolute figure. Its determination entails a variety of accounting decisions—most notably the time period for which that figure will be computed. . . .

Because the Court finds that Calcagni has failed to demonstrate fraud, the parole evidence rule applies. The Court may not consider any evidence for the purpose of contradicting the initial purchase price and principal terms as stated in the writings, and must therefore conclude that those writings express the final agreement of the parties.

III. Applying the Adjustment Clause.

Because no fraud or misrepresentation occurred, the court applied the adjustment clause to the $3,403,843 initial purchase price. After analyzing several theories relating to various adjustments to offset against the purchase price, the Court concluded that Calcagni was entitled to an adjustment of $95,181.81 in the purchase price.

■ ■ ■

IV. Revocation of Corporate Charter

After the close of evidence, Calcagni moved for judgment as a matter of law, pursuant to Super. R. Civ. P. Rule 52, on the grounds that TST was no longer an active corporation. It argues, therefore, that it lacks the capacity to sue on its promissory note in C.A. No. 06-4986. It is undisputed that in October 2005, the Secretary of State revoked TST's articles of incorporation for failure to file its annual report. See G.L. 1956 § 7-1.2-1310(a)(3) (providing for such revocation).

When a revocation has occurred, "the authority of the corporation to transact business in this state ceases." Section 7-1.2-1311(b). However, a corporation whose articles of incorporation have been revoked "nevertheless continues for five (5) years after the date of the . . . revocation for the purpose of enabling it to settle and close its affairs, to dispose of and convey its property, to discharge its liabilities, and to distribute its assets, but not for the purpose of continuing the business for which it was organized." Section 7-1.2-1325. This statute modifies the common law rule, which provided that the dissolution of a corporation "marked the death of its corporate existence and, in the absence of statutory provisions to the contrary, terminated that existence for all purposes whatsoever." *Theta Props. v. Ronci Realty Co.*, 814 A.2d 907, 912 (R.I. 2003). Under these statutes, a corporation continues to exist for the "limited purposes of winding up the business and of defending lawsuits or filing any claims related to the business." Id.

While conceding that TST is within the five-year "wind up" period, Calcagni argues that TST is continuing its business and is not winding up as contemplated by § 7-1.2-1325. For example, it contends that it still owns one waste disposal truck and is servicing one account. However, the Court will not engage in an inquiry of the principals' subjective intent in order to determine whether TST actually intends to wind up its business.

Rather, the Court notes that the "shareholders, directors, and officers" of a revoked corporation "have power to take any corporate or other action that is appropriate to carry out the purposes" of the wind-up provision in § 7-1.2-1325. The Court finds that bringing suit upon a defaulted note is an appropriate action for the purpose of settling and closing a corporation's affairs, and may even be a necessary step in distributing its remaining assets to creditors and shareholders. Therefore, even if the principals do not intend to wind up the business as the law requires, the Court will not prevent their suit from proceeding.

If the principals are operating the business without authority, perhaps they are running the risk of incurring personal liability arising from those endeavors. Perhaps there exists some other remedy for such unlawful actions. See, e.g., § 7-1.2-303 (allowing for certain classes of individuals, which do not include corporate creditors, to assert the lack of capacity or power as a defense, and also allowing the attorney general to seek an injunction of unauthorized transactions). However, the Court finds that the revocation should not prevent TST from maintaining its suit on the promissory note.

V. Default under the Note

TST seeks judgment on the Note because it claims that TSTRI, LLC is in default. Similarly, it seeks judgment against Peter Calcagni on his personal guaranty of the Note.

A. Findings of Fact

As one would expect, failure to remit the required installment payments is an event of default under the Note. Upon default, TST was required to give notice of default and a ten day opportunity to cure. TST gave such notice on August 7, 2006. On August 28, 2006, TST accelerated the entire amount of indebtedness owed on the Note.

The Court accepts the payment schedule found in [a trial exhibit] as evidence of when payments were made. Those payments are allocated based upon Schedule A to TST's Post-Trial Memorandum of March 23, 2007. This schedule reveals that Calcagni fell behind on the required payments, and ceased making all payments after July 28, 2006. After accounting for principal, regular interest, default interest, and late fees through March 23, 2007, Calcagni allegedly owes $2,174,941.47 on the accelerated Note. This amount includes default interest for the period from May 2006 to August 2006.

B. Acquiescence

Calcagni argues that TST acquiesced to the late payments on the Note.

The evidence reveals that between November 2005 and August 2006, the parties communicated in an attempt to reconcile the various amounts owed by each party to the other. Therefore, the Court will not permit default interest to be charged prior to the August 7, 2006 notice of default, because the Court finds that TST acquiesced to those defaults. However, when the efforts to reconcile failed, and Calcagni made no further payments, TST was within its rights to demand in August 2006 that Calcagni cure the default, and to accelerate the Note when the default was not cured.

C. Right to Set Off Indebtedness

Calcagni claims a right to set-off the various purchase price adjustments against his indebtedness on the Note. [H]owever, many of his claimed adjustments were erroneous. The purchase price adjustment of approximately $95,000 provides no excuse for non-payment, as Calcagni's overdue payments far exceeded that amount at the time TST invoked its default and acceleration rights. Moreover, the Note did not provide a right to set-off such debts against the required payments under the Note. Therefore, the Court finds that TST properly found TSTRI, LLC to be in default under the terms of the Note.

The entire amount due should be calculated as evidenced by Schedule A, but excluding default interest for May 2006 through August 2006. Therefore, without any regard to the effect of the adjustment clause, the total amount due on the accelerated Note is $2,142,766.10.

VI. Type of Relief

Calcagni has described its request for relief as an adjustment to the purchase price and the principal due under the Note, and it is clear that some adjustment is warranted. It is not clear, however, whether Calcagni asks this Court to reform the price terms contained in the Agreement and Note, or simply to award damages against TST in the amount of the adjustment.

Following the asset sale, in July 2005, TST conveyed a security interest in the Note to Rhode Island Resource Recovery Corporation (Resource Recovery), which held the Note under a pledge agreement. Based on its status as a holder of that security interest, Resource Recovery moved to intervene in this litigation prior to trial, and this Court denied its motion.

After trial, in March 2007, Resource Recovery gave notice to TST that it had defaulted under its obligations, that it was accelerating its loan with TST, and was assuming the rights of TST under the Note. Therefore, it seems that Resource Recovery is now the owner of the Note in question.

Any reformation of the Note's principal term, reducing the amount owed, would certainly be opposed by Resource Recovery. Therefore, the Court will conduct such further proceedings as are necessary to address whether reformation, damages, or perhaps some other remedy is appropriate as a result of the Court's conclusion that Calcagni is entitled to an adjustment to the purchase price. As the net effect of these actions is approximately $2 million owed to TST, the Court will order that the remaining funds presently held in escrow be released. However, the parties are directed to address whether those funds should be paid to either TST or Resource Recovery. Finally, in light of the indebtedness now owed by TSTRI, LLC to TST, the parties are directed to address TST's request that TSTRI, LLC be placed into receivership. The resolution of these questions is necessary before judgments may enter in either of the actions pending before the Court.

VII.

■ ■ ■

[T]he Court finds that TST is entitled to damages against TSTRI, LLC on Count I of its complaint in the amount of $2,142,766.10 arising from default under the Note. Similarly, it is entitled to damages against Peter Calcagni on his guarantee of that Note. The Court will not enter judgment against We Dispose, LLC, however.

■ ■ ■

Counsel may present orders consistent with the foregoing, which shall be settled after due notice to counsel of record.

Review Questions

1. Black represents a company that makes computers. Black tells the White Manufacturing Company that installation of a system for cost control and the rental of the machines to process them will save the company $6,000 per month. As a result, the White Manufacturing Company rents the machines and installs the system. After six months of operation, it is determined that the system is losing money for the company at the rate of $3,000 per month. The White Manufacturing Co. sues Black's employer for the $18,000 lost by use of the system to date and the $36,000 that would have been saved if Black's claim had been correct. White Manufacturing Co. charges fraud. How much can White Manufacturing Co. recover? On what basis?

2. List the elements required for fraud. Which of these elements are required for misrepresentation?

3. Gray, an engineer for a manufacturer of die-cast parts, inadvertently used the weight of the parts rather than the labor cost in calculating the bid price. As a result the price bid per part was about 25 percent above the cost of the raw die-cast material rather than about 80 percent above the raw material cost (which would have resulted from proper bid calculation). The manufacturer to whom the bid was submitted immediately accepted. Gray's employer claims that he should be allowed to avoid the contract since an error was made in the calculation. Will the court be likely to uphold the contract or dissolve it?

4. Distinguish between duress and undue influence.

5. Brown, a machine operator in Green's plant, lost her left hand in the course of her employment. Because of the character of the work, the state workers' compensation laws do not apply. Brown threatens to sue Green for $80,000 for damage to her ability to earn a living for the remainder of her working life. Brown claims, with apparent good reason, that the machine should have been guarded at the point where the injury occurred. Green offers to pay $20,000 to Brown if Brown will drop the court action. Brown agrees, and Green pays her $5,000 on account. It soon becomes apparent that this is all Green intends to pay. Brown starts a court action for the remaining $15,000. Green sets up as a defense that the contract resulted from duress. Who would win the court action? Why?

6. The defendants won in the case of *McMullen v. Joldersma*. Suppose the plans were complete for the M-37 bypass and the financing was in place when the McMullens bought the party store. How would this change the outcome of the case?

7. As explained in *Deann Thomas v. Lewis Engineering, Inc.* what is the exception to the privity rule as it relates to professionals?

8. What are the two exceptions to the parole evidence rule as described in *Tri-State Trucking of Rhode Island, LLC v. Tri-State Trucking, Inc.?*

9. Consider the case Turja v. Turja in chapter 6. What facts presented could support a successful claim that the son, Stephen, was guilty of fraud, undue influence, and duress? Are all the elements met? If not, what additional facts would need to be found?

Consideration

HIGHLIGHTS

- Four Essentials of Consideration
  1. Must have value.
  2. Must be legal.
  3. Must be possible at the time the contract is made.
  4. Must be either present or future.

Consideration is the price (in terms of money, goods, or services) given by one party to a contract in exchange for the other party's promises. Consideration may take the form of a benefit to one of the parties. Consideration also can consist of a detriment to a party or a forbearance by a party. Whatever form it takes, if the price is freely bargained by the parties, it is construed by the courts as their consideration.

Generally, the courts will consider the question of whether the consideration was "sufficient" to serve as a basis for a binding contract. In the vast majority of situations, the promises exchanged constitute sufficient consideration; the sum of $10 may suffice. An example of insufficient consideration is a promise to perform an undisputed legal duty already owed to the other party.

Although the courts will consider the question of the sufficiency of the consideration, the courts very rarely consider the "adequacy" of the consideration. **Adequacy** refers to the relative values of the promises, goods, or services exchanged. Usually, the courts are not concerned with whether someone made a wise bargain; the courts only consider whether the bargain amounted to an enforceable contract.

ESSENTIALS OF CONSIDERATION

The consideration given by each party to a contract must meet certain requirements.

1. **The consideration must have value.** Except for money given in satisfaction of a debt—the amount of which has been acknowledged by both parties and is not in dispute—the values exchanged by the parties need not be equal. Usually it is assumed that in the free play of bargaining, the parties have arrived at what each believes to be a reasonable consideration. If the consideration exchanged appears grossly unfair, this disparity in value may be used to establish a case of fraud or other lack of reality of agreement, but inequality of value usually is not sufficient proof by itself.

   Black promises an uncle, White, that she will not smoke until attaining the age of 22. White, in turn, promises to pay Black $5,000 for not smoking until Black is 22. Each consideration has value. Black's consideration is a forbearance of a legal right to smoke before age 22. White's consideration is the $5,000. It may seem to an outsider that $5,000 is a high price to pay for the value received. Or it might seem to others that no price would be sufficient to pay for giving up the use of tobacco. However, in this contract, Black and White set the price—he value of Black's forbearance of a legal right to them was $5,000. A court would not question their determination of values.

2. **The consideration must be legal.** The courts usually do not enforce a contract in which the consideration given by one of the parties to the contract is contrary to an established rule of law. If, in the previous example, a statute pronounced the act of smoking before age 22 a crime, Black's forbearance of smoking would not have been valid consideration; legally, Black couldn't smoke anyway. To be valid consideration, Black's forbearance would have to be forbearance of something that Black was lawfully entitled to do. Similarly, if it is a positive act that is promised as consideration, the act must be lawful. If Black's consideration had been a promise to use illegal narcotics or, for that matter, not to use them, the contract would not be enforced.

3. **The consideration must be possible at the time the contract is made.** If the impossibility of performance was known to either or both of the parties at the time of the offer and acceptance, the courts ordinarily do not grant damages for nonperformance. The situation is a little more difficult, however, if the contract calls for performance that later becomes an impossibility. Occasionally, the subsequent destruction of an item essential to the contract, through no fault of either party, renders performance of the contract impossible. Impossibility of performance is discussed further in Chapter 13.

4. **The consideration must be either present or future.** A present contract cannot be supported by a past consideration. Suppose that a number of students decide to paint several classrooms. After the painting has been completed, the dean calls the students together, compliments them, and then promises to give them each a $50 credit toward tuition next year. The only possible consideration for the dean's promise is the painting, a past act. Hence, there is no consideration.

An exception to this rule may exist in certain situations where it appears that a gift was made. For the past consideration to support a contract, it must not have been intended as a gift when it was given to the other party. If the past act or forbearance was not intended as a gift but was done at the other's request, then an **unliquidated obligation** (that is, an obligation of some undetermined amount) was created by it. A present promise to pay for such a past act or forbearance is binding on the promisor. The unliquidated obligation has become liquidated. Past relationships between the parties concerning similar acts or forbearance often explain their intent either to make a gift or to create an unliquidated obligation. But valid past considerations form a rather minor exception to the general rule.

The vast majority of contracts generally look to the future. The exchange of a future promise for a similar promise has virtually unquestioned standing at law. If Green orders 10,000 die castings of a particular design to be delivered by August 31 and agrees to pay for them on delivery, both parties' exchanges of consideration are to take place in the future. Either this or present payment for future delivery sets up an obligation that the other party must carry out or risk action for breach of contract.

Seal of Consideration

In ancient times an impression was made on a piece of clay or (later) a blob of wax attached to a document. The seal took the place of the signature. Certain documents of exceptional importance or high security were required to be sealed. The seal came to have such significance that the common law courts would not inquire whether any consideration existed for a contract if the contract bore a seal. The formality of the act of sealing the document made it legal and binding.

In private transactions the seal is no longer an impression made on wax. Now it may be composed of the word seal or any other notation intended by the parties as a seal, following the signature. Many government institutions (such as the courts) have formal seals. When a notary notarizes a document, the notary usually adds the notary's seal. In most states, the use of private seals to formalize contracts has no real legal effect. Most states have abolished the old common law rules on seals. However, many states still require that certain documents (such as deeds for real property) be notarized; as noted, this usually involves the notary's seal.

> There are 3 theories of consideration:
> 1. Bargain theory
> 2. Injurious reliance theory
> 3. Moral consideration theory.

Theories of Consideration

Certain theories of consideration form the basis of court opinions in contract cases. A primary one is the **bargain theory,** which in effect holds that the parties have bargained in good faith for the consideration to be received. Each of the parties has determined the value of the consideration. If the agreement to exchange consideration was voluntary and intentional, the values involved—regardless of how dissimilar they may appear—are not subject to question.

The bargain theory is the basis of court decisions in which it appears that the parties negotiated freely. But what happens when one party deals from a position of great strength with a much weaker party? Is it fair when a dominant party may dictate terms on a "take-it-or-leave-it" basis to one who is practically forced to take it? Should a landlord, for example, be allowed to hold a tenant to a very one-sided contract that the tenant was forced to take because she had no alternative?

Many cases have indicated that gross inequities of this nature may not be enforced. For example, if the contract wording would require the weaker party to assume responsibility for not only the weaker party's own negligence but also the dominant party's negligence, such terms may not be given their literal effect. These adhesion contracts have gradually come to be less enforceable. The courts seem to favor a balancing of the equities in such situations rather than blindly following the contract language.

Another theory is the **injurious reliance theory,** which is based on the principle of estoppel. Courts occasionally enforce promises that are not supported by consideration under a doctrine referred to as **promissory estoppel.** (See *Pickus Construction and Equipment v. American Overhead Door* at the end of the chapter for an analysis of the elements required to prove promissory estoppel. Note that all the elements have to be proved.) Essentially, this doctrine prevents a promisor from asserting the absence of consideration as a defense in situations in which a person relied on the promisor's promise to his detriment. Where one of two innocent parties must suffer, so the reasoning goes, the party whose act occasioned the loss should bear that loss.

Suppose Gray promises to contribute $10,000 to a charitable organization's building fund. In reliance on Gray's promise, the charitable organization contracts with others for a structure to be built. Gray's promise to contribute will be enforced in the courts if this is necessary to pay for the building.

A third theory, the **moral consideration theory,** is generally confined to those cases in which a preexisting debt has been discharged by law and the debtor, subsequent to the termination of the obligation to pay the debt, agrees to pay it anyway. The basis of the argument here is that though the obligation to pay the original debt has been relieved by law, the debt still exists. The debtor, in reaffirming this obligation to pay the debt, has transformed the moral obligation to pay into a legal duty to pay. This moral obligation is sufficient consideration to sustain a court action to recover on the debt. Cases of this nature often arise from bankruptcy proceedings and creditor's compositions. See Chapter 13 for more on compositions.

Mutuality

Generally, both parties must be bound to a valid bilateral contract or neither is bound. The promises constituting consideration in a contract must be irrevocable. If either party to a contract can escape the duties under it as desired, there is no contract between the parties. The contract must fail for lack of consideration and mutuality.

Legal Detriment

Consideration for a contract may consist of a legal detriment. **Legal detriment** means that the promisor agrees, in return for the consideration given by the other party, either to give up some right that is lawfully the promisor's or to do something that the promisor might otherwise lawfully avoid doing. Agreeing, for consideration in the form of a discount in price, to restrict this year's purchases of raw material requirements to a particular source is an example of legal detriment. A company agreeing to so restrict its purchases would be giving up its legal right to choose to purchase its raw materials from any other source for the year's time.

Forbearance

Forbearance, or a promise to forbear or give up a legal right, is sufficient consideration to support a contract. One instance of forbearance occurs when one party gives up the right to sue another. A person has a legal right to resort to the courts to have a claim adjudicated. For such forbearance to sue to constitute consideration, the claim on which the suit would be based must be a valid one. The party that forbears must have at least reasonable grounds on which to bring court action. If the claim is reasonably doubtful, or would be a virtual certainty in favor of the plaintiff, the promise to forbear is sufficient consideration.

- Common in charity
- equity cases - "not fair" otherwise

In a small minority of cases, some courts have extended this reasoning even further, allowing as sufficient consideration forbearance to sue when only the party who would be plaintiff thinks, in good faith, that he or she has a valid claim. However, if a person brings a court action maliciously (for example, only as a nuisance to the defendant), forbearance of court action fails to constitute consideration on which a contract may be based.

One type of case in which forbearance of court action often occurs is that in which the amount of a debt is in dispute. If, in a compromise, the disputing parties agree on the amount to be paid, this agreement is binding on them. Each has given up the right to have the court decide the correct amount of the debt. In fact, most court cases are settled by compromise before they ever go to trial.

It is considered to be a different situation when the amount of debt is certain and due or past due. When such is the case, the acceptance by the creditor of a lesser sum may not discharge the complete debt. The creditor may bring an action at law later for the remainder of the debt. However, time has value. Prepayment of a lesser sum may, by agreement, discharge a fixed debt. Also, payment by means other than money may discharge a liquidated debt.

Pre-existing Duty

The law allows you to collect only once for what you do. If a person already has a duty to perform in a certain way, she cannot use that same performance as consideration in another contract. For example, assume that the house next door is on fire. The city's fire department is attempting to extinguish the blaze. An offer by you to pay a firefighter $500 extra to spray water on your roof to keep the fire from spreading to your house would not create a valid contract on acceptance. The firefighter is already paid for doing everything to keep the fire from spreading. The firefighter can give nothing in exchange for the $500 that has not already been paid for.

An exception exists to the rule that consideration must not be something that one already is bound to do. Assumptions are usually made before parties enter into a contract. The actual conditions may not exist in the manner that the parties assumed. When this is found to be true and the performance of the resulting contract is thereby made more difficult, the added difficulty may be treated as consideration to support a modification of the contract price.

If, under these conditions, the parties agree to added compensation, the courts will often enforce the added payment. To be enforceable, however, the difficulty must be something that could not be reasonably anticipated by normal foresight. (See *Plante v. Moran Cresa v. Kappa Enterprises, LLC* at the end of the chapter.)

Black, a contractor, agrees to build a structure for White for a fixed price. Test borings have shown the soil to be mainly clay. However, during excavation for the foundation and basement, Black discovers that the test borings somehow missed a substantial rock layer. If White agrees to pay extra because of the necessity of excavating through the rock, the modified agreement should be enforceable. This is true even though the original contract did not require White to make such payments if unusual conditions were met. White might successfully refuse to pay extra, but once White has agreed to pay, White is bound to do so. If the price increase had been based on a factor normally considered foreseeable, such as increases in labor or materials costs, there probably would not have been lawful consideration. Such contingencies could be anticipated and covered in the original price; such changes would not support added consideration.

PICKUS CONSTRUCTION AND EQUIPMENT v. AMERICAN OVERHEAD DOOR, 761 N.E.2d 356 (App. Ill. 2nd District, 2001)

Study terms: Reliance, promissory estoppel, custom, trade practice

Defendant, American Overhead Door (AOD), appeals from a judgment of the circuit court of Lake County entered in favor of plaintiff, Pickus Construction & Equipment (Pickus Construction). Following a bench trial, the trial court awarded plaintiff $35,079. Defendant contends that the trial court's judgment was contrary to the manifest weight of the evidence. For the following reasons, we reverse.

■ ■ ■

Background

Plaintiff is a construction company that was engaged as the general contractor on a project known as the Wheaton Public Works Project (Wheaton Project). James Pickus has been the company's vice-president of operations since 1990. In this position, he is involved in bidding and reviewing contracts between plaintiff and its various clients and subcontractors. Defendant is a company that is in the business of supplying and installing overhead doors. Larry Hooker has been its president for over 30 years. At the time of the events that form the basis of this suit, Trevor Murphy was employed by defendant as a salesman.

Plaintiff received plans for the Wheaton Project early in September 1997. Bidding on the project was to close on September 26, 1997. The parties dispute whether plaintiff invited defendant to submit a bid or defendant learned of the project from an independent source. . . . The specifications for the Wheaton Project required the use of Fimble doors or a comparable alternative. Fimble is a manufacturer of overhead doors.

On September 22, 1997, plaintiff received defendant's first bid on the Wheaton Project by fax. In this bid, defendant proposed to supply 18 doors, of various sizes and types, at a cost of $76,895. The doors were classed as types A, B, C, and D. These different classifications apparently referred to the doors ' sizes. Fourteen doors were type A, two were type B, one was type C, and one was type D. The bid did not indicate from which manufacturer defendant intended to procure these doors. This bid also included 18 operators. "Void" was written across this bid because it was superceded by later bids.

Plaintiff received a second bid from defendant on September 24, 1997, at 8:41 a.m. in the amount of $69,575. This bid consisted of three pages; however, only the first and third appear in the record. On the first page, type A doors are listed, but no quantity is shown. Type C doors are omitted. Thus, only three doors, two type B and one type D, are listed. Pickus testified that the second page of the bid listed two or three additional doors. Eighteen operators were also included in this bid. This bid has "void" written over it. At 8:57 a.m. of the same morning, defendant faxed plaintiff a corrected first page that listed the quantity of type A doors as 13. All bids contained a statement indicating that, because of potential price increases by suppliers, defendant would not hold prices for more than 30 days and a signed purchase order must be tendered.

On September 26, 1997, the date bidding on the Wheaton Project was to close, plaintiff received four additional bids from other overhead door companies. The lowest of these was from a company called Door Systems in the amount of $113,500. Pickus stated that, after receiving these bids, he became concerned about defendant's bid because it was considerably lower and telephoned Murphy to express his concern. Pickus testified to the following conversation with Murphy. He advised Murphy that he believed defendant's bid was erroneous. . . . They went over the plans to insure defendant's bid complied with them. Murphy assured Pickus that the bid was correct. Pickus also inquired as to whether the bid had been sent to other general contractors, and Murphy said that it had. Pickus told Murphy that he felt the bid was too low and that Murphy should advise all of the general contractors to whom it had been sent of this problem. Pickus explained that, if he did not use the bid and the other general contractors did, plaintiff would be at a competitive disadvantage relative to the other general contractors. Murphy stated the bid was correct and would be honored. Pickus informed Murphy that plaintiff would be using the bid and if it was awarded the project, it would forward a subcontract to defendant. Murphy replied, "Fine, I look forward to working with you" and added, "Good luck." Defendant disputes whether this conversation occurred. Plaintiff was awarded the project later the same day.

On October 1, 1997, defendant faxed plaintiff a new bid, lowering the price to $66,000. Plaintiff contends that there was some contact between plaintiff and defendant in the 30 days following the date defendant's last bid was received. Defendant disputes this contention. Outside of defendant's October 1, 1997, fax, the record does not divulge the nature of any such contact. It is undisputed that plaintiff did not forward a signed purchase order or any other writing indicating the acceptance of defendant's bid during this period.

Plaintiff contacted defendant early in November 1997 to set up a meeting to award defendant the contract for the overhead door system. The meeting took place on November 17, 1997. Pickus was called into the meeting as it was nearing completion. Murphy told Pickus that he had not figured the specified manufacturer, Fimble, into the bid and that defendant was no longer willing to enter into a contract based on its bid. Murphy stated that the door on which defendant had based its bid was equal or superior to the door identified in the plans and specifications for the project and that he would be able to get the architect to approve it. The architect, however, did not do so.

Subsequently, defendant submitted additional proposals using different doors. All were rejected by the architect. Finally, on May 13, 1998, defendant submitted a proposal using Fimble doors at a cost of $122,800. Pickus then spoke with Hooker about the matter. Hooker told Pickus that the only way it would perform the job was if plaintiff sent defendant a contract reflecting the May 13 proposal. Plaintiff declined and instead entered into a contract with Door Systems at a cost of $114,500.

Plaintiff then filed this action, seeking damages on a promissory estoppel theory. After a bench trial, the court found for plaintiff. The trial court found that plaintiff had established a prima facie case of promissory estoppel. The court held that defendant's bid was not ambiguous despite not mentioning the manufacturer, as it was based on the plans and specifications for the Wheaton Project. . . . The court acknowledged that defendant's last bid, of which the second page was not made a part of the record, included only 16 doors, although the plans called for 18. The court reduced damages proportionately as a result. Regarding the 30-day written confirmation provision in defendant's bid, the court observed that, in the construction industry, this was not a firm rule. Accordingly, the trial court awarded plaintiff $35,079. Defendant now appeals.

Analysis

In order to succeed on a claim of promissory estoppel, a plaintiff must show (1) that the defendant made a promise unambiguous in its terms, (2) that the plaintiff relied on the promise, (3) that this reliance was expected and foreseeable from the defendant's position, and (4) that the plaintiff's reliance on the promise was detrimental. Defendant contends that plaintiff failed to demonstrate both the existence of an unambiguous promise and that plaintiff's reliance was foreseeable to defendant. . . .

Turning to defendant's first argument, we conclude that the trial court's determination that an unambiguous promise was made is not contrary to the manifest weight of the evidence. Defendant points to two sources of potential ambiguity. First, defendant asserts that the bid upon which plaintiff relied was for 16 doors, while the Wheaton Project required 18 doors. We find this argument unpersuasive. The trial court based its award on defendant's having forwarded a bid to plaintiff for 16 doors. . . . Defendant does not explain how its bid, taken as a bid for 16 doors as the trial court took it, was in some way ambiguous.

Second, defendant argues that its bid was ambiguous because it did not specify the manufacturer of the doors it proposed to use. The trial court ruled that no ambiguity existed sufficient to preclude the application of promissory estoppel because defendant's bid was based on the plans and specifications for the Wheaton Project, which clearly set forth what constituted an acceptable door. . . .

Defendant asserts that evidence to the contrary exists. According to defendant, it was undisputed at trial that, when a subcontractor intended a particular manufacturer, the manufacturer was specified in the subcontractor's bid. In support, defendant points to three bids from other subcontractors that did name specific manufacturers and testimony from Hooker that defendant's bid did not. Such evidence is wholly insufficient to establish a trade practice or custom. If plaintiff should have known to interpret defendant's bid in light of the custom to which defendant alludes, defendant failed to introduce any substantial evidence of this at trial.

Some support for defendant's position can be found in the fact that the bid sheet contained a space to indicate that the bid was based on plans and specifications. This space was left blank. However, the weight of this evidence is undermined by the fact that defendant's later bids, including its final one using Fimble doors, also contained no indication in the space provided that the bid was based on any plans or specifications. . . . The trial court's decision on the ambiguity issue finds support in the record, and, although evidence to the contrary exists, it is not so compelling as to render the trial court's decision erroneous. We will not substitute our judgment for that of the trial court in such circumstances.

■ ■ ■

We now turn to defendant's next argument. Defendant contends that plaintiff's reliance upon its bid was not reasonable and hence not foreseeable from defendant's point of view. Defendant advances two theories on this point. First, defendant asserts that, because its bid was so much lower than the others plaintiff received, plaintiff knew it was erroneous and thus should not have relied on it. Second, defendant points to plaintiff's failure to forward a signed purchase order or some other writing to it within 30 days, as was required in the bid documents. According to defendant, this failure to comply with a condition required by the bid renders any subsequent reliance unreasonable.

We find defendant's first contention unpersuasive. It is generally true that a general contractor that receives a bid that is substantially lower than other bids it receives is put on notice that the bid may be erroneous. . . . In the present case, Pickus, when confronted by defendant's low bid, contacted defendant and expressed his concern that the bid was erroneous. Murphy stated that the bid was accurate, that defendant would honor it, and that it had been submitted to other general contractors. Murphy's assurances were sufficient to dispel any impression held by Pickus that the bid had been made in error. Under these circumstances, we cannot find that the trial court's decision was against the manifest weight of the evidence simply because defendant's bid was substantially lower than the others received by plaintiff.

Second, defendant asserts that plaintiff's failure to forward a written purchase within 30 days, as specified on the face of the bid, renders any subsequent reliance unreasonable. It is undisputed that no such order was tendered to defendant. The trial court rejected this argument, finding that the 30-day provision was not a firm rule in the construction industry and that "generally, things go beyond that time frame." Defendant contends that there is no evidence in the record to support this finding.

Defendant's observation is correct. In its brief, plaintiff points to no basis for this finding, and our reading of the record reveals none as well. In fact, the only evidence in the record addressing the existence of a trade practice regarding the 30-day provision is contrary to the trial court's finding. Hooker explained that defendant included this provision in the bid because its suppliers would only hold their prices for 30 days. Hooker added that it is a custom of the industry that contracts be in writing and that subcontractors order no material until they receive something in writing. Defendant also points out that two other door companies included similar provisions in their bids; however, this fact provides weak support, at best, for the existence of this trade practice, as the inclusion of such provisions in these bids says nothing as to whether they were typically enforced. Plaintiff submitted no contradictory evidence from which the existence of a trade practice could be inferred.

The party asserting the existence of a trade practice bears the burden of proving its existence. Plaintiff submitted no evidence of a trade practice indicating that such provisions like the one in defendant's bid were routinely disregarded. Accordingly, the trial court erred in finding that such a trade practice existed. Absent this trade practice, plaintiff cannot establish that it was reasonable for it to rely on defendant's bid, after the bid was to expire on its own terms, without taking the specified action to consummate the contract.

Plaintiff relies on Illinois Valley Asphalt, Inc. v. J.F. Edwards Construction Co., 90 Ill. App. 3d 768, 45 Ill. Dec. 876, 413 N.E.2d 209 (1980), in arguing that its failure to respond within 30 days does not preclude its promissory estoppel claim. That case is distinguishable. Illinois Valley Asphalt did involve a factual situation similar to the present case, where a general contractor responded to a subcontractor's bid in writing after the date the subcontractor fixed in its bid. However, after the bidding had opened, a representative of the general contractor told a representative of the subcontractor that, if the general contractor was awarded the contract, it would award the subcontract to the subcontractor. At trial, evidence was presented that a trade practice existed that such a communication was sufficient to award the subcontract. In that case, the plaintiff presented evidence that established a trade practice demonstrating that it was reasonable to rely on the defendant's bid. In the present case, no such evidence exists.

Plaintiff points out that a conversation occurred between Murphy and Pickus similar to the one the court relied on in Illinois Valley Asphalt. As noted above, in Illinois Valley Asphalt it was shown that the conversation was sufficient to award the subcontract to the subcontractor considered in the context of a prevailing trade practice. The existence of a trade practice is a question of fact. We cannot assume the existence of a similar trade practice absent evidence to support its existence merely because such a trade practice was proved in Illinois Valley Asphalt. Hence, this similarity between the two cases provides no support for plaintiff's position.

In light of the foregoing, the judgment of the circuit court of Lake County is reversed.

Reversed.

HUTCHINSON, P.J., and BOWMAN, J., concur.

PLANTE & MORAN CRESA, L.L.C. v. KAPPA ENTERPRISES, L.L.C., et al., 2006 U.S. Dist. LEXIS 38135

Study terms: Pre-existing duty, unjust enrichment, consideration, fixed-fee

Opinion and Order Granting Defendant's Motions for Summary Judgment

Before the court are Kappa Enterprises' motion for summary judgment, filed March 30, 2006, and Flagstar Bank's motion for partial summary judgment, filed April 3, 2006. Plaintiff submitted a combined response on April 10, 2006; Flagstar filed a reply on April 17, 2006. The court heard oral argument on June 8, 2006, and took the matter under advisement. For the reasons set forth below, Defendants' motions are granted.

Background Facts

Plaintiff, Plante & Moran Cresa, L.L.C., offers project management services to property owners to help them design and construct buildings; such services typically include overseeing the architect, land acquisition, contractor bidding, and construction. Plaintiff provided project management services to Defendant Kappa Enterprises, L.L.C., with respect to the development of a hotel in Southfield, Michigan. Plaintiff and Defendant Kappa entered into a contract on December 9, 1999, which stated that Plaintiff would provide project management services for the hotel design and construction in exchange for a lump sum payment of $189,000. (contract). Plaintiff apparently obtained the contract by convincing Kappa that its original choice intended to charge too much. Defendant Kappa paid Plaintiff in installments in accordance with the contract, beginning on January 15, 2000, and continuing until

February 15, 2001.[1] The contract, which was drafted by Plaintiff, does not provide for a termination date or for any additional payment to Plaintiff should unforeseen delays occur.

According to Plaintiff, the hotel project took significantly longer than Plaintiff expected. One of Plaintiff's representatives, Paul Rivetto, testified that he expected the hotel to be designed and built within twelve months, with a completion date around the end of 2000. Plaintiff claims to have submitted a schedule to Defendant Kappa, but neither party has been able to produce it. The design and construction of the hotel was delayed for various reasons. The hotel was substantially complete in July 2002 when a sprinkler head burst and other water leaks caused flooding. The flooding damaged carpeting and drywall, which had to be ripped out and replaced. The flood remediation work took approximately 18 months and ended in January 2004.

Plaintiff contends that Kappa owes it an additional $105,000 for project management services it provided after the flood. According to Plaintiff, such services are outside the scope of the parties' fixed-fee agreement. Plaintiff further claims that Kappa agreed to pay the additional amount, but never followed through.

Defendant Kappa contends that the flood remediation work is within the scope of the parties' flat fee agreement and that it does not owe additional payments. Kappa denies that it agreed to pay any additional amount and argues that such an agreement is barred by the preexisting duty rule. Kappa further argues that Plaintiff's unjust enrichment claim should be dismissed.

Defendant Flagstar Bank has also filed a summary judgment motion on the same grounds. Flagstar is a defendant because it has an interest in the hotel property, upon which Plaintiff has asserted a lien.

Law and Analysis

Essentially, Defendants argue that the alleged oral agreement to pay $105,000 above the contract price is invalid for lack of consideration. "An essential element of a contract is legal consideration." *Yerkovich v. AAA*, 461 Mich. 732, 740, 610 N.W.2d 542 (2000). "Under the pre-existing duty rule, it is well settled that doing what one is legally bound to do is not consideration for a new promise." *Id. at 740-41*. The pre-existing duty rule "bars the modification of an existing contractual relationship when the purported consideration for the modification consists of the performance or promise to perform that which one party was already required to do under the terms of the existing agreement." *Id. at 741*.

Plaintiff contends that the pre-existing duty rule does not apply here, due to the unforeseen circumstances of the flood. Plaintiff cites a 1941 Sixth Circuit case for this proposition. See *Grand Trunk Western R. R. Co. v. H. W. Nelson Co., Inc.*, 116 F.2d 823, 834 (6th Cir. 1941) (applying Michigan law).

> The general rule that a promise to pay a construction contractor additional compensation for performance of a contract which he is under obligation to perform is invalid because [it is] without consideration, but the exception is equally well recognized that where, during the prosecution of the work, some unforeseen and substantial difficulties in its performance occur which were not known or anticipated by the parties when the contract was entered into and which casts upon the contractor additional burdens not contemplated and the contractee promises the contractor extra pay or benefits if he will complete the contract, such an agreement is valid and enforceable. *United States v. Cook*, 257 U.S. 523, 526, 42 S. Ct. 200, 66 L. Ed. 350, 57 Ct. Cl. 595; *Scanlon v. Northwood*, 147 Mich. 139, 110 N.W. 493.

---

[1]Kappa paid all of the contract price except for $5,000, which Plaintiff seeks to recover here, along with additional damages. The parties do not appear to dispute that Kappa owes Plaintiff $5,000.

Grand Trunk, 116 F.2d at 834.

The issue here is whether the project management services related to the flood remediation were within the scope of the parties' written agreement. Exhibit A to the agreement outlines the scope of services Plaintiff was to provide. These services included preliminary design and site selection, design and construction, contractor bidding, and project close-out. Specifically, Plaintiff agreed to "perform job observations as necessary to evaluate construction progress, adherence to project plans and specifications, and *assist in resolving field problems* and disputes in the most economical and expeditious manner possible." *Id.* (emphasis added). The parties clearly anticipated that problems would arise in the construction of the hotel and that Plaintiff would assist in resolving those problems. The work involved in providing project management services during the flood remediation phase was thus contemplated by the agreement and was within the scope of the services Plaintiff promised to provide.

The court finds that the flood was not an "unforeseen difficulty" that would allow Plaintiff to seek additional payments beyond its fixed fee and defeat the application of the pre-existing duty rule. Plaintiff drafted this agreement, agreed to a fixed fee, and did not include any language regarding the anticipated time period for the project or what would constitute "extra work." Rather, Plaintiff specifically agreed to assist in resolving "field problems," which are presumably common in construction projects. Plaintiff assumed the risk that this project would take longer or involve more hours than it anticipated—that is the essence of a fixed-fee agreement. The court finds that the "additional work" performed by Plaintiff was covered by the parties' written agreement, and that any alleged modification of that agreement is barred by the pre-existing duty rule.

In addition, the court will dismiss Plaintiff's unjust enrichment claim. "A contract will not be implied under the doctrine of unjust enrichment where a written agreement governs the parties' transaction." *King v. Ford Motor Credit Co.,* 257 Mich. App. 303, 327, 668 N.W.2d 357 (2003). In this case, a written agreement governs the parties' transaction. Accordingly, Plaintiff cannot recover under an unjust enrichment theory.

Therefore, IT IS HEREBY ORDERED that Defendant Kappa Enterprises L.L.C.'s March 30, 2006 Motion for Summary Judgment is GRANTED.

IT IS FURTHER ORDERED that Defendant Flagstar Bank's April 3, 2006 Motion for Partial Summary Judgment is GRANTED.

Review Questions

1. Does a conditional promise, such as a promise to resell a piece of property if you are able to buy it, constitute a legal consideration? Why or why not?

2. Black agreed to pay White, a member of the state house of representatives, $1,000 to do everything "reasonable and lawful" to defeat a bill that was to be presented to the legislature. Can White collect the $1,000 whether the bill passes or not? Why or why not?

3. Black hires White Automation to build a piece of automatic machinery for $150,000. During the time that the machinery is under construction, union pressures result in a wage increase for the electrical suppliers. As a result, electrical equipment costs more than anticipated. White demands $8,000 more for the machinery and Black agrees to pay. After installation, Black refuses to pay more than $150,000 for the machine. White sues. What is the result? Why?

4. Gray owed Brown $600 on an old debt. Gray refused to pay the debt several times, claiming lack of funds each time. Brown finally attempted to collect the debt at a time when Brown knew Gray had sufficient money. Gray offered to pay $400 and give Brown a wrist watch worth about $30 if Brown would consider the debt paid in full. Brown accepted. Later Brown brought an action for the remainder of the debt. Can Brown get it? Why or why not?

5. In *Pickus Construction v. American Overhead,* what were the defendant's arguments against the court's finding of promissory estoppel against them? How did the court respond?

6. In the usual employment contract between an engineer and his or her employer, how much notice is required for termination of the employment by either party?

7. The defendants in *Plante & Moran Cresa v. Kappa Enterprises, LLC* argued that the alleged oral agreement to pay $105,000 above the contract price is invalid for lack of consideration. According to the defendants and the court, what is the name of the rule that applies here? What should the plaintiff have done in order for this rule not to apply?

Lawful Subject Matter

■ Contracts based on illegal subject matters are generally void. Categories of illegal subject matter include those that are
1. Contrary to statutes,
2. Contrary to common law and
3. Contrary to public policy.

The necessity that any contract have a lawful purpose is fundamental. Generally, courts treat contracts based on illegal subject matter as void. Subject matter considered by the courts to be illegal may be classified in three categories: contrary to statutes (federal, state, local); contrary to common law; or contrary to public policy. More generally, illegal consideration may consist of any act or forbearance, or promise to act or forbear, that is contrary to law or morality or public policy. Ignorance by the parties as to what constitutes illegality cannot make valid a contract that is void because of its illegality.

INTENT

The intent of the parties at the time the contract was made often determines its legality. If both parties intended the contract to be performed illegally, it usually will not be enforced by the courts. This is true even though actual performance was lawful. If only one of the parties intends to perform illegally, the courts will usually uphold the contract.

A very large number of contracts are capable of being performed either legally or illegally. A contract to drive a truck, for instance, may be performed either according to the law or in violation of it. If there is no evidence of an intent by both parties to perform the contract illegally, the truck-driving contract would be valid. Generally, the contract must be incapable of being performed in a legal manner for the courts to declare it void on the basis of illegality.

Knowledge that the subject matter of a contract is intended for later unlawful use by one of the parties usually will not void the contract. However, where the subsequent unlawful use involves a heinous crime (treason or murder, for instance), an exception is made to this rule. An action to collect for the sale of a gun where the seller knows the buyer intends to commit murder with the gun would not be successful. If the seller had no knowledge of the intended use of the gun and complied with the law in all other respects in the sale, however, the contract of sale would be valid and enforceable.

CONTRACTS CONTRARY TO STATUTE OR COMMON LAW

If the subject matter of a contract is contrary to an existing statute, the contract is usually void. The statute concerned may be a federal or state statute or a local ordinance. It is not necessary for the contract to violate the wording of the statute; the contract may be legally void if it conflicts with the implied meaning of the statute or the intent of the legislature when the law was passed.

Law Passage Following Contract Formation

If a continuing contract is formed and later a law is passed making its performance unlawful, the entire contract is not necessarily rendered void. Performance of the contract subsequent to the passage of the statute cannot be recovered for in the courts, but the contract can be enforced for the parties' performance prior to passage of the statute.

> A *continuing contract* is a contract the performance of which will take place over some length of time.

Shortly after the United States entered World War II, Congress passed several regulations outlawing the sale of various items and restricting the sale of others. In a case resulting from these restrictions where a new car sales agency had leased premises for the sole purpose of display and sale of new cars (so stipulated in the lease), it was held that because of the government restrictions on the sale of new cars, the lease contract was terminated. If the lease had been written so that the use of the premises had not been restricted to the sale of new cars, the governmental restrictions would have had no effect.

Type of Statute

In considering the legality of a contract's subject matter, the courts sometimes distinguish between laws passed for the protection of the public and laws that have as their primary purpose the raising of revenue. If a contract violates a statute passed for the protection of the public, it is treated as void and unenforceable. If the contract violates a revenue statute, however, it is usually enforceable, subject to penalties for avoidance of the statute. An excellent example is found in the state licensing laws, such as those for business establishments and professional people. If the purpose of the licensing law is to protect the public and safeguard the lives, health, property, and welfare of citizens (for instance, licensing of professional engineers), a contract for professional services made by one who is not licensed in the state will not be enforced by the courts. If the primary purpose of the law is to collect revenue only (for instance, state gasoline taxes), the courts usually enforce the contract and impose a penalty on the violator.

Harm to Third Person

If a third person would be harmed by a crime or tort in connection with the performance of a contract, the subject matter of the contract probably will be held unlawful. A court will not punish for the commission of a crime or tort on one hand and support a contract to commit such a crime or tort on the other. Where the contract is an inducement to commit a crime or tort, but the commission of such crime or tort is not necessarily required to perform the contract, the courts will examine the strength of the inducement and the general nature of the contract. Life insurance may be an inducement to murder, or fire insurance an inducement to commit arson. For this reason, an insurable interest often must be shown before insurance can be obtained. A contract, in the performance of which a party to the contract must breach an existing contract, is usually unenforceable in the courts.

Contracts That Restrain Trade

Our economy is based on the principles of free enterprise and freedom of competition. Under the Sherman Antitrust Act, the Clayton Act, and state statutes similar in scope, contracts that tend to create a monopoly or maintain price levels or in other ways restrain trade may be unlawful. The penalty for a violation of these statutes is up to triple the amount of damages shown, plus payment of the court costs and reasonable attorney's fees for the plaintiff's attorney.

An example of an agreement that would probably be unenforceable as a violation of the antitrust laws would be an agreement by two competitors to fix prices at a set minimum or to divide the market. Suppose two concrete manufacturers together have 90 percent of the business in a given state. They agree not to underbid each other in several counties, and each agrees not to bid at all for jobs in several other counties. Such an agreement would violate the antitrust laws and should be unenforceable. Other business practices, however, are not so clear-cut. Should it be unlawful for a computer software manufacturer to require its customers to also buy maintenance services for the software? If so, are both the maintenance services agreement and the underlying software license agreement unenforceable? The law is not always clear in these areas.

In connection with their employment, many people sign written agreements in which they agree that, for some period of time after the termination of their employment, they will not compete against their former employer. Such **noncompetition** clauses (sometimes called **covenants not to compete**) are considered restraints of trade, and were strictly construed against enforcement by the common law. In some states, such covenants are prohibited in certain employment relationships. Generally, such noncompetition provisions are enforceable to the extent necessary to protect the legitimate business interests of the employer, such as the employer's trade secrets or good will. If the provisions are reasonable as to the restrictions on territory, scope of the former employee's activity, and the time period of the restrictions, then the provisions often will be upheld and enforced. (see *Modern Environments, Inc. v. Stinnett* at the end of this chapter)

Such provisions are also common in agreements involving the sale of a business. Despite the seeming restraint of trade present in such agreements, the courts will protect the purchaser of a business and the "good will" that goes with it. By such an agreement the seller states that the seller will not compete with the buyer of the business. Typically, the agreement states that the seller agrees not to enter into the same type of business in the same market area for a given period of time. If the restrictions about type of business, market area, and time are all reasonable, the restrictions probably will be upheld in court. It is only when these restrictions are unreasonable that a court will hold them to be in violation of the law.

Engineers are often asked to sign preprinted employment agreements that detail their obligations as employees. Besides noncompetition clauses, such employment agreements often include provisions dealing with the employee's duties regarding confidentiality and inventions. Occasionally, such forms provide that all inventions of an employee—whether or not they relate to the employer's business and whether or not they are made during business hours and at work— belong to the employer. Some forms provide that any inventions made by an employee within a year or so after the employment relationship ends are presumed to belong to the employer. A few states have statutes that limit the enforceability of such provisions; in some other states, such provisions are considered unenforceable as contrary to public policy. In still other states, however, such contracts can be enforced.

Usury

Most states have passed laws limiting the rate of interest that may be charged for the use of money. Considerable variation exists in the attitudes of the various states when these laws are violated. In some states, when usurious interest is charged, the courts will not aid in collection of principal or interest; in other states, all interest may be forfeited as a result of usury; and in still other states, collection of principal and maximum allowable interest results.

Wagering Contracts

In most states wagering contracts are illegal. It is often difficult for the courts to determine whether a particular contract is a legitimate business transaction or a wager. The test used by the courts in making their determination is centered about the creation of the risk involved. Is a risk created for the purpose of bearing that risk? If so, the contract is deemed a wager. In a dice game, on the first throw, a total of seven or eleven on the two dice pays off for the holder of the dice. The two bettors have created a risk of loss by placing their bets before the dice are thrown, with the score on the dice determining who shall lose the money. In a business contract calling for future delivery of a commodity for the payment of a present price, risk of loss resulting from a change in price of the commodity between the present and the delivery date is assumed by at least one of the parties. Has a risk been created for the purpose of assuming it? The courts answer no, providing that future delivery is actually intended. Manipulations in the stock and commodity markets are sometimes criticized on the grounds that the final result of the contract is, and was originally intended by the parties to be, a transfer of money rather than commodity.

Other Types of Violations

A contract to withhold evidence in a court case is probably unlawful; so are contracts that tend to promote litigation. This is one reason why contingency fees for attorneys or others connected with a trial are sometimes frowned on. If a would-be plaintiff stands to lose nothing in a court case, this is an incentive to undertake the litigation. Contracts having the effect of compounding a crime are against public policy; for example, tax evasion agreements are unlawful.

A large category of unlawful contracts is included in those that either restrain marriage or promote divorce. The courts generally do everything within their power to promote matrimony and to discourage its dissolution.

Contracts that are immoral in subject matter or tend to promote immorality are sometimes declared by the courts to be unlawful and against public policy.

CONTRACTS CONTRARY TO PUBLIC POLICY

Contracts must conform to the common law and the applicable statutes and regulations to be lawful and enforceable. However, these are not the only limitations imposed by the courts on a contract's subject matter. In addition to the requirement that contracts conform to the written and unwritten laws of the community, the courts require conformity to public policy. **Public policy** is rather difficult to define because it is continuously changing. It must change to conform to changing ideas and changing technology.

Generally, those contracts that would be held contrary to public policy are those that, if enforced and thereby encouraged, would be injurious to society in some way. It is not necessary that someone, or the public in general, be injured by the performance of the contract. Courts have held that a contract may be unenforceable because of an evil tendency found to be present in it. (See *Allied Erecting & Dismantling, Co., Inc. et al. v. USX Corporation* at the end of the chapter.)

Unenforceability because of the "evil tendency" of a contract is particularly characteristic of contracts in which the judgment or decision of public officials might be (or has been) altered by the contracts. In fact, any contract that tends to cause corruption of a public official may be subject to censure in the courts. This brings up a rather sensitive problem with which the courts often are faced: What acts of lobbying are to be condoned?

- The end sought to be accomplished by the lobbying must be a lawful objective—one which, if accomplished, would improve the public welfare, or at least would do it no harm.
- The means used must be above reproach. Generally, a lobbyist must be registered as such; his or her lobbying practices must not involve threats or bribery or secret deals.
- Courts generally treat contingent fees based on the passage of legislation as being against public policy.

Assume a consultant is hired for an engineering design job, with payment to depend on his ability to obtain the acceptance of the design by political officials. The consultant, being human, might be tempted to go beyond what is considered right and ethical to obtain the fee. In the court's eyes, contingent fees are an inducement to the use of sinister and corrupt means of gaining the desired objective. Although the court may in certain cases overlook the presence of contingent fees, it nearly always detracts from the case presented by the proposed recipient of the fee. Contingency fees are allowed for attorneys in certain cases. The courts allow such fees based on the idea that, by providing an incentive to the lawyer, more lawyers will make their services available to persons who otherwise could not afford the lawyer's fees.

Fraud or Deception

A contract that has the effect of practicing fraud or deception on a third person is against public policy. Black agrees to pay White $1000 if she (a prominent nuclear engineer) will recommend

Black to Gray for a job as a nuclear engineer. White knows nothing of Black's qualifications for the job. Even though White wrote the recommendation and Black succeeded in getting the job, it is likely that White could not get the $1,000 by court action. The contract would be void as against public policy.

In contrast, payment for a recommendation from an employment agency would be quite enforceable. The employment agency is in the business of recommending people for jobs for a fee. Thus, those who hire employees through an employment agency have knowledge of the usual arrangement; those who hire based on individual recommendations have a right to assume that such recommendations are given freely and without prejudice.

Breach of Trust or Confidence

Contracts that breach a trust or a confidential relationship are against public policy and will not be supported in a court. Agency is one such fiduciary relationship—the agent acts for his principal in dealing with others. White, as Black's agent, contracts with Green for the purchase of steel. Unknown to Black, White is to receive a 5-percent kickback from Green for placing the order with Green. Even though the steel purchase occurred, the 5-percent kickback agreement could not be enforced in court.

Sources of Public Policy

Public policy is found in the federal and state constitutions, in statutes, in judicial decisions, and in the decisions and practices of government agencies. Lacking these, the court must depend on its own sense of moral duty and justice for its decisions.

EFFECT OF ILLEGALITY ON CONTRACTS

Generally, no action in law or in equity can succeed if it is based on an illegal agreement. That is, court actions based on agreements that are illegal, immoral, against public policy, intended to prompt the commission of a crime, or forbidden by statute probably will be unsuccessful. Because defining "public policy" is sometimes difficult, it is also difficult to predict whether a given agreement will be held to violate the "public policy." The defense of illegality is allowed, not as a protective device for the defendant but as a disability to the plaintiff. The public is better protected against dishonest transactions if the court places this stumbling block in the path of those attempting to avoid the law. Suppose a member of a gang of thieves, on being cheated out of a share of the loot, went into court to try to force a split. What should the court do as to the division of the spoils? The court would contend that the agreement to split is void, being based on a crime. The public interest is best served if the court does everything in its power to discourage the crime. Thus, as far as the loot-splitting is concerned, the case would be dismissed.

Public interest, then, is the determining factor in such cases. The court's decision is based on its opinion of what would be the greatest aid to the public. If the public interest would be promoted better by a decision for the plaintiff, the court will so decide. Consider a confidence game in which the victim is placed somewhat afoul of the law by an agreement with the confidence men (as is usually the case). Swallowing pride and guilt alike, the victim asks the court to order the return of the money taken. The court may decide to do just that as a discouragement to continued practice of such con games on the public.

Disaffirmance of an illegal contract while the illegal portion of the agreement is still executory often allows the person disaffirming the agreement to recover the lawful consideration previously given by that person. A person paying money to another to have a crime committed ordinarily may disaffirm the contract so that the crime may be stopped and thus obtain restoration of the money. Here again, the public interest is the determining factor, with the court acting as the representative of the public interest.

MODERN ENVIRONMENTS, INC. v. JOHNETTA R. STINNETT 263 Va. 491; 561 S.E.2d 694 (2002)

Study terms: Covenant not to compete, legitimate business interest

In this appeal, an employer contends that the trial court erred in holding that a covenant not to compete prohibiting a former employee from being employed in any capacity by a competitor is over-broad and unenforceable.

Johnetta R. Stinnett worked as a salesperson for Modern Environments, Inc. (Modern) from 1995 until the fall of 2000. Modern is in the business of selling and installing office furniture. In April 2000, Stinnett signed an employment agreement with Modern that contained a one year non-compete clause. Within one year after leaving Modern's employ, Stinnett accepted employment with a company that was a competitor of Modern. Modern notified Stinnett and her new employer by letter that Stinnett's employment with the competitor violated the non-compete clause and that legal action would be instituted unless Stinnett terminated her new employment.

Stinnett filed a declaratory judgment action seeking a declaration that the non-compete provisions of her employment contract were unenforceable because they were over-broad and contrary to public policy. Modern responded by filing a demurrer and cross-bill seeking an injunction against Stinnett's further employment in violation of the non-compete clause. Following briefing and argument of counsel, the trial court entered an order reciting that "for reasons stated on the record," the restrictive covenants in the employment agreement are "over-broad and unenforceable as a matter of law." We awarded Modern an appeal.

This Court evaluates the validity and enforceability of restrictive covenants in employment agreements using well settled principles. First, covenants in restraint of trade are not favored, will be strictly construed, and, in the event of an ambiguity, will be construed in favor of the employee. Second, the employer bears the burden to show that the restraint is no greater than necessary to protect a legitimate business interest, is not unduly harsh or oppressive in curtailing an employee's ability to earn a livelihood, and is reasonable in light of sound public policy. Finally, each case must be determined on its own facts.

The provision of the employment agreement at issue states:

Employee agrees that for as long as Employee remains employed by the company, and for a period of one (1) years [sic] after Employee's employment with the Company ceases, Employee will not (i) directly or indirectly, own, manage, operate, control, be employed by, participate in or be associated in any manner with the ownership, management, operation, or control of any business similar to the type of business conducted by the company or any of its affiliates (a "competing business"), which competing business is within a fifty (50) mile radius of the home office or any business location or locations of the Company or any of its affiliates at which Employee worked.

Modern asserts that the trial court erred by declaring that the covenant not to compete was facially over-broad because it prevented the former employee from working in any capacity for a competitor of her former employer.

This assertion is based on Modern's position that this Court has previously enforced identical or similar language in other employment agreements and has not held such language to be over-broad. . . .

[T]his Court did not limit its review to considering whether the restrictive covenants were facially reasonable. The Court examined the legitimate, protectable interests of the employer, the nature of the former and subsequent employment of the employee, whether the actions of the employee actually violated the terms of the non-compete agreements, and the nature of the restraint in light of all the circumstances of the case. The language of the non-compete agreement was considered in the context of the facts of the specific case. In no case did the Court hold that the language contained in the restrictive covenant at issue was valid and enforceable as a matter of law under all circumstances.

■ ■ ■

Based on this review, we reject Modern's assertion that our holdings in these prior cases require the conclusion that as a matter of law the language at issue is reasonable and not over-broad. A conclusion that the restrictive covenant is a reasonable restraint must rest on other grounds. However, no other grounds were advanced by Modern in this case.

Modern's sole contention is that it met its burden of showing that the restrictive covenant is reasonable because the time and geographic limitations are reasonable and because Stinnett is only prohibited from working for Modern's competitors. Other than the bald statement in its brief that the non-compete agreement "is reasonable and no greater than necessary to protect Modern's legitimate business interests," Modern offers neither argument nor evidence of any legitimate business interest that is served by prohibiting Stinnett from being employed in any capacity by a competing company.[3] In the absence of any justification for imposing the instant restraint on an employee's ability to earn a livelihood, Modern has not carried its burden of showing that the restrictive covenant at issue is reasonable and no greater than necessary to protect a legitimate business interest.[4]

Accordingly, we will affirm the judgment of the trial court.

Affirmed.

ALLIED ERECTING & DISMANTLING, CO., INC., et al. v. USX CORPORATION, 249 F.3d 191 (USCA 3rd Cir. 2001)

Study terms: Public policy, "evil tendency", "last look" provision, Sherman Act

Previous litigation between appellant, Allied Erecting and Dismantling, Co., Inc., and appellee, USX Corporation, was settled on the eve of trial. In this suit, Allied claims that USX violated several provisions of that settlement agreement. The District Court granted summary judgment in favor of USX, and Allied appeals. . . . As fully explained below, we will reverse the District Court's judgment in two respects, affirm the balance, and remand the cause to the District Court.

---

[3]The record establishes that Stinnett was a salesperson while employed by Modern and that she was subsequently employed by a competitor of Modern, but is silent as to the nature of Stinnett's employment with her new employer.

[4]Although the case originated as a declaratory judgment proceeding, Modern, as the employer, had the burden to produce evidence of reasonableness because the contract is in restraint of trade, . . . and the success of its cross-bill depended on its ability to demonstrate the reasonableness of the clause.

I. Background

On the eve of trial for the first action brought by Allied against USX Corporation, the two entities entered into a court-supervised Settlement Agreement. Allied claimed that it suffered sixty-six million dollars in damages as a result of contracts it entered into with USX to dismantle several of USX's steelmaking facilities. The Settlement Agreement, which is at the heart of this controversy, provided: (1) USX would pay Allied eight million dollars; (2) Allied would be granted all dismantling projects at USX's Fairless Works plant; (3) Allied could bid in good faith as a primary bidder on USX's subsequent dismantling projects; (4) Allied would be granted "last look" rights for a period of seven years to equal or better the most acceptable bid received by USX for any dismantling projects; (5) Allied would be awarded dismantling contracts at USX's Ambridge Works, Saxonburg Works, and McDonald Works.

1. Allied's Section V Claim

Allied's first two claims arise from Section V of the Settlement Agreement, which reads:

Except as to any dismantling work at USX's South Works facility in Chicago, Illinois, Allied shall be invited to bid in good faith as one of U.S. Steel Groups' primary bidders on any dismantling work, of whatever nature or type at any steelmaking facilities, or former steelmaking facilities, owned by the U.S. Steel Group consistent with and pursuant to specification and performance standards developed and issued by U.S. Steel Group for such work and, for a period of seven (7) years from the date of this Settlement Agreement and General Release, Allied shall be offered an opportunity to equal and/or to better the most acceptable bid received by the U.S. Steel Group for any such further dismantling activity. If, within ten (10) days of receipt of bids, Allied offers in writing to perform such work on such terms which are equal to or better than the bid otherwise most acceptable to the U.S. Steel Group, when, in such event, the work shall be awarded to Allied, provided, however, that Allied is then able to meet U.S. Steel Group performance standards then in effect and, further, that Allied has not been adjudicated to be in default under any dismantling contract with U.S. Steel Group then in effect at the time of such bidding.

Under this "last look" provision, USX issued Allied and other dismantlers specifications for projects up for bid. The third parties then bid on the projects subject to the condition that "Purchaser [USX] reserves the right to reject any or all bids." No third party was told that Allied held the right to review and match their final bids. For the first few sessions, Allied formulated and submitted earnest bids on the projects. Formulating a bid for projects such as these can be cost and labor intensive, and Allied later determined that instead of participating in the bidding process it would merely review the best bid offered to USX and decide if it wanted to take the job at that price.

Allied claims that USX violated the terms of Section V through its relationship with Allied's archrival, Brandenburg Industrial Services Company, Inc. Because the litigation soured Allied and USX's working relationship, USX awarded most of its dismantling projects thereafter to Brandenburg. As a result of this close working relationship, Brandenburg prepared most of the specifications for the projects on which Allied held the "last look" right.

Because Brandenburg prepared these specifications, Allied argued, it held an "unquestioned and substantial advantage over the other bidders for this work." Within one bid, Brandenburg offered to forgive $379,500 that USX owed for developing the project specifications if USX granted the project to Brandenburg. Allied was unable to compete with an offer that included debt forgiveness, and therefore claims that such dealings between USX and Brandenburg materially varied the terms of the project specifications and violated Allied's last look rights.

■ ■ ■

II. Discussion

A. The Enforceability of Section V

The District Court "declared [Section V] a nullity and unenforceable as against public policy as practiced to date." Allied contends that Section V does not offend the common law consensus and therefore cannot be voided as against public policy. We agree.

"It has long been settled," the District Court stated, "that a court will not become an aid in the enforcement of contractual provisions where to do so would violate public policy." The Court recited 19th and early 20th century caselaw, beginning with *Veazey v. Allen,* 66 N.E. 103, 173 N.Y. 359 (1903):

There are . . . phases of public policy which are as enduring and immutable as the law of gravity. One of them is that, as applied to the law of contracts, courts of justice will never recognize or uphold any transaction which in its object, operation, or tendency is calculated to be prejudicial to the public welfare. That sound morality and civic honesty are cornerstones of the social edifice is a truism which needs no reinforcement by argument. It may therefore be taken for granted that whenever our courts are called upon to scrutinize a contract which is clearly repugnant to sound morality and civic honesty, they need not look for a well fitting definition of public policy, nor hesitate in its practical application in the law on contracts.

"The rule is," the Court continued as it cited the Pennsylvania Supreme Court in *Kuhn v. Buhl,* 251 Pa. 348, 96 A. 977, 984 (1916), "that courts having in their view public interests, will not lend their aid to the enforcement of an unlawful contract." The Court drew its direct authority from *Pittsburgh Dredging and Constr. Co. v. Monongahela and Western Dredging Co.,* 139 F. 780, 784 (Circuit Court, W.D.Pa. 1905), which stated that "viewed from that standpoint of morals, square dealing, and commercial integrity, combinations for collusive, misleading, biddings, wherever made, cannot be approved; yet to enforce rights based on an agreement to make such bids is to make the law an active agent to accomplish such deceptive purposes."

Regarding Section V of the settlement Agreement, the Court found that the evil tendency of the contract was or would be to perpetrate a fraud on the third-party bidders and to deny one bidder on each project the natural consequence of the bidder's endeavors . . . This potential fraud has been and would continue to be perpetrated on innocent third parties by using the court and the confidentiality of the settlement agreement to keep this practice undisclosed as to both the victims and the public. The injury to competition may not be immediate in any particular project, but if this arrangement were to be carried into effect repetitively, over the seven-year period, it is clear that it would have an injurious effect on competition by denying certain of those third parties what should otherwise have been awarded to them for their honest work and labor in formulating the bids.

Stating that Section V "contains the tendency to work fraud on the innocent third party bidders, repugnantly distorts the natural consequences of bona fide competition, and uses the court to shelter this state of affairs from detection," the Court held that the "provision is, indeed, void as against public policy." Further, the Court refused to shift the blame for the illegality of Section V to USX because Allied could have genuinely participated in the bidding process and demanded that USX disclose Allied's last look right to the other bidders. The Court was unwilling to find that Allied was victimized by this provision because it concluded that Allied chose not to participate in the bidding process on several of the projects, unfairly benefitted from having others shoulder the expense of drafting the specifications, and unscrupulously viewed the bids of its rivals without their consent. The Court also refused to allow its supervisory approval of the Settlement Agreement to validate Section V, concluding that it was Allied's failure to genuinely participate in the bidding and inform other bidders of its last look rights that rendered it infirm.

When ruling on the grounds of public policy, a court must speak for a "virtual unanimity" that can "be found in definite indications in the law." We cannot find such a consensus. To the contrary, the state of the law on this issue is entirely unclear, as "last look" and "first refusal" rights are typically found unproblematic in a variety of contexts. Considering the wide acceptance of last look rights, the District Court's policy analysis contradicts common law consensus and is therefore unsupportable. In addition, Section V was an element of the court-supervised agreement to settle the previous litigation. The principles of this agreement were met with approval by the trial judge presiding over that litigation, and this authorization further militates against the argument that there exist a virtual anonymity against last-look rights.

■ ■ ■

The District Court loosely referred to antitrust principles in its opinion, and Allied argues on appeal that because Section V constitutes neither a per se nor a rule of reason violation of the Sherman Act, it cannot violate public policy.

Because we have already held that, irrespective of the Sherman Act, Section V does not violate general principles of public policy, we need not discuss Allied 's Sherman Act argument. We note in passing, however, that we do not believe that Section V violates the Sherman Act. We ruled in Sitkin, a case involving a similar last look right exercised by a dismantling company, that the parties "desired to find the market price rather than influence the market price," and that, therefore, the contract did not violate the Sherman Act. We believe that Allied and USX had similar intentions in drafting Section V, and, therefore, we are confident that there has been no Sherman Act violation.

In sum, we conclude that the District Court erred by finding Section V unenforceable as contrary to public policy. We will remand for the District Court to determine if USX must pay damages to Allied for breach of contractual obligations under Section V. . . .

E. USX's Counterclaims

Because we are reversing the District Court's conclusion that Section V of the Settlement Agreement is unenforceable, we will reinstate only USX's counterclaims that were dismissed as a direct result of the District Court's enforceability ruling. In light of our holding that Section V is not void as against public policy, the District Court must now address USX's counterclaim that Allied breached Section V by failing to submit initial bids on certain projects.

III. Conclusion

In conclusion, we will reverse and remand the Section V and [and other] issues to the District Court, affirm the District Court's order granting summary judgment on the fraudulent inducement claim, and reinstate USX's relevant counterclaims.

Review Questions

1. Black, an engineer, agrees to act as an expert witness for White in a court case. In return, Black is to receive a $50,000 fee plus expenses for the expert witness services. Is the contract lawful? Why or why not?

2. In question 1's example, assume that Black agrees to accept $50,000 if White wins, or payment for expenses only if White loses. Is the contract lawful? Why or why not?

3. Gray is a process engineer for Brown. Green has made the low bid on equipment for a manufacturing operation that Gray had begun setting up. The equipment meets the specifications as well as equipment proposed by other bidders. To improve the chance for acceptance of Green's bid, Green offered to give Gray 2 percent of the bid price if Green's equipment is used. If the equipment is bought, can Gray get the 2 percent? After receiving the 2-percent offer, what if anything, should Gray have done about it?

4. Explain why an insurance contract is not a wagering contract.

5. What is public policy? In *Allied Erecting & Dismantling, Co., Inc. v. USX,* the lower court had very strong arguments in support of its opinion that the Settlement Agreement's Section V was against public policy. Do you think that there were additional reasons why the court of appeals refused to find the same?

6. Why are contingency fees looked on with disfavor by courts?

7. In *Modern Environments v. Stinnett,* the former employee was technically in violation of the employment agreement she had signed. The lower court and the Virginia Supreme Court determined that the non-compete agreement was over-broad, and therefore not reasonable. How would you modify the language of the agreement in order to meet the reasonableness standard set forth by the court?

Statute of Frauds

CHAPTER 11

HIGHLIGHTS

- There are two types of contracts most often encountered that are subject to the Statute of Frauds:
  1. Promises that cannot be performed within a year
  2. Contracts for the sale of goods in excess of $500

An oral contract is generally just as enforceable as a written contract, but a written one has at least one major advantage—its terms are easier to ascertain. The relationships of the parties are set forth for interpretation by the individuals concerned and by a court of law if the need should arise. Although it is true that in most circumstances an oral contract is as enforceable as a written one, it is sound practice to put into writing any contract of more than a trivial nature.

ENGLISH STATUTE OF FRAUDS

In England, the courts were faced with many cases concerning oral agreements. Instances of perjured testimony were frequent as the contending parties attempted to prove or disprove the existence or terms of contracts. In 1677, the English Statute of Frauds was passed as "An Act for the Prevention of Frauds and Perjuries" to relieve the courts of the necessity of considering certain types of contracts unless their terms were set forth in writing and signed. The statute did not require a formally drawn instrument. The only writing required was the minimum needed to establish the material provisions of the agreement. Generally, the writing had to fulfill all of the following criteria:

- Reasonably identify the subject of the contract
- Indicate that a contract has been made between the parties
- State the essential terms of the contract
- Be signed by or on behalf of the party to be charged

The writing was not required to be all on one instrument, but if it was not, a connection between the various instruments usually had to be apparent from the documents themselves. The statute did not require a signature from both parties. Only the person sought to be charged with the contractual duties has to have signed. The signature itself could consist of initials, rubber stamp, ink or pencil, or anything else intended by the party to constitute identification and assent. In addition, the signature could appear anywhere on the document.

The English Statute of Frauds consisted of several sections, but only the sections numbered 4 and 17 are of major importance to us today. These two sections have become law in all of the states in the United States, with only minor modifications.

Fourth Section

According to the fourth section, "no action shall be brought" on certain types of contracts unless the agreement that is the basis for such action is in writing and signed by the defendant or the defendant's agent. The following types of contracts are subject to the requirement of a written and signed instrument:

1. Promises of an executor or an administrator of an estate to pay the debts of the deceased from the executor's or administrator's own estate
2. Promises to act as surety for the debt of another
3. Promises based on marriage as a consideration
4. Promises involving real property
5. Promises that cannot be performed within a year

Seventeenth Section

The seventeenth section sets forth the requirements for an enforceable contract having to do with the sale of goods, wares, or merchandise. It says that unless one of the following three actions is done to secure such a transaction where the consideration is at least ten pounds sterling (£10), the contract will be unenforceable at law:

1. Part of the goods, wares, or merchandise must be accepted by the buyer.
2. The buyer must pay something in earnest toward the cost of the goods.
3. Some note or memorandum of the agreement must be made and signed.

The monetary minimum in the United States is $500 in each of the states upon their adoption of the Uniform Commercial Code (see figure 11.1).

Promises that cannot be performed within a year and contracts for the sale of goods in excess of $500 are the two types of contracts most commonly encountered.

Uniform Commercial Code § 2-201.

§ 2-201 Formal Requirements; Statute of Frauds.

(1) Except as otherwise provided in this section a contract for the sale of goods for the price of $500 or more is not enforceable by way of action or defense unless there is some writing sufficient to indicate that a contract for sale has been made between the parties and signed by the party against whom enforcement is sought or by his authorized agent or broker. A writing is not insufficient because it omits or incorrectly states a term agreed upon but the contract is not enforceable under this paragraph beyond the quantity of goods shown in such writing.

(2) Between merchants if within a reasonable time a writing in confirmation of the contract and sufficient against the sender is received and the party receiving it has reason to know its contents, it satisfies the requirements of subsection (1) against such party unless written notice of objection to its contents is given within 10 days after it is received.

(3) A contract which does not satisfy the requirements of subsection (1) but which is valid in other respects is enforceable

(a) if the goods are to be specially manufactured for the buyer and are not suitable for sale to others in the ordinary course of the seller's business and the seller, before notice of repudiation is received and under circumstances which reasonably indicate that the goods are for the buyer, has made either a substantial beginning of their manufacture or commitments for their procurement; or

(b) if the party against whom enforcement is sought admits in his pleading, testimony or otherwise in court that a contract for sale was made, but the contract is not enforceable under this provision beyond the quantity of goods admitted; or

(c) with respect to goods for which payment has been made and accepted or which have been received and accepted (Sec. 2-606).

Only applies to goods, not services

Figure 11.1

CONTEMPORARY APPLICATION OF THE STATUTE OF FRAUDS

Promises Requiring More Than a Year to Perform

The legal interpretation of a year is important here. Generally, the time starts to run from the time of making the agreement, not from the time when performance is begun. An <u>oral</u> contract, then, to work for another for a year, starting two days from now, would be unenforceable under the English Statute of Frauds. In most jurisdictions it is held that parts of days do not count in the running of time. If a contract is made today, time will start to run on that contract tomorrow.

The statute refers to contracts "not to be performed within the space of one year of the making thereof." The court interpretation of this statement is that it means contracts that, by their terms, cannot be performed within one year. White promises orally to pay Black $80,000 if Black will build a certain house for White. No time limit is set on the construction. The fact that actual construction took place over a two-year period would not bring the contract under the Statute of Frauds. If the contract could be completed within one year no matter how improbable this may be, an oral contract for its performance is binding. By this reasoning, a contract to work for someone "for life" or to support someone "for life" is capable of being performed in one year and need not be in writing to be binding.

If the performance of an oral contract is to take place in less than a year and is extended from time to time by increments of less than a year in such a way that performance continues for more than a year, such a contract is valid. For instance, an oral lease contract for nine months might be continued orally at the end of the period to run for another nine months for a total of eighteen months.

Suppose that an oral contract cannot be fully performed within a year. If one of the parties nonetheless fully performs, is that party unable to enforce the contract against the other? The majority of courts treat full performance by a party as removing the contract from the effect of the one-year requirement of the Statute of Frauds, even though the party took more than a year to perform. Accordingly, most courts would enforce such a contract. Some courts, however, hold that one party's full performance still does not remove the contract from the written requirement of the Statute of Frauds.

In situations where the Statute of Frauds applies but there has been at least a partial performance, most courts allow the party who has partially performed to recover the value of the services rendered. In some situations, one party may be estopped (that is, precluded) from asserting the Statute of Frauds as a defense if the other party has already performed in whole or in part. Courts are usually careful in such situations. They will refuse to allow the Statute of Frauds to be used as a defense by someone who uses it to secure an unfair advantage and thus "defraud" the other party to the agreement. In short, the courts refuse to allow the Statute of Frauds as a defense when doing so helps one party perpetrate a fraud.

Sale of Goods

As noted, a large majority of the states have adopted the Uniform Commercial Code (UCC), which governs the sale of goods within those states. The UCC drafters adopted the rule of the seventeenth section of the English Statute of Frauds in slightly altered wording.

If a transaction involves goods of a value great enough to be governed by the Statute of Frauds in the particular state, an oral agreement will be valid only if

1. some of the goods are accepted by the buyer, or
2. the buyer pays part of the purchase price in earnest, or
3. the buyer signs some note or memorandum as to the terms of the sale, or
4. the party to be charged with the contract formally admits that a contract was made, or
5. the goods are to be specially made, are not suitable for sale in the ordinary course of the seller's business, and the seller has taken substantial steps to begin performance of the contract. Payment of part of the purchase price may be made in money or in anything of value to the parties.

See *Davis v. Crown Central Petroleum Corporation* at the end of the chapter.

What actually constitutes goods under the UCC has led the courts into some difficulty on occasion. It is apparent that if the goods already exist, a contract for their sale will not involve services. However, if the contract is for the purchase of goods not now in existence but to be made by the seller, there is a question whether this is not a contract for services to be performed. If the contract is primarily for services rather than goods, the UCC (and the accompanying UCC Statute of Frauds) will not apply.

Even when the UCC applies, however, there are still exceptions to the UCC Statute of Frauds. For example, the UCC provides the previously noted exception for an oral contract for the sale of goods that are to be specially made for the buyer and are not suitable for sale by the seller in its ordinary course of business. Such an oral sales contract is enforceable if the seller has begun making the goods or has made commitments for procuring the goods. Black wishes to purchase a gate to match a very old wrought iron fence around his house. Black contracts orally with White to have such a gate specially made. The contract will be enforced. This, of course, is a logical rule, because the gate probably would be unsalable to anyone else.

Another exception to the UCC rule is if the party against whom enforcement of the contract is sought admits in pleadings, in testimony, or in court that a contract was made. Thus, testimony from a buyer who agrees that a contract was made but denies its enforceability for one reason or another removes the availability of the UCC Statute of Frauds as a defense (although the buyer may indeed have other defenses). If a written confirmation is sent by one merchant to another and the receiving party knows its contents, the confirmation can serve as the writing and satisfy the UCC Statute of Frauds if the receiving party fails to object within 10 days after receiving it.

OTHER ACTIVITY REGULATED BY THE STATUTE OF FRAUDS

Promises of Executors or Administrators

In law an **executor** is a person appointed in a will by the **testator** (the person who made the will) to execute, or put in force, the terms of the instrument. An administrator serves a somewhat similar function where the deceased died without a will (**intestate**). The administrator is appointed by a court to collect the assets of the estate of the deceased, pay the estate's debts, and distribute the remaining estate to the heirs—those persons entitled to it by law.

If the executor of an estate agrees to pay the debts of the deceased out of the executor's own estate, the executor is acting as a surety for the debt of another. In effect, the executor is saying: "If the estate of the deceased is insufficient to pay you, I will pay the debt." This situation is covered under surety contracts, which are described next.

Promises to Act as Surety

A promise to act as a surety for another's debts is quite similar to an executor's promise to pay the debts of the deceased. The requirement that a contract to answer for the debt or default of another person must be in writing covers all types of guaranty and surety contracts. Lending institutions frequently require either collateral or a responsible cosigner as security for a loan. If White wishes to borrow $1,000 from Black Loan Company, a cosigner may be required for the loan. If Gray is to act as surety, agreeing to pay if White does not pay, Gray's agreement to do so must be in writing to be enforceable against Gray.

It is appropriate here to distinguish between primary and secondary promises. An oral **primary promise** (also called an **original promise**) is enforceable but an oral **secondary promise** (also called an **collateral promise**) is not. Wording of the promise can be quite important, but the apparent intent of the promisor when the promise is made is even more important. If the wording and other facts make clear an intent such that "If White does not pay, I will," the promise to pay is a secondary promise. However, if the circumstances show the intent to be that "I will pay White's loan," the promise is a primary one. The primary promise is enforceable against the promisor if it is made either orally or in writing. Even though a "primary" promise may be for someone else's debt, the court views such a promise as an independent obligation that can be enforced, even if made orally. Similarly, when someone agrees to make good any loss sustained by the plaintiff if the plaintiff acts as surety for a third party, the promise is primary in nature and enforceable even though the promise was oral. This comes closer to an undertaking of indemnity or insurance than surety.

Indemnity (insurance) contracts need not be in writing. Under such contracts, no liability is held to exist until an obligation arises between the insured and some third party. In an indemnity agreement, one person agrees to pay another's debts to third persons. For example, your insurance company agrees to pay your debts to persons injured in an automobile accident as a result of your negligence. The difference between a promise to act as a surety and to insure or indemnify you, however, is that the promise of indemnity is made to the debtor (that is, the insurer makes its promise to the insured), not to the creditor.

Promises in Consideration of Marriage

Marriage is said to be the highest consideration known to law. This provision of the Statute of Frauds applies particularly to situations in which the agreement to marry is based on consideration such as a prenuptial agreement.[1] If Ms. White requires that Mr. Black waive all spousal rights in her property in the event of divorce as a condition for marriage, such an agreement would have to be in writing. The statement that contracts in which marriage is to be a consideration must be in writing does not mean that when a man and woman simply agree to get married, with neither giving up anything but their unmarried status, such an agreement would have to be in writing. Oral promises to marry are actionable under the common law if one of the parties attempts to breach the promise to marry the other.

Real Property Transactions

Transactions involving "lands, tenements, and hereditaments," that is, **real property,** must be in writing to be enforceable at law. Real property law is treated more thoroughly in Chapter 19. It is necessary here, however, to distinguish between real and personal property. Real property is anciently defined as consisting of land and those things permanently attached to it, or immovables. Personal property, then, includes the movables, or things not firmly attached to the land. These definitions comprise only a part of the distinction currently applied by the courts.

Minerals in the soil, water rights, and trees on the land, for instance, are usually considered by the courts as real property, requiring a writing for their sale. An exception exists, however, in dealing with these natural fruits of the land. When the contract contemplates immediate severance or removal of these things, an oral contract to such effect will be enforceable.

Black orally sells White a stand of timber to be cut and sold by White, with Black to receive payment as cutting proceeds. If cutting is to begin immediately, the oral contract is enforceable; if it is to begin 10 months from now, it is unenforceable.

However, things such as cultivated crops, which result from human effort, usually may be the subject of an enforceable oral contract. For instance, a contract to sell the fruit in an orchard as it ripens would be enforceable under the Statute of Frauds, even though made orally. Lease

---

[1] See *Bloomfield v. Bloomfield*, 723 NYS 2d 143, 281 A.D. 2d 301 (1st Dept. 2000).

contracts involve an interest in land but, for most purposes, are considered personal property. An oral lease for a year or less is valid, but a lease must be in writing if it is to run longer than a year.

Part performance of an oral contract for the purchase of real property can influence a court to disregard the Statute of Frauds. It is only in an unusual case, however, that the courts will enforce such an oral contract. For example, suppose that the buyer set about building a new house on the land. The extent of improvements made by the buyer is so vast and material as to make it unjust to hold the oral agreement unenforceable.

EFFECT OF THE STATUTE

In some states, the Statute of Frauds is phrased so as to make contracts void unless in compliance with it. In most states, however, a contract that fails to comply with the Statute of Frauds is merely voidable. In these latter states, then, the Statute of Frauds operates as a defense to the enforcement of the contract. If the Statute of Frauds is not raised as a defense, the court will enforce the contract.

Even if the Statute of Frauds applies, a party that has performed may still be able to recover something. When one party has performed under such an oral contract, that party has probably enriched the other party. A court of equity may recognize the obligation created by such a performance and enforce payment by the other party under a quasi-contract theory of recovery. Generally, when the result would be inequitable or grossly unfair, the courts will not allow the statute to stand as a defense.

When the oral contract that should have been written is either completely executory or completely executed, the courts often will not consider enforcing it. The parties, generally, are left where they are found.

DAVIS v. CROWN CENTRAL PETROLEUM CORPORATION
483 F.2d 1014 (4th Cir. 1973)

Study terms: Statute of frauds, oral agreement, preliminary injunctive relief, balance of the hardship, mutuality, definiteness of duration

These two cases, combined on hearing because of their similarity of issues, are unfortunate by-products of the current oil shortage. Both plaintiffs and defendant are victims of the shortage in one form or another. They may properly be termed independents in their own particular type of operation. The plaintiffs, both citizens of North Carolina, on the one hand, are small independent oil dealers with their main operations in North Carolina. The defendant, on the other hand, is a refiner, dependent almost entirely on producers for its supply of crude oil. It has for some years been selling its product to the plaintiffs. In anticipation of the oil shortage, which all parties in the industry apparently foresaw, discussions were had among the parties as to future supplies. It is contended by the plaintiffs that the defendant agreed to supply them with certain fixed quantities of gasoline. As the energy crisis deepened, the suppliers of the defendant reduced drastically its supply of crude oil. It accordingly proceeded to allocate on a lower percentage its deliveries to its contract customers and to notify customers such as the plaintiffs, whom it denominated noncontract customers, that it would make no further sales to them. These actions followed that notification. The plaintiff Davis filed his action originally in the Western District of North Carolina, seeking injunctive relief against what he claimed was a breach of contract involving irreparable injury and damages for violation of the Sherman Act. Jurisdiction was based on diversity of citizenship and federal question. The other action was first instituted in the State Court and removed to the District Court for the Middle District of North Carolina. The plaintiff in this action sought similar relief to that demanded in Davis, and federal jurisdiction was predicated on similar grounds.[2] Preliminary injunctive relief was sought in both cases by the plaintiffs and were granted in both instances by the District Court having jurisdiction of the actions. It is from these grants of injunctive relief that the defendant has appealed in each case. Since there is similarity both of facts and issues in both cases, we ordered the appeals argued together and we shall dispose of both together. We reverse in both cases.

In granting temporary injunctive relief, the District Court in each instance made specific Findings of Fact. The defendant contends that these Findings show on their face that the temporary injunctive relief was improvidently granted. Specifically, it attacks the Finding made by the District Court in each case that the plaintiffs were likely to prevail eventually on the merits. Such a finding is necessary for the granting of preliminary injunctive relief. . . . This basic contention presents the issue on appeal.

In support of its conclusion that the plaintiffs were likely to prevail, the District Court in each case found that there was an oral agreement between the plaintiff and the defendant whereby the defendant was to supply the plaintiffs with their gasoline requirements. . . .

In the Davis case, the District Court found there was an agreement between the plaintiff and the defendant "for the purchase of goods vastly exceeding $500.00," which was "not written nor evidenced by any writing" and which, though obligating defendant to supply the plaintiff as much 1,200,000 gallons per month, did not obligate the plaintiff "to purchase that quantity nor any set quantity of gasoline." Although it recognized that the defendant was suffering hardship too, on account of the oil crisis, the District Court found that "(T)he balance of hardship favors the relief sought by the plaintiff" and "(T)hat the public interest will not be harmed by the issuance of an injunction." It accordingly ordered the defendant to "continue to supply plaintiff with at least 300,000 gallons of gasoline products per month . . . pending further orders of this Court. . . ."

The defendant denied in both cases the existence of any valid agreement. In support of this position, it asserts the alleged agreements are lacking in mutuality and definiteness of duration. Moreover, it argues that if there were any agreement, it was void as violative of the controlling Statute of Frauds. It is conceded by all parties that the North Carolina law is controlling on these several contentions.

We find it necessary to consider only the contention based on the Statute of Frauds. The Findings of Fact in each case make it clear that any agreement between the plaintiffs and the defendant was within the North Carolina Statute of Frauds. That statute, unlike its counterpart in other jurisdictions where its scope is merely remedial, "affects the substance as well as the remedy." . . . On the basis of the present Findings of Fact in each case, the statute would be a complete bar to a ruling that there was a valid contract or agreement between the plaintiffs, on the one hand, and the defendant, on the other. It is true, as the plaintiffs have argued, that in exceptional cases, courts of equity will find an estoppel against the enforcement of the statute, but such an estoppel can arise in North Carolina only "upon grounds of fraud" on the part of him who relies on the statute. . . . The District Court in neither case made a finding of "fraud" on the part of the defendant and, absent such finding, there can be no estoppel. Nor, on the facts in the record before us, would it appear that any such finding would have been in order. In *United Merchants & Mfrs. v. South Carolina El. & G. Co.* . . . it was stated:

> "A mere failure or refusal to perform an oral contract, within the statute, is not such fraud, within the meaning of this rule, as will take the case out of the operation of the statute, and this is ordinarily true even though the other party has changed his position to his injury."

---

[2]The state court clearly lacked jurisdiction of the action under the Sherman Act but, whether jurisdiction existed after and as result of removal to Federal Court, *see Freeman v. Bee Machine Co.*

After all, as the District Court in one of the cases observed, the plight of the defendant was not substantially different from that of the plaintiffs. It was experiencing hardships which forced it to take action it obviously did not relish.

The claim of the plaintiffs is appealing and our sympathies are with them. As the District Courts indicated, it is equitable in periods of scarcity of basic materials for the Government to inaugurate a program of mandatory allocations of the materials. This, however, is a power to be exercised by the legislative branch of Government. The power of the Court extends only to the enforcement of valid contracts and does not comprehend the power to make mandatory allocations of scarce products on the basis of any consideration of the public interest. Public reports indicate that Congress is cognizant of the problem presented by these actions and is giving active consideration to the establishment of a program of mandatory allocations that would apply from the oil producer down to the oil retailer. It is earnestly hoped that Congress will take the necessary steps to establish such controls. For that control, however, the parties must look to the Congress and not to the Courts.

The District Courts were in clear error in finding on the record before them that there was the reasonable likelihood that the plaintiffs would prevail on the merits. For that reason, the granting of injunctive relief during the pendency of the actions was improper in both cases and the injunctions are hereby vacated.

Reversed.

MICHAEL CURTIS v. MICHELE ANDERSON
106 S.W.3d 251, 2003 WL 1832257 (Tex.App.-Austin)

Study terms: Statute of frauds, oral agreement, conversion, conditional gift role

Opinion

This is an appeal from a summary judgment in a suit brought by appellant Michael Curtis to recover a diamond ring from appellee Michele Anderson after Curtis terminated the couple's engagement. Curtis sued for breach of an oral agreement and conversion, and the trial court granted Anderson a summary judgment. Curtis appeals arguing that Anderson was not entitled to summary judgment because the ring was a conditional gift, and Anderson's possession of the ring became an unlawful conversion when Anderson refused to return the ring. We will affirm the judgment of the trial court.

Factual Background

In the summer of 2000, Curtis and Anderson became engaged to be married. Curtis gave Anderson a ring. Approximately six or eight weeks later, the engagement ended. Anderson refused to return the ring to Curtis. Curtis alleges that at the time that he gave her the ring, Anderson agreed that if the wedding was called off she would return the ring.

The only summary judgment evidence before the trial court were excerpts from Curtis's deposition. Concerning the agreement to return the ring, Curtis testified that "we had a mutual understanding that I clearly stated and she accepted that if I did not—if we did not become married that I would retain—retain the stone." He admitted that the "mutual understanding" was not reduced to writing. When asked who "broke off the engagement," he testified, "I did. . . . I did it for several reasons. One is that I felt like she had some sexual hang-ups. I felt that she had some previous general issues with men, and she also had a very volatile temper." Based on this record, the trial court granted summary judgment in favor of Anderson.

Discussion

Anderson's sole ground for seeking summary judgment was that Curtis could not prevail because the statute of frauds prohibits the enforcement of any alleged oral agreement concerning return of the ring. Curtis argues by his first issue that the statute of frauds is not applicable; he claims the case is governed instead by the conditional-gift rule.

According to Curtis, the ring was a conditional gift, and because the contingent condition of marriage was not met, the gift was not completed and the ring should be returned to him. In his second issue, he contends that he presented sufficient evidence to establish the elements of a tort claim for conversion.

Statute of Frauds

The application of the conditional-gift rule assumes that there is no binding agreement between the parties about ownership of the engagement ring should the marriage not occur. If a binding agreement between the parties exists, then application of the conditional-gift rule is not appropriate. Curtis contends that when he and Anderson became engaged, Anderson agreed that she would return the ring if they did not marry. He testified that their "mutual understanding" was an express agreement, but their "mutual understanding" was not reduced to writing.

In response, Anderson asserted in her motion for summary judgment that Curtis's contract and conversion claims could not prevail because of the statute of frauds found in section 1.108 of the family code, which prohibits enforcement of oral agreements in consideration of marriage. According to Anderson, even if she agreed to return the ring, her promise is unenforceable because it is not in writing.

In 1997, the legislature added section 1.108 to the family code. It states:

> A promise or agreement on consideration of marriage or nonmarital conjugal cohabitation is not enforceable unless the promise or agreement or a memorandum of the promise or agreement is in writing and signed by the person obliged by the promise or agreement.

Id. This statutory provision has yet to be interpreted by any court. We must decide whether section 1.108 encompasses agreements between engaged parties regarding the disposition of engagement gifts should the engagement fail. . . .

Although section 1.108 was obviously intended to apply to prenuptial agreements, its plain language is broad enough to include Anderson's alleged promise to return Curtis's ring. . . . We hold that Curtis's allegation that he and Anderson had expressed their "mutual understanding" that the ring would be returned if the marriage did not occur comes within the scope of this statute as written. Therefore, to be enforceable any such agreement must be in writing.

Conditional-Gift Rule

In the absence of an enforceable agreement, we turn to the conditional-gift rule. . . . Texas courts have held that the rule operates to require that the ring be returned to the donor if the donee is at fault in terminating the engagement. See *McLain v. Gilliam*, 389 S.W.2d 131. . . . The court in *McLain* expressed the rule as follows:

> A gift to a person to whom the donor is engaged to be married, made in contemplation of marriage, although absolute in form, is conditional; and on breach of the marriage engagement by the donee the property may be recovered by the donor. 389 S.W.2d at 132.

In this case, Curtis as donor judicially admitted to terminating the engagement. He does not contend that Anderson was at fault in ending the engagement. . . . Thus, this case involves the opposite situation than that involved in McLain . . . ; here, the donor was responsible for breaching the promise to marry.

This is a case of first impression in Texas. We have found no Texas case in which the donor was responsible for terminating the engagement. We have examined cases in other jurisdictions, and it appears that most courts apply a conditional-gift rule in adjudicating ownership of engagement gifts when the marriage fails to occur. . . .

Texas courts, including this Court, have applied the fault-based conditional-gift rule when a donee breaks the engagement. We believe that the same rule should apply when the donor defaults. We hold that absent a written agreement a donor is not entitled to the return of an engagement ring if he terminates the engagement.

Conversion Claim

Curtis also complains that the trial court erred in granting summary judgment on his conversion claim. He argues that Anderson's retention of the ring once he demanded its return transformed her continued possession of the ring into conversion. The tort of conversion is defined as the (1) unauthorized and wrongful exercise of dominion and control over (2) another person's property (3) to the exclusion of or inconsistent with the rights of the plaintiff.

A fundamental element of a conversion cause of action is that the plaintiff be the owner of or legally entitled to immediate possession of the property allegedly converted. In order to prevail on his conversion claim, Curtis had to establish his ownership or superior right to possession of the ring in question. Curtis could not make this showing under either the parties' purported agreement or the conditional-gift rule. As a result, Anderson was entitled to judgment on this claim as a matter of law.

Conclusion

We hold that when an agreement between an engaged couple as to the disposition of engagement gifts is not in writing, any dispute arising over ownership when the engagement is broken is subject to the fault-based conditional-gift rule. We affirm the summary judgment granted by the trial court.

LINDSAY v. MCENEARNEY ASSOC. 260 Va. 48, 531 S.E. 2d 573 (2000)

Study terms: Statute of frauds, oral modification, exclusive right, real estate, accord and satisfaction

The primary issue we consider in this appeal is whether a contract that must be in writing pursuant to the statute of frauds, Code § 11-2, may be modified by a parol agreement.

■ ■ ■

Lindsay executed a written "Exclusive Right to Represent Buyer Agreement" with McEnearney Associates, Inc., which gave McEnearney Associates the exclusive right to represent Lindsay in any real estate transactions for the purchase of a home. The written agreement required Lindsay to pay McEnearney Associates a commission of three percent of the sales price of any residential real property that Lindsay purchased between August 13, 1997, and November 30, 1997.

Subsequently, Lindsay signed a sales contract for the purchase of property located at 2343 South Nash Street in Arlington County. She negotiated the price and terms of the sale "on her own," and she purchased the property without the assistance of McEnearney Associates. Lindsay claimed that she entered into an oral agreement with Loretta Connor, McEnearney Associates' sales associate. Pursuant to the terms of that parol agreement, Lindsay claims that she was required to pay to McEnearney Associates a commission of one percent rather than three percent because Lindsay found the property that she purchased without the assistance of McEnearney Associates. She closed on the property, and a statement shows that McEnearney Associates received a commission of one percent. In response to requests for admission, Lindsay admitted that she did not sign a written modification of the "exclusive right to represent buyer agreement."

B.

McEnearney Associates filed a warrant in debt against Lindsay in the Arlington County General District Court and alleged that Lindsay breached the written agreement. Lindsay alleged in the general district court that McEnearney Associates breached the agreement, and she asserted that the written contract had been orally modified. . . . The general district court entered a judgment in favor of Lindsay, and McEnearney Associates appealed the judgment to the circuit court for a trial *de novo* as permitted by Code §§ 16.1-106 and -107. Lindsay did not appeal the general district court's judgment denying her counterclaim.

McEnearney Associates filed a motion for summary judgment in the circuit court and asserted that it was entitled to judgment because the "exclusive right to represent buyer agreement" required that Lindsay pay McEnearney Associates a three percent sales commission in the event that Lindsay purchased residential property, that Lindsay purchased and closed upon such property, but she only paid a one percent commission to McEnearney Associates. Continuing, McEnearney Associates asserted that the written "exclusive right to represent buyer agreement" could not be modified orally. Responding, Lindsay asserted in the circuit court that the contract between the parties had been modified orally and that she had a viable defense of accord and satisfaction that could not be defeated by the statute of frauds. The circuit court granted summary judgment in favor of McEnearney Associates, and Lindsay appeals.

II.

Code § 11-2, often referred to as the statute of frauds, states in relevant part:

"Unless a promise, contract, agreement, representation, assurance, or ratification, or some memorandum or note thereof, is in writing and signed by the party to be charged or his agent, no action shall be brought in any of the following cases:

■ ■ ■

"7. Upon any agreement or contract for services to be performed in the sale of real estate by a party defined in § 54.1-2100 or § 54.1-2101. . . ."

Code § 54.1-2100 defines real estate broker as

"any person or business entity, including, but not limited to, a partnership, association, corporation, or limited liability corporation, who, for compensation or valuable consideration (i) sells or offers for sale, buys or offers to buy, or negotiates the purchase or sale or exchange of real estate. . . ."

Code § 54.1-2101 defines real estate salesperson as "any person, or business entity of not more than two persons unless related by blood or marriage, who for compensation or valuable consideration is employed either directly or indirectly by, or affiliated as an independent contractor with, a real estate broker, to sell or offer to sell, or to buy or offer to buy, or to negotiate the purchase, sale or exchange of real estate, or to lease, rent or offer for rent any real estate, or to negotiate leases thereof, or of the improvements thereon."

Lindsay argues that the statute of frauds does not apply to the written contract she signed with McEnearney Associates or, alternatively, that the statute of frauds has no application here because the contract has been fully performed. We disagree.

The purposes of Code § 11-2 are to provide reliable evidence of the existence and terms of certain types of contracts and to reduce the likelihood that contracts within the scope of this statute can be created or altered by acts of perjury or fraud. . . . Its primary object was to prevent the setting up of pretended agreements and then supporting them by perjury. There is further a manifest policy of requiring contracts of so important a nature as the sale and purchase of real estate to be reduced to writing since otherwise, from the imperfection of memory and the honest mistakes of witnesses, it often happens either that the specific contract is incapable of exact proof or that it is unintentionally varied from its original terms."

We held in *Heth's v. Wooldridge's,* 27 Va. (6 Rand.) 605, 609-11 (1828), that the statute of frauds rendered unenforceable an oral modification of a written contract for the sale of land. Explaining our holding, we stated that the parol modification of an agreement required to be in writing by the statute of frauds would permit "the very mischiefs which the statute meant to prevent." *Id.* at 610.

The written contract that Lindsay executed with McEnearney Associates falls within the scope of Code § 11-2 because the contract is an agreement for services to be performed in the sale of real estate by a real estate broker and a real estate salesperson. Applying our established precedent, we hold that when, as here, a contract is required to be in writing pursuant to Code § 11-2, any modification to that contract must also be in writing and signed by the party to be charged or his agent.

It is true, as Lindsay observes, that in certain circumstances written contracts, even those that contain prohibitions against unwritten modifications, may be modified by parol agreement. This principle, however, does not apply to an agreement which must be in writing to satisfy Code § 11-2.

We find no merit in Lindsay's contention that the statute of frauds does not apply to her agreement with McEnearney Associates because the agreement has been fully performed. The agreement has not been fully performed because, as Lindsay admitted in the circuit court, she failed to pay McEnearney Associates three percent of the sales price of the real property she purchased.

Lindsay argues that the circuit court erred by ruling that the statute of frauds barred the presentation of her evidence of an accord and satisfaction as a defense to McEnearney Associates' breach of contract claim. We disagree.

We have discussed the following principles of accord and satisfaction which are equally pertinent here:

"'Accord and satisfaction is a method of discharging a contract or cause of action, whereby the parties agree to give and accept something in settlement of the claim or demand of the one against the other, and perform such agreement, the "accord" being the agreement, and the "satisfaction" its execution or performance.'

"'The thing agreed to be given or done in satisfaction must be offered and intended by the debtor as full satisfaction, and accepted as such by the creditor.'

"'Thus an accord and satisfaction is founded on contract embracing an offer and acceptance. The acceptance, of course, may be implied, and as a general rule, where the amount due is unliquidated, i.e., disputed, and a remittance of an amount less than that claimed is sent to the creditor with a statement that it is in full satisfaction of the claim, or is accompanied by such acts or declarations as amount to a condition that if accepted, it is accepted in full satisfaction, and the creditor accepts it with knowledge of such condition, then accord and satisfaction results.'"

Virginia-Carolina Elec. Works, Inc. v. Cooper, 192 Va. 78, 80-81, 63 S.E.2d 717, 718-19 (1951) (citations omitted).

We hold that the statute of frauds does not permit Lindsay to establish a defense of accord and satisfaction. The purported contract that constitutes the accord is predicated upon the existence of an oral modification of Lindsay's written "exclusive right to represent buyer agreement" with McEnearney Associates. Yet, as we have already stated, Code § 11-2 requires that any modifications to that contract must be in writing. Thus, Lindsay cannot assert the defense of accord and satisfaction based on a contract that violates the statute of frauds. Approval of this defense in these circumstances would permit Lindsay to circumvent the statute of frauds.

In view of our holdings, we need not consider Lindsay's remaining contentions. Accordingly, we will affirm the judgment of the circuit court.

Affirmed.

Review Questions

1. Why is a written contract better than an oral one?

2. Name three ways in which the seventeenth section of the English Statute of Frauds differs from the fourth section.

3. What is the difference between surety and insurance or indemnity?

4. Black orally contracts with White to build a brick garage for Black for $5000. When the garage is finished Black refuses to pay the price, claiming that the garage is an addition to real property and, therefore, should have been in writing to be enforceable. Can White collect? Why or why not?   *ucc 3b*

5. Green orally contracts with the Gray Die Shop to build a punch press die for $2,000. The production for which Green was going to use the die is canceled and Green refuses to accept and pay for the die. Gray claims that the contract was for a service and, thus, Gray is entitled to payment. Green points out that dies are the usual product of the shop and, therefore, that the contract had to be in writing to be enforceable. Who is right? Why?

6. In *Davis v. Crown Central Petroleum Corporation,* suppose Crown Central had arranged binding contracts with all its outlets to supply them based on their "needs." What recourse would it have if its supply of crude oil were suddenly cut in half?

7. Would you enter into an exclusive buyer agreement after reading the *Lindsay v. McEnearney* case? Why or why not? How could potential buyers protect themselves when confronted with such an agreement? What should Lindsay have done?

8. Compare and contrast *Lindsay v. McEnearney* with *Melvin v. West* in Chapter 7.   *written*   *oral*

exam

9. Refer to *Curtis v. Anderson* where the court found that Curtis could not recover the ring given to Anderson upon their engagement. Consider the question of whether Curtis would have asked Anderson to sign an agreement at the time of the engagement. Is that the court's intention? *Y*

★ oral allowed in TX, oral not allowed in VA
Highlight statute of frauds
IRAC

Third-Party Rights

CHAPTER

12

HIGHLIGHTS

■ An important concept regarding third-party rights is that of *privity* where only those who are parties to a contract can enforce its terms, that is, only a person who is in privity with another may enforce contract rights.

■ Contract rights can be assigned only if it is allowed under the contract or there is no evidence that the rights cannot be assigned.

■ While rights can be assigned, duties are delegated. When both rights and duties are given to a third party, there has been a *novation*.

A contract is a voluntary, intentional, and personal relationship. The parties to the contract are the ones who determine the rights and obligations to be exchanged. Ordinarily, only those who are directly involved have rights stemming from the contract. In law, the relationship between parties to a contract is known as **privity of contract;** generally, only a person who is in privity with another may enforce any rights under the contract.

A strict interpretation of the privity-of-contract principle would prevent many common transactions of considerable value and convenience in our economy. Indeed, the early common law that developed in England during the Middle Ages refused to recognize attempts to assign a contract right. These rules developed when wealth was essentially either land or tangible items (such as horses, food, gold, or the like). In today's economy, wealth is instead embodied in intangibles like bank accounts, stocks and bonds, intellectual property, accounts receivable, and so forth. Therefore, the courts recognize three major exceptions:

1. Rights assigned by a party to a contract to a third party are enforceable by the third party.

2. Rights arising from third-party beneficiary contracts are enforceable by the beneficiary.
3. Rights arising under a manufacturer's warranty of its products are often enforceable by persons besides the one who actually bought the product.

ASSIGNMENT

Probably the most common form of assignment occurs when an indebtedness that is not yet due is assigned to another for value. Suppose White Company sells industrial machinery. In payment for some machinery, the White Company received $5,000 from Black Company on delivery of the machinery and also the promise of Black Company to pay another $10,000, six months after delivery. That promise to pay is often embodied in a separate written instrument called a note. If the White Company finds itself in need of funds, it may be able to sell this right to payment from Black to someone else (possibly a local bank). The bank would pay something less than face value (known as **discounting**) and then take over Black's note with the same rights White had.

An assignment involves at least three parties. The **obligor** (or **debtor**) is the party to the original contract who now finds that because of the assignment, the obligation is now owed to someone not in privity of contract with him. The **obligee** (or **creditor,** or **assignor**) is the person to whom the obligor originally owed the duty to perform (see Figure 12.1). Now, because of the assignment, the right to that performance has been assigned to another. The assignee is the one to whom the duty has been assigned and is the person to whom the obligor now owes the duty of performance. (See *Summit v. Commonwealth Plumbing* at the end of the chapter.)

Most contract rights are assignable if there is no stipulation to the contrary in the contract. Many contracts state that the rights and obligations involved cannot be assigned by one party without the other party's consent. Generally, an assignment in violation of a contract provision renders the assignment voidable at the option of the obligor. Such anti-assignment clauses are not necessarily enforceable. For example, the Uniform Commercial Code (UCC) states that a right to receive damages for breach of a contract for the sale of goods can be assigned despite an agreement to the contrary.

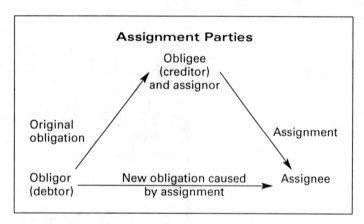

Figure 12.1

Assignment Parties

Obligee (creditor) and assignor

Original obligation

Assignment

Obligor (debtor)

New obligation caused by assignment

Assignee

There are situations in which contract rights are not freely assignable (even absent a contract clause that prohibits an assignment). Generally, a contract right cannot be assigned if any of the following applies:

■ The assignment would materially change the other party's duty.
■ The assignment would materially increase the other party's burden or risk under the contract.
■ The assignment would materially impair the other party's chance of obtaining performance of the contract in return.

What amounts to a "material" change or increased burden obviously involves many shades of gray and may vary from situation to situation.

Contract rights to personal services and services based on the skill of an individual are treated exceptionally, and it seems right that they should be. If Black contracts to perform personal services for White, it should not be possible to force Black into a choice of either performing those services for a third person (to whom White has assigned such rights) or breaching the contract. The same reasoning applies where the rights involved are in the nature of a trust or confidence. For example, suppose Black Machinery Co. obtains a license to certain patents and trade secrets of White. If the license agreement requires Black Machinery Co. to maintain all of the trade secrets in confidence (as such agreements usually do), is it fair to allow Black Machinery Co. to assign the license agreement to Green Co.? The answer will be "no" in many situations.

Rights and Duties

If you can transfer your rights under a contract to someone else, can you also transfer your obligations or duties? The terminology used is important. Rights are assigned. Duties, however, are delegated. This distinction is helpful in considering the legal issues involved. (Unfortunately, not all lawyers and courts use these terms correctly.)

In contrast to rights, duties generally cannot be freely delegated (in the absence of the obligee's consent). At least, one cannot relieve himself of liability for nonperformance or poor performance by delegating the duties to someone else. Even if the duties are delegated to another, the person delegating the duties is still responsible for their performance. The delegating person is essentially a guarantor of the performance. Black hires White to build a structure. White subcontracts the plumbing to Gray. White's subcontract with Gray does not relieve White of White's responsibility for Black's structure, including the plumbing. But White is only responsible as a guarantor if White was the one to deal with Gray. White's responsibility is relieved if Black makes the plumbing arrangements with Gray.

Assignment and delegation differ from **novation.** In a novation, both rights and duties are effectively transferred. The distinction depends on the number of parties involved. (For a more thorough explanation, see Chapter 13.) If all three parties agree to the substitution of one party for another, the effect is a new agreement by means of a novation. Assignment of rights requires only an agreement between assignor and assignee. The obligor or debtor must, of course, be informed of the assignment if it is to be binding on the obligor. Also, the obligation must be made no greater or more burdensome as a result of the assignment—the obligor may not be forced to do more than the obligor originally agreed to do.

Assignee's Rights

The assignee (the third party) acquires no better right than the assignor had. The assignee's right to performance is subject to any fault the obligor could have found with the right in the hands of the original obligee. In other words, if there is a defense available to the obligor (the obligee's fraud or misrepresentation, for instance), the obligor may use this defense against the assignee's claim, just as the obligor could have against the original obligee. If the obligor exercises such a claim, the assignee is left with only an action against the assignor (who is also the obligee) for whatever consideration the assignor has given up. If the obligor has made partial payment to the assignor or has a counterclaim against the assignor, the right obtained by the assignee may be subject to these claims.

Black has a $6,000 claim against White for installation of a machine. Black assigns this claim to Gray in return for operating cash. If the installation proves to be substantially less than a proper performance, requiring White to spend $1,000 more to put it in operating condition, this amount might be used as a set-off against the $6,000 claim. As a result, Gray would get only $5,000 from White and have to look to Black for the other $1,000.

An assignee may have rights against the assignor for a breach of any express (or implied) warranties made by the assignor. Generally, there are several implied warranties in an assignment. The assignor warrants (guarantees, according to common parlance) that there is a valid claim, that the assignor has a right to assign it, that there are no defenses available against it except as noted in the assignment, and that any document delivered is genuine and not a forgery. If any of these is not as warranted, the assignee has a right of action against the assignor. However, the assignor does not warrant that the obligor will be able to perform. If the obligor becomes bankrupt, for instance, the assignee may have to settle for considerably less than the amount bargained for. The assignee takes the same risk here as taken in any claim where the assignee becomes the obligee.

A gratuitous assignment creates no rights. If an assignment is made gratuitously (the assignor getting nothing back for it), the assignee gets nothing but a promise of a future gift. Such a promise is unenforceable. Until the assignee actually gets something, the gratuitous assignment may be recalled and avoided at will by the assignor.

Notice

A debtor cannot be charged with nonpayment of the debt if the debtor innocently pays a creditor who has assigned the obligation. To make the assignment legally effective against the debtor, there must be **notification.** When an assignment of rights is made, the assignee obtains, along with the rights, a practical duty to notify the obligor of the assignment. If such notice is not given and the obligor pays the debt to the original creditor, the obligor's debt is discharged. The assignee is then left with an action for recovery against the assignor. If notice has been given to the obligor and the obligor still pays the assignor, such payment does not discharge the obligor's debt.

It is possible, though unlawful, to make subsequent assignments of the same right. The results for the respective assignees are somewhat controversial. Some court holdings are based on the argument that after the first assignment the assignor had nothing more to assign, so the assignee received nothing. Consequently, subsequent assignments are null and void. Another view is taken by other courts. According to them, the first assignee to notify the obligor has priority as to the claim. The argument here is that only proper notification completes the assignment. The losers, in either instance, have only their actions against the assignor (who may have since disappeared). Some assignments of contract rights are subject to statutory filing requirement. For example, the UCC provides that, as between two assignees (each of whom paid value for the assigned rights) of the same rights, the first who files the appropriate notice with the appropriate county clerk or state agency (as provided for by the statute) is the prevailing party.

Wages

The assignment of future earnings for a present debt is governed by statute in most states. These statutes vary considerably, some making such assignments void. Many set a maximum amount (for example, 25 percent) that may be assigned from expected wages. In states where wage assignments are upheld, the right to lawfully assign future wages usually depends on present employment. One must be presently employed to make an effective assignment of expected wages. Generally, if a person is employed, that person can assign his or her wages, even though it might be argued that the employment may be terminated at the option of the employer. However, an unemployed person cannot assign expected wages from hoped-for employment, even though the hopes may be well founded.

THIRD-PARTY BENEFICIARY CONTRACTS

If Black and White make a contract that will benefit Gray, what right does Gray have? The nature of the contract, the intent of the parties, and prior indebtedness existing between them all have bearing on Gray's rights. Courts are not in complete agreement in their holdings in such cases. Given the same set of facts, a case may be decided quite differently in two different jurisdictions. Only in insurance contracts is the law regarding third-party beneficiaries likely to be applied uniformly.

Insurance

Probably the most common type of third-party beneficiary contract is the life insurance contract. There are two possibilities relating to the beneficiary's right to recover following death of the insured. First, if the insured has purchased insurance from an insurer, the insured has the right to name anyone as beneficiary. When the insured dies, the beneficiary has a right of action to collect the amount of the policy from the insurer. White contracts with the Black Insurance Co. to insure White's own life, agreeing to pay the stipulated annual premiums, and naming Gray as beneficiary. When White dies, Gray can collect.

The second situation differs from the first in that it is the beneficiary who purchases the insurance. Gray contracts with the Black Insurance Company to insure White's life. Gray agrees to pay the annual premiums and is to be beneficiary. For Gray's right as beneficiary to exist, Gray must be able to show an insurable interest in White's life; Gray must risk a loss that would occur on White's death. In other words, there must be some anticipated benefit to Gray resulting from White's continued existence. An insurable interest is present where the insured is a member of the beneficiary's immediate family. It does not extend to outsiders unless there is an economic tie involved, as between creditor and debtor or partners in an enterprise. Sometimes a business will obtain "key man" insurance for its president or other key employees. Presumably, if such employees are key to the business' continued existence, then the business should have an insurable interest.

An insurable interest in property extends to those who have an ownership or lien interest in the property, or merely possession of it with the attendant risk of loss. Both mortgagor and mortgagee, for instance, have an insurable interest in real estate being purchased under a mortgage contract. There is a limit to the insurable interest created by an economic tie, however; usually the courts consider the justification for insurance to be limited to the amount of economic benefit expected or the risk of loss involved.

The time when the insurable interest must exist differs in the two types of insurance. In life insurance it must be shown to have existed at the time when the insurance contract was made. A later change in the relationship does not serve to terminate the insurance. In property insurance, however, the insurable interest often must exist when the loss is suffered. An example of a third-party beneficiary insurance case is *Frazier v. Twentieth Century Builders* at the end of this chapter.

Donee Beneficiary

An unsealed promise of a future gift is generally worthless as far as enforceability is concerned. That is, it is worthless if the donee tries to enforce it against the donor—as a contract it is unenforceable for lack of consideration. A donee beneficiary has a right to enforce a contract created to benefit that person. Gray is a third-party beneficiary if Black and White agree that White will pay Gray $500. Gray has an enforceable legal claim against White. However, Gray would not have a right of action against Black, who is Gray's benefactor.

Creditor Beneficiary

Assume that Gray is to receive $500 in payment of an obligation Black owes to Gray. If Black and White make a contract whereby Black's consideration is to be paid for by White's paying $500 to Gray, Gray is a creditor beneficiary of that agreement. As such, Gray may enforce the claim against either Black or White. As a creditor beneficiary, Gray is in a stronger position than as a donee beneficiary. Gray may collect from either the promisor or the promisee. Gray can sue the promisee on the previous debt, or Gray can elect to force the promisor to perform.

Incidental Beneficiary

Many contracts are made that indirectly benefit third parties. Failure of the promisor to perform properly does not necessarily give an incidental beneficiary the right to take legal action. Generally, if there was no intent by the contracting parties to benefit the third party, the third party has no legally enforceable interest in the contract. Brown hires White (a landscape architect) to landscape Brown's estate. Gray, who lives next door, will incidentally benefit as a result of the landscaping. However, Gray has no legally enforceable interest in the completion of the landscaping. Gray was not a party to the contract, and whatever benefit Gray might have received was merely incidental to the contract's primary purpose.

Now suppose Brown is a manufacturer and supplier of parts for the automotive industry. Brown contracts with Green, a builder of automation equipment, for Green to build and install an automatic machine. Gray supplies most of Green's steel. Although Gray would be likely to benefit from the automation contract, the agreement is not for Gray's benefit, and Gray could not enforce it.

Contracts in which the government is a party are sometimes considered to be third-party beneficiary contracts. This is the case when it is reasoned that citizens are to benefit from contracts made by the governing body, and that they have an interest in these contracts because they pay taxes and thereby acquire rights to benefits. If a small segment of a community is to receive benefits and possibly pay a special assessment toward them, the argument for enforcement by a citizen is even stronger. However, the right of a citizen to take action as a third-party beneficiary to enforce a government-made contract is largely a matter of local statute.

Intended Beneficiary

A modern rule reformulates the above traditional categories of donee beneficiaries, creditor beneficiaries, and incidental beneficiaries. Under this rule, an "intended beneficiary" may enforce the contract. To be an intended beneficiary, the third party must show that recognizing a right to performance is appropriate to give effect to the parties' intentions and either

1. The performance of the promise will satisfy an obligation of the promisee to the beneficiary, or
2. The promisee intended to give the third party the benefit of the promised performance.

WARRANTIES

Traditionally, any promises or warranties made by one party to a contract benefitted only the other party and persons in privity with that party. However, the requirement of privity has eroded over time. Today, a seller's warranties about the goods sold often flow to persons besides the purchaser of the goods. The courts recognize that a number of people will use a product like a car. Essentially, the courts have allowed persons besides the buyer to sue for injuries received due to a breach of warranty. These concepts are discussed in greater detail in the chapter on Product Liability.

The Uniform Commercial Code adopted three alternatives that govern the **flowdown** of the warranties provided by sellers of goods:

1. The seller's warranty extends to any person in the family or household of the buyer or who is a guest if it is reasonable to expect that such a person may be affected by the goods and that person suffers personal injuries due to a breach of the warranty.
2. The second alternative expands the first to include all persons who may reasonably be expected to be affected by the goods and who are personally injured due to a breach.
3. The third alternative extends the second to include persons who may reasonably be expected to be affected by the goods and who are "injured" by a breach (that is, the injury need not be a personal injury). Different states have adopted different alternatives, thus leading to a variety of rules.

SUMMIT PROPERTIES PARTNERSHIP, L.P., ET. AL. v. COMMONWEALTH PLUMBING SERVICES, INC., ET. AL.
At Law No. 25992 (Circuit Court of Loudoun County, June 25, 2002)

Study terms: Privity, subrogation, insurance, indemnification

This is an action for damages caused by a fire. Summit Properties Partnership, L.P. (Summit Properties) is the owner of a tract of land upon which apartment houses were to be erected by its general contractor, Summit Apartment Builders, Inc. (Summit Builders). Summit Builders entered into a subcontract with Commonwealth Plumbing Services, Inc. (Commonwealth Plumbing), to provide plumbing services in connection with the construction of the apartments. Kathryn Osborne operated the business known as Commonwealth Plumbing. It is claimed that an employee of Commonwealth Plumbing ruptured a gas line while performing plumbing work and that the resulting explosion caused a fire that destroyed the building under construction and damaged other aspects of the apartment complex.

Summit Properties had obtained a builder's risk insurance policy on the property with Travelers Indemnity Company (Travelers). It is alleged that Summit Properties has been paid $1,225,000.00 by Travelers as a result of their fire loss.

Plaintiffs, Summit Properties, Summit Builders, and Travelers seek recovery against the defendants as a result of the losses they have allegedly sustained by reason of the fire and payment to the insured. Summit Properties seeks recovery of $252,000.00 for lost income suffered but not reimbursed by Travelers. Summit Builders asserts a claim for the $25,000.00 deductible provided for in the builder's risk policy. It is the contention of Travelers that it is entitled to recover the $1,225,000.00 paid to its insured.

Defendants have demurred to the motion for judgment. They have previously craved and been granted oyer* of the written contracts that form the basis of plaintiffs' claims for damages. Accordingly, copies of the subcontract and builder's risk insurance policy have been produced by the plaintiffs and are to be read with the pleadings and considered by the Court in ruling upon the instant demurrer.

The centerpiece of the demurrer filed by the defendants is their contention that the provisions of the contract of indemnity limit the exposure of the defendant subcontractor to third party claims against either the owner or general contractor. Thus, it is argued, since Travelers subrogation claim can be based on no greater rights than those of the owner, it may not recover against Commonwealth Plumbing for claims paid that have their aegis in the negligent actions of Commonwealth Plumbing's own agent.

Such a narrow reading of the subcontract is neither justified by the plain words of the agreement between Summit Builders and Commonwealth Plumbing or cases relied upon by the defendants. It is the opinion of the Court that, under the express terms of the contract, the plumbing subcontractor is subject to liability for its agents' negligence whether third parties, subcontractor, or owner initiates those claims. Although reference is made to parties obtaining insurance, the express terms of the contract do not purport to shift recovery for loss to the builder's risk policy.

Travelers issued a builder's risk policy to Summit Properties on July 12, 1999 in connection with the apartment project that is the subject of this suit. The policy provides insurance during the construction phase of the project, including the time during which Commonwealth Plumbing would be on the job.

On July 22, 1999, Summit Builders entered into an agreement with Kathryn A. Osborne as president of Commonwealth Plumbing to provide plumbing services identified in exhibits attached to the contract. As a part of the contract, the parties agreed as follows:

[2]0. Indemnification

To induce SABI (Summit Builders) to enter into this Subcontract Agreement and for ten dollars (10.00) and other good and valuable consideration, the receipt of which are hereby acknowledged by Subcontractor, Subcontractor agrees to indemnify and hold SABI and the Owner and their agents and employees wholly harmless from and against any damages, claims, demands, suits, losses, and expenses (including but not limited to reasonable attorneys' fees and court costs arising out of or resulting from the execution of the work provided in this Subcontract Agreement or occurring in connection therewith, whether caused in whole or in part by Subcontractor or any direct or indirect employee or any of them or anyone for whose acts any of them may be liable (including Subcontractor's failure to comply with the terms of this Subcontract Agreement), excluding liability for the sole negligence of SABI or its agents or employees.

21. Insurance

Subcontractor shall purchase and maintain at Subcontractor's own cost insurance with the minimum limits of liability as specified below:

A. Commercial General Liability. . . .
B. Automotive Liability . . .
C. Worker's Compensation . . .

[*Oyer* is a term used by the Commonwealth of Virginia that technically means "read aloud", or in this case, the defendants had been granted a hearing regarding the written contracts.]

The insurance certificate shall provide for notice of cancellation to SABI thirty (30) days   prior to cancellation or lapse of coverage, and the certificate shall include that the above provision has been included. This certificate shall be furnished at the time this Subcontract Agreement is signed.

22. Builder's Risk Insurance

Subcontractors (sic) hereby acknowledges that the builder's risk insurance, as provided by the Owner, does not cover the loss of Subcontractor's trailers, scaffolding, forms, supplies or tools or other losses of Subcontractor which are contained within the deductible amounts of such insurance. All such uninsured losses of Subcontractor's property shall be the sole responsibility of Subcontractor.

■ ■ ■

32. Governing Law

This contract shall be construed and interpreted under the laws of Maryland.

Travelers asserts that it is entitled to assert its subrogation claim against the defendants based upon the terms of its contract of insurance and the indemnification provisions of the subcontract. Commonwealth Plumbing and Ms. Osborne contend that it was the intent of the parties that the risk of loss by fire was to reside with the builder's risk insurance carrier. To support this assertion, defendants draw the attention of the court to other provisions of the contract and the law of the State of Maryland.

Maryland courts recognize that,

[w]here parties to a business transaction mutually agree that insurance will be provided as a part of the bargain, such agreement must be construed as providing mutual exculpation to the bargaining parties who must be deemed to have agreed to look solely to the insurance in the event of loss and not to liability on the part of the opposing party . . . where . . . and insured has entered into a contract which is intended to substitute insurance for personal liability, the insurer is bound by such agreement and may not sue the third party under a theory of subrogation."

It has been observed that, "[f]ire insurance covers the property loss sustained regardless, generally speaking, of its cause. Insurance against negligence indemnifies the negligent person as to his liability to another." Thus, agreements that seek to shift risk of loss by fire to an insurer differ from those that may require indemnification. In the instant case, the subcontract agreement contains provisions relating to insurance coverage as well as indemnification.

Although the provisions of the builder's risk insurance expressly limit the liability of the insurer to pay for the subcontractor's losses, neither such policy nor the terms of the subcontract admit of an agreement to shift to the policy losses sustained by the owner as a result of the negligent acts of Commonwealth Plumbing. While, under Maryland law, the parties might shift any fire loss caused by their negligence to the policy of insurance, they struck no such bargain in this case. The clear and unambiguous terms of the indemnification agreement permit recovery for damages arising out of or resulting from the execution of work by the subcontractor. To enforce indemnification between the parties to the contract, as apposed to third party claims, does not offend the public policy of the Commonwealth.

Consistent with the allocation of risk among the parties to the subcontract, a commercial general liability policy was to be obtained by the subcontractor to protect against third party claims arising out of the project. Builder's risk insurance would provide protection for damages to its property caused by the negligence of the parties to the instant contract. However, the owner and contractor might also look to Commonwealth Plumbing in the event of loss occasioned as a result of the subcontractor's negligence.

Counsel for the defendants has argued that there are, in general, three types of indemnification agreements recognized by the Maryland courts. He notes that all three involve an indemnitor, indemnitee, and third party entitled to payment. However, the provisions of the contract and the intent of the parties would control whether such liability for indemnification would be limited to third party claims. In the instant case, the Court is of the opinion that the provisions of the indemnification agreement are sufficiently broad to include not only claims by third parties, but claims against the subcontractor as well.

The right of Travelers, as subrogee, can rise no higher than those of its insured. Conversely, the recovery by Travelers is dependent upon the assignment of the owners right to recover against the subcontractor. No such assignment has been pled. Accordingly, the demurrers of the defendants to Traveler's claims will be sustained with leave to replead. . . .

The demurrer to the claims of Summit Builders and Summit Properties is overruled.

Counsel for the defendants are requested to draw an order consistent with this opinion to which counsel may note their exception.

FRAZIER, INC. v. 20TH CENTURY BUILDERS, INC. 198 N.W.2d 478 (Neb. 1972)

Study terms: Consequential damages, negilgence, insurance, wrongful denial of coverage

This appeal is from the sustaining of a motion for summary judgment in a third party action. The principal action was a mechanic's lien foreclosure brought by Frazier, Inc., against Majors, Inc., et al., in which Omaha Poured Concrete Company was subsequently included as a defendant. Majors filed a cross-petition against Frazier and Omaha Poured, alleging damages as the result of the negligence of Frazier in the installation of the plumbing system that resulted in the collapse of a cement floor poured by Omaha Poured. Omaha Poured was given leave to file a third party complaint against Transamerica Insurance Company as a third party defendant, to recover under the terms of an insurance policy providing coverage to Omaha Poured for consequential damages. We affirm.

Frazier, as a subcontractor under Omaha Poured, installed the plumbing in a building being erected for Majors. The plumbing was installed with removable sleeves around the floor drain to permit the installation of a poured concrete floor. After the completion of the building, Majors moved its plastic moulding machines onto the concrete floor and started operations. These machines required the discharge of large quantities of water through the drain. After the first day's operation, the concrete floor pulled away from the wall, buckled, and collapsed. It was then discovered that the sleeves had not been properly installed, and after they were removed the water, instead of running off through the drain, went through an unsealed gap into the ground under the floor, undermining it, causing the floor to sink. Notice was immediately given to Transamerica who investigated the damage. Omaha Poured replaced the floor at its own expense. The present third party action involves the amount paid to settle the consequential damages sustained by Majors because of the collapse of the floor, and does not include any part of the cost of replacing the floor.

Omaha Poured was granted leave to file its third party action against Transamerica January 22, 1969. The petition filed January 30, 1969, alleged that Transamerica is obligated to defend Omaha Poured, and to pay any sums of money that may be awarded against Omaha Poured as the result of consequential damages sustained by Majors. Omaha Poured alleged it complied with all of the conditions precedent of said policy, but Transamerica refused to provide coverage for said incident which falls within the terms and conditions of the policy, and has refused to defend Omaha Poured. On June 5, 1969, the issues raised by the third party petition were ordered separated for trial purposes from the issues raised by the principal pleadings.

On July 14, 1969, the day the main action was called for trial, Omaha Poured made a settlement with the other defendants, and judgment was entered against it for Majors' consequential damages in the amount of $14,298.22. The agreement included a covenant that the judgment creditor would not execute against the personal assets of Omaha Poured for more than $1,500 of the judgment, and would thereafter look to the Transamerica policy for all remaining sums thereunder.

Immediately previous, and on the day judgment was entered, Majors amended its cross-petition to include a special allegation of negligence against Omaha Poured. No notice was given Transamerica of this amendment, but a copy of the judgment entry was mailed to it the same day the judgment was entered.

The trial court sustained Omaha Poured's motion for summary judgment, specifically finding that there was no substantial controversy as to the following material facts: (1) Transamerica knew of the collapse of the floor shortly after it occurred and knew that the potential loss and claim by Majors, Inc., would include not only damage to the floor itself but the very consequential damages which Majors, Inc., ultimately sought by the instant litigation. (2) Pursuant to investigation by representatives of Transamerica, it unequivocally and unconditionally denied any coverage on the policy of insurance for the incident involved herein. (3) Transamerica refused to accede to the demand of Omaha Poured that it take over and defend the crosspetition filed against Omaha Poured by Majors, Inc. (4) Omaha Poured, without the knowledge or consent of Transamerica, entered into a compromise settlement and agreement with Majors, Inc., resulting in a consent judgment against Omaha Poured for the consequential damage sustained by Majors, Inc., with an agreement that no more than $1,500 of the judgment would be collected from Omaha Poured personally. (5) The amendments to the crosspetition of Majors, Inc., just prior to the entry of the judgment were made without the knowledge or consent of Transamerica. (6) The evidence is sufficient to support a finding of legal liability by Omaha Poured to Majors, Inc., for the consequential damage loss to Majors, Inc., resulting from the collapse of the floor, in the amount of the judgment.

The trial court made the following specific findings of law: First, the policy in question affords coverage to Omaha Poured for the consequential damage claim asserted by Majors, Inc. Exclusions raised by Transamerica in support of its denial of coverage do not apply. Second, once Transamerica denied coverage to Omaha Poured in the manner in which it did, Transamerica breached its contract with Omaha Poured, relieving Omaha Poured of any obligation under the contract to notify or deal with Transamerica any further regarding the claim.

Third, the allegations contained in the cross-petition of Majors, Inc., prior to the amendments made on July 14, 1969, together with all the pleadings in the case, were sufficient to compel Transamerica to assume the defense of Omaha Poured, and its refusal of Omaha Poured's demand that it do so constituted a breach of its contract with Omaha Poured.

Fourth, Transamerica's breach of its contract with Omaha Poured, by denying coverage and refusing to defend its insured, entitled Omaha Poured to enter into the most favorable settlement and consent judgment possible, which it did.

Fifth, the consent judgment, supported by the evidence, was obtained in good faith and without fraud or collusion. Therefore, having declined to defend the action when called upon to do so, Transamerica may not again litigate the issues that resulted in the judgment.

Only one of the court's findings of fact set out above is questioned in Transamerica's brief, that being No. (2). . . . Dwight Stephens, Transamerica's claim representative who first investigated the claim, stated in his affidavit that after completing his investigation he informed Omaha Poured's representative that no coverage was afforded to the insured for the collapse of the floor for the reason that the aforementioned policy did not provide coverage for completed operations and products' liability, and that he thereafter reported his findings to his superior, Harold Pace. The affidavit of Harold Pace states that coverage was denied under said policy shortly after May 27, 1967, for the reason that coverage was not afforded thereunder to the insured for completed operations and products liability coverage. There can be no question that Transamerica at all times was insisting that the damage was for completed operations and products liability coverage and it was not covered under the policy. Transamerica overlooked the fact that its policy did insure Omaha Poured for consequential damages and that the action brought by Majors, Inc., was for consequential damages resulting from the incident. Nowhere in Transamerica's brief does it challenge, by assignment of error or discussion, the finding of the trial court that the Transamerica policy in question affords coverage to Omaha Poured for the consequential damage claim asserted by Majors, Inc. Nor, incidentally, does Transamerica challenge in any manner the trial court's conclusion that as a matter of law the "exclusions raised by third party defendant in support of its denial of coverage" do not apply.

■ ■ ■

As to Transamerica's third assignment of error, when we pierce the allegations of the pleadings there is no question the material issues are those of law and not of fact.

As to Transamerica's fourth assignment of error, there is not the slightest evidence of fraud or collusion. The settlement was consummated on the morning of the scheduled trial of the main action. Omaha Poured, left to its own resources, at the last moment made a settlement to avoid increased liability. Omaha Poured must have realized, as would any reasonable person in its position, that prudence required it to protect its assets as best it could. There was no question about its liability for consequential damages. There is not the slightest merit to this assignment.

Nor is there merit in Transamerica's fifth assignment of error. Transamerica investigated the incident and denied coverage. It further refused to tender a defense for Omaha Poured. Exhibit 84 to the deposition of Transamerica's claim manager, is a letter from attorneys for Omaha Poured to Transamerica under date of January 10, 1969, reminding it of two previous denials of coverage and of a previous letter dated August 31, 1967, to the effect that Omaha Poured was taking steps to protect itself from the effects of the wrongful denial of coverage. The letter of January 10, 1969, gave Transamerica the trial date of the main action, again tendered the defense of Omaha Poured, and advised that in the event it again declined to provide a defense, a third party petition would be filed. To hold that Omaha Poured had any additional obligation to keep Transamerica informed of the developments would be absurd. Transamerica's sixth assignment of error misconstrues the decisions of this court on the point in question. In *National Union Fire Ins. Co. v. Bruecks*, . . . we said: "It would seem that the obligation to defend a suit for an insured should be determined on the basis of whether the petition filed against him attempts to allege a liability within the terms of the policy." Clearly, the inference from the crosspetition and the prayer before the amendment included Omaha Poured, and sufficiently inferred liability to require action on the part of Transamerica. If there was the slightest question, certainly the proceedings culminating in the filing of this third party action, which occurred six months before the amendment in question, would have removed any doubt. Transamerica's seventh assignment of error is predicated on its assertion that affidavits offered into evidence, notarized by an attorney of record, are improper and should be excluded. We direct counsels' attention to section 25-1245. R.S. Supp., 1969, which provides: "An affidavit may be made in and out of this state before any person authorized to take depositions, and must be authenticated in the same way. An attorney at law who is attorney for a party in any proceedings in any court of this state shall not be disqualified as the person before whom the affidavit is made by reason of such representation."

In view of previous comment, Transamerica's eighth assignment of error requires no further discussion. The trial court correctly sustained the motion for summary judgment. The judgment is affirmed.

Affirmed.

Review Questions

1. What is meant by privity of contract?

2. Black is purchasing a house under a real estate mortgage held by the White Mortgage Company. Under the terms of the mortgage, Black is obligated to make a payment to White Mortgage Company each month. Part of the payment is for interest, another part reduces the mortgage balance, and a third part is deposited in an escrow account to pay for taxes and insurance. Black sells his house to Gray, with Gray agreeing to take over the mortgage payments. If Gray defaults, can White Mortgage Co. take action to recover from Black?

3. Why must the obligor be notified of an assignment?

4. What is meant by insurable interest?

5. In *Summit v. Commonwealth Plumbing,* the court granted leave to Traveler's insurance to replead its claim. It found that Traveler's may be entitled to recovery, but could not receive it for what reason?

6. In the case of *Frazier, Inc. v. 20th Century Builders, Inc.,*. if the insurer reimbursed the owner for the damages caused by collapse of the floor, would the insurer then have a right to take action for reimbursement against the subcontractor, Frazier, Inc., based on Frazier's alleged negligence? Why did Majors amend its cross-petition just before the entry of the agreed judgment to include an allegation of negligence on the part of Omaha Poured?

Performance, Excuse of Performance, and Breach

CHAPTER 13

HIGHLIGHTS

- Performance is often hindered or helped by the terms of a contract. These terms may include:
  1. Conditions precedent
  2. Architect's approval
  3. Satisfaction
  4. Completion date
  5. Substantial performance

- Performance may or may not be excused if one of the following occurs:
  1. Death or illness
  2. Destruction of an essential element of the contract
  3. Unexpected hardship
  4. Commercial frustration

Contracts, as you have observed, are made every day and are often quite ordinary. The terms of the agreements are usually carried out with no dispute. When you complete your obligations under a contract (such as by paying the purchase price of this book), you have performed under the contract. This chapter deals with the exceptional situations, where a party has bargained away more than intended, or where performance became more difficult than expected. Public attention is called to the exceptional disputes such as publicized court cases covered in newspaper articles. It must be kept in mind, however, that these cases are the exception to the general rule, and that the vast majority of contracts are performed satisfactorily and with benefit to both parties.

Many contracts are clearly worded and the meaning of a contract is not disputed. In this chapter, we shall see how the parties are held to their bargain, and what constitutes performance, or a real offer to perform. We shall see when the law will discharge a person's obligations or, in effect, excuse that person's performance. In Chapter 14, we will examine the remedies available if a contract is breached by one of the parties to it.

PERFORMANCE

Theoretically, a party to whom a contractual obligation is owed has a right to precise performance of the obligation. Anything short of that performance constitutes a breach of the agreement. However, performance or non-performance is not always as self-evident as it might appear. The nature of the obligation undertaken determines the required performance. As noted earlier in Chapter 9, Consideration, when a liquidated amount of money is owed, payment of a lesser amount will not satisfy the obligation. Where the contract requires the construction of an office building, however, exact performance may not always occur and cannot seriously be expected. This is particularly true where research and development are to be part of the performance of the contract. Such is the case in many engineering contracts.

Conditions

Frequently contracts are written so that the apparent intent of the parties is that performance by one party will precede performance by the other. For instance, one might contract to have a machine built and installed in a plant, with payment to be made when the installation is complete. A condition such as the installation would constitute a **condition precedent** in a contract. If the parties to a contract agree that, on the occurrence of some future event, an obligation will come into existence, the event is a condition precedent. One common type of condition precedent is obtaining an architect's certificate. The passage of time until the completion date is also often a condition precedent.

A frequent source of confusion is the distinction between a **condition precedent** and a **condition subsequent.** A condition subsequent is a condition that, when it occurs, ends an existing contract. Assume, for example, that White's house is insured by the Black Insurance Company. The insurance policy states that if the house is ever left vacant for a period of 30 days or more, the insurance will no longer be in effect. Such a conditional event is used as a condition subsequent.

Architect's Approval

When an architect's approval is required as a condition precedent to the owner's duty to pay the contractor, the courts usually enforce the contractual requirement. It must be recognized, however, that architects are human too, and they are capable of human failings. The architect may have died or may be ill when the certificate is required, or the architect may unreasonably or fraudulently refuse to issue the certificate. Under such extreme circumstances the court ordinarily dispenses with the requirement of the architect's certificate. If there is any sound reason for the architect's objections, however, the court will probably enforce the requirement.

Satisfactory Performance

Where the purchaser of a certain performance must be satisfied, two possibilities exist. The nature of the contract may be such that the only test of satisfaction is the personal taste of the buyer. Black, an artist, agrees to paint White's portrait to White's satisfaction for $5,000. White may never be satisfied, even though to a third person the portrait appears to be a perfect likeness. White could conceivably state, after each submission of the portrait for approval, that "it just isn't me" and require Black to continue. If the court determined the contract to be binding (the consideration of Black's performance might be termed illusory), Black would have to continue to paint or else breach the contract.

The nature of the contract may be such that, even though satisfaction of the buyer is specified, it provides for mechanical or operational suitability. In this case, if performance is such that it would satisfy a reasonable person, the court will deem the condition satisfied. Black agrees to install an air conditioning system for White, to White's satisfaction. After the installation, if White is not satisfied, Black may still be able to collect by proving that the air conditioning system will do all that a reasonable person could expect it to do.

Completion Date

Time limits are frequently stated in contracts. Where performance is to take place by a certain date and it actually extends beyond that date, any of several situations may result. If the time for performance is not important, there may not even be a breach of the agreement. Generally, time is important. Thus, many contracts include provisions for **liquidated damages** that state an amount to be paid for each day's delay in performance beyond the date specified in the contract. If the amount specified is a reasonable estimate of the difficult-to-calculate damages that would be caused by delay, the provision should be enforced.

If the specified amount of the liquidated damages is excessive, the court may hold that the clause is really a **penalty clause** and will refuse to enforce it. If there is no liquidated-damages clause and performance runs beyond the time agreed on, the party aggrieved can sue the other party for damages due to the delay. Usually, either actual or liquidated damages merely serve to reduce the price paid for the work. Ordinarily, nothing but an unreasonably late performance gives the buyer sufficient grounds to terminate the contract and pay nothing. Where performance is not unreasonable, the buyer must accept the performance, but at a price reduced by the damages suffered. See *Eatherly Construction Co. v. HTI Memorial Hospital* at the end of the chapter.

In certain circumstances, the time of performance is critical in a contract. In this event, a clause to the effect that "time is of the essence of the contract" is usually included. In such instances the courts generally will allow termination of the contract for late performance.

Substantial Performance

Many building contracts or contracts in which machinery or equipment is to be built and installed are performed in a manner that deviates from the exact terms of what is specified. Some flaw in such a project can be found by looking long enough and thoroughly enough. If variations from specifications were not allowed, few projects would be undertaken and fewer yet would be paid for. As it is, if a building contract calls for a concrete floor of a uniform five-inch thickness, it is not likely that the resulting floor will be precisely five inches thick throughout. However, if the result accomplished approximates closely the result specified, the contractor can still recover the contract price; the performance has been substantially the same as that specified.

If gross inaccuracies occur or substitutions are found to have been made that do not reasonably satisfy the specifications, the claim is no longer substantial performance. Just where the line is drawn between substantial performance and an outright breach is a question of fact, and is usually submitted to a jury.

Cases involving substantial performance frequently result in the performer receiving the amount for which the parties contracted, less the other party's damages. The extent of damages is usually determined by the value of the performance as rendered compared with the value of the performance as specified, or the cost of additional work to complete the performance properly.

Substantial performance assumes that the performer has performed the contract in good faith and with the intent that the performance would agree with the standards specified. If this is not the case—if the performer willfully abandons the performance or the work is unreasonably poor— then there has been a breach of the contract and not a substantial performance.

Impossibility of Performance

The law is not entirely clear or settled with regard to impossibility of performance. In certain cases involving death or illness of the promisor, definite statements can be made. In other cases, particularly where performance has turned out to be much more difficult than was anticipated, general conclusions can be drawn from the majority of decisions, but numerous exceptions exist.

Generally, when a person undertakes a contractual obligation, that person assumes certain risks. One can hedge against such risks in various ways. Contract clauses can state that if certain things occur, performance will be excused. A contractor can add enough money to the price to cover "contingencies." You can purchase insurance against some of the risks you assume. In fact, almost the only hazard you cannot cover in some way is liability for your performance if the public is injured by it. For example, injuries resulting from the collapse of a public building caused by the contractor's negligence could allow recovery against the contractor by either the injured person or the insurance company. Presumably, the contractor will have its own liability insurance to cover such claims.

There are at least four categories under impossibility of performance, when nonperformance is due to

1. Death or illness,
2. Destruction of an essential to the contract,
3. Unexpected hardship, and
4. Commercial frustration.

Death or Illness.   Ordinarily, the death or illness of a party to a contract does not discharge that contract. If a person dies or becomes incapacitated by illness, that person's estate or those appointed to act for that person must take over and complete the obligations. Only where a contract is such that personal services are involved will death or incapacity due to illness serve as a lawful excuse for nonperformance. For instance, death or illness of a free-lance consulting engineer would discharge the engineer's remaining obligations to her clients.

Destruction of an Essential to the Contract.   **Destruction of an essential to a contract** means that something is destroyed without which the contract cannot be performed. Black hires the White Construction Company to build an addition to Black's plant. Before the work is begun, the plant is destroyed by fire without fault of either party. White Construction Company's obligation is terminated. If the White Construction Company had begun work and were, say, half finished,

White's obligation to complete the structure would be ended, but White could collect for the work the company had completed, in addition to any materials that had been accepted by Black. If, as a third possibility, the contract had been for the building of a structure by itself (not an addition to an existing structure), and if the work again were half finished when the building was destroyed, White Construction Company's obligation would not be ended and the contractor would have to rebuild. One cannot make an addition to a structure that no longer exists, but you can build a separate structure even though the first attempt to do so was destroyed.

Unexpected Hardship.   As previously indicated, one who contracts to perform in some way runs the risk that conditions may not remain as they are when the contract is entered into, and that conditions may not be as they seem. It is not an uncommon experience to have a materials price increase, energy costs or a wage increase cut deeply into the profit margin. If the cause of hardship is anything the contractor reasonably could have anticipated, the courts generally will not relieve the contractor of the contractual duties. It is only where the difficulties that have arisen are of a nature such that no one reasonably could have anticipated them that the court may, in some way, either relieve the burden or lighten the load on the contractor. Under such circumstances a subsequent contract with the owner, whereby the contractor is to receive additional money for performance, may be enforceable in court. Or the court may enforce a subsequent contract to give the contractor more time in which to perform. It should be remembered, however, that these rulings are exceptions; if the difficulty was foreseeable the law often gives no relief. A subsequent contract based on a foreseeable difficulty that did arise (for example, a materials price rise) may be void for lack of consideration as to the increase in the contract price.

Commercial Frustration.   The doctrine of commercial frustration is often treated in the same manner as the destruction of an essential to a contract. In the United States courts, the result is the same. **Commercial frustration** commonly results when a contract is made to take advantage of some future event not controlled by either party. The event is then called off and, as a result, the purpose of the contract is said to have been frustrated. Black leases a concession stand from White for a certain week during which an athletic event is to be held. The athletic event is called off (or moved to a different location). The courts would allow Black to avoid the lease contract. Similarly, if a law were passed preventing such an event, Black probably would not be held to the lease.

Prevention of Performance.   It is implied in every contract that each party will allow the other to perform his or her obligations. If one party prevents the other's performance of the contract, that party discharges the other party's obligation. The interfering party also has breached the contract and may be sued for damages. Black sells White some standing timber, giving White a license to use a private road to the timber. Black prevents White from using the road. White's obligation is terminated, and White may sue Black for damages.

Waiver.   A **waiver** consists of voluntarily giving up a right to which one is legally entitled. To waive a right, the party waiving it must first know that he or she is entitled to it. That person must also intend to give up the right. If Black purchases a machine from White according to a detailed specification, Black can expect to receive the described machine. If the actual machine received varies significantly from the description, Black has a right to refuse to accept it. If, with knowledge of the differences involved, Black agrees to accept it for a slightly lower price, then keeps the machine and uses it, Black probably has waived the right to return the machine to White and get one according to the specifications.

Modifications by Agreement

If two parties have a right to make a contract by an agreement between them, it is only reasonable that they could also agree to disagree. If no rights of a third person are involved, the parties may discharge their contract by mutual agreement without performance in several ways discussed below.

Renunciation.   The parties may agree merely not to be bound by the terms of the original agreement, or they may make a new contract agreement involving the same subject matter, thus discharging the old contract. The original agreement itself may specify some event, the occurrence of which will end the contractual relation (that is, a condition subsequent). Both of the parties may ignore their rights under the contract, each going about their respective business in such a manner that a waiver of performance may be implied from their actions. When a contract is discharged by such methods, the release of one party constitutes the consideration for the release of the other.

Accord and Satisfaction.   Accord and satisfaction occur when a party agrees to accept a substitute performance for the one to which that party was entitled. Ordinarily this occurs when there has been a breach of performance by one party, giving the other party a right to sue for damages. In common terminology, this is the "settlement out of court" one frequently hears about in connection with both contract and tort cases.

To be effective as a discharge, both accord *and* satisfaction must have occurred. (Recall *Lindsay v. McEnearney Associates* in Chapter 11.) **Accord** refers to a separate agreement substituted for the original one. **Satisfaction** occurs when the conditions of the accord have been met, that is, the performance of the "accord." As a practical matter, most disputes never reach court. Most cases that reach court are settled by one side paying some amount to the other.

Novation.   A **novation** replaces one of the parties to a contract. For a novation to be legally effective, all the parties to a contract must agree to it. As a simple illustration, assume that Black owes White $100. White owes Gray $100. If the three parties agree that Gray will collect the $100 debt from Black, a novation has occurred that completely relieves White of White's obligation to Gray. Black no longer owes White $100, but Black has a legally enforceable obligation to pay Gray. Common examples of novation occur when a person buys a house or a car from another, substituting himself or herself as a mortgagor and agreeing to make the loan payments to the lending institution (with consent by the lending institution to the substitution).

Arbitration.   A court action for damages is sometimes impractical because of the time required for it, the cost involved, or some other reason. In many states and in federal court actions it is possible to substitute a procedure known as arbitration for a court action. **Arbitration** is a procedure in which a dispute is submitted to an impartial umpire or board of umpires whose decision on the matter is final and binding. The legality of the procedure depends upon the statutes of the state in which the suit arises or the Federal Arbitration Act. In common law, arbitration has no standing; even if a decision were rendered by an arbitrator, the case might still be taken to court and the arbitration would have no effect. Many of the states and the federal government have seen in arbitration a means of relieving crowded court dockets, and have passed laws setting up the procedure and giving an arbitration decision almost the same force as a court judgment. In these jurisdictions almost any controversy in which damages are requested can be submitted to arbitration—not just contract cases, but tort cases and even property settlements following divorces.

Arbitration has several inherent advantages when it is compared to court proceedings. Probably the main advantage is found in specialization—the disputing parties decide among themselves what person is to act as judge and jury. This allows them to select someone who has a specialized knowledge of the field involved—someone who would not have to be educated on the general technical principles before deciding the case. Often this results in a more equitable decision than a judge and jury might render in court.

A second advantage of arbitration is the speed of the procedure. Court dockets are quite crowded; it is not unusual for a year to pass before a particular case comes up, and delays as long as five years occur in some jurisdictions. In the intervening period, witnesses may die or move away, and memories dim. Arbitration may afford an immediate solution.

Cost-saving is a third advantage that may result from arbitration. In addition to the saving of whatever monetary value may be attached to waiting time, the cost of the procedure itself is often less than court costs.

> Advantages of arbitration:
> 1. Subject matter specialization
> 2. Time savings
> 3. Cost savings

Arbitration is not bound by the evidence rules encountered in a court of law. Whether this is an advantage or a disadvantage is questionable and would depend largely on the case. However, if the arbitrator considers hearsay testimony, for instance, as desirable in determining an issue, such testimony can be taken.

In jurisdictions where arbitration is used, the proceeding is conducted as an extrajudicial action of a court. Questions of law may be submitted to the court for determination, and the final arbitration award is enforced by the court.

There are three main legal requirements for arbitration.

1. The parties must agree to arbitrate—when the agreement to arbitrate took place is of little matter as long as the parties did agree at some time prior to the arbitration.
2. A formal document known as a submission must be prepared by the parties and given to the court. The **submission** is roughly a combination of the complaint and reply required in a court case; it presents the issue to be decided.
3. The arbitrator(s) must be impartial and disinterested parties. If these requirements are met, the arbitrator's award will bind the parties.

The popularity of arbitration as a means of settling disputes has increased considerably in recent years. Probably the main reason is the efficiency of the procedure. It is conceivable that laws requiring arbitration in certain types of civil cases may be passed to further relieve the courts of burdensome cases in civil disputes.

Tender of Performance

Tender of performance, if refused, may discharge the obligations of the party tendering performance. Three conditions, however, generally must be present in a lawful tender of performance:

1. The party offering to perform must be ready, willing, and able to perform the obligation called for.
2. The offer to perform must be made in a reasonable manner at the proper time and place according to the contract.
3. The tender must be unconditional.

Not only is the obligation discharged if such a tender is refused, but the party who wrongfully refused the tender has breached the contract and may be sued by the tendering party. There is one notable exception to the general rule by which a tender of performance discharges an obligation. If the tender is an offer to pay a debt that is due and payable in money, the debt is not discharged by refusal of the payment. However, there are at least two rather important effects:

1. The accrual of interest is stopped.
2. Any liens used to secure the debt are discharged.

If the debt is payable in money, an offer to pay with anything other than legal tender may be refused by the creditor; the creditor is under no duty to accept a check, for instance. If the offer to pay is made before maturity of the debt, the creditor also need not accept. In such cases, if the creditor rejects the offer to pay for the reasons indicated, there has been no tender of performance.

Anticipatory Breach

A breach of contract usually results from a failure to perform one or more contractual obligations as agreed and at the proper time. However, a contract may be breached before the time of performance has arrived. If the party who is to perform notifies the other party that he cannot or will not perform when the time comes to do so, that party is said to have repudiated the contract. Because the repudiation occurred before the time for performance, that party has committed an **anticipatory repudiation** (or **anticipatory breach**) of the contract. Such an anticipatory repudiation gives the other party several choices:

■ The nonrepudiating party may accept the repudiation and immediately sue the other party for whatever damages may have been caused.
■ The nonrepudiating party may treat the repudiation as inoperative, await performance, and (if there is no performance) hold the breaching party responsible for the resulting damages.
■ The nonrepudiating party may accept the repudiation and obtain performance from someone else if it is possible to do so.

To constitute an anticipatory repudiation, a party must demonstrate a distinct and unequivocal refusal to perform as promised. Such a refusal may be shown by words or by conduct. Often, there is no express statement by a party to the effect of "I will not perform as promised." Hence, the courts often consider the circumstances to see whether the statements or conduct really amounted to a repudiation. If the nonrepudiating party continues to urge the other to perform, for example, the court may hold that urging performance demonstrates that the other's conduct was not a clear and absolute refusal to perform and thus was not an anticipatory repudiation.

Suppose Black Construction Company agrees, as general contractor, to build a structure for White. Gray is hired as subcontractor to do the electrical work. A month before the electrical work is to be undertaken, Gray informs Black that Gray cannot do it because of other commitments. At this point, if there were no rule as to anticipatory breach, Black would be in quite a dilemma. If Black obtained the electrical work from someone else and Gray subsequently had a change of heart, Gray could demand to be allowed to perform and sue Black if denied the opportunity to do so. If Black hired a second subcontractor and then Gray returned and performed the work, the second subcontractor could sue. If Black waited until Gray's time for performance was past and the contract was breached, thus causing Black to be late with the contract with White, then White could sue Black.

The rules relating to anticipatory repudiation give Black a way out. Black can hire another electrical contractor to do the work without fear that Gray will be able to take successful action against Black. Before Black actually hires another subcontractor, however, Gray has the right to withdraw the repudiation, thus resuming Gray's obligations under the contract. Generally, the nonrepudiating party should be careful to be sure that the other party has clearly expressed a refusal to perform before taking action.

Anticipatory repudiation does not apply to the payment of a debt. Although a debtor may notify a creditor that the debt will not be paid when due, the creditor must wait until the duty to pay has actually been breached before taking action. The creditor, of course, is not likely to be placed in a dilemma similar to that in which Black, in the foregoing example, would be in the absence of the rules relating to anticipatory repudiation.

BY OPERATION OF LAW

Certain laws have been passed and rules developed to provide for contract discharge as a matter of law. Chapter 10, Lawful Subject Matter, discussed the result of a change in legislation and how it relates to discharging contracts. Here we will consider alteration of the contract, the statute of limitations, bankruptcy, and creditor's compositions.

Alteration of the Contract

If a contract is intentionally altered by one of the parties, without the consent of the other, the obligations of the other party under the contract may be discharged. A party cannot be held to the other's later changes in the contract terms if that party never consented to the changes. Black uses White as surety to secure a $500 loan from Gray, dated July 1 and due September 1. During August Black concludes that the debt cannot be paid when due. Black asks Gray for a loan extension to October 1, to which Gray agrees. On October 1, Gray, discovering that Black has moved without leaving a forwarding address, turns to White as surety.

White's surety agreement would not be enforceable under these circumstances unless White had consented to the extension. Generally, when one party makes a material alteration of a written contract without the other's consent, the contract is rendered unenforceable. However, if the nonconsenting party later finds out about the alteration and then continues to treat the contract as in force (such as by continuing to perform under the contract), then the alteration may be said to have been ratified and accepted. If the alteration is ratified, the contract will be enforceable even by the party who made the alteration.

Statute of Limitations

Each of the states has adopted a statute (or statutes) that limits the time during which a lawsuit can be brought for a breach of contract. Many of the states specify a certain length of time for oral contracts, a longer time for written contracts, and a still longer time for contracts under seal. The state of Florida, for instance, specifies four years for oral contracts and five years for written contracts. Texas, however, has a general rule of four years for all contracts, written or oral. A party who has a right of action on a contract must take such action within the time limits stated in the statute of limitations; otherwise, the party will lose the right to take such action. The time is figured from the date the cause of action accrued, which is usually when the contract was breached. However, there is the possibility of renewal whenever the debt is acknowledged, such as by part payment. Under most such statutes, time does not continue to run while the person who has breached is outside the state.

Suppose Black orally hires White to add a roof to a structure. The contract is made March 1, 2005. White finishes performance on March 31, 2005, but is never paid. If the statute of limitations states five years for this type of contract, White has until March 31, 2010, to commence a court action for recovery. In some states, if Black made a part payment or in some other way acknowledged the debt on, say, April 15, 2007, the time for White to take action would not expire until April 15, 2012. Similarly, if Black left the state for a year, White would have until March 31, 2011, to begin the suit. If a suit is not begun by the dates mentioned, however, White loses the right to take action for a recovery from Black.

Although the legal duty to perform a contract may be discharged by the statute of limitations, it may be reinstated by the debtor. An act or promise by the debtor by which the debtor resumes the obligation will revive the debt and give the contract new life under the statute. If Black made a partial payment for the roof in 2013, the statute of limitations would start to run again. Much the same is true of bankruptcy, discussed below. Reacknowledgment by the debtor of a debt discharged in bankruptcy serves to reinstate its legal life despite the discharge.

Statutes of Repose

As you have learned, most contract-based causes of action can only be brought by parties to a contract, otherwise known as *privity*. Until 1957, professional engineers were protected from liability if an injured party was not in privity with the engineer. *Inman v. Binghampton Housing Authority*, 3 NY 2d 137, 143 N.E. 2d 895, changed all that. In that case, the court found that despite lack of privity, design professionals, could be found liable for defects.

As discussed above, a statute of limitation dictates until when a cause of action may be filed. In many states, causes of action may be filed upon discovery of a latent defect in design or construction under the "discovery rule." This rule exposes design professionals and construction firms to liability well beyond the time anticipated when contracted to do the work which meant that the statute wouldn't start to accrue until the defect was discovered, regardless of how many years had passed since building completion. The intent of the discovery rule is to prevent companies from avoiding liability when defects in design or construction are found regardless of when the building was constructed or whether these companies had entered into a contractual relationship with an injured party. Statutes of repose were the response to this seemingly perpetual liability scheme. Statutes of repose come into play and start to accrue from the date of a building's completion or date of occupancy, thereby limiting the reach of discovery rules and statutes of limitation. If a party suffers losses due to defects in design or construction before the relevant statute of repose has run, that party may seek recovery from the design or construction firm. However, once the statute of repose has run, no one can bring a suit against the design or construction firm if defects are found. Most states have statutes of repose.[1]

> Bankruptcy filings are down. In 2005, there were 1,637,254 bankruptcy filings. In 2006, there were 1,484,570, a change of 9.3%.

---

[1]See *http://www.acec.org/advocacy/statute_of_repose_limitations.pdf* and *http://www.aia.org/SiteObjects/files/statuteofreposecompendium.pdf* for additional information.

Bankruptcy

When a person owes more than that person can pay, should the law help or leave the person to fend for himself? Should one creditor be allowed to receive payment for that creditor's entire debt at the expense of other creditors? These and similar questions have been debated by legislatures since the problem of bankruptcy was first recognized. The stigma that has historically been attached to bankruptcy has been lifted significantly, as reflected in the high number of personal bankruptcy filings. According to the Administrative Office of the U.S. Courts, these filings make a large demand on court resources while at the same time court budgets are not increasing significantly in response.

Under the United States Constitution "Congress shall have Power . . . To establish . . . uniform Laws on the subject of Bankruptcies throughout the United States".[2] Debt was the greatest single cause of imprisonment at the time of the American Revolution. Inability to pay a debt was a prison offense. In fact, forgiving a debtor's obligations and allowing the debtor to begin again with a clean slate is a somewhat recent innovation in the law. Bankruptcy proceedings for the purpose of paying off creditors are not new, but it is only recently that the debtor could receive a discharge of the debtor's obligations in such an action.

Three federal bankruptcy laws were passed and repealed after very short lives before the Bankruptcy Act was passed in 1898. The current law, commonly called the Bankruptcy Code, was enacted in 1978 and has been amended as needed over the years.[3] For this discussion, the Bankruptcy Code can be found under 11 U.S.C. United States Code, Title 11, Bankruptcy. According to the Administrative Office of the U.S. Courts, the present Code has two main purposes:

1. To give an honest debtor a "fresh start" in life by relieving the debtor of most debts, and
2. To repay creditors in an orderly manner to the extent that the debtor has property available for payment.

The legal machinery enacted to carry out these functions is quite lengthy and complicated. The purpose of this text will be adequately served if we consider only an abbreviated version of the process and some of the key concepts (see Table 13.1).

The U.S. federal courts have exclusive jurisdiction over bankruptcy cases. All 94 federal judicial circuits handle such cases. Appeals from the federal bankruptcy court may be taken to the district court, then to a U.S. circuit court of appeals and, finally, to the U.S. Supreme Court.

To Whom the Law Applies.   The Bankruptcy Code applies to all entities (individuals, businesses, municipalities, railroads) with one major exception. Financial institutions such as banks, savings and loans, credit unions, and the like generally have separate statutes to govern their insolvency. A special bankruptcy chapter, number 13, is designed primarily to give relief in cases of consumer debt. Two chapters, 7 and 11, are designed primarily to give relief in the form of liquidation and reorganization, respectively. Chapter 9 of the act handles municipalities with debt problems.

The law distinguishes between voluntary and involuntary bankruptcy. Almost anyone who has the capacity to make a contract may become bankrupt voluntarily. The prospective voluntary bankrupt is not even required to be insolvent, although as a practical matter bankruptcy would seem pointless otherwise.

---

[2]United States Constitution, Article I, Section 8.

[3]Enacted By Pub. L. 95-598, Title I, Sec. 101, Nov. 6, 1978, 92 Stat. 2549.

[4]*http://www.uscourts.gov/bnkrpctystats/bankrupt_ftable_dec2006.xls* last viewed 8/15/07.

TABLE 13.1 Bankruptcy Chart

	Chapter 7	Chapter 11	Chapter 13
Who can file?	Involuntary/Voluntary	Involuntary/Voluntary	Voluntary*
Purpose?	Liquidation to pay debt	Reorganization	Consumer debt
Benefits?	Voids existing debts and prevents beginning or continuation of actions based on discharged debts.	Debtor is allowed to continue business	Assets are not sold and debtor is allowed to develop own payment plan.
How?	1. Debtor's/Creditor's petition 2. Order for relief 3. Trustee appointed 4. Collection and selling of debtor's assets	1. Meeting of interested parties 2. Creation of a plan setting forth a means of continuing business 3. Confirmation	1. Creditor's petition 2. Order for relief 3. Trustee appointed 4. Debtor has to devise method of payment 5. Affirmation by interested parties and court
When?	As soon as practicable	Once plan is confirmed, the debtor is discharged.	3 years. Debtor is discharged after all payments have been made.

The Estate. Bankruptcy contemplates rehabilitation of the debtor. Depriving the debtor of all property so that the debtor winds up requiring aid from a government welfare program isn't rehabilitation. Therefore, certain real and personal property cannot be taken. Each state has "homestead laws" that specify certain minimums, but the Bankruptcy Code also lists some exemptions. Various forms of income and benefit exceptions are also found in nonbankruptcy parts of the U.S. Code. For example, Social Security and certain other retirement benefits are exempt from bankruptcy, as are veteran's benefits.

The property of the debtor's estate, for bankruptcy purposes, consists of all the debtor's assets that are not exempt. In addition to the obvious types of property mentioned above, this includes patent rights, copyrights, trademarks, liens held by the debtor on property owned by a third party, and community property. Generally, the estate rights are limited to those the debtor had at the beginning of the case.

Voluntary Bankruptcy. As stated earlier, almost anyone can become bankrupt voluntarily. Each chapter under which a debtor files has different procedures. The Supreme Court has recently adopted updated bankruptcy procedures that each federal judicial circuit uses as guidelines. Features of Chapters 7, 11, and 13 are discussed briefly below.

Chapter 7 Bankruptcy. The objective of a Chapter 7 bankruptcy is to liquidate the debtor's available assets to pay off the creditors. Debtors may be individuals or businesses. When a business files under Chapter 7, it is usually because it cannot be run profitably and it cannot reorganize. When a debtor files for Chapter 7 bankruptcy, a trustee is appointed. The trustee is tasked with liquidating the debtor's assets in an orderly fashion. The trustee collects the debtors assets and sells those that are non-exempt. The trustee then distributes the proceeds to creditors. Once the process is begun, the debtor cannot dismiss the Chapter 7 case.

Chapter 11 Bankruptcy. Generally, under Chapter 11, the debtor (usually a business) is allowed to continue whatever business there is until there is a court order to the contrary. Chapter 11 bankruptcies are usually filed by a business when it is having trouble with cash flow or the business is not going well. The debtor may continue to use, acquire, or dispose of property in the same manner as if no petition had been filed. Of course, counter-measures are available in case of abuse of the creditors' rights by the debtor. For example, the court may appoint an interim trustee to preserve the estate until the creditors' meeting determines what is to be done with it. The debtor in a Chapter 11 bankruptcy is responsible for developing a plan for how the debtor will pay debts.

Chapter 13 Bankruptcy.   The title of Chapter 13 is "Adjustment of Debts of an Individual with Regular Income." The main problem treated by Chapter 13 is consumer indebtedness. It is available to a debtor on a voluntary bankruptcy basis.

A trustee is appointed by the bankruptcy court to perform most of the same duties as those in Chapter 7, except for the collection and selling of the debtor's assets. It is up to the debtor to devise a means of paying off the debts. This, of course, may be done by assigning a part of the debtor's wages to this purpose or by assigning a portion of earnings from self-employment or a business. Whatever the source of income, courts usually allow the debtor broad freedom to choose an acceptable plan and operate within it, as long as it appears that the past due debts to creditors will be paid off.

The period of operation of the plan is nominally limited to three years, but it may be extended to five years with the court's permission. Creditors cannot force the debtor into a plan unacceptable to the debtor, for the same reasons they can't force a debtor into a Chapter 13 bankruptcy in the first place. (They could, of course, force the debtor into a Chapter 7 bankruptcy.) Only the debtor knows the debtor's family requirements, family plans, and other such personal information. Secured creditors in a Chapter 13 bankruptcy may be paid off in many ways, one of which is for the debtor to return goods on which a lien was placed.

Confirmation of the debtor's plan requires only the affirmation of the parties with an economic interest in the action, as well as the court's blessing as to the plan's legality. One of the court's confirmation considerations is that the plan's contemplated payoff to the unsecured creditors is at least as much as they would have realized under a Chapter 7 liquidation.

Involuntary Bankruptcy.   Involuntary bankruptcy is confined to Chapters 7 and 11 of the Act, and there are restrictions about who may be driven into bankruptcy. Specific exceptions are made for farmers, charitable corporations, insurance companies, and financial institutions indicated earlier. Such debtors are protected from involuntary bankruptcy. Involuntary bankruptcy is usually begun by three or more creditors, holding a total of $5,000 or more of unsecured obligations, jointly filing a petition in a bankruptcy court. If there are fewer than 12 creditors, only one creditor with a secured claim of $5,000 is needed to start the bankruptcy proceeding. If the debtor is a partnership, one or more of the partners may file.

Discharge.   Under Chapter 7 and Chapter 13, discharge of the debtor's obligations is limited to individuals. The reason for the Chapter 7 limitation is that even though a corporation or partnership may go through the Chapter 7 liquidation, what is left after bankruptcy is nothing but a hollow shell. The effect of the debtor's discharge in bankruptcy is to void existing debts and enjoin the beginning or continuation of any action based on discharged debts.

Certain debts, however, are not discharged by bankruptcy, and some acts by the debtor may prevent the discharge of obligations. Generally, the following debts are not dischargeable in bankruptcy:

- Tax obligations
- Debts and continuing obligations for alimony, maintenance, and child support
- Liability for fraud or willful or malicious injury

- Certain debts for fines, penalties, or forfeitures for the benefit of a governmental unit
- Debts resulting from fraud, embezzlement, or larceny while acting in a fiduciary capacity
- Debts for educational loans less than five years past due
- Debts remaining after a previous bankruptcy proceeding
- Debts not listed or listed too late to be included with the bankruptcy

Discharge of the debtor's remaining obligations is generally forthcoming if the debtor has dealt honestly and fairly with the creditors during the bankruptcy action. Most of the reasons for denying discharge are based on actual or reasonably suspected deceit by the debtor. Under Chapter 7, discharge will be denied for the following reasons:

- The debtor is not an individual.
- The debtor transferred, destroyed, or concealed some of the debtor's property within a year prior to the filing of the bankruptcy petition, intending to thereby defraud the creditors.
- The debtor failed to keep or preserve accounts of transactions.
- The debtor made fraudulent statements in the bankruptcy examination regarding the estate or the debtor's financial affairs.
- The debtor failed to explain satisfactorily any losses of assets.
- The debtor refused to obey a lawful court order during the proceedings.
- The debtor was granted a discharge under Chapter 7 or Chapter 11 (or their predecessors) in a case commenced within six years prior to the date the petition was filed.
- The debtor was granted a discharge under Chapter 13 within the past six years, unless the debtor had paid at least 70 percent of the amounts of the unsecured claims in the prior case.

Generally, confirmation of a Chapter 11 plan discharges the debtor. The main exception is when the Chapter 11 action essentially amounts to the Chapter 7 liquidation. In such an instance, the Chapter 7 limitations just noted would apply. Under a Chapter 13 bankruptcy, the court is required to discharge the debtor after all payments under the plan have been made. However, there is a noticeable reluctance of courts to confirm plans that do not contemplate the payment of at least 70 percent or so of the amounts of claims by unsecured creditors.

Once the debtor has been through the bankruptcy mill and has received a discharge, the debtor's contractual obligations are ended. New property acquired, and new transactions and business dealings, are free from interference by former creditors. The debtor is given new economic life.

Creditors' Compositions

One alternative to filing under Chapters 7 and 11 of the Bankruptcy Code is a Creditors' Composition. It is often to the advantage of both the creditors and the debtor to avoid bankruptcy proceedings. There are many costs of bankruptcy and each cost reduces the assets to be divided. It is costly, for instance, to pay the trustee to maintain and then dispose of the property involved. There is a much less expensive procedure available. The creditors' composition accomplishes almost the same thing as bankruptcy. It discharges the debtor's obligations. Each participating creditor gets some return on the creditor's account receivable. The procedure is informal but binding. It is not necessary for all the creditors to join in the composition; two or more are sufficient. If only one creditor is involved, it is not a creditor's composition and the remainder of the debtor's obligation is not discharged, as pointed out in chapter 9, "Consideration."

For example, Black owes White $1,000, Gray $2,000, and Brown $3,000. Black has $3,000 cash available plus various other assets but cannot pay all the debts and remain solvent. Black meets with White, Gray, and Brown, telling them of the situation. The creditors are faced with the possibility of bankruptcy proceedings where, after the costs are paid and Black's assets sold for whatever price they may bring, the creditors may get $0.25 for each dollar of debt. As an alternative, the creditors may choose to divide Black's cash assets in any way they see fit, $0.50 for each dollar being one such possibility. The creditors may agree, instead of taking a straight percentage, to divide the assets in some other way that is satisfactory to each. They might agree, for instance, that White will receive $600, Gray $1,000, and Brown $1,400. There are several explanations of the consideration involved in a creditor's composition. Perhaps the most common one holds that the consideration received by each creditor for giving up the right to sue for the remainder of the debt is found in the forbearance of the same right by the other creditors. Several states have statutes that specify the conditions and procedures for creditors' compositions. Where such statutes exist, examination of the consideration involved is unnecessary.

EATHERLY CONSTRUCTION COMPANY v. HTI MEMORIAL HOSPITAL d/b/a MEMORIAL HOSPITAL, 2005 Tenn. App. LEXIS 575

Study terms: Liquidated damages, penalty, retainage, reasonable estimation, public policy, contract documents, instructions to bidders, reformation of contract, equitable relief, mutual mistake

Opinion

This is a breach of contract action arising from the construction of a water line and pumping station for a new hospital. Eatherly Construction Company filed suit to recover $35,250 for installation of 705 linear feet of twelve-inch ductile iron pipe that was omitted from its bid and $10,000 of retainage funds withheld by the owner, HTI Memorial Hospital Corporation. The hospital denied Eatherly's claims and filed a counterclaim to recover *inter alia* liquidated damages of $500 per day for each day the construction was delayed beyond the agreed completion date and attorney fees. The trial court summarily dismissed Eatherly's bidding error claim of $35,250 and the hospital's claim for attorney fees. Following a bench trial, Eatherly was awarded the $10,000 retainage while the hospital's claim for liquidated damages was dismissed. Both parties appealed. We affirm.

Opinion

HTI Memorial Hospital Corporation (Memorial) planned to construct a new acute care hospital, Skyline Medical Center. Part of the massive construction project required the construction of a new pumping station and the installation of 705 linear feet of a twelve-inch water line under an existing interstate highway.

Memorial circulated a Request for Bids for construction of the water line to a number of contractors. Eatherly was one of several contractors to bid on the water line. It submitted a bid of $432,035 for the water line on November 11, 1999. Its bid however contained a $35,250 error. The specifications called for the installation of 705 linear feet of 12 inch ductile iron pipe. Eatherly priced that part of the work at $50 per unit price (per linear foot). Eatherly correctly entered $50 in the "Unit Price" column of its bid; however, it failed to write the extended price of $35,250 in the "Extension" column on the bid form. The bid form at issue, entitled "Water Main Take-Off/Unit Price Schedule," comprised two pages and some sixty pre-printed lines of items and quantities stated thereon. Of course, the columns for the UOM unit prices and extension prices appeared as blank lines. When the bid was completed by Eatherly and submitted to Memorial, the bid form was in the following format:

Water Main Take-Off/Unit Price Schedule

Item	Qty.	UOM Unit Price	Extension
12/12/12 Tee	1 EA	0	0
120 Gate Valve	1 EA	1,800.00	1,800.00
16/12 Reducer	1 EA	0	0
16" D.I.P. Pipe	1140 LF	64.00	72,960.00
12" D.I.P. Pipe	705 LF	50.00	1
Site Restoration			
Asphalt Paving	LS	24,500.00	24,500.00
Conc. Paving	LS	2,500.00	2,500.00
Landscaping, Seed & Sod	LS	500.00	500.00
	Total Bid		$432,035.00

The mathematical product that should have been entered in the extension column for 705 linear feet of ductile pipe was $35,250. As a result of the error in the extension column, the total bid of $432,035 for the water line work was $35,250 less than Eatherly intended.

After submitting the bid for construction of the water line, Memorial requested Eatherly to submit an additional bid, for construction of a pumping station. As requested, Eatherly submitted a bid, via letter dated November 30, 1999, for the pumping station in the amount of $126,425. Unfortunately, Eatherly did not notice nor correct its prior $35,250 error for the water line when it submitted the bid for the pumping station. Thus, Eatherly's original bid of $432,035 for the water line was not corrected. The aggregate total of Eatherly's bid for the pumping station and the water line, with the $35,250 error embedded therein, was $566,836.

Eatherly's bid for the pumping station and the water line (hereinafter the "Work") was the lowest bid submitted for the Work. The next lowest bid was $593,710. Memorial awarded the contract for the Work to Eatherly because it was the "lowest responsive, responsible bidder."

The parties entered into a contract for the Work (the Agreement) on February 7, 2000. The Agreement is comprised of numerous documents which are referred to as the "Contract Documents." . . . Contract documents relevant to the issues presented include the Invitation for Bid, Instruction to Bidders, General Conditions, Specifications, Eatherly's bid, Notice of Award and Notice to Proceed. The Work to be performed by Eatherly was identified in Article 1 of the Agreement as the "Old Due West/New Due West Water Main/Pump Station."

Memorial issued the Notice to Proceed on February 11, 2000. Eatherly commenced work on the water line and pumping station shortly thereafter. Eatherly did not submit a request for payment until after it had substantially completed the installation of the 705 linear feet of water line under Interstate 65. The $35,250 bidding error was discovered by Eatherly while it was preparing its first request for payment. Eatherly informed Memorial of the $35,250 error when it submitted its first request for payment, and requested that its compensation be increased accordingly. Memorial declined the invitation to pay more than the stipulated sum required of it in the Agreement. Thereafter, a number of differences arose resulting in the commencement of this litigation on April 25, 2001 when Eatherly filed suit against Memorial alleging the contract price should be increased because of the error in the bid. Eatherly also sought to recover $10,000 which it contended Memorial wrongfully retained. Memorial filed an answer denying liability. Memorial also filed a counterclaim contending it was entitled to $27,500 as liquidated damages due to Eatherly's failure to complete construction on time, and attorney fees. Eatherly answered the counterclaim denying it was liable to Memorial on any of the claims.

Following discovery, the parties filed competing motions for partial summary judgment. Memorial sought a monetary judgment against Eatherly for liquidated damages and attorney fees and dismissal of Eatherly's claim of $35,250 for the ductile iron pipe. In its competing motion for partial summary judgment, Eatherly sought dismissal of Memorial's claims for liquidated damages and attorney fees and a $35,250 judgment against Memorial for the ductile iron pipe.

In June 2003 the trial court granted in part and denied in part the competing motions for summary judgment. . . .

Both parties appeal. . . .

Standard of Review

The issues were resolved in the trial court upon summary judgment. . . . We consider the evidence in the light most favorable to the non-moving party and resolve all inferences in that party's favor.

■ ■ ■

The interpretation of a contract is a question of law. Issues as to interpretation and application of unambiguous contracts are likewise issues of law, the determination of which enjoys no presumption of correctness on de novo appellate review. Therefore, the trial court's interpretation of a contract is not entitled to a presumption of correctness under Tenn. R. App. P. 13(d) on appeal. . . .

The cardinal rule for interpretation of contracts is to ascertain the intention of the parties and to give effect to that intention consistent with legal principles. . . .

The court, at arriving at the intention of the parties to a contract, does not attempt to ascertain the parties' state of mind at the time the contract was executed, but rather their intentions as actually embodied and expressed in the contract as written. . . .

Liquidated Damages

The trial court dismissed Memorial's claim for liquidated damages pursuant to a Tenn. R. Civ. P. 41.02(2) motion of Eatherly at the close of Memorial's case in chief. The trial court held the liquidated damages provision was a penalty and, therefore, was unenforceable. That ruling was based upon the specific finding that Memorial failed to establish that the liquidated sum, $500 a day, was a reasonable estimation of the damages it would likely suffer in the event of a delay. Ruling from the bench, the trial judge stated:

> I didn't hear anything with regards to how this contract was negotiated and that figure was calculated. The term adhesion contract was inserted in one of the closing arguments. I'm not willing to step up and say that this was an adhesion contract, but I will state that there is a lack of evidence as to how this provision was negotiated and what was the foresight that these individuals had when they entered into it as to what might be the potential harm.

Mr. Dennis said that it's insignificant at $500 a day as to the damages that they might sustain for a breach, and I would concur that $500 a day would be insignificant if I had some evidence to tie it to. And it might mean that there are other facts that could have been inserted into the proof. It may be that everybody was working so hard that you didn't sustain necessary harm to compensate you.

For all these reasons, the Court finds that the liquidated damages provision is a penalty and is unenforceable under existing case law in this state. Therefore, the motion is granted.

■ ■ ■

The liquidated damages provision at issue pertained to the timely completion of the installation of a pumping station and water lines to service the Skyline Medical Center. The liquidated damages provision is set forth in the agreement under the subtitle "Contract Time." It provides:

ARTICLE 3. CONTRACT TIME

■ ■ ■

3.2 Damages for Delay. The OWNER [Memorial] and Contractor [Eatherly] recognize that the Project has highly important environmental and economic significance and is required for compliance with Federal and State Regulations, that time is of the essence of this Agreement, and that the OWNER will suffer direct financial loss if the project is not completed as described in Article 3.1, and if the Work is not finally completed within 120 days after the date of substantial completion as set above. Accordingly, Contractor agrees to pay OWNER as liquidated damages the amount of $500.00 per day for each day that the project is not substantially completed after the period stated in Article 3.1 and the additional amount of $500.00 per calendar day for each day that the project has not been finally completed after the date as set forth above. . . .

Eatherly was to complete the Work by June 10, 2000. Memorial contends that Eatherly did not complete the Work until August 12, 2000; thus it is entitled to $500 for each of the 63 days for a total liquidated damage of $31,500. Eatherly denies liability on two grounds. It challenges the validity of the liquidated damages provision and contends the Work was completed in a timely manner because it was granted additional time to complete the Work.

Damages are deemed "liquidated" if parties to a contract agree in advance on the amount of damages to be recovered for compensation upon the occurrence of a particular defaulting event. The term "liquidated damages" is defined by case law as:

[A] "sum stipulated and agreed upon by the parties at the time they enter their contract, to be paid to compensate for injuries should a breach occur." *V.L. Nicholson Co. v. Transcon Inv. & Fin. Ltd., Inc., supra*; *Kimbrough & Co. v. Schmitt*, 939 S.W.2d 105, 108 (Tenn. Ct. App. 1996), perm. app. denied (Tenn. 1996). The stipulated amount represents an estimate of potential damages in the event of a contractual breach where damages are likely to be uncertain and not easily proven. *V.L. Nicholson*, 595 S.W.2d at 484.

Guiliano [v. Cleo, Inc., 995 S.W. 2d 88] at 96-97. The purpose of liquidated damages is to provide a means of compensation in the event of a breach where "damages would be indeterminable or otherwise difficult to prove." *Id.* at 98. By stipulating to the damages that might reasonably arise from a breach, the parties estimate the potential damages likely to be sustained by the non-breaching party. . . . If, however, the stipulated amount is unreasonable in relation to the potential or estimated damages, then it will be treated as a penalty. If the provision is found to be a penalty, then it is unenforceable as against public policy.

Determining the propriety of the parties' agreement as to liquidated damages requires an assessment of two competing interests. One is the freedom of parties to bargain for and to agree upon terms. The other is limitations set by public policy. Generally, parties to a contract are free to agree upon other terms that may not seem desirable to outside observers. In that respect, courts should not interfere in the contract, but should carry out the intentions of the parties and the terms bargained for in the contract, unless those terms violate public policy.

When parties agree to a liquidated damages provision, it is generally presumed that they considered the certainty of liquidated damages to be preferable to the risk of proving actual damages in the event of a breach.

■ ■ ■

Tennessee follows what is called the "prospective approach" when addressing the propriety of a liquidated damages clause. Under the prospective approach, courts must focus on the intentions of the parties based upon the language in the contract and the circumstances that existed "at the time of contract formation." Those circumstances include:

> Whether the liquidated sum was a reasonable estimate of potential damages and whether actual damages were indeterminable or difficult to measure at the time the parties entered into the contract." *Id.* at 100-101; *see also V.L. Nicholson,* 595 S.W.2d at 484. If the provision satisfies those factors and reflects the parties' intentions to compensate in the event of a breach, then the provision will be upheld as a reasonable agreement for liquidated damages.

Guiliano, 995 S.W.2d at 100-101.

■ ■ ■

Here, the trial court specifically found there was no proof in the record to support a finding that the liquidated sum agreed upon was a reasonable estimate of the damages to be incurred by Memorial. The best, if not only evidence offered by Memorial to support its contention that $500 per day is reasonable is that the parties "agreed" to the amount.[12] While the fact the parties "agreed" to the amount is relevant, and it is a factor to be considered in order to determine whether the amount was a reasonable estimate at the time the parties entered into the contract, that evidence—the parties' agreement—standing alone does not preponderate against the trial court's specific finding to the contrary.

In the trial court, Memorial had the burden to establish that the liquidated sum was a reasonable estimate of potential damages. See *Guiliano,* 995 S.W.2d at 100-101. The trial court found Memorial failed to meet that burden by making the specific finding that the evidence was inadequate to establish the liquidated sum agreed upon was a reasonable estimate. As a consequence of the trial court's finding, Memorial must establish on appeal the evidence preponderates against the trial court's finding of fact. See Tenn. R. App. P. 13(d). Specifically, Memorial must establish that the evidence supports another finding of fact with greater convincing effect. The evidence in the record does not preponderate against that finding. Accordingly, we affirm the decision to dismiss Memorial's counterclaim for liquidated damages.

■ ■ ■

---

[12]The only other "evidence" offered by Memorial in support of the "reasonableness" of the $500 per day value were the conclusory statements by Mike Dennis and John Massey. The essence of their testimony substantiated only the difficulty ascertaining actual damages in the event of construction delays. Their testimony provided no guidance as to how the figure $500 per day was ascertained or its reasonableness.

The $35,250 Error

Eatherly contends the trial court erred by summarily dismissing its claim of $35,250 for the installation of 705 feet of twelve-inch ductile iron pipe, the price for which Eatherly failed to include in its bid. This claim is based upon two provisions in the contract documents. One, Eatherly contends that Memorial had a contractual duty to tabulate the bid—after it was submitted by Eatherly—to assure its accuracy and that it failed to do so. Two, Eatherly contends the contract documents provide that the bids are controlled by the unit prices stated therein, not the extension prices, and the "sum to be paid" is to be derived by "correctly extending the unit prices and then correctly adding the correctly extended figures." Memorial counters, stating the contract was for a stipulated sum which is not subject to modification absent agreement of the parties. In support of this argument, Memorial primarily relies on paragraph 4.1 of the Agreement which reads: "The OWNER shall pay CONTRACTOR a stipulated sum for performance of the Work in accordance with the Contract Documents on the basis of the prices indicated on the Bid Form." We find the contract documents support Memorial's position and the trial court's decision.

The court's task is to ascertain and give effect to the contracting parties' intentions by fairly construing the terms of their written agreement. . . .

If the contract is unambiguous, we must enforce it according to its terms. . . .

It is undisputed that the 705 linear feet of ductile iron pipe was to be provided and installed by Eatherly. It was itemized in the bid Eatherly submitted and it was expressly required of Eatherly in the contract documents. Eatherly admits the omission of the extension price of $35,250 was due to its oversight; however, it contends that of the contract documents required Memorial, along with the Engineering and Utility District, to tabulate the bid after it was submitted to assure its accuracy and that Memorial failed to do so. This contention is based upon a section the Bid Proposal Form which, Eatherly contends, places a duty upon Memorial to tabulate and correct any errors in Eatherly's bid. In pertinent part, the Bid Proposal Form provides:

> After Bid Proposals are received, tabulated and evaluated by the Owner, the Engineer and the Utility District, and the successful bidder for each package has been determined, said bidder will be contacted to determine any scope duplications or omissions.

The trial court did not accept Eatherly's contention that the above placed an affirmative duty on Memorial to identify and correct bidding errors by Eatherly. Neither do we. The section above contemplates that bid proposals will be tabulated and evaluated after they are received; however, the language therein does not place an affirmative contractual duty on Memorial, as Eatherly contends, to identify and correct Eatherly's errors or be liable for the consequences.

Eatherly also contends the contract documents, specifically the Instructions to Bidders, requires errors, such as that committed, be corrected, no matter when the error is ascertained, based upon the specified unit price, the extension price and the total bid price notwithstanding. This contention is based upon section 14.5 of the Instructions to Bidders. It provides:

> Discrepancies between the indicated sum of any column of figures and the correct sum thereof shall be resolved in favor of the correct sum. Discrepancies in the extension of the Unit Price times the estimated quantity for any line item shall be resolved in favor of the correct extension.

The primary fallacy with Eatherly's argument is the timing of its application. Eatherly contends it can be applied whenever the error is ascertained. Particularly, Eatherly contends it can and should be applied post contract and post performance of the Work. We disagree.

The section upon which Eatherly relies is in the Instructions to Bidders. After reviewing the contract documents, it becomes obvious that this section of the contract documents is for the limited purpose of correcting errors in bids prior to Notice of Award and prior to executing the Agreement. It is obviously a mechanism to correct known errors in bids in order to award the contract for the work to the lowest responsive, responsible bidder. Section 14.5 has no application to ascertainment of errors after the parties enter into the Agreement, as is the case here.

■ ■ ■

Reformation of Contract

As an alternative to the contract theories discussed above, Eatherly contends it is entitled to reform the contract, as an equitable remedy, and an award in the amount of $26,974, which represents the difference between Eatherly's total bid and that of the next lowest bidder. Specifically, Eatherly claims it is entitled to equitable relief because Memorial knew or should have known Eatherly made a mistake in its bid.

The trial court dismissed Eatherly's claim for reformation of the contract finding *inter alia* no mutuality of mistake, that the mistake was the unilateral mistake of Eatherly, and the requested relief could not be granted without prejudice to Memorial. We find no error with the trial court's decision.

A mutual mistake will provide a basis for equitable relief if the proof is clear and convincing. A contract, which by reason of a mutual mistake in its execution does not conform to the real agreement of the parties, may be reformed when the mistake is established by clear and convincing proof. The Chancellor's finding of no mutual mistake comes to this court with a presumption of correctness unless the preponderance of the evidence is otherwise.

Mutual mistake has been defined as "one common to both parties to a contract, each laboring under the same misconception . . . respecting a material fact, the terms of the agreement, or the provisions of the written instrument designed to embody such agreement." 17 C.J.S., CONTRACTS, § 144. A mistake is remediable only if it relates "to a fact which constitutes, or goes to, the very essence, or basis, of the contract. . . ." *Id.* Our cases have recognized that equitable relief on the ground of mutual mistake is available "not only when the mistake is expressly proved, but also when it is implied from the nature of the transaction." *Robinson v. Brooks*, 577 S.W.2d 207, 208 (Tenn. Ct. App. 1978), quoting from 12 C.J.S., CANCELLATION OF INSTRUMENTS, § 27b.(1).

■ ■ ■

Eatherly's burden of proof was to establish by clear and convincing evidence that Memorial knew or should have known of the error by Eatherly. *Commercial Standard Ins. Co.*, 245 S.W.2d at 778. Eatherly failed to satisfy this burden of proof. Accordingly, we affirm the trial court's dismissal of Eatherly's claim for reformation of the contract.

In Conclusion

The judgment of the trial court is affirmed and this matter is remanded with costs of appeal assessed against HTI Memorial Hospital Corporation.

YALE DEVELOPMENT CO. v. AURORA PIZZA HUT, INC.
420 N.E.2d 823 (III. App. Ct. 1981)

Study terms: Anticipatory breach, willingness and ability to perform, liquidated damages, compensatory damages

This case involves an appeal and cross-appeal from a judgment against defendant, Aurora Pizza Hut, Inc., finding it breached a written contract but limiting plaintiff's recovery to $1,000 under the terms of a liquidated damages clause. The sole contention of plaintiff, Yale Development Co., on appeal is that the liquidated damages clause should not be enforced and that it should be able to recover compensatory damages in excess of the amount stipulated. Defendant, in its cross-appeal, argues that its termination of the contract did not constitute an anticipatory breach and, in the alternative, that if such action did amount to a breach, the plaintiff is not entitled to recover because of its subsequent inability to perform under the terms of the contract. The parties executed a contract on September 19,1975, in which defendant agreed to pay $90,000 in exchange for property located at Route 53 and Butterfield Road in unincorporated Du Page County.

The sale was specifically made contingent upon plaintiff being able to obtain a zoning change and liquor license which would allow the construction and operation of a Pizza Hut restaurant. Although time was made of the essence, no specific time was mentioned in the contract.

Plaintiff filed an application for rezoning with the Du Page County Zoning Board of Appeals on December 14, 1975. After a public hearing, the Zoning Board of Appeals recommended denial of plaintiff's petition, and on February 23, 1976, the County Board of Du Page County concurred in the denial. Plaintiff then filed a complaint in the circuit court seeking a declaratory judgment that the existing zoning was void and that the intended use of the premises be permitted. A bench trial was conducted on October 7, 1976, and the case was taken under advisement on January 5, 1977, after trial briefs were submitted. On February 24, 1977, before the court issued a decision, defendant sent plaintiff a letter purporting to terminate the contract "inasmuch as the city (sic) has denied our right to rezone the property." Subsequently, the court denied rezoning of the property, of which the court below took judicial notice in its written reasons for decision. Plaintiff then filed the present action alleging that defendant was in breach for repudiating the contract prior to the time performance was due. The trial court held for the plaintiff but entered an order limiting plaintiff's recovery to $1,000 under the terms of the liquidated damages clause. This appeal and cross-appeal followed, and we reverse on the basis of the defendant's cross-appeal. We find the second issue raised by defendant's cross-appeal to be dispositive of the case at bar but turn first to the question of breach.

Defendant's first contention on cross-appeal is that its good faith termination of the contract did not constitute an anticipatory breach under these facts and circumstances since plaintiff was unable to obtain the rezoning of the subject property within a reasonable period of time. We disagree. Defendant's notice of termination came at a time when the plaintiff had an action for declaratory relief to rezone the property pending before the circuit court. Defendant reasons that the rezoning had not been accomplished in the 17-month period between the time of execution and the notice of termination and that there was no reason to believe the pending action would be successful when proceedings before the Zoning Board of Appeals and the County Board were not. Although the contract did not specify the time for performance, the law will imply a reasonable time. . . . The intention of the parties controls what time is reasonable, and the court must look to the surrounding circumstances to discover the intention of the parties. . . . A reasonable time for performance is such time as is necessary to do conveniently what the contract requires. . . . Under the facts and circumstances of the present case we do not view a 17-month period as being unreasonable when the necessity of rezoning was contemplated by both parties. The record also indicates that the parties had had previous similar dealings which required rezoning actions and that two of such dealings required periods in excess of two years to be resolved. In light of these previous contracts and the period of time typically involved in obtaining a rezoning, we cannot say that a 17-month period was manifestly unreasonable and not within the intention of the parties. The trial court, therefore, correctly viewed the February 24 letter as an anticipatory breach.

Under the law of anticipatory repudiation, when one party to a contract repudiates his obligations before the time performance is due, the other party may elect to treat the repudiation as a breach and sue immediately for damages. . . . This leads us to defendant's second contention and the one we find dispositive of the case at bar. Defendant contends that, assuming the letter of February 24 is regarded as an anticipatory breach, the plaintiff cannot maintain an action on the contract due to his subsequent inability to fulfill the condition precedent of obtaining the rezoning. We agree. Although it was not necessary for plaintiff to actually tender performance since an anticipatory breach excuses the non-repudiating party from further performance, . . . we hold that a plaintiff suing on an anticipatory breach must still show a willingness and ability to perform had not the breach occurred. While it appears that no Illinois court has yet had the opportunity to consider this precise question, Illinois courts have consistently held that where both parties to a contract are in default, there can be no recovery by either against the other. . . . It has also been held that where one party has put it beyond his power to perform, although no demand or tender is necessary to allow the other to recover for breach, the plaintiff must show that he was ready and willing to perform the contract on his part. Similarly, in *Nation Oil Co. v. R.C. Davoust Co., Inc.,* . . . the court held that a party seeking to recover for breach of contract must show that he has performed or offered to perform his own obligation under the contract or that such performance was excused. Although none of these cases involve anticipatory breach, we consider their basic principles to be applicable to the case at bar.

As these cases demonstrate, if plaintiff had elected to wait until the time for defendant's performance to sue, instead of bringing suit immediately on the anticipatory breach, it would have been required to show a tender or offer of performance on its part, or that its performance was excused. Due to plaintiff's inability to procure a rezoning, it could not have tendered its performance and thus could not have maintained an action for breach at the time when defendant's performance was due. It would be anomalous to allow plaintiff to put itself in a better position by suing immediately on an anticipatory breach and thus avoid the necessity of proving its tender of performance. We therefore hold that where a plaintiff sues on an anticipatory breach prior to the time when defendant's performance is due, he must show an ability and willingness to perform his part of the contract within the time specified in the contract.

Indeed, this seems to be the law in those jurisdictions which have considered the question. For example, in *Ufitec, S.A. v. Trade Bank and Trust Co.,* . . . the question before the court was whether a holder of a draft drawn against a letter of credit issued by the defendant bank could sue the bank for an anticipatory repudiation due to the bank's act of revoking the letter of credit before it had expired. The court concluded that since the holder never complied with the letter of credit, nor had it shown that it could ever comply during the term of the letter of credit, it could not maintain an action for anticipatory breach. The court stated, "(a)n anticipatory breach, in a proper case, may excuse one from performing a useless act, but it does not excuse one from the obligation of proving readiness, willingness, and the ability to have performed the conditions precedent."

The problem is also discussed in the third edition of Williston's Treatise on Contracts:

The difficulty is not peculiar to cases of supervening illegality, but is involved in every other case where, after an anticipatory breach, supervening impossibility occurs which would in any event prevent and excuse performance of the contract. The situation differs, also, in only a slight particular, hereafter referred to, from one where it appears after the anticipatory breach that, although there was no legal excuse, the return performance could not or would not have been rendered to the repudiator.

In each of these cases, if all the facts could have been known or foreseen at the time of the repudiation, no cause of action on the contract would have arisen.

It is a practical disadvantage of the doctrine of anticipatory breach that an action may be brought and perhaps judgment obtained before the facts occur which prove that there should have been no recovery. Fortunately, in most cases, this evidence, though not available at the time of the repudiation, becomes available before judgment can be obtained.

It seems clear that if the evidence thus becomes available the plaintiff can recover no substantial damages, and in the case of supervening illegality which is not due to the defendant's fault, there seems no reason to allow even nominal damages.

The loss should rest where chance has placed it; and the same should be true in case of any supervening excusable impossibility. Still more clearly, there should be this result if the facts show that the plaintiff could not or would not have performed for reasons which would not be a legal excuse. The fact that a right of action has already arisen should not preclude the defense. Failure of consideration after a cause of action has arisen bars recovery. . . .

Similarly, comment a of section 277 of the Restatement of Contracts makes the following rule applicable to anticipatory breach:

Where in promises for an agreed exchange a promisor commits a breach, and a right of action for the breach arises, the right of action is extinguished if it appears after the breach that there would have been a total failure to perform the return promise, even if the promisor had not justified the failure by his own breach or otherwise. . . .

Since the plaintiff has failed to obtain rezoning of the subject parcel it cannot demonstrate an ability to perform and, hence, is not entitled to recovery under the contract. The judgment of the trial court is accordingly reversed.

Reversed.

In Re: Plaintiff: J.A. WALKER CO., INC. v. Defendants: CAMBRIA CORPORATION, et al., 159 P.3d 126; 2007 Colo. LEXIS 447

Study terms: Arbitration agreements, mechanics' liens, fraudulent inducement, incorporation by reference

This opinion is a companion to our decision announced today in *Ingold v. AIMCO/Bluffs, L.L.C. Apartments*, No. 06SA240, 159 P.3d 116, 2007 Colo. LEXIS 445 (Colo. May 29, 2007). In *Ingold*, we held that the former version of the Colorado Uniform Arbitration Act distinguishes between two types of allegations of fraudulent inducement. Allegations of fraudulent inducement specifically directed to an arbitration agreement, including an arbitration provision in a contract, must be resolved by the trial court. Fraudulent inducement allegations directed more broadly to a contract as a whole, of which an arbitration agreement is only a part, must be resolved in arbitration.

In this case, we hold that the current version of the Colorado Uniform Arbitration Act, sections 13-22-201 to -230, C.R.S. (2006) (the "CUAA"), recognizes the same distinction between fraudulent inducement allegations. Thus the trial court—not an arbitrator—must resolve allegations that a party was fraudulently induced specifically into entering an arbitration agreement. Here, Petitioner J.A. Walker Company, Inc. ("Walker") directs its fraudulent inducement allegations specifically to the arbitration agreement relied upon by Respondent Cambria Corporation ("Cambria"), not to the parties' contract as a whole. The trial court ordered the parties to arbitrate their dispute, but the arbitration order is unclear whether the trial court resolved Walker's fraudulent inducement challenge to the arbitration agreement. We therefore make the rule to show cause absolute so that the trial court can "proceed summarily to decide," pursuant to section 13-22-207(1)(b), Walker's fraudulent inducement challenge directed specifically to the agreement to arbitrate.

I.

■ ■ ■

In April 2004, 450 Seventeenth, LLC ("450") entered into a construction contract (the "Prime Contract") with Cambria for improvements to property owned by 450 in downtown Denver. Cambria executed a separate contract (the "Subcontract") in May 2005 with Walker, a subcontractor, for the placement and finish of structural concrete at the project site.

While the Subcontract itself does not contain an explicit arbitration provision, it incorporates by reference the dispute resolution procedure detailed in the Prime Contract. The Prime Contract incorporates by reference a document known as the General Conditions of the Contract for Construction (the "General Conditions"). The General Conditions provides that "[a]ny Claim arising out of or related to the Contract . . . shall . . . be subject to arbitration."

Walker alleges that it requested but never received a copy of either the Prime Contract or the General Conditions. Walker further alleges that when it asked Cambria whether arbitration was required by the Prime Contract, Cambria assured Walker that it was not. Walker alleges that it relied on those representations when it executed the Subcontract with Cambria.

Walker subsequently filed suit seeking both payment for work that it allegedly performed and foreclosure on its mechanics' liens. In its answer to Walker's complaint, Cambria asserted that the trial court lacked jurisdiction because Walker's claims were subject to the mandatory arbitration provision of the General Conditions, incorporated by reference into the Prime Contract and applicable to Cambria by operation of the Subcontract. On similar grounds, 450 moved to compel arbitration. Walker objected, claiming it was not bound by the arbitration provision because Cambria fraudulently induced it into entering the Subcontract with assurances that the dispute resolution procedure did not include arbitration. . . .

The trial court compelled arbitration and stayed the remainder of the case. In its order, which was based on the parties' briefing and "the file in this matter," the trial court made no mention of Walker's allegations that it was fraudulently induced into agreeing to arbitrate its dispute with Cambria. Instead, the trial court held that "the arbitration provisions incorporated into the Subcontract Agreement between the Plaintiff and Defendant Cambria Corporation through the General Conditions of the prime construction project are enforceable and valid in accordance with the Uniform Arbitration Act, C.R.S. § 13-22-201, et seq." We issued a rule to show cause to consider the trial court's order compelling arbitration.

II.

Walker argues that the trial court erred in compelling arbitration because it has alleged that it was fraudulently induced into agreeing to arbitrate. We hold that section 13-22-206 requires the trial court to resolve allegations of fraudulent inducement, like Walker's, that are directed specifically to an agreement to arbitrate. Because it is unclear from the trial court's order whether it resolved this issue, we make our rule to show cause absolute and remand the case to the trial court for its determination of whether the parties' arbitration agreement was fraudulently induced.

A.

Colorado law favors the resolution of disputes through arbitration. . . . The current version of the CUAA applies to agreements, like the Subcontract, entered into after August 4, 2004. See § 13-22-203(1).

Interpreting the former version of the CUAA, we held in *Ingold* that the arbitrability of an allegation of fraudulent inducement depends upon whether the allegation is directed specifically to the agreement to arbitrate or more broadly to the contract containing the arbitration agreement. A fraudulent inducement claim directed specifically to the arbitration agreement is a challenge to "the existence of the agreement to arbitrate," § 13-22-204(1), C.R.S.

(2003), and therefore must be resolved by the trial court under the statute. A fraudulent inducement claim directed to a contract as a whole—of which the arbitration agreement is only a part—is to be decided by the arbitrator, not the trial court. As explained in *Ingold,* the United States Supreme Court drew the same distinction between fraudulent inducement claims under the Federal Arbitration Act in *Prima Paint Corp. v. Flood & Conklin Mfg. Co.,* 388 U.S. 395, 87 S. Ct. 1801, 18 L. Ed. 2d 1270 (1967).

Cambria argues that the current version of the CUAA does not distinguish between fraudulent inducement allegations in the way we recognized in *Ingold.* According to Cambria, the mere showing of an unambiguous arbitration agreement is sufficient to demonstrate that "an agreement to arbitrate exists" under *section 13-22-206(2),* and therefore, the arbitrator must decide all other issues concerning the enforceability of the agreement, including allegations of fraudulent inducement. We disagree.

Section 13-22-206 of the current version of the CUAA recodifies the statutory distinction between challenges to arbitration agreements that we recognized in *Ingold* and that the United States Supreme Court recognized in *Prima Paint.* As under the former version of the CUAA, the current version empowers the trial court to determine "whether an *agreement to arbitrate* exists or a controversy is subject to an *agreement to arbitrate.*" § 13-22-206(2) (emphasis added). As we explained in *Ingold,* a fraudulent inducement allegation directed specifically to the arbitration agreement is a challenge to the existence of the agreement to arbitrate. This is different from a situation where a party alleges fraudulent inducement of a contract as a whole. Under both the former and current version of the statute, the arbitrator must decide "whether a contract containing a valid agreement to arbitrate is enforceable." § 13-22-206(3) (emphasis added). Cambria offers no reason for why the current version of the CUAA should be interpreted differently from the former version of the CUAA.

In this case, Walker claims it is not bound by the arbitration provision because Cambria fraudulently induced it into entering the Subcontract through assurances that the dispute resolution procedure did not include arbitration. These allegations of fraudulent inducement are specifically directed to the arbitration agreement, and consequently, must be resolved by the trial court. See § 13-22-206(2). Here, it is unclear whether the trial court resolved Walker's allegations of fraudulent inducement. Its order states that "the arbitration provisions incorporated into the Subcontract . . . are enforceable and valid," but it does not specifically address Walker's contention that the arbitration agreement was fraudulently induced. . . .

Cambria argues that this court can decide Walker's challenge based on the evidence before us. It points out that Walker is a sophisticated contracting party and that there is evidence undermining Walker's allegations of fraudulent inducement. Much of Cambria's argument echoes our teaching that a party—particularly a sophisticated party—claiming fraudulent inducement faces a formidable challenge in proving its claim if the contracting parties "have access to information that was equally available to both parties and would have" dispelled the alleged fraud. *M.D.C./Wood, Inc. v. Mortimer,* 866 P.2d 1380, 1382 (Colo. 1994). . . .

[The court declined to reach the merits of Walker's fraudulent inducement claims and left the decision to the trial court on remand.]

III.

We hold that under section 13-22-206 of the current version of the CUAA, allegations of fraudulent inducement directed to the arbitration clause itself are to be resolved by the trial court, while allegations challenging the validity of the contract as a whole are to be decided by the arbitrator. Since it is unclear whether the trial court considered Walker's fraudulent inducement challenge, we make the rule to show cause absolute and remand the case for the trial court to "summarily decide" this issue in accordance with section 13-22-207(1)(b).

Review Questions

1. Black Tool and Die Company agreed to make a punch press die for White for $10,000. A one-month delivery time was agreed on. Black was ready to begin work on the die when White called and told Black to hold up until further notice. White then shopped around in an attempt to find a lower price for the die. White could not find a better price and called Black about two weeks later to tell Black to go ahead on the die, but Black refused, saying that its work schedule was now such that the die could not be completed within six months. White claims breach of contract and threatens to sue. What is the likely outcome of the case? Why? *Black showed tender of performance. White modified contract without agreement*

2. Why is it necessary for courts to recognize an anticipatory repudiation in connection with contracts involving a structure that is to be built and installed?

3. Black, under a contract with White, built an automatic assembly machine to assemble drive mechanisms for automobile window regulators. The contract calls for a machine capable of producing 1,500 assemblies per hour. The resulting machine ran at a speed that would easily produce 1,500 assemblies per hour. However, its longest run since installation a month ago has only been about two minutes before it jammed. Frequently it will run only one or two pieces before stopping. The cause of jamming is slight variations in the dimensions of the component parts of the window regulator assemblies. The parts are manufactured by White in the same manner they have been produced for many years. Black had access to unlimited quantities of the parts while the machine was being built. According to the contract, Black's performance was finished when the installation of the machine was completed. White has paid $162,000 of the $180,000 agreed price of the machine. Black demands payment of the remainder. White claims a right to retain part or all of the $18,000 to compensate him for efforts spent in making the machine work. Has Black substantially performed? Can Black recover the remaining $18,000? Why or why not?

4. Green leased a building near two metal working plants for a period of five years. The lease specified no restrictions for use of the building. Green set up a tool and die shop that operated profitably for about a year when the first plant left and relocated in another city. Shortly thereafter the second plant was liquidated in bankruptcy. Green wants to avoid the lease, claiming commercial frustration in that the remaining tool and die work is insufficient to be profitable. Is Green bound by the lease or can Green get out? Why?

5. In *Eatherly Construction Co. v. HTI Memorial Hospital*, did the court find that the amount of $500 per day for construction delays was unreasonable? What did this court need in order to determine whether the liquidated damages clause was reasonable?

6. In *Yale v. Pizza Hut*, the defendant was held to have committed an anticipatory breach of the contract. Why, then, did the court refuse damages to the plaintiff? *Tender of perf*

7. According to *J.A. Walker Co., Inc. v. Cambria Corporation, et al.*, who can decide whether a party was fraudulently induced into entering an arbitration agreement? An arbitrator or a trial court? Why?

Remedies

HIGHLIGHTS

- Remedies can be enforced by statute through several methods including:
  - Writ of execution
  - Garnishment
  - Attachment proceedings

as well as pursuant to a mechanic's lien.

- Remedies for breach of contract can take several forms:
  - Damages—where there can be monetary award for contract breaches and may include punitive damages;
  - Restitution—where the plaintiff can only recover the thing or the value of the thing given under the contract;
  - Equity—where the plaintiff has no adequate monetary remedy, the court can grant other equitable remedies.

We have already considered the requirements for the formation of a lawful contract. Chapter 13 examined the discharge of the obligations imposed on the parties by such contracts. We have observed that not all contractual obligations are discharged as the parties originally intended. When a party fails to perform as promised, a **breach of contract** has occurred. Each party to a contract has a right to obtain proper performance for the rights or performance given up. If this right is not satisfied, the law affords a remedy. The extent and type of remedy afforded is determined by the nature and extent of the breach.

We have defined a **contract** as an agreement enforceable at law. The enforceability aspect of the definition sets a contract apart from other agreements. Saying that contract rights are enforceable at law implies that there must be remedies available for their breach. An old equity maxim is that "For

223

every right, a remedy." This pertains to contract rights as well as to other rights the law recognizes and protects.

Legal and equitable remedies exist to enforce a right, to prevent the violation of a right, and to compensate for an injury. Chapters 21 and 22 consider **tort** laws. Generally, this body of law defines the circumstances in which an injury due to another's actions or inactions will be compensated. Probably the most common remedy sought and obtained in both contract and tort is monetary damages. However, there are numerous instances in which damages will not afford an adequate or complete remedy. For this reason, other remedies have developed. Such remedies as restitution, specific performance, injunction, rescission, and reformation are examples of remedies that may be available under particular circumstances. In addition, a court of equity, with its historical origin based on unusual remedies, can combine and select remedies or, if necessary, invent new ones to fit new circumstances.

This chapter considers only the more common remedies of damages, restitution, specific performance, and the injunction. One or a combination of these remedies will be appropriate in nearly any case in which an engineer is likely to be involved.

DAMAGES

Damages are the compensation in money awarded by a court to one who has suffered a loss, detriment, or injury. Most breach-of-contract cases involve some sort of claim for damages. When the breach is not of a nature such that compensation is really justified, the injured party may still win the case and be awarded nominal damages (such as an amount of one dollar). An award of nominal damages merely means the court has recognized that there was an invasion of a technical right of the plaintiff. Of course, as with any other award, the loser will probably be assessed the court costs.

Compensatory Damages

The usual reason for undertaking a damage action is to obtain compensation for an injury to one's person, property, or rights. To obtain compensatory damages it is necessary to prove that a right existed and was violated; in addition, the amount of damages suffered must be established with reasonable certainty. If the amount of damages cannot be reasonably established, the result is likely to be either an award of nominal damages or else no award of damages at all. The amount of compensation to which the plaintiff may be entitled usually is a question for juries.

The basic purpose of awarding compensatory damages is to make the plaintiff whole for any losses suffered by the defendant's wrongful conduct. The approach used to determine the compensatory damages awarded may vary depending on whether the dispute involves a contract right or a tort.

The **contract approach** attempts to compensate the plaintiff for his or her lost expectations. This approach does so by compensating for any out-of-pocket costs to the plaintiff and other damages such as profits missed as a result of the contract breach. The objective of this method is to place the plaintiff in the position that the plaintiff would have enjoyed if the contract had not been breached.

Of course, the damages claimed must be directly connected to the contract breached if the plaintiff is to be compensated. Black has a contract to build automation equipment for White. Gray states that if the automation works properly for White, Gray will be interested in a similar installation. Black may even have submitted preliminary plans and drawings to Gray. White then breaches the contract with Black. Black could collect compensation for the costs incurred on White's contract, plus the profits Black could reasonably expect from White. However, Black probably could get nothing from White to cover Black's anticipated profit from Gray; Black's claim probably would be considered too speculative.

The **tort approach** attempts to compensate the plaintiff for the plaintiff's actual losses. The plaintiff's out-of-pocket costs are considered, and lost profits (or lost wages for an individual) also may be considered. The results, in terms of money damages, can be quite different, depending on which theory is followed. However, the results also can be quite similar; a tort case can look very much like a contract case, and damages for a breach of contract can appear to be the same as those for a tort. It is not unusual to see someone pursue a recovery under both contract and tort theories in the same case. How can this happen? If you're injured in a car accident, you may decide to sue the manufacturer for breach of warranty and negligence, as well as for strict products liability. All of these types of actions are discussed later.

Types of Compensatory Damages.

It is worth noting that different types of compensatory damages exist. More importantly, however, different rules apply to different types of damages with respect to the question of what damages can be recovered in a given case. **General damages** are damages that logically and naturally flow from the wrongful conduct or breach of contract.

Each kind of legal injury has its own kind of associated general damages. For example, the general damages a buyer of land might recover due to the seller's fraud may be different from the general damages one might recover due to injuries received in a car wreck. Usually, general damages need not be "foreseeable." In other words, it does not matter that the defendant did not, or even could not be expected to, foresee the nature and extent of the plaintiff's injuries. The courts essentially presume such foreseeability with respect to what are classified as general damages. Moreover, a plaintiff usually is not required to "specially plead" (that is, specifically include a claim for) general damages. Thus, when a plaintiff prepares the complaint to be filed against the defendant, the plaintiff's general allegation of damages will be enough.

As opposed to general damages, a plaintiff usually needs to specifically plead **special damages.** Special damages consist of those damages resulting from a party's particular circumstances. Courts usually consider lost profits as a type of special damages. To recover such damages, the plaintiff must specifically plead in some detail the facts relating to the damages in the complaint. Moreover, special damages must be foreseeable to be recovered. (See *Lewis Jorge Construction Management, Inc. v. Pomona Unified School District* at the end of the chapter.)

Suppose White Manufacturing Co. agrees with Black Inc. to buy a surface mounting system for use by White in manufacturing circuit boards to be included in personal computers. The system fails to work. Before White discovers the problems, the system ruins about $1,000 worth of White's inventory. Because the system fails to work, White also cannot deliver 20,000 circuit boards it promised to Green Computers. As a result, White cannot obtain the $20,000 profit it would have made on the contract with Green. It appears likely that the $1,000 would be general damages, whereas the $20,000 would be considered special damages. Should White be able to recover the $20,000 from Black? Assuming that the amount of lost profits can be shown with reasonable certainty and it is clear that White would have made the profits "but for" Black's breach of contract, the issue becomes one of foreseeability. If White had told Black, prior to the execution of their contract, that White's contract with Green would yield White $20,000 in profits, and had further told Black that White needed the system to perform the contract with Green, then White's lost profits seem likely to be considered "foreseeable" and therefore recoverable.

Exemplary (Punitive) Damages

In certain cases the courts allow a party to recover more damages than the reasonable compensation for the injury or wrong suffered. Where a person's rights were violated or an injury caused under circumstances such as fraud, malice, oppression, or other despicable conduct by the defendant, **exemplary damages** (also called **punitive damages**) may be awarded. Such damages have a primary purpose of punishing the defendant for the reprehensible conduct and making the defendant an example to deter others from similar conduct. Secondarily, exemplary damages are added compensation for the shame, degradation, or mental anguish suffered by the plaintiff.

Exemplary damages are much more likely to be allowed in tort cases than in contract cases. At common law, exemplary damages could not be recovered for a breach of contract, no matter how willful the breach was. Exemplary damages are governed generally by the state laws on damages, and the relevant principles vary from state to state. Large awards of punitive damages often grab headlines and generate a considerable amount of debate. For several years, the U.S. Supreme Court has left limitations on punitive damage awards to individual state legislatures, reflecting the Court's hesitancy to impose a standard that could be applied nationwide when the standard may only be applicable to a particular state.

In 1996, the Supreme Court was presented with a case so compelling in its final result, that the Court took it on. This case was *BMW of North America v. Gore* where at the trial level, BMW was faced with a punitive damage award several times over the amount in compensatory damages.[1] In this case, BMW was found to have sold 983 refinished cars as new. Mr. Gore bought one of them and upon finding that the car had been repainted, began legal action. The jury awarded Gore $4,000 in compensatory damages and $4 million in punitive damages. The Court set down a three-prong test to determine whether a punitive damages award rose to the level of being unconstitutional, specifically, that it was against the 8th Amendment's prohibition against excessive fines and cruel and unusual punishment. The 8th admendment reads, "Excessive bail shall not be required, nor excessive fines imposed, nor cruel and unusual punishments inflicted." This led to two other Supreme Court decisions expanding on the three-prong test, resulting in a balanced approach to be used by the lower courts when they are presented with similar disparities between actual or compensatory damage awards and punitive damage awards.[2] More recently in *Philip Morris USA v. Williams,* 127 S. Ct. 1057; 166 L. Ed. 2d 940; 2007 U.S. LEXIS 1332, which cited the *BMW* case, the Supreme Court held that a judge cannot use a punitive damages verdict to punish a defendant directly for harm to non-parties under the taking of property without due process theory.

> The Supreme Court's punitive damages test measures:
> 1. degree of reprehensibility on part of defendant;
> 2. difference between the actual or potential harm suffered and the punitive damages award;
> 3. difference between civil penalties awarded in comparable cases and punitive damage award in instant case.

Liquidated Damages

It has become almost standard practice to provide some form of remedy such as liquidated damages in the wording of engineering contracts.

Liquidated damages are damages that are liquidated in the sense that they are a known, predetermined amount or measure of damages, such as damages of $100 per day for each day a

---

[1] 517 U.S. 559, 116 S.Ct. 1589 (1996).

[2] *Cooper Industries, Inc. v. Leatherman Tool Group,* 532 U.S. 424 (2001) and *State Farm Mutual Automobile Insurance Co. v. Campbell,* 2003 WL 1791206 (S.Ct. 2003).

contractor is late in completing the contract. Such provisions are enforceable in court only if they are made in good faith by the parties and are a reasonable estimation of the uncertain amount of the damages caused by delay. In *R.P. Wallace, Inc. v. U.S.* at the end of the chapter, R.P. Wallace sought remission of liquidated damages assessed against it by the Navy.

If a court views the contract provision as one intended to secure performance rather than provide for just compensation, the court is likely to view the clause as a "penalty." If considered a penalty, the contract clause will not be enforced.

Duty to Mitigate Damages

If there were no recourse to the courts to recover damages when your rights were invaded, you would certainly make every effort to keep your damages as small as possible. The law imposes a duty by requiring the plaintiff to **mitigate** (that is, minimize) the damages suffered by taking reasonable steps to minimize or avoid additional damages. If one is injured, every reasonable effort to keep the injury to a minimum should be taken. For instance, if an employment contract (to run for a certain period of time) is breached by the employer, the employee must actively seek work elsewhere. If the employee is successful, the difference between the two salaries (assuming the original job paid more) will be awarded; if the employee is unsuccessful after a reasonable effort to find subsequent employment, the total lost pay may be recovered. If the employee does not make a reasonable effort to find subsequent employment or refuses suitable work, the employee's damages may be considerably diminished. However, the employee probably need not take work for which he or she was not suited (for instance, an experienced engineer would be unlikely to be criticized for refusing employment as a farmhand); neither would the employee be required to move a great distance.

To further illustrate the concept of mitigation, suppose Black is a manufacturer of appliance parts, particularly chrome-plated ones. Black has a contract with White whereby White is to supply Black with nickel, at a stated price, for use in the copper-nickel-chromium plating process. During the life of the agreement, White raises the price of the nickel supplied, thus breaching the contract. Black could try to find an alternate source of nickel or pay White's increased price. Black might even use the increased price as an excuse to cease manufacturing appliance parts for Black's customers, relying on White's breach to cover any losses Black might sustain.

Either of the first two alternatives might be considered reasonable as an attempt to mitigate damages. Black probably cannot simply breach Black's contracts with the appliance manufacturers and pass along to White the damages assessed against Black. Neither could Black maintain an action for lost profits if Black ceased manufacturing parts on this basis. Black's damages suit probably will get only the difference between the contract price and the price actually paid for the nickel.

It might appear in the mitigating situation just described that the equitable remedy of specific performance would be fair. Such is usually not the case, however, unless a statute exists to make it available. Without a statutory provision, specific performance probably would be denied on the basis that monetary damages would be a sufficient remedy. Black could conceivably obtain the same quality and quantity of nickel from other suppliers, with the difference in price being the only source of damages.

Economic Loss Rule

The **economic loss rule** is a convention used by courts to distinguish losses that may sound in tort *or* contract. As discussed above under Compensatory Damages, the contract approach and the tort approach may result in a similar recovery, e.g., money. While most losses can be assigned a

[handwritten margin note, rotated: If suing under contract, must sue for contract remedies & cannot sue for tort remedies]

monetary value, the effectiveness of the economic loss rule lies in the ability of the court to determine whether the risks of loss should have been articulated more clearly in the contract, or if the injury suffered was not foreseeable by the parties, and, therefore, could not be negotiated into the contract. If remedies are available in the contract, the court will apply those remedies to the case. (See *Arguro Alejandre, et al. v. Mary M. Bull* at the end of the chapter. In a concurring opinion to this case, Judge Chambers suggests that when one reads the phrase "economic loss," replace the phrase with "commercial loss.")[1]

[handwritten note: economic loss rule - suing party must have contract w/ defendant in order to sue for purely economic losses; cannot sue on a tort theory. Economic loss - damages for disappointed expectations, relating to a product that does not perform as promised (lost profits, delay damages, loss of benefit, reduced value)]

RESTITUTION

Restitution is generally similar to an award of money damages. Awards based on restitution are usually made in money. Such awards are based on the plaintiff's having parted in good faith with consideration and the defendant's having breached a duty owed the plaintiff. The difference between restitution and damages lies in the purpose and amount of the award. Restitution only restores what is lost or the value of the thing given. This remedy focuses on the value received by the defendant that would be unjustly retained unless returned to the plaintiff. There is no attempt, as in a damages action, to compensate the plaintiff for what the plaintiff expected to gain (such as the plaintiff's lost profits). In effect, the defendant is required to return whatever consideration was received from the plaintiff. The courts, however, do not adhere strictly to this rule. They will not apply it where, by so doing, they offer a shield to the defendant for the defendant's wrongdoing. For example, if the defendant has made a profit through wrongful conduct, the profits of the defendant may be awarded to the plaintiff. The injured party generally is required to return what was received where it is reasonably possible to do so.

EQUITABLE REMEDIES

Specific performance and injunction are the principal equitable remedies. Neither remedy may be used where an adequate remedy at law (for example, damages, restitution, or a statutory remedy) is available. However, either remedy may be used in conjunction with damages where damages alone would be an insufficient remedy. When either an injunction or specific performance would require extensive supervision of the court for enforcement, an attempt should be made to find a different remedy. In *Advanced Marine Enterprises, Inc., et al. v. PRC Inc.*[2] at the end of the chapter, the chancery court was able to order both legal and equitable remedies. When, for instance, specific performance of a contract to perform computer maintenance services is requested, it is likely that the court would deny the request. Such a remedy seems likely to require a fair amount of court supervision; hence, the court would look for a more appropriate remedy.

---

[1] For further discussion, read *O'Connor, et al. v. Hertz, et al.*, 2005 Cal. App. Unpub. LEXIS 10732, where the plaintiffs sought recovery from an architect for failure to keep construction costs within the budget under professional negligence theory.

[2] See also Appendix B, Case No. 00-9, Employment Offer of Employment By Vendor.

Specific Performance

Simply put, the remedy of specific performance takes the form of a court ordering a party to perform the specific action required by a contract. The most common, though not exclusive, use of the remedy of specific performance occurs when a unique piece of property is involved. A piece of land, such as a city lot or a farm, is unique. So is an original painting by an old master or a tailor-made piece of automation equipment.

Courts, since ancient times, have considered land as unique (the extension of this concept to other types of property is of more recent origin). If a contract to sell a particular piece of land is breached by the seller, sufficient money damages might be awarded to allow neighboring property to be purchased. However, no two pieces of land have the same location, and it is likely that there would be other tangible and intangible differences; for instance, the view from one location may give a sense of security or be more appealing than the view from another location. Also, consider the uniqueness of the neighbors adjoining different lots.

Courts usually will not require specific performance of personal service contracts. However, a court might be persuaded to prohibit a person who has contracted to perform services to another from performing the same services for anyone else.

Injunction

Originally, **injunctions** were only prohibitive in nature ("thou shalt not"). Now, in most jurisdictions, an injunction may be either prohibitive or mandatory ("thou shalt"). An order for specific performance can be viewed as a type of **mandatory injunction.** Even where injunctions must be prohibitive, it is possible to write what is, in effect, a mandatory injunction. In one case a tenant, enraged at his landlord, piled garbage on the front lawn of the tenant house before leaving. Though a mandatory injunction ordering the tenant to remove the garbage could not be issued in that state, a prohibitive injunction did the job just as well. The tenant was prohibited from allowing the garbage to remain on the lawn at the former residence.

The injunction is often used where irreparable injury to property is imminent—not just "possible," but very probable. The probable damage must also be a type of damage that could not be satisfactorily repaired. Damage to one's good will or reputation, for example, is generally considered irreparable. (Recall the elements necessary for an injunction in Chapter 3, Development of Law under the Unusual Remedies section.)

In addition to a **permanent injunction** entered after a trial as part of the court's final judgment, courts can enter interim orders that amount to injunctions controlling a person's conduct pending a trial or final resolution of the case. In emergency situations, courts can issue **temporary restraining orders.** Such orders are a type of injunction and are usually limited in duration to a matter of days or weeks. Pending a final trial on the merits, a court can also issue a **preliminary injunction.** Usually, preliminary injunctions are entered only after a hearing; unlike a temporary restraining order, a preliminary injunction usually remains effective up to the entry of a court's final judgment. The purpose of a temporary restraining order or preliminary injunction is usually to preserve the "status quo" pending a full trial.

The speed and convenience of the injunction as a means of enforcing a law often has appeal to the litigants (the parties to a lawsuit). If a particular law can be made to call for an injunction or a "cease and desist" order to be issued when the law is violated, the time and expense of a jury trial are often avoided. It is necessary only that the order be issued and probable violators informed. Any further violation would be a contempt of court, possibly resulting in a jail sentence or a fine. The speed and simplicity of the injunction are often appealing to the party filing a lawsuit. Generally, however, the courts view injunctions as extraordinary remedies that are available only in limited circumstances.

ENFORCEMENT OF REMEDIES

A remedy without enforcement would be meaningless. The law must "have teeth" if it is to be effective. There are three common means of enforcing court awards against the loser in a suit at law: execution, garnishment, and attachment. Enforcement of equitable remedies usually takes the form of contempt of court.

After a judgment is rendered, the loser is expected to comply with that judgment. When security has been posted, the award may be deducted from it. When no security has been pledged and the loser does not comply with the court's order (by paying the amount of damages awarded), the other party may return to court for an order to confiscate property to satisfy the judgment.

Such an order is a **writ of execution.** The writ is addressed to the sheriff or other law enforcement officer, and gives the officer the right to seize as much of the loser's property (both real and personal) as may be necessary to satisfy the judgment. The property so obtained is then sold at an execution sale, the proceeds being used to satisfy the award of damages, with any remainder going back to the loser.

Garnishment is the legal mechanism used to obtain the loser's property that is held by or under the control of a third party. Notice is given to the third party to turn over the judgment debtor's property in satisfaction of the obligation and to not pay a debt to the judgment debtor until the judgment is paid. *Obtain wages from defendant's employer*

Attachment is a legal mechanism used when the defendant is not within the jurisdiction of the court, but some of the defendant's property is available. The attachment process usually takes place before the court proceedings to ensure that the plaintiff, if successful in court, will be paid the amount of the judgment. Because the seizure of the property occurs before a trial is held, the plaintiff is usually required to post a bond to protect the defendant. An attachment prevents the defendant from removing property from the court's jurisdiction prior to the court's judgment. If the defendant wishes to remove the attachment, the defendant can post a counter bond in an amount sufficient to cover the plaintiff's claim.

Execution, garnishment, and attachment are all limited by statutes in the various states. The homestead laws that limit creditors' rights also protect the loser in a damages action. Also, many state laws severely restrict garnishment, particularly when the wages of the head of a household are concerned.

Another device that should be considered as security or enforcement is the **mechanic's lien.** It is a lien against not only the structure built, improved, or worked on, but also generally against the land on which the structure rests. Generally, such a lien may be established by any unpaid laborers, contractors, subcontractors, materialmen, or others having a hand in the work involved. The mechanic's lien does not exist under common law or equity; instead, it is a creation of state statutes. The statutes tend to be similar and tend to emphasize benefits to laborers, materialmen, and subcontractors rather than those to contractors, engineers, or architects. The reason is that the latter group is much more likely to be in position to pursue a remedy for nonpayment by means of breach of contract. For a further discussion of mechanics' liens, see Chapter 16.

Contempt of court is the principal means of enforcing equitable remedies such as injunctions or an order requiring specific performance. The extent of the enforcement is pretty much within the court's discretion. Generally, an order duly issued by a court that has been properly served on the persons to whom the order is directed must be obeyed by those persons. Any violation may amount to contempt. The court has considerable discretion, however, in assessing an appropriate fine or penalty. A court may order the offending party to serve time in jail, pay damages caused by the contemptuous (or contumacious) conduct, or set a daily fine for each day the contempt (violation of the order) continues. The remedy chosen may be coercive (that is, to try to force compliance with court's earlier order) or punitive (to punish for the violation of the order). In addition, the contempt proceedings may be criminal in nature or civil in nature. Contempt procedures can vary widely.

LEWIS JORGE CONSTRUCTION MANAGEMENT, INC. v. POMONA UNIFIED SCHOOL DISTRICT et al., 102 P.3d 257 (Cal. 2004)

Study terms: Special damages, general damages, lost profits, bonding capacity

A school district terminates a construction contract when the contractor, four and a half months after the promised due date, still has not finished the project. The contractor's bonding company then hires another firm to complete the project, but it suspends then later reduces the amount of bonding for the contractor. The latter successfully sues the school district for breach of contract, recovering in damages some $3 million for potentially lost profits, which the contractor claimed it would have earned on prospective construction contracts it never won because of its impaired bonding capacity. The Court of Appeal concluded that those potential profits were a proper item of general damages in this action for breach of contract. We disagree.

I.

In 1994, the Pomona Unified School District (District) solicited bids for building improvements at Vejar Elementary School. The District awarded the contract to Lewis Jorge Construction Management, Inc. (Lewis Jorge), the low bidder at $ 6,029,000. Although the contract originally provided for completion in December of 1995, heavy rains delayed work, and the parties agreed to a revised completion date of January 22, 1996. That date came and went, but the project remained unfinished.

The District withheld payments to Lewis Jorge for work completed in April and May, 1996. On June 5, the District terminated the contract with Lewis Jorge and made a demand on the contractor's surety to finish the project under the performance bond the surety had provided for Lewis Jorge. The surety then hired another contractor to complete the school project for $164,000. That contractor completed the project between early July and mid-September, 1996.

Lewis Jorge sued the District, alleging it breached the contract by declaring Lewis Jorge in default and terminating it from the construction project. The complaint sought damages and alleged six causes of action. The first, alleging breach of contract, and the second, alleging breach of an implied warranty of sufficiency of the plans and specifications for the project, are both contractual claims naming the District as a defendant. Causes of action three through five—alleging nondisclosure of material facts, inducing breach of contract, and negligence—named a district employee as a defendant. The sixth cause of action sought equitable indemnity against both the District and the employee for claims against Lewis Jorge by its surety and its unpaid subcontractors. Lewis Jorge did not plead as special damages the profits it claimed to have lost on future contracts.

Lewis Jorge, in turn, was sued by a number of its subcontractors for nonpayment of their past due bills.

At trial, Lewis Jorge presented evidence from its bonding agent that in June 1996 it had a bonding limit of $10 million per project, with an aggregate limit of $30 million for all work in progress. By mid-1997, the only sureties willing to provide Lewis Jorge with bonding imposed a limit of $5 million per project, with an aggregate limit of $15 million, a reduction of its bonding capacity to the level its surety had imposed in the early 1990's. Sometime in 1998, Lewis Jorge ceased bidding altogether and eventually closed down.

Lewis Jorge sought to prove the extent of its lost future profits on unidentified construction projects, using as the relevant period the date of the District's breach to the date of trial, and relying on its profitability during the four years preceding the breach. Robert Knudsen, a financial analyst who specialized in calculating lost profits claims, projected that Lewis Jorge had lost $95 million in gross revenue for future contracts that, based on its past history, it would likely have been awarded. Historically, Lewis Jorge had realized a profit of about 6 percent of revenue. Knudsen calculated lost profits on unidentified projects at $4,500,000, which discounted to present value came to $3,148,107.

The jury returned special verdicts in favor of Lewis Jorge, finding the District liable for $362,671 owed on the school construction contract, of which $143,755 was attributable to the District's "breach of warranty as to the fitness of its plans or specification" (the complaint's second cause of action). It awarded $3,148,197 in profits Lewis Jorge did not realize "due to the loss or reduction of its bonding capacity." Having found the District's employee negligent, the jury found him and the District jointly and severally liable for $3,510,868.

The District and its employee appealed. Lewis Jorge also appealed, raising issues that are not material here and were rejected by the Court of Appeal. Although the Court of Appeal reversed the judgment against the District's employee, and reversed awards against the District for prejudgment interest and contractual attorney fees (Civ. Code, § 1717), it rejected the District's claim that the award to Lewis Jorge of $3,148,197 for potential profits on future projects was an improper component of *general damages* for breach of contract. The Court of Appeal granted the District's petition for rehearing on that question; after receiving additional briefing, it concluded that "the lost profit damages sought by Lewis Jorge were in the nature of general damages, [] not special damages as claimed by the District."

We granted the District's petition for review to resolve whether general damages for breach of a construction contract include potential profits lost on future contracts that a contractor does not win when, as a consequence of the property owner's breach, the contractor's surety reduces the contractor's bonding capacity. We later solicited and received briefing from the parties on the related issue of whether an award of lost potential profits would have been proper here as special damages.

II.

(1) Damages awarded to an injured party for breach of contract "seek to approximate the agreed-upon performance." (*Applied Equipment Corp. v. Litton Saudi Arabia Ltd.* (1994) 7 Cal.4th 503, 515 [28 Cal. Rptr. 2d 475, 869 P.2d 454] (*Applied*).) The goal is to put the plaintiff "in as good a position as he or she would have occupied" if the defendant had not breached the contract. (24 Williston on Contracts (4th ed. 2002) § 64:1, p. 7.) In other words, the plaintiff is entitled to damages that are equivalent to the benefit of the plaintiff's contractual bargain.

(2) The injured party's damages cannot, however, exceed what it would have received if the contract had been fully performed on both sides. (Civ. Code, § 3358.) This limitation of damages for breach of a contract "serves to encourage contractual relations and commercial activity by enabling parties to estimate in advance the financial risks of their enterprise." (*Applied, supra,* 7 Cal.4th at p. 515.)

(3) Contractual damages are of two types—general damages (sometimes called direct damages) and special damages (sometimes called consequential damages). (24 Williston on Contracts, *supra*, § 64.1, pp. 11–12; 3 Dobbs, Law of Remedies (2d ed. 1993) § 12.2(3), pp. 39–42.)

A. General Damages

(4) General damages are often characterized as those that flow directly and necessarily from a breach of contract, or that are a natural result of a breach. Because general damages are a natural and necessary consequence of a contract breach, they are often said to be within the contemplation of the parties, meaning that because their occurrence is sufficiently predictable the parties at the time of contracting are "deemed" to have contemplated them.

B. Special Damages

(5) Unlike general damages, special damages are those losses that do not arise directly and inevitably from any similar breach of any similar agreement. Instead, they are secondary or derivative losses arising from circumstances that are particular to the contract or to the parties. Special damages are recoverable if the special or particular circumstances from which they arise were actually communicated to or known by the breaching party (a subjective test) or were matters of which the breaching party should have been aware at the time of contracting (an objective test). Special damages "will not be presumed from the mere breach" but represent loss that "occurred by reason of injuries following from" the breach. Special damages are among the losses that are foreseeable and proximately caused by the breach of a contract.

California follows the common law rule that an English court articulated some 150 years ago in *Hadley v. Baxendale* (1854) 156 Eng.Rep. 145. After Hadley's mill shut down because of a broken crankshaft, he entered into a contract to have a new one built. When the builder asked Hadley to send him the broken shaft to use as a model, Hadley took it to Baxendale, a common carrier, for delivery to the builder. Baxendale did not deliver until seven days later. Hadley then sued Baxendale for lost profits for that period. Hadley's lost profits, the court held, were not recoverable, because he had failed to inform the carrier that the mill would be shut down until delivery of the new shaft. (*Id.* at p. 151.) Because the special circumstance—the mill's inoperability without a mill shaft—was not communicated to Baxendale, he did not assume the risk of compensating Hadley for mill profits lost as a resulting of Baxendale's late delivery of the mill shaft.

(6) *Hadley* did not expressly distinguish between general and special damages. But such a distinction flows naturally from that case; hence the rule that a party assumes the risk of special damages liability for unusual losses arising from special circumstances only if it was "advised of the facts concerning special harm which might result" from breach—it is not deemed to have assumed such additional risk, however, simply by entering into the contract.

■ ■ ■

III.

Here, the Court of Appeal affirmed the jury's award to Lewis Jorge of $3,148,197 in *general damages*, based on profits Lewis Jorge did not earn on future unidentified contracts because its surety had reduced its bonding capacity after the District's termination of the construction contract. The Court of Appeal concluded that such potential profits were recoverable as general damages because they followed "from the breach in the ordinary course of events" and were a "natural and probable consequence." The Court of Appeal found it significant, as did the trial court, that the contract at issue, like much of Lewis Jorge's business, was a public contract that required bonding.

The Court of Appeal reasoned: When the contract was formed, the District knew of its own bond requirements, and it knew that public works contractors must provide bonds to secure their performance. Because impaired bonding capacity "has long been recognized as a direct consequence of an owner's breach of a construction contract," the Court of Appeal concluded that the District should have known that breaching the contract and resorting to the surety to complete the project could impair Lewis Jorge's ability to obtain bonds without which it could not bid on other public contracts. Accordingly, the Court of Appeal held that the potential profits Lewis Jorge lost on contracts it did not win after the District's termination of the school construction contract were general damages attributable to the District's breach.[3]

---

[3]The District advances various public policy arguments in urging us to preclude lost future profits as a component of general damages when the hiring party is a public entity and especially when, as here, it is a school district. Lewis Jorge responds that because public contracts require bonding, profits lost on potential projects because of impaired bonding capacity after an owner's breach of a public contract will always be general damages. Whatever the merits of these arguments, we need not base our holding on the circumstance that the contract was a public contract or that a public school district was the breaching party. For bonding, although it is statutorily required for most public contracts, is also commonly imposed under contracts between private parties for larger construction projects.

(8) The Court of Appeal, however, failed to consider a threshold inquiry. If the purpose of contractual damages is to give the nonbreaching party the benefit of its contractual bargain, then the first question is: What performance did the parties bargain for? General damages for breach of a contract "are based on the value of the performance itself, not on the value of some consequence that performance may produce." (3 Dobbs, Law of Remedies, *supra,* § 12.4(1), p. 62.) Profits "'which are the direct and immediate fruits of the contract'" are "'part and parcel of the contract itself, entering into and constituting a portion of its very elements; something stipulated for, the right to the enjoyment of which is just as clear and plain as to the fulfillment of any other stipulation.'" (*Shoemaker v. Acker* (1897) 116 Cal. 239, 245 [48 P. 62].)

(9) Unearned profits can sometimes be used as the measure of general damages for breach of contract. Damages measured by lost profits have been upheld for breach of a construction contract when the breaching party's conduct prevented the other side from undertaking performance.

(10) Lost profits from collateral transactions as a measure of general damages for breach of contract typically arise when the contract involves crops, goods intended for resale, or an agreement creating an exclusive sales agency. The likelihood of lost profits from related or derivative transactions is so obvious in these situations that the breaching party must be deemed to have contemplated them at the inception of the contract.

We are not aware of any California authority involving a construction contract that has upheld an award of general damages against a breaching owner for profits unearned on unidentified contracts the contractor did not get when its bonding was impaired as a result of the contract breach. Lewis Jorge, nevertheless, urges us to permit such recovery, citing a Montana decision, *Laas v. Mont. Hwy. Comm'n et al* (1971) 157 Mont. 121 [483 P.2d 699]. The Montana court's . . . decision in *Laas* appears to represent a singular instance of upholding lost profits on future construction projects as an item of general damages for breach of a construction contract, a holding that has not been followed in a published opinion outside Montana in the 33 years it has been on the books.

The only California decision upholding damages for a contractor's lost profits on future contracts it did not win because its bonding capacity was impaired arises not, as here, from a construction contract but from a contract to provide future bonding. (*Arntz Contracting Co. v. St. Paul Fire & Marine Ins. Co.* (1996) 47 Cal.App.4th 464, 489 [54 Cal. Rptr. 2d 888].). . .

Applying these rules to the school construction contract here, we cannot say that the parties' bargain included Lewis Jorge's potential profits on future construction projects it had not bid on and been awarded. Full performance by the District would have provided Lewis Jorge with full payment of the contract price. Certainly, Lewis Jorge anticipated earning a profit on the school contract with the District, but that projected profit was limited by the contract price and Lewis Jorge's costs of performance. If Lewis Jorge's bid accurately predicted its costs, the benefit of its contractual bargain for profits was capped by whatever net profit it had assumed in setting its bid price.

The District's termination of the school contract did not directly or necessarily cause Lewis Jorge's loss of potential profits on future contracts. Such loss resulted from the decision of CNA, Lewis Jorge's surety at the time of the breach, to cease bonding Lewis Jorge.

■ ■ ■

(14) Having here concluded that profits Lewis Jorge might have earned on future construction projects were improperly awarded as general damages, we now decide whether those lost potential profits were recoverable as special damages. Lost profits, if recoverable, are more commonly special rather than general damages, and subject to various limitations. Not only must such damages be pled with particularity, but they must also be proven to be certain both as to their occurrence and their extent, albeit not with "mathematical precision." (*Berge v. International Harvester Co.* (1983) 142 Cal. App. 3d 152, 161 [190 Cal. Rptr. 815].) "When the contractor's claim is extended to profits allegedly lost on *other* jobs because of the defendant's breach" that "claim is clearly a claim for special damages." (3 Dobbs, Law of Remedies, *supra,* § 12.4(3), fn. 12, p. 71.) Although Lewis Jorge did not

plead its lost future profits as special damages, the issue of their availability as special damages was presented to the jury, and at oral argument the District expressly stated that it was not relying on that pleading omission.

Although a few cases state that a contractor suing for breach of contract may recover as special damages any profits it might have earned on other unawarded construction contracts, such damages are frequently denied as too speculative. . . .

California . . . has not upheld as special damages a contractor's unearned profits after breach of the construction contract. . . .

(15) At trial, Lewis Jorge presented evidence that its bonding capacity was reduced by its surety after the District's termination of the contract. But Lewis Jorge did not establish that when the contract was formed the District could have reasonably contemplated that its breach of the contract would probably lead to a reduction of Lewis Jorge's bonding capacity by its surety, which in turn would adversely affect Lewis Jorge's ability to obtain future contracts. As the evidence at trial disclosed, Lewis Jorge's bonding agent, who had obtained the construction bonds from CNA, anticipated that CNA's suspension of Lewis Jorge's bonding capacity would only be temporary. . . . But the breaching party "is not required to compensate the injured party for injuries that it had no reason to foresee as the probable result of its breach when it made the contract."

Evidence at trial established that the owner's terminating a contract might or might not cause the contractor's surety to reduce its bonding capacity. As the District pointed out at oral argument, when it signed the contract it did not know what Lewis Jorge's balance sheet showed or what criteria Lewis Jorge's surety ordinarily used to evaluate a contractor's bonding limits. Absent such knowledge, the profits Lewis Jorge claimed it would have made on future, unawarded contracts were not actually foreseen nor reasonably foreseeable. Hence they are unavailable as special damages for the breach of this contract.

(16) To summarize: It is indisputable that the District's termination of the school construction contract was the first event in a series of misfortunes that culminated in Lewis Jorge's closing down its construction business. Such disastrous consequences, however, are not the natural and necessary result of the breach of every construction contract involving bonding. Therefore, as we concluded earlier, lost profits are not general damages here. Nor were they actually foreseen or foreseeable as reasonably probable to result from the District's breach. Thus, they are not special damages in this case.

Disposition

The judgment of the Court of Appeal must be modified to read: "The judgment against Christopher Butler is reversed; the award of prejudgment interest is reversed; the award of attorney fees is reversed; and the award of $3,148,197 for lost profits is reversed. In all other respects, the judgment is affirmed. The matter is remanded to the trial court for an award of prejudgment interest consistent with the opinion of the Court of Appeal." As modified that judgment is affirmed.

R.P. WALLACE, INC. v. THE UNITED STATES, 63 Fed. Cl. 402 (2004)

Study terms: Delays, liquidated damages, contract specifications, government contracts, critical path method, *United States v. Spearin*

At least since the shock of exploding fireworks caused the scales to fall on poor Helen Palsgraf, courts have applied temporal and spatial limitations on the availability of tort damages, invoking familiar doctrines such as proximate cause and forseeability. See *Palsgraf v. Long Island R. Co.,* 248 N.Y. 339 (1928). Similar principia have evolved in contract cases and, not surprisingly, have wended their way into the regulatory and decisional fabric of Federal procurement law, particularly in assessing responsibility for delays in the completion of a contract, and especially in considering the validity of liquidated damages assessed for delay.

The latter topics are the subject of this action, which springs from a contract between plaintiff, R.P. Wallace, Inc. ("R.P. Wallace"), a construction firm, and defendant, the United States acting through the Department of the Navy ("the Navy"), for renovation and repair work to a New Orleans, Louisiana naval facility. When, in its view, this contract was not completed on time, the Navy assessed liquidated damages against plaintiff. Following trial in New Orleans, plaintiff seeks remission of those liquidated damages, as well as compensable damages, for what it contends was government-caused delay and disruption.

I. Findings of Fact

Based on the record, including the parties' stipulations, the court finds as follows:

On September 28, 1993, the Navy and plaintiff entered into contract number N62467-91-C-7226 ("the contract") for various repairs to be made to Building 58 at the Naval Support Activity in New Orleans, Louisiana. Under the contract, plaintiff was responsible, *inter alia,* for re-roofing the building, removing an elevator penthouse, painting the interior and exterior, and replacing all the existing exterior windows with thermal break insulating glass aluminum frame units. Morton-Verges, Architects (Morton-Verges), the Navy's architect and engineer for this project, prepared the contract's plans and specifications, and was responsible for reviewing and commenting on all submittals made by the contractor.

Because Building 58 is a historic structure, the contract specifications called for replacement windows that conformed to the building's historic appearance and, therefore, differed from standard office-building windows in several respects. The contract drawings provided various details regarding these windows, including specifics on their panning and muntins. Detail 3/4 of sheet 4 of the contract drawings specified the design for the aluminum panning of the window jambs, requiring it to be convex. Additionally, the specifications required that the replacement windows use historically accurate "true" muntins that could support a "design pressure" of 65, indicating an ability of the glass to withstand wind loads of almost 160 miles per hour.[2] Further, Section 08520, P1.4.6.a of the contract required R.P. Wallace to submit as a sample "one full-size window with muntins of type proposed for use, complete with [American Architectural Manufacturers Association] label, glazing, hardware, anchors, and other accessories." Regarding installation of the windows, General Note 16 of sheet 1 of the contract drawings provided that the "Contractor will not be allowed to remove more windows than can be replaced in the same day."

The contract required R.P. Wallace to commence performance within ten days of the award, and to finish no later than 165 days after the award (15 calendar days for the mailing of the award and the contractor's submission of required bonds, and 150 calendar days to prosecute the work). Based upon the September 28, 1993, award date, the original completion date thus was March 12, 1994. Section 0101, P 1.9 of the contract provided that "if the Contractor fails to complete the work within the time specified in the contract, or any approved extension, liquidated damages shall be assessed the Contractor in the amount of $200.00 for each calendar day of delay."[4]

On November 3, 1993, the parties conducted a preconstruction conference, at which, according to the minutes, the Navy recommended a "tentative start date" of "early December 1993." . . .

---

[2]Emphasizing that the muntins had to be capable of bearing this pressure, paragraph 2.1 of the window specifications incorporated, by reference, American Architectural Manufacturers Association publication 101, section 1.5.2 of which states that "structural members shall be designed to withstand the full design load for the project site."

[4]The contract also incorporated, by reference, the liquidated damages provisions of section 52.212-5 of the 1984 version of the FAR. In relevant terms, these provisions are essentially identical to the language quoted above.

Notwithstanding, as of January 3, 1994—over ninety days into the expected 165-day contract period and sixty days after the preconstruction conference—plaintiff had neither provided any of the required submittals nor performed any work at the Building 58 site. On that date, plaintiff notified the Navy that it would provide the required window submittals on January 7, 1994. . . . Despite plaintiff's prior assurances, January 7, 1994, passed without plaintiff providing the Navy with the window submittals.

On February 7, 1994, the Navy issued a cure notice to R.P. Wallace pointing out that "with only 34 days remaining before the contract completion date . . . you have made only minimal progress toward completion" and listing a dozen submittals that remained outstanding. This notice closed by indicating that "the Government considers your failure to perform a condition that is endangering completion of this contract. Therefore, unless this condition is cured within 10 days after receipt of this notice, the Government may terminate for default." . . .

On February 28, 1994, two weeks prior to the scheduled completion date, plaintiff finally provided its window submittal. On March 11, 1994, plaintiff sent the Navy a letter indicating that all outstanding submittals had been provided and citing, but not identifying, "several factors which have contributed to the delay in completion of the contract." While this letter indicated that "we will be addressing these factors under separate cover shortly," there is nothing in the record indicating that such a document was ever prepared or sent to the Navy. . . . On March 16, 1994—four days after the original scheduled completion date for the project—the Navy rejected the window submittal for four reasons: (i) the sample proposed a concave, rather than a convex, panning, as required by the contract; (ii) plaintiff proposed using false muntins instead of the true muntins called for by the contract; (iii) the sill was not properly sloped, as specified in the contract; and (iv) the submittal was not a full-sized window sample.

On March 24, 1994, Alenco, the contractor that had created the window submittal for R.P. Wallace, sent plaintiff a letter discussing the grounds for the Navy's rejection. It advised plaintiff that it could revise the submittal to include the convex panning required by the contract, noting that it owned "two extrusion presses, which allows us to design and extrude our own profiles in house." Alenco, however, indicated that it was impossible to build an aesthetically acceptable window using true muntins that complied with the contract's wind resistance requirements, emphasizing "we can offer the Navy either the historic type false muntin as we have already submitted, or we can offer a true muntin. However, we can no longer offer a combination of both, nor can anyone else." Regarding the muntins, this letter further adumbrated:

> What it boils down to . . . is that the Navy can have either the structural integrity offered by a bulky true muntin, or they may have the atheistic [sic] value of a false muntin, but not both! The Navy will not find another manufacturer capable of performing this feat, or able to present a test report to substantiate it.

This is the first clear evidence in the record that plaintiff was aware of the defects in the contract's window specifications. On March 30, 1994, Alenco again wrote plaintiff, reemphasizing that true muntins of a size necessary to meet the wind resistance requirements would not be "the most pleasing to the eye."

On March 31, 1994, plaintiff delivered another window submittal to the Navy, albeit not a full-sized sample, together with a letter from Alenco explaining why it was impossible to build the windows as specified. At this time, plaintiff requested a variance from the contract specifications regarding muntins. . . .

On April 21, 1994, the Navy advised R.P. Wallace that it had approved the variance for the muntins, contingent upon the receipt and approval of a sample window. Describing the rationale for this decision, the parties have stipulated, as follows:

> The Navy's specifications for the windows were defective in that they included two incompatible requirements: (1) the windows had to meet current wind load requirements to remain intact through a hurricane, and, at the same time, (2) were required to be duplicates of the original windows they were replacing, i.e., to have full muntins between each pane of glass in a multi-pane window. The penetration of the window

by true muntins would have weakened the glass to such an extent that it could not meet the wind load requirement. A solid pane for the entire window could meet the wind load requirement; however, the muntins required to maintain the wind load requirement with individual panes would have been much larger than the original muntins, rendering them unacceptable for aesthetic reasons.

On April 26, 1994, plaintiff wrote the Navy highlighting the defective specification and seeking relief from anticipated liquidated damages, asserting that "the delay encountered is as a result of defective specifications and further is clearly from unforeseeable causes beyond the control and without fault or negligence of the contractor." In a response dated April 29, 1994, the Navy wrote that "the government is in agreement that the design is defective, and that an equitable time adjustment must be made." This letter requested that plaintiff apply for a time extension listing "the specific delays (i.e., provide description and dates), the number of days lost by reason of such delays, that such delays impacted the critical path of the project and were not concurrent with one another." It further stated that "the government will reduce the amount of liquidated damages for the number of days of the government estimated time extension."

On May 6, 1994, R.P. Wallace delivered the approved window submittal to Alenco for production. At this point, Alenco began to take steps to fabricate the windows. On May 23, 1994, plaintiff received a response from the Navy to one of its invoices in which the latter had deducted anticipated liquidated damages. That same day, plaintiff wrote the Navy asserting that the "delay is clearly and indisputably the result of a defect in contract specifications" and that it should "not be subjected to liquidated damages as a result of same." Notwithstanding, on May 26, 1994, the Navy sent plaintiff a letter in which it expressed its intention to assess liquidated damages based upon a revised contract completion date of May 1, 1994. On June 6, 1994, Alenco received the custom extrusion dies needed to create the convex panning, conducted tests and promptly sent samples to R.P. Wallace. For reasons unexplained, R.P. Wallace did not approve the custom panning profiles and release the windows for manufacturing until July 6, 1994.

On August 19, 1994, plaintiff delivered the Navy a letter dated August 17, 1994, in which it contested the deduction of liquidated damages from a prior invoice. In this letter, plaintiff asserted:

> As has been well documented in writing, by both the contractor and the Navy, the delay we have encountered is a direct result of a defect in contract specifications. As a consequence of our job progress being delayed, as a result of the defective specifications, it is my understanding that we will not be subject to liquidated damages.

On August 22, 1994, the Navy notified R.P. Wallace by letter of its intent to extend the contract by 36 days in connection with the window submittals. . . .

On September 20, 1994, Alenco delivered the first of several batches of windows to the work site, at which point, following the Navy's approval, R.P. Wallace began to install them. On October 21, 1994, the Navy executed modification no. P00004, unilaterally modifying the contract to grant plaintiff a time extension of 36 calendar days in compensation "for any and all window delays for the period between the return of the initial window submittal (16 March 1994) and approval of the resubmittal (26 April 1994)." The adjustment (together with two other unrelated adjustments totaling 19 days) resulted in a final contract performance date of May 6, 1994.

Following the receipt of a last batch of windows on January 9, 1995, plaintiff finished installing the windows on January 11, 1995, at which time the Navy took beneficial occupancy of Building 58.[9] The Navy assessed plaintiff liquidated damages of $200 per day for the entire period from the modified completion date of May 6, 1994, to

---

[9]Although plaintiff had originally projected that it would take 28 days to complete the window installation, it actually took 113 days to complete this task.

January 10, 1995—a total of 250 days—resulting in $50,000 in liquidated damages. On June 7, 1995, plaintiff delivered to defendant a claim for compensation in the amount of $130,402 for delays allegedly caused by the defective specifications, as well as a claim for remission of the entire $50,000 in liquidated damages. Both claims asserted that all the contract delay was directly attributable to the defective window specifications. Unpersuaded, on or about December 1, 1995, the Navy denied both claims. While conceding that the design specifications were defective, the Navy asserted that the 36-day equitable adjustment properly compensated plaintiff for the delays caused by the defect and contended that other factors within plaintiff's control—notably, what it claimed was plaintiff's failure to perform the preliminary project work for the project in a timely manner—caused the additional delay. It also found that plaintiff had failed to provide evidence that any additional delay resulted from the Navy's conduct.

On April 24, 1996, plaintiff filed its complaint in this court, seeking $130,402 in delay damages, $50,000 in remitted liquidated damages, interest, costs, and attorney's fees. On January 12, 1998, R.P. Wallace amended its complaint to reduce its claim for delay damages to $65,753.65. On August 15, 2001, plaintiff filed a motion for partial summary judgment. This case was transferred to the undersigned on July 24, 2002, and on October 16, 2002, the court granted, in part, plaintiff's motion for partial summary judgment. Specifically, the court held that defendant would be bound by its concession that both the contract's painting specifications and aluminum window specifications were defective, but that genuine issues of fact precluded granting summary judgment on the issue whether the defective specifications proximately caused delay in completion of the contract.

Trial in this case was held from May 5 to 7, 2003, in New Orleans. . . .

II. Discussion

Having failed to complete the renovations of Building 58 by the appointed date, plaintiff was assessed $50,000 in liquidated damages under . . . the contract. That section provided that "if the Contractor fails to complete the work within the time specified in the contract, or any extension, liquidated damages shall be assessed the Contractor in the amount of $200.00 for each calendar day of delay." Plaintiff seeks remission of these damages, as well as recovery of delay damages allegedly caused by the Navy.

A. Background Legal Principles

As a threshold matter, the court must focus on whether plaintiff's delay in completing the contract was "excusable" within the meaning of FAR § 52.249-10(b), a provision incorporated into the contract. That section states that a contractor shall not be charged with damages, if "the delay in completing the work arises from unforeseeable causes beyond the control and without the fault or negligence of the Contractor." See *Sauer, Inc. v. Danzig*, 224 F.3d 1340, 1345 (Fed. Cir. 2000). . . .

The excusable delay provisions of FAR § 52.249-10 are silent as to how a contractor should demonstrate the existence and extent of such delay. Nonetheless, the decisional law is well-settled that—

> When a contractor is seeking extensions of contract time, for changes and excusable delay, which will relieve it from the consequences of having failed to complete the work within the time allowed for performance, it has the burden of establishing by a preponderance of the evidence not only the existence of an excusable cause of delay but also the extent to which completion of the contract work as a whole was delayed thereby.

Morganti Nat'l, Inc. v. United States, 49 Fed. Cl. 110, 132-33 (2001). . . .

Thornier issues are posed by concurrent or sequential delays—the first occurring where both parties are responsible for the same period of delay, the second, where one party and then the other cause different delays *seriatim* or intermittently. . . . Summarizing the law on this point, the Federal Circuit, in *Essex Electro Engineers,*

224 F.3d at 1295 (internal quotations omitted), recently reiterated that a contractor "generally cannot recover for concurrent delays for the simple reason that no causal link can be shown: A government act that delays part of the contract performance does not delay the general progress of the work when the prosecution of the work as a whole would have been delayed regardless of the government's act."

[In *United States v. United Engineering & Constructing Co.*, 234 U.S. 236 (1914)]... the Supreme Court concluded that the contractor could not be held liable for liquidated damages under the original contract, reasoning that "when the contractor has agreed to do a piece of work within a given time, and the parties have stipulated fixed sum as liquidated damages . . . in order to enforce such payment the other party must not prevent the performance of the contract within the stipulated time; and that where such is the case, and thereafter the work is completed, though delayed by the fault of the contractor, the rule of the original contract cannot be insisted upon, and liquidated damages measured thereby are waived." *Id.*

■ ■ ■

It remains to apply these principles to the facts found here, focusing initially on the remission of liquidated damages and, ultimately, to the extent necessary, on the claimed delay damages.

B. Is Plaintiff Entitled to a Remission of Liquidated Damages?

Although not listed in FAR § 52.249-10, defective specifications plainly can give rise to excusable delay. See *United States v. Spearin*, 248 U.S. 132, 135-36 (1918) (discussing the rule at common law). R.P. Wallace contends that the Navy's defective window specifications delayed and disrupted its performance, entitling it to remission of all the liquidated damages asserted against it. While defendant does not contest that the defective specifications were harmful, it vigorously asserts that, on the record before the court, plaintiff has not shown that it is entitled to more than the 36 days of excusable delay already granted in modification P00005 of the contract. Tracking the approach employed by the experts who testified in this matter, in analyzing these assertions, the court will examine three distinct periods in the contract's performance—the first leading up to the approval of the window submittal, the second while the approved windows were being fabricated, and the last primarily involving the installation of the windows.

The first period in question—the submittal phase—began with the award of the contract and ended with the Navy's approval of the second window submittal on April 21, 1994, a period of approximately 210 days. For this period, plaintiff provided evidence that it was delayed in finding a window manufacturer because the contract originally called for a window that was essentially unavailable and could not be manufactured—one in which the historical type of true muntins specified could not reasonably accommodate the wind resistance requirement. . . . As such, based upon the record, the court thus concludes that, beyond the 36 days already allowed by defendant, plaintiff has established that, for the submittal period, it is entitled to an additional period of excusable delay totaling 21 days.

■ ■ ■

The second period in question—the window fabrication and delivery phase—began with the approval of the window submittal and ended with the delivery of the majority of the windows approximately 110 days later, on September 20, 1994. Plaintiff does not seriously argue that the defective specifications directly delayed the fabrication of the windows—indeed, the record indicates that the lion's share of the manufacturing delay was attributable to fabricating the convex aluminum panning on the windows, a feature of the original specifications that was not defective. Rather, R.P. Wallace contends that once its efforts to meet the original window specifications led it to Alenco and the specifications then were determined to be defective, it was inalterably committed to Alenco and "stuck" with whatever delivery schedule the latter could provide. The factual flaw of this argument, however, is that plaintiff produced no evidence, beyond the self-serving statements of its officers, that, once the defective specifications were corrected, it had no reasonable choice but to subcontract with Alenco. . . .[20] Yet, there is no indication that plaintiff even considered going elsewhere. Making matters worse for plaintiff, the record reveals delays in the fabrication process that are indisputably attributable to plaintiff, including a one-month delay—from June 6, 1994, to July 6, 1994—for plaintiff to approve the sample panning that Alenco had extruded.

Because it builds upon the supposed foundational delays encountered in the fabrication phase, plaintiff's case is weakest for the final period involved here—the installation phase—which began with the arrival of the windows and ended approximately 16 weeks later, on January 11, 1995, when the windows were completely installed. While plaintiff initially estimated that it could install all 74 replacement windows in 28 days, it actually took 113 days, or 85 days longer. Again, plaintiff does not assert that the defective specifications directly caused this delay. Rather, it contends that: (i) it had originally anticipated installing the windows while it reroofed the building; (ii) by the time the windows were received, that task had been completed; and (iii) as a result, there were fewer individuals on site to assist with the installation of the windows. But, this assertion makes no sense for several reasons.

First, the record reveals no critical path relationship between the roofing and window installation tasks, as plaintiff used different subcontractors for each of these purposes. Second, daily progress reports indicate that plaintiff's installation subcontractor sometimes had adequate staff in place to install multiple windows, but more often did not—indeed, for entire weeks it had no staff at the site at all. . . . Accordingly, the record does not support plaintiff's excusable delay claim for the installation period.

Finally, a point about notice. As to all the claims in question, defendant argues that plaintiff failed to comply with that portion of FAR §52.249-10 which requires the contractor to notify the contracting officer in writing within 10 days from the beginning of any excusable delay, stating the cause of that delay. Research reveals no case that has really interpreted this provision. . . .

Under the court's interpretation of the notice clause, plaintiff appears to have provided timely notice to the contracting officer for the delay associated with the submittal phase. . . . Accordingly, in the court's view, the notice requirement of FAR § 52.249-10 was met in the case of the 21 additional days delay the court has found to be excusable.

The situation, however, is far different as to plaintiff's claims with respect to the fabrication and installation periods. According to plaintiff, it knew that it would be unable to install the windows on a timely basis as early as April 21, 1994, when it authorized Alenco to proceed with the fabrication of the aluminum panning and the fabrication of the windows. At this time, plaintiff knew that the windows would be fabricated in approximately 22 weeks, rather than the originally-projected six to eight weeks. Yet, it did not notify the Navy of this until August 19, 1994, when a letter dated August 17, 1994, was provided to Lt. Turner—even though on May 26, 1994, Lt. Turner had indicated the Navy's intention to assess liquidated damages based on the assumption that the fabrication and installation of the windows would take eight weeks and 31 days respectively. Moreover, the August 17, 1994, letter did not reveal that fabrication delays were continuing and could cause additional delays in the installation of the windows. Unlike the situation with the defective specifications, then, plaintiff's failure to alert the Navy to the problems being experienced in fabricating and installing windows prejudiced the Navy by preventing it from reviewing the situation and taking appropriate steps to minimize the projected delays. As such, plaintiff's failure to provide proper notice as to the delays being encountered in fabricating and installing the windows provides a second, independent basis upon which to deny any further excusable delay for these phases of the project.

---

[20]Mr. Wallace testified that once the window submittal was approved, he did not go to another window manufacturer, because "Alenco was the only game in town." While this may have been true prior to the time the defective muntin specification was discovered and corrected, there is no indication that such was the case after the specifications were changed. . . .

C. Is Plaintiff Entitled to Delay Damages?

The court need not tarry in respect of the remaining issue here—whether plaintiff is entitled to delay damages for the 21 days that the court has concluded constitute a further period of excusable delay. In fact, such damages are not awardable for at least two reasons. First, there is clear indication that during the three-week period in which plaintiff had difficulty in identifying a window supplier, there was concurrent delay. Among other things, during the fall of 1993, when plaintiff encountered the delay in question, it failed to meet a number of contract deadlines, including that for submitting at least a dozen administrative and material submittals. Moreover, it also appears that during this same period, delay occurred because of the death of Mr. Wallace's grandmother. Accordingly, plaintiff cannot recover damages for this period because "prosecution of the work as a whole would have been delayed regardless of the government's act." *Essex Electro Engineers,* 224 F.3d at 1295 (internal quotations omitted). Second, to cinch matters, plaintiff, at trial, failed to provide evidence from which this court can authenticate and validate its allegedly recoverable expenses. Instead, plaintiff was satisfied to rely on attachments to its complaint and various analyses conducted by defendant rejecting the lion's share of its claimed expenses. In the court's view, such evidence is inadequate to establish a basis for recovery.

III. Conclusion

Based on the foregoing, the court finds that plaintiff is entitled to remission of liquidated damages for 21 days, which, at $200 per day, amounts to a recovery of $4,200. While the court recognizes that this may be somewhat a Pyrrhic victory, the short of it is that, in all other respects, plaintiff's claims simply are unsupported by the record or the law. The Clerk shall enter an appropriate judgment. No costs.

IT IS SO ORDERED.

ARTURO ALEJANDRE, et al. v. MARY M. BULL, 153 P.3d 864 (Wash. 2007)

Study terms: Economic loss rule, fraudulent misrepresentation, negligent misrepresentation

Petitioner Mary M. Bull sold a house to the respondents, Arturo and Norma Alejandre. The Alejandres subsequently learned the septic system was defective and sued Ms. Bull for fraudulently or negligently misrepresenting its condition. The trial court dismissed the Alejandres' claims after they rested their case, determining as a matter of law that the Alejandres had failed to prove their claims and that the claims are barred by the economic loss rule. The Court of Appeals reversed, concluding that sufficient evidence was presented in support of the claims and that the economic loss rule did not apply because the parties did not contractually allocate risk for fraudulent or misrepresentation claims.

We reverse the Court of Appeals. Under Washington law, the defective septic system at the heart of plaintiffs' claims is an economic loss within the scope of the parties' contract, and the economic loss rule precludes any recovery under a negligent misrepresentation theory. . . .

Facts

Ms. Bull owned a single family residence that was served by a septic system. The year before she put the house up for sale, Ms. Bull noticed soggy ground over the septic system. She hired William Duncan of Gary's Septic Tank Service to pump the tank. She also contacted Walt Johnson Septic Service, which emptied the tank and patched a broken pipe leading from the tank to the drain field. In April 2000, Ms. Bull applied for a connection to the city sewer, but when she learned there was a $5,000 hook-up fee she abandoned the idea.

Ms. Bull placed her home on the market in June 2000. In September 2001, Ms. Bull and the Alejandres entered into an earnest money agreement for the sale of Ms. Bull's home to the Alejandres. This agreement contained Ms. Bull's representation that the property was served by a septic system and her promise to have the septic tank pumped prior to closing. The earnest money agreement contained an addendum providing, among other things, that the sale was contingent on an inspection of the septic system. . . . The addendum also provided that if the buyer disapproved of any inspection report, the buyer had to notify the seller and state the objection. If the seller did not receive such notice, the inspection contingency would be deemed satisfied.

As provided in the earnest money agreement, a septic tank service (Walt's Septic Tank Service) pumped the tank, and the Alejandres received a copy of the bill. The bill stated on it that the septic system's back baffle could not be inspected but there was "[n]o obvious malfunction of the system at time of work done." In addition, prior to closing Ms. Bull provided the Alejandres with a seller's disclosure statement as required by RCW 64.06.020.[1] The disclosed that the house had a septic tank system which was last pumped and last inspected in the fall 2000 and that "Walt Johnson Jr. replaced broken line between house and septic tank," and she answered "no" to the inquiry whether there were any defects in operation of the septic system.[2] Ms. Bull also disclosed that she was aware of changes or repairs to the system. The Alejandres reviewed the disclosure statement with their agent and then signed the section of the disclosure statement headed "BUYER'S WAIVER OF RIGHT TO REVOKE OFFER." The Alejandres thus acknowledged, as expressly explained in the disclosure statement, their duty to "pay diligent attention to any material defects which are known to Buyer or can be known to Buyer by utilizing diligent attention and observation.

Also prior to closing, the Alejandres' lending bank required an inspection of the property. The resulting inspection report stated that its purpose was to notify the client of all defects or potential problems. The report indicated that the septic system "Performs Intended Function" and stated that "everything drains OK."

On December 10, 2001, the sale closed. The Alejandres moved into the house a week later. In January 2002, the Alejandres smelled an odor inside their home. They also heard "water gurgling like it was coming back up." They noticed a foul odor outside the home as well, which they believed came from the ground around the septic tank, which they said was soggy. In February, they hired William Duncan of Gary's Septic Tank Service. Mr. Duncan told the Alejandres that he could pump the tank but could not fix the problem because the drain fields were not working. He also told the Alejandres that he had told Ms. Bull that the drain fields were not working and that she needed to connect to the city's sewer system.

The Alejandres subsequently hired another company to connect to the city sewer system. During this work, the company discovered that the baffle to the outlet side of the septic system was gone, thus allowing sludge from the septic tank to enter the drain field and plug it.

The Alejandres sued Ms. Bull for fraud and misrepresentation, claiming costs and damages totaling nearly $30,000. After the plaintiffs rested their case, Ms. Bull moved for judgment as a matter of law. The court granted the motion, ruling that the economic loss rule bars the Alejandres' claims and that they failed to present sufficient

---

[1] Chapter 64.06 RCW contains requirements for sellers of residential real property to make certain disclosures unless the buyer has expressly waived the right to receive the disclosure statement or the sale is exempt from the disclosure requirements under RCW 64.06.010. If the buyer does not waive the right, the buyer can, in the buyer's sole discretion, rescind the earnest money agreement within three business days after receipt of the disclosure statement.

[2] The Alejandres maintain that Ms. Bull knew that the disclosure statement was wrong in stating that the tank was last pumped in the fall, rather than in May 2000, and that a broken pipe was replaced between the house and the tank, rather than between the tank and the drain field. At trial, Ms. Bull was unsure why she said "fall" rather than "May" and she testified that the line was not broken between the house and the tank. The Alejandres also maintain that Ms. Bull failed to disclose that she had to do her laundry outside the home because of the failed system. Ms. Bull says she did not disclose that she did her laundry outside the home for one month because the problem with the system was taken care of by the time she filled out the disclosure form.

evidence in support of their claims. The court entered judgment in favor of Ms. Bull and awarded her attorney fees as provided for in the parties' purchase and sale agreement.

The Alejandres appealed. The Court of Appeals reversed, holding that the Alejandres presented sufficient evidence to take their claims to the jury and that the economic loss rule does not apply because the parties' contract did not allocate risk for fraudulent or negligent misrepresentation claims.

Analysis

When reviewing a trial court's decision on a motion for judgment as a matter of law, the appellate court applies the same standard as the trial court and reviews the grant or denial of the motion de novo. *Davis v. Microsoft Corp.*, 149 Wn.2d 521, 531, 70 P.3d 126 (2003).

Ms. Bull maintains that the Alejandres' tort claims are precluded by the economic loss rule, as the trial court ruled.

The economic loss rule applies to hold parties to their contract remedies when a loss potentially implicates both tort and contract relief. It is a "device used to classify damages for which a remedy in tort or contract is deemed permissible, but are more properly remediable only in contract. . . .

■ ■ ■

"Tort law has traditionally redressed injuries properly classified as physical harm." *Stuart v. Coldwell Banker Commercial Group, Inc.*, 109 Wn.2d 406, 420, 745 P.2d 1284 (1987). . . . In general, whereas tort law protects society's interests in freedom from harm, with the goal of restoring the plaintiff to the position he or she was in prior to the defendant's harmful conduct, contract law is concerned with society's interest in performance of promises, with the goal of placing the plaintiff where he or she would be if the defendant had performed as promised.

■ ■ ■

In short, the purpose of the economic loss rule is to bar recovery for alleged breach of tort duties where a contractual relationship exists and the losses are economic losses. If the economic loss rule applies, the party will be held to contract remedies, regardless of how the plaintiff characterizes the claims. Washington law consistently follows these principles. The key inquiry is the nature of the loss and the manner in which it occurs, i.e., are the losses economic losses, with economic losses distinguished from personal injury or injury to other property. If the claimed loss is an economic loss and no exception applies to the economic loss rule, then the parties will be limited to contractual remedies.

■ ■ ■

The Alejandres maintain that the economic loss rule does not apply in the context here, i.e., the sale of a residence. However, as Ms. Bull contends, in this state the economic loss rule applies to tort claims brought by homebuyers. And, as in other circumstances, where defects in construction of residences and other buildings are concerned, economic losses are generally distinguished from physical harm or property damage to property other than the defective product or property. The distinction is drawn based on the nature of the defect and the manner in which damage occurred. In *Stuart*, 109 Wn.2d at 420-22, and in *Atherton*, 115 Wn.2d 506, we declined to recognize any tort cause of action for negligent construction because the plaintiffs in each of these cases presented no evidence of personal or physical injury resulting from the manner in which the condominium complexes in each case were constructed and instead sought only economic damages.

Here, the injury complained of is a failed septic system. Purely economic damages are at issue. There is no question that the parties' relationship is governed by contract. Thus, unless there is some recognized exception to the economic loss rule that applies, the plaintiffs' claim of negligence cannot stand because they are limited to their contract remedies. No exception to the economic loss rule has been established.

The plaintiffs allege that Ms. Bull made negligent misrepresentations about the condition of the septic system contrary to the duty of due care under the *Restatement (Second) of Torts* § 552 (1977).[4] Both *Berschauer/Phillips* and *Griffith* hold that although Washington recognizes a tort claim for negligent misrepresentation under the *Restatement (Second) of Torts* § 552, this claim is not available when the parties have contracted against potential economic liability.

Accordingly, the Alejandres' reliance on § 552 and what must be proven under it is foreclosed by our precedent. Because the parties' relationship is governed by contract and the loss claimed is an economic loss, the trial court correctly concluded that plaintiffs' negligent misrepresentation claim must be dismissed.

■ ■ ■

In accord with the overwhelming weight of authority from other jurisdictions and under our decision in *Berschauer/Phillips,* the economic loss rule applies regardless of whether the specific risk of loss at issue was expressly allocated in the parties' contract.

■ ■ ■

The plaintiffs also assert a claim of fraudulent concealment. . . . [h]owever, the fraudulent concealment claim fails because, as the trial court ruled, the Alejandres failed to present sufficient evidence to support the claim. Under *Obde,* 56 Wn.2d 449, and similar cases, the vendor's duty to speak arises (1) where the residential dwelling has a concealed defect; (2) the vendor has knowledge of the defect; (3) the defect presents a danger to the property, health, or life of the purchaser; (4) the defect is unknown to the purchaser; and (5) the defect would not be disclosed by a careful, reasonable inspection by the purchaser. The Alejandres failed to meet their burden of showing that the defect in the septic system would not have been discovered through a reasonably diligent inspection. . . . A careful examination would have led to discovery of the defective baffle and to further investigation.

Next, insofar as the Alejandres have asserted common law fraud theories, they have failed to present sufficient evidence of the . . . elements of fraud. In particular, they have failed to present sufficient evidence as to the right to rely on the allegedly fraudulent representations about the condition of the septic service. . . . As explained, the Alejandres were on notice that the septic system had not been completely inspected but failed to conduct any further investigation and, indeed, accepted the findings of an incomplete inspection report. Having failed to exercise the diligence required, they were unable to present sufficient evidence of a right to rely on the allegedly fraudulent representations.[6]

Accordingly, the trial court correctly determined, as to the Alejandres' fraudulent conveyance and fraudulent representation theories, that Ms. Bull was entitled to judgment as a matter of law under CR 50 because the Alejandres failed to present sufficient evidence in support of these theories.

■ ■ ■

---

[6]The Alejandres urge the court to hold that the economic loss rule does not apply to claims of fraud in the inducement, and they argue their fraud claims are claims of fraud in the inducement. We are aware that some courts recognize a broad exception to the economic loss rule that applies to intentional fraud. Other courts recognize a limited exception to the economic loss rule for fraudulent misrepresentation claims that are independent of the underlying contract (sometimes referred to as fraud in the inducement) but only where the misrepresentations are extraneous to the contract itself and do not concern the quality or characteristics of the subject matter of the contract or relate to the offending party's expected performance of the contract. We need not address the question whether any or all fraudulent representation claims should be foreclosed by the economic loss rule because we resolve the Alejandres' fraudulent representation claims on other grounds.

Conclusion

In this case involving the sale of a residence with a defective septic system, we hold that the economic loss rule applies and forecloses the buyers' claim that the seller negligently misrepresented the condition of the septic system. The buyers' claim of fraudulent conveyance is not subject to the economic loss rule. However, the buyers failed to present sufficient evidence on this claim and on their claims of fraudulent misrepresentation to take these issues to the jury. The trial court properly dismissed all of the claims under CR 50 at the close of the plaintiffs-buyers' case.

We reverse the Court of Appeals and reinstate the trial court's judgment, including the award of attorney fees and costs, and we award attorney fees and costs to Ms. Bull for the appeal and this discretionary review, as provided for in the parties' contract.

Chambers, J. (concurring in result)—I agree with the majority in result but write separately to suggest a different analytical approach to the economic loss rule.

Like the majority, I would reject Arturo and Norma Alejandre's negligent misrepresentation claim. Once the economic loss rule is applied, this negligent misrepresentation claim is revealed to be a breach of contract claim, not remediable in tort. Like the majority, I would hold that the contract does not control the Alejandres' fraudulent concealment claim. In this state, fraud is not a contract claim. Like the majority, I would hold that the Alejandres failed to present sufficient evidence on all of their claims and that the trial court properly dismissed them.

Unlike the majority, I do not believe that the best approach to the economic loss rule is to find it bars recovery for any undefined economic loss between parties whose relationship is governed by contract unless an exception applies. The economic loss rule is a misnomer, and the majority mistakes the name of the doctrine for its function.

Instead, I would approach the economic loss rule in light of what it is: a tool we use to ensure that tort is tort, contract is contract, and that each comes with its own remedies. The distinction between tort and contract matters because our society has made the rational choice to limit contract remedies to the typically efficient remedies laid out by the specific contract signed by the parties or provided by background contract and commercial law. Our society has also made the rational choice that tort remedies should make the victim whole and thus often include significant consequential damages, such as pain and suffering, which are generally inappropriate for mere breaches of contract. Only after I determined that there was a potential question as to whether a suit filed in tort should instead sound in contract would I examine whether the loss was, rightly understood, an "economic loss"—that is to say, a "commercial loss," properly addressed in contract law. In this case, either approach achieves the same results.

The majority aptly recites the relevant facts. . . .

The Economic Loss Rule

Bull argues, and the trial court agreed, that the Alejandres' claims are barred by the "economic loss rule." Claims for breach of contract and some tort claims, especially products liability claims, often bear great similarity to one another. Tort remedies are often, perhaps always, significantly larger than contract remedies. It appears to me that the economic loss rule is a response to the risk that the tort remedies available in products liability law, if applied in contract law, could gut it. One way we have prevented the death of contract is through the economic loss rule. It prevents one party to a contract from rewriting the damage provisions after a breach by styling the case in tort. As the inimitable Judge Richard A. Posner put it:

> The insight behind the [economic loss rule] doctrine is that commercial disputes ought to be resolved according to the principles of commercial law rather than according to tort principles designed for accidents that cause personal injury or property damage. A disputant should not be permitted to opt out of commercial law by refusing to avail himself of the opportunities which that law gives him.[8]

---

[8]Over the years, the economic loss rule has been applied in cases where there was no privity of contract between the parties. This is because there are types of injuries for which the law gives no remedy, and injuries to third parties stemming from someone else's breach of contract are often (though not always) of that type. Properly used, the economic loss rule can be a useful tool to tell us if the claim is also of that type. None of this is before us today.

I say the rule is unfortunately named because describing the "loss" as economic is not particularly helpful and can be positively misleading. Again, as Judge Posner quite aptly noted:

> It would be better to call it a "commercial loss," not only because personal injuries and especially property losses are economic losses, too—they destroy values which can be and are monetized—but also, and more important, because tort law is a superfluous and inapt tool for resolving purely commercial disputes. We have a body of law designed for such disputes. It is called contract law.

"Economic loss" (for which I suggest we read in our heads "commercial loss") includes "'the diminution in the value of [a] product because it is inferior in quality and does not work for the general purposes for which it was manufactured and sold.'" Christopher Scott D'Angelo, *The Economic Loss Doctrine: Saving Contract Warranty Law from Drowning in a Sea of Torts*, 26 U. Tol. L. Rev. 591, 592 (1995). It "is called in law an 'economic loss,' to distinguish it from an injury to the plaintiff's person or property (property other than the product itself), the type of injury on which a products liability suit usually is founded." *Miller*, 902 F.2d at 574.

Thus, merely because a loss can be expressed in economic terms, it is not necessarily an "economic loss" triggering application of the unfortunately named "economic loss rule." See *Miller*, 902 F.2d at 575. I recognize that this is the sort of linguistic perversity that gets lawyers laughed at. Nonetheless, I point it out because I fear the majority may be misunderstood as holding that we *start* from the position that any damage, or at least any property damage, that can be expressed in a dollar figure is presumptively an economic loss and the economic loss rule will keep a case out of tort (whether or not a contractual relation that could give rise to relief exists) unless an exception applies. While in most cases, that will get us to the right type of law, it is a needlessly complicated way to approach the problem.

In my view, we should start by recognizing that the "economic loss rule" is the analytical tool we use to determine whether a dispute implicates tort or contract law in those cases that could potentially sound in either.[9] Once the choice is made, then the applicable measure of damages, either in tort or contract, may be applied. This approach empowers contracting parties to negotiate remedies just like any other contract term, while recognizing the tort duties of care all owe to all. It also gives appropriate deference to the legislature's decision to impose different statutes of limitations in different areas of law. Especially in products liability, an area of law that had its origins in contracts before finding its home in tort, the economic loss rule can be very helpful in deciding whether a particular "products liability" case is really a breach of contract claim in disguise.

We have used the economic loss rule in residential purchase and sale disputes already. Often in the real property context, the breach of contract is revealed when the property suffers damage. Property damage often invokes tort remedies, but "[i]ncidental property damage, however, will not take a commercial dispute outside the economic loss doctrine; the tail will not be allowed to wag the dog." *Miller*, 902 F.2d at 576.

This demonstrates why understanding "economic loss" to mean "commercial loss" would be helpful. When a piece of property is bought that is *worth* less because of a property defect, that is easily understood to be a commercial loss. But that *exact* same damage caused by a trespass or nuisance, or occurring in a products liability context, may well sound in tort. In those cases, the "loss," properly understood, is not a commercial loss and does

---

[9]As the Pennsylvania District Court noted: In general, the economic-loss doctrine "prohibits plaintiffs from recovering in tort economic losses to which their entitlement flows only from a contract." "The rationale of the economic loss rule is that tort law is not intended to compensate parties for losses suffered as a result of a breach of duties assumed only by agreement." Compensation for losses suffered as a result of a breached agreement "requires an analysis of damages which were in the contemplation of the parties at the origination of the agreement, an analysis within the sole purview of contract law." "In order to recover negligence, 'there must be a showing of harm above and beyond disappointed expectations evolving solely from a prior agreement. A buyer, contractor, or subcontractor's desire to enjoy the benefit of his bargain is not an interest that tort law traditionally protects.'" *Factory Mkt.*, 987 F. Supp. at 395-96.

not arise from a breach of contract. While the loss can be expressed in economic terms (and what cannot be in these days?), the damage should properly be understood to be a property damage, potentially giving rise to relief in tort.

While negligent misrepresentation *may* sound in tort, see *Restatement (Second) of Torts* § 552 (1977), in this case, the claim falls under the contract these parties signed. The remedies available to the parties are controlled by the contract between the parties.

Turning briefly to whether there is potential relief in contract, under the inspection addendum to the contract, the Alejandres were authorized to inspect the septic system and required to notify Bull that they found it unsatisfactory within 10 days. Fairly read, that contractual language put a duty of due diligence on the Alejandres to take steps to protect themselves and to anticipate that Bull might not have complete knowledge of the workings of an underground system. Thus, it was the Alejandres' duty, under the purchase and sale agreement, to exercise due diligence and to satisfy themselves that the septic system was acceptable. If, upon a reasonably diligent inspection, they discovered the septic system was not in good working order, their remedy under the purchase and sale agreement was to rescind the contract or seek other contract remedies. I conclude under the facts of this case that the contract controls, and this claim properly sounds in contract, not tort. To recover, they must prove that the contract they signed was breached. They have not done so. I agree with the majority that the economic loss rule takes this case to contract and, under the contract, they have no claim.

Fraud

The majority is correct that the economic loss rule does not preclude the Alejandres' fraudulent concealment claim. Whether we see this as an exception to the economic loss rule or simply that we recognize that in this state, being defrauded is a dignitary injury, not a commercial one, we reach the same result. I concur with the majority that the Alejandres have not submitted sufficient evidence to go to the jury with this claim or on their common law fraud theories. I also agree with the majority that Bull is entitled to her attorney fees.

Conclusion

A house was purchased with a defective septic system. I do not wish to minimize the significant injury the Alejandres have suffered because of this. The cost of repair was close to a third of the purchase price of the house. I too would be outraged and looking for someone to sue.

But the Alejandres' claim for negligent misrepresentation was primarily a claim for a commercial loss, stemming from an alleged breach of contract. To recover in tort, "'there must be a showing of harm above and beyond disappointed expectations evolving solely from a prior agreement.'" *Factory Mkt.*, 987 F. Supp. at 396. They have not shown this, nor have they proved breach of contract. While a fraud claim is not barred by the economic loss rule, they have not submitted sufficient evidence to take theirs to the jury. I concur with the majority in result.

ADVANCED MARINE ENTERPRISES, INC., ET AL. v. PRC INC. 256 Va. 106; 501 S.E.2d 148 (1998)

Study terms: Temporary restraining order, equitable remedy, punitive damages, business conspiracy, non-competition, goodwill, tortuous interference with contract, confidential and proprietary information

In this appeal, we consider issues in a chancery proceeding involving both equitable and legal claims arising from an alleged business conspiracy and breach of an employment agreement.

PRC Inc. (PRC) is a Delaware corporation that, among other things, provided marine engineering services under contract to the United States Navy. Included in those services was "shipbuilding support" that PRC rendered to the Naval Sea Systems Command (NAVSEA). Advanced Marine Enterprises, Inc. (AME), a Virginia corporation engaged in the business of marine engineering, also provided services under contracts with the Navy, including NAVSEA.

PRC requires every new employee to sign a uniform Employment Agreement as a condition of employment. The Employment Agreement obligates PRC employees to protect PRC's proprietary information and to refrain from disclosing such information to individuals outside the company. The Employment Agreement also contains a non-competition provision, which provides in relevant part:

> Employee agrees not to compete with PRC for a period of eight months following termination of employee's employment, by rendering competing services to or, with respect to such services, solicit any customer of PRC for whom Employee performed services while employed by PRC, within 50 miles of a PRC office.

At various times during 1995, due to the loss of certain marine engineering contracts, PRC informed some of its marine engineering employees that they should look for other employment. On December 13, 1995, PRC announced that the company would be sold to Litton Industries, Inc. (Litton).

In November 1995, prior to the announcement of the sale, C. Michael Pirrera, a senior manager in PRC's marine engineering department, contacted AME and inquired whether AME would be interested in employing all seven managers from PRC's marine engineering department (PRC Managers). When AME expressed interest in hiring the PRC Managers, AME and the PRC Managers, led by Pirrera, formed a plan (the Plan) under which AME would attempt to hire every employee in the PRC marine engineering department.

Under the Plan, AME agreed to make secret job offers to all employees in PRC's marine engineering department. These employees would be required to resign on the same day, December 29, 1995, without notice to PRC, despite PRC's requirement that employees provide two weeks notice of their intent to leave PRC's employ. . . .

AME knew about the terms of PRC's Employment Agreement before implementing the Plan. AME was aware that it faced a potential lawsuit by PRC to enforce the Employment Agreement, and that PRC could assert other causes of action against AME, such as tortious interference with contract. After projecting the nature and amount of damages that might result from a lawsuit by PRC, AME decided that the benefits of the Plan outweighed the potential consequences of a lawsuit.

To implement the Plan, some of the PRC Managers developed a "matrix" describing how the PRC Managers would obtain the business of PRC's marine engineering department. This "matrix" included detailed confidential and proprietary information about PRC's workload, the value of certain work, and the amount of government funding available for each job in PRC's marine engineering department. . . .

Based on employee information supplied by the PRC Managers, AME prepared "offer" letters to each of the PRC Managers and other marine engineering employees. . . . The "offer" letters included a provision in which AME agreed to indemnify and hold harmless each PRC employee against any claim, demand, damage, or injury asserted by PRC in connection with the employee's employment by AME. . . .

On December 20, 1995, Pirrera learned that rumors of the Plan might "leak out " to PRC. In response, Pirrera sent an e-mail message to the other PRC Managers on December 21, 1995, which stated:

> Subject: Execute
>
> Gentlemen:
>
> Based on yesterday's events we need to do the following:
>
> With the exception of the highest risk team members (i.e., people we are absolutely sure will blab), talk to the rest of the team today.
>
> Determine task backlogs immediately.
>
> Back up computer files immediately.
>
> Transfer files to client sites immediately.
>
> Remember gentlemen, we got to this point as a team and we will see this through as a team. Let's roll!
>
> Mike

On December 29, 1995, the whole group of 26 managers and employees from PRC's marine engineering department submitted letters resigning their employment with PRC, effective immediately. Before leaving PRC and without PRC's knowledge or consent, many of the PRC Managers and other PRC employees copied their client files and sent the files to client sites so that the files would be available once the employees began working at AME. . . .

In January 1996, PRC filed a bill of complaint against AME, two AME executives, the former PRC Managers, and the other former PRC marine engineering employees. Among other things, the bill of complaint contained a request for a temporary restraining order to prevent the defendants from using or disseminating PRC's confidential and proprietary information and from soliciting or performing services for their former PRC customers. The chancellor entered the temporary restraining order on January 2, 1996, but later modified its terms to exclude AME's business with governmental entities.

As amended, the bill of complaint also asserted both legal and equitable claims for relief. The five Counts relevant to this appeal are: 1) breach of fiduciary duty (Count I); 2) intentional interference with contractual relations (Count II); 3) intentional interference with prospective business and contractual relations (Count III); 4) specific performance and breach of the Employment Agreement (Count IV); and 5) violation of Code § 18.2-499 (Count VII).

The matter was tried before a chancellor, who heard testimony from two expert witnesses and 41 other witnesses. PRC presented the testimony of Mark Bleiweis, a certified public accountant, who is an expert in the area of damage calculation in contract disputes. Bleiweis estimated that, of the several types of economic damage suffered by PRC in the loss of its marine engineering unit to AME, the largest amount of damages resulted from lost goodwill. Bleiweis defined goodwill as the excess of the sales price of a business over the fair market value of the business' identifiable assets.

■ ■ ■

Bleiweis estimated that PRC sustained $925,123 in goodwill damages from the loss of its marine engineering unit to AME. Using the sale of AME to Nichols, Bleiweis estimated that PRC's lost goodwill damages were $841,965.

■ ■ ■

Bleiweis estimated that the present value of the expected lost profits was $265,655, based on the revenues that the former employees' labor would have generated for PRC. However, he testified that these damages were included in his estimate of lost goodwill.

[AME's expert witness]Edward H. Ripper, a certified public accountant, as an expert in government contract accounting claims and valuation. Ripper testified that Bleiweis' conclusions were "substantially overstated," "highly speculative," and contained many "calculation errors." . . .

On June 19, 1996, the chancellor found in favor of PRC on all counts at issue in this appeal, stating, "I think the method by which the [PRC Managers]] elected to do this was covert, surreptitious, violated civil duties, [and] was absolutely wrong." . . .

Ruling from the bench, the chancellor awarded $1,245,062 in compensatory damages on each of Counts I, II, III, and VII. . . . Under Code § 18.2-500, the chancellor then trebled the $1,245,062 compensatory damage award entered on Count VII. Thus, the total amount of non-punitive damages awarded was $3,735,186.

The chancellor awarded punitive damages in the amount of $1,000,000 against AME, noting that he might be required to reduce that amount to $350,000 under Code § 8.01-38.1. . . .

The chancellor took under advisement AME's argument that the award of treble damages under Code § 18.2-500 was subject to the punitive damages ceiling fixed by Code § 8.01-38.1. . . .

On Count IV, based on breach of the non-competition clause of the Employment Agreement, the chancellor enjoined certain PRC Managers and employees for seven and one-half months from performing services for and soliciting work from those NAVSEA jobs for which each manager or employee provided services while employed by PRC. The chancellor stayed the injunction pending resolution of this appeal, and ruled that if the damages he awarded are approved on appeal, the injunction will be dissolved.

After a hearing on several post-trial motions, the chancellor entered the final decree on June 18, 1997, about one year after the trial. . . .

On appeal, AME argues that the chancellor erred in (1) ruling that AME violated Code § 18.2-499, (2) enforcing the non-competition covenant of the Employment Agreement, (3) his calculation of PRC's lost goodwill and profits and his determination that PRC met its burden of proving damages, [and] (4) awarding punitive and treble damages. . . .

III. "Goodwill" Damages and Sufficiency of Evidence of Damages

AME contends that the chancellor erred in accepting PRC's evidence of damages, including its evidence of lost goodwill and profits. . . . Specifically, AME asserts that the chancellor failed to consider that PRC's marine engineering department made a profit of only $45,108 in 1995, and that the price for the sale of PRC to Litton did not change after the departure of the PRC Managers and employees. We disagree with AME's arguments.

■ ■ ■

After hearing detailed testimony from Bleiweis, PRC's expert, and Ripper, AME's expert, the chancellor accepted Bleiweis' methodology and evidence of damages. We cannot say, as a matter of law, that the chancellor's determination was plainly wrong.

In determining PRC's damages for lost goodwill, the chancellor accepted Bleiweis' variation of the market value approach, a frequently-used method for computing goodwill damages that is based on the difference between the price a business would sell for and the value of its non-goodwill assets. . . .

Since the chancellor's findings regarding PRC's lost profits and damages for lost goodwill are supported by credible evidence, we will not disturb those findings on appeal.

For the same reasons, we find no merit in AME's contention that the damage award was excessive as a matter of law. Although the record shows that the price for the sale of PRC to Litton did not change after the departure of the PRC Managers and employees, Bleiweis emphasized that the departing group had goodwill value for purposes of maintaining the customer relationships necessary for contract retention. As stated above, the chancellor based the award of damages on his acceptance of Bleiweis' testimony, which constituted credible evidence in support of that award.

IV. Punitive and Treble Damages

AME argues that a chancellor in equity may not award punitive damages because any award of damages in equity is limited to compensating an injured party to make it "whole." AME contends that punitive damages are in the nature of a penalty and extend beyond mere compensation. AME also asserts that treble damages are punitive in nature and, thus, are likewise unavailable in a court of equity. We disagree with AME's arguments.

When a court of equity acquires jurisdiction of a cause for any purpose, the court may retain the entire cause to accomplish complete justice between the parties. Thus, the chancellor may hear legal claims and enforce legal rights by applying remedies available only at law. . . .

If a chancellor decides to retain jurisdiction over legal claims, the chancellor acts "as a substitute for the court of law."

PRC's bill of complaint sought both equitable and legal remedies. AME could have moved to transfer its legal claims to the law side of the court under Code § 8.01-270, where it would have been entitled to a jury trial, but chose not to proceed in this manner. Thus, AME cannot now complain that the chancellor improperly awarded legal relief to PRC. We also observe that the chancellor awarded compensatory, punitive, and treble damages under the various legal claims, not under any equitable claims. Therefore, we conclude that the chancellor acted within his discretion in awarding legal relief on the law claims before him. . . .

We also disagree with AME's contention that the chancellor erred in awarding treble damages. Code § 18.2-500(a) provides in relevant part:

> Any person who shall be injured in his reputation, trade, business or profession by reason of a violation of § 18.2-499, may sue therefor and recover three-fold the damages by him sustained, and the costs of suit, including a reasonable fee to plaintiff's counsel; and without limiting the generality of the term, "damages" shall include loss of profits.

This subsection explicitly allows an award of treble damages on proof of the cause of action provided under Code § 18.2-499. Nevertheless, AME asserts that treble damages may not be awarded in equity because Code § 18.2-500(b), which sets forth the equitable relief available for business conspiracy claims brought under the statute, does not specifically state that treble damages may be awarded in a chancery case. Code § 18.2-500(b) provides in relevant part:

> Whenever a person shall duly file a bill in chancery in the circuit court of any county or city against any person alleging violations of the provisions of § 18.2-499 and praying that such party defendant be restrained and enjoined from continuing the acts complained of, such court shall have jurisdiction to hear and determine the issues involved, to issue injunctions pendente lite and permanent injunctions and to decree damages and costs of suit, including reasonable counsel fees to complainants' and defendants' counsel.

We conclude that this provision does not preclude an award of treble damages in a law claim heard in chancery. . . .

AME next contends that the chancellor's award of both punitive and treble damages was duplicative. Although AME concedes that the chancellor awarded punitive and treble damages under separate counts of the bill of complaint, AME argues that the conduct underlying the claims is the same and, therefore, that the chancellor erred in awarding both types of damages. We disagree with AME's argument.

The awards of punitive and treble damages were based on separate claims involving different legal duties and injuries. The chancellor awarded punitive damages under Counts I, II, and III, for breach of fiduciary duty, intentional interference with contractual relations, and intentional interference with prospective business and contractual relations. The award of treble damages was limited to the business conspiracy claim of Count VII.

■ ■ ■

Under the plain language of Code § 8.01-38.1, the limitation of $350,000 applies only to an award of "punitive" damages. If the General Assembly had intended for an award of treble damages to be subject to this limitation, it would have included an express reference to such damages in the statutory language. . . . In the absence of such a reference, we will not construe the plain language in a manner that amounts to holding that the legislature meant other than what it actually stated.

■ ■ ■

Affirmed in part, reversed in part, and remanded.

Review Questions

1. Black hired White Automation to build a special machine to be used by Black in the manufacture of automobile door handles. The door handles were to be sold to Gray Motor Company. The price of the special machine was to be $200,000. Shortly before work was to begin on the special machine, Gray canceled the order for door handles and Black immediately canceled the contract for the special machine. Does White have a right to resort to legal action? If so, against whom and for how much?

2. What is the purpose of (a) nominal damages, (b) compensatory damages, (c) exemplary damages?

3. Distinguish between general damages and special damages.

4. Green bought a new car from Brown Motor Sales for $30,000. A few days later, Green attempted to pass another vehicle, and the steering linkage locked when Green turned the wheels to the left. The resulting crash destroyed the car and sent Green to the hospital. Green's hospital bill amounted to $8,800; during the first month away from Green's job, Green received full pay from her employer ($3000 per month), but in the next two months she did not. When she returned to work with a 20-percent disability, she was asked to take a job paying $2,000 per month because of her inability to perform her former job. Green is 45 years old. Examination of the wrecked automobile showed that one joint in the steering linkage appeared too tight and that there was no grease fitting at the joint and, apparently, there never had been one even though a hole had been drilled and tapped for the fitting. Does Green have a right of action against anyone? If so, for how much? Based on what theory?

5. Why should punitive damages be awarded to a plaintiff in a fraud action but denied when the action is based on a breach of contract?

6. In *Lewis Jorge Construction Management, Inc. v. Pomona Unified School District,* the plaintiff sought recovery of its lost future profits due to the loss of its bonding capacity. The court found that an award under general damages was improper and considered whether Lewis Jorge could recover under special damages for lost profits. What was the court's conclusion?

7. In *R.P. Wallace v. The United States,* the court found that the plaintiff was only entitled to a recovery of $4,200.00 for *liquidated damages.* Why wasn't the plaintiff entitled to "delay damages?"

8. Judge Chambers in a concurring opinion to *Arturo Alejandre, et al. v. Mary M. Bull,* states that the economic loss rule is a tool to ensure what?

9. In *Advanced Marine,* would it have been reasonable for the chanecry cout to think that the $350,000 limit applied to an award of treble damages? What was the "good will" that was lost?

Sales and Warranties

CHAPTER
15

HIGHLIGHTS

■ There are at least three issues under the topic of sales and warranties that are of particular concern to engineers:
1. Engineers are often responsible for making recommendations for the goods and services to be purchased.
2. Timely delivery of goods and services should be insisted upon.
3. Purchasing goods, suggesting appropriate services, and all the varied decision-making that goes along with these tasks should be considered along with the scheduled dates for project completion.

The building of any structure, whether a productive machine, a building to house it, a road, or a bridge, requires the purchase of many things. In a large organization, a purchasing department buys what the engineer specifies; in a small firm, the engineer may need to do some of the buying. In either case, the engineer's job includes at least effective recommendation of the goods and services to be purchased and a vital concern with their adequacy and delivery. A manufacturing engineer or process engineer may not be responsible for making the actual purchase; nevertheless, the engineer remains very much concerned with the items to be bought, because they will be components of the final structure or process. In addition, a target date for completion must be met, and delay in receipt and installation of components may be quite costly. From this standpoint, the engineer must be concerned also with the transportation of the items purchased, and even with the financial arrangements involved.

When things go wrong—when a vendor claims that a product will perform according to the engineer's specifications, but the product fails to do so, for example—does the engineer have any recourse? Such problems do arise, and engineers will benefit from some understanding of their rights and responsibilities regarding sales and warranties.

The term **sale** can refer to a transfer of real property, personal property, or even intangible property. As it is used in this chapter, however, it refers only to the transfer of ownership of tangible personal property, for a price usually stated in money. In other words, we are here concerned only with contracts for the sale of goods. The law that governs such transactions is set forth in practically every state by the Uniform Commercial Code (UCC).

OWNERSHIP

Two concepts basic to the sale of goods are **ownership** and **risk of loss.** Because a sale involves the transfer of ownership, it can be important to know who owns the goods at a particular time. Generally, only the party who owns the goods can lose them, but this is where the risk-of-loss concept comes in. When goods are stolen or destroyed by fire, flood, or other catastrophe, the one who loses (at least, initially) is the party who had the risk of loss.

The UCC allows the parties a great deal of freedom to determine when title (or ownership) passes from seller to buyer and when the risk of loss passes from seller to buyer. The parties can have risk of loss pass at the same time as title passes. However, there is no requirement that they do so. In most situations, a buyer or seller can obtain insurance to cover possible losses or damage to the goods. Thus, buyer and seller may negotiate about when and how risk of loss and title are transferred and about which party pays for insurance (or, if no insurance is purchased, who bears the risks). Consequently, the time when title and risk of loss passes from seller to buyer and the shifting of ownership risks are important.

Title

For title to pass from seller to buyer, the parties must identify the goods as those to which the contract refers. This identification can be made at any time and in any manner explicitly agreed to by the parties. The goods do not have to be in a deliverable state before they can be identified. Also, despite the identification of goods (and the passing of title), the risk of loss remains on the seller until the risk is passed, either as agreed by the parties or pursuant to the terms of the UCC (some of which are discussed below).

Generally, title to goods passes in any manner and on any conditions explicitly agreed to by the parties. In the absence of an explicit agreement, title passes to the buyer at the time and place at which the seller completes performance with reference to the physical delivery of the goods.

For example, Black, in need of 20 cooling fans, visits White's warehouse, where White has about 100 such fans stored. After examining a few sample fans, Black agrees to take 20 of them at an agreed price per fan. At this point there is no sale: White still owns the entire 100 fans, and should anything happen to them, the entire loss is White's. Now assume that Black asks White to separate out 20 fans. Black intends to go get a truck to pick up the fans. If no other material thing remains to be done, Black has title to the 20 separated fans. Suppose now, however, that the parties agreed to have White deliver the fans to Black. Since it appears to be their intent to have title pass on delivery, title would not pass until delivery had occurred. Of course, if the fans were not yet in existence and Black purchased them merely by a description or a sample, title could not pass until the fans were made or at least started and tagged or appropriated for (or somehow "identified" as relating to) Black's contract.

Risk of Loss

F.O.B., F.A.S., C.I.F., and C. & F. are abbreviations commonly used in contracts to describe the terms concerning shipment of goods from seller to buyer (see Figure 15.1). Among other considerations, such terms specify the point at which the risk of loss passes from seller to buyer. For instance, **F.O.B.** San Francisco (which stands for free on board) indicates that when the seller has delivered goods into a carrier's possession in San Francisco with instructions that they be delivered to the buyer, the risk of loss passes to the buyer. Until the goods reach this point, however, the seller suffers if the goods are damaged or lost. **F.A.S.** generally means the same thing with respect to delivery by ship. The term F.A.S. San Francisco, then, would mean that the seller is required to deliver the goods alongside the vessel in the manner usual in that port or on a dock designated by the buyer.

C.I.F. and C. & F. are similar to the concepts of F.O.B. and F.A.S., but they require the seller to do more than just hand the goods over to the common carrier. In a **C. & F.** (cost and freight) sale, the price includes the cost of the goods and freight to the named destination; moreover, the seller must obtain a negotiable bill of lading for the goods, load the goods, and send all the documents plus an invoice to the buyer. A **C.I.F.** (cost, insurance, and freight) contract requires the seller to do everything required in a C. & F. sale in addition to insuring the goods. The buyer must pay the price of the goods (which, of course, includes shipping costs and any other extras) when the documents arrive. In addition, the buyer has no right to wait for the goods or inspect them prior to making payment; instead, the buyer must make payment against tender of the required documents. If the goods are lost or damaged in transit, the buyer has no action available against the seller. The buyer, however, would probably recover under the insurance policy and, failing that, the buyer can file a legal action against the carrier.

Despite the standard definitions given above, the buyer and seller can use these abbreviations in their contracts and still negotiate their own terms. That is, if the wording of their agreement makes it clear that they intended another meaning (for instance, giving the buyer the right to approve of the goods before paying for them), such will be the court's interpretation. The point is that, where these terms are used and not modified by some other agreement, the UCC definitions dictate their meaning.

Security Interest

In our economy, buying on credit accounts for a large number of sales. People (and businesses) buy goods and services now and agree to pay later. But the simple promise to pay later leaves something to be desired. When "later" comes, the buyer may have many debts and very few assets. Without some added security, the seller may be reduced to the status of creditor in bankruptcy and receive only a few cents for each dollar of the debt. In response to these concerns, arrangements have been devised to improve the seller's position (and willingness to sell on credit). Under the UCC the arrangement is known as a security interest. When a seller has a **security interest** in the goods, the seller retains certain rights to the goods sold until the buyer has paid for them.

Consider Black, who is in need of a refrigerator and lacks funds sufficient to pay the entire cash price. Having located the refrigerator of choice, Black arranges to purchase the refrigerator by paying $90 immediately and agreeing to make a series of 10 monthly payments of $90 each to the Brown Appliance Company. To "secure" Black's promise, Brown wants to retain a security interest. To fortify Brown's rights, Brown may want to file a financing statement regarding the security interest. Among other things, a financing statement should include the names and

U.C.C. - ARTICLE 2 – SALES, PART 3.
GENERAL OBLIGATION AND CONSTRUCTION OF CONTRACT

§ 2-319. F.O.B. and F.A.S. Terms.

(1) Unless otherwise agreed the term F.O.B. (which means "free on board") at a named place, even though used only in connection with the stated price, is a delivery term under which

    (a) when the term is F.O.B. the place of shipment, the seller must at that place ship the goods in the manner provided in this Article (Section 2-504) and bear the expense and risk of putting them into the possession of the carrier; or

    (b) when the term is F.O.B. the place of destination, the seller must at his own expense and risk transport the goods to that place and there tender delivery of them in the manner provided in this Article (Section 2-503);

    (c) when under either (a) or (b) the term is also F.O.B. vessel, car or other vehicle, the seller must in addition at his own expense and risk load the goods on board. If the term is F.O.B. vessel the buyer must name the vessel and in an appropriate case the seller must comply with the provisions of this Article on the form of bill of lading (Section 2-323).

(2) Unless otherwise agreed the term F.A.S. vessel (which means "free alongside") at a named port, even though used only in connection with the stated price, is a delivery term under which the seller must

    (a) at his own expense and risk deliver the goods alongside the vessel in the manner usual in that port or on a dock designated and provided by the buyer; and

    (b) obtain and tender a receipt for the goods in exchange for which the carrier is under a duty to issue a bill of lading. . . .

§ 2-320. C.I.F. and C. & F. Terms.

(1) The term C.I.F. means that the price includes in a lump sum the cost of the goods and the insurance and freight to the named destination. The term C. & F. or C.F. means that the price so includes cost and freight to the named destination.

(2) Unless otherwise agreed and even though used only in connection with the stated price and destination, the term C.I.F. destination or its equivalent requires the seller at his own expense and risk to

    (a) put the goods into the possession of a carrier at the port for shipment and obtain a negotiable bill or bills of lading covering the entire transportation to the named destination; and

    (b) load the goods and obtain a receipt from the carrier (which may be contained in the bill of lading) showing that the freight has been paid or provided for; and

    (c) obtain a policy or certificate of insurance, including any war risk insurance, of a kind and on terms then current at the port of shipment in the usual amount, in the currency of the contract, shown to cover the same goods covered by the bill of lading and providing for payment of loss to the order of the buyer or for the account of whom it may concern; but the seller may add to the price the amount of the premium for any such war risk insurance; and

    (d) prepare an invoice of the goods and procure any other documents required to effect shipment or to comply with the contract; and

    (e) forward and tender with commercial promptness all the documents in due form and with any indorsement necessary to perfect the buyer's rights. . . .

Figure 15.1

addresses of the debtor and the secured party, and a description of the property covered by the security interest. Most often, a seller of goods or a lender usually prepares the financing statement by completing a blank form. The place of filing differs from state to state; Brown may want to file it in the county where Black lives and in the secretary of state's office. If Brown were to keep the refrigerator pending full payment, the filing of the financing statement would not be required, because Brown's interest is a special type of security interest. It is a **purchase money security interest** (that is, Brown loaned Black the "purchase money" for the collateral) in a consumer good.

Default.   Still considering Black's refrigerator purchase, suppose that Black makes eight of the ten required payments but cannot make the last two. In other words, Black defaults. The possibility of this happening was, of course, Brown's reason for obtaining the security interest in the first place. So now what can Brown do about it?

When a debtor defaults, the secured party (in this case, Brown) can regain possession of the goods. In other words, Brown can repossess Black's refrigerator. However, Brown has certain obligations, and Black has certain rights even after default and repossession. If the secured party elects to repossess (other methods of enforcing the security interest may be appropriate), that party must proceed in a commercially reasonable manner. For example, if the goods are resold, the secured party must use the proceeds to pay the debtor's obligation and the reasonable costs of repossession and then turn any remainder back to the debtor, but not profit. The secured party is not obligated to give any profit made from the sale of repossesed goods.

However, repossession and resale constitute only one of several possible remedies available to the secured party. The remedy the secured party will elect to use depends to a considerable extent, of course, on the nature of the goods involved. One alternative is simply to obtain a court judgment against the debtor in default, using the security agreement as proof of the debt. The debtor, then, retains possession of the goods, and the secured party can use the powers of the court to collect the debt. Another alternative is to repossess the goods and lease them, using the rental payments to satisfy the balance due and costs. The secured party may sometimes even repossess the goods and keep them to satisfy the debt, but the debtor must be notified that the secured party intends to do this. The UCC attempts to balance the rights of both parties. Brown, as secured party, has a variety of remedies if Black defaults on the refrigerator payments. On the other hand, Black has a right to expect fair treatment from Brown if, for some reason, it becomes impossible for Black to make the payments.

Sale on Approval and Sale or Return

A device frequently used to sell goods is to place them in the prospective buyer's hands for a period of time. Of course, this method can be highly effective because the consumer often finds it difficult to surrender possession at the end of the trial period. This technique can also be highly effective with merchants; they can take goods to sell and then return them if they do not sell. The UCC addresses two distinct types of such sales activity.

In a **sale on approval,** the buyer is given possession of the goods to use or inspect for a period of time. While the goods are in the prospective buyer's hands, the title to them rests with the seller. Thus, any loss of the goods is the seller's loss, with, of course, the right to recover if the prospective buyer caused the loss. Approval may be either express or implied. Expression by the buyer of a willingness to take title to the goods constitutes approval. If a time limit is stated and the goods are held beyond the time limit, the buyer's approval may be implied. If no time limit was stated but the goods are held beyond a reasonable period, approval also may be implied. If the buyer uses the goods as his or her own, exceeding what would be considered a reasonable trial, the buyer's acceptance also may be implied. Until the buyer has registered approval, however, the prospective buyer in a sale on approval arrangement is merely a bailee of the seller's goods. A **bailee** is one who possesses property belonging to another (the **bailor**) with that person's acquiescence.

A sale in which the seller gives the merchant or dealer the right to return the goods at the merchant's option is a **sale or return** transaction. In such sales, the risks of ownership of the property pass to the merchant. Frequently, such sales take the form of sales to a merchant "on consignment" or "on memorandum." Return of the goods revests title and ownership risks in the seller. If a time limit is stated and the goods are not returned in that time, the sale to the merchant becomes final. If no time limit is expressly agreed to by the parties, the merchant has a reasonable time in which to return the goods.

SALES CONTRACTS

The law of contracts sets forth certain rules for the formation, interpretation, and discharge of contracts. If the parties to a contract agree on a specific provision, that provision controls the contract. However, if the parties do not agree to a specific provision, the UCC's rules probably will supply the applicable rule of law. In short, the UCC fills any gaps in the parties' agreement.

According to the law of contracts, the offeror determines the terms of the acceptance (even, perhaps, the means of acceptance). Any alteration of the offer by the offeree in the intended acceptance constitutes a counteroffer. The formation of contracts, both at common law and under the UCC, was considered in Chapter 7.

Many contracts that would have been considered unenforceable for a lack of certainty are enforceable under the UCC. For example, the parties may choose to determine the quantity of product to be delivered later, or they may choose to base the price on a future market quotation or a later arrangement by the parties. Such contracts would probably be considered questionable as "gentlemen's agreements" (or agreements to later agree) under the common law of contracts. But under the UCC, such agreements probably have the legal status of fully enforceable contracts.

The main purpose of the UCC is to reflect the intent of the parties more accurately. If it is apparent that the parties intended to make a contract, the law attempts to enforce the essence of the agreement even if some terms are missing or indefinite. Generally, the parties retain a great deal of freedom to structure their relationship. Where the contract terms are unclear (either because of ambiguities in the language used or the parties' failure to consider certain issues), the UCC attempts to provide rules that supply or define the contract's terms.

WARRANTIES

A warranty (or guarantee) provides added assurance to a buyer. The common-law courts, early on, adopted the concept of privity of contract. Under this view, only a party to the contract could sue on the warranty. Thus, if you bought a car and a family member was injured while driving it, your family member could not sue for a breach of warranty, such as that the car's steering system was defective. As discussed in Chapter 22, Products Liability, that concept has eroded in terms of who may receive the benefits of a warranty. In this section, however, we will discuss the two major types of warranties: express warranties and implied warranties. (See Figure 15.2, Limited Warranty Example.)

Limited Warranty Example

DRILLS-4-ALL warrants the Drill-4-U ("Drill"), to be free from defects in material and workmanship existing at the time of manufacture and appearing within three (3) years from the date of original purchase. If such a defect appears during the warranty period, we will (at our sole option) repair or replace the defective unit with no charge for service or parts, provided that the unit is delivered at the customer's expense to one of the authorized service centers listed.

This warranty does not cover claims resulting from misuse, failure to follow instructions on installation and use, neglect, use of unauthorized attachments, commercial use, use on a current or voltage other than specified on the brewer, or unauthorized service during the warranty period on conditions otherwise covered by warranty.

THIS WARRANTY SHALL BE EXCLUSIVE AND SHALL BE IN LIEU OF ANY OTHER EXPRESS WARRANTY, WRITTEN OR ORAL, INCLUDING BUT NOT LIMITED TO ANY EXPRESS WARRANTY OF MERCHANTABILITY OR FITNESS FOR A PARTICULAR PURPOSE. THE DURATION OF ANY IMPLIED WARRANTIES, INCLUDING BUT NOT LIMITED TO ANY IMPLIED WARRANTIES OF MERCHANTABILITY OR FITNESS FOR A PARTICULAR PURPOSE, IS EXPRESSLY LIMITED TO THE THREE YEAR PERIOD OF DURATION OF THIS LIMITED WARRANTY.

Some states do not allow limitations on how long an implied warranty lasts, so the above limitation may not apply to you.

THE CUSTOMER'S EXCLUSIVE REMEDY FOR BREACH OF THIS WARRANTY OR OF ANY IMPLIED WARRANTY OR OF ANY OTHER OBLIGATION ARISING BY OPERATION OF LAW OR OTHERWISE SHALL BE LIMITED AS SPECIFIED HEREIN TO REPAIR OR REPLACEMENT, AT OUR SOLE OPTION. IN ANY EVENT, RESPONSIBILITY FOR SPECIAL, INCIDENTAL AND CONSEQUENTIAL DAMAGES IS EXPRESSLY EXCLUDED.

Some states do not allow the exclusion or limitation of incidental or consequential damages, so the above limitation or exclusion may not apply to you.

This warranty gives you specific legal rights, and you may have other rights that vary from state to state.

Figure 15.2

Express Warranties   shared + clear

A **warranty** is a promissory statement. In making a warranty, the vendor makes a contract somewhat similar to insurance. The vendor essentially agrees to assume a risk that would normally be borne by the buyer. Here, it is important to distinguish between statements that may be fraudulent and warranty statements. You recall that fraud involves a false representation of a fact—something in the past or present. Warranty, on the other hand, has to do with either the present or the future.

When Black sells White an automatic screw machine, for example, and tells White that it has just been overhauled, when in truth it has not been, such a statement is fraudulent. If Black, on the other hand, promises to repair the machine if it should break down in the first year White uses it, this statement constitutes a warranty. Black is promising to take over White's risk of repairing the machine during the year after White's purchase.

Statements having to do with the present, however, are a little harder to distinguish. A statement by Black that all the collets and pushers on the machine have been replaced and are new would be a statement regarding fact; if untrue, the statement could be the basis of fraud. In contrast, a statement that the machine is in such a condition that it will be useful in the manufacture of White's product would constitute a warranty. Another distinction between a warranty and a fraudulent representation is that a warranty becomes part of a contract. In cases where a false representation has been made to induce a sale, the buyer may elect to pursue either a remedy for fraud or a remedy for breach of warranty.

Opinions.   A warranty is a statement of fact, not one of opinion or judgment. With few exceptions, a statement made as the seller's opinion cannot constitute a warranty. A statement by the vendor that the merchandise is "first rate," "the best," or "superior quality" is usually construed as sales talk, or "puffing." The courts have long adhered to the idea of *caveat emptor* ("let the buyer beware"). According to this logic, buyers are free to inspect goods before buying them and are free not to buy if the seller does not allow them to inspect or if they find something wrong.

The recent tendency in the law, however, is to place more responsibility upon the seller. It might even be called *caveat vendor*. This trend is based upon the notion that many statements a vendor makes, in effect, relieve the buyer of the duty to ascertain the value of the goods. An affirmation of value having this effect has been held to be a warranty in some recent decisions. Also, many states have adopted statutes aimed at certain deceptive trade practices. Under such statutes, representations may be actionable even though the representations fail to rise to the level of a warranty.

Description or Sample.   Contracts for the sale of goods often involve samples of the goods the buyer is to receive or descriptions of them. The use of such descriptions or samples generally constitutes an express warranty of what the buyer is to get. Because the buyer normally does not have an opportunity to examine the goods when a description or sample is used, the seller would have a chance to substitute inferior goods. The warranty is meant to prevent this substitution. Accordingly, the UCC provides that any description of the goods (or any sample or model) that is made part of the "basis of the bargain" creates an express warranty that the goods will conform to the description (or the sample or model, as the case may be).

Implied Warranties   *implied by behavior*

In addition to the seller's express warranties, there may be implied warranties. These implied warranties automatically exist in all sales contracts unless they are expressly disclaimed by the parties. (See *Eastern Steel Constructors v. The City of Salem* at the end of this chapter).

Good Title.   In a contract to sell goods, there is an implied warranty that the buyer will receive good title, guaranteeing that there are no rights or liens on the goods other than those of which the buyer is made aware. In other words, no other person has valid claim to the merchandise in question. Specific exceptions to the warranty of good title exist in the form of sheriff's sales, certain auctions, mortgagee's sales, and the like. Such sales are often authorized by law, but they include no assurances regarding prior owners or lien holders.

With a related warranty, the seller guarantees that the goods are free of any rightful claim of a third person by way of infringement or the like. Such a warranty applies to sellers who regularly deal in the goods sold. Thus, a seller of an automated materials-handling system impliedly warrants that the system does not infringe someone else's patents. However, if the seller builds goods in accordance with the specifications furnished by the buyer, the buyer indemnifies the seller against any claim against the seller that might arise out of such specifications.

Fitness for Described Purpose.   The seller is assumed to be more familiar with the goods than the purchaser. For example, a buyer makes a purchase by describing to the seller what the goods are to do, with the seller selecting and then supplying the goods for that purpose. Because the buyer relies on the seller's skill or judgment to select appropriate goods, the goods come with an **implied warranty of fitness** for the buyer's purpose (as he or she described it to the seller). This warranty is not applicable, however, if the buyer orders goods according to a trade name or trade specification. When the buyer orders in such a way, it makes no difference if the seller knows of the intended purpose and believes the goods will not fit the purpose.

For example, Black requires a punch press for a blanking operation on a production line being set up. Black's calculations (in error) show a requirement of a 40-ton press. White, with full knowledge of the purpose intended, is called upon to supply the press and does so. On the first day of operation, the punch-press crank breaks, and Black brings an action on the implied warranty of fitness for the purpose of the press. Since Black specified the press, Black cannot recover for any breach of such a warranty. If, on the other hand, Black had asked White to supply a press for the blanking operation, allowing White to determine the required press size, the fitness of the press for Black's purpose would have been impliedly warranted.

Merchantability.   All goods sold by a merchant are subject to an implied warranty of **merchantability.** To be merchantable, the goods must at least

1. Pass without objection in the trade under the contract description;
2. Be of fair to average quality within the description (in the case of fungible goods);
3. Be fit for the ordinary purposes for which such goods are used;
4. Be of even kind, quality, and quantity within each unit and among all units involved;
5. Be adequately packaged and labeled as may be required by the agreement;
6. Conform to the promises or statements of fact made on the container or label.

In the above example, if it could be shown that the punchpress crankshaft had not been properly heat treated and, therefore, could not withstand a 40-ton force, Black might recover for breach of the implied warranty of merchantability.

Disclaimers

Parties to a sales contract are free to contract in any lawful manner they choose. Frequently, sales contracts contain disclaimers—clauses that, in effect, state that there are no warranties, express or implied, and that the buyer takes the goods at his or her own risk. As long as public policy is not seriously involved, such clauses are usually lawful and binding. A contract provision that the buyer takes the goods "as is" or "with all faults" will be held to relieve the seller of all implied warranties. Of course, even in an "as is" sale, the merchandise sold must be what it is purported to be. That is, sale of a vertical milling machine indicates that what is bought will constitute a vertical milling machine even though the term "as is" is used in the sale.

Generally, any such disclaimer of the implied warranty of merchantability must mention merchantability and, if in writing, it must be "conspicuous." This means that the disclaimer must stand out from the rest of the writing and attract the attention of the reader for it to be effective as a disclaimer. In other words, if one wishes to exclude implied warranties, the seller cannot hide such wishes from the buyer at the time of the sale.

Moreover, warranties are construed to be cumulative. Thus, the buyer's rights under any implied warranties are added to the buyer's rights under any express warranties. Generally, the courts try to construe disclaimers to be reasonably consistent with any express warranties; in cases of doubt, however, courts often construe disclaimers of warranties narrowly (i.e., in favor of the buyer).

PERFORMANCE AND BREACH

The parties' obligations in a contract are for the seller to transfer and deliver the goods and for the buyer to accept and pay for the goods in accordance with the contract. In other words, the seller is to package and ship goods that conform to the contract (i.e., meet or exceed the warranties applicable) so that the goods arrive at the time and place specified by the agreement. Assuming that the seller meets these obligations, the buyer is obligated to accept the goods and to pay the contract price. In the vast majority of such sales contracts, the parties perform without a problem. The rest of this chapter considers the occasions when a party does not.

The UCC requires that the seller "tender delivery" by placing conforming goods at the buyer's disposal. The manner, time, and place for tender are to be determined by the parties' agreement and the relevant provisions of the UCC. Unless otherwise agreed, the buyer has a right to inspect the goods in a reasonable manner before accepting or paying for them.

If the goods conform to the contract, the buyer is obligated to accept them and pay for them. If the goods received by the buyer are nonconforming, however, the buyer has several options. The buyer may

1. Accept the goods sent and pay for them at the contract price, subject to the buyer's right to sue for the cost of the nonconformance of the goods;
2. Reject the entire shipment of goods; or
3. Accept any conforming commercial unit or units and reject the rest.

If the buyer rejects the goods, he or she must inform the seller within a "seasonable" time after rejection and state the reason for rejection. The seller may then instruct the buyer as to disposition of the goods. If no instructions as to the disposition of the goods are forthcoming, the buyer is simply required to act reasonably according to the nature of the goods and the circumstances. If the buyer rejects the goods but the time allowed for the seller's performance has not yet passed, the seller may notify the buyer of the seller's intention to "cure." A seller may then cure by making a delivery of conforming goods within the contract time for performance.

A buyer may accept the goods by expressly telling the seller that he or she accepts them. Probably more often, however, the buyer accepts the goods by continuing to retain possession of them and to use them after a reasonable opportunity to inspect them has come and gone. If the buyer accepts nonconforming goods, the buyer must notify the seller of the nonconformity within a reasonable time after the buyer discovered (or should have discovered) the nonconformity. A failure to give such notice usually will bar the buyer from any remedy.

In some situations, a buyer may accept nonconforming goods because of the seller's representations about fixing any problems. A buyer may revoke an acceptance of a lot or commercial unit whose nonconformity "substantially" impairs its value if the acceptance was made either (1) on the reasonable assumption that its nonconformity would be cured, but the nonconformity has not been "seasonably" cured; or (2) without discovery of the nonconformity, and such acceptance was reasonably induced either by the difficulty of discovery or by the seller's assurances. Thus, if a seller assured that a nonconformity would be fixed (but did not fix the nonconformity) or assured that a defect or problem was really not a nonconformity, then the buyer may revoke the acceptance. If the defect is latent and cannot be reasonably detected by normal inspection techniques, a buyer may also have a right to revoke an earlier acceptance.

Suppose Green ships the Brown Company 10,000 zubit components known as ZIP-2s. Brown Company's receiving inspection shows the lot of ZIP-2s to consist of about 20 percent defective units and, therefore, to be unacceptable. Brown Company rejects the entire shipment of 10,000 ZIP-2s and notifies Green. Green may require the entire lot of 10,000 to be returned or may ask Brown Company to sort the 10,000. If the nonconformance could shut down a production line, the idea of mitigation of damages would probably require Brown Company to sort and use the good ones. In any event, Green should be liable for any added costs caused by the nonconformance of the shipment. If the ZIP-2s are nonconforming, Green probably will be liable for Brown Company's inspection costs. On the other hand, Green would not be liable for such inspection costs if the goods entirely conformed to the contract specification.

BUYER'S REMEDIES

As you recall from Chapter 14, Remedies, if the seller breaches the contract, the buyer has a right to take legal action. The parties may anticipate the buyer's damages, however, in a "liquidated ① damages" clause in the contract. As previously noted, such a clause may be enforced. If considered a penalty, however, the clause will not be enforced.

If the seller breaches the contract by failing to deliver the goods or by repudiating the contract, the buyer has several options available. First, the buyer may procure substitute goods from ② another source. The buyer's damages, then, would reflect the difference between the costs of cover (i.e., the price for the cover goods) less the contract price, together with any costs resulting from the delay involved or the costs of obtaining the goods from another source. Second, the buyer may seek damages based on the difference between the market price for the goods at the time the buyer learned of the seller's breach less the contract price for such goods. Additional damages, such as lost profits, also may be recoverable.

The UCC also provides the buyer with the right to replevin (or recover possession of) the goods if, after reasonable effort, the buyer cannot procure substitute goods or if the circumstances indicate that such effort will be unavailing (see Figure 15.3). Finally, if the buyer can establish that the goods are unique, the buyer can resort to the equitable remedy of specific performance. In effect, the buyer can seek an order requiring the seller to deliver the goods to the buyer.

If the buyer rightfully rejects or revokes acceptance, the buyer has a security interest in the goods to the extent of any part payments or expenses reasonably incurred in their inspection, receipt, transportation, or the like. At that point, then, the buyer's rights are those of the secured party.

What if the seller delivers the goods as required, but the goods are nonconforming? Damages for breach of contract may take the form of a set-off or deduction from the price to be ③ paid for the goods. If making a partial payment may leave the buyer in breach for failure to pay the contract price, the buyer must notify the seller of his or her intention to do so. The buyer's right to take legal action based on the seller's breach eventually ends if the buyer does nothing about it. The statute of limitations for such claims specifies this length of time as no longer than four years. Similarly, the seller's time to sue a buyer for breach is four years. In many states, statutes allow the parties to shorten this time limit in their agreement; such statutes, however, usually set a minimum time, such as one year, to which the parties can agree as a shortened statute of limitations.

U.C.C. - ARTICLE 2 – SALES, PART 7. REMEDIES

§ 2-711. Buyer's Remedies in General; Buyer's Security Interest in Rejected Goods.

(1) Where the seller fails to make delivery or repudiates or the buyer rightfully rejects or justifiably revokes acceptance then with respect to any goods involved, and with respect to the whole if the breach goes to the whole contract (Section 2-612), the buyer may cancel and whether or not he has done so may in addition to recovering so much of the price as has been paid

  (a) "cover" and have damages under the next section as to all the goods affected whether or not they have been identified to the contract; or

  (b) recover damages for non-delivery as provided in this Article (Section 2-713).

(2) Where the seller fails to deliver or repudiates the buyer may also

  (a) if the goods have been identified recover them as provided in this Article (Section 2-502); or

  (b) in a proper case obtain specific performance or replevy the goods as provided in this Article (Section 2-716).

(3) On rightful rejection or justifiable revocation of acceptance a buyer has a security interest in goods in his possession or control for any payments made on their price and any expenses reasonably incurred in their inspection, receipt, transportation, care and custody and may hold such goods and resell them in like manner as an aggrieved seller (Section 2-706).

§ 2-703. Seller's Remedies in General.

Where the buyer wrongfully rejects or revokes acceptance of goods or fails to make a payment due on or before delivery or repudiates with respect to a part or the whole, then with respect to any goods directly affected and, if the breach is of the whole contract (Section 2-612), then also with respect to the whole undelivered balance, the aggrieved seller may

  (a) withhold delivery of such goods;

  (b) stop delivery by any bailee as hereafter provided (Section 2-705);

  (c) proceed under the next section respecting goods still unidentified to the contract;

  (d) resell and recover damages as hereafter provided Section 2-706);

  (e) recover damages for non-acceptance (Section 2-708) or in a proper case the price (Section 2-709);

  (f) cancel.

Figure 15.3

SELLER'S REMEDIES

If the seller delivers goods to the buyer and those goods conform to the buyer's specification, the seller expects to be paid for them. The buyer, however, might refuse to accept the goods, might repudiate the contract, or might not pay the price. The UCC provides the seller with a set of remedies in such cases (see Figure 15.3). One rather obvious remedy is to sue the buyer for damages. Of course, the seller also may simply cancel the contract if the buyer breaches. The seller has other alternatives, too, and they are discussed below.

Withholding Delivery

If the buyer breaches the whole contract, the seller has a right to retain the goods and withhold delivery. If the goods are in the hands of a common carrier or other bailee when the seller has the right to withhold delivery, the buyer's breach allows the seller to stop delivery of the goods by notifying the carrier or bailee. However, the seller's right to stop delivery automatically ceases when the goods are received by the buyer, when a bailee of the goods (except a carrier) acknowledges to the buyer that the bailee holds the goods for the buyer, or when any negotiable document of title is negotiated to the buyer.

A problem arises when the goods are being manufactured for the buyer and are to be shipped in a series of shipments. When the buyer's breach occurs, the goods may be in various stages of completion. In such situations, the seller must decide whether to finish any work in process. Depending on the nature of the goods, they might conceivably be finished and sold to another buyer. If the first buyer is the only customer for them, however, further work on the inventory would seem pointless.

Seller's Resale

If the buyer breaches the contract, the seller has a right to resell the goods to another, either by public or private sale. The UCC imposes certain conditions on such sales to protect the buyer. For example, the terms, method, manner, time, and place of resale must be commercially reasonable. If the sale is public, even the seller may purchase the goods. Any buyer at the resale takes the goods free of any claim by the original buyer.

If the resale yields a lower price than the original buyer was to pay (i.e., the contract price), the seller has a right to recover from the buyer the difference between the resale price and the contract price, plus incidental costs. However, if resale yields a profit for the seller, the original buyer has no claim on that profit.

Unless the parties have agreed otherwise, when one party breaches a contract, the other's right to compensatory damages usually includes lost profits. In other words, the goal is to place the non-breaching party in the position he or she would have enjoyed had no breach occurred. Since most commercial transactions are undertaken for a profit motive, claims for lost profits are fairly common.

> ## EASTERN STEEL CONSTRUCTORS, INC. v. THE CITY OF SALEM, A MUNICIPAL CORPORATION, et al., *209 W. Va. 392; 549 S.E.2d 266; 2001 W. Va. LEXIS 3*

Study terms: Specifications, economic loss doctrine, privity, implied warranty, duty of care, third party beneficiary, warranty, professional negligence

Eastern Steel Constructors, Inc., a contractor, appeals an order of the Circuit Court of Harrison County granting summary judgment in favor of Kanakanui Associates, a design professional, as to Eastern Steel Constructors' claims for professional negligence, implied warranty of plans and specifications, and as a third-party beneficiary to a contract between Kanakanui Associates and the City of Salem, West Virginia. The circuit court rejected the claim for professional negligence based upon its conclusion that, because Eastern Steel Constructors' sought only economic damages, this cause of action could be maintained only as a cause of action in contract. With respect to the implied warranty claim, the circuit court reasoned that absent a contract between the parties, there was no duty owed. Finally, the circuit court found that Eastern Steel Constructors' was not a third-party beneficiary of the contract between Kanakanui Associates and the City of Salem. After reviewing the parties briefs, the record submitted on appeal, and the relevant law, we find that a contractor may assert a negligence cause of action against a design professional seeking purely economic damages even in the absence of privity of contract, that there exists an implied warranty of plans and specifications that inures to a contractor in the absence of a contract, and finally, that Eastern Steel Constructors' failed to establish any evidence supporting its third-party beneficiary claim.

I. Factual and Procedural History

The City of Salem, West Virginia (hereinafter "Salem"), a defendant and third-party plaintiff below and an appellee herein, entered into a contract with Kanakanui Associates (hereinafter "Kana-kanui"), also a defendant below and appellee herein, under which Kanakanui was to provide engineering and architectural services for certain improvements to Salem's existing sewer system, including the design of a new sewage treatment plant to be built under one construction contract, and of two sewer lines to the new plant that were to be built under two additional construction contracts. Kanakanui created particular documents, namely plans and specifications, to be used to solicit bids from interested construction companies in connection with the three separate contracts, and to be further used by the successful bidders in constructing the project. After the completion of the bidding process, Eastern Steel Constructors, Inc. (hereinafter "Eastern"), plaintiff below and appellant herein, was awarded a contract for the construction of one of the sewer lines to the new plant.

Eastern contends that after beginning construction on the project, it experienced significant delays caused by sub-surface rock conditions and existing utility service lines that had not been disclosed in the documents prepared by Kanakanui. Kanakanui submits that under the contract between Salem and Eastern, Eastern was "to be responsible for the installation of the facilities *regardless of the type, nature, or quantity of subsurface conditions, including rock, on the Project.*" (Emphasis added).

As a result of the delays encountered in the project, Eastern asserts, it incurred substantial actual and consequential damages. In addition, Eastern maintains that Kanakanui failed to properly administer and manage the project, which, according to Eastern, caused it further financial damage.

Eastern subsequently filed tort actions against both Salem and Kanakanui for its damages. Eastern's complaint contained three allegations that involved Kanakanui: (1) that Kanakanui had been negligent in its provision of construction engineering services, consultation, project inspection, project management, and project administration; (2) that both Salem and Kanakanui breached an implied warranty of plans and specifications; and (3) that the Eastern was entitled to damages as a third-party beneficiary of the contract between Salem and Kanakanui.

Kanakanui responded with a motion for summary judgment pursuant to Rule 56(b) of the West Virginia Rules of Civil Procedure. Following a hearing on Kanakanui's motion, the Circuit Court of Harrison County, by order entered November 2, 1999, granted the motion. In reaching its conclusion that summary judgment should be awarded in favor of Kanakanui, the circuit court found that Kanakanui's motion presented two issues of law. First, whether the execution and rendering of a contract between an engineer/architect and an owner for the design, plans, and specifications of a project imports a duty from the engineer/architect to a contractor constructing part of the project under a separate contract with the owner. If the answer to this legal question is affirmative, the circuit court observed, then subsequent questions of whether such a duty conveys to the contractor a right to maintain actions against the engineer/architect for negligence in performing its contract obligations and for breach of an implied warranty of plans and specifications must be addressed. The second legal issue identified by the circuit court was whether a construction contractor hired by an owner has a third-party beneficial interest arising from a contract between an engineer/architect and the owner, to which the construction contractor was not a party, such that the construction contractor may institute an action against the engineer/architect for its failure to properly render the services contracted. After identifying these issues, the circuit court concluded, as to the first issue, that the prevailing law in West Virginia limits the recovery of a building contractor to an action for economic damages against the owner as an action in contract only, and that there is not a duty owed by the engineer/architect to the building contractor regarding the plans, drawings and specifications, the adequacy or inadequacy of any or all of them and for the administration of the contract, under the engineer/architect's contract to and for the owner.

With regard to the second issue, the circuit court concluded "that the building contractor is precluded by *West Virginia Code* § 55-8-12 from maintaining any claim as a party with a beneficial interest in the contract between the Owner and the Engineer/Architect because the contractor was not specifically identified and the action instituted was in tort." It is from this order that Eastern now appeals.

II. Standard of Review

It is well established that our review of a grant of summary judgment is de novo. . . .

III. Discussion

A. Design Professional's Liability to Contractor For Purely Economic Damages Resulting from Professional's Negligence

The first question raised in this appeal is whether there exists in West Virginia a cause of action sounding in negligence whereby a construction contractor may recover damages for purely economic losses from a design professional (e.g. architect or engineer) in the absence of a contract between the contractor and the design professional. The trial court concluded that such a cause of action may be pursued only as an action in contract law. We disagree.

This Court previously addressed the question of whether a claim for negligence may lie in the context of the construction industry where there is no contract between the parties to a dispute in the case of *Sewell v. Gregory*, 179 W. Va. 585, 371 S.E.2d 82 (1988). . . .

For the purposes of the instant case, then, Sewell instructs us that Eastern may properly assert a cause of action for negligence against Kanakanui if it can be established that Kanakanui owed a duty of care to Eastern. With regard to the existence of a duty of care, the *Sewell* Court further held, in Syllabus point 3,

> the ultimate test of the existence of a duty to use care is found in the foreseeability that harm may result if it is not exercised. The test is, would the ordinary man in the defendant's position, knowing what he knew or should have known, anticipate that harm of the general nature of that suffered was likely to result?

179 W. Va. 585, 371 S.E.2d 82. The Sewell Court concluded that it was foreseeable to the contractor when he constructed the house that there would be subsequent purchasers. Therefore, the Court reasoned, the contractor had "a common law duty to exercise reasonable care and skill in the construction of a building . . . [and a] subsequent homeowner can 'maintain an action against a builder for negligence resulting in latent defects which the subsequent purchaser was unable to discover prior to purchase.'" *Id. at 588.* In reaching its conclusion, the *Sewell* Court was not, however, required to address the general rule precluding economic damages in a cause of action, such as the case at bar, where negligence is claimed in the absence of either physical injury, property damage or a contract.

More recently, in a case that did involve a plaintiff seeking purely economic damages as a result of the defendant's negligence, this Court conducted an elaborate review of the determination of the existence of a duty of care. See *Aikens v. Debow, 208 W. Va. 486, 541 S.E.2d 576, 2000 W. Va. LEXIS 118* (No. 27376 Nov. 6, 2000). The *Aikens* Court began by noting that the question of whether a duty exists is a question of law for the court to resolve: "the determination of whether a defendant in a particular case owes a duty to the plaintiff is not a factual question for the jury; rather the determination of whether a plaintiff is owed a duty of care by a defendant must be rendered by the court as a matter of law." Syl. pt. 5, *Aikens*. In defining the proper considerations for ascertaining the existence of a duty, we observed in *Aikens* that, in addition to the primary question of foreseeability of risk in discerning the existence of a duty, consideration must also be given to "'the likelihood of injury, the magnitude of the burden of guarding against it, and the consequences of placing that burden on the defendant.'" *Aikens*, 208 W. Va. 486, 541 S.E.2d 576, 2000 W. Va. LEXIS 118, *12. . . .

Having established the existence of a well settled general rule against permitting recovery in negligence for purely economic damages, however, the *Aikens* Court acknowledged that a minority of jurisdictions have permitted such recovery "under certain limited circumstances." 208 W. Va. 486, 541 S.E.2d 576, 2000 W. Va. LEXIS 118, *34. After a thorough review of case law from jurisdictions strictly adhering to the general rule of no economic recovery, as well as that from jurisdictions that have developed and applied exceptions to that general rule in order to permit economic recovery, we expressed "our belief that a hybrid approach must be fabricated to authorize recovery of meritorious claims while simultaneously providing a barrier against limitless liability." *Aikens*, 208 W. Va. 486, 541 S.E.2d 576, 2000 W. Va. LEXIS 118, *45. We went on to explain that "the common thread which permeates the analysis of potential economic recovery in the absence of physical harm is the recognition of the underlying concept of duty. Absent some *special relationship*, the confines of which will differ depending upon the facts of each relationship, there simply is no duty." *Id.* (emphasis added). The Court further explained that the existence of a special relationship will be determined largely by the extent to which the particular plaintiff is affected differently from society in general. It may be evident from the defendant's knowledge or specific reason to know of the potential consequences of the wrongdoing, the persons likely to be injured, and the damages likely to be suffered. Such special relationship may be proven through evidence of foreseeability of the nature of the harm to be suffered by the particular plaintiff or an identifiable class and can arise from contractual privity or other close nexus.

While the economic loss asserted in the *Aikens* case involved a disruption to commerce, and the case did not involve the construction industry, we nevertheless adhere to our belief, expressed in that opinion, that recovery of economic damages should be allowed in certain meritorious claims when an adequate barrier against limitless liability, such as the existence of a special relationship, can be identified:

> Where a special and narrowly defined relationship can be established between the tortfeasor and a plaintiff who was deprived of an economic benefit, the tortfeasor can be held liable. In cases of that nature, the duty exists because of the special relationship. The special class of plaintiffs involved in those cases were particularly foreseeable to the tortfeasor, and the economic losses were proximately caused by the tortfeasor's negligence.

Aikens 208 W. Va. 486, 541 S.E.2d 576, 2000 W. Va. LEXIS 118, *46. To this end, we note that, for reasons similar to those we expressed in *Aikens*, numerous courts have allowed the recovery of economic damages by a contractor for the negligence of a design professional where there was no contract between the two and where there was no physical injury or property damage.

■ ■ ■

The *Guardian* Court [*Guardian Construction Co. v. Tetra Tech Richardson, Inc.*, 583 A.2d 1378, 1381 (Del. Super. Ct. 1990)] ultimately relied on the *Restatement (Second) of Torts § 552* (1977), as well as case law from other jurisdictions, to conclude that, under the circumstances of the case before it, the contractor's action in negligence was not barred by a lack of privity notwithstanding the fact that the damages sought were purely economic. The Court explained:

> Modern legal authority supports the proposition that if, in the course of its business, [a design professional] negligently obtained and communicated incorrect information specifically known and intended to be for the guidance of [contractors], and if it is specifically known and intended that [the contractors] would rely in calculating their project bids on that information, and if [the contractors] rely thereon to their detriment, then [the design professional] should be liable for foreseeable economic losses sustained by [the contractors] regardless of whether privity of contract exists.

Guardian, 583 A.2d at 1386. The Court further observed the close nexus between the negligence and the economic harm suffered: "the use of the information negligently supplied was not an indirect or collateral consequence . . . it was the end and aim of the transaction." *Id.*

Another like case is *Donnelly Construction Co. v. Oberg/Hunt/Gilleland*, 139 Ariz. 184, 677 P.2d 1292 (1984). In Donnelly a construction contractor sued a design professional for negligence. The contractor claimed that it had relied upon plans, specifications, and a site plan prepared by the design professional to calculate its bid on a project. The contractor was awarded the project based upon its bid amount. However, once construction began, the contractor discovered that the aforementioned documents contained substantial errors that resulted in increased costs to the contractor. In overturning a dismissal granted by the trial court, the Supreme Court of Arizona found that the absence of a contract between the contractor and the designer did not preclude the contractor's negligence action as privity was not required to maintain an action in tort. . . .

The Court elaborated:

> Design professionals have a duty to use ordinary skill, care, and diligence in rendering their professional services. . . . When they are called upon to provide plans and specifications for a particular job, they must use their skill and care to provide plans and specifications which are sufficient and adequate. . . . This duty extends to those with whom the design professional is in privity, . . . and to those with whom he or she is not. . . .

Id. Finally, the Court concluded that it was foreseeable that the contractor, who was "hired to follow the plans and specifications prepared by [the design professional], would incur increased costs if those plans and specifications were in error." *Donnelly*, 139 Ariz. at 187-188, 677 P.2d at 1295-96.

We are persuaded by our prior analysis in *Aikens v. Debow*, and the foregoing authority from other jurisdictions allowing contractors to assert negligence causes of action to recover economic damages in the absence of contractual privity, consequently we expressly hold that a design professional (e.g. an architect or engineer) owes a duty of care to a contractor, who has been employed by the same project owner as the design professional and who has relied upon the design professional's work product in carrying out his or her obligations to the owner, notwithstanding the absence of privity of contract between the contractor and the design professional, due to the special relationship that exists between the two. Consequently, the contractor may, upon proper proof, recover purely economic damages in an action alleging professional negligence on the part of the design professional. . . .

Having established that a design professional owes a duty of care to contractors, we endeavor to give some definition to that duty. We note that the exact nature of the specific duty owed by a design professional *may* be impacted by provisions contained in the various contracts entered among the parties (e.g. the contract between the owner and the design professional, and the contract between the owner and the contractor), provided that such contractual provisions do not conflict with the law. In addition, the duty of care may be further defined by rules of professional conduct promulgated by the agencies charged with overseeing the specific profession of which a defendant is a member. West Virginia Rules of Professional Conduct for Architects, 1A W. Va. C.S.R. § 2-1-9 et seq. (1998). Consequently, we hold that when a special relationship exists between a design professional and a contractor, the specific parameters of the duty of care owed by the design professional to the contractor must be defined on a case-by-case basis. However, in general, the duty of care owed by a design professional to a contractor with whom he or she has a special relationship is to render his or her professional services with the ordinary skill, care and diligence commensurate with that rendered by members of his or her profession in the same or similar circumstances.

Having determined that a contractor may indeed maintain a cause of action for negligence seeking purely economic damages against a design professional where there is no privity of contract between the two, we conclude that the circuit court erred in granting summary judgment in favor of Kanakanui as to Eastern's claim of professional negligence.

B. Design Professional's Liability to Contractor for Implied Warranty of Plans and Specifications

We next address the circuit court's award of summary judgment in favor of Kanakanui as to Eastern's claim for breach of an implied warranty of plans and specifications. As with Eastern's claim for professional negligence, the circuit court concluded that a claim for breach of implied warranty could be pursued only as an action sounding in contract. We again disagree. . . .

Due to the special relationship that exists between a design professional and a contractor, which is discussed in the preceding section of this opinion, we believe a similar conclusion is warranted in the case of an implied warranty of plans and specifications. While, in a technical sense, the plans and specifications are prepared for the owner of a project, a design professional nonetheless knows that they will be relied upon by contractors vying for the project, and ultimately will be further relied upon by the contractor who is hired to perform the actual work. Furthermore, errors and inadequacies in the specifications will forseeably work to the financial detriment of the contractor. Consequently, an innocent contractor should be protected by a warranty, and design professionals thereby held accountable for their work.

Following existing West Virginia precedent . . . we hold that a design professional (e.g. an architect or engineer) providing plans and specifications that will be followed by a contractor in carrying out some aspect of a design, impliedly warrants to the contractor, notwithstanding the absence of privity of contract between the contractor and the design professional, that such plans and specifications have been prepared with the ordinary skill, care and diligence commensurate with that rendered by members of his or her profession.

Because we conclude that a contractor may pursue a claim for the breach of an implied warranty of plans and specifications in the absence of privity, we find that the circuit court erred in granting summary judgment to Kanakanui as to Eastern's claim for breach of implied warranty.

C. Design Professional's Liability to Contractor as a Third-party Beneficiary to Contract Between Design Professional and Project Owner

Finally, we address the circuit court's award of summary judgment in favor of Kanakanui as to Eastern's third-party beneficiary claim. The circuit court concluded that Eastern was precluded from pursuing this claim as it was not a third-party beneficiary of the contract between Salem and Kana-kanui as defined in West Virginia Code § 55-8-12 (1923) (Repl. Vol. 2000), which states:

> If a covenant or promise be made for the *sole* benefit of a person with whom it is not made, or with whom it is made jointly with others, such person may maintain, in his own name, any action thereon which he might maintain in case it had been made with him only, and the consideration had moved from him to the party making such covenant or promise.

(Emphasis added).

The foregoing statute expressly allows a person who is not a party to a contract to maintain a cause of action arising from that contract only if it was made for his or her "sole benefit." We have repeatedly applied this statute and have consistently given force to the "sole benefit" requirement.

■ ■ ■

Eastern has failed to direct this Court to any language in the contract between Kanakanui and Salem that either expressly or impliedly declares an intent that the contract was for Eastern's sole benefit. While it is clear that the contracting parties knew the contract would result in professional work product by Kanakanui that would ultimately be *relied* upon by a construction contractor building the project, it is equally clear that the contract itself was for the benefit of the contracting parties. Consequently, we find no error in the circuit court's grant of summary judgment on this ground.

IV. Conclusion

For the reasons stated in the body of this opinion, the November 2, 1999, order of the Circuit Court of Harrison County is affirmed insofar as it granted summary judgment in favor of Kanakanui as to Eastern's claim as a third-party beneficiary, reversed insofar as it granted summary judgment for Kanakanui as to Eastern's claims for professional negligence and implied warranty, and remanded for additional proceedings not inconsistent with this opinion.

Affirmed in part, Reversed in part, and Remanded.

PLEASANT v. WARRICK 590 So. 2d 214, 25 ALR5th 922 (Ala. 1991)

Study terms: Conversion, negligence, secured party, repossession

Ingram, Justice.

The plaintiff, E.L. Pleasant, sued Rodney Warrick, John Deere Industrial Equipment Company, and Deere Credit Services, Inc. ("Deere Credit"), for conversion, negligence, and wantonness for destruction of a logging skidder, which he contends had been wrongfully repossessed. All of the defendants filed timely motions for summary judgment supported by affidavits and deposition excerpts, contending that Pleasant was in default at the time of the repossession and that the logging skidder had been lawfully repossessed. The trial court entered summary judgment for the defendants.

The dispositive issue on appeal is whether the trial court erred in granting the defendants' motions for summary judgment.

The record, in pertinent part, reveals the following: Pleasant, a timber subcontractor, purchased a John Deere 440C logging skidder from Warrior Tractor and Equipment Company, Inc., an independent John Deere dealer. Pleasant traded a used John Deere 440B skidder and received a credit of $7,117.21 against the total purchase price of $25,268.24, leaving a balance due of $18,151.03. The skidder was used as collateral for a security agreement related to the $18,151.03 balance. Pleasant agreed to make 36 consecutive monthly payments of $721.66 each. The security agreement was transferred or assigned to defendant Deere Credit for administration and collection.

The security agreement, which was signed by Pleasant, contained the following pertinent language:

"This contract shall be in default if I (we) shall fail to pay any installment when due. . . . In any such event Lender may take possession of any Goods in which Lender has a Security Interest and exercise any other remedies provided by law, and may immediately and without notice declare the entire balance of this contract due and payable. . . .

". . . Waiver or condonation of any breach or default shall not constitute a waiver of any other or subsequent breach or default."

The evidence is not in dispute that Pleasant was behind with his payments almost from the beginning. In fact, Pleasant admitted that his account was in default and that Deere Credit had sent him past-due notices of his default. At the time of repossession, which was only 20 months after Pleasant had purchased the skidder, Pleasant was approximately 6 months past due in making the payments pursuant to the security agreement. The record shows that Deere Credit, through its agent, Rodney Warrick, made every reasonable effort toward working out Pleasant's default in order to bring his account current. There was evidence, although disputed, that, instead of a full payment every 30 days, Pleasant was permitted to make partial payments at more frequent intervals. Pleasant contends that this constituted a modification of the original agreement and that, therefore, there was no default. Although we find that Deere Credit may have tried to accommodate Pleasant in an effort to bring Pleasants' account current, we do not find that this accommodation constituted a modification. Nevertheless, even if we concluded that there was a modification, Pleasant admitted that he did not make all of the partial payments. Further, the record reveals that two of the partial payments were made with checks that were dishonored for insufficient funds. The record does not reveal that Pleasant brought his account current at any time.

The record further reveals that on September 1, 1989, Warrick met with Pleasant concerning his account, which was then over $4,000 past due. Warrick had spoken with Pleasant's wife the previous night and had told her that he was coming to see Pleasant about his past-due account. Warrick testified that he obtained directions to the skidder and that he then drove to its location and verified its identity. Pleasant testified that Warrick did not tell him that he was going to repossess the skidder, only that he was going to check the condition of it, and that he would then come to Pleasant's home later that afternoon to get some money Pleasant would have for him.

When Warrick found the skidder, he testified that he attempted to obtain a truck to haul the skidder. When no truck was available, Warrick asked for and received permission to park the skidder at a local dealership over the week-end. When Warrick returned to the skidder, he testified that he checked the oil, the water in the radiator, and the fuel. He then repossessed the skidder; he started its engine and drove it down the highway toward the dealership where he would leave it. No one else was present when he drove the skidder away.

On his way to the dealership, the unexpected occurred. Warrick testified that he heard the engine backfire and saw steam coming from the engine. Assuming that the skidder had simply overheated, he parked the skidder, locked the brakes, and removed the ignition key. He walked back to his vehicle, and as he drove past the skidder, he saw smoke and found the skidder to be on fire. He testified that he believed the fire was beyond his control, so he called the fire department.

The origin of the fire is unknown, and Pleasant did not introduce any evidence concerning the cause of the fire. The investigation report from the fire department lists the cause as "unknown." Warrick testified that the skidder was steaming, but that it was not on fire, when he parked it to return to his car.

Warrick reported the fire loss to the insurance company, and Pleasant filed and insurance claim and proof-of-loss statement. The claim was approved, and proceeds form the insurance policy were forwarded to Deere Credit. Deere Credit retained the amount equal to Pleasant's debt ($13,589.51) and forwarded a check for the balance of $3,598.26 to Pleasant, which Pleasant cashed.

In reviewing a summary judgment, this Court uses the same standard as that of the trial court to test the sufficiency of the evidence. . . . The party moving for a summary judgment must make a prima facia showing that there is no genuine issue of material fact. . . . The burden then shifts to the nonmoving party to show by substantial evidence the existence of a genuine issue of material fact. . . .

The applicable law regarding secured transactions and repossessions is well settled in Alabama. Ala. Code 1975, sect. 7-9-503 provides:

"Unless otherwise agreed a secured party has on default the right to take possession of the collateral. In taking possession a secured party may proceed without judicial process if this can be done without breach of the peace. . . ."

This section allows the secured party, after default, to take possession of collateral without judicial process if possession can be accomplished without risk of injury to the secured party or to any innocent bystanders. . . . The secured party may repossess collateral at his own convenience and is not required to make demand for possession or have the debtor's consent prior to taking possession. . . .

Here, we find the evidence to be undisputed that Pleasant was in default on his account, The record is clear that his payment history over the 20-month period prior to the repossession was irregular and that his account was never brought up to date. The record is undisputed that Pleasant failed to meet his regular payment schedule, failed to make promised payments on his arrearage, and remitted checks that were not honored because of insufficient funds. The record is further undisputed that at the time of repossession, Pleasant's account was over $4,000 in arrears. Indeed, the record is clear that Pleasant was in default and that pursuant to sect. 7-9-503, the defendants were entitled to take possession of the skidder.

The record is also undisputed that the defendants did not commit any breach of peace in obtaining the skidder. There was no evidence of any actual or constructive force used when the skidder was repossessed. In fact, no one other than Warrick was present when the repossession took place. Nor do we find that Warrick repossessed the skidder through fraud or trickery. . . . Pleasant contends that Warrick "tricked" him into informing Warrick of the location of the skidder. However, in view of the payment history of this case, we do not agree. Pleasant was well aware of his $4,000 arrearage, as well as the fact that there was a strong possibility that the skidder would be repossessed.

In view of the above, Warrick's repossession of the skidder was lawful; i.e., Pleasant was in default and no breach of the peace occurred while taking possession of the collateral. Therefore, we find that Pleasant's claims of conversion, negligence, and wantonness are without merit. . . .

The judgment in this case is due to be affirmed.

AFFIRMED.

Review Questions

[Handwritten note in left margin: "White is able to modify agreement because White imposed conditions on self + it was not material to buyer"]

1. Brown, in Seattle, Washington, ordered a machine from the White Company in Detroit, Michigan. The contract stated that the machine was to be shipped F.O.B. Detroit. At the bottom of the sheet the note "we will deliver the machine to you in Seattle at our cost" was written in long-hand and signed by White. En route between Detroit and Seattle the machine was destroyed in a cyclone. There was no insurance. Whose machine was lost? Why? Who pays for or bears the costs of the loss of the machine? Why? *[handwritten: White + White]*

2. Black bought an automatic machine from White to replace a machine that had previously required two employees to operate. In selling the machine to Black, White pointed out the savings in the cost of the employees' time. White also stated that "Productive capacity will be doubled" with the new machine. Hourly production increased even more than double—1,100 pieces per hour against 500 pieces per hour previously. However, there was a problem involved. The machine seldom ran an hour without breaking down. Whenever it broke down, either one or two machine service personnel would be called upon to fix it. After a month or so, one person was ordered to stand by for breakdown whenever the machine was scheduled to run. Weekly production on the automatic machine is only slightly greater than previous weekly production on the hand-operated machine. Black has charged White with fraud. Can Black recover on this basis? Why or why not? Is Black stuck with the machine? What defense does White have? *[handwritten: yes, or breach of warranty; No implied warranty; statement is true; No sale on approval]*

3. Distinguish between express and implied warranties. See Figure 15.2 Limited Warranty Example where it says some of its terms may be void (duration of implied warranty and exclusion or limitation of incidental or consequential damages) under some state statutes. What are the provisions in your state?

4. How does the existence of a security interest protect a creditor from debtor default?

5. What rights does the debtor have in a security interest arrangement?

6. Distinguish between sale on approval and sale or return.

7. What is required of the seller to sell something as is?

8. What remedies are available to a seller if the buyer refuses to pay for goods the buyer has accepted?

9. Compare and contrast *Eastern Steel Constructors v. The City of Salem* with *Arguro Alejandre, et al. v. Mary M. Bull* in Chapter 14 with regard to the economic loss rule.

10. Referring to the case of *Pleasant v. Warrick:* a. What would be both the legal and practical implications of repossessing the logging skidder if Pleasant had paid half of the outstanding balance and then was three weeks behind with a payment? b. Suppose the skidder had not been destroyed by fire and Deere Credit had been able to realize only $10,000 upon the resale of the used skidder. How much of the $10,000 would be paid to Pleasant?

Engineering Contracts

PART

3

The general public is well aware of construction contracts that determine how buildings are to be erected, bridges built, and highways created. The media often report the stories of such construction projects and refer to the underlying contracts. Legal problems and entanglements concerning contracts in the civil engineering, architectural engineering, and construction engineering fields occur frequently enough that special courses in contract writing are commonly included in those curricula. The need for such education in mechanical, industrial, electrical, chemical, and other engineering curricula may be less apparent. However, all engineers seem to be involved in writing specifications and interpreting contract documents at some time during their careers. Some, especially those who rise in management, spend their entire working lives dealing with contracts. Thus, all engineers need to be aware of laws regarding ownership, independent contracting, and agency, and of contract documents that spell out work to be done and the responsibilities of the parties to contracts.

Contracting Procedure

CHAPTER 16

HIGHLIGHTS

- In general, parties to construction contracts include the owner, engineer, and contractor. Similarly, parties to engineering services contracts may be the client, engineer, and subcontractors. The most common methods of payment under these contracts are:
  - Lump sum
  - Cost-plus percentage
  - Cost plus fixed fee
  - Guaranteed maximum price
  - Unit price

Before delving into the contracting procedure, several terms and relationships between contracting parties need to be discussed. Although the following discussion focuses on construction contracts, most (if not all) of the discussion applies equally to other large projects, such as systems or software development.

PARTIES TO A CONTRACT

The parties to a construction contract generally are the owner, the engineer or architect, and the contractor. The owner is the party for whom the work is to be done. He or she is the one to whom the others look for payment for services. The owner has the final authority in questions as to what is to be included or left out of a project. The owner may be a private individual, president of a corporation, chair of a board of directors, or a public official charged with the responsibility for the project.

As the terms will be used here, engineer and architect are virtually interchangeable. At present, considerable controversy (including court cases) exists over the meanings of the terms and the work to be considered the proper field of each. No attempt will be made to add to that controversy here. It will be assumed that the person is properly employed, whether an architect or engineer. Such a person acts as an agent of the owner. Specifically, that person furnishes the technical and professional skill necessary in the planning and administration of the project to accomplish the owner's purpose. That person is the owner's designer, supervisor, investigator, and adviser. Most state laws require such a person to be registered as a professional. This person may be a consultant or an employee of the owner. Generally, this is the person with whom the contractor deals directly.

The third party to the contract, the contractor, is the one who undertakes the actual construction of the project. The contractor furnishes the labor, materials, and equipment with which to complete the job. The term **contractor** is frequently further broken down into general contractor and subcontractor.

The **general contractor** agrees to accept the responsibility for the complete project and frequently undertakes the major portion of the project. **Subcontractors** are hired for particular specialties by the general contractor and, in effect, work for the general contractor while completing the portions of the project for which they have been hired. The general contractor retains responsibility for the entire project, even when the wiring, the plumbing, and the roofing are done by various subcontractors.

TYPES OF CONSTRUCTION

Independent contractors construct much of the capital wealth of our country. Most of the buildings, machines, bridges, utilities, production facilities, and so forth responsible for our standard of living were built under contract. There are two major types of such construction—public and private. The two types differ significantly in the motives and relationships involved with each.

Differences :- Regulation
— Public Scrutiny

Private Works

Private projects are those that are undertaken for an individual or a company. The restrictions imposed upon the parties are those of contract law in general. The agreement may be achieved by advertising and then choosing the best bid, or by direct negotiation without advertising. The acceptance may be oral or in writing; in fact, the entire contract could be oral if the parties so choose.

The profit motive is the usual reason for construction of private works. The owners believe the project will give them a desirable return on investment. However, this need not necessarily be the reason for the project. In private works, the reason could be nothing more than a personal whim of the owner. The desirability of the project is not open to question by the public, as long as no one is harmed by it.

Public Works

The motive for construction of **public works** is public demand or need. Financial return sufficient to justify the investment is frequently of less than primary importance. Money for public works comes either from direct payments from available funds, such as tax receipts, or from loans, such as bonds.

Voluminous statutes usually govern the letting of public-works contracts. These laws apply certain restrictions to the contracting procedure. Generally, formal advertising and bidding are required to ensure competition among the bidders. The "lowest responsible bidder" gets the job. Usually the means of advertising and the length of time the advertisement is to run are specified. Changes in plans or specifications often cannot be made after the award of the contract, even though such changes might be beneficial. Acceptance of a bid may not be effective until it has been ratified by a legislature or a legislative committee.

Once a contractor's bid on a public project has been accepted, the amount the contractor will get is fixed. The official in charge of the project cannot agree to pay more, regardless of the credibility of the contractor's claim.

The purpose of the restrictions in letting public projects is, of course, to prevent dishonesty, collusion, and fraud among bidders and between bidders and public officials. Sometimes the restrictions seem cumbersome and unfair; sometimes the results are less than might have been accomplished without them. The results in general, however, are beneficial to the public.

PROJECT DELIVERY SYSTEMS

Design-Bid-Build

This is the traditional construction industry building delivery model where the owner enters into separate contracts with the contractor, who is responsible for building the structure, and the architect or other design professional, who is responsible for drafting the building design. The model provides a lot of control for the owner who has overall approval authority. In addition, in the event of a loss or other liability related to the project, the owner can seek recovery from either or both the architect and contractor. However, because there is lack of privity as well as potential conflicts of interest between the contractor and architect, projects can easily become stalled often requiring the owner to pay the cost.

Design-Build

The design-build model attempts to remedy the conflict scenario presented by the traditional design-bid-build process be eliminating the contractual conflict between the contractor and architect. Under this model, the owner contracts directly with a design-build entity which provides construction and design services. The design-build entity can be a joint venture between a contractor and an architectural firm or designer or a business entity with design and builder capabilities in-house. This model differs from the design-bid-build model significantly from a liability perspective because the owner has only one entity to look to for recovery.

Multi-Prime

For both the design-bid-build and design-build models, the owner usually enters into a relationship with one contractor, often called the "prime contractor" who subcontracts most if not all of the work to others. The owner, in a multi-prime model scenario, enters into several contracts with several contractors performing separate, but often overlapping tasks. In this instance, the owner hires a prime contractor to manage the work of the other contractors, but is not responsible for performing the work in its entirety. When liability issues arise, the owner will have to seek recovery from several potentially responsible parties, which may cause significant delays to the full performance of the project.

Construction Management

Under a construction management delivery model, a construction manager is introduced into the mix. Here, the owner enters into separate agreements with the architect and the contractor, but where the architect would usually serve as the owner's representative on the work site, the construction manager steps in to monitor progress. The construction manager usually holds one of two possible roles: 1) construction manager as an agent of the owner or 2) construction manager at risk, where the construction manager enters into subcontracting relationships with others in order to complete the work.

Integrated Project Delivery

This model, while widely adopted in Australia where it is called "alliancing" or "partnering", is gaining some currency in the United States and elsewhere. Under this delivery model, all interested parties agree beforehand how risks and rewards will be allocated and how disputes will be handled. One purpose of this model is to eliminate delays caused by disputes and conflicts of interest. Similar models are prevalent in the manufacturing industry, however, due to the nature of construction, e.g., the lack of economies of scale, the manufacturing industry lessons don't apply as readily to the construction industry. Encouragingly, advances in software and applied technology are moving this delivery method forward. Building Information Modeling, or "BIM" software supports a more integrated and cooperative approach to building design and construction which may aid in the adoption and use of this model in the future.

CONSTRUCTION, MANUFACTURING, AND ENGINEERING SERVICES

Construction projects, the manufacture of goods for a market, and engineering services are similar. All are concerned with creating a finished product or deliverable. All must deal with economic considerations in acquiring supplies, material and labor. All aim for efficient management in an effort to show a profit. All are concerned with schedules that must be met if penalties of one kind or another are to be avoided. Quality must be maintained and costs must be minimized.

The main distinction among these stems from the location of the final product. In most construction, whether it is a building or a piece of productive equipment, the place where the product is built is the place where it will stay. A construction product is usually custom-made, built to specific requirements, and not to be reproduced. Such a product does not lend itself to the economies of standardization of method, as do the products of most manufacturing companies. Nor is such a product exclusively the often intangible product of engineering services. In the following sections, note where you think there are distinctions from the construction process and manufacturing or engineering services.

DIRECT EMPLOYMENT OR CONTRACT

After the decision has been made by the owner to undertake a construction project, the question frequently arises as to whether the work should be done by the available staff, by contract, or by a combination of both. The decision involves many factors; two very important ones are the relative size of the job and the skill of the staff. Many jobs are too large for the present staff or too small to be submitted to an outside contractor, so the decision is clear. But many projects of intermediate size could be completed either way. It is with these intermediate-size jobs that the present discussion is concerned. The owner (or engineer) should consider a number of benefits and disadvantages in choosing the best procedure.

Cost

Certain savings in cost are apparent when direct employment is used:

1. The owner does not have to pay the contractor's profit margin if the owner does the work.
2. The amount added in by the contractor for contingencies is saved if no contingencies arise. The owner who expects to save costs in this area is taking a risk. If the owner's staff is inexperienced, the odds are not good. On the other hand, an experienced contractor can often see and avert incidents that would otherwise constitute contingencies.
3. The cost of making multiple estimates is avoided if the owner's staff undertakes the job.

Flexibility

A project undertaken by direct employment has more flexibility. If it becomes necessary to make changes while the job is under way, the owner has only to order the changes made. When a contractor is hired to build according to plans and specifications, and work is under way, change proposals will usually meet with resistance. Contractors are interested in completing the project as soon as possible so that they can get paid and go on to the next job. They will be tempted to charge heavily for the delay caused by a change.

Subsequent Maintenance

A machine or other structure is more readily repaired by the original builder than by maintenance personnel who have had no experience with it. Therefore, the issue of subsequent maintenance might induce the owner to use as many of his or her own people as possible.

Grievances

Labor problems are usually somewhat reduced by using present crews. The possibility of disputes about what a given set of workers can or cannot do under a labor contract exists, but the likelihood of these and other disputes (such as grievances) arising can be minimized by using direct employees. The individuals forming the nucleus of the crew, at least, have probably learned to work with each other and with others employed by the owner in different jobs. (For further discussion about this issue, see Chapter 24, Labor.)

Specialization

The great and sometimes decisive advantage of hiring a contractor for a project is that the contractor is a specialist. The contractor's specialty is in labor, supervision, and procurement of materials. It is probably this factor, more than any other, that gives the contractor an advantage.

Public Relations

The public relations programs of many large concerns tip the scales in favor of hiring contractors to undertake jobs for them. This factor plays a big part when a company sets up an operation in a new community. To establish itself in a favorable light, the company may hire local people to set up its facilities, even though it has a staff capable of doing the job.

PAYMENT ARRANGEMENTS

An owner may pay for work undertaken by a contractor in four basic ways. In addition to these basic payment arrangements, a nearly infinite number of variations and combinations of them exists, each adjusted to a given project. Here, the features of each basic arrangement will be considered primarily from the owner's standpoint.

Lump-Sum Contract

If an individual purchases a product manufactured by a company, that person usually knows the price before making the agreement to purchase. The price of an automobile, for example, is arranged between the dealer and the buyer before the sale is actually made. The **lump-sum contract** arrangement gives the owner the same assurances as to the price to be paid for the job. This aspect of the lump-sum contract appeals to people, and it is the main advantage of the arrangement from the owner's standpoint.

The fixed-price, or lump-sum, contract has several inherent disadvantages, however. Probably the main one is the antagonistic interests of the owner and the contractor. Once the contract is signed, the contractor's main interest is in making a profit on the job and in doing it as quickly as possible. The contractor has an incentive to do no more than the minimum requirements set forth in the plans and specifications. With an unscrupulous contractor, the results may be shoddy work and a poor structure. Even a responsible contractor may be tempted to cut corners if contingencies start eating into the profit margin or if the contractor is already losing on the job. The owner's interest, on the other hand, is to obtain the best possible structure for the agreed price.

Under a lump-sum contract, changes can be quite costly. The change represents an impediment to the contractor's speedy completion of the job, and the contractor is in the position to dictate the cost of changes. The owner has hired the contractor to do the job, and the contractor is on the premises with equipment to render the service called for in the contract. The owner's alternative to paying the contractor's price for a change is to wait for completion of the job and then hire someone else to make the change. This is often very costly.

A lump-sum contract requires that considerable time and money be spent by the contractor in examining the site, estimating, and drawing up and submitting a bid before the job can start.

The delays may run to months or even years, and the cost of delay can also be considerable. If the job is to be started as quickly as possible, the lump-sum contract is not recommended. In addition, if the work required is at all indefinite or uncertain, the lump-sum contract should not be used. The contractor will have to add in a sufficient amount to cover the uncertainty to show a profit.

The greater the degree of uncertainty, the greater the probable spread in bidders' proposals. If the equipment were a piece of automation with only the raw material and the end product known, the highest bid might be four or five times that of the lowest. Each contractor would try to hedge as well as possible against a large number of unknowns and gamble that the amount submitted would result in a successful bid.

Lump-sum contracts are used commonly and successfully, but their successful use is generally restricted to situations in which unknowns are at a minimum. They are not appropriate

1. Where uncertainties exist
2. Where a speedy start is necessary—emergency work, for instance or
3. Where ongoing plant operations will interfere with the contractor's work.

Where such circumstances exist, one of the two basic "cost-plus" types of contract can be used.

Cost-Plus-Percentage Contract

The most rapid means of starting a project is through the use of the cost-plus-percentage pricing arrangement. In an emergency, an owner may have to start work immediately, while the plans for the completed structure are still being drawn up. Using a cost-plus-percentage arrangement, today's email can result in action today.

With the cost-plus-percentage arrangement, the owner usually pays all the contractor's costs plus an added percentage (often 15%) of these costs as the contractor's profit. Cost refers to the cost of materials and services directly connected with the owner's structure. Such items as the contractor's overhead, staff salaries, and the like are usually excluded unless direct connection to the project can be shown.

A major advantage of the cost-plus-percentage contract is its flexibility. Changes may be made readily when the owner desires them. Because the cost to the owner for making changes includes profit to the contractor, the contractor has little reason to object to them.

One disadvantage of this approach is that the risks involved in construction are assumed by the owner. In fact, the owner does not know with any certainty what the project will cost until it is completed. On the other hand, if no adverse conditions (contingencies) arise, the benefit goes to the owner.

Another disadvantage is that the contractor's profit is tied to costs, so the costs may be quite high. Gold-plated doorknobs may show up, for example. The contractor has what amounts to an incentive for dishonesty—a financial reward for running up costs. The most honest of contractors (or people in any profession, for that matter) would be tempted to be inefficient under the circumstances. Even if the contractor pursues the completion of the project in the best interests of the owner, efficiency is not assured. Supervisors and workers are not likely to put forth outstanding effort for their employer if they will gain nothing as a result. The need for the owner to police the project is apparent. It is because of this disadvantage that the federal government looks with disfavor on the cost-plus-percentage contract.

Cost-Plus-Fixed-Fee Contract

The remedy for the main undesirable feature of the cost-plus-percentage contract is found in the cost-plus-fixed-fee arrangement. Here, the contractor is paid a fixed fee as the profit, and the owner picks up the tab for all the contractor's costs of undertaking the project. The amount of the fixed fee is subject to negotiation between owner and contractor, or bid by the contractor based upon estimates of the cost of the completed project. The advance settlement of the amount of the fixed fee precludes an immediate start on the project, but the time required is not nearly as great as that necessary for a lump-sum contract.

The cost-plus-fixed-fee contract is probably the best basic arrangement from the standpoint of all parties. The contractor has no incentive to run up costs or work inefficiently. In fact, with a view toward maximizing profit in a given period of time, the contractor has an incentive to hasten the completion of the project to obtain the fee.

However, less flexibility exists with this arrangement due to the contractor's desire to complete the job rapidly. A proposal by the owner for a major, time-consuming change in the project is likely to be met with objections and a request for more money. But if the contractor is required to spend a longer time in pursuit of the fee, it is fair to expect that fee to be increased.

As in the cost-plus-percentage contract, the owner runs the risk of contingencies, but here the owner is in a somewhat better position. Under the cost-plus-percentage arrangement, the occurrence of contingencies tends to increase the contractor's ultimate profit. Such contracts, then, have built-in incentives to bring about contingencies or, at least, not to actively avoid them. In contrast, in the cost-plus-fixed-fee arrangement, the desire for early completion is an incentive for the contractor to avoid contingencies if possible.

Unit-Price Contract

Certain kinds of structures may be conveniently built under the unit-price type of arrangement. This scheme applies best where a large amount of the same kind of work must be done. In the building of a road, for instance, the main elements are excavating, filling, and pouring concrete. The contract would specify so much per cubic yard of excavation (plus an extra amount if rock formation is encountered), so much per cubic yard of fill, and so much per cubic yard of concrete. The price per unit includes the contractor's cost per unit plus an amount for contingencies, overhead, and profit. The engineer or architect usually estimates the quantities required ahead of time for the benefit of both owner and contractor. The actual quantities may be considerably different from the estimate, but the contractor is paid according to the actual quantities required.

Usually the unit-price arrangement is used in conjunction with a lump-sum contract. The combination is usually quite beneficial, largely because it is flexible, reflecting the realities of most projects. In nearly every job, for example, there are some elements (such as clearing, grading, and cleaning up the site) that do not lend themselves to unit pricing.

With a unit-price contract, much of the contractor's uncertainty is relieved, and this is reflected in a lower estimate for contingencies. The risks not assumed by the contractor, however, must be borne by the owner, and the owner has no precise knowledge of the cost of the job until its termination.

Just as is true with a lump-sum contract, the job cannot begin immediately under a unit-price arrangement. The contractor requires time to examine the premises and prepare an estimate for the owner. Normal bidding procedure is usually used in the award of such contracts. (See the *D.A. Elia Construction Corporation v. New York State Thruway Authority* case in Chapter 17 where the parties entered into a unit price contract and the claimant tried to collect money for additional work.)

Variations

The types of payment arrangements described above are the four basic forms of construction contracts. In addition to these, many variations, or hybrids, exist—many of them tailored to particular needs of projects. A common variation is the guaranteed maximum price, or GMP, payment structure, where the parties agree upfront to a maximum price. A hybrid of the lump sum and cost-plus agreements, a GMP payment structure includes the contractor's anticipated profit in the price. The advantage of the GMP arrangement is that the owner's risk is lessened because the price is known at the beginning, e.g., the owner pays for the labor, materials, and equipment costs plus the contractor's profit. However, an unsophisticated contractor may be at a disadvantage because unknown costs may eat away at the anticipated profit already identified in the contract. Nevertheless, a GMP may be the best payment structure when the parties are sophisticated negotiators and the project is large enough to encourage the parties to enter into such an arrangement. Another common variation involves the addition of an incentive system of some sort. Profit-sharing and percentage-of-cost saving are examples. Another kind of variation involves the addition of a kind of reverse liquidated damages clause. That is, the contractor is offered a fixed amount per day additionally if the contractor finishes ahead of schedule.

Another variant form of contract is the **management contract**. In such a contract, the owner hires a contractor, not necessarily to undertake the work with the contractor's own organization, but to oversee the job, often hiring others to do the work. Frequently the contractor's duties include managing work to be done with the owner's labor force.

The contractor under a management contract still holds independent contractor status. The contractor agrees to produce a result, but the contractor's actions do not bind the owner under a management contract as an agent would bind the principal. The management contract is, therefore, distinct from the agency relationship, which exists between an owner and the engineer or architect. Usually the services of an engineer or architect are not used.

STAGES OF A PROJECT

As you may recall from Chapter 1, there are three broad project stages:

1. Conception of the idea,
2. Reduction of the idea to practice, and
3. Refinement and oversight.

In discussing the development of a project in the following sections, consider the lump-sum or the unit-price type of contract arrangement, since each requires extensive preliminary work. In addition, pay attention to potential legal pitfalls in each phase.

Conception of the Idea

A project starts as an idea or a dream for the future. If it is soundly conceived, planned, and developed, the idea may be realized. The first step toward realization is a feasibility study.

Feasibility Study.   Regardless of how beneficial a project may seem to its originator, it is nearly always advisable to conduct a preliminary investigation. Although wars and other emergencies may preclude such an investigation, a state of emergency is somewhat unusual and presumably won't prevent most preliminary investigations. The purpose of the feasibility study is to answer several preliminary questions:

1. What is the cost of undertaking the project? This cost is usually determined on the basis of either annual cost or present worth (or capitalized cost if perpetual service is contemplated).
2. Can the objective be attained better (or cheaper) in some other way? The comparison is made on the basis of annual cost or present worth.
3. What benefits will result from the use of the various alternatives? In most private projects and some public projects, the ratio of economic benefit to cost must be sufficient to justify the project. In many public works, a crying need may push consideration of economic benefit into the background.

The preliminary investigations and reports are usually not intensive. Some schemes are shown to be obviously impractical after only minimal data are obtained. Where the scheme appears to be profitable, however, the expenditure of considerable time and money on the feasibility study may be justified. Surveys and investigations of such elements as markets, sources of raw material and labor, costs of transportation, and applicable laws and ordinances may become necessary.

Even when a tentative decision has been made to go ahead with a project based on reports that show it will pay off, it is usually worthwhile to investigate cheaper ways of achieving the same goal. Most such studies can be undertaken at a reasonable cost. However, the more closely balanced the evidence, the more detailed must be the study. Investigations about alternatives can save money in the long run, but wisdom must dictate the point at which a study should be terminated and a decision about the project made. Feasibility studies do not complete projects. Thus, when the benefits of a course of action (or inaction, as the case may be) are clear, the feasibility study has usually served its purpose.

Legal Considerations.   When a business decides to conduct its own feasibility study, or if it is required by the proposal, there are several legal issues that should be addressed at this stage. For instance, if a hiring company uses another company in the business of preparing feasibility reports, who bears the burden of liability if the hiring company relies to its detriment on the feasibility study results? What happens if the study is conducted by in-house engineers? Another consideration could be changes in the legal environment; for example, statute regulations may have changed since the study was conducted. How do you keep track of those changes?

Reduction to Practice—Design

Once the feasibility study is conducted and the project is approved, the development team endeavors to reduce the idea to practice. This often means creating a design or model of the project.

Normally, all designs are complete in some detail before work is started. A complete feasibility study usually includes sketches of possible layouts. The design, however, includes functional requirements, layouts, and dimensions. Preparation of the drawings is a part of the engineer's or architect's task. When the drawings and specifications are completed, they combine to give the basic information on the project.

The design drawings provide the owner and engineer with a picture to use in talking to others about the project. A copy of the design drawings of the proposed construction ordinarily must be filed with the proper authorities to obtain a building permit.

Legal Considerations.   Once the design is created, the legal ramifications may be difficult to determine. As an example, consider architectural designs. Architectural designs are protected by copyright law. If a Company A hires Company B to draft the architectural designs, who owns them? This can be easily resolved by contract, but the rights to architectural designs are sometimes not included in the design agreement. Of course, if the design is created by in-house staff, it is owned by the employer company, unless indicated otherwise in the employment agreement. (For further discussion see Chapter 20, Intellectual Property.)

Refinement and Oversight

The next stage in the project is discovering potential flaws or bottlenecks in the project design and refining it. At this point, the project may be ready to be put out to bid. The development team needs to consider what legal arrangements have to be made and prepare contract documents.

Legal Arrangements.   If such arrangements have not already been made, land must be obtained for the project by purchase or by exercise of eminent domain if the project is a public one. Access to railroad sidings, highway connections, and utility services must all be considered. If a building is being constructed, zoning is likely to be a consideration, and building codes may need to be satisfied.

Another legal matter is the engineer's or architect's right to practice in the state. State licensing laws generally require that a registered professional engineer sign construction plans or issue them under his or her seal. All members of the engineering staff need not be registered. If a member of the staff who is registered takes responsibility for the plans, the law is satisfied.

Of increasing importance are citizenship requirements or the immigration status of employees. In some instances, work permits or security clearances must be issued.

If the project is to be undertaken for the public, strict adherence to laws governing such projects is necessary. The order and appropriation for the project must be passed before the work is undertaken. In many types of public construction, minimum standards must be met.

Preparation of Contract Documents.   During the preliminary stages of a project, it is necessary to prepare contract documents that will guide the remainder of the project. One of the main purposes of these documents is to set forth the relationships between the parties. If this purpose is to be accomplished adequately, the documents must be prepared with great care and skill. The engineer's task of preparing these documents often seems dull and routine—an obstacle to be overcome to get to something more interesting. Still, careful attention to this task prevents many future controversies, and it may make controversies that do arise easier to settle.

The pieces that make up a construction contract generally include

1. The advertisement,
2. The instructions to bidders,
3. The proposal,
4. The agreement,
5. The bonds,
6. The general conditions, and
7. The specifications and drawings.

Each component of the documents that constitute a construction contract has a name or heading. The parts do not bear the same names in all contracts, and the material covered under the headings is not always the same. For example, what is covered in "general provisions" in a federal government contract might be covered in "general conditions" or "information for bidders" in a state or private contract. Certain instances of overlap may be noted. Ideally, a subject should be thoroughly treated in only one section, with reference made to this section whenever the occasion arises again in the documents. In larger projects, however, duplication of coverage is not unusual.

The American Institute of Architects has attempted to standardize contract documents. It has drawn up and copyrighted documents for use in building construction contracts. The federal government and most state governments also have standardized contract components; the same is true of most large companies. Engineers will need to become familiar with the particular terms and usages of a new employer.

Advertisement.   The advertisement is usually the last contract document prepared, but it is the first one seen by the contractor. The primary purpose of the advertisement is to obtain competitive bidding on the project. In public contracts, a minimum of three bidders on any project may be required, and a special authorization may be needed if an award is to be made otherwise. In private work, there is no legal requirement to advertise. Contract awards are often made to contractors with whom the owner has successfully dealt on previous occasions. Many companies consider competitive bidding to be desirable, however, and they have established formal procedures to require it.

A second purpose of the advertisement is to attract appropriate bidders—those interested in the type of work involved. To accomplish this end, the wording of the advertisement should give a clear, general picture of what is to be done. The advertisement should not attract those who have insufficient capacity or who are not interested in the type or location of the work.

The title of the advertisement is important, too. It must attract attention, and it should state very generally what is to be built. The phrase "Notice to Contractors" or "Call for Bids" together with a phrase generally describing the type of structure, such as "Elementary School," serves these functions well. A glance attracts the qualified contractor and prompts the contractor to read further.

The information to be included in the advertisement varies with the type of project contemplated, but the following facts should usually appear:

- The kind of job.
- Location of the project—construction equipment is rather costly to move.
- The owner and the engineer—previous contracts with an owner or engineer may influence a contractor to submit a bid.
- Approximate size of the project.
- The date and place for receipt of bids, and the date and place of opening and reading bids.
- How an award is to be made (for example, "to lowest responsible bidder"), whether the owner reserves the right to reject any and all bids, and when the contract is to be signed.
- Notice of any deposits, bid bonds, or performance bonds to be required.
- Procedure for withdrawing bids if withdrawal is to be allowed.
- Where copies of the contract documents may be obtained.
- Any special conditions.

Finally, the advertising medium should be carefully selected. Public contracts must be advertised in local newspapers. For private contracts, local newspapers and trade journals make good advertising media. Whatever the medium, however, brevity best serves the purposes of advertisement.

Instructions for Bidders.   Contractors who remain interested in the project after reading the advertisement are offered the opportunity to obtain more information, usually from the engineer. It is customary to require a deposit (usually from $50 to $500) from the contractor when the contractor obtains a set of the contract documents.

The "Instructions for Bidders" or "Information for Bidders" may be compiled as an independent document. Probably just as frequently, however, this information section is the first portion of the specifications. Variations do occur. For example, items included in the instructions for bidders in one project may be found in the general conditions document or in the general provisions section of the specifications in another project.

Much of the information contained in the instructions is the same as that given in the advertisement, but it appears in expanded form—the "what," "where," "when," "how," and "for whom" are given in more detail. Additional inclusions in the instructions vary considerably from one project to another, but the most common are listed below:

■ A requirement that the bidders follow the proposal form.
■ A statement as to whether alternative proposals will be considered.
■ The proper signing of bids (e.g., the use of a power of attorney).
■ Qualification or prequalification of bidders. This is a requirement to satisfy the owner that the contractor has the necessary skill, financial means, staff, and equipment to do the job.
■ Discrepancies. Provision is usually made for review with the engineer of any discrepancies appearing in the contract documents.
■ Examination of the site. An invitation is normally extended to bidders to examine the site, with a warning that no additional compensation will be allowed as a result of conditions of which the bidders could have informed themselves.
■ Provision for return of required bid deposits to unsuccessful bidders.

The Proposal.   The proposal is a formal offer by the contractor to do the required work of the project. Acceptance of the proposal often completes the formation of the contract, so considerable care must be exercised by the engineer in drawing it up.

Proposal forms are usually standardized for all bidders on a job. If the proposal is to be broken down into components (as in a unit-price contract), all bidders must break it down in the same way so that their proposals can be compared. Even if only one lump-sum bid is requested, all bidders must use the same proposal form so that bids can be considered on an equivalent basis. The following elements may be contained in a proposal form:

■ The price or prices for which the contractor agrees to perform the work involved.
■ The time when work is to start and when it is to be finished.
■ A statement that no fraud or collusion exists. Of particular concern is the possibility of fraud or collusion among bidders or between any bidder and a representative or employee of the owner.
■ A statement that the bidder accepts responsibility for having examined the site.
■ A proffer of any required bid bond or guarantee and agreement to furnish whatever other bonds may be required, as well as an agreement to forfeit these bonds or portions of them under the conditions stipulated.
■ Acknowledgement that the various other documents are to become parts of the contract.
■ A listing of subcontractors if required.
■ Signatures of contractor and witnesses.

Agreement.   To a lawyer, the agreement probably would consist of all the various documents. However, the parties often use the term **agreement** for the short "outline" of the transaction. The term contract is preferred when one is referring to the complete array of contract documents.

The agreement may be one or two pages and describes the work to be done, largely by reference to the other contract documents. It also states the owner's consideration (the price to be paid for the work) and the means and time of payment. Although they are included in the other contract documents incorporated by reference, a few terms are usually repeated in the agreement. If time of completion is an essential element of the project, that date and amount of liquidated damages for a failure to complete on time are mentioned. If there is to be a warranty of the work by the contractor for, say, a year after completion, that is usually included. Any amounts to be held back from payments to the contractor are also set forth.

Bonds.   A **bond** gives the owner financial protection in case of default by the contractor. Three types of bonds are commonly used in contracting—the bid bond, the performance bond, and the labor and material payment bond. Each protects the owner by assuming a risk the owner would otherwise normally run in hiring a contractor to undertake the work. See *Alton Ochsner Medical Foundation v. HLM Design Of North America, Inc. et al.,* as an example of how poor bond management affected liability for poor performance.

The Bid Bond.   A **bid bond** assures the owner that the bidder will sign the agreement to do the work if that bid is the one accepted. In theory, the amount of the bid bond covers the owner's loss if the lowest responsible bidder fails to sign the agreement and the owner must then turn to the next higher bidder. Usually, however, the owner does not require a bid bond as such. Instead, the ower requires a deposit that may be satisfied by a bid bond, a certified check, or some other security posted by the contractor to assure that the owner will be reimbursed in case of a failure to sign. The chosen bidder forfeits the bid security only if he or she fails to agree to the project. It is returned to the bidder after the agreement is made. The bid securities of unsuccessful bidders are, of course, returned to them.

The Performance Bond.   Through the **performance bond,** the owner's risk that the contractor may fail to complete the project is passed on to a surety company. The risk involved may be very slight—most contractors are in business to stay—but there are still risks. The contractor may meet with insurmountable obstacles. Unforeseen price rises, strikes, fires, floods, storms, or unsuspected subsoil conditions, for example, can financially ruin even the best-backed contractors.

If the contractor is faced with such disaster, insolvency may result, and if there is no performance bond, the owner may be left trying to obtain blood from a stone. However, if a performance bond has been required, completion of the structure is guaranteed by the surety. Many surety companies maintain facilities that can be pressed into service on such occasions; more frequently, the surety company hires another contractor to complete the job.

The cost of a performance bond will, of course, be passed on to the owner in the price to be paid for the work. This price varies, but in many situations will be around 1 percent of the bid price for the contract. For this price, the owner not only gets risk protection but a preselection of contractors. Surety companies are quite choosy about the contractors they agree to back. The irresponsible and the "fly-by-nights" are poor risks, and a reputable surety company will not back them. The requirement of a performance bond, then, allows the owner to select from the best.

The Labor and Material Payment Bond.   The labor and material payment bond offers protection against labor and material liens, known generally as **mechanics' liens.** (See *Britt Construction, Inc. v. Magazzine Clean, LLC.* at the end of this chapter) The lien laws vary considerably from state to state. In any state, however, if a contractor fails to pay for the labor or material, a lien may be obtained against the property that the labor or material was used to improve. In addition to the labor and material supplier's liens, the contractor, subcontractor, engineer, or architect may secure payment by recording a lien. Based upon the assumption of the owner's honesty and integrity in dealing with the contractor and engineer or architect, discussion here will focus on labor liens and material suppliers' liens.

Mechanics' liens secure payment for anything connected with the improvement of real estate. The security is an encumbrance upon the property that makes it more difficult to sell or mortgage. Technical procedures and delays, specified in lien laws, must be followed to remove the encumbrances. Owners may protect themselves from such encumbrances in two ways:

1. By withholding sufficient funds from payments to the contractor to pay for labor and materials if the contractor fails to do so, and
2. By requiring the contractor to obtain a labor and material payment bond.

To withhold part payment, the owner must know the amount that he or she may need later. This usually requires the owner to get a sworn statement from the contractor as to the outstanding bills before making payment to the contractor. Some state lien laws require such a sworn statement if a payment bond has not been provided in the contract. The procedure can become somewhat cumbersome, however. Requirement of a labor and material payment bond from the contractor is much simpler for the owner. According to the bond, the surety company assumes liability for the contractor's unpaid bills.

When the federal government is owner in a project, a performance bond and a labor and material payment bond are automatically required. The Miller Act (passed in 1935) requires such bonds for federal contracts exceeding $100,000 (40 USC § 270d-1, as revised January 22, 2002, 48 CFR 28.102-2b, as revised October 1, 2002).

General Conditions.   In a set of construction contract documents, the document that sets out the general relationships between the parties is usually referred to as the **General Conditions.** This document deals with rights, obligations, authority, and responsibility of the parties. Topics covered differ from contract to contract, but the following are almost invariably present.

- A statement near the beginning attests to the unity of the contract documents—meaning that a requirement in one document is just as binding as if it appeared in all of them. Such unity is necessary, since the specifications for even a small project would reach tremendous size if each specification had to be followed by all conditions pertaining to it.
- The right of inspection (of the site) by the owner or the owner's representative at any time is usually reserved. The documents usually are written so that the owner "may" inspect the site but is not required to do so. It is common also to allow inspection by public officials as a general condition.
- Conditions for terminating the contract by the owner and by the contractor are set forth. Bankruptcy of the contractor and failure by the owner to make scheduled payments to the contractor when due are usually made conditions for termination.
- The insurance program for the project is outlined. Provision is often made that worker's compensation will cover medical costs and partial wage payments in case of injury to a worker as a result of the worker's employment by the contractor. The owner's protective liability insurance protects the owner from contingent liability if the contractor's operations should injure anyone. Fire insurance and vehicle liability are among other types of insurance commonly required.
- The extent or limits of authority should be spelled out clearly. For example, the engineer or architect is to make decisions on the work involved in the project. That person usually must then be given the right to stop the work if it appears necessary to do so.

- Provision is often made for arbitration as a final step in any dispute between the owner and the contractor.
- A requirement is usually included that the contractor keep the site clean as work progresses and that the contractor clean up the site thoroughly when finished.

Specifications and Drawings.   The specifications and drawings are sufficiently important to warrant an entire chapter, Chapter 17. What is useful to note here is that precision and clarity are critical for successful contract bids as well as contract performance.

Final Considerations.   Almost invariably the owner is under considerable pressure to get things started soon after the decision has been made to undertake a project. Advertising and bidding procedures require time, however, and to shorten the time allowed for the initial stages is usually an act of folly. Prospective bidders need time to investigate the work site and to plan their work. If they do not have enough time for the investigation and planning stages, many unknowns will remain. Generally, the less the bidder knows, the higher will be the bid price. There are few fears greater than fear of the unknown, and fear of unknowns in a proposed project drives the price up. Owners need time also, to carefully consider the quality of the bidders and their bids.

Collusion.   Most owners, engineers, and contractors are honest people. There are some of each, however, who will eagerly forget ethics and the law to seize an unfair advantage. Unfair advantages take many forms, and there are many "shades of gray" between fair, honest practices and unethical or illegal ones. The time for receiving proposals, for example, should be fixed and inflexible. It is unfair to give one contractor a longer time to prepare a proposal. Similarly, the opening of the proposals should take place at the time specified. Also, if an owner or engineer gives preference to one contractor over another in other ways, either by giving the contractor added vital information privately or by writing out specifications, that preference constitutes an unfair advantage. If the owner unfairly helps a contractor, it will almost certainly be resented by the others, even if it's not illegal.

Collusion occurs occasionally when a limited number of responsible bidders bid on a project. The bidders get together and decide who will take the job and for what price; the other bidders then submit prices higher than that submitted by the "successful" bidder. The result of these and other forms of collusion is higher cost to the owner.

The existence of such practices is the main reason for governmental restrictions in letting public contracts. Collusion is, of course, against public policy and the Sherman Anti-Trust Act. It is punishable by both fines and imprisonment. In addition, up to triple damages can be recovered in a civil action by those injured.

Lowest Responsible Bidder.   A contract award for a public project normally goes to the **lowest responsible bidder.** Many factors enter into the determination of the contractor who is to be awarded the contract. If prequalification has been required, the decision is based upon price alone. The nonmonetary factors used to determine the lowest responsible bidder are those that indicate a reasonable likelihood of completing the contract. The chief factors considered include the following:

- The contractor's reputation. Such things as shoddy work, constant bickering for extra payments, and an uncooperative or lackadaisical attitude are causes for caution. Rejection of a bidder on the basis of reputation often leads to arguments, but each person builds a reputation and has to live by it.
- The contractor's finances. Inadequate finances or credit can be a source of trouble in a project. Even though a bond is required of the contractor, finding a substitute would cause delay, and the project may suffer.

- The contractor's experience. If the project is a new field for the contractor, the contractor may have to experiment, whereas an experienced contractor would automatically know what to do. Of course, the work of an inexperienced contractor might be better because of a new approach or new ideas. The best guide is the contractor's past performance. The size of the project is also an important consideration. A contractor's success on a $40,000 job does not ensure success on a million-dollar project.
- The contractor's equipment. Most contracting work requires expensive, often specialized equipment. Without access to the proper equipment, the contractor may be doomed to failure despite good intentions. A trip by the engineer to examine the contractor's machinery and equipment may be recommended.
- The contractor's staff. The skill of the contractor's staff members in their areas of specialization can mean the success or failure of an enterprise. Purchasing of materials and supplies, for instance, can add to profit or be an excessive addition to cost. The same or similar items are often sold for widely divergent prices by different suppliers. A highly skilled purchasing agent will know how to get the best prices.

Disqualification of a bidder is a rather serious step, and the engineer should be sure of the reasons before taking such action. Charges of favoritism or debates over qualifications compel the engineer to overlook any personal feelings about a particular bidder. With all bids, the owner's right to reject any and all proposals should be reserved for use if needed.

CONTRACTOR'S COSTS

What is the major reason why contractors lose money on jobs? Probably the best answer is that they bid too low. And why would a contractor bid too low? Often, the contractor has overlooked one or more of the elements of a proposal. The contractor must consider five basic components in his or her proposal to "make out" on the job. These are

1. Direct labor
2. Materials
3. Overhead
4. Contingencies
5. Profit.

Direct Labor

In putting together the elements necessary for any kind of structure, be it a special machine or a building, direct labor is an important item. The cost-per-hour for toolmakers, carpenters, brick layers, and other specialists adds up to a significant amount. The contractor can usually anticipate and estimate this direct payment cost with reasonable accuracy. Other costs tied to direct labor, however, are often overlooked.

Fringes.   Fringe benefits include payments (other than direct wages) by an employer caused by the presence of the worker on the payroll and those imposed on the employer by a governmental requirement or by collective bargaining. Many payments are fringes according to this definition. Some examples follow:

- Worker's compensation. To conform with a state's worker's compensation law, the employer usually pays an insurance company or a state agency an amount set in accordance with the amount of the employer's payroll. The amount paid by the employer is determined either from a rate manual or according to the employer's injury experience. These costs can run from a few cents per $100 on office employees to nearly $20 per $100 of payroll on certain types of construction workers. Such costs can vary from state to state.
- Social Security. Employer's payments to Social Security are equal to those deducted from the worker's wage. When the Social Security Act was first passed, these payments were little more than a trivial annoyance; now they constitute a significant cost to the employer.
- Holiday pay and vacation pay. A recent collective bargaining goal has been to add each employee's birthday as a holiday for that employee. Although these payments are made for time when no work is performed, they are still costs of direct labor.
- Medical insurance and life insurance. It is the policy of many employers to share the cost of these benefits with their employees.
- Overtime premiums and shift premiums.
- Retirement funds. Such funds will later supplement an employee's Social Security benefits.
- Stock-buying plans and profit-sharing plans. In a stock-buying plan, the employer usually helps the employee become a part owner in the company. Frequently, deductions are taken from the employee's wages and then the employer adds to the employee's investment a few years later. Turnover is reduced if the employee must wait for the employer's addition to the investment. Profit-sharing plans work in many ways—one way being the division of company profits after the first, say, $100,000 of profit in a particular year. The increased incentive and company loyalty that result from such plans cannot be denied—but neither can the added cost.

Availability and Transportation of Personnel.   With increased automation, it has become more and more possible—and even desirable—to locate plants without reference to the location of the labor supply. Structures are being built in the frozen regions near the Arctic Circle and the Antarctic and in desert regions. The contractor who undertakes such installations has the problem of getting people to go to such places and work on the projects. Often this requires that relocation expenses and housing be furnished. In less spectacular instances, a contractor often overlooks these costs. For example, in many locations there may be an abundance of people but a shortage of specialists. Getting specialists there may entail payment by the contractor for daily time in transit or for relocation.

Materials

One of the largest costs of a project is the cost of the materials that go into it. Purchasing can make or break a contractor. There is, in effect, an added profit for the contractor who, because of a knowledge of sources for materials and supplies, is able to make fortunate procurements throughout the project.

The cost of materials and, thus, profits can also be affected by prompt payment of invoices. It is common practice among suppliers to offer discounts for prompt payment—for example, 2 percent off for payment within 10 days, net in 30 days. Although 2 percent on any one bill may not amount to much, the accumulation can be significant, especially over the life of a long and costly project.

Availability and Transportation of Materials.   Local availability of materials and supplies is an important consideration. Apparent local availability may not be sufficient because contractors can run into very high local prices. Usually, it pays a contractor to investigate and make sure of local supplies before bidding on work in unfamiliar settings.

One story runs something like this: The Black Construction Company, of a neighboring state, was the successful bidder on a contract, bidding in competition with local contractors. The bid was made on the basis of cement, mortar, sand, and gravel being locally available (suppliers in the area were quite plentiful). Black set up to begin operations on the site, but when Black Company tried to buy concrete ingredients, it found that they were "earmarked" for its former bidding competitors. As a result, Black had to transport cement, mortar, sand, and gravel several hundred miles, resulting in a loss on the contract.

Storage.   The owner often allows the contractor to use the owner's storage facilities if such facilities are available. In many contracts, however, such facilities are not available or are inadequate. The successful bidder should include storage costs in planning the bid.

Overhead

These are the costs that many small contractors or those new in the field may tend to overlook; as a result, they seem to "make out" on their contracts but lose money in the long run.

The contractor's overhead consists of many things: maintenance of an office, equipment maintenance, taxes, supervision, depreciation on equipment, utilities, and others. A discussion of all the items that normally constitute overhead would be beyond the scope and purpose of this book. Only one, the cost of estimating, will be discussed here because it is so often overlooked.

The costs of investigation, planning, and estimating usually run somewhere around 2 percent of the bid price on a project. Two percent does not seem too much, but very few contractors get every job on which they bid. The average is probably about one successful bid out of every six or so submitted. In that case, the cost of making a successful bid is substantial. It is worth noting that this cost constitutes one of the wastes of competitive bidding practices. There have been instances in which owners solicited bids with no real intention of going through with their projects.

Contingencies

Very few projects run smoothly from start to finish. It is rare that every day is a weather-working day or that delays do not occur for one reason or another. To cover the unexpected events, the contractor usually adds an item known as **contingencies** into the bid. The cost of this item runs anywhere from 5 percent to 15 percent of the total, depending upon the likelihood that hazards will occur. If liquidated damages are a prospect, contingencies may be quite high. If all work to be done is clearly and completely shown, and the probability of hazards is small, the amount for contingencies will be reduced.

Profit

People who work for others as independent contractors do so in an endeavor to make a profit. Contractors are entitled, as is anyone else, to be paid a reasonable return on the projects they undertake. The percentage of profit varies with such things as the size of the job, competition, and other factors, but there must be profit if the contractor is to remain in business.

ALTON OCHSNER MEDICAL FOUNDATION v. HLM DESIGN OF NORTH AMERICA, INC. ET AL. 2002 U.S. Dist. LEXIS 24252 (USDC ED La. 2002)

Study terms: Bonds, certificate for payment

Before the Court is the motion of defendant Federal & Deposit Company of Maryland ("F&D") for summary judgment. For the following reasons, the Court DENIES F&D's motion.

I. Background

In 1987, Alton Ochsner Medical Foundation retained HLM to design and oversee the construction of a five-story multi-purpose building. On October 1, 1991, Broadmoor, LLC contracted with Ochsner to serve as the contractor. Broadmoor agreed to furnish all labor and materials, guaranteed that the work would be free from defects, and agreed to correct all nonconforming work at its own expense. Broadmoor also agreed to indemnify Ochsner from any losses, expenses, damages, or claims arising out of any damage to property in connection with Broadmoor's work.

Also on October 1, 1991, Broadmoor, as principal, and F&D, as surety, executed three bonds in the amount of $22,488,000 each, naming Ochsner as obligee: a Bond for Private Work, a Performance Bond, and a Labor and Materials Bond. The Private Work Bond and the Performance Bond make no reference to one another. The Private Work Bond provides both for performance and payment protections, and it does not provide a time limit for instituting a lawsuit:

> WHEREAS, the Principal [Broadmoor] has entered into a certain written contract the 1st day of October, 1991, with the Obligee [Ochsner] for Multi-Purpose Building for Alton Ochsner Medical Foundation . . . which contract is hereto attached and made a part hereof.
>
> NOW, THEREFORE, THE CONDITION OF THIS OBLIGATION IS SUCH, That if the above bounden Principal [Broadmoor] shall truly and faithfully perform said contract according to its terms, covenants and conditions, and shall pay all subcontractors, journeymen, cartmen, workmen, laborers, mechanics and furnishers of material, machinery or fixtures, jointly as their interests may arise, then this obligation to be null and void; otherwise to remain in full force and effect.
>
> It is expressly understood and agreed by and between the parties to this Bond that the same is given in accordance with Louisiana Revised Statutes of 1950, Title 9, Chapter 2, Sections 4801 to 4855 inclusive.

The Performance Bond, on the other hand, provides that any suit on the bond must be filed no later than two years from when the final payment under the construction contract between Ochsner and Broadmoor falls due:

> WHEREAS, the Principal [Broadmoor] has by written agreement dated October 1, 1991, entered into a contract with Owner [Ochsner] for Multi-Purpose Building for Alton Ochsner Medical Foundation . . . in accordance with drawings and specifications . . . which contract is by reference made a part hereof, and is hereinafter referred to as the Contract.
>
> NOW, THEREFORE, THE CONDITION OF THIS OBLIGATION is such that, if Contractor [Broadmoor] shall promptly and faithfully perform said Contract, then this obligation shall be null and void; otherwise it shall remain in full force and effect.
>
> Any suit under this bond must be instituted before the expiration of two (2) years from the date on which final payment under the Contract falls due.

In January of 1994, before the project was substantially completed, Broadmoor notified Ochsner of cracking in three of the pile caps in the building 's foundation constructed by Broadmoor. Ochsner repaired the cracking when Broadmoor and HLM refused to do so. As of April 5, 1995, the project was substantially completed and Broadmoor submitted to Ochsner its Certificate for Payment. Ochsner made its final payment to Broadmoor in August of 1996. In April of 1996, Ochsner became aware of additional cracking in these and other pile caps. In July of 2000, Ochsner learned that the cracking had worsened. After Broadmoor and HLM again did not repair the cracking despite Ochsner's requests, Ochsner filed this lawsuit on May 30, 2001. Ochsner claims losses in excess of $4,000,000.

In its Complaint, Ochsner names F&D as a defendant and sues F&D under the Private Work Bond. Regarding F&D, the Complaint states:

> On or about October 1, 1991, Broadmoor, as principal, and F&D, as surety executed a Bond for Private Work ("Bond") wherein Ochsner was named as obligee. Pursuant to the Bond, F&D bound itself to Ochsner for Broadmoor's true and faithful performance of the Ochsner/ Broadmoor Contract in accordance with its terms, covenants, and conditions.

Later in its Complaint, Ochsner also states:

> Pursuant to the Bond, F&D bound itself to Ochsner for Broadmoor's true and faithful performance of the Ochsner/ Broadmoor Contract in accordance with its terms, covenants and conditions.

■ ■ ■

> F&D is liable to Ochsner for all losses and damages incurred by Ochsner as a result of the Ochsner/ Broadmoor Contract.

Nowhere in its Complaint does Ochsner make any claim upon the Performance Bond or the Labor and Materials Bond.

In its motion for summary judgment, F&D argues that the two-year suit limitation clause in the Performance Bond renders Ochsner's suit against it untimely, as the suit was filed more than two years since the project was substantially completed. Ochsner responds that it has not brought suit under the Performance Bond, but rather under the Private Work Bond, which does not contain the two-year time limit. In addition, Ochsner asserts that the two-year limit in the Performance Bond is invalid. The Court DENIES F&D's motion for the following reasons.

II. Discussion

A. Legal Standard

Summary judgment is appropriate when there are no genuine issues as to any material facts, and the moving party is entitled to judgment as a matter of law. The moving party bears the burden of establishing that there are no genuine issues of material fact.

If the dispositive issue is one on which the nonmoving party will bear the burden of proof at trial, the moving party may satisfy its burden by merely pointing out that the evidence in the record contains insufficient proof concerning an essential element of the nonmoving party's claim. The nonmovant may not rest upon the pleadings, but must identify specific facts that establish a genuine issue exists for trial.

B. The Private Work Bond

F&D argues that Ochsner's lawsuit is time-barred under the Performance Bond. Ochsner sued under the Private Work Bond, not the Performance Bond. In fact, Ochsner makes no mention of the Performance Bond in its Complaint.

The Private Work Bond contains both performance and payment protections, as evidenced by its language:

> NOW, THEREFORE, THE CONDITION OF THIS OBLIGATION IS SUCH, That if the above bounden Principal [Broadmoor] shall truly and faithfully perform said contract according to its terms, covenants and conditions, and shall pay all subcontractors, journeymen, cartmen, workmen, laborers, mechanics and furnishers of material, machinery or fixtures, jointly as their interests may arise, then this obligation to be null and void: otherwise to remain in full force and effect.

But it contains no clause providing a time limit on lawsuits. Ordinarily, the prescriptive period on a contract between an owner and a surety is ten years. F&D does not even refer to the Private Work Bond in its brief, much less explain why it does not govern the claims asserted against it. Accordingly, F&D's motion for summary judgment is DENIED.

BRITT CONSTRUCTION, INC. v MAGAZZINE CLEAN, LLC, ET AL., 271 Va. 58 (2006)

Study terms: Mechanic's liens, statutory construction

In this appeal, we consider whether Code § 43-4 requires that a general contractor, as a condition of perfecting a mechanic's lien, contemporaneously file with the memorandum of lien a "certification" that a copy of the memorandum has been mailed to the property owner.

In February 2003, Magazzine Clean, L.L.C. (Magazzine Clean) hired Britt Construction, Inc. (Britt) as the general contractor for construction of a commercial car wash facility on Magazzine Clean's property in Loudoun County. As a result of disputes between the parties during the construction process, Britt recorded 12 separate memoranda of mechanic's liens against Magazzine Clean's property.

Britt recorded the memoranda of liens in Loudoun County between June 18, 2004 and October 14, 2004. However, Britt did not mail copies of these memoranda of liens to Magazzine Clean, nor did Britt file certifications of such mailings at the time of filing its memoranda. Instead, Britt waited until December 17, 2004 to record certifications of mailing for each of the 12 memoranda previously filed.

Magazzine Clean initiated this suit by filing a petition to invalidate the mechanic's liens pursuant to Code § 43-17.1[2] Magazzine Clean argued that none of the mechanic's liens met the perfection requirements contained in Code § 43-4 because Britt did not mail copies of the memoranda of mechanic's liens to Magazzine Clean, nor did Britt file

Sub-contractors cannot sue magazzine directly because they do not have a contract, called privity.

---

[2]Code § 43-17.1 provides that: "Any party, having an interest in real property against which a lien has been filed, may, upon a showing of good cause, petition the court of equity having jurisdiction wherein the building . . . is located to hold a hearing to determine the validity of any perfected lien on the property. After reasonable notice to the lien claimant and any party to whom the benefit of the lien would inure and who has given notice as provided in § 43-18 of the Code of Virginia, the court shall hold a hearing and determine the validity of the lien. If the court finds that the lien is invalid, it shall forthwith order that the memorandum or notice of lien be removed from record."

certifications of mailing along with the memoranda. As amended by the General Assembly in 2003, Code § 43-4 states in relevant part that:

> A general contractor . . . in order to perfect the lien given by § 43-3 . . . . shall file a memorandum of lien at any time after the work is commenced or material furnished, but not later than 90 days from the last day of the month in which he last performs labor or furnishes material, and in no event later than 90 days from the time such building . . . is completed, or the work thereon otherwise terminated. . . . A lien claimant who is a general contractor also shall file along with the memorandum of lien, a certification of mailing of a copy of the memorandum of lien on the owner of the property at the owner's last known address. . . .[3]

After considering the parties' briefs and arguments, the circuit court granted Magazzine Clean's amended petition and invalidated the liens. The circuit court held that the mechanic's liens were invalid because Britt did not file certifications of mailing along with the memoranda of liens. Britt appealed from the circuit court's decree.

Britt argues that the provision in Code § 43-4 directing a general contractor to file a certification of mailing is not a requirement for perfection of the general contractor's mechanic's lien. Britt asserts that the statute's only requirement for perfection of such a lien is the timely filing of the memorandum of lien, and that the certification of mailing need only be filed in order for a property owner to be deemed to have notice of the lien. Thus, Britt maintains that the statutory directive for filing a certification of mailing is merely a note provision that should be construed liberally.

In support of its argument, Britt notes that two other statutes, which address liens of subcontractors and persons performing labor or furnishing materials for a subcontractor, expressly require as a condition of perfecting a lien that written notice of the lien be given to the owner. Britt contends that because Code § 43-4 does not contain similar express language, the General Assembly did not intent to impose such a requirement in this statute. We disagree with Britt's arguments.

We Consider the language of Code § 43-4 under basic rules of statutory construction. We examine the statute in its entirety and determine the General Assembly's intent from the plain and natural meaning of the words used in the statute.

When statutory language is unambiguous, we are bound by the plain meaning of that language. *Williams v. Commonwealth*, 265 Va. 268, 271, 576 S.E. 2d 468, 470 (2003) Therefore, when the General Assembly has used words of a definite import, we cannot give those words a construction that amounts to holding that the General Assembly meant something other than that which it actually expressed.

We further observe that when a statute has been amended, there is a presumption that the General Assembly intended to effect a substantive change in the law. Thus, we will assume that a statutory amendment is purposeful, rather than unnecessary.

Because the mechanic's lien statutes are in derogation of the common law, the statutory requirements regarding the existence and the perfection of a mechanic's lien must be strictly construed. *Carolina Builders Corp. v. Cenit Equity Co.*, 257 Va. 405, 410 (1999). A mechanic's lien must be perfected within the specific time frame and in the manner set forth in the statutes, or the lien will be lost.

Applying these principles, we conclude that the certification of mailing requirement of Code § 43-4 is plain and unambiguous. The statute expressly requires that a general contractor "file along with" the memorandum of lien a certification that the general contractor has mailed a copy of the memorandum of lien to the owner at the owner's last known address.

---

[3]The General assembly amended Code § 43-4 to include the certification of mailing requirement effective July 1, 2003.

By using the word "file," the General Assembly made its intention clear thai the certification of mailing is not merely a notice provision. Moreover, in requiring that the certification be filed "long with" the memorandum of lien, the statutory language directs thai the memorandum of lien cannot be filed alone without the certification of mailing, and that both documents must be filed in order to perfect the lien.

Britt's contrary argument is unavailing because it would permit a general contractor to mail a copy of its memorandum of lien and to file its certification of mailing at a time of the general contractor's own choosing. This result would render the plain language of the statutory amendment meaningless and would undermine the clear intent of the amendment to prevent a general contractor from filing undisclosed liens against an owner's property. . . .

Britt did not file the required certifications of mailing along with its memoranda of liens but waited more than two months after filing the final memorandum of lien to record the certifications. Thus, Britt's actions clearly demonstrate its failure to comply with the certification requirement of Code § 43-4.

Our conclusion regarding the plain meaning of Code § 43-4 is not altered by Britt's observation that this statute, unlike Code §§ 43-7 and -9, does not expressly state that written notice to the owner is a condition of perfecting a mechanic's lien. When statutory language is plain and unambiguous, we will not look to other provisions of the Code to interpret that statute. Thus, the fact that the General Assembly chose to use different language in stating a perfection requirement in those other statutes cannot alter the plain language of Code § 43-4, which requires that a general contractor "file" its certification of mailing "along with" its memorandum of lien.

For these reasons, we will affirm the circuit court's judgment.

Affirmed.

Review Questions

1. Describe the relationships between the parties to a construction contract.

2. How do public projects and private projects differ?

3. In what ways are contracting and manufacturing similar? How do they differ?

4. If you planned to build a two-car garage on your lot, what factors would you consider in deciding whether to attempt to build the garage yourself or hire a contractor to do it? How does this differ from an industrial situation in which a company is trying to decide whether to build a special machine or hire someone else to build it?

5. Summarize briefly the advantages and disadvantages of: (1) a lump-sum contract, (2) a cost-plus-percentage contract, (3) a cost-plus-fixed-fee contract, and (4) a combination unit-price and lump-sum contract.

6. In *Alton Ochsner Medical Foundation v. HLM Design of North America, Inc. et al.,* what should defendant F&D have done to ensure that the suit limitation clause would apply to the Private Work Bond?

7. What purposes are served by a feasibility study? When should it end?

8. The vice president has asked you to head a feasibility study for Big Construction Project because you took a class in Engineering Law. This is your first major project and you are a recent hire. While you are anxious to give a good impression to your superiors, you don't have any substantive experience conducting such a study. What steps should you take to ensure that the feasibility study is: (a) is comprehensive, (b) addresses other business considerations, and (c) limits the risk of losing your job?

9. Why do governments strictly regulate competitive bidding practices? What formal procedures are required in your state?

10. Locate in a local newspaper an "advertisement" or "call for bids" for a construction project. What details does the advertisement give its reader?

11. What are mechanic's lien laws? How do your state's statutes call for such liens to be established and removed?

12. What are the purposes of each of the three types of bonds commonly required of contractors?

13. According to *Britt Construction, Inc. v. Magazzine Clean, LLC,* why must mechanic's lien statutes be strictly construed? What did the court conclude with regard to the certification of mailing requirement of the statute?

Specifications

HIGHLIGHTS

- Contract problems may be avoided (or at least minimized) if specifications are put in writing rather than being given orally for two reasons:
  1. They create a permanent record for resolving disputes, and
  2. They assure planning.

As the term is used here, **specifications** refer to the description of work to be done or things to be purchased. Specifications are most frequently heard of in building construction, but they exist in every area of engineering. The definition above is broad enough to include many things we encounter daily in our dealings with others. If, for example, you hire painters to paint your house and tell them the house is to be white with gray trim, you have made a specification. If you take your car to an auto mechanic and tell the mechanic to do whatever is necessary to remove the "grind" from the transmission, you have made a specification. A person ordering four 1/4″ stove bolts, 3″ long, from a hardware store clerk is using a specification. The specifications mentioned so far have been oral, but they are just as much specifications as the voluminous written ones for such a construction project as the Big Dig in Boston.

Building construction is seldom undertaken without specifications; process engineers, project engineers, accountants, and purchasing agents toil over specifications for machinery and equipment. Very few government purchases are made without reference to written specifications. In fact, the federal government only proceeds without such specifications in isolated instances of purchases of services. Generally, if the purchase involved is of sufficient monetary value so that a breach of the specified conditions could cause the purchaser substantial injury, the specifications should be written. Two main reasons exist for putting specifications in writing rather than giving them orally: 1) They create a permanent record for resolving disputes, and 2) they assure planning.

The written record of specifications defines the duties of the parties. The specification then becomes part of the contract. Later, it may be interpreted in court or by an arbitration board if a dispute arises between the parties. Lawsuits on some construction projects have been undertaken years after completion of the projects. Without a permanent record of the rights and obligations of the parties, the result can be utter confusion. Written specifications can also frequently prevent costly lawsuits because they act as a ready reference in controversies between the parties.

The second reason for putting specifications into writing, the planning reason, is based in both practicality and efficiency. It is easy to specify something orally without giving it much thought. When it becomes necessary to reduce the specification to written form, however, the writer becomes more cautious, examining his or her reasons for making a particular specification. The result, in specification writing, is usually a more efficient purchase. The likelihood that the thing specified will do the job required at a lower overall cost is improved greatly by the writing.

SPECIFICATION WRITING

A specification must communicate to the reader what is required by the writer. Its job is essentially quite simple. As such, it can be best accomplished in simple language. Complex sentence structure and complicated wording can often be interpreted more than one way. With simple structure and simple wording, ambiguity is much less likely. It is best to avoid jargon, abbreviations, and symbols, unless they are of a recognized standard and fully understood in the trade.

Specifications, together with the drawings, show the contractor what is to be done. Each supplements the other. In addition, the specifications indicate the relation between the parties in greater detail than does the agreement. The working drawings show the work to be accomplished, but, frequently, not the quality standards required. The quality standards are usually more conveniently stated in the specifications.

Style

Specifications should be written with great clarity. A specification should not contain flowery language or complicated legal terms. Although they must be brief and terse, the specifications still must be complete. The writer should strive to be brief up to the point where further brevity may be accomplished only by sacrificing some of the meaning in the specification.

Literature written to be read for entertainment frequently refers to a subject in terms of "he . . . hers," and other such pronouns. The use of such words adds considerably to the ease and enjoyment of reading such material. The vast majority of readers will understand which subject the author is referring to. Those who do not will usually pass over the sentence without hesitation to maintain interest in the story.

The writer of anything that is likely to be interpreted in a court of law, however, cannot afford to risk this kind of open-endedness. Where a misinterpretation is at all likely, the subject should be repeated. Consider the sentence: "In case of controversy between the contractor and the inspector, he shall immediately refer the question to the engineer." Who is to refer the question? The contractor could say, "I thought the inspector was to contact the engineer," and vice versa. Replacement of the word "he" with "the contractor" or "the inspector" clears up any doubt.

Precise Wording

Our language consists of many words with similar meanings. It includes general words and specific words. The word **property,** for instance, has a very broad meaning, including all items of personal possessions as well as real estate. Many other terms have broad meanings and should be used very carefully in the wording of specifications. The words used should indicate what is to be accomplished so precisely that no doubt can exist. Commonly used words are often not exact enough in meaning. For instance, it is quite common to hear someone say that something should have an addition on either side of it, as in "light on either side of the gateway," when what is intended is that the addition must be present on each side, as in "a light on each side of the gateway."

Many other words are commonly used interchangeably, such as *any* and *all, amount* and *quantity, malleable* and *ductile,* and *hardness* and *rigidity,* for example. The inexactness of meaning in oral conversation usually causes no problems, but in specification writing the result can be trouble, measured in dollars. There are few instances in the English language where one word means exactly the same thing as another. Care in wording is definitely called for where the writing may have to be reviewed in court.

Reference Specifications

With the wealth of old specifications, texts on specification writing, and standards published by various organizations, it is a rare specification today that is entirely original. As a practical matter, this is desirable. Previous specifications have been tested, errors removed, and the language improved. If the rewriting for the new specification is carefully done, other potential problems may be discovered and eliminated. In addition, the writer is less likely to overlook something in the present specifications if he or she can follow a pattern. However, a word of caution is necessary. Wholesale copying and pasting of paragraphs and requirements from previous specifications can result in problems. While two specifications may be similar in many ways, it is unlikely that any two will be exactly alike. If the writer does not exercise caution in adapting paragraphs and clauses from old to new, certain features of the new job are likely to be different from what was intended. For example, features of the new job could be completely omitted from the new specification.

Many organizations have written standards for various items of equipment or elemental components, making it unnecessary for the specification writer to do more than incorporate these standards by reference. The United States government, for one, has published a vast array of standard specifications. (See Appendix E.) The writing of a specification where such standards exist consists largely of determining the appropriate elements and listing the standards in logical order.

The advantage of such standardization is quite obvious. A specialist in one element of the entire assembly has specified the appropriate quality to be used. No person can be a specialist in everything. The specification resulting from the combined efforts of several specialists is quite likely to be better than it would be if the total specification were to be written by any one of them.

ARRANGEMENT

The specification usually consists of at least two parts, one general and the other specific and detailed. The first part is known by different names (such as **general conditions, general provisions,** or **special conditions**) in various specifications, and the information included is not well standardized. Most contracts also include a document, previously discussed, known as the **General Conditions,** which deals with the basic rights and responsibilities of the parties. To avoid confusion with the more basic document, we will here use the term general provisions as the name for the general portion of the specifications.

General Provisions

In the relationships between the parties to a construction contract, there is a vast middle ground between the provisions in the agreement and the general conditions, and the detailed provisions in the specifications. The general conditions are broad enough to apply to any contract work the owner may want and to any contractor he or she may hire.

In contrast, the general provisions part of the specifications pertains to a particular contract or type of contract. It sets the owner's policy as to control of the work, the scope and quality of the work, and any special requirements or precautions. It is here that answers are found to many of the day-to-day questions that arise between the parties. To cover these questions, the general provisions must consider many topics. A few of the topics appear frequently enough to justify consideration here, these being representative of the contents of a general provisions section.

Schedule of Work. A topic that is almost certain to be covered (usually by a requirement for consultation) is the schedule of work. It is poor policy, from both a legal and a cost standpoint, to rigidly control the details of the work to be accomplished. However, there are occasions when the owner must exercise some control to protect the public interest (as in highway construction) or to dovetail the project with the work undertaken by other contractors.

Usually a work schedule is called for in the specifications, to help the engineer and contractor plan the work to the best advantage of the owner. Such a work schedule is made by showing graphically the starting and completion dates for the component parts of the project. The work schedule resembles the Gantt Charts so commonly used in production scheduling in industry. Such schedules are frequently used in a progressive payment scheme (so much per week or month according to accomplishments during the period). The schedule allows the parties to make a realistic estimate of progress. The amount to be held back may also be specified in the general provisions.

Changes and Extra Work. Very few large contracts are completed without changes being made during the course of the work. Questions of payment for such changes can cause severe disagreement if no provision has been made to anticipate them. A provision for changes and extra work is usually made in either the general conditions or the general provisions portion of the specification. Just as there are four basic means of paying for a contract, there are also four means of paying for extra work:

- Lump sum
- Cost plus percentage
- Cost plus fixed fee
- Unit price

The cost-plus-percentage and unit-price types of contracts present little or no problem as to compensation for changes or added work. The extra work is compensated at the fixed rate in the unit-price contract; the open-endedness of the cost-plus-percentage contract also accommodates changes easily. (See *D.A. Elia Construction Corporation v. New York State Thruway Authority* at the end of the chapter.) In the lump-sum and cost-plus-fixed-fee contracts, however, the parties must decide in advance how extra work will be compensated. The contractor is already on the job; it is usually inconvenient and costly to get someone else to take on the added work. If the contractor is unscrupulous and the engineer has not provided adequately for changes, the contractor can make these changes costly. The engineer who gives careful attention to the contract provision regarding payment for changes and extra work may find the time well spent.

Other Suppliers.   Work or materials to be supplied by others (either by the owner or other contractors) should be shown in the specification. Frequently this information takes the form of one or more **right of supply** clauses, in which the owner reserves the right to supply motors or other components (often salvaged from worn-out machines). When more than one contractor is to work on a project, the areas of responsibility must be delineated clearly in the specification. This precaution is necessary for avoiding conflict and assuring that someone will be responsible for the details of each item; in other words, it will avoid both overlaps and gaps in the contracted work.

Drawings.   The drawings are an integral part of a construction contract. The following statement is commonly included in the general provisions of the specification: "The Contract Documents are complementary, and what is called for by any one shall be binding as if called for by all." The drawings and specifications, particularly, supplement each other as everyday working documents. Occasionally a conflict between a drawing and the specifications becomes apparent.[1] To resolve such conflicts, the general provisions frequently state that "in case of difference between drawings and specifications, the specifications shall govern."

Some provision is usually made for submission of working drawings and sketches for the engineer's approval during the course of the construction. Such details as the number of copies of drawings, even size and type of paper to be used, are sometimes specified.

Information Given.   Usually, considerable information pertinent to the proposed project is passed along to the bidders and thus, eventually, to the contractor. Some information may be such that the owner and the engineer do not want to warrant its completeness or the indications given by it. This is commonly true of test borings, for instance, where a defect in the subsoil might not show up in the samples that are taken. To protect the owner in such circumstances, the general provisions usually state that neither the owner nor the engineer will warrant that the information given shows the entire picture. Generally, contractors are held responsible for having made their own examinations of the site.

Services Furnished.   When an addition to an existing structure is made, or when machinery or equipment is installed, certain of the services necessary are commonly supplied by the owner. Such services as water, compressed air, electricity, and crane service to unload equipment are frequently furnished by the owner. It is in the owner's interest to mention the availability of such services in the specification, to give contractors a truer picture of the costs they will not have to bear.

Receipt and Storage of Materials.   By similar reasoning, if the contractor is to be allowed to use the owner's receiving and storage facilities, this should be made clear in the specification.

---

[1] In *Centex Construction Company, Inc. v. The United States* at the end of the chapter the specifications state: "Anything mentioned in the specifications and not shown on the drawings, or shown on the drawings and not mentioned in the specifications, shall be of like effect as if shown or mentioned in both." This clause is one of the reasons for the controversy in this case.

Such facilities add considerably to contractors' costs if they have to supply them. The costs will be passed along to owners who do not supply such facilities.

Wage Rates. Particularly in government contracts, the wages paid by the contractor are important. Several federal statutes (and usually similar state statutes) regulate wages, hours, and even sources of materials on public contracts. By way of example, many federal contracts cite the Davis-Bacon Act,[2] the Buy American Act,[3] and the Federal Labor Standards Act[4] in their specifications. But the wages paid to workers are of importance even when the owner is a private party. Wages and working conditions poorer than those to which the area is accustomed can cause strikes and other labor problems. Projects are sometimes delayed and even destroyed as a result of labor strife. To avoid trouble of this nature, many specifications for private contracts provide that wage rates and working conditions must be equal to or better than those prevailing in the trade or locality.

Safety. A requirement such as the following, as to safety and accident prevention, usually appears in the specifications: "The contractor shall, at all times, exercise reasonable precautions for the safety of employees in the performance of this contract, and shall comply with all applicable provisions of federal, state, and municipal safety laws and building construction codes."

Detail Specifications

The second major division of the specifications gives the details of the work to be undertaken. These **detail specifications,** together with the engineer's drawings, state how the job is to be done and what is to be accomplished. In writing the general provisions portion, the engineer has access to guideposts and instructions. When writing the detail specifications, however, the engineer has fewer guideposts and must rely on his or her writing skills and knowledge of what is to be accomplished.[5] Detail specifications are concerned with the materials and work quality in the finished project.

Materials. When someone buys a lathe or a punch press, he or she is really purchasing the ability to turn, bend, blank, or perform some other operation on materials. All a possession can do is provide a service of some kind. Much the same is true in the purchase of materials for a construction project. The materials purchased must render a service. But it is up to the engineer to

---

[2]Davis Bacon Act, 40 USC 276a relating to rate of wages for laborers and mechanics employed by contractors and subcontractors on public buildings.

[3]Buy American Act, 41 USC §10a, providing preferential treatment for domestic sources of un-manufactured articles, manufactured goods, and construction services.

[4]Federal Labor Standards Act, 29 USC §207, anyone who is engaged in commerce or in the production of goods for commerce for a workweek longer than forty hours must receive compensation at a rate not less than one and one-half times the regular rate at which employed. See also Contract Work Hours And Safety Standards, 40 USC §328 the wages of every laborer and mechanic employed by any contractor or subcontractor in performance of work on any contract of the character specified in section 329 of Title 40 must be computed on the basis of a standard workweek of forty hours, and for excess work, the rate cannot be less than one and one-half times the basic rate of pay.

[5]As you'll read in *Performance Abatement Services, Inc. v. Lansing Board of Water and Light,* the court had neither patience nor sympathy with the drafter who seemed to have little knowledge about the drawings included in his specifications.

obtain the best service available for the owner at the most favorable cost. Thus, two factors normally oppose each other in the selection of material: cost and service. In many instances, the same or similar service can be rendered by two quite different materials (or pieces of equipment or machines). Transportation, storage, and inspection costs join other costs to be weighed against the service offered by a particular material. Part of an engineer's "stock in trade" is a knowledge of various means of obtaining services for the employer.

Materials are commonly called for in specifications according to established standards. SAE 1090 steel, for instance, indicates a steel with 90 points of carbon that hardens with proper heat treatment and has some unique physical properties. The various physical properties of materials, such as strength, elasticity, conductivity, and appearance, are important in determining the service they will render in a specific application. Very often, specifications are written with "or equal" clauses; for example, "Electrical controls to be XYZ *or* equal." The or *equal* means "not literally exact equality." A more precise way of stating the meaning might be *or equivalent*. With this, as with other contract language, the courts interpret wording according to trade usage.

Work Quality.   A person who takes an automobile to a mechanic for a valve job usually isn't interested in the order of removal and replacement of the screws, nuts, and bolts. Similarly, an owner is rarely interested in how a particular result is accomplished, providing the result is satisfactory. It is usually far better to specify the results to be accomplished rather than the process by which the results are to be accomplished. For instance, it is much better to specify the compressive strength of concrete a week after pouring than to specify the quantities of cement, sand, and broken stone or gravel and the method of mixing.

There are exceptions, of course, to the principle of specifying results only. A particular method may interfere with the rights of others or with other contractors (e.g., using blasting in place of air hammers). Usually, however, it is sufficient in specifications to require that the work quality be equal to the best available without going into the details of the method. Language like the following might be used: "All sheet-metal work shall be performed and completed in accordance with the best modern sheet-metal practice, and no detail necessary therefor shall be omitted, although specific mention thereof may not be made either in these specifications or on the drawings."

Specifications and the Owner-Contractor Relationship

Generally, there are two possible relationships between the owner and the contractor. The relationship of owner-independent contractor indicates that the contractor has been hired to produce a particular result. The employer-employee relationship prevails where the owner (employer) or the owner's agent, the engineer, supervises the work of the contractor too closely. (For further information on the owner-independent contractor relationship, see *Vizcaino v. Microsoft Corp.,* 120 F. 3d 2006 (9th Cir. 1997).)

The owner has two main advantages in retaining the owner-independent contractor relationship. Probably the primary one is that the contractor retains liability for his or her acts. There are many cases on record where, because the supervision was too close or the specifications made the contractor a mere employee of the owner, the owner was held directly liable for the acts of the contractor. Statements like the following make the contractor sound like an employee of the engineer (and, hence, an employee of the owner): "The contractor shall begin and continue work on whatever parts of the project the engineer shall direct, at whatever time the engineer shall direct." Such statements make it easy for a court to find against the owner and hold the owner liable for the actions of the contractor. The second advantage to maintaining the owner-independent contractor relationship (and specifying results rather than means of attaining them) involves liability as well. A contractor who follows a specified procedure cannot be held responsible if the result proves faulty. On the other hand, if the contractor agrees to produce a specified result, with the means of accomplishment left to the contractor, he or she must produce the result. A contractor who fails to produce the specified results may become the target of a damage action, or may be required to assure repairs.

D.A. ELIA CONSTRUCTION CORPORATION v. NEW YORK STATE THRUWAY AUTHORITY 289 A.D.2d 665; 734 N.Y.S.2d 295 (NY 2001)

Study terms: Bid documents, unit price, fraudulent misrepresentation

This action arises out of a contract between claimant and the State Thruway Authority (hereinafter the State) for the repair of four concrete piers supporting the Castleton-on-Hudson Bridge. Claimant was paid the unit prices specified in the contract for the number of cubic yards of concrete actually placed and the number of gallons of epoxy bonding compound estimated in the project's bid documents. As a result of having to repair more but smaller areas of the piers than were identified in the bid documents and the State's alteration of the repair procedure, claimant brought this action seeking compensation for additional work and costs, including the cost of additional gallons of bonding compound. Finding, after trial, that claimant had agreed to the changed repair procedure, that the increased number of repaired areas did not reflect a qualitative change in the work and that the extra bonding compound had not been "incorporated into the work," the Court of Claims dismissed the claim. Claimant appeals.

Absent proof of fraudulent misrepresentation or qualitative alteration of the work, the express terms of a unit price contract govern the parties' rights and obligations . Seeking to avoid this result, claimant contends that the State misrepresented the extent of the repairs and that the Court of Claims erred in finding no qualitative change in the contemplated work.

To rely on fraudulent misrepresentation, claimant had to prove that the additional work performed resulted directly from the State's concealment or nondisclosure of material facts either known to it or within its possession. Claimant's vice president, Daniel Elia, testified that its bid had assumed that the bid documents were based on an accurate field inspection of the piers by means of recent hammer soundings. The bid documents, however, do not make this representation but, rather, warn that only major repair areas are indicated and that they are approximations subject to actual field conditions. Claimant's own expert witness, Douglas Pressley, testified that the bid documents' depiction of a small number of large areas indicated that the piers had not been hammer sounded. Since the contract advised bidders to conduct their own inspections and claimant failed to do so, it could not reasonably rely on the bid documents for anything more than an approximation of the repair areas.

In the alternative, claimant argues that the contract's exculpatory clauses should not be enforced because it did not have enough time to thoroughly inspect the piers. Although this Court has stated that such clauses will not be given effect if an inspection would not have revealed the inaccuracy of the contract's representation, this principle is inapplicable here because there is no evidence that a sufficiently thorough inspection could not have been made before claimant submitted its bid.

Claimant's proof at trial also failed to prove a qualitative change in the nature of the work. Although significantly more areas required repair, the total area repaired and the amount of concrete used was comparable to the original estimates. Thus, claimant established only that the work was more difficult than expected, a circumstance that is plainly insufficient to free it from the unit price contract. Nor are we persuaded that the Court of Claims erred in rejecting claimant's argument that the State's unilateral alteration of a repair procedure amounted to a qualita-

Expert should not testify knowing that testimony will contradict his employer's position. Discuss testimony w/ employer & lawyer beforehand.

tive change in the work. The record supports the court's finding that the procedure was revised before the contract was executed and, thus, was within claimant's contemplation. For these reasons, we conclude that the Court of Claims properly dismissed the claims for extra work.

We similarly conclude that sufficient evidence supports the finding by the Court of Claims that the additional gallons of bonding compound were not "incorporated into the work" as that phrase is used in the parties' contract. The contract provided that any ambiguities concerning its specifications were to be determined with finality by the State. The project's resident engineer, Carl Niemann, testified that one gallon per 50 square feet was determined to be a satisfactory standard for what would be considered as "incorporated into the work." Niemann also testified that the additional gallons of compound were wasted by claimant due to its short shelf life, improper mixing, adherence problems and application inaccuracies. The only evidence offered to contradict Niemann, who was at the job site on a daily basis and for 60 hours a week, was the testimony of Elia, who admittedly was at the site about twice a month. Since the Court of Claims credited Niemann's testimony in this regard and its assessment of credibility is to be given considerable deference, we find ample evidence supporting the court's findings.

Ordered that the judgment is affirmed, without costs.

CENTEX CONSTRUCTION COMPANY, INC. v. THE UNITED STATES 49 Fed. Cl. 790; 2001 U.S. Claims LEXIS 128 (US Ct. Fed. Cl. 2001)

Study terms: Specifications and drawings, typical

The basic issue in this case is whether the contract in question, to construct an addition to a Veteran's Administration medical facility, required the installation of channel bracing in stud walls with door openings. If it did not, then plaintiff is entitled to additional compensation for ultimately having to install those braces; if the contract did so provide, then plaintiff is entitled to no compensation. After careful consideration of the briefs filed and the oral argument, and for the reasons discussed below, the court concludes that the contract required the channel bracing and, therefore, GRANTS defendant 's motion for summary judgment.

I. Statement of Facts

On or about September 30, 1993, the Department of Veterans Affairs ("VA") entered into Contract No. V101DC0086 with Centex Construction Company, Inc. ("Centex" or "plaintiff") to construct a clinical addition and Spinal Chord Injury Center ("SCI Center") at the VA Medical Center in Dallas, Texas. Section 1.45 of this contract, entitled "SPECIFICATIONS AND DRAWINGS FOR CONSTRUCTION (FAR 52.236-21) (APR 1984)," contained a standard Federal Acquisition Regulation (FAR) clause, which provided, in pertinent part:

> (a) The Contractor shall keep on the work site a copy of the drawings and specifications and shall at all times give the Contracting Officer access thereto. Anything mentioned in the specifications and not shown on the drawings, or shown on the drawings and not mentioned in the specifications, shall be of like effect as if shown or mentioned in both.

The VA contract contained two sets of drawings—one set pertaining to the SCI Center and the other to the clinical addition. Architectural Drawing No. 74-71R, pertaining to the SCI Center, and Architectural Drawing No. 2-240R, pertaining to the Clinical Addition, both were entitled "Door Schedules and Details." Architectural Drawing No. 74-71R includes detail 74-71/04 entitled "Elevation of Framing at Door Opening," which applies to metal door frames and requires the installation of 3/4 inch channel bracing. Architectural Drawing No. 2-240R identifies this same detail as 2-240/04 and also requires the installation of 3/4 inch channel bracing.

Centex subsequently subcontracted various work to Cleveland Construction, Inc. ("Cleveland"), including the installation of interior door frame systems in the SCI Center. The subcontract required, in pertinent part:

34. STATEMENT OF WORK AND SUBCONTRACT PRICE

B. INCLUSIONS. In addition to the foregoing, it is further understood and agreed that this Subcontract also includes the furnishing and installation of the below listed items regardless of whether or not they are in the above specification section(s), or any other specification section, or shown on the plans:

4. The receiving, unloading, and storing of all hollow metal frames as directed by the Contractor's project office. Subcontractor will distribute, install and grout (if required) all frames, including elevator frames, and provide all wall framing and bracing as required to keep the frames plumb and square.

On November 14, 1995, Centex and Cleveland employees met with the VA's senior resident engineer, Paul Newman, to discuss several outstanding issues regarding the VA contract. At this meeting, Cleveland asserted that, because detail 74-71/04 was not described as a "typical" detail on the contract drawing, it was therefore not part of the VA contract. In his response dated December 14, 1995, Mr. Newman stated "the 3/4" channel as wall bracing shown by 74-71/04 is in the drawings. Because the detail 74-71R/04 is in the contract you are required to install the wall bracing."

On December 21, 1995, Centex again disputed, in writing, Mr. Newman's assertion that the VA contract required the installation of the door support material at the hollow metal door frames. On January 22, 1996, Mr. Newman responded to Centex's letter, writing "our position is that details is on the drawings, so it is part of the contract and therefore must be incorporated into construction." On February 7, 1996, Centex and Cleveland employees again met with Mr. Newman to discuss detail 2-240/04. At that time, the VA directed Cleveland to install the door opening bracing in the clinical addition in accordance with the contract detail. On February 21, 1996, Cleveland informed Centex that it would install channel bracing at all door openings at the clinical addition pursuant to detail 2/240-04, but added that it would submit a proposal for the added cost associated with the installation.

On December 10, 1997, Centex submitted a claim to the VA's contracting officer in the amount of $82,777, asserting that it was entitled to this amount as a result of the "extra contractual requirements" imposed by the VA in connection with the installation of channel supports at hollow metal door frames. On March 24, 1998, the contracting officer denied Centex's claim. On March 23, 1999, Centex filed suit in this court. On October 25, 2000, defendant filed its Motion for Summary Judgment. Subsequently, this case was reassigned to the undersigned judge. On May 15, 2001, this court heard oral argument in this case.

II. Discussion

Summary judgment is appropriate when there is no genuine dispute as to any material fact and the moving party is entitled to judgment as a matter of law. Such is the case here.

Contract interpretation is a matter of law and, as such, is amenable to disposition on summary judgment. In interpreting a contract, the court's examination begins—and in this particular case, ends—with the plain language used in the contract. The court must interpret the contract as a whole to give reasonable meaning to all its parts and to avoid "conflict or surplusage of its provisions." *Granite Const. Co. v. United States*, 962 F.2d 998, 1003 (Fed. Cir. 1992).

In the instant case, the drawings attached to the contract clearly required the installation of 3/4 inch channel bracing when metal door frames were utilized. However, this requirement was not reflected in the specifications of the contract. As recited above, however, a standard clause required by the FAR (48 C.F.R. § 52.236-21(a)) was included in the contract, which indicated, in pertinent part, that—"anything . . . shown on the drawings and not mentioned in the specifications, shall be of like effect as if shown or mentioned in both." This clause has consistently been interpreted to require a contractor to comply with drawings even where details depicted therein are not listed in the specifications. Accordingly, the inclusion of this clause in the contract at issue obliged plaintiff to comply with details of drawings as if they were in the specifications, thereby requiring plaintiff and it subcontractor to install the channel bracing in question.

Plaintiff, however, claims that, notwithstanding this standard clause, it is unreasonable to hold it to every minute detail in the voluminous drawings attached to the contract, particularly because such details ordinarily would be expected to be found in the specifications. "This lack of information," it contends, "would lead a reasonable contractor to conclude that channel bracing was not required." But, this contention ignores the fact that plaintiff signed a contract that indicated otherwise. Indeed, even assuming that a reasonable contractor would act in the fashion that plaintiff claims—far from apparent, in this court's view—it remains that any notion that the scope of the specifications and drawings clause is limited by some overarching concept of commercial reasonableness, essentially permitting a contractor to ignore certain details in the drawings, was soundly rejected by the Court of Claims in *Unicon Management Corp.,* supra. In that case, a contractor sought compensation for installing steel plate covers on certain floors, a requirement that was indicated in the drawings, but not in the specifications. Judge Oscar Davis, writing on behalf of a unanimous court, rejected a similar argument to that made by plaintiff here, stating that: "if they examined the plans and specifications carefully they could not have helped notice the drawings which specifically embodied the . . . requirement. If they were not aware of this fact they should have been." 375 F.2d at 806. The court concluded that if "plaintiff did not study the plans and specifications before bidding, it cannot complain that the [Armed Services] Board [of Contract Appeals] and this court strive, in accordance with the established canon, to read the relevant contract provisions together rather than at odds." Id. at 807. Based upon *Unicon,* this court concludes that the fact that the contract in question contained numerous drawings does not excuse plaintiff from being familiar with those drawings and complying with the actual contract it executed.

[handwritten margin note: Not a valid defense]

Plaintiff also claims that it was not obliged to install the bracing because it was neither identified as "typical" in the drawings nor highlighted elsewhere in the contract. But, again, plaintiff fails to anchor this proposition to anything in the contract's language. The details 71-74/04 and 2-240/04, incorporating the bracing requirement, clearly were shown on the drawings and nothing in the contract limits its requirements to those identified as typical—indeed, if that were true, very few of the details in the drawings would be binding as most appear not to have been labeled "typical." Other documents in the record confirm that both plaintiff and its subcontractor understood that they were required to implement details in the drawings whether or not they were labeled "typical," without any expectation of additional compensation. Moreover, while excerpts of depositions in the record indicate that the agency contracting officials expected a contractor to incorporate into the construction details in the plans that were identified as typical or otherwise highlighted in the contract, there is no indication in these depositions or elsewhere in the record that those officials believed that only in those circumstances was a contractor required to comply with a detail listed in the drawings. In short, this deposition testimony does not suggest any variation in practice from the plain terms of the contract. . . .

III. Conclusion

For the foregoing reasons, the court GRANTS defendant's motion for summary judgment. The Clerk is hereby ordered to dismiss plaintiff's complaint.

PERFORMANCE ABATEMENT SERVICES, INC., v. LANSING BOARD OF WATER AND LIGHT, ET AL 168 F. Supp. 2d 720; 2001 U.S. Dist. LEXIS 13765 (USDC WD MI 2001)

Study terms: Specifications and drawings, general requirements, changes to the work, voluntary dismissal, innocent misrepresentation, express warranty, implied warranty, reformation, mutual mistake

This matter is before the Court on Defendants SCS Group, L.C.'s (hereafter "SCS") and Lansing Board of Water and Light's (hereafter "BWL") Motion to Dismiss Certain Claims without Prejudice, SCS's Motion for Partial Summary Judgment against Plaintiff Performance Abatement Services, Inc. (hereafter "PAS"), BWL's Motion for Partial Summary Judgment against PAS, PAS's Motion for Partial Summary Judgment against BWL, and PAS's Motion for Partial Summary Judgment as to SCS's counter-claims. . . .

I. Procedural History

This suit was filed by PAS on May 21, 1998 against Defendants SCS and BWL. Thereafter, SCS counter-claimed against PAS and cross-claimed against BWL; BWL then cross-claimed against SCS.

On December 4, 1998, the Magistrate Judge assigned this matter approved PAS's request to amend its Complaint. An Amended Complaint was thereafter filed which included claims against a bonding company—International Fidelity Insurance Company (hereafter "Fidelity"). . . .

Facilitative mediation was conducted in this matter in December 2000. The mediation did not result in a complete settlement between the parties, but did result in a partial settlement agreement between SCS, BWL, Fidelity and a second bond company (Deerfield Insurance Company) entered into on January 24, 2001. . . .

As part of the previous motion practice, the Court discussed in its previous decisions the effect of the partial settlement. The Court previously concluded that those settlements were helpful in advancing this litigation. The Court also concluded that those settlements do not directly yield a conclusion that PAS is entitled to payment by SCS for unpaid work—since the payments were made for a variety of purposes including the responsibility of SCS to defend and indemnify BWL from further suit by PAS.

These instant Motions were filed by the parties between May 1, 2001 and May 11, 2001.

II. Factual Background

A. Bid History and Expert Opinions

This tangled story involves the aspiring hopes of the City of Lansing's Board of Water and Light to remodel the Ottawa Station Development Project, a former power plant, by removal of various boilers, by abatement of asbestos and other dangerous substances, and by remodeling of portions of the plant. BWL informed potential bidders of the Project in 1995 and 1997 and conducted public walk-through meetings with potential bidders who wished to view the plant. As part of the bid process, BWL designed its "Specification," which informed bidders of the "nature and scope" of the work. The Specification did so in part by attaching very general construction drawings and by attaching asbestos laboratory test results (i.e., samples were taken randomly from insulation throughout the plant and tested to determine if the insulation contained asbestos). The Specification, however, further indicated that BWL had not conducted a field survey for asbestos, that asbestos was located in most of the plant, and that bidders were responsible to survey asbestos themselves.

BWL's Specification was designed by Michael Ring. Michael Ring testified at deposition that he conducted the 1995 pre-bid walk-through with bidders and at the time informed bidders (as stated in the Specification) that further drawings of equipment and insulation were available for review at BWL's offices. His testimony contradicts the testimony of Edward Champagne (PAS's bidder) who has testified that he explicitly asked Ring for insulation plans and was told by Ring that no such plans existed. PAS has also pointed out that Ring's statement that he disclosed to all bidders that additional drawings were available for review at BWL's offices is questionable in that no additional bidders reviewed the drawings "on file" during the bidding process.

There were, nevertheless, many such plans and drawings to review in BWL's offices and the plant. The parties' several experts have examined the many plans and have asserted many and diverse opinions concerning the plans and conduct of the employees of the parties in the bidding process. . . .

B. Contract, Specification and General Requirements

BWL awarded the contract to SCS in the fall of 1997 (which award assumed BWL's use of PAS as the asbestos subcontractor). Article I of the Contract stated that SCS would perform the services and provide the materials described in the attached Specification in exchange for payment of $2,686,000.00. It also indicated that the "Specification, General Requirements, Drawings and Addenda, if any, are made a part of this contract and are incorporated by reference as if fully stated."

The Specification of BWL (which invited the bids) provided in pertinent part:

17. Owner's Information.

The Owner will furnish to the Contractor any design, manufacturer, operation or maintenance information available . . . There is no guarantee as to the completeness of the information.

The existing plant and equipment drawings are available for review at the Board of Water and Light. Copies can be made available as required to the successful bidder.

The Specification, § V, P 2 described the scope of work as including: the removal of energy production equipment (two 225,000 lb./hour steam boilers, three 275,000 lb./hour steam boilers, three 25 Megawatt steam turbines, one 4000 Kilowatt steam turbine and other assorted equipment specifically excluding a 2500 Kilowatt steam turbine and other property listed in section 8 of the Specification) and the "removal and disposal of all hazardous materials, fluids and solids."

The Specification made the following statements as to asbestos:

1. General

The Contractor should be aware that Ottawa Station was an operating power plant from 1940 to about 1990. Most hazardous materials have been removed, except as noted in this Section. However, this list should not be considered complete. The Contractor should be aware that there may be other materials present but not specifically specified herein. Also, there may other small amounts of various hazardous materials in the equipment, piping, and tanks even though the materials were removed. . . .

2. Asbestos

No asbestos survey has been performed. It is the responsibility of the Contractor to determine the extent of the asbestos to be removed. Asbestos thermal and electrical insulation and other materials are in most areas of the plant. The Contractor shall remove and dispose of all asbestos materials in all areas of the plant, including all areas which contain equipment which will remain property of the Owner. . . .

3. Refractory

The Owner has randomly tested the refractory in the boilers for asbestos. The results of the tests are attached to this specification. The sample numbers on the test results correspond to locations on the equipment and refractory in the plant. Locations 44 through 74 are marked on the equipment. There may be other asbestos containing refractory which has not been identified, and the Contractor shall take all necessary precautions during performance of the work.

Following the Specification document in the bid package was a legal document, General Requirements for Construction and Erection ("General Requirements"), which included 63 sections pertinent to the bidding and completion of this project. Pertinent sections of the General Requirements are explained below.

Section 6 of the General Requirements required bidders complete a site inspection:

6. EXAMINATION OF SITE

Bidder shall fully inform themselves regarding access to the site and work area, the conditions to be met at the site where the work will be done, and on other relevant matters concerning the work to be performed. Failure to take this precaution will not relieve the successful bidder from furnishing all material and labor necessary to complete the Contract without additional cost to the Owner.

Bidder also should examine drawing prints and other data if mentioned as being on file and available for reference of Bidder at [the] offices of [the] Engineer in Lansing, Michigan.

Section 10 of the General Requirements provided that the successful bidder assumed the risk of the work:

10. RISKS OF THE WORK

The Contractor shall carry on the work at his own risk and responsibility until it is fully completed and accepted by the Owner.

The work is to be made complete including any implied [work] that may have been omitted in the description of said work, but the use of which is implied or necessary, and shall be deemed to be included in this Contract. Such implied work shall be provided by the Contractor as if the same had been stated specifically without any additional charge to the Owner.

The mention of any specific responsibility or liability of the Contractor in this, or any part of the Contract Documents, shall not be construed as a limitation or restriction upon the general responsibility or liability imposed on the Contractor.

Sections 32 and 33 further described the drawings and specifications of the Contract, including the intended legal effect of their incompleteness:

32. DRAWINGS AND SPECIFICATIONS

The drawings, accompanying these specifications, are construction drawings portraying the scope and intent of the work, the layout and structure of the coordinate physical parts, and the general arrangements of apparatus and equipment necessary to fulfill the purposes of the project. . . .

33. ENGINEER'S DRAWINGS AND SPECIFICATIONS

Any work indicated on the drawings and not particularly described in the specifications, or specified and not indicated on the drawings, shall be included by the Contractor, and the omission from both drawings and specifications of express references to any detail of work necessary and obviously intended shall not relieve the Contractor from furnishing the same without additional cost to the Owner.

The work shall be made complete including any implied [work] that may have been omitted in the description of said work, but the use of which is implied, and shall be deemed to be included in this contract. Such implied work shall be provided by the Contractor as if the same had been stated specifically without any additional charge to the Owner.

Also included within the General Requirements were sections relating to changed conditions and requests for additional payment:

46. CHANGES IN THE WORK

The Owner, without invalidating the Contract, may order extra work or may make changes by altering, adding to, or deducting from the work, the Contract Sum being adjusted accordingly. All such work shall be executed under the conditions of the original contract, except that any claim for extension of time caused thereby shall be adjusted at the time of ordering such change.

In giving instructions, the Engineer shall have authority to make minor changes in the work, not involving extra cost, and not inconsistent with the purposes of the work; but otherwise, except in an emergency endangering life or property, no extra work or change shall be made unless in pursuance of a written order from the Engineer stating that the Owner has authorized the extra work or change, and no claim for an addition to the contract sum shall be valid unless so ordered.

47. CLAIMS FOR EXTRA COSTS

If the Contractor claims that any instructions by drawings or other conditions involve extra cost under this contract, he shall give the Engineer written notice of such claim and in no event proceed with the work unless in pursuance of a written order from the Engineer stating Owner has authorized the change or in the event of an emergency endangering life or property. No such claim shall be valid unless so made.

C. PAS's Subcontract with SCS

Subsequent to the entry of the Contract, BWL and PAS entered into a Subcontract dated January 7, 1998. The Subcontract was bonded by St. Paul on or about January 13, 1998. The Subcontract required PAS to complete those duties of the Contractor which were listed on Schedule A. The Subcontract also required PAS to assume the duties of SCS as to the scheduled work: "SUBCONTRACTOR . . . shall be bound to CONTRACTOR to assume the same obligations and responsibilities which CONTRACTOR must assume to Owner." In exchange, SCS was required to pay PAS the sum of $1.298 million dollars.

Schedule A listed a variety of activities relating to asbestos removal, foremost of which was the obligation for the "complete removal and legal disposal of all asbestos containing materials (ACM) on the site." The Schedule also included the obligations to do "all work required to enclose, access, remove and transport ACM on the site" and "all testing, notifications, clearances, final cleaning and reports for ACM on the site." The Schedule excepted from removal flyash but only if the flyash was tested and certified by PAS as not containing ACM. (Id.) The Subcontract thus differed from SCS's prior proposal for flyash abatement—which had another company (Aqua Tech) responsible for flyash cleaning (instead of removal by PAS).

This Subcontract also contained a Schedule B—which was an Amendment to the original contract language. The Amendment expressed at paragraph 7 that "neither party shall be liable for special, incidental or consequential damages of any kind." However, it narrowed the paragraph 7 limitation at paragraph 5 so as to give PAS a greater right to recover these damages against SCS when caused by delay:

5. Notwithstanding anything in the contract Documents to the contrary, including but not limited to Section 24, in the event Subcontractor is delayed in its work or is otherwise required to accelerate or resequence it work for reasons other than the fault of Contractor or others under Contractor's control, then Subcontractor shall be entitled to additional compensation to the same extent as Contractor. If, however, Subcontractor is delayed in its work or is otherwise required to accelerate or resequence its work due to the fault of the Contractor or others under Contractor's control, then Subcontractor shall be entitled to additional compensation regardless of Contractor's right of recovery.

Furthermore, the Subcontract contained the following paragraphs pertinent to PAS's duties under the Subcontract:

6. Even if specific work set forth in the Contract Documents relating to SUBCONTRACTOR'S Work is not described in this Subcontract and/or every item necessary to perform SUBCONTRACTOR'S work may not be mentioned in the Contract documents, SUBCONTRACTOR shall still perform all work necessary to finish its scope of work, including all work normally construed to come within the scope of its activities as required of CONTRACTOR in the Contract Documents.

7. SUBCONTRACTOR . . . warrants that it has received and reviewed all plans and specifications relating to its work. . . .

23. No alterations shall be made in the SUBCONTRACTOR'S Work except upon written direction by CONTRACTOR. CONTRACTOR may, at any time, by written order and without notice to SUBCONTRACTOR's surety, make changes in the work contracted for and SUBCONTRACTOR shall proceed with the work as directed. If the changes cause an increase or decrease in the cost or time for performance, an equitable adjustment shall be requested in writing by SUBCONTRACTOR. Nothing hereon shall excuse SUBCONTRACTOR from proceeding with the prosecution of work as changed. SUBCONTRACTOR is required, however, to submit a list of pending change orders with its monthly invoice. Failure to comply with this requirement will relieve CONTRACTOR of its obligation to adjust the contract sum or time for performance. No charges for premium time will be honored by CONTRACTOR unless signed work tickets have been obtained from CONTRACTOR.

24. SUBCONTRACTOR agrees to make any claims to CONTRACTOR for damages or additional compensation based on alleged extra work delays or changed conditions in the same manner as prescribed in the Contract Documents for similar claims by CONTRACTOR to Owner. CONTRACTOR shall not be liable to SUBCONTRACTOR for any claim until it is allowed by Owner. ANY PAYMENT OF ADDITIONAL MONEYS FROM CONTRACTOR TO SUBCONTRACTOR ARE CONTINGENT UPON PAYMENT FROM OWNER TO SUBCONTRACTOR. SUBCONTRACTOR hereby expressly waives any other right to damages or additional compensation. SUBCONTRACTOR shall also make no claim for delays or damages which it claims was caused by other subcontractors on the Project.

28. SUBCONTRACTOR shall coordinate and check his work with all drawings. Where discrepancies are discovered, SUBCONTRACTOR shall be responsible to immediately notify CONTRACTOR in writing and request a clarification. If notification is not given, SUBCONTRACTOR shall be responsible for the required correction. Moreover, if SUBCONTRACTOR encounters any conditions upon which it may base a claim for extra compensation or time, it shall be its duty to give written notice to CONTRACTOR prior to commencing work involving this area. If no notice is given, SUBCONTRACTOR shall be fully liable for any and all expense, loss or damage resulting from this condition.

34. THIS SUBCONTRACT SHALL, WHEN ACCEPTED BY SUBCONTRACTOR CONSTITUTE THE ENTIRE AGREEMENT BETWEEN THE PARTIES AND SHALL SUPERSEDE ALL PROPOSALS, CORRESPONDENCE AND/OR ORAL AGREEMENTS BETWEEN THE PARTIES.

D. Discovery of Asbestos and Correspondence

Of course, "additional" asbestos was in fact discovered in the boiler areas of the plant in early 1998 (on or about February 25, 1998) by a foreman for PAS in the course of his work. . . . Details concerning the "additional" asbestos are found in the correspondence of Edward Champagne to Robert Greenlees of SCS. In addition to these "changed conditions," PAS has made other requests for payments regarding various conditions some of which are apparently unrelated to the asbestos abatement issues.

Discovery of added asbestos at the plant prompted PAS's project manager (Champagne) to write SCS on March 2, 1998 to inform SCS that PAS had discovered a second wall containing asbestos in the small boilers (boilers 1 and 2). Champagne indicated in the letter that the discovery was a "changed condition" and that PAS would not proceed further without a written order.

SCS apparently responded to this letter by indicated to Champagne, in a letter of March 30, 1998, that it deemed the "hidden asbestos" as covered by the Subcontract. . . . Many other letters were also exchanged between the parties concerning other "additional asbestos" and other "changed conditions" for which PAS was seeking additional payment.

III. Standard for Voluntary Dismissal

Defendants and Cross-Claimants SCS, Fidelity and BWL have moved to voluntarily dismiss without prejudice their claims against one another in accordance with their earlier partial settlement. This Motion is made pursuant to Federal Rule of Civil Procedure 41(a)(2). . . . This Rule is intended to allow for the liberal dismissal of claims, even after a counter-claim is filed, subject to the court approval to determine that no party is prejudiced thereby.

In its Grover decision, the Sixth Circuit identified four factors which a court should consider in determining whether to grant a voluntary dismissal: (1) the amount of time and effort the defendant has incurred in preparing for trial; (2) any lack of diligence on the part of the plaintiff in prosecuting the action; (3) the plaintiff's failure to explain the need for a dismissal; and (4) whether the defendant has filed a motion for summary judgment. Grover by Grover, 33 F.3d at 718.

IV. Analysis of Dismissal Motion

It is obvious and elementary from the above legal analysis that the Rule 41(a) analysis is intended to prevent legal prejudice to a party who has previously opposed a claim to be dismissed. In the instant case, the claims to be dismissed are not lodged by Plaintiff PAS, but rather concern SCS, BWL and Fidelity. PAS believes that its interests will nevertheless be impacted because the dismissal "will affect PAS's claims that are pass-through claims through SCS against the Board" and because "Fidelity has raised as a defense that PAS cannot be paid for its claims unless the Owner pays SCS."

Upon review of the factors stated in Grover, the Court determines that PAS will suffer no legal prejudice and that the Motion to Dismiss Certain Claims without Prejudice will be granted.

V. Standards for Summary Judgment

The instant cross-motions for summary judgment are brought pursuant to Federal Rule of Civil Procedure 56. Under the language of Rule 56(c), summary judgment is proper if the pleadings, depositions, answers to interrogatories and admissions on file, together with affidavits, if any, show that there is no genuine issue as to any material fact and that the moving party is entitled to judgment as a matter of law. . . .

Credibility determinations, the weighing of the evidence, and the drawing of legitimate inferences are jury functions.

VI. Summary Judgment Analysis

This brings the discussion to the issues raised in SCS's Motion for Partial Summary Judgment against PAS, BWL's Motion for Partial Summary Judgment against PAS, PAS's Motion for Partial Summary Judgment against BWL, and PAS's Motion for Partial Summary Judgment as to SCS's counter-claims. The Court will discuss the Motions in that order.

1. Count V—Breach of Contract

SCS argues that is entitled to summary judgment as to Count V because the Count assumes that SCS had an obligation to reimburse PAS for the removal of additional, undisclosed asbestos and that this assumption is factually and legally mistaken. PAS contests this argument on several grounds.

First of all, PAS observes that Count V relates to a variety of damages other than the "additional asbestos removal." Indeed, a variety of other damages are pleaded at Paragraph 71 of the Second Amended Complaint. These other damages relate to items separate from the asbestos removal—such as delay and losses allegedly caused PAS by SCS's "wrongful stop order" (letter) of May 20, 1998, water loss, Dowtherm evacuation, enclosure repairs, flooding, small boiler enclosure fire, and elevator shutdown. . . . Therefore, at best Defendant SCS's argument can only be granted in part—with respect to claims for additional payment respecting the removal of asbestos allegedly not contemplated by PAS.

Second, the issue of asbestos abatement must, because of the standard pertinent to summary judgment issues, be resolved in PAS's favor. While considerable pages could be devoted to an analysis of the contract language used, most of which favors SCS, this is unnecessary in light of the holdings in [several Michigan cases]. Those cases require as part of Michigan law that a party which submits a project for bids has a legal duty to inform bidders of the nature and scope of work to be performed by giving bidders all relevant information on the project in its possession. Failure to comply with this duty has been treated as a breach of contract or implied warranty. Furthermore, the above cases do not permit a city or its contractor to circumvent the implied duty by contract language requiring the bidder to make its own inspections, requiring the bidder to assume the risk of non-discovery, or disclaiming the accuracy of the information.

SCS has raised any number of arguments in its various briefs that the Court should limit the holdings in the above cases so as not to apply to the current dispute.

In this case, there is testimony from PAS's bidder that he inquired of Michael Ring of BWL at a pre-bid meeting whether there were more detailed drawings of insulation and was told by Ring that there were not such drawings. At that time, Ring admitted to knowing of the drawings and admitted knowing that they showed insulation which might detail asbestos, but has denied (at least implicitly) that Champagne ever posed those questions to him. The factual issue of which testimony is to be believed is complicated and interesting and one for the jury to decide.[6] As such, summary judgment must be denied as to Count V.

---

[6]On the one hand, Champagne's statement that Ring told him that there were no such drawings seems to be an all too convenient recollection intended to preserve his self-interest; it would have been particularly irrational for Ring to have said this when he knew such drawings were lying about in public view in BWL's offices and the plant. On the other hand, Ring's statements in his Affidavit of July 18, 2001 which denies the supposed conversation with Champagne are couched in "remembrance" rather than "fact"—reflecting either undue care in framing affidavits or undue experience in lying. (See Ring Affidavit of July 18, 2001 P 3, stating that Ring "did not recall discussing the . . . drawings . . . with Champagne. . . . However, I do not believe that I would have told or represented to . . . Champagne that drawings showing insulation did not exist. . . .") Absent the unexpected (that Ring had this important conversation and has simply forgotten it), one of the two is lying under oath. This Court's opinion of perjurers is not far off that expressed by Dante in Canto XXX of the Inferno (of the Divine Comedy): they belong in the Eighth Circle of Hell consumed by a fever of their own making.

Count VI—Innocent Misrepresentation   *SCS for judgment against PAS*

Count VI of the Second Amended Complaint seeks damages for innocent misrepresentation by SCS. PAS has alleged that the Specification, drawings, and bid package which SCS passed on from BWL to PAS understated the extent of asbestos abatement. PAS further alleges that this false information caused it to make an artificially low bid, which inured to SCS's benefit.[7]

Under Michigan law, to prevail on a claim of innocent misrepresentation, a plaintiff must show that a false and material misrepresentation was made, that the plaintiff detrimentally relied upon the misrepresentation, and that the injury inures to the benefit of the person making the representation. . . .

The evidence relied upon by PAS essentially consists of the affidavits and deposition testimony of Edward Champagne. Champagne states that he relied upon the bid architectural drawings in making measurements of the asbestos to be abated.

On the evidence of record, a reasonable finder of fact could not find in PAS's favor because PAS did not detrimentally rely upon misrepresentations of SCS in bidding the project. Therefore, summary judgment shall enter in SCS's favor as to Count VI.

Count VII—Breach of Express Warranty   *SCS for judgment against PAS*

Count VII of PAS's Second Amended complaint seeks relief for breach of express warranty relating to the drawings and testing information attached to the Specification, which PAS asserts was expressly warranted by the statement in the General Requirements that the "drawings, accompanying these specifications, are construction drawings portraying the scope and intent of the work. . . . PAS deems that the warranties were breached in that there was asbestos in areas of the plant not disclosed on the drawings and the testing. SCS deems that there were no express warranties in the Subcontract and Contract in light of the particular language used.

The snippet of language cited above by PAS is the only source of express contract language which could at all be construed as providing an express warranty to PAS as to the accuracy of the drawings. It is noteworthy that this language does not reference the testing data and cannot be reasonably construed as providing a promise as to the testing data. However, even the notion that there was an express warranty as to the drawings falls away once it is understood what was included in the drawings and what other contract language was used in the contract documents. As for the drawings themselves, it is evident from the previous discussion that they were not understood by the person preparing them (Michael Ring) nor the person relying upon (Edward Champagne) as even addressing the issue of the location of asbestos within the plant except in a very general way. Since skilled parties on both side of this conflict regard the drawings as "useless" from the standpoint of asbestos disclosure, a jury could not reasonably conclude that the drawings made express warranties about the location of asbestos in the plant nor the boilers.

Furthermore, these conclusions are strengthen by the fact that SCS/BWL made clear in the governing contract language that no such warranties were made and that PAS was assuming the risk of non-disclosure. Section 3, paragraph 17 of the Specification provided that "there is no guarantee as to the completeness of the information" and further indicated that plant and equipment drawings were available for review at BWL. Section VI, paragraph 2 of the Specification further provided that "no asbestos survey has been performed. It is the responsibility of the contractor to determine the extent of the asbestos to be removed. Asbestos thermal and electrical insulation and other materials are in most areas of the plant." . . .

---

[7]As to the allegation of reliance upon false testing data, the only evidence suggesting reliance by PAS are the affidavits of Edward Champagne, which state the bare legal conclusion that Champagne relied upon the testing, but do not detail when, if at all, Champagne examined the testing data and how, if at all, it was used in advance of the bidding.

This Court's interpretation of the express contract language is determined as a matter of law. The instant contracts, which contained integration clauses, are unambiguous as to whether express warranties were to be extended to the testing data and drawings.

In light of the nature of the documents at issue, the multiple contract references which proclaimed loudly that warranties were not made, and the pertinent law, there are no genuine issues of material fact since a jury could not reasonably conclude that SCS had made an express promise as to the drawings and testing which has been breached. Accordingly, summary judgment shall issue on Count VIII in favor of SCS.

5. Count IX—Implied Warranty *SCS for summary judgment against (PAS)*

Count IX of the Second Amended Complaint states a claim against SCS for breach of an implied warranty under the Subcontract that the underlying drawings, Specification and test results were adequate for their intended purpose of determining the quantity of asbestos to be abated. SCS argues in its Motion for Summary Judgment under the federal case of *United States v. Spearin*, 248 U.S. 132, 136, 63 L. Ed. 166, 39 S. Ct. 59 (1918) that no implied warranty attaches to a construction contract specifications unless the specifications relate to the design or construction of a project. Since the instant project involved "performance specifications" to eliminate "all asbestos" and not design specification, SCS concludes that summary judgment is warranted. SCS also argues that any implied warranties were disclaimed in the contract language. PAS's answer to this argument is that Michigan law does not recognize the rule announced in Spearin and that the contract language disclaimers herein were ineffective.

PAS is correct that the authorities cited by SCS interpret either federal laws or the laws of states other than Michigan. Since this case involves Michigan law, the decisions of its Supreme Court are deemed controlling. As this Court has already mentioned with respect to Michigan law, Michigan law implies in every construction contract a duty to inform bidders of all material information pertinent to the bid. . . . As such, summary judgment will be denied SCS on Count IX.

6. Count XI—Reformation/Mutual Mistake *SCS for summary judgment against PAS*

Count XI of the Second Amended Complaint seeks reformation of the Subcontract because of mutual mistake so as to require SCS to compensate PAS an additional $118,000 for the removal of flyash from the plant. SCS argues that PAS has not submitted evidence of mutual mistake in support of this claim.

This Subcontract explicitly required the removal of all asbestos-containing materials ("ACM"). It specifically excluded flyash, but if and only if the flyash was "tested and certified by PAS to be non-ACM."

Reformation on the ground of mutual mistake under Michigan law requires proof that both parties were mistaken as to the contract terms or that one party was mistaken and the other party acted fraudulently or inequitably. In this case, PAS has filed insufficient evidence to create a genuine issue of material fact as to mutual mistake. SCS is entitled to summary judgment on Count XI as a matter of law.

VII. Conclusion

In accordance with this Opinion, SCS and BWL's Motion to Dismiss Certain Claims without Prejudice will be granted; SCS's Motion for Partial Summary Judgment shall be granted in part and denied in part; BWL's Motion for Partial Summary Judgment will be granted in part and denied in part; PAS's Motion for Partial Summary Judgment against BWL will be denied, and PAS's Motion for Partial Summary Judgment as to counter-claims of SCS will be granted. An Order and Partial Judgment shall issue consistent with this Opinion.

Review Questions

1. Why should specifications be written rather than oral?

2. What are the inherent advantages of standardized specifications?

3. What danger is there in requiring a contractor to perform according to a particular method?

4. Why is the owner interested in the wages the contractor pays to employees?

5. What are the inherent advantages and disadvantages in copying portions of old specifications into new ones?

6. Notice that the expert witness in *D.A. Elia Construction Corp. v. New York State Thruway Authority* testified that the bid documents indicated that the piers to be repaired had not been hammer sounded. Recall in Chapter 4 an engineer's obligations as an expert witness. If you were Mr. Pressley, how could you have avoided the conflict between your testimony and that of the vice-president, Daniel Elia (your client)? Would your recommendation comply with the NSPE Code of Ethics?

   — Don't testify
   — Discuss testimony w/ VP in advance
   — Help client prepare case

7. Interpreting from *Centex Construction Company, Inc. v. The United States,* to the extent that plaintiffs thought that the specifications were not "typical," do you think that the court would have supported any assertion that what the government wanted was not a "trade practice"?

8. In *Performance Abatement Services, Inc. v. Lansing Board of Water and Light,* the court granted summary judgment for Count VII, Breach of Express Warranty to SCS because "a jury could not reasonably conclude that SCS had made an express promise as to the drawings and testing which has been breached." What additional facts, not mentioned in this case, could support the PAS claim that there was an express promise?

9. Compare and contrast *Performance Abatement Services* with *D.A. Elia Construction Corporation.*

   Summary judgment
   Breach of contract
   Breach of warranty
   ~~Settlement agreed~~
   → Earlier stage
   Lump sum contract

   Appeal
   Sought compensation for add'l cost
   ~~br~~
   → Later stage
   Unit price contract

   → specs are similar → did not detail scope, only general in purpose
   → contract laid risk on contractor
   → Both contracts asked contractor to perform testing
   → owner-independent contractor relationship
   scope + level of effort were not clear to both parties

Agency

HIGHLIGHTS

- The agency relationship usually involves 3 people:
  1. Principal
  2. Agent
  3. Third Parties

- A relationship can be established by:
  - Agreement
  - Ratification
  - Estoppel
  - Necessity

When a person's duties and desired objectives become too numerous to handle, he or she can usually delegate some of those duties to others. This delegation of duties may take one of three forms:

1. The employer-employee relationship,
2. The owner-independent contractor relationship, or
3. The agency relationship.

In satisfying the normal requirements of a job, an engineer (whether a consultant or an employee) must act as an agent at least part of the time for the person who hired him or her. The engineer's rights and liabilities while so engaged are dictated by the law of agency. **Agency** can be defined as a <u>consensual</u> fiduciary relationship between two persons by which one (the principal) has the right to control the conduct of the other (the agent), who in turn has the authority to affect the principal's legal relationships with others.

An agent represents a principal in dealing with other persons. Herein lies the distinction between the agency relationship and the employer-employee relationship. An agent is often an employee for many purposes, but special rights and duties are involved in an agency relationship. A lathe operator in a plant, for example, is an employee. If the operator deals with others as a representative of the employer, the operator becomes an agent.

The intent of the parties and the degree of control exercised by the owner determine whether a relationship is one of owner-independent contractor or employer-employee. As indicated, the parties must agree to have one act as an agent for another. If one person decides to negotiate a price for a piece of land prior to receiving any authority to do so, there is no agreement and no agency (although, as discussed later, the parties may subsequently decide to create the agency relationship). The degree of liability of the employer (or owner) for another's conduct is, in turn, determined by the relationship between the employer and the other person whose conduct is at issue. For example, an employer may be held liable for tortious injuries caused by employees acting within the scope of their employment and for harm to them from any injuries that arise from their employment. Although a principal may be liable for an agent's actions or for harm done to an agent while he or she was working for the principal, a principal's liability for injury to, or injury caused by, an independent contractor is usually much more limited. In many instances, particularly worker's compensation cases, a supposed owner-independent contractor relationship was held to be an employer-employee relationship because of the degree of control exercised by the employer (owner).

Consider this example: Black, a manufacturer, hires White as a time-study engineer in Black's plant. In setting production standards and making methods changes, White is Black's employee. White is soon promoted to process engineer and is charged with installing an automatic machine being purchased from Gray Automation. When White deals with Gray Automation and other outsiders, White acts as Black's agent. White fills a dual role, then, for while White acts as an agent in dealing with others, White remains Black's employee. In installing the machine, Gray Automation is an independent contractor unless either Black or White exercise an excessive degree of control over Gray Automation.

Agency involves three people:

1. The principal, the person who is represented by the agent and who is the source of the agent's authority;
2. The agent, the person who represents the principal; and
3. The persons with whom the agent deals in the name of the principal.

CREATION OF AGENCY

The agency relationship may be created in any number of ways. Such relationships are commonly created in four situations:

1. By agreement,
2. By ratification,
3. By estoppel, or
4. By necessity.

The responsibility and authority of the agent differs slightly in each.

Agreement

As you may recall from Chapter 6, the parties to an agreement must be competent. In agency, the principal is the party to be bound to a third party so the principal must be competent to contract. The agent may be a gray-haired man of 60, but if his principal is a minor, the contract is voidable at the minor's option. From this, there arises a practical desirability of investigating the principal before contracting with him or her. Not only could an incompetent principal avoid a contract made by his or her agent, but the principal might also avoid the contract with the agent.

Though competency of the principal is of major concern to third parties and agents, competency of the agent is not. Almost anyone may be an agent; it is the principal who is bound. A minor agent contracting with a third party would bind the adult principal to the contract.

For an agency-by-agreement relationship to exist, both parties must intend to create a relationship that amounts to agency. The parties may not consider the relationship to be agency when they enter into it, but if the result amounts to an agency, it will be so construed. The means the parties use to express their intentions to form an agency are ordinarily unimportant, as long as the ideas are exchanged. An agency contract is much the same as any other contract in this respect. The intent may be either expressed or implied; the agreement may be written or oral. The agency agreement is not necessarily a contract, but it usually is. A contract requires consideration, but a simple agreement may create a valid gratuitous agency.

Consider the following example: Black owns a truck; White does not. Black, without being promised any compensation, agrees to transport a machine for White from a freight depot to a machine shop across town. In doing this, Black must deal with others. Even though Black is to be paid nothing, Black's actions are controlled by White when Black picks up, transports, and unloads the machine. Thus, an agency was formed. Now suppose that Black goes alone, but White telephoned ahead to tell the freight depot that Black was to pick up the machine. Again, the parties have formed an agency.

As noted, an agreement to act as another's agent ordinarily may be either oral or written. However, for certain purposes, a written or a sealed instrument may be required. Where the instrument that binds the principal requires a seal from a notary public, the agent's authority usually must be in writing and notarized also. Probably the most common example is the power of attorney, which establishes the agent as an attorney-in-fact for the principal. Where a transaction requires a public recording, such as in the sale of real estate, any powers of attorney involved are also recorded. Another form of written agency is the corporate proxy. By the use of a proxy, a stockholder appoints some particular person to vote the shareholder's shares of stock in a particular way.

Ratification

If a person purports to act on another's behalf and contracts with a third person, but had no authority to do so, the person for whom the "agent" purports to act is not bound to the contract. Much the same is true if an actual agent exceeds the authority given by the principal. In either of these circumstances, the principal may agree to be bound by the terms of the contract. Agreeing to do so is referred to as **agency by ratification,** and it cures any defect in a lack of authority. Such a ratification by the principal is retroactive, that is, the ratification goes back to the time the contract was made. The effect of ratification is the same as if the principal had previously retained an agent to act in that particular manner.

For example, suppose Black hires White as a salesman to sell the company's products. White finds a buyer, Gray, for a used milling machine that Black has wanted to sell for some time. White, without delay, contracts in Black's name to sell the machine to Gray. If Black no longer wishes to part with the machine or, possibly, has contracted to sell it to another, Black will not be bound to the agreement with Gray. If Black does not ratify White's agreement with Gray, White will be personally liable for any harm to Gray resulting from a breach of the contract. If Black ratifies the agreement, Black will be held to the contract just as if White had been given specific orders to sell the machine. However, the fact that White lacked the authority to sell the machine does not allow Black to ratify only a portion of the contract. Black must ratify all or nothing. Meanwhile, Gray is not bound to the agreement until Black ratifies. If Gray finds that White acted without authority, he may withdraw before Black's ratification.

There are at least four requirements for a valid ratification:

1. A principal must exist when the supposed agent acts. A corporation, for instance, cannot make a binding ratification of contracts made in the corporation's name before it was formed. New contracts will be required if the corporation is to be bound.
2. The person acting without authority must act as an agent. If the person acts on his or her own behalf, subsequent ratification by another will not create enforceable obligations between the third party (the supposed principal) and an outsider.
3. The principal must be aware of the facts when the principal ratifies the contract. A principal's actions that would imply ratification have little effect unless the principal knows the facts. Of course, if the principal does not investigate the details when the principal has a duty to do so, the principal's negligence may be interpreted to support the conclusion that the principal did know or should have known the facts.
4. The principal must intend to ratify. Intent and ratification may be interpreted from the principal's actions after the principal obtains knowledge of the transaction and the relevant facts. If the principal does nothing after being informed of the transaction, for example, ratification could be implied from the principal's inaction. In such a situation, the principal is said to have ratified the transaction through acquiescence.

Estoppel

Agency by estoppel arises where one person appears to have the authority to act for another and, despite a lack of real authority, does act in the name of the other. Apparent authority is one key to this concept. If the principal acts in such a way that another person appears to be the principal's agent (thus, in effect, deceiving the third party), the principal is then estopped (i.e., prevented) from denying that the other person is an agent. The second key for an agency by estoppel is that the third party must have acted in reliance upon the supposed agent's apparent authority. The third party must, of course, have dealt with the supposed agent to prove agency by estoppel.

Necessity

An **agency by necessity** may occur as a result of an emergency. This situation occurs rather infrequently, however. If, to save the principal from some disaster, an employee must deal with others without an opportunity to obtain authorization, an agency by necessity is created. Generally, such an agency has some relationship to what the agent's normal duties involve.

Where a dependent binds a guardian to pay for necessaries, agency is sometimes said to exist. The most common holding, however, is that the guardian had a duty to support, and binding the guardian to such contracts is a result of this duty, rather than agency.

AGENT'S AUTHORITY

The **agent's authority** comes from the principal. It consists of express orders or directions the principal gives in addition to authority that may reasonably be implied. **Implied authority** is based on previous dealings between the parties, or local or trade customs. If none of these control the situation, the extent of implied authority is the extent necessary to accomplish the purpose of the agency.

If the agent contracts beyond the scope of his or her authority, the principal is not bound. From this, an obvious burden is placed upon the third party to determine whether the agent has authority to enter into a given contract. The third party is safe in dealing with the agent if he or she can obtain evidence of the agent's mission. Implied authority necessary to accomplish the purpose is included despite a principal's instructions to an agent to the contrary.

For example, Black is a buyer for the White Manufacturing Company. Black's present assignment is to buy a six-station automatic indexing table, drive, and base, to be used in machining small die castings. Black has been specifically told not to buy any tooling with the machine. Nevertheless, Black contracts with Gray for a machine with tooling that, he is told, may be reworked for the die castings. The contract for the machine and the tooling probably will be binding, because Black's specific orders to buy the machine might well be taken to imply that he also had a right to purchase tooling for it. If Gray knew of Black's assignment, he would not have a duty to go further and determine any unusual restrictions that might have been imposed on Black.

A third type of authority is **apparent authority.** The reasoning of apparent authority runs so close to that of agency by estoppel as to make the two nearly indistinguishable. Apparent authority exists when, by some act or negligence on the part of the principal, an agent either appears to be clothed with more authority than he or she really has, or a person who is not an agent is made to appear as if he or she were one. A person usually creates apparent authority by conduct or statements that give the impression that he or she is authorized.

Agents are chosen for their particular capabilities and fidelity. The agent is the one the principal trusts. Therefore, the agent generally cannot freely delegate his or her authority to another. If the agent hires a sub-agent, the principal has no liability to the sub-agent for wages or other benefits and is not liable for the sub-agent's acts. The rule is not without exception; if part of the agent's express or implied task is hiring others for the principal, those so hired work for the principal. Obviously, when a principal hires a corporation as an agent, that corporation must hire individuals to perform its duties.

AGENT'S DUTIES

Agency is a fiduciary relationship; that is, the principal's trust and the agent's loyalty are implied. The term **fiduciary** comes from Roman law and is used to describe a person who has a duty to act primarily for another's benefit. A fiduciary generally has duties of trust and confidence that are owed to the person for whom the fiduciary acts. The law enforces these qualities in the relationship; a breach can give rise to a cause of action. Furthermore, the agent is personally liable for the results of any disloyalty. (See *Adam D. Sokoloff et al. v. Harriman Estates Development Corp.,* et al at the end of the chapter.)

Obedience

Memo

An agent owes the principal a duty of strict obedience in all ordinary circumstances. Disobedience constitutes a breach of the agency agreement. It is not the agent's function to question or judge the wisdom of the principal's orders; it is the agent's function to do everything in his or her power to obey them. Of course, when the principal outlines a general purpose to be accomplished, the agent may be required to use judgment and discretion in working out the details. Still, the purpose to be accomplished is not open to question. Direction by the principal is implicit, even in a gratuitous agency (that is, one in which the agent acts without compensation), once the agent has begun to perform as an agent.

Of course, strict obedience is limited to lawful and reasonable acts. The agent need not, for example, follow instructions of an unlawful nature. Neither would an agent be expected to accomplish an impossibility. In an emergency situation, an agent may have the right to fail to obey instructions strictly. The reasoning here is the same as that in authority of necessity: If the agent's failure to follow instructions will save the principal from disaster, the agent's right to disobey is apparent. In such situations, the agent is to act as the principal would direct if the principal knew of the emergency.

Care and Skill

The agency relationship normally implies that the agent will use ordinary care and skill in carrying out his or her duties. The test of whether or not the agent has done so is the test of the reasonably prudent person. That is, has the agent acted as a reasonably prudent person would be expected to act in like circumstances? Negligence in following the principal's orders may make the agent liable for payments to the principal; in addition, the agent may have to pay for losses suffered by the principal that could be reasonably anticipated from the agent's failure to follow instructions properly.

If the agent professes to be a specialist (for example, a consultant in some professional field), the standard of skill expected is that which is normally attributed to such a person. The standard is the same whether the person really is such a specialist or not; if the agent fails to perform as a specialist would be expected to perform, the agent is liable. For example, Black, a manufacturer, hires White—an engineering consultant on conveyors—to design, recommend, and oversee the installation of a monorail conveyor system in Black's plant. If, because of very poor planning and design, the conveyor must be removed shortly after its installation, Black may have an action against White. If Black's action is to be successful, he must prove (usually by expert testimony) that anyone possessing the knowledge and skill normally possessed by an engineering consultant specializing in conveyors would have made a more effective design. Damages could include a refund of the consultant's fee plus the cost of the improper installation and the cost of its removal.

To Act for One Principal

Memo

Conflict is relieved if disclosed

An agent has a duty to act for and accept compensation from only one principal. An agent could not, for instance, reasonably represent both buyer and seller in a sales contract. The buyer's interest and the seller's interest are at opposite poles; each desires to get the best possible deal. If the agent represents more than one party to a contract, the transaction is voidable at the option of either principal. The agent is relieved of the responsibility to act for only one party if the parties are told of the multiple relationships, and acquiesce. An agent's interest that is adverse to a principal's interests is allowable if the principal knows of the interest, is fully informed of all the relevant facts, and continues the relationship in spite of it.

The agent's compensation should come only from his or her principal. Hence, an agent is not allowed to secretly profit from his or her actions on behalf of the principal. If the agent acquires property in his or her own name that should go to the principal, the agent will be deemed to be holding the property as bailee. Similarly, the agent cannot contract with himself or herself as the third party without fully disclosing the facts to the principal and gaining the principal's consent. Without such disclosure and consent, the principal has the right to avoid the transaction. Kickbacks or secret commissions from third parties are, of course, against public policy.

Loyalty

Memo

The essence of agency is the identity of the agent with his or her principal's purpose. Generally, an agent has a duty to act solely for the interests of the principal within the agent's authority. The agent is often in a position to gain personally from information he or she acquires. To use the information to add to the agent's personal fortune is an act of disloyalty. Recovery by the principal for such misuse of information is possible.

At first blush, the duty of loyalty seems very similar to that of the duty to act for one principal. As with the duty to act for one principal, an agent can only act for one party and cannot behave in a way that is adverse to a principal's interest. The duty of loyalty goes beyond a single transaction. First, it requires the agent not to use or disclose confidential information, except for the principal's benefit. Second, it continues even after the termination of the agency. Thus, although the agent may compete with a former principal after termination of the agency, the agent may not compete by using any of the principal's trade secrets, such as secret customer lists, blueprints, or the like. (See Appendix B, Case No. 01-2 Conflict-Of-Interest: Third Party Developer.)

Accounting

The agent has a duty to account for all money or property involved in agency transactions. Further, the principal's property must be kept separate from the agent's property. If commingled money or property belonging to both principal and agent is lost, the agent must make good the principal's loss. If the property were kept separate and the loss occurred through no fault of the agent, only the principal would lose.

The agent has no right to use the principal's property as his or her own without consent to do so. Using the principal's property without the principal's consent is the same in agency as elsewhere—it is conversion.

Closely akin to the agent's duty to account to the principal for money or property is the agent's duty to report information to the principal. This duty often involves relating offers, financial information, or other details encountered in the course of business negotiations or transactions. Generally, notification to an agent has the same effect as notice to a principal. In either case, the principal is charged with possession of the knowledge.

> According to a survey conducted in 2005, the average first-year cost estimate to comply with the Sarbanes-Oxley Act is almost $3 million for 31,000 hours of work plus additional audit fees of $823,200, or an increase of 53% in costs.[1]

---

[1]Zhang, I.X., Simon, W. E., *Economic Consequences of the Sarbanes-Oxley Act of 2002,* http://w4.stern.nyu.edu/accounting/docs/speaker_papers/spring2005/Zhang_Ivy_Economic_Consequences_of_S_O.pdf (last viewed 8/29/07)

Accounting practices will come under increasing scrutiny since the passage of the Sarbanes-Oxley Act (15USC § 78). This law, passed in October 2002 in response to the effects of Enron and other publicly held companies that reported false information to their shareholders and the SEC, imposes more stringent financial reporting compliance and auditing guidelines upon public companies. The full range and breadth of this act is being tested, as everyone from the chief operating officer to maintenance services may be required in some measure to participate in compliance programs.

PRINCIPAL'S DUTIES

Most agency agreements are contractual in nature, and each party has a duty to live up to the agreement. The principal does not, however, have to pay for disloyal service. Neither will the principal have to pay if payment is contingent upon the agent's success (as is the case with commissions for sales) and the agent's efforts do not meet with success.

Payment

In the usual agency contract, the means and amount of payment are stipulated. If such a stipulation is not made, however, the agent is entitled to reasonable payment for his or her services. If principal and agent are not close friends or relatives, and there is no other reason to suggest that the agent acted for nothing, an unliquidated obligation of payment by the principal to the agent exists. A gratuitous agency is, of course, an exception to this rule.

Payment of an agent on a commission basis presents some special problems. When, for example, has the agent earned the commission? What happens if the principal accepts an order through the agent to sell to a third party, but the principal and third party cannot reach an agreement on the terms? Does the principal have to pay if he or she deals directly with the third party? Such problems arise not only in agencies to sell a company's products, but also in real estate and other agencies.

Generally, and barring an agreement to the contrary, the agent has performed and is entitled to a commission when the agent has achieved the desired result, such as when the agent has found a buyer and has contracted with the buyer for the principal. A real estate agent (or broker) ordinarily does not have the power to contract for the principal. Therefore, the role of the real estate agent is only to find a buyer who is ready, willing, and able to buy the property at the owner's price. With these conditions satisfied, the principal has an obligation to pay the agent the agreed commission.

If, after the agent has performed the obligation, the principal and the third party do not complete the transaction, the principal is still bound to pay the agent. It matters not that the principal will no longer profit by the transaction.

The agency agreement often states whether a commission is to be paid to the agent if principal and third party deal directly. An **exclusive agency** for the sale of real estate, for instance, requires that the principal pay the real estate agent's commission regardless of who sells the realty. Without the exclusive agency feature, the owner can sell the realty to another and be free of the obligation to pay the agent. The same is generally true of an agent who sells a product. If the sale is made without his or her services, the agent has done nothing to earn a commission and is not entitled to it. (Recall *Melvin v. West* in Chapter 7, and *Lindsay v. McEnearney Associates* in Chapter 11.)

Expenses

The principal is legally bound to pay an agent's expenses. To constitute an obligation of the principal, the expenses must, of course, be connected to the purpose of the agency. Thus, the cost of travel, meals, and overnight hotel or motel accommodations connected with an agent's trip to sell a principal's products should be paid by the principal. Similar costs incurred by the agent on a pleasure trip with his or her family ordinarily would not be covered by the principal.

Indemnity

memo

While the agent has a duty to follow the principal's orders, the principal also has a duty to indemnify the agent if the result injures someone and the agent has to pay for the injury. Of course, the agent is not required to perform unlawful acts; he or she is prohibited in the same manner as anyone else from committing a crime. If, however, a tort or crime is committed by the agent innocently following a principal's instructions, the principal is liable for the result. Both principal and agent are liable to the third party for acts committed out of and in the course of the agent's employment, but if the agent has to pay, he or she can recover from the principal. The agent alone is responsible for acts not connected with his or her employment.

Black, for example, is an agent of the White Machinery Company. His function is to answer customers' complaints and thus make the sales of the machines permanent. He is a troubleshooter. As such, he is entitled to compensation according to his agreement with the White Machinery Company. He is also entitled to payment for legitimate expenses in connection with his job. If, while instructing someone in the proper use of a machine, an injury should occur, White Machinery Company would be liable.

THIRD PARTY RIGHTS AND DUTIES

So far, we have considered the rights and liabilities of two of the agency parties—the principal and the agent. But the third party also has a stake in the relationship. Questions frequently arise concerning the extent to which an agent's transactions bind the principal and a third party, and the extent to which the third party can rely on the agent's claims. These and other questions are considered below.

Duty to Question Agent's Authority

We have noted that the agent has the authority given by the principal. The agent also has authority common to other agencies of a similar nature. Third parties are safe in relying upon the agent's authority to this extent once they have established that they are dealing with an agent of an existing principal. If the third party knows nothing of either the principal or the agent from previous contacts, he or she should determine by some objective means whether the principal and the agency relationship actually exists.

Obviously, if the principal is nonexistent or the principal exists but there is no agency, the third party may part with something of value in good faith and get nothing for it; a person cannot be bound as a principal simply because someone claimed to be his or her agent. Under such circumstances, the third party would be left with an action against the agent only, and it is likely that he or she would be hard to find.

Determination of Who Should Collect

Does the agent have the right to collect from a third party? Usually not, barring specific authority, or trade or local practice to the contrary. Ordinarily, the third party must give the consideration directly to the principal.

Black, for example, represents the White Manufacturing Company. He obtains an order from Gray for a quantity of his principal's product and receives part payment for the goods. Black is not seen again. Who bears the loss of the part payment? It depends upon whether Black had express or implied authority to receive payment. If no authority to receive payment can be found, Gray is the loser to the extent of the payment. The situation would be different if Black had brought the goods along with him; under this circumstance, the right to collect can be implied. If Gray has doubts about Black's authority to collect, Gray ignores them at his or her risk.

Somewhat akin to the agent's authority to receive payment is the authority to sign a negotiable instrument in a principal's name. Authority to do this must be expressly given, to have binding effect.

Transactions Binding Agent

In the normal course of affairs, the third party has no action available against the agent in a transaction. If the agent has acted within the scope of his or her express and implied authority, it is the principal and the third party who are bound. It is possible, however, for an agent to act as surety for the principal or to contract with the third party in such a way that it is the agent rather than the principal who is bound. Ordinarily an agent identifies the principal and discloses the fact that he or she is agent for the principal. If the agent however, merely agrees to something in his or her own name or signs as "Black, agent," indicating no principal, he or she may be held personally liable by the third party.

Tort

If, while in the course of a principal's business, the agent commits a tort against a third party, the third party may charge either the agent or the principal with the act. Successful action against the agent then gives the agent a right to recover from the principal, since the principal has a duty to indemnify the agent. An exception to this rule appears to exist where the principal is a minor and the agent an adult. Here the agent must bear the loss.

The third party can engage in tortious conduct as well. For example, if the third party, without cause, brings about the principal's discharge of the agent, the third party has committed a tort. In fact, the rule is more general than this. Anyone who maliciously causes another to lose an employment relationship (without any legal privilege to do so) has probably committed a tort, and an action will lie against him or her.

Undisclosed Principal

Normally third parties are aware that they are dealing with an agent of a known principal. Such is not always the case, however. The agent may not reveal that he or she is working for any principal, thus allowing the third party to assume that the agent is the party to be bound. Or the agent may reveal that he or she represents another without naming the principal (a partially disclosed principal).

In either case, where the agent has not disclosed the principal to the third party, the third party may elect to hold either principal or agent to the contract. If the third party elects to hold the agent, and the agent has acted within the scope of his or her authority, the agent has a right to be indemnified by the principal. Either the principal or the agent may hold the third party to the transaction, but the principal's right to do so is superior to the agent's.

The enforceability of **undisclosed-principal transactions** appears to be counter to the concept that a contract must be entered into voluntarily and intentionally by the parties. However, the legality and enforceability of such contracts are well established. It is, in effect, an exception to the general rule of contracts. An undisclosed-principal contract will not be enforced, however, where the third party, either expressly or by implication, makes clear his or her intent to deal exclusively with the agent.

Consider this example. White Manufacturing Company wishes to expand its operation into another section of the country. It retains Black to purchase land for the expansion without revealing the company's name. (Such purchases are sometimes undertaken to keep local land prices from soaring.) Black contracts with Gray to buy 200 acres of suitable land. Gray can elect to hold Black to the contract or, when the principal is revealed, hold the White Manufacturing Company to the contract. Gray is bound to the contract unless he has either expressed or implied his intent to deal exclusively with Black. If Black, under questioning by Gray, were to deny the existence of a principal, it would be grounds for fraud, making the contract voidable at Gray's election.

TERMINATION

The rules for winding up an agency agreement are about the same as those for winding up any employment agreement, except that the third party must be considered. The usual contract of employment is oral and terminable at the option of either employer or employee, often times called **employment-at-will.** Not all agency contracts are this simple.

By Law

Death, insanity, or bankruptcy of either principal or agent automatically terminates an agency relationship. Death or insanity of the principal is effective even if the agent is not aware of the event. That is, if an agent deals with another after the principal dies or becomes insane, but before the agent is informed of it, the transaction is not binding. If the agency has been created for a specific purpose, destruction of an element essential to the accomplishment of the purpose ends the agency. Similarly, passage of a law that makes the purpose of the agency unlawful terminates the agency.

By Acts of the Parties

An agency created to accomplish a specific purpose or to last for a stated time is generally not terminable at the option of the parties without possible repercussions. If an agency is created to accomplish a particular purpose, it ends when the purpose is accomplished. If a time limit for the agency is set, it ends when the time runs out.

Where an agency has been created for a purpose or to last a certain time, neither the principal nor the agent may unilaterally terminate the agency without the other's agreement. Both parties may, of course, agree to disagree before the contract is finished. Revocation of the agency by the principal terminates the agency, but if it is done without just cause or agreement, the principal is likely to be held liable for payment to the former agent for the remainder of the term for which the agency was to run. Similarly, renunciation of the principal by the agent ends the agency, but the principal may be allowed recovery. As indicated earlier, disloyalty by the agent would be just cause for early termination by the principal. In any case, termination by a unilateral act of either principal or agent does not become effective until the other party is informed of it.

Agency with an Interest

If the agent has an interest in the subject matter of the agency, the relationship cannot be terminated by an act of the principal. The term **interest** refers to more than just the agreed compensation for the agent's services. Essentially, this term implies part ownership or an equity in the subject matter. This issue might come up in a partnership venture, for example; in most instances, one partner could not very well fire another.

Notice

When an agency is terminated, third parties should be notified. If the agency is terminated by law, notification is considered to have taken place, since death, insanity, or bankruptcy would be a matter of public record. When termination takes place by acts of the parties, however, there is a particular necessity to inform those who have dealt with the agent. If, in ignorance of the dissolution of the agency, a third party deals with an agent as he or she had dealt with the agent before, the principal will be bound. The reason is the agent's apparent authority to act for the principal. Because it is the principal who runs the risk of being bound by the ex-agent's actions, a careful principal will make sure that third parties who have dealt with the former agent receive notice that the agency relationship has ended.

Notification to those who have previously dealt with the agent prevents this liability, but notification is not effective until the third parties receive it. Thus, a notice in a newspaper or trade journal may not be effective notification to third parties.

PARTNERSHIPS AND CORPORATIONS

Each partner in a partnership is an agent for the partnership. Generally, the rules of agency apply to partnerships. As long as a transaction takes place in the normal course of business of the partnership, an agreement by one partner binds all partners to the contract. Transactions beyond the normal scope of the partnership generally require approval of all the partners. Just as notice given to an agent is the virtual equivalent of notice given to the principal, notice given to any partner is also notice to the partnership.

A distinctive feature of a partnership is the agreement of the partners to share the profits. As you might guess, the partners also share the liabilities. Thus, each partner remains liable for the debts of the other partners resulting from their activities on behalf of the partnership. For this reason, many businesses are organized as corporations. In a corporation, the business is owned by the shareholders. The shareholders' liability is limited to their investment in the corporation. If White and Black are partners, for example, and White commits a tort while acting on behalf of the partnership, then White, Black, and the partnership can be held liable. If, however, Black is simply a shareholder in White Corporation, then Black's personal liability is usually limited to the loss of his investment in White Corporation due to a suit against White and the White Corporation.

Another distinctive feature of corporations is that they can "live" forever. A partnership ends when a partner leaves by death or agreement. A corporation's directors, officers, and employees may come and go, but the corporation remains a distinct and unchanging legal entity.

> *Usually* is the key word when discussing a shareholder's personal liability. If a shareholder owns sufficient shares to be a controlling shareholder, e.g., whose votes can determine board of directors membership, this shareholder can be subject to liability above personal investment.[2]

ADAM D. SOKOLOFF ET AL. v. HARRIMAN ESTATES DEVELOPMENT CORP., ET AL. 96 N.Y.2d 409; 754 N.E.2d 184; 729 N.Y.S.2d 425 (CA NY 2001)

Study terms: Duty of loyalty, agency, specific performance, third-party beneficiary theory, replevin

On this appeal, we review the dismissal on the pleadings, pursuant to CPLR 3211 (a) (7), of plaintiffs' cause of action seeking specific performance of an alleged contract. The facts as alleged in the complaint and other averments submitted in opposition to the motion to dismiss are as follows. In March 1998, plaintiffs purchased land in the Village of Sands Point, Nassau County, in contemplation of building a new home on the property. For a total of $65,000, defendant Harriman Estates Development Corp., a residential contractor, offered to provide plaintiffs with certain pre-construction services, including furnishing an "architectural and site plan/landscape design" and assisting them in obtaining a building permit. The offer was set forth by Harriman in a March 12, 1998 letter, which established a payment schedule and requested payment of a $10,000 retainer fee. Plaintiffs accepted the offer by paying Harriman the retainer fee. Thereafter, following several meetings between plaintiffs, Harriman and defendant Frederick Ercolino, an architect, the architectural plans were finalized, filed with the Village and approved.

Although plaintiffs paid Harriman a total of $55,000 for the architectural plans and other services, and tendered the remaining balance due under the terms of their agreement with Harriman, Harriman and Ercolino refused to allow plaintiffs to use these plans to build their home. After plaintiffs rejected Harriman's offer to build the home for an estimated cost of $1,895,000 (a sum significantly greater than Harriman's earlier estimates), Harriman for the first time informed plaintiffs that the architectural plans could not be used to construct the house unless it was hired as the builder. Harriman predicated its claim to the exclusive use of the plans on the terms of a contract it had entered into with Ercolino in May 1998 for the "Sokoloff Residence."

Plaintiffs then brought this action against Harriman and Ercolino for specific performance of the "contract dated March 12, 1998" (the first cause of action) and for replevin of the architectural plans (the second cause of action).

---

[2]See *Monica A. Beam, derivatively on behalf of Martha Stewart Living Omnimedia v. Martha Stewart, et al.* at the end of the chapter.

With respect to the first cause of action, seeking specific performance, plaintiffs alleged that Harriman was acting as their agent in procuring architectural drawings and plans from Ercolino, that the plans were unique and based upon a design conceived by them and that they had no adequate remedy at law. Plaintiffs requested an order directing Harriman and Ercolino to permit them to use the architectural plans. In their second cause of action, for replevin, plaintiffs alleged that they owned the architectural plans "by reason of being a third-party beneficiary" of the contract between Harriman and Ercolino.

Harriman moved to dismiss the complaint pursuant to CPLR 3211 (a) (7) for failure to state a cause of action. Supreme Court granted the motion in part by dismissing the cause of action for replevin, leaving intact plaintiffs' cause of action for specific performance. On Harriman's appeal from Supreme Court's failure to grant the motion to dismiss in its entirety, the Appellate Division reversed, dismissed the specific performance claim and severed the action against Ercolino. The court reasoned that even plaintiffs' first cause of action was barred by a provision in the Harriman-Ercolino contract stating that "nothing contained in this Agreement shall create a contractual relationship with or a cause of action in favor of a third party against either the Client [Harriman] or Architect." We granted leave to appeal and now reverse.

On a motion to dismiss pursuant to CPLR 3211, we must accept as true the facts as alleged in the complaint and . . . accord plaintiffs the benefit of every possible favorable inference. . . . Applying these principles, we conclude that plaintiffs adequately alleged a cause of action against Harriman for specific performance.

Plaintiffs' first cause of action was not predicated on a third-party beneficiary theory and therefore was not barred by the contractual provision cited by the Appellate Division. To be sure, the complaint alleged that plaintiffs owned the architectural "plans by reason of being a third-party beneficiary of a contract between [Harriman and Ercolino]." However, that third-party beneficiary theory was interposed only in support of plaintiffs' second cause of action, which was not before the Appellate Division and is not before us now. By contrast, in the first cause of action—the only cause of action at issue at the Appellate Division and before this Court—plaintiffs seek specific performance of their contract with Harriman and allege that Harriman was acting as their agent in procuring architectural drawings and plans from Ercolino. The contractual provision barring third-party actions is irrelevant with respect to plaintiffs' cause of action for specific performance of their contract with Harriman because plaintiffs are themselves parties to that contract.

Harriman argues that plaintiffs nonetheless are not entitled to specific performance because their claim is improperly predicated on an "invoice" that does not have the "status of a contract." This contention lacks merit. Plaintiffs alleged in their complaint that Harriman offered to provide them with an architectural design and other services for $65,000 and that they accepted that offer. They attached to their complaint a copy of the March 12, 1998 letter from Harriman, which sets forth a payment schedule for the proposed work and states that Harriman "started the architectural and site plan/landscape design process" and that plaintiffs' "retainer for these services is required at this time." Plaintiffs further alleged that the architectural and other services were completed and that they paid $55,000 and tendered the remaining $10,000 balance to Harriman. At this pleading stage of the litigation, we cannot conclude as a matter of law that the March 12, 1998 letter does not represent and memorialize a binding, bilateral agreement under which Harriman agreed to procure architectural plans and other services for plaintiffs and, for that, plaintiffs agreed to pay Harriman $65,000.

We also reject Harriman's assertion that specific performance is an inappropriate remedy because the architectural plans are not unique and a dollar value can be placed on the purchase of replacement plans. In general, specific performance will not be ordered where money damages "would be adequate to protect the expectation interest of the injured party" (Restatement [Second] of Contracts § 359 [1]. Specific performance is a proper remedy, however, where "the subject matter of the particular contract is unique and has no established market value" (*Van Wagner Adv. Corp. v S & M Enters.*, supra, at 193).

The decision whether or not to award specific performance is one that rests in the sound discretion of the trial court. In determining whether money damages would be an adequate remedy, a trial court must consider, among other factors, the difficulty of proving damages with reasonable certainty and of procuring a suitable substitute performance with a damages award (see, Restatement [Second] of Contracts § 360). Specific performance is an appropriate remedy for a breach of contract concerning goods that "are unique in kind, quality or personal association" where suitable substitutes are unobtainable or unreasonably difficult or inconvenient to procure (see, id., comment c).

In this case, plaintiffs have alleged that "the architectural plans and drawings are unique in that they are based upon a design conceived by the plaintiffs," and that without specific performance they "would have to change their requirements" as to the design of their new home. These allegations are sufficient to withstand a motion to dismiss for failure to state a cause of action. Whether money damages would adequately compensate plaintiffs for loss of these allegedly unique architectural plans is a matter to be resolved at a later stage, not on a motion to dismiss the complaint.

Harriman's final contention is that it has an absolute defense to specific performance of the alleged March 12 contract because the contract it entered into with Ercolino stipulates that the plans "shall only be used by Harriman Estates Development Corp. for their one time use at Harriman Estates . . . and . . . shall not be transferred or sold to others except" with the written consent of, and the payment of appropriate compensation to, Ercolino. Harriman essentially maintains that, although plaintiffs never entered into a contract with it to construct their house, it has the right to withhold the architectural plans it allegedly agreed to procure on plaintiffs' behalf unless plaintiffs hire it as their construction contractor. Harriman's reliance on its contract with Ercolino for this proposition is misplaced.

On Harriman's motion to dismiss, we of course accept as true plaintiffs' allegation that Harriman was acting as plaintiffs' agent when it entered into the contract with Ercolino to prepare the architectural design for their home. A person who enters into a contract with another to perform services as an agent "is subject to a duty to act in accordance with his promise" (Restatement [Second] Agency § 377). Moreover, fundamental to the principal-agent relationship "is the proposition that an [agent] is to be loyal to his [principal] and is 'prohibited from acting in any manner inconsistent with his agency or trust and is at all times bound to exercise the utmost good faith and loyalty in the performance of his duties'" (*Western Elec. Co. v Brenner*, 41 NY2d 291, 295).

Agents "must act in accordance with the highest and truest principles of morality" (Elco Shoe Mfrs. v Sisk, 260 NY 100, 103) and, as fiduciaries, are forbidden from engaging in "many forms of conduct permissible in a workaday world for those acting at arm's length" (*Meinhard v Salmon*, 249 NY 458, 464). It thus follows that an "agent must not seek to acquire indirect advantages from third persons for performing duties and obligations owed to [the agent's] principal" (*Brenmer Indus. v Hattie Carnegie Jewelry Enters.*, 71 AD2d 597, appeal withdrawn 56 NY2d 648). If "an agent receives anything as a result of his violation of a duty of loyalty to the principal, he is subject to a liability to deliver it, its value, or its proceeds, to the principal" (Restatement [Second] Agency § 403; see also, id. § 407).

Under those guiding principles, Harriman, as the alleged agent of plaintiffs, would not be entitled to rely on the Ercolino contract to defeat plaintiffs' claim for specific performance of Harriman's promise to furnish them with the plans. Accepting plaintiffs' allegations as true, Harriman could be found to have breached its duty of loyalty to plaintiffs by entering into a stipulation with a third party preventing plaintiffs, its principals, from using the architectural plans it procured for them unless they also agreed to Harriman's terms for building their home. As plaintiffs' agent, Harriman would not have been acting in plaintiffs' best interests when it agreed with Ercolino to place such a restriction on their right to use the architectural plans. Under those circumstances, it should not be permitted to invoke a contractual provision entered into with a third party in breach of its fiduciary duties as a legal ground for withholding its own consent to plaintiffs' use of the architectural plans.

Accordingly, the order of the Appellate Division should be reversed, with costs, and the motion of defendant Harriman Estates Development Corp. to dismiss the first cause of action of the complaint against it denied.

Order reversed, etc.

[handwritten: ↑ on behalf of all shareholders]

[handwritten margin note: Board of Directors are all friends of M.S. so they have loyalty to M.S. As a result, a demand to BoD would be futile. Grounds for suit w/o demand]

MONICA A. BEAM, derivatively on behalf of MARTHA STEWART LIVING OMNIMEDIA, INC. v. MARTHA STEWART, et al. 833 A.2d 961; 2003 Del. Ch. LEXIS 98 (Del. Ct. Chancery 2003)

Study terms: Derivative suit, "duty to monitor", duty of loyalty, corporate opportunity doctrine, demand futility

Monica A. Beam, a shareholder of Martha Stewart Living Omnimedia, Inc. ("MSO"), brings this derivative action against the defendants, all current directors and a former director of MSO, and against MSO as a nominal defendant. The defendants have filed three separate motions seeking (1) to dismiss Counts II, III, and IV under Court of Chancery Rule 12(b)(6) for failure to state claims upon which relief may be granted; (2) to dismiss the amended complaint under Court of Chancery Rule 23.1 for failure to comply with the demand requirement and for failure adequately to plead demand excusal; or alternatively (3) to stay this action in favor of litigation currently pending in the U.S. District Court for the Southern District of New York. This is the Court's ruling on these motions.

I. Factual Background

Plaintiff Monica A. Beam is a shareholder of MSO and has been since August 2001. Derivative plaintiff and nominal defendant MSO is a Delaware corporation that operates in the publishing, television, merchandising, and internet industries marketing products bearing the "Martha Stewart" brand name.

Defendant Martha Stewart ("Stewart") is a director of the company and its founder, chairman, chief executive officer, and by far its majority shareholder. MSO's common stock is comprised of Class A and Class B shares. Class A shares are traded on the New York Stock Exchange and are entitled to cast one vote per share on matters voted upon by common stockholders. Class B shares are not publicly traded and are entitled to cast ten votes per share on all matters voted upon by common stockholders. Stewart owns or beneficially holds 100% of the B shares in conjunction with a sufficient number of A shares that she controls roughly 94.4% of the shareholder vote. Stewart, a former stockbroker, has in the past twenty years become a household icon, known for her advice and expertise on virtually all aspects of cooking, decorating, entertaining, and household affairs generally.

Defendant Sharon L. Patrick ("Patrick") is a director of MSO and its president and chief operating officer. The amended complaint reports that in 2001, MSO paid Patrick a salary of $700,000, a $280,000 bonus, and granted her options for 130,000 Class A shares. She also serves as the secretary of M. Stewart, Inc., which is described in the complaint as "one of Stewart's personal companies." Prior to Patrick's employment at MSO, she was a consultant to the magazine, Martha Stewart Living, and developed extensive experience in the media, entertainment, and consulting businesses. Patrick is also a longtime personal friend of Stewart.

Defendant Arthur C. Martinez ("Martinez") has been a director of MSO since January 2001. Martinez is the former chairman of the board of directors and chief executive officer of Sears Roebuck and Co . . . A March 2001 article in Directors & Boards reported that Patrick and Stewart both consider Martinez to be "an old friend." Also, Martinez was recruited to serve on MSO's board by then-board member Charlotte Beers ("Beers"), another "longtime friend and confidante" of Stewart.

Defendant Darla D. Moore ("Moore") has been a director of MSO since September 2001, when Beers resigned and Moore replaced her. . . . Moore, too, is reported to be a longtime friend of both Stewart and Beers, as evidenced by a 1996 Fortune magazine article highlighting the close friendship among the three women and by the amended complaint's report of Moore's attendance at a wedding reception in 1995, which was attended by both Stewart and Samuel Waksal and hosted by Stewart's lawyer, Allen Grubman. . . .

The amended complaint states that compensation paid to MSO's directors includes all of the following:

- $20,000 as an annual retainer;
- $1,000 for each meeting attended in person;
- $500 for each meeting attended telephonically; and
- $5,000 annually for serving as chairman of any committee.

Twenty-five percent of directors' fees are paid in shares of MSO's Class A common stock, with the remaining 75% payable either in Class A shares or cash at the choice of the director. In addition, MSO has a stock option plan for the directors.

The plaintiff seeks relief in relation to three distinct types of activities. The first involves the well-publicized matters surrounding Stewart's alleged improper trading of shares of ImClone Systems, Inc. ("ImClone") and her public statements in the wake of those allegations. The second relates to the private sale of sizeable blocks of MSO stock by both Stewart and Doerr in early 2002. The third challenges the board's decisions with regard to the provision of "split-dollar" insurance for Stewart.

A. Stewart's ImClone Trading

The market for MSO products is uniquely tied to the personal image and reputation of its founder, Stewart. MSO retains "an exclusive, worldwide, perpetual royalty-free license to use [Stewart's] name, likeness, image, voice and signature for its products and services." In its initial public offering prospectus, MSO recognized that impairment of Stewart's services to the company, including the tarnishing of her public reputation, would have a material adverse effect on its business. The prospectus distinguished Stewart's importance to MSO's business success from that of other executives of the company noting that, "Martha Stewart remains the personification of our brands as well as our senior executive and primary creative force." In fact, under the terms of her employment agreement, Stewart may be terminated for gross misconduct or felony conviction that results in harm to MSO's business or reputation but is permitted discretion over the management of her personal, financial, and legal affairs to the extent that Stewart's management of her own life does not compromise her ability to serve the company.

Stewart's alleged misadventures with ImClone arise in part out of a longstanding personal friendship with Samuel D. Waksal ("Waksal"). Waksal is the former chief executive officer of ImClone as well as a former suitor of Stewart's daughter. More pertinently, with respect to the allegations of the amended complaint, Waksal and Stewart have provided one another with reciprocal investment advice and assistance, and they share a stockbroker, Peter E. Bacanovic ("Bacanovic") of Merrill Lynch. Bacanovic, coincidentally, is a former employee of ImClone. When MSO made its initial public offering of Class A common stock in 1999, Bacanovic sold $10 million of the shares to Stewart's friends and family members, including Waksal and members of his family. Waksal has encouraged and assisted Stewart not only with regard to the purchase of ImClone stock but also has persuaded Stewart to invest in a venture fund he started, Scientia Health Group, and has helped Stewart and her friend, Beers, to become private investors in Cadus Pharmaceutical Corp. prior to its initial public offering in 1996. . . . The speculative value of ImClone stock was tied quite directly to the likely success of its application for FDA approval to market the cancer treatment drug Erbitux. On December 26, Waksal received information that the FDA was rejecting the application to market Erbitux. The following day, December 27, he tried to sell his own shares and tipped his father and daughter to do the same. Stewart also sold her shares on December 27. That day she was traveling with another friend, Marianna Pasternak ("Pasternak"), when Stewart spoke with Bacinovic's assistant, Douglas Faneuil, and sold all of her ImClone shares. The next day, December 28, Pasternak's husband, Bart Pasternak, also sold 10,000 shares of ImClone. After the close of trading on December 28, ImClone publicly announced the rejection of its application to market Erbitux. The following day the trading price closed slightly more than 20% lower than the closing price on the date that Stewart had sold her shares. . . . After barely two months of such adverse publicity, MSO's stock price had declined by slightly more than 65%. In August 2002, James Follo, MSO's chief financial officer, cited uncertainty stemming from the investigation of Stewart in response to questions about earnings prospects in the future.

B. Private Sales of MSO Stock

In January 2002, Stewart and the Martha Stewart Family Partnership sold 3,000,000 shares of Class A stock to entities designated in the amended complaint as "ValueAct." In March 2002, Kleiner, Perkins, acting through its general partner, Doerr, sold 1,999,403 shares of MSO to ValueAct.

C. Split-Dollar Insurance

MSO provides Stewart with what is commonly termed a "split-dollar" insurance policy. The premiums for such policies are paid entirely or in large part by the employer. . . .

Some have suggested that split-dollar insurance constitutes an interest-free loan to the employee on whose behalf the policy is purchased. If so, such policies could run afoul of recent federal legislation banning loans to corporate executives and directors.[3] A spokesperson for MSO stated that, in light of the questions raised regarding the nature and propriety of split-dollar insurance policies for executives, MSO's lawyers and accountants would be evaluating the advisability of paying the premiums in the future and would make a determination before the next premium payment came due in February 2003.

II. Analysis

A. Motions to Dismiss Counts II, III, and IV—Court of Chancery Rule 12(b)(6)

In ruling on a motion to dismiss under Rule 12(b)(6), the Court considers only the allegations in the amended complaint, and any documents incorporated by reference therein. For this purpose, the Court accepts as true all well-pled factual allegations contained in the amended complaint, but conclusory statements—those unsupported by well-pled factual allegations—are not accepted as true. . . .

1. Count II—Failure to Monitor Stewart's Personal Activities

Count II of the amended complaint alleges that the director defendants and defendant Patrick breached their fiduciary duties by failing to ensure that Stewart would not conduct her personal, financial, and legal affairs in a manner that would harm the Company, its intellectual property, or its business.

The "duty to monitor" has been litigated in other circumstances, generally where directors were alleged to have been negligent in monitoring the activities of the corporation, activities that led to corporate liability. Plaintiff's allegation, however, that the Board has a duty to monitor the personal affairs of an officer or director is quite novel. That the Company is "closely identified" with Stewart is conceded, but it does not necessarily follow that the Board is required to monitor, much less control, the way Stewart handles her personal financial and legal affairs.

In *Graham v. Allis-Chalmers Manufacturing Co.*, the Delaware Supreme Court held that "absent cause for suspicion there is no duty upon the directors to install and operate a corporate system of espionage to ferret out wrongdoing which they have no reason to suspect exists." . . .

First, plaintiff does not allege facts that would give MSO's Board any reason to monitor Stewart's activities before mid-2002 when the allegations regarding her divestment of ImClone stock became public. Second, the quoted statement from Graham refers to wrongdoing by the corporation. Regardless of Stewart's importance to MSO, she is not the corporation. And it is unreasonable to impose a duty upon the Board to monitor Stewart's personal

---

[3]See Sarbanes-Oxley Act of 2002, Pub. L. No. 107-204 § 402(a), 116 Stat. 745, 787-88 (codified at 15 U.S.C. § 78m(k) (2003)).

affairs because such a requirement is neither legitimate nor feasible. Monitoring Stewart by, for example, hiring a private detective to monitor her behavior is more likely to generate liability to Stewart under some tort theory than to protect the Company from a decline in its stock price as a result of harm to Stewart's public image.

Even if I accept that the board knew that Stewart's personal actions could result in harm to MSO, it seems patently unreasonable to expect the Board, as an exercise of its supervision of the Company, to preemptively thwart a personal call from Stewart to her stockbroker or to fully control her handling of the media attention that followed as a result of her personal actions, especially where her statements touched on matters that could subject Stewart to criminal charges. Plaintiff has not cited any case to support this new "duty" to monitor personal affairs. Since the defendant directors had no duty to monitor Stewart's personal actions, plaintiff's allegation that the directors breached their duty of loyalty by failing to monitor Stewart because they were "beholden" to her is irrelevant. Count II is dismissed for failure to state a claim.

2. Count III—Stock Sales by Stewart and Doerr

Count III of the amended complaint alleges that Stewart and Doerr breached their fiduciary duty of loyalty, usurping a corporate opportunity by selling large blocks of MSO stock to ValueAct. Defendants Stewart and Doerr are essentially in the same position with respect to Count III. The basic requirements for establishing usurpation of a corporate opportunity were articulated by the Delaware Supreme Court in *Broz v. Cellular Information Systems, Inc.*:

[A] corporate officer or director may not take a business opportunity for his own if: (1) the corporation is financially able to exploit the opportunity; (2) the opportunity is within the corporation's line of business; (3) the corporation has an interest or expectancy in the opportunity; and (4) by taking the opportunity for his own, the corporate fiduciary will thereby be placed in a position [inimical] to his duties to the corporation.

In this analysis, no single factor is dispositive. Instead the Court must balance all factors as they apply to a particular case. For purposes of the present motion, I assume that the sales of stock to ValueAct could be considered to be a "business opportunity." I now address each of the four factors articulated in Broz.

a. Financial Ability of MSO to Exploit the Opportunity

The amended complaint asserts that MSO was able to exploit this opportunity because the Company's certificate of incorporation had sufficient authorized, yet unissued, shares of Class A common stock to cover the sale to ValueAct. Defendants do not deny that the Company could have sold previously unissued shares to ValueAct. I therefore conclude that the first factor has been met.

b. Within MSO's Line of Business

An opportunity is within a corporation's line of business if it is "an activity as to which [the corporation] has fundamental knowledge, practical experience and ability to pursue." Because I have already determined that MSO had sufficient authorized but unissued shares available and because no special expertise is required to issue stock, I find that the "ability to pursue" prong is met. The question then becomes whether selling its own stock is an activity as to which the Company has fundamental knowledge and practical experience.

Plaintiff states that the Company's line of business is "creating 'how-to' content and domestic merchandise for homemakers and other consumers." Nevertheless, the Court recognizes that raising capital is a fundamental activity in which businesses are often engaged. MSO made its initial public offering in October 1999. . . . By definition, a company's issuance of its stock does not generate income. For the foregoing reasons, I therefore conclude that the sale of stock by Stewart and Doerr was not within MSO's line of business.

c. MSO's Interest or Expectancy in the Stock Sales

A corporation has an interest or expectancy in an opportunity if there is "some tie between that property and the nature of the corporate business." . . . Here, plaintiff does not allege any facts that would imply that MSO was in need of additional capital, seeking additional capital, or even remotely interested in finding new investors. Had MSO wished to do so, it had a readily available, liquid market in which to accomplish that aim, as MSO's Class A stock is traded on the New York Stock Exchange.

I fail to see any connection between the potential sale of stock to ValueAct and the nature of MSO's business. . . . In the absence of specific allegations indicative of corporate interest or expectancy, I must conclude that this factor of the Broz test has not been met.

d. Whether the Stock Sales Placed Stewart and Doerr in a Position Inimical to Their Duties to MSO

"The corporate opportunity doctrine is implicated only in cases where the fiduciary's seizure of an opportunity results in a conflict between the fiduciary's duties to the corporation and the self-interest of the director as actualized by the exploitation of the opportunity." Given that I have concluded that MSO had no interest or expectancy in the issuance of new stock to ValueAct, I fail to see, based on the allegations before me, how Stewart and Doerr's sales placed them in a position inimical to their duties to the Company. Were I to decide otherwise, directors of every Delaware corporation would be faced with the ever-present specter of suit for breach of their duty of loyalty if they sold stock in the company on whose Board they sit.

Additionally, Delaware courts have recognized a policy that allows officers and directors of corporations to buy and sell shares of that corporation at will so long as they act in good faith. . . . In the absence of allegations based on well-pled facts that call into question the propriety of Stewart and Doerr's sales at the time they made them, it is impossible to infer that they acted in bad faith in selling their shares or placed themselves in a position inimical to their duties to MSO. Given the allegations before me, the fourth factor of the Broz test is not met.

On balancing the four factors, I conclude that plaintiff has failed to plead facts sufficient to state a claim that Stewart and Doerr usurped a corporate opportunity for themselves in violation of their fiduciary duty of loyalty to MSO. Count III is dismissed in its entirety under Rule 12(b)(6) for failure to state a claim upon which relief can be granted.

3. Count IV—Split-Dollar Insurance

Count IV alleges, "putting aside whether the approval of such policies . . . [was] proper, [the defendants] have failed to act responsibly and address the impropriety of the Company's payment of split-dollar insurance policy premiums." Although it is possible that such policies ultimately will be held to violate certain provisions of Sarbanes-Oxley, the amended complaint fails to state a claim under Delaware law. Plaintiff does not allege that the premiums paid by MSO were unlawful, that MSO failed to disclose the existence of the policy, or that the Board failed to take action once the governing law changed. To the contrary, plaintiff alleges merely that the split-dollar policies might be held to be unlawful, then admits that the Board properly disclosed the existence of the policy in its April 1, 2002 Definitive Proxy Statement, and further concedes that the Company's lawyers and accountants were looking into whether the policy should be unwound. . . .

Although I do not have a developed record before me at this stage of the litigation, plaintiff here has conceded that MSO's Board has considered and is considering, with the assistance of expert advisors, the potential ramifications of continuing the split-dollar insurance policy. Two key elements necessary to proving a breach of fiduciary duty claim in this instance would be that (1) the directors knew or should have known that a violation of the law was occurring and, (2) "the directors took no steps in a good faith effort to prevent or remedy that situation." Giving full weight to the facts pled in the amended complaint, I must conclude that plaintiff is unable to meet this test, and has thus failed to state a claim on which relief can be granted.

Plaintiff also asserts in Count IV that the policy was improper, noting that "additionally, and under any circumstance, the conduct allege [sic] herein does not deserve any such reward." The complaint is unclear as to whether this is a separate, albeit oblique, allegation, challenging the Board's business judgment when taking out the policy in the first place, or whether it calls into question the continuation of premium payments. Nevertheless, such a vague and conclusory statement is inadequate to sustain Count IV, and it is therefore dismissed.

B. Motions to Dismiss the Amended Complaint—Court of Chancery Rule 23.1

Defendants have moved to dismiss the amended complaint under Court of Chancery Rule 23.1 for failure to make demand upon MSO's board of directors or adequately to plead why demand would be futile. . . .

Plaintiff concedes that demand was not made but asserts that demand would be futile because the board of directors is incapable of acting independently and disinterestedly in evaluating demand with respect to plaintiff's claims. As a practical matter, because Counts II through IV are dismissed for failure to state a claim, I need only determine whether demand would be futile with respect to Count I. Count I alleges that Stewart breached her fiduciary duties to MSO and its shareholders by selling (perhaps illegally) shares of ImClone in December of 2001 and by public statements she made regarding that sale. . . . Demand is required if, in view of all the particularized allegations in the complaint and drawing all reasonable inferences in favor of the plaintiff, there is no reasonable doubt of the ability of a majority, here four of the six directors, to respond to demand appropriately.

1. Inside Directors: Stewart and Patrick

Defendants do not suggest that either Stewart or Patrick should be considered disinterested, independent, and able to consider demand on Count I without interference of improper extraneous influences. Still, Rule 23.1 places the burden on the plaintiff to specify in the complaint the reasons why demand would be futile.

The amended complaint alleges that in her sales of ImClone stock and subsequent statements on the subject, Stewart not only breached her fiduciary duties to MSO and its shareholders but may have committed criminal acts. It seems that these allegations indicate a significant likelihood that Stewart may be both civilly and criminally liable with respect to these actions. This is sufficient to raise a reasonable doubt of Stewart's disinterest in the challenged acts. I find that Stewart must be considered incapable of appropriately considering demand with respect to Count I.

Patrick is the president, chief operating officer, and a director of MSO. Her 2001 compensation included $980,000 in salary and bonuses along with a grant of options to purchase 130,000 Class A shares of MSO. Based on the magnitude of compensation described flowing to Patrick from her work at MSO, I find that Patrick has a material interest in her own continued employment. . . . This raises a reasonable doubt whether Patrick can evaluate and respond to demand on Count I without being influenced by improper consideration of the extraneous matter of how pursuit of that claim would affect Stewart's interests.

2. Outside Directors: Martinez, Moore . . .

Stewart has overwhelming voting control of MSO, controlling over 94% of the shareholder vote. It is reasonable to infer that she can remove or replace any or all of the directors. This ability does not by itself demonstrate that Stewart has the capacity to control the outside directors, but is not without relevance to whether there is a reasonable doubt of the outside directors' independence of Stewart. . . . [I]t is not obvious from the allegations that such compensation would be sufficient to entice any of the outside directors to ignore fiduciary duties to MSO and its shareholders. Nor does plaintiff suggest that the outside directors have a history of blindly following Stewart's will or even accepting her recommendations without adequate independent study and investigation. . . .

The allegations regarding the friendship between Moore and Stewart are somewhat more detailed, yet still fall short of raising a reasonable doubt about Moore's ability properly to consider demand on Count I. . . . On the facts pled, however, I cannot say that I have a reasonable doubt of Moore's ability to properly consider demand. . . .

In sum, plaintiff offers various theories to suggest reasons that the outside directors might be inappropriately swayed by Stewart's wishes or interests, but fails to plead sufficient facts that could permit the Court reasonably to infer that one or more of the theories could be accurate. Evidence to support (or refute) any of the theories might have been uncovered by an examination of the corporate books and records, to which the plaintiff would have been entitled for this purpose. . . . Armed with such information, plaintiff (and this Court) would be in a much better position to evaluate whether there exists a reasonable doubt of the outside directors' resolve to act independently of Stewart. It appears, however, that plaintiff made no such investigation, instead relying largely, if not solely, on information from media reports to support the assertion that demand would be futile.

It is troubling to this Court that, notwithstanding repeated suggestions, encouragement, and downright admonitions over the years both by this Court and by the Delaware Supreme Court, litigants continue to bring derivative complaints pleading demand futility on the basis of precious little investigation beyond perusal of the morning newspapers. This failure properly to investigate whether a majority of directors fairly can evaluate demand may lead to either (or both) of two equally appalling results. . . . If the facts to support reasonable doubt could have been ascertained through more careful pre-litigation investigation, the failure to discover and plead those facts still results in a waste of resources of the litigants and the Court and, in addition, ties the hands of this Court to protect the interests of shareholders where the board is unable or unwilling to do so. This results in the dismissal of what otherwise may have been meritorious claims, fails to provide relief to the company's shareholders, and further erodes public confidence in the legal protections afforded to investors. . . .

I would be remiss, though, if I failed to point out that with a bit more detail about the "relationships," "friendships," and "inter-connections" among Stewart and the other defendants or with some additional arguments as to why there may be a reasonable doubt of the directors' incentives when evaluating demand with respect to Count I, there may have been a reasonable doubt as to one or all of the outside directors disinterest, independence, or ability to consider and respond to demand free from improper extraneous influences. Nevertheless, on this pleading, no such doubt is raised. The defendants' motions to dismiss the amended complaint for failure adequately to plead demand futility are granted with respect to Count I. . . .

Demand → shareholders must make demand to Board of Directors to fix problem (ie, give them a chance) before suing in court

III. Conclusion

For the reasons stated above, (1) the defendants' motions to dismiss Counts II, III, and IV for failure to state a claim are GRANTED; (2) the defendants' motions to dismiss the amended complaint for failure to make demand or adequately to plead that demand is excused are therefore moot as to Counts II, III, and IV; (3) the defendants' motions to dismiss the amended complaint for failure to make demand or adequately to plead that demand is excused are GRANTED as to Count I; and (4) the defendants' motion to stay further proceedings in this action pending the resolution of the Federal Action is mooted by the dismissal of all counts in the amended complaint.

IT IS SO ORDERED.

MOUNDSVIEW IND. S.D. NO. 621 v. BUETOW & ASSOC.
253 N.W.2d 836 (Minn. 1977)

Study terms: Plans and specifications, on-site inspections, acts or omissions, duty to supervise, "clerk of the works"

Buetow & Associates, Inc., (Buetow) entered into an agreement with Moundsview Independent School District No. 621 (Moundsview) to perform architectural services. Buetow agreed to prepare plans and specifications for an addition to a school as well as to provide general supervision of the construction operation. After the completion of construction, a windstorm ripped a portion of the roof off the school, allegedly due to the failure of a contractor to adequately fasten the roof to the building. The trial court granted Buetow's motion for summary judgment based on Buetow's contract with Moundsview, which provided that Buetow was not responsible for the failure of a contractor to follow the plans and specifications. We affirm.

In August 1968, Moundsview retained Buetow to prepare plans and specifications for an addition to an elementary school. At the time of the execution of the agreement, Moundsview had the option of requiring Buetow to provide (1) no supervision, (2) general supervision, or (3) continuous onsite inspection of the construction project by a full-time project representative referred to as a "clerk of the works." Moundsview elected to have Buetow provide only a general supervisory function, the specific language of the contract enumerating the requirements as follows:

> The Architect shall make periodic visits to the site to familiarize himself generally with the progress and quality of the Work and to determine in general if the Work is proceeding in accordance with the Contract Documents. On the basis of his on-site observations as an Architect, he shall endeavor to guard the Owner against defects and deficiencies in the Work of the Contractor. The Architect shall not be required to make exhaustive or continuous on-site inspections to check the quality or quantity of the Work. The Architect shall not be responsible for construction means, methods, techniques, sequences, or procedures, or for safety precautions and programs in connection with the Work, and he shall not be responsible for the Contractor's failure to carry out the Work in accordance with the Contract Documents.

The contract further provides:

> The Architect shall not be responsible for the acts or omissions of the Contractor, or any Subcontractors, or any of the Contractor's or Subcontractor's agents or employees, or any other person performing any of the Work.

Buetow prepared plans and specifications requiring the placement of wooden plates upon the concrete walls of the building. The plates were to be fastened to the walls by attaching washers and nuts to one-half inch studs secured in cement. During the 79-week construction period, the president of Buetow made 90 visits to the construction site in performance of Buetow's general supervisory obligation.

On May 19, 1975, a severe windstorm blew a portion of the roof off the building causing damage to the addition and to other portions of the school. It was discovered that the roof had not been secured by washers and nuts to the south wall of the school as required by the plans and specifications.

Moundsview brought an action for damages caused by the roof mishap against Buetow, the general contractor, and the roofing subcontractor. In response to an interrogatory from Buetow requesting Moundsview to state all facts upon which it relied to support its allegations against Buetow, Moundsview replied:

> Defendant Buetow failed to properly supervise the roof construction, failed to supervise and discover the missing nuts and studs and take proper corrective action.

Thereafter, Buetow made a motion for summary judgment, basing its motion upon the affidavit of one of its officers which stated that Buetow did not observe during any of its construction site visits that the washers and nuts had not been fastened to the studs on the south wall. The motion was also accompanied by the architect's contract which the parties entered into and Buetow's interrogatories and Moundsview's answers thereto. Moundsview did not file a responsive affidavit to oppose the motion.

The trial court granted Buetow summary judgment, accompanying its decision by memorandum which states:

> Since the contract of the architect did not require detailed supervision by the architect of the construction project, and since the architect was not contractually liable, as a matter of law, for the acts and omissions of the general contractor or any subcontractor, the architect is entitled to summary judgment under the principle enunciated in *J & J Electric, Inc. v. Moen Company*. . . .

Moundsview appeals from the judgment entered pursuant to the order for summary judgment.

The issue presented for consideration is whether there exists a genuine issue of fact in this case that will preclude the entry of summary judgment dismissing the complaint against Buetow.

1. Initially, we note that the rule in Minnesota is that a party cannot rely upon general statements of fact to oppose a motion for summary judgment. Instead, the nonmoving party must demonstrate at the time the motion is made that specific facts are in existence which create a genuine issue for trial. . . .

   The general statements included within Moundsview's complaint and the equally general answers to Buetow's interrogatories are insufficient to create a genuine issue of fact to successfully oppose a motion for summary judgment. Thus, since Moundsview failed to present any specific averments of fact in opposition to the motion for summary judgment, our review of the case is limited to a consideration of the contract between the parties.

2. Moundsview argues that Buetow breached its duty of architectural supervision by failing to discover that a contractor had failed to fasten one side of the roof to the building with washers and nuts as required by the plans and specifications. It is the general rule that the employment of an architect is a matter of contract, and consequently, he is responsible for all the duties enumerated within the contract of employment. . . . An architect, as a professional, is required to perform his services with reasonable care and competence and will be liable in damages for any failure to do so. . . .

Thus, consideration of whether Buetow breached a duty of supervision requires an initial examination of the contract between the parties to determine the parameters of its supervisory obligation. The argument that Buetow breached its duty to supervise would be more persuasive had Moundsview contracted for full-time project representation rather than mere general supervision. An architect's duty to inspect and supervise the construction site pursuant to a contract requiring only general supervision is not as broad as its duty when a "clerk of the works" is required. The mere fact that Buetow received additional compensation for performing the general supervisory service does not serve to expand its responsibilities to an extent equivalent to the duties of a full-time project representative. Moundsview cannot be allowed to gain the benefit of the more detailed "clerk-of-the-works" inspection service while in fact contracting and paying for only a general supervisory service.

Thus, the question of whether Buetow breached its duty to supervise the construction project is to be determined with reference to the general supervisory obligation enumerated in the contract. The contract provided that the architect" . . . shall not be responsible for the Contractor's failure to carry out the Work in accordance with the Contract Documents." When this section is read in conjunction with the section which provides that "(t)he Architect shall not be responsible for the acts or omissions of the Contractor, or any Subcontractors, or any of the Contractor's or Subcontractors' agents or employees or any other persons performing any of the Work," it is apparent that by the plain language of the contract an architect is exculpated from any liability occasioned by the acts or omissions of a contract. The language of the contract is unambiguous. The failure of a contractor to follow the plans and specifications caused the roof mishap. By virtue of the aforementioned contractual provisions, Buetow is absolved from any liability, as a matter of law, for a contractor's failure to fasten the roof to the building with washers and nuts.

Thus, based upon the language of the architect's contract, Buetow was entitled to summary judgment. . . .

Affirmed.

Review Questions

1. Identify the following persons as agents, employees, or independent contractors according to the usual duties involved in their work.
   a. Research chemist working for chemical company
   b. Free-lance consulting engineer in the labor relations field
   c. TV repairman for local electronics store—on house call
   d. Troubleshooter for steel company (keeps steel sold to customers by recommending proper treatment of a particular heat of steel)
   e. Engineering vice-president for local company
   f. Dentist

2. How may the agency relationship be created?

3. Brown is a process engineer for White Manufacturing Company and is about to recommend the purchase of certain machinery and equipment. The Green Equipment Company is one prospective supplier. On a recent trip to the Green Company, Green offered Brown a new station wagon if Green was chosen as the equipment supplier. Brown has always considered herself to be quite ethical, but she is also human, and the station wagon sounds tempting. Neglecting the ethical aspects of the situation, Brown is still faced with certain legal and practical problems. What are Brown's rights if, after recommending Green as supplier, Green fails to produce the station wagon? What can happen to Brown if White Manufacturing finds out about the deal?

4. Gray, engineer for Black, White, and Company, was sent to observe an automation installation at a plant some 50 miles away. On the return trip, he approached an intersection and applied his brakes. His car hit a patch of ice. As a result, he hit another car, injuring its occupants, both cars, and himself. Who is liable for injuries to the other car and its occupants? Who is liable for injury to Gray and Gray's car?

5. In *Adam D. Sokoloff et al. v. Harriman Estates Development Corp., et al.,* the plaintiffs are looking for the equitable remedy of specific performance (see Chapters 3 and 14). Why do you think they are seeking this remedy rather than money damages? *[handwritten: They want the plans, not the money to make new plans]*

6. In *Monica A. Beam v. Martha Stewart,* the court seemed willing to consider the plaintiffs' *[handwritten: ?]* allegations regarding whether outside directors were inappropriately swayed by Stewart's wishes favorably, but cannot do so. Why? What should the plaintiffs have done?

7. Consider the case of *Moundsview v. Buetow & Assoc.* and respond to the following.
   a. Does the summary judgment in favor of Buetow mean that Moundsview cannot recover for the windstorm damage to its school roof? *[handwritten: Moundsview could recover from contractor " " " not recover from Buetow because β·spolation of evidence]*
   b. Rewrite the contract requirements in such a manner that Buetow is responsible for such disasters in the factual situation described in the case. *[handwritten: Use "continuous onsite inspection..."]*

Property

PART

4

Most engineers work for other people—either for a private enterprise or for the public (by working for a local, state, or federal government). Engineers use other people's property in their work; hence, they must observe other people's rights. Many situations require the engineer to deal simultaneously with an employer's property and that of others, possibly including his or her own. Effectively handling property and property rights in these circumstances requires some knowledge of property law. Property is either tangible or intangible. Because of the obvious differences between tangible property, such as an automobile, and intangible property, such as a patent on a new computer chip, the rights and responsibilities relating to each may be treated differently. What is confusing for many is distinguishing between the "thing" embodying the intangible rights and the "thing" itself representing personal property. As an example, let's say Blue has bought a bottle of the Coca-Cola soft drink from a local grocery store. It is now Blue's *personal property*. If someone steals the drink, it will be considered theft of personal property. However, Blue has not purchased, nor can Blue recover for, the *intangible property rights* or often called *intellectual property rights* manifested in the bottle, specifically, the copyrights to the drawings or decoration on the bottle, the trade secret protected content, or any trademarks that are embodied in the bottle design. Those rights remain with the Coca-Cola Company. For purposes of this section, intangible property rights and intellectual property rights may be used interchangeably. Since property rights are fundamental in our legal system, you will notice that legal terminology regarding tangible property rights creeps into discussions of intangible property rights.

Tangible Property

CHAPTER 19

HIGHLIGHTS

■ Real Property can be acquired by
  ■ will or inheritance,
  ■ purchase,
  ■ gift, or
  ■ legal action.

■ Personal property can be acquired by
  ■ original acquisition, for example, by possession or creation;
  ■ procedure of law, such as by death, foreclosure, or bankruptcy; or
  ■ acts of others, such as by will, gift, or contract.

All of us have things that we consider to be ours: our clothing, books, writing instruments, watches, perhaps a home in the suburbs. These things are our **tangible property.** However, the word **property** can be used in two senses. In its usual sense, it denotes **things** owned by a person. In a broader sense, however, it also refers to the rights involved in ownership. These rights, known as **property rights,** signify dominion over the things owned. That is, the right to use and to exclude others from using the things we own, the right of control over and enjoyment of them, and the right to dispose of them.

Tangible property may be further classified as real or personal. In the discussion of the statute of frauds in Chapter 11, we distinguished between real and personal property. Generally, **real property** has been defined as land and anything firmly attached to it. **Personal property** is all property other than real property, such as goods, chattels (defined momentarily), choses in action (also defined below), money, and accounts receivable or other evidences of debt. Sometimes, however, personal property is treated as real property; this is called **mixed property.**

REAL PROPERTY

Real property has been defined as land and anything firmly affixed to it. When an engineer becomes a party to building a road or renovating a manufacturing plant, he is concerned with real property. As a citizen in a community, the engineer will either own real property or lease it. As an investor, an engineer may speculate in real estate, since such speculation has become nearly as popular a sport as speculation in the stock market. Because engineers deal with real property in both their personal and professional lives, we will consider here the transfer (or conveyance) of real property and some of the rights and duties created.

We may buy real property, use it pretty much as we please, and transfer it to others with very few restrictions involved. Only when title is threatened or, perhaps, when prescriptive rights (that is, the right to continue using another's property that arises out of longstanding use of that property) are exercised against real property do we become aware that any limitations exist.

Evolution of Property Rights

Our property laws involving real property developed over the centuries in England. When William the Bastard won the Battle of Hastings in 1066, he set about establishing control over England. He accomplished this feat partly by asserting absolute claim to all English land, then parceling it out to those who had helped and supported him. Tied to the land granted by the king was the obligation to serve him.

Under this feudal system, then, the land belonged to the Crown. The right to hold realty depended upon military service and fealty to the king, and was theoretically terminable at his option. Land was parceled out to the gentry who, in turn, divided it up among their servants. Originally, when a landholder died, another person was appointed to take his place. Very early this practice was replaced by provisions that tended to ensure that the property would remain in the family of the grantee or tenant, provided the heir was able to meet the military obligations entailed. The system of estates developed in England is essentially the system that generally remains in place there today.

Kinds of Estates

Estates in real property can be classed as freehold, less than freehold, and future estates.

Freehold. A **freehold estate** is an estate of undetermined duration in real property. It may be an estate in fee simple or it may be an estate for life. An **estate in fee simple** is the highest real property estate known to law. The holder of an estate in fee simple has the right to complete use and enjoyment, as long as this use does not harm another, and the right to transfer the estate to anyone he or she chooses. Upon death, the holder's estate will be distributed according to his or her will, or according to law if the holder dies intestate.

A **life estate** is an estate of undetermined duration and, therefore, a freehold estate. The life upon which the term of the estate depends is usually that of the holder of the estate, but it could be that of anyone else. An estate that is terminable upon some contingency other than death but is, in some way, dependent upon the duration of a person's life is treated the same as a life estate. For example, Black gives an estate to a young widow, terminable when she remarries. She might die without remarrying, and this would also terminate the estate. It is, therefore, treated in law the same as a life estate, but with marriage as an added contingency.

The rights in a life estate are not so complete as they are in a fee simple estate. The holder of a life estate may not sell it to another. Though the holder is allowed to use and enjoy the real property, he or she may not destroy its value. For instance, the holder may not sell the topsoil or remove ornamental trees, although he or she could cut and sell ripe timber.

Less than Freehold.   An estate in real property that is to run for a fixed or determinable time is **less than freehold.** In law it is considered as personal property. Thus, a 10-year lease or a grant of property to run "as long as the property is used for educational purposes" would be less than a freehold estate. By contrast, a lease for "99 years, renewable forever" would be a freehold estate, since the duration is undetermined.

Future Estate.   A **future estate** is an estate that someone will have when a future event occurs. According to Gray's will, for example, White obtains Grayacre (Gray's estate) for the duration of White's life and, upon White's death, Grayacre will pass to Black. Black thus gets the estate left to White when White dies. Black has a future estate in the property involved.

TRANSFER OF REAL PROPERTY

Real property and real property rights are transferred in four major ways:

1. By will or inheritance,
2. By sale,
3. By gift, and
4. By legal action.

We will consider the documents required in these transfers and the principles that govern them in the following pages.

Will

Originally, the word **will** indicated a disposal of real property only; another document, a testament, disposed of the testator's personal property. Thus the use of the phrase **last will and testament** came to be popular when the testator wished to combine the two functions in one document. By common and legally accepted usage, the term **will** today indicates a document that provides for the disposition of both real and personal property. Below, we will consider some aspects of will-making.

Age.   Under common law, anyone of sound mind and 21 years of age can make a valid will in the United States. Most states, however, have passed laws that reduce the age requirement.

Mental Capacity.   A person without his or her proper mental faculties (that is, someone who is insane) cannot make a valid will. The law does not require a towering intellect as a testator, however. The law requires only that the testator have (a) sufficient mental capacity to comprehend his or her property, (b) capacity to consider all persons to whom he or she might desire to leave property, and (c) understanding that he or she is making a will. In these requirements, nothing prohibits an eccentric person from making a will. A person physically or mentally ill may make a will.

Even an insane person could, in his or her rational moments, make a valid will. (See *Turja v. Turja* in Chapter 6.)

Who May Inherit.   Inheritance is not limited to relatives of the deceased under a valid will. Almost anyone may inherit. Municipalities, universities, and charitable organizations often have benefited from the terms of wills. The law, however, usually prevents inheritance by the murderer of the testator.

Similarly, the testator has a right to disinherit as he or she chooses within the limits of the state statutes. Still, complete disinheritance of the testator's husband or wife cannot be done successfully in most states, and provision for any minor children may be required.

Types of Wills.   In addition to the ordinary written will with witnesses, two other types of wills are recognized in some states. The first, the **holographic will,** is one that is written entirely in long-hand by the testator. Usually no witnesses are required. Where the holographic will is recognized—in about half the states—it has the same standing as any other will.

The second type, a **nuncupative will,** is an oral will. Such a will usually must be made before a certain number of witnesses, and they, in turn, must reduce it to writing shortly thereafter. The testator cannot will real property to another orally. In fact, in the states where nuncupative wills have legal standing, a limit is usually placed on the amount of personal property that may be willed orally.

Witnesses.   A witness to a will should be a disinterested party capable of being a witness in any judicial proceeding. Generally, a minor can act as a witness. A person who stands to gain or lose by the will, however, may be incompetent as a witness. For this reason, it is usually preferred that a witness have no stake in the property being disposed of by the will.

Essentials of a Valid Will.   Excluding holographic and nuncupative wills, there are four common requirements for a valid will:

1. To pass the testator's real and personal property along to others, the will must be in writing. The law does not require any special kind of writing, such as typing or long-hand, so long as the will is written. Neither is there a requirement as to the material on which the writing appears. A will chiseled in stone or etched on glass could be as legally binding as one drawn up on a form prepared by an attorney.
2. A valid will must be signed by the testator and notarized. The signature normally appears at the end of a will, and its validity may be open to question if it appears elsewhere.
3. Wills (except holographic) must be witnessed. State laws usually require either two or three persons to attest the signing of a will. The will must be signed in the presence of the witnesses. Depending upon the state's law, the witnesses may or may not need to sign in the presence and sight of each other and of the testator.
4. A will must be published. In connection with wills, the word **publication** means something different from its ordinary sense. Publication of a will occurs when the testator declares that this is his or her last will and testament. The witnesses must know that it is a will being signed; they need not necessarily know the terms of it, only that it is a will.

Codicils.   A codicil is a supplement to a will, such as an amendment. It is used to explain, modify, add to, or revoke a part of an existing will. If there is any question as to the date of the making of the will, the time when the last codicil was drawn is the effective date of the will. The making of a codicil to a will requires the same formality as is required in making a will.

Probate.   After the death of the testator, the will may be presented for probate to a probate court. State statutes determine the next steps, but the state laws follow a general pattern. Opportunity is given to question the validity of a will. It may be held invalid for fraud, undue influence, improper execution, forgery, mistake, or incapacity of the testator. If the validity is unchallenged or any challenges attempted are unsuccessful, it is admitted to probate (i.e., received by the court as a valid statement of the testator's intent). If the will is successfully challenged, a previous will may be reinstated or, if no previous will exists, the result is the same as intestate death.

In drawing up a will, it is customary to name someone as **executor** or **executrix.** Upon the death of the testator and probate of the will, the executor is called upon to carry out the terms of the will. This usually must be done under bond unless the testator has specifically exempted the executor from bond. If no will was left or if no executor was named in the will, the court will appoint an administrator or administratrix. The functions of an administrator are similar to those of an executor. However, if the decedent left no will, the administrator must follow state laws for distributing property following intestate death. In many situations, family members are named as executor or executrix. This naming does not prevent them from also being a beneficiary.

Deed

Conveying real property by sale or gift requires a **formal transfer.** Each state has jurisdiction over the real property within its boundaries. Each has set forth the formalities required to convey ownership from one person to another. Although the statutes vary from state to state, there is a general uniformity in the requirements.

Kinds of Deeds.   Two kinds of deeds are in common use in the United States today: warranty and quit-claim. On certain occasions, either may be used, but the better title is obtained in a warranty deed.

A **warranty deed** warrants that the title obtained by the grantee is good. In any deed, the grantee gets only the title that the grantor has to give; but if a grantee's title under a warranty deed is ever successfully attacked, the grantee may recover any damage suffered from the grantor. By granting a warranty deed, the grantor usually warrants three things:

1. That he or she has good title and the right to convey it;
2. That the property has no encumbrances other than those mentioned in the deed; and
3. That the grantee and his or her heirs or assigns will have quiet, peaceful enjoyment of the property conveyed.

A **quit-claim deed** transfers title but does not warrant it. In effect, it is the conveyance by the grantor of whatever title he or she may have to the property. Such a deed might be used where inheritance of the property by several members of a family sometime in the past has left a clouded title.

Essentials of a Deed.   For a deed to be valid, it must usually include several essential elements. The deed must:

- Name the grantor and grantee as well as any consideration involved
- Describe or otherwise identify the property
- Use words of conveyance transferring ownership from the grantor to the grantee
- Be signed, sealed (in some states), witnessed, delivered and accepted by the grantee

Description of property within a city is likely to be by lot number and plat. Rural property may be described according to metes and bounds, or in a few jurisdictions, by the Torrens System of sections and fractions. In interpreting a deed, the court will endeavor to carry out intent of the parties even when there is an error in the description. Corner markers or monuments and natural landmarks may show this intent better than descriptions, since these can be seen by the parties. Thus, the presence of such a marker may cause the court to disregard the technical description.

Recording.   The law requires any proceeding that affects real estate to be recorded as a notice to the public. Thus, to have full force or standing at law, a deed, mortgage, lien, attachment, or other encumbrance must be filed at the local recorder's office or registry of deeds.

The recording of a deed does not pass title to the property; the making of the deed takes care of that. Still, it is essential to record the deed, because the states have various "recording" statutes that adopt and give effect to the idea that a subsequent buyer may obtain good title. Thus, if the grantor were to make a second deed to a second grantee fraudulently, the first grantee may lose the property if the second grantee purchased the property in good faith and without knowledge of the first grant. By recording the deed, the first grantee essentially tells the world of the first grant. Because the grantee is at risk, it is the grantee who makes sure the deed is recorded.

> Recording statutes may be "race", "race-notice", or "notice" which indicate priority as to deed notice and recording requirements as they relate to subsequent bonafide purchasers for value.

Title Search and Title Insurance.   A title search involves following the changes in title to a piece of property from the initial grant from the state to the present. The result of the search should show an unbroken chain; a break is cause for suspicion and further search. For example, a will leaving the property to more than one person may be questioned. Any encumbrance on the property is also questioned, to determine whether it has been cleared up. Of course, a search may have to stop at some point. If, for example, because of destruction of records or some other reason, ancient title cannot be cleared, it will usually be certified despite the void.

In most communities title insurance may be purchased to warrant title to real property. The title insurance company will search the title and issue insurance for a one-shot fee based on the outcome of the search.

Mortgages

Most buyers of real property do not have sufficient assets to pay cash for their real estate; therefore, they must borrow the money. They may borrow the money on a personal loan or a note, but the problem of securing the loan exists. For example, Black borrows $100,000 from White on a note, the money to be used to buy a house and lot. If Black, at some future time, cannot pay an installment on the note, White may obtain a judgment in court for the remainder of the note. However, under most state statutes, Black's homestead and much of his personal property would be protected from execution or attachment. Thus, White would have little real security.

In contrast, a mortgage offers the lender substantially greater security. A real estate mortgage is a contract between the **mortgagor** (the borrower) and the **mortgagee** (the lender). The mortgagor borrows funds from the mortgagee, perhaps for the purchase of real estate, promising to return the money with interest, and offering the real estate as security.

Mortgage Theories.   Mortgages began under common law as **defeasible conveyances.** That is, the mortgage took the form of a deed from mortgagor to mortgagee. The mortgagor could defeat the deed (get it back) by paying the loan on which the mortgage was based in the time specified. The mortgagor who missed a payment, however, had nothing. Any default gave the mortgagee absolute right of ownership. The results seemed rather harsh, and the treatment has since become more lenient.

In equity jurisdiction, where mortgage foreclosures are normally handled, certain mortgagor's rights have come to be recognized. Where prior common law considered the mortgagee as owner, allowing the mortgagee to collect rents and profits from the property, the mortgagor now usually has the rights of ownership. A missed payment no longer terminates forever the mortgagor's right; usually the mortgagor has a certain time in which to redeem the property under a right known as **equity of redemption.**

In equity, the view is taken that the mortgage is security for a loan—that it is nothing more than a lien. This lien theory now constitutes the accepted reasoning on mortgages in the majority of the states. Some states, however, still hold to the older common law ideas in modified form, under the **title theory of mortgages.**

Formality.   Since a real estate mortgage represents an interest in real property, it must be in writing, pursuant to the state's Statute of Frauds, discussed in chapter 11. The form of a mortgage, even in lien theory states, is similar to that of a deed, and the same formal requirements usually pertain to both. The mortgage usually must be signed, acknowledged, witnessed, and recorded. When the mortgage is satisfied, this too must be recorded. Recording of the mortgage and its discharge serves as notice to the public of this type of property encumbrance.

Mortgagor's Rights.   The presence of a mortgage on real property does not, of course, prevent reasonable use of the property by the mortgagor or disposal of it subject to the mortgagee's rights. The mortgagor may use and enjoy the mortgaged property in whatever way he or she wishes, as long as the use does no harm to another, including the mortgagee.

The mortgaged property may be willed to another, sold, or given away. If the grantee takes the property merely subject to the mortgage, he or she will not be held to have assumed personal liability. In other words, in case of default, a deficiency judgment might still be obtained against the original mortgagor. On the other hand, if the grantee assumes the mortgage, he or she should be held to replace the mortgagor in all respects. Hence, a seller of property is usually careful to be sure that he or she has no further obligation if the buyer later defaults.

Mortgagor's Duties.   Security is the reason for a mortgage. While the mortgagor is owner of the mortgaged property, his or her right of ownership must be somewhat restricted. The mortgagor cannot, for example, tear down all buildings, sell off the trees and topsoil, and then allow the mortgagee to take over the worthless remainder. To do so would diminish the mortgagee's security and constitute waste. If the mortgagee's security is so threatened, he or she probably has reason to institute foreclosure proceedings.

As the loan balance declines, the mortgagor's right to unlimited use and disposal of the property increases, since less security is required to protect the mortgagee's interest.

The mortgagor usually must pay all taxes, assessments, and insurance on the mortgaged property. Unpaid taxes and assessments become liens, and endanger the mortgagee's security. Insurance on the property protects mortgagee and mortgagor alike. In case of near total destruction, the mortgage balance is usually paid first, and any remainder goes to the mortgagor.

The mortgagor must, of course, make the mortgage payments when they are due. Nonpayment constitutes default and gives the mortgagee the right to institute foreclosure. Generally, the mortgagee need not accept early payment of the balance due; he or she has a right to the interest contracted for. The standard FHA and Veteran's Administration mortgages have provisions for early payment, but many other mortgages do not.

Mortgagee's Rights.   The mortgagee generally may assign the mortgage note, together with the mortgage, to a third person, who then has the same rights as the mortgagee, or "stands in the mortgagee's shoes."

In case of default in payment or a diminishing of the security, the mortgagee may foreclose. Generally, foreclosure proceedings begin with the filing of a bill in equity. The bill outlines the mortgagee's rights and the mortgagor's breach of the agreement. If foreclosure is allowed, the court will appoint a master to sell the mortgaged property. The purchaser gets a deed to the property. Court costs are paid first from the proceeds, then from the mortgage balance. If anything from the sale remains, it is returned to the mortgagor.

If the foreclosure sale of the property does not return enough to pay off court costs and the mortgage, the court may issue a **deficiency decree.** Such a decree holds the mortgagor personally liable for the unpaid balance of the obligation. Enforcement by execution or attachment will be likely to follow. The courts lately have exhibited some reluctance to issue deficiency decrees, however. This is especially true where the mortgagee loaned the money for the purchase of real property. The reasoning goes this way: The mortgagee loaned the money on the security of the realty and not on the mortgagor's personal credit. In other words, the mortgagee must have evaluated the risks involved and the mortgagor's security when the loan was made. Recovery, then, should be limited to the price this property will bring; part of the value of the interest charged is payment for risk.

Mortgagee's Duties.   Although the mortgagee may assign the mortgage note, together with the mortgage to a third person, the mortgagee must notify the mortgagor of the assignment if the mortgagor is required to pay the assignee. The mortgagee must, of course, notify the mortgagor of the assignment if the mortgagor is required to pay the assignee. If the mortgagor, lacking knowledge of the assignment, pays the original mortgagee, the mortgagor will diminish the amount of the note by the amount of the payment.

Land Contracts

An arrangement sometimes used in the purchase of real property is the **land contract.** It resembles a mortgage quite closely under the title theory. According to a land contract, the purchaser, in addition to making a down payment (if any), agrees to make a series of payments. When the buyer reduces the balance to some agreed amount, frequently half the purchase price, the seller will deed the property to the buyer and take a first mortgage.

Land contracts typically include the right of the seller to declare all payments due immediately if the purchaser misses a payment or makes a late one. This, of course, has the effect of forcing forfeiture by the buyer if he or she misses one payment, since the buyer can rarely pay the entire balance in such a situation. If the buyer gives back the land upon default, the seller takes it back with no problem of foreclosure proceedings and public resale. The seller is still owner and merely takes over the property.

This removal of the buyer and repossession of property by the seller with no balancing of the equities involved is known as **strict foreclosure.** Strict foreclosure is allowed in connection with both mortgages and land contracts where the prospective buyer has acted very improperly or is insolvent. State laws govern the handling of land contracts as well as mortgages. Where the buyer has acted in good faith under a land contract and has made only a slight default, the equities will usually be balanced in some manner—but the buyer must ask for such relief. If the buyer merely returns the land to the seller, the buyer gives up any chance to recoup a part of the loss; in other words, the buyer's equity is cut off.

Eminent Domain

A private party has a lesser right to possession and use of property than does a public body. The government's right to take private property for public use is known as **eminent domain** (or **condemnation**). The right of eminent domain may be exercised by the state, a municipality, or another public entity, but the right is not limited to these bodies. Quasi-public enterprises or private businesses whose functions serve the public at large (e.g., railroads or power companies) also may have this right.

However, the United States Constitution prohibits a taking of private property without just compensation. Compensation is usually made according to an assessment of the market value of the property taken. (See *Fairfax v. Virginia Department of Transportation* in this chapter).

Dedication

When land is required for public use such as a road or school playground, many owners will donate land for the purpose. Such donations are known as **dedications.** Although dedications are usually made expressly by the owner to the public officials involved, they may also be implied. For example, if public use of private property is made continuously for a period of 20 years or longer, dedication may be conclusively presumed.

No formality is required in the offer to dedicate a piece of property to public use. Similarly, no formality is required for acceptance. To complete a dedication, however, there must be some kind of acceptance. When the offer is expressly made, it is usually answered by expressed acceptance. Acceptance may be implied, however, from public activities, such as maintaining the dedicated property. Public maintenance of a privately owned but publicly used road, for instance, would indicate acceptance of the road.

It may seem unnecessary to determine whether a particular piece of property has been dedicated to the public use. The question of tort liability, however, makes title determination important. If someone is injured because of a large hole in the road, the question arises as to who owned the road and who, therefore, had the duty to maintain it.

Adverse Possession

Title to real property may be acquired by adverse possession. Legal requirements generally make it quite difficult to acquire title to land in this manner; however, if the requirements are met, a new title to the land is issued to its possessor. That is, adverse possession results not in a transfer of present title, but in an entirely new title being issued.

The right to take title by **adverse possession** results from the theory in law that doubt and uncertainty as to title to anything should be removed. Reasoning from this, owners should be reasonably diligent in defending their rights.

There are four general requirements to take title by adverse possession:

1. Possession must be open, and it must be notorious, actual occupation. The possessor must occupy the realty in the same manner as one might expect of the true owner.
2. The possession and occupation must be adverse to the interests of the owner. Thus, a tenant or lease-holder could not obtain title to the realty by adverse possession.
3. The adverse possession must be continuous over the statutory period required. For example, if the state statute requires 20 years (as a large number do), two 10-year periods separated by a period when the property was occupied only by the true owner would not suffice.

4. There must be either a claim of right or color of title by the possessor. **Claim of right** is interpreted from acts of the possessor such as improving the land or fencing it. **Color of title** is some symbol of claim of ownership that is in some way defective. Suppose that a former owner deeded the property to two different purchasers. Each would have color of title due to the deeds. Payment of taxes on the property possessed is required by some statutes to show claim of right or color of title.

For further discussion of these elements, see *Abood v. Johnson* in this chapter.

Prescription

Gaining **prescriptive rights** is just about the same as gaining title from adverse possession. The only major difference in most states is that prescriptive rights do not give a person ownership of real property. Instead, prescription deals in rights involved with real property, particularly easements. As with adverse possession, the use of the property must be open and notoriously adverse to the owner. For example, Black, for many years (a sufficient number according to the state statute), has crossed White's land to get to his own. Black has established an easement by prescription. He does not own the path across White's property, but he has the right (called an easement) to continue to cross it in the same manner in which he is accustomed to crossing it.

PERSONAL PROPERTY

The term **chattels** is often used synonymously with personal property. **Chattels personal in possession** are the tangible items (e.g., a watch, a truck, or a machine in a factory). **Chattels personal in action,** commonly known as **choses in action,** are intangible rights arising from a tort or contract—the right to goods contracted for or the right to recover for injuries suffered in an automobile collision, for example. A **chattel real** is an interest in real property, such as a 10-year lease.

Personal property also includes intangible property rights. For example, patents, copyrights, and trademarks are considered personal property. These property interests are considered in Chapters 20 through 23.

Acquisition

A person may lawfully obtain ownership of personal property by

1. Original acquisition,
2. A procedure of law, or
3. Acts of other persons.

Original Acquisition. Unowned things in their natural state become the property of the first person to obtain possession of them. Most things are owned by someone today, but, for example, the possibility of reducing wild animals to personal property still exists. Obtaining ownership by taking possession of such an animal with the intention of becoming its owner is known as **acquiring title by occupancy.**

Generally, property that one creates by mental or physical efforts belongs to the creator unless there is an agreement to transfer it to another. Books, inventions, trade names, and other such creations are of this nature. Even where no formal agreement exists, however, the results of such mental or physical efforts may belong to the creator's employer.

Property may be acquired by **accession**—by adding to other property. Generally, an owner of property owns what the property produces and what is added to it. A new windshield of a car or a gear in the transmission, for example, becomes the property of the owner of the automobile. This is particularly true where the addition becomes an integral, built-in part of the whole in such a way that it is not readily detachable. Even where the innocent purchaser of stolen property adds value to it, he or she is merely adding value to property belonging to another.

For example, White buys a car from Gray who, unknown to White, has stolen the car from Black. White adds a new motor, transmission, and paint job to the car. Later, Black locates the car. Black is entitled to regain possession of the car in its improved state. Probably there is no other place where the law adheres so strictly to the principle of *caveat emptor* (the buyer beware) than in the purchase of stolen property.

Accession also applies to a natural increase of purchased property. Here, for example, Black sells White a mare. Shortly after the sale, a foal is born. White is the owner of both the mare and the foal.

Procedure of Law.   Property may be distributed according to certain legal procedures. Examples of such procedures include: Distribution via an intestate death, Mortgage foreclosure, Judicial sale, and Bankruptcy.

Intestate. When a person dies and has not left a will, his death is termed **intestate.** The various states have, by statute, declared how property shall be distributed in case of such intestate death. Such statutes are called the **laws of descent.** These statutes vary considerably as to who will inherit the estate. If no relatives of the deceased can be found, the property will go to the state—in legal terms, it **escheats** to the state. Even when a person leaves a will, that person is limited somewhat in the way she may leave an estate to heirs. A person may not, according to most state laws, leave a spouse or minor children destitute by willing an entire estate to strangers.

Mortgage Foreclosure. Other property-related statutes provide for **mortgage foreclosure** when a mortgagor defaults. Although **chattel mortgages** commonly specify that the mortgagee will take and sell the mortgaged property in case of default, a court procedure is usually possible if such provision has not been made.

Judicial Sale. Sales of property may also be undertaken to satisfy a judgment of a court. If the loser fails to voluntarily pay the judgment, the winner can have certain of the loser's property seized (within limits stated in the laws of each state) and then sold to satisfy the judgment.

Abandoned Property. When abandoned property is found by the police or sheriff's department in a community, it is kept for a statutory period of time and then sold at public auction. Such sales usually must be advertised and public, with the property going to the highest bidder.

Bankruptcy. Under bankruptcy procedures, a trustee may be appointed by the court with the duty to convert the assets of the bankrupt into money. The trustee may take over the property with the right and duty to sell it. Generally the buyer of such property does not get any better title than the seller had.

In judicial sales, sales of abandoned property, and bankruptcy sales, title is usually not warranted by the seller. The seller sells by virtue of a legal right or duty to do so; the buyer assumes the risk that title may not be good.

Consider this example. Black steals White's car and abandons it in a neighboring town. The local police department holds it for the required period of time and then sells it to Gray at a public auction. Later, White finds his car in Gray's possession. White can claim and get his car.

Acts of Other Persons.   A person may lawfully acquire title to personal property from others by will, gift, contract, confusion, or abandonment. A person may also acquire possession of property if it is lost or mislaid by another.

The subject of wills was discussed briefly above and will be discussed further under real property. It is sufficient to note here that testators who comply with the law may leave their property to whomever they wish.

> A gift may be recoverable when the property was stolen from another, or when the donor anticipated impending bankruptcy, or when the donor was dying and diminished the property that would go to his or her heirs.

If a person acquires property by gift, title to the property follows possession. For example, Black promises White a gift of $500. At this point, White has nothing. The promise of a future gift, either oral or in writing, is unenforceable, since it is unsupported by consideration. Of course, if the written promise of a gift were signed and sealed, a consideration may be imputed under common law, and the promise would be enforceable, in those states that still recognize the validity of seals. As soon as Black actually gives the $500 to White, however, it becomes White's property, and Black loses any claim to it. Under a few exceptional circumstances, the gift may be recoverable, particularly when a third party has rights in the gift.

The subject of acquisition of personal property by contract is covered in Chapter 15, Sales and Warranties. Good title to personal property is warranted in any sale unless there is a disclaimer of the warranty in the contract.

Property may be acquired by what is known as **confusion,** primarily when fungible goods are involved. **Fungible goods** are goods any unit of which is replaceable by any other unit—for example, grain of a particular type, crude oil, or screws in a bin. Such goods are usually sold by weight or measure. If fungible property of two or more owners is mixed together so that the identity of each owner's property is lost, each owner owns an undivided share of the confused mass. For example, after harvest, Black stores 500 bushels of wheat with 700 bushels of wheat belonging to White in a common granary. Each owns an undivided share of the 1,200 bushels of wheat. Destruction of a part of the mass will be shared by each party on the basis of that party's contribution to the total.

If confusion of goods results from the tortious act of one of the parties, the innocent party will be protected. If ownership by the **tortfeasor** (that is, the person who committed the tort) cannot be determined, the innocent party becomes owner of the total.

Abandoned property is unowned. It becomes the property of the first person to take possession of it. Taking possession of abandoned property is about the same as taking possession of something that has never been owned.

Lost and mislaid properties give rise to some legal problems. In each case, the owner has unintentionally parted with possession of his or her property. In each case the owner still owns the property even though it is no longer in his or her possession. The finder of lost property has a right to the property against all persons except the true owner. By contrast, the holder of mislaid property has possession of it as a bailee—in other words, he or she is holding it to give to the owner.

The distinction between lost and mislaid property is derived largely from the circumstances in which it is found. If the property is found in such a location that it is apparent that the owner intentionally placed it there and then inadvertently left it, it is mislaid. A purse left on a store counter would be mislaid; if it were found on the floor, it would have been lost.

In many states, the problems involved in lost and mislaid property have been cleared up by statute. The requirements of the statutes usually are met by advertising the property in a local newspaper. If no one claims the property within a certain time after the publication of the advertisement, the finder obtains title to the property.

Bailment
— leave car at mechanic
— leave clothes at dry cleaner
↳ not the same as bail from jail

A relationship that closely resembles property ownership is that of **bailment.** The bailment relationship occurs when personal property is left by the **bailor** (the owner of the property) with a second person, the bailee.

Distinguishing bailment from similar relationships requires careful definition. Generally, a bailment is made up of three elements:

1. Title to the property remains with the bailor.
2. Possession of the property is completely surrendered by the bailor to the bailee.
3. The parties intend the return of the bailor's property at the end of the bailment.

Notice the similarity between a bailment and a sale or a trade with a slight delay in it, as in this example. Black stores a spare conveyor at White's warehouse for an agreed period and a fee. Black still has title to the conveyor. White has possession, and the same conveyor is to be returned to Black. Therefore, it is an instance of bailment. With certain other types of property, however, an inherent difficulty exists. If, instead of a conveyor, Black were to store grain in a common granary, he might not expect to get back the identical grain that he stored. The same might be true of animals in a herd and a few other instances where the owner does not expect that the identical property will be returned. Courts are not uniform in all jurisdictions in their holdings under such circumstances. Generally, however, if the owner is not to receive back the identical thing given, it is not bailment. It is held that title passed with possession and that title to other similar goods will be passed back later. In other words, such a transaction represents a sale.

It becomes necessary to find out who owns what when one party or the other goes bankrupt, or a writ is issued pursuant to a judgment against someone's property. Property being held for a bailor by a bailee generally cannot be successfully taken for the bailee's debt. It is possible, however, if a judgment were to be issued against the bailor, to obtain the property from the bailee.

Duty of Care.   The person entrusted with the property of another has a duty to care for it. Under a particular set of circumstances, the degree of care may be great, ordinary, or slight. Two primary considerations determine the necessary degree of care: (a) the nature of the property involved, and (b) the purpose of the bailment. As to the nature of the property, it is obvious that a person should take greater care of a new computer than of a used anvil. The bailment relationship benefits someone—the bailor, the bailee, or both. If the bailment is to benefit the bailor only, the bailee need exercise only **slight care** in protecting the bailed property. The bailee is liable only if he or she has been grossly negligent. Such a bailment might occur as a result of the owner requesting a friend to care for his or her property gratuitously.

If property is borrowed for the benefit of the bailee (as one would borrow a neighbor's lawn mower), **great care** is required of the bailee. The property must be returned in the form in which it was borrowed. About the only damage for which the bailee would not be responsible would be that resulting from an act of God—such as destruction by a cyclone.

Probably the most common form of bailment occurs as a benefit to both bailor and bailee. Whenever the bailor pays the bailee to take goods and alter them in some way, or pays the bailee just to store them and then return them, both parties benefit. Thus, **mutual-benefit bailments** would occur in the following situations: one party leaves a car at a garage with orders to fix the transmission; one party asks another to transfer a machine from Cleveland, Ohio, to Fort Worth, Texas; and one party stores an unused machine at another's warehouse during a slack period.

When the bailment is to benefit both parties, the bailee is required to use at least **ordinary care**. Ordinary care means the care that a person would be likely to use in preserving his or her own property. The bailee is liable for damage resulting from his or her negligence. If the bailee has used the requisite amount of care in preserving the bailor's property, the bailee will not be held liable for damages. Loss of the property or damage to it, then, will follow title and be borne by the bailor.

Bailee's Right and Duties.   The bailee's right to possession of the bailed property is second only to the bailor's right. The bailee may sue a third party to recover the property if necessary. This rule reflects a general presumption of property law: The person in possession of the property is its owner.

If the bailee gives the bailed property to someone other than the owner or the owner's agent, and it is thereby lost, the bailee is liable. However, if the bailee retrieves the property and returns the property to the person who gave it to him or her originally, assuming that that person is still the owner, the bailee should not be liable, even though the property left the bailee's hands.

The bailor has a duty to disclose any known defects in the bailed property that might harm the bailee or the bailee's employees. If harm results from a failure to disclose such defects, the bailor may be held liable for tort. For example, consider what a bailor of steel drums containing toxic chemicals or radioactive waste should disclose to the warehouse acting as a bailee.

Negligence Liability.   Most bailments are contracts. It is possible, therefore, for the bailee, by contract clauses, to remove any or all liability for negligence, but only if he or she is a private bailee. For a public or quasi-public bailee—a hotel or trucking company that offers its services to the public at large, for instance—to make such a contract stipulation probably would be unlawful, especially if the stipulation conflicted with statutes governing such relationships. Such a bailee will be liable for negligence when serving the public, regardless of contract clauses to the contrary. If a private bailee insists on eliminating liability, the bailor has only two choices—to do business with the bailee on those terms or do business elsewhere.

↳ parking garage not responsible for lost/stolen/damaged property

Fixtures

A **fixture** is personal property that becomes attached to real property. By the attachment, the personal property becomes a part of the real property. The concrete and other building materials that are worked into a plant become real property. Similarly, a heating unit or a television aerial becomes a fixture when it is attached to a house.

Generally, ownership of a fixture goes to the owner of the real estate to which it is attached. There are so many exceptions to this generality, however, that the principle might be restated this way: Unless something appears to the contrary, the owner of the real property also owns the fixture. In *Manderson & Associates, Inc. v. Gore,* at the end of the chapter, the intent of the parties determines whether the fixture is part of the real property. The main condition to the contrary is the intention of the parties when the fixture was attached. If it appears that both parties intended that ownership of the fixture should not go to the owner of the real property, the original owner will have a right to remove it. For example, Black Construction Company undertakes the building of a structure for the White Company. A small building, complete with plumbing and lighting, is erected on the premises as a superintendent's office. Although the superintendent's office is firmly fixed to the ground, it may be removed at the end of the project.

The relationship of landlord and tenant often involves the determination of ownership of fixtures. When real property is leased or rented, the tenant normally may install personal property (e.g., machines or conveyors) and then take them when he or she leaves. If the tenant fails to remove and take a fixture, ownership goes to the owner of the land. For example, Black rented a house from White. Requiring hot water, which White's house did not have, Black bought a suitable water heater from Gray Appliance Store, paying 10 percent down. Black installed the water heater, hooked it up to the plumbing, used it for a month, and then moved to another state. Gray Appliance tried to get the water heater back, but could not, since it was now part of White's real property. Gray's only available action is against Black, and he may be hard to reach.

FORMS OF OWNERSHIP

Just as there are many types of property, there are many forms of legal ownership of property. Because engineers may become involved, either personally or professionally, in transactions involving property (whether a house or a patent), the following discussion outlines different types of property ownership.

Trusts

Titles to both real and personal property may be involved in a trust relationship. **Trusts** involve at least two people—the **trustee,** who either holds or sells property, and the **beneficiary,** who is to benefit from the trust. There are two titles to trust property. The trustee has legal title, with the right to sell or otherwise use the property involved. The beneficiary has equitable title to the property, since the trustee is required to handle the property involved for the benefit of the beneficiary. Equitable title is regarded by equity as the real ownership, even though legal title is vested in someone else.

Rights in Common

Ownership of property by several persons can arise under at least five situations:

1. Partnership,
2. Joint tenancy,
3. Tenancy in common,
4. Tenancy by the entireties,
5. Community property.

In each situation, the rights of more than one person are involved in any property dealings.

Partnership.   A **partnership** is an association of two or more persons who agree to carry on a business as co-owners for profit. People often join their assets to more effectively carry on an enterprise. Each partner has rights in the partnership property and in other property acquired by the partnership, and each has attendant liabilities. Every partner has a right to act as agent in the business of the firm, thus adding to or disposing of assets in the transactions for which the partnership exists.

As noted in chapter 18, partners in an enterprise have what is known as unlimited liability for the debts of the partnership; each stands to lose some or nearly all of his or her personal fortune if the enterprise folds. However, for many purposes, the property of individual partners is separate from that of the partnership. When a solvent partnership is dissolved—possibly because of the death of a partner or agreement to dissolve—each partner has a claim to a share of the partnership assets, but no claim upon the property individually owned by other partners. If a partnership becomes bankrupt, the firm's creditors have first claim upon the partnership assets; the creditors of the bankrupt partner have first claim against his or her individual property.

Joint Tenancy.   A **joint tenancy** is created by a will, deed, or other instrument naming two or more parties as joint tenants. Under joint tenancy, each tenant has an equal, undivided interest in the property and the right to use the property. A joint tenant may not exclude the other joint tenants from it. The right of survivorship is a main feature of joint tenancies. In a joint tenancy with **rights of survivorship,** a joint tenant's interest automatically passes upon death to the surviving joint tenants.

A joint tenant may sell his or her share in the property to another, but the buyer then becomes a tenant in common (defined below) with the remaining joint tenants. The buyer is a tenant in common, but the remaining joint tenants are still joint tenants. Since a right of survivorship acts before a will does, a joint tenant cannot successfully leave an interest in the property to his or her heirs.

Tenancy in Common.   A **tenancy in common** is about the same thing as a joint tenancy, except that it includes no right of survivorship. A tenant in common may leave an interest in the property to heirs or sell it to someone. The result is merely a substitution of one or more tenants in common.

Each tenant in common has an undivided part interest in the whole property. Each is entitled to a proportionate share in possession, use of, and profits from the property. If one tenant pays property costs, say taxes, that tenant has the right to contribution from the others.

Tenancy by the Entireties.   **Tenancy by the entireties** might be thought of as a special case of joint tenancy. The relation is created by conveying to husband and wife in a conveyance that states "by the entireties." Neither husband nor wife can destroy the relationship without consent of the other. This type of ownership applies only to spouses.

Community Property.   Some nine of our states have a somewhat exceptional treatment of property owned by husband and wife. In those states, property acquired by a couple after their marriage is known as **community property.** Each has an equal share in it. Upon the death of either party, the other is entitled to at least half the community property, or to the entire amount if there is no will to the contrary.

As noted, community property is property acquired after marriage. It is possible, however, for either husband or wife to have and acquire separate property either before or after marriage. The property that each has when they marry remains "separate" property. Also, if one of the two acquires property after marriage by gift, will, bequest, or descent, it is that spouse's separate property. Property obtained through a trade of separate property remains separate from community property.

Real Property Leases

A **lease** is a contract by which a tenant acquires less than a freehold interest, or something less than complete rights in real property owned by another, the landlord. The lease itself is a chattel real, that is, personal property. Although a lease is a contract, it involves an interest in land, so it is treated somewhat exceptionally.

Creation and Characteristics.   The original statute of frauds considered leases as real property transactions and required them to be in writing. The statute of frauds has been changed in various states, however, so that an oral lease contract to run for a year or less (three years or less in some states) is binding.

A lease creates the relationship of lessor and lessee or, more commonly, landlord and tenant. The relationship created is not the same as that between a roomer and proprietor of a rooming house, or a hotel owner and a guest. It involves more than these. Tenants are placed in possession of the property to use it as they please (within the limits of the lease agreement), and as long as they abide by law and the lease, their rights to use and enjoyment are about the same as that of ownership.

The provisions of the lease contract bind the parties. This, of course, is fundamental—but it should be noted that if the lease is written, any oral provisions not reduced to writing are probably worth nothing. For example, Black leases a building from White for the manufacture of boat trailers. White orally promises to rewire the building, but the written lease is silent about the wiring. If White does not rewire the building, Black is likely to be in the market for some extension cords. Black might be able to use the oral promise to show fraud in the creation of the contract, but this would be about the only value of the oral promise.

Those covenants expressed in the lease agreement will be adhered to strictly in a court interpretation. If the Black and White lease above contained a statement that Black "agreed to return the building in as good condition as when received, save for normal wear and tear and natural decay," such clause would hardly sound ominous. However, if the building burned down, Black might find himself replacing it or paying for it under this clause in the lease.

A lease runs for a definite period of time or is terminable at will by either party. It is this feature that makes it something less than a freehold estate. There are four general types of leases:

1. A lease for a definite period of time; whether 1 week or for 99 years, it is still a lease;
2. A lease from year to year, month to month, or week to week;
3. A tenancy at will; and
4. A tenancy at sufferance.

A lease for a definite period of time, say two years, needs little explanation. The tenant's rights end with the passage of time. A lease from month to month might arise from the expiration of a lease that was taken for a definite period of time. A tenant who continues to hold the property with the consent of the landlord after the original lease expires has a lease of this nature. The rent periods in the original lease dictate how long the tenant's new lease right will last (e.g., if the expired lease was to run a year at a fixed amount per month, the new lease runs from month to month).

When the tenant is given possession in such a way that a lease would be presumed and yet no term is called out, the tenant is said to have a **tenancy at will.** It may be terminated at any time by either party.

A **tenancy at sufferance** occurs when the tenant remains in possession of the property without the landlord's consent after expiration of the lease. The landlord may terminate a tenancy at sufferance at any time.

Unless law or the lease prohibits it, a tenant's rights under a lease contract usually may be assigned to another. The result is known either as **assignment** or **sublease,** depending upon how much of the lease contract was assigned. If the entire remainder of the lease is assigned to another, it constitutes an assignment; if the lessee fails to assign all rights under the lease—he or she assigns only part of them—it is a **sublease.** In a sublease, the original lessee still has property rights in the lease.

Consider the Black-White lease again. Black, finding the manufacture of boat trailers quite seasonal and rather unprofitable, assigns the remaining term of the lease to Gray. If Black has reserved nothing to himself and has assigned the full remainder of the term, Gray is now bound by the terms of the original lease. If Black has assigned to Gray only a part of the building, or has assigned to him only two of the remaining, say, four years under it, this would constitute a sublease. In such a situation, Gray would not be bound by the terms of the original lease, but by the terms of the new one between himself and Black.

Landlord's Rights.   The landlord is, of course, entitled to the agreed compensation for the use of the premises. The landlord also has the right to come peaceably upon the premises for purposes of collecting the rent when it is due. In certain states, he or she has the statutory right to exercise a lien against the tenant's personal property if other efforts to collect the rent fail. As a last resort, the landlord may obtain an eviction order against the tenant, thus removing the tenant, if the rent is not paid when due. If the lease only designates a patch of ground, as many do, the destruction of a building there will not reduce the amount of rent, even though the leased property may become untenantable as a result. At the end of the term, the leased premises must be returned to the landlord in substantially the same condition as when leased. Though the landlord is not allowed to interfere with the tenant's enjoyment of the property, the landlord may, after notice, inspect the premises for waste. Also, the landlord has a right to come onto the premises for purposes of repairing damage.

The landlord has available an action for waste against the tenant if waste can be shown. But the landlord's available action does not end with the tenant. The landlord still owns the land; she or he is **remainderman,** meaning that she or he is said to have a **reversionary interest** in the property. The landlord, therefore, has a right to prevent third persons from injuring the property or obtaining easements upon it. The law will support an action by the landlord for recovery or preventive relief as the case may be.

Tenant's Rights.   The tenant has a right to the property she or he has leased, free from interference by the landlord. The tenant also has a right to the appurtenances on the property, such as buildings, if such was the intent of the lease. Whether the leasehold will serve the tenant's purposes or not is beside the point, unless fraud or concealment can be proved. Here the rule of *caveat emptor* applies: If the property would not serve the tenant's purposes, she or he should not have leased it. The tenant, of course, is at liberty to use the property for any purpose she or he wishes, as long as the use does not violate the law or a provision of the lease.

A tenant often must improve the premises to suit them to her or his purpose. When the term of the lease expires, who owns the improvements? The answer depends upon the extent of improvements and the intent of the parties when the improvements were made. The extent of improvements depends upon two factors: the purpose of the lease and the length of its term. Generally, the landlord is entitled to the return of the property in substantially the same form it was when she or he leased it. A lease to run 100 years, for example, would allow a great deal more alteration than a 1-year lease. If the stated use of the premises or restrictions in the lease make changes obviously necessary, agreement to those changes will be implied. Many leases provide that the landlord owns all improvements. This would show the parties' intent and the landlord would indeed own the improvements.

Taxes and assessments are normally paid by the landlord. If the tenant must pay real estate taxes or assessments to retain the leased property, she or he has a choice of two remedies: (1) pay the agreed rent and maintain a damage action against the landlord, or (2) set off the payments against the rent. Of course, the landlord could shift responsibility for taxes and assessments to a tenant as part of the lease agreement.

Liability.   The tenant is liable for injuries to her or his employees, guests, or invitees to nearly the same extent as an owner of the property. The tenant has a duty to such persons to keep the premises in a reasonably safe condition.

Unless the law or a covenant in the lease requires it, the landlord is generally under no duty to repair the premises. It follows, then, that the landlord was usually not held liable to outsiders for injuries sustained by them. In the past, it was only rarely that the landlord was held liable for injury to a tenant. When a defective condition of the premises was known to the landlord, and the landlord did not reveal it to the tenant, the landlord was held liable. However, there seems to be a recent trend in which landlords are being found liable more often.

Easement - *utility companies*

An **easement** is an interest in land. It gives a person a right to do something with the real property of another or a right to have another avoid doing something with her or his property. An easement is heritable, assignable, and irrevocable. These features distinguish an easement from a **license,** which is the revocable and unassignable permission or authority to use the property of another. A valid license may be given orally.

Consider this example. Black and White own adjoining property. Black has secured written permission to cross White's land. The writing states that the right to cross White's land pertains to "Black, his heirs or assigns." Such a grant would be held an easement, since it is capable of being assigned to another. If Black sells his land to Gray, the easement may be transferred to Gray in the sale. A license, on the other hand, could not be transferred.

There are several types of **natural easements.** An owner of a building, for example, owns a right (in the form of an easement) to prevent a neighbor from excavating in such a way as to cause her or his building to tend to fall into the hole. This is known as the **right of lateral support.** For further example, if Black sells White a piece of property that is completely surrounded by Black's property (there is no other means of access except by air travel), White has an implied natural easement across Black's property. This is also known as a **way of necessity.** Although the idea did not become popular in the United States, a natural easement to light and air developed in English law: No structure could be built that interfered greatly with the natural light available to a neighbor.

Easements, other than natural, are created by grant or prescription. The grant may be in the form of a deed of the easement right itself, or of a covenant in the deed that transfers the real property. All manner of easement rights are created by grant or prescription—roads, power lines, gas lines, or sewers may cross land under such easements. Even raising the water level and inundating part of someone's land by damming a stream may involve an easement.

WATER RIGHTS

A person who owns a piece of real property ordinarily owns the things on it and under it as well. From this, the conclusion might logically be drawn that water on and beneath the land belongs to the owner of that land. Actually, the rights may or may not extend to ownership of the water, depending upon the jurisdiction. Water rights are handled in several ways, ranging from individual ownership to state ownership.

The question of water ownership has lately become more and more pressing, as our population has increased considerably, bringing with it an increased demand by each individual for water. In some areas, such as southern California, water rights are hot political issues. As public demand for water increases still further, we are likely to see continued legislation in this field. Although laws and court cases lack uniformity regarding water rights, some generalities may be stated.

Boundaries

The extent of real property is often limited by a watercourse or a body of water. Under common law, the defining of property limits in such a manner depends upon the nature of the body of water.

If, for example, a nonnavigable stream separates the property of two **riparian owners**—owners of property bordering on a stream or other body of water—each owns to the center of the stream channel, or to the "thread of the stream," as it is known. Shifting the stream to a new channel does not change the rights of the two owners. The property line remains as before if the channel change comes about suddenly or in such a way that the old channel may continue to be identified.

In contrast, riparian owners are entitled to additions to their property that come about gradually as a stream adds to one shore or another. If, as a result of these natural accretions, the channel gradually changes, the dividing line of the properties will also change; the property of one riparian owner will be extended at the expense of the neighbor across the stream.

The owner of property bordering on water affected by the tides generally owns only to the high-water mark. The **foreshore** (the land between the high-water mark and low-water mark) is public property, belonging to the state. Land bordering upon navigable lakes or streams is generally owned to the low-water mark. In addition, the owner usually has the right to build a pier extending to the line of navigability. The difficulty of establishing a fixed property line by the concepts of high-water or low-water marks has led some of the states to establish riparian property lines by other means. Lines so established are not so apt to fluctuate with droughts or floods.

When a stream divides two states, state ownership does not follow any general pattern. One state may own all of the stream or none of it; in other cases, the center of the stream or the channel may be the dividing line.

Riparian Rights and Duties

A riparian owner generally has the right to reasonable use of water bordering his or her property. Reasonable use is a little difficult to define, however, and each controversy over the right to use water must be decided on its own merits. Generally, it means that the owner may use the water as long as the use has no significant effect upon the quantity, quality, and velocity of the stream. The owner has a duty not to pollute the water that his or her property borders. For instance, the owner cannot dump garbage or sewage into a stream bordering the property. This dumping would constitute an unreasonable use, since it would change the quality of the water with respect to the downstream riparian owners. Domestic use of water (for drinking or bathing) is held to be more important than either agricultural or industrial uses.

Underground Water

Issues concerning water and the rights to its use are not limited to the rivers and lakes on the surface of land. Who, for example, owns water in the soil (percolating water) or underground rivers? Generally, the right to use and the duty not to pollute extend to underground streams as well as surface streams.

A large quantity of water is present beneath the soil as water with no appreciable direction of flow. The results of tapping this source of water are often quite unpredictable. The owner of land generally has the right to drill a well and capture a quantity of this water for his or her own use. However, if in so doing the owner lowers the level of the water table so that a neighbor cannot get water, an injury is apparent, and the neighbor's cause may be actionable. Where a watershed supply is quite limited, the court will have a tough time assigning the rights to the water.

Irrigation has been the salvation of many areas of our country. If water must be pumped from a watershed, however, the lowered level of the water table may harm neighboring communities. Not all the water returns to the soil; much is lost by evaporation from the ground surface and through the leaves of the plants fed. The states have varying rules that govern the extent of a landowner's rights to such underground water.

A property owner often cannot raise the level of groundwater and do harm. The damming of a stream, for example, could result in flooding nearby. Similarly, rules often restrict a landowner's ability to cause water due to rain or the melting of snow to drain onto another's property.

Prior Appropriation

Our western states have far more serious water problems than the eastern states. The common law rights mentioned above seem satisfactory where water is plentiful, but other rules have developed in the West.

The rights of **prior appropriation** (roughly—first come, first served) and **prescription** predominate in the West. Under this view, the first of two or more persons to appropriate water for his or her own use has the superior right to continued use. Continued use of water for an extended period of time, even by a nonriparian owner, gives one the right to further use. If a person wishes to appropriate a large amount of water for his or her own use, that person must first obtain a permit to do so. In this way, use of the water is controlled for the benefit of the public.

MANDERSON & ASSOCIATES, INC. et al. v. GORE
193 Ga. App. 723; 389 S.E.2d 251 (Ga.App. 1989)

Study terms: Limited partnership, closely held company, fiduciary duty, personal property, chattel, real property, "intent of the parties", unity of title

Opinion

Case No. A89A1294 is an appeal by Manderson & Associates et al. asserting two enumerations of error of the trial court's final order and judgment. Case No. A89A1295 is an appeal by James W. Gore, asserting six enumerations of error, of the same final order and judgment, the subsequent order correcting the final order and judgment, and of orders granting defendants' motion to dismiss jury demand and denying plaintiff's motion for summary judgment.

Lewis M. Manderson, Jr., and James W. Gore became business associates in an outdoor advertising business. Manderson, the principal owner and primary shareholder, sold the business at a profit. Gore shared in these profits, as he held company stock under a profit sharing plan. Manderson, Gore and another Manderson employee, named Hubbert, subsequently entered a new business venture to acquire interest in three other outdoor advertising businesses. Three limited partnerships were organized to own, operate and eventually sell each of the three businesses. Manderson & Associates, Inc. (M &A) was formed to be the general partner of each of the three limited partnerships. Apparently, the parties contemplated that M&A would serve as general partner for investment limited partnerships engaged in the outdoor advertising business. Manderson, Gore, and Hubbert also purchased limited partnership interests in one of the businesses.

As they apparently intended to invest in other business ventures, the men formed a second corporation, MGH Management, Inc. (MGH). The primary business purpose of MGH was to provide management services for all limited partnerships for which M&A served as general partner.

Manderson was the president and majority stockholder of each corporation. Gore and Hubbert were employees of MGH, as M&A apparently had no employees. Gore also was an officer and director for MGH and M&A. Gore was vice-president in charge of finance and accounting; he was the top financial officer, and also was in charge of finding further business ventures.

To preserve the closely-held nature of the companies, each shareholder executed certain employment agreements and stock transfer agreements. The stock transfer agreements required each man to transfer his share back to the corporation upon termination of his employment. Each stock transfer agreement contained a formula for computation of the stock's purchase price. According to trial testimony, Manderson became concerned that Gore did not have the organization's best interests at heart. In January of 1986, two of the three outdoor advertising plants were sold and Gore had located no other ventures. In the early summer of 1986, Gore located a potential investment in a plastic jug manufacturing company. Gore then proposed, contrary to the parties' prior agreement, that he and Manderson be vested with equal ownership in the jug company. Manderson rejected this proposal. Manderson subsequently heard a rumor that Gore was planning to leave. He then asked Gore to present his plan for staying with the company, and repeated this request during subsequent meetings. On September 19, 1986, Manderson officially terminated Gore 's employment with the company, in part asserting that Gore refused to provide such a plan. Gore contests this assertion, claiming that he provided a general plan to Manderson, and that he merely wanted to "slow down." Appellants demanded that Gore present his stock for sale to the company. Gore did not do so, and disputed the stock price due and owing him.

Gore subsequently brought suit asserting a breach of fiduciary duty by the majority stockholders and requesting declaratory judgment of the parties' rights and obligations under the agreements. Assuming the agreements are enforceable, which Gore contests, he seeks the amount due under the agreements in addition to damages caused by Manderson's alleged breach of fiduciary duty. The case was tried without jury. The trial court denied Gore's fiduciary duty claim, construed paragraph 8 (A) (1) of the stock transfer agreements in a manner more favorable to Gore, and awarded pre-judgment interest to Gore.

■ ■ ■

8. Appellant asserts that the trial court erred in concluding that MGH-owned billboards are personal property rather than real property.

The trial court supported its legal conclusion by finding inter alia that the billboards can be readily moved from site to site with proper equipment. This finding of fact is adequately supported by the trial record and will not be disturbed.

Evidence of record reflects that MGH owned approximately 50 billboards of three different types, and some boards were bolted to concrete foundations. The superstructure of certain of the boards weighed approximately 25,000 to 30,000 pounds. The typical board is 45-50 feet high. The boards could be moved with proper equipment. Except for one or two instances, the boards were placed on land leased by the outdoor advertising limited partnerships who had leased the boards directly from MGH. The leases each contained a clause authorizing removal of the boards from the property on expiration of the lease and on certain other circumstances. When boards with concrete foundations were removed, in all instances except one when the foundation was removed, the concrete foundations were abandoned and left on the land. The boards were treated by MGH as real property for certain tax purposes and as personal property for other tax purposes.

As appellant Gore has not raised in his enumerations of error any appellate issue regarding the concrete foundations on which certain of these billboards were affixed, that question is not before us for review.

In Georgia the law governing the title and disposition of land is exclusively subject to the laws of the State where it is situated. Likewise, "[t]he question of whether property is a part of real estate is to be determined according to the law of the jurisdiction in which the real estate is located." 36A CJS 593, Fixtures, § 1. Accordingly, the law of Georgia, and not the law of Alabama, is dispositive of the issue before us as to all billboards located on Georgia land on the determination date of the MGH stock transfer agreement. Examination of the record reveals that on the determination date certain of the billboards may have been located on land in other states, for example in Pennsylvania or South Carolina. As appellant Gore has failed to provide any citations of authority or argument to address the correctness or incorrectness of the trial court's ruling under the laws of states other than Georgia and Alabama, any appellate issue concerning such billboards is deemed abandoned. Further, appellate briefs do not direct this court to any place in the record where it is shown that any billboards were located in Alabama on the "determination date" of the MGH stock transfer agreements. Accordingly, the parties have no standing to complain of our determination to resolve this issue solely by applying Georgia law. "It is not the function of this court to cull the record on behalf of a party. . . ." *Armech Svc. Co. v. Rose Elec. Co.,* 192 Ga. App. 829, 830 (386 S.E.2d 709) (1989).

Under Georgia law various factors should be considered in determining whether an article of personalty has become a part of the real property to which it has been actually or constructively annexed. "Varying weight, according to the circumstances of the case, may be given to the determinative factors." 36A CJS 592, supra at § 1. "'Whether an article of personalty connected with or attached to realty becomes a part of the realty, and therefore such a fixture that it can not be removed therefrom, depends upon the circumstances under which the article was placed upon the realty, the uses to which it is adapted, and the parties who are at issue as to whether such an article is realty or detachable personalty.'" *Aquafine Corp. v. Fendig &c. Advertising Co.,* 155 Ga. App. 661 (1) (272 S.E.2d 526); *State of Ga. v. Dyson,* 89 Ga. App. 791, 793 (81 S.E.2d 217); *Consolidated Warehouse Co. v. Smith,* 55 Ga. App. 216 (1) (a) (189 SE 724). There must be some form of active or constructive annexation of personalty to the realty for the chattel to become a part of the realty. *State of Ga. v. Dyson,* supra at 793; 12 EGL, Fixtures, § 4. Moreover, some cases have required that there be unity of title as a prerequisite for the creation of a fixture. Inherently included within these factors is that of intent of the parties.

It is the intent of the parties vested with ownership and use of the chattel to be annexed "as to whether the chattel is to become a permanent part of the realty" which is the primary test "in determining whether or not it becomes a fixture." 12 EGL, Fixtures, § 3; see generally, 36A CJS, Fixtures, § 2 a; see *Pease &c. Realty Trust v. Gaines,* 160 Ga. App. 125, 129 (2) (a) (286 S.E.2d 448). In certain circumstances, intent can be the controlling factor. In *Sawyer v. Foremost Dairy Prods.,* 176 Ga. 854, 861 (169 SE 115), the court held "[i]t seems to be well settled that whether fixtures have become attached to the realty as a part thereof, and not merely for incidental, transitory use, depends upon the intention of the parties vested with the ownership or use thereof." Id.; accord *Aquafine Corp. v. Fendig &c. Advertising Co.,* supra at (1).

"'Where it is doubtful, under all the circumstances, whether the article in question is personalty or is a fixture, the doubt is to be solved by the [factfinder].'" *Aquafine Corp.,* supra at 661. Normally, the question of intent is for the factfinder.

Considering all relevant factors, in light of the operative circumstances, we are satisfied that the billboards are personalty rather than real property under Georgia law.

Appellant Gore asserts that the trial court erred in relying upon the unity of title requirement (see *State of Ga. v. Dyson,* supra), MGH 's tax treatment of the structures, and certain ground lease provisions which designated structures as trade fixtures. Regarding unity of title, one noted authority has found the rule to be "vague" and "seldom applied in practice." Ga. Real Estate Law, supra at § 10-7. We need not resolve these questions, however, as the trial court's ruling was correct, and accordingly will not be reversed notwithstanding the reasons therefore given. . . .

Judgment affirmed.

FAIRFAX v. VIRGINIA DEPARTMENT OF TRANSPORTATION 247 Va 259, 29 ALR5th 759 (1994)

Study terms: Trust agreement, beneficiary, condemnation, restricted use, indenture agreement, fair market value

In this appeal of a condemnation award, we consider the appropriate basis for valuing land subject to a trust agreement providing that ownership of the land vests in a church if it is used for purposes other than a park.

In 1970, David Lawrence executed a trust agreement dedicating a 639-acre parcel to be used as a park. The trust agreement names Fairfax County Park Authority (FCPA) as the beneficiary in possession of the trust so long as the property in the trust is used as a public park dedicated to the memory of Ellanor C. Lawrence. If the property is used for any other purpose, title to the land passes to the trustees of the St. John's Episcopal church, who then own the land free from any restrictions. In the event of condemnation, the trust agreement requires the FCPA to contest the condemnation proceedings in "every fashion reasonably possible," but condemnation of a portion of the park land does not pass the remaining property to the Church.

In 1988, the Virginia Department of Transportation (VDOT) filed a certificate of condemnation for approximately 13 acres of the park property to expand and improve existing roadways. The certificate subsequently was amended to reduce the taking to 2.6497 acres in fee simple, 1.0584 acres for a temporary easement, and .3685 acre for a drainage easement.

VDOT filed a motion in limine seeking to establish the criteria for valuation of the condemned park property. The trial court ruled that the measure of compensation was the property's fair market value as restricted by the trust. The parties waived a hearing before the commissioners and presented evidence to the trial court of the value of the condemned property restricted to use as a park.

VDOT's appraisal witness, Edward S. Williams, III, testified that the value of the property used as a park or an open space was $2,125 an acre based on four sales he considered comparable. Clyde A. Pinkston, the appraisal witness presented by FCPA, testified that there was no market for park land and, therefore, the property, if restricted to use as a park, had no market value. If used for residential purposes, Pinkston testified that the condemned property would have a market value of $125,000 an acre.

The trial court, treating the land as park land, held that there was a market for the property as park land and accepted Williams's appraisal of $2,125 an acre. The court entered a final order setting the condemnation award of $6,450. We awarded FCPA an appeal.

We first consider FCPA's contention that the trial court erred in holding that the value of the condemned land must include consideration of its restricted use. In reaching this conclusion, the trial court acknowledged that cases in other jurisdictions have taken contrary positions. Some jurisdictions value the land based on the restricted use while others disregard such restrictions. . . . The trial court elected to follow those cases which required consideration of the restricted use because of their factual similarities to this case. In each instance, the land taken comprised only a small portion of the condemnee's parcel. In our opinion, this factor—the amount of land taken relative to the amount left the owner—has little relevance to determining just compensation due as measured by fair market value, the standard mandated in condemnation proceedings in this commonwealth. . . . Therefore, we must look to other grounds for determining the appropriate treatment of the use restriction in valuing the land taken in this condemnation proceeding.

In seeking support for their respective positions, the parties noted that we have not previously determined whether use restrictions such as those present in this case should be applied in determining the amount of the condemnation award. We have held, however, that similar restrictions should not be taken into account when valuing land for taxation purposes.

In a series of cases the Richmond, Fredericksburg and Potomac Railroad Company (RF&P) challenged the assessments of its railroad yard in Alexandria, the Potomac Yard. . . . The assessments were attacked on various grounds, but as pertinent here, RF&P sought reduction in the appraised value because the land could only be used as a railroad yard. Furthermore, as in the instant case, price of the land was subject to an indenture agreement that provided that if the land was used for purposes other than maintenance or construction of railroad tracks, ownership of the land would revert to the United States government. We rejected on various grounds RF&P's argument that the restricted use as a railroad yard and the indenture agreement should be considered in determining the market value of the property. . . .

As relevant here, we held that it is the fair market value of the land, not the value of the land **to the owner,** which is subjected to taxation, Therefore, the market value of the land is derived by considering the various uses to which the land is susceptible, not just those uses to which a particular owner may be restricted. If, however, the land is so committed to a particular use that it cannot be put to another use economically, we held that, under those circumstances, it is appropriate to take the committed use of the land into consideration when determining the market value of the land. . . .

While these principles establish the circumstances under which use restrictions should be considered in calculating the fair market value of property in the context of real property taxation, we see no reason why they should not be applied in the context of the instant case. Fair market value of land is used not only for taxation purposes, but, as we have said, it is the prescribed method for determining the amount of "just compensation" due in condemnation proceedings. . . . To adopt one set of principles for determining the fair market value of real property in a condemnation proceeding and another set to make the same determination for taxation purposes could result in a single parcel of land having more than one fair market value. Such a result would be inconsistent and inequitable and is unnecessary.

These principles are consistent with the condemnation jurisprudence of this Commonwealth. Condemnation is an _in rem_ proceeding and, while the land is valued from the point of view of an owner rather than the condemnor, the value established is not the value to the owner personally. . . . A determination of a particular owner's loss relative to that of others is only undertaken in the second step of the condemnation proceeding in which the condemnation award is allocated among those with interests in the property. . . .

Finally, there is no evidence here that the condemned land was so committed to use as a park that it was not economically feasible to put the land to other uses. In fact, the trial court held that the highest and best use of the property was for residential purposes. There were no legal impediments to that use. Nor are future improvements required to adapt it for residential use. . . . Therefore, applying the principles stated above, we find that fair market value of the property condemned in this case should be calculated without regard to the use restrictions placed on it by the trust agreement. In light of our disposition of this issue, we need not address the other errors assigned by FCPA.

Accordingly, we will reverse the order of the trial court and remand the case for further proceedings consistent with this opinion.

Reversed and remanded.

ABOOD v. JOHNSON 200 N.W.2d 20 (Neb. 1972)

**Study terms:**   Adverse possession, indefeasible title, conveyance

This is a boundary line dispute between adjoining landowners. The court found generally in favor of the plaintiffs, fixed the boundary line, and awarded damages to the plaintiffs for a fence destroyed by the defendant. The plaintiffs

have appealed, contending that the boundary line fixed by the court did not include all the land enclosed by the fence. Defendant's cross-appeal contends that the line should be the lot line.

Plaintiffs were the owners of Lot 6, and the defendant was the owner of Lot 7, both located in Section 12, Township 8, Range 16, Buffalo County, Nebraska. The quarter-section line running north and south in Section 12 was the lot line between the two lots. Lot 6 was east of the line, and Lot 7 was west of it. The Platte River formed the north boundary of both properties, and a county road formed the south boundary. The distance from north to south was approximately 1,500 feet. The plaintiffs first acquired an interest in Lot 6 in 1926 and became the sole owners in 1953. Defendant acquired Lot 7 in 1947. The allegations and evidence of the plaintiffs were that a boundary line fence had been in existence for at least 50 years. It extended the entire distance of the lot line but was located some distance west of it. Plaintiffs' evidence fixed the location of the boundary line fence 61 feet west of the lot line at the south, 49.2 feet west of the line at a point near the center of the properties, and 28 feet west of the line at the north end of the properties.

The defendant contends that the correct boundary was the lot line; that there was no fence which was a boundary line fence; and that any use which plaintiffs had made of Lot 7 was with permission. Defendant has cross-appealed from that portion of the decree which fixes any part of the boundary line at any point other than the platted lot line.

Evidence convincingly established the existence of a fence between the two properties for more than 50 years. The fence was located at a point west of the quarter-section lot line, but the exact distance west of the line at particular locations in prior years was not surveyed or measured. The evidence was undisputed that the plaintiffs had dug and established an irrigation well approximately 40 feet west of the lot line on the south end of the properties and had used it continuously since 1957. The evidence was also undisputed that the plaintiffs had planted and cultivated crops on much of the south half of the disputed area. The evidence was convincing that plaintiffs had built and maintained a silo or ensilage bed at a point some 700 feet north of the south line of the properties. The west edge of the ensilage bed was 49.2 feet west of the lot line.

The evidence is undisputed that in 1967, the plaintiffs and defendant jointly removed approximately 400 feet of fence at the north end of the property to permit gravel operations to continue moving easterly along the south side of the Platte River and from defendant's to plaintiffs' property. In April of 1971, the defendant used a bulldozer and destroyed the fence on the remainder of the property. At the trial, the evidence established that there were some broken fence posts at the south edge of the property 61 feet west of the lot line. There was a fence post on the west side of the ensilage bed approximately 700 feet north of the south line and 49.2 feet west of the lot line. There was a broken concrete fence post some 400 feet further north. A line connecting these three specific locations and extended north for the remaining distance of approximately 400 feet ends 28 feet west of the lot line at the north border. Plaintiffs' evidence was that the fence had been located on that line. There is no evidence that a fence was ever located on the lot line at any point.

The court found generally in favor of the plaintiffs, fixed the boundary line, and directed the preparation of a survey accordingly. The boundary fixed by the court commenced 43.7 feet west of the platted lot line on the south, rather than 61 feet as shown by the plaintiffs' evidence; the next point was some 740 feet north, and at that point the line fixed by the court and the line established by plaintiffs' evidence were identical, both being 49.2 feet west of the lot line. From that point, the boundary fixed by the court ran generally north but toward the east where it joined the lot line at a point some 400 feet further north. The boundary then continued to the north end of the properties on the quarter-section lot line. The court also awarded damages to the plaintiffs for the fence destroyed by the defendant.

A thorough review of the record confirms the court's general findings in favor of the plaintiffs. The boundary line fixed by the court, however, does not fully conform to the evidence. The trial court did not indicate the basis upon which the boundary was fixed, nor make specific explanatory findings. The court determined that the dividing line on the north 400 feet of the property should be the quarter-section lot line rather than any fence line. Comments in the record indicate that determination was made because the plaintiffs and defendant had jointly removed the fence in that area for the sole purpose of removing gravel from both properties in 1967.

We believe this case is controlled by *McCain v. Cook*. . . . It is the established law of this state that, when a fence is constructed as a boundary line between two properties, and parties claim ownership of land up to the fence for the full statutory period and are not interrupted in their possession or control during that time, they will, by adverse possession, gain title to such land as may have been improperly enclosed with their own. See also *Ohme v. Thomas*. . . .

"After the running of the statute, the adverse possessor has an indefeasible title which can only be divested by his conveyance of the land to another, or by a subsequent disseisin for the statutory limitation period. It cannot be lost by a mere abandonment, or by a cessation of occupancy, or by an expression of willingness to vacate the land, or by the acknowledgement or recognition of title in another, or by subsequent legislation, or by survey." *McCain v. Cook*, supra.

Under the evidence here, the plaintiffs had acquired title to the property long before 1967, either by boundary line acquiescence or adverse possession, or both. Plaintiffs' title could be divested by nothing short of a validly executed deed, by adverse possession, or by other legal means not pertinent here.

The county surveyor who testified on behalf of both parties also prepared exhibit 57, describing the boundary fixed by the court. The surveyor also prepared a survey reflecting the fence line as established by plaintiffs' evidence. The survey was exhibit 50, but it was not admitted into evidence because a copy of it had not been furnished to the defendant within the time specified in the pretrial order. The surveyor did testify as to the location of the various points used as the basis of that survey and that he had observed the fence posts at the points shown. Exhibit 50 designates the line on which the dividing line fence was located in reference to the north-south quarter-section line. Exhibit 50 may not be complete insofar as a legal survey description of the property line is concerned, but subject to any such technical completion, it designates the boundary line established by the plaintiffs' evidence.

The decree of the district court was generally correct, but it should be modified by substituting the division line shown in exhibit 50 for the line shown in exhibit 57 and described in paragraph 2 of the journal entry. As so modified, the decree is affirmed.

Affirmed as modified.

Review Questions

1. Distinguish between
    a. personal property and real property
    b. sale and bailment
    c. warranty deed and quit-claim deed

2. In what ways may personal property be lawfully acquired?

3. Are the following things real property or personal property?
    a. A gas-operated water heater.
    b. A bird bath.
    c. Flowers and shrubs in a flower bed.
    d. A satellite dish mounted on a roof.
    e. A window-mounted air conditioner.

4. What are the requirements for a valid will?

5. Why must real estate transactions be recorded?

6. In what ways are a mortgagee's interests in real property similar to those of a landlord?

7. How are water rights controlled in your state?

8. For what purposes other than condemnation (as in *Fairfax v. Va. Dept. of Transp.*) is real estate valuation appraisal performed? *Tax*

9. According to the logic in *Fairfax v. Va. Dept. of Transp.*, on what occasions, if any, should the market value of land not be determined by the various uses to which it is susceptible, but only to its restricted uses?

10. In *Manderson & Associates, Inc. v. Gore*, why did Gore want the billboards to be considered real property rather than personal? *Value of shares would likely increase*

11. Return to the requirements for adverse possession and compare these requirements with the court finding for the plaintiff in *Abood v. Johnson.*

Intellectual Property

CHAPTER
20

HIGHLIGHTS

■ There are four major bodies of law that govern intellectual property:
1. Federal patent laws
2. Federal trademark laws
3. Federal copyright laws
4. Trade secret laws

The phrase **intellectual property** is used to describe the types of intangible property that result from the creative exercise of the mind.

In this chapter, we will consider different legal systems as they apply to patents, trademarks, copyrights, and trade secrets. As we do, keep in mind that each legal system of protection is different, and the rights and subject matter protected by each legal system may differ and, sometimes, overlap.

PATENTS

Engineers often find themselves in fortunate positions to invent things or to improve upon others' inventions. They are particularly favored by training in science and by the nature of engineering jobs for this endeavor. Inventions generally involve the application of scientific principles to practical problems with a practical solution as the objective. The relationship between invention and the training offered by engineering curricula seems quite obvious. Because of the engineer's training and talents, the engineer should be alert to the possibilities of making inventions, of obtaining protection for them, and of exploiting them.

Suppose that Mr. Black, an engineer, has just invented an entirely new type of internal combustion engine after many months of effort. The engine will run on a most readily available resource and promises to have a longer life than even the best competitors. What should Black do to secure for himself the fruits of his creative efforts? A number of systems of legal protection are available, depending on the particular embodiment of his ideas that Black wishes to protect. For example, a patent might protect the engine, while trade secret protection might be useful for the method or process used to manufacture the engine. Probably the best course would be for him to contact an expert on intellectual property, such as a patent attorney or a patent agent. But even when the task of obtaining a patent is delegated to another, there is still considerable information Black should have. What, for instance, may be patented? What is the nature of a patent? Who is entitled to one? What protection does it afford? What are trade secrets and copyrights? What protection do they provide? How does one obtain them? What could prevent an inventor from getting a patent? How long will it take, and what delays can be anticipated? This section attempts to answer these and a host of other questions that are likely to arise.

History

Prior to the adoption of the U.S. Constitution, patents were issued by the individual colonies, but a patent issued by one colony secured rights only in that colony. If someone in another colony could obtain the essence of the patent, he was free to make any use of it that he might desire. It is for this reason, among others, that the U.S. Constitution provides that Congress shall have the power "to promote the Progress of Science and useful Arts, by securing for limited Times to Authors and Inventors the exclusive Right to their respective Writings and Discoveries" (United States Constitution, Article 1, Section 8, Clause 8).

The first United States patent law was enacted in 1790. Since then there have been numerous changes of the law and its administration. Interestingly enough, the adoption of patent laws has been a consistent concern of Americans since the colonial days. For example, the Republic of Texas enacted patent laws shortly after its independence from Mexico in 1846. Similarly, the Confederacy adopted its own patent laws and issued its own patents while the Civil War raged.

The Patent Office (the official name is the United States Patent and Trademark Office of the Department of Commerce, but is most commonly called the Patent Office, PTO, or USPTO) is now headed by a Director of Patents, and is part of the U.S. Department of Commerce. Congress continues to modify the laws in an effort to make sure the patent laws keep up with the lightning speed of discovery, as well as harmonize our laws with those of other nations.

What Is Patentable?

An idea is not patentable, but the machine, process, or thing into which it has been incorporated may be. That is, a physical law or principle, such as Newton's law of gravity, no matter how beneficial it is likely to be, is not patentable as long as it remains an idea or principle. Its physical embodiment might be patented, however, if the other tests of patentability are met.

There are three general categories of patents:

1. Utility Patents (35 USC §101)
2. Design Patents (35 USC §171)
3. Plant Patents (35 USC §161)

These categories are divided into classes. Debate continues over the proper legal scope of these categories, as well as whether the legal scope of the categories is appropriate as a matter of public policy. While it may appear to be otherwise, ethical issues, such as stem cell research and cloning, are not a bar to patentability. However, ethical issues may be a bar to exploitation of the invention in the market.

The Patent Right

One meaning of the word **patent** is "open or disclosed, obvious or manifest." These terms apply to the rights issued to an inventor by the Patent Office, because the inventor must make a full and complete disclosure of his invention. No material feature or component may be withheld. The revelations must be such that a person skilled in the field could, by use of the patent, duplicate the thing patented.

> A relatively new class of patent, Business Methods Patent, is the subject of considerable controversy. The business method classification 705 covers computer implemented business inventions. In 1999, pursuant to *State Street Bank & Trust Co. v. Signature Financial Group, Inc.*, 149 F.3d 1368, Congress enacted the American Inventors Protection Act, 35 USC 273 (Defense to infringement based on earlier inventor) to protect innocent infringers who had inadvertently used a patented business method. See *eBay Inc., et al. v. MercExchange, LLC* at the end of the chapter.

The **patent right** is often regarded as a contract between the government and the inventor. This "contract" can only be formed with a natural person as the inventor, as opposed to a copyright or trademark, where either can be issued to a business. In consideration for disclosing the invention, the government gives the inventor the right to exclude others from making, using, or selling his invention. The right extends for 20 years from the date of issue [35 USC §154(a)(2)], after which the patent enters the public domain. A patent only gives the right to exclude others from making, using, or selling the invention—not the right to make, use, or sell it. The reason is to avoid conflict with state or other federal laws that might prohibit making, using, or selling such things as those covered by the patent. For example, a statute might prohibit, making, using, or selling things in a particular manner proposed by the inventor. For instance, you might invent a truly fantastic insecticide, but the EPA and FDA may have something to say about its manufacture, sale, and use. Also, the inventor should not be able to impose upon the prior rights of another or of the public in general merely because he has obtained a patent for his invention. For instance, if use of the invention injures another, the presence of the patent should not allow the injury to continue.

A description of those who may obtain a patent is neatly summarized in Title 35 of the United States Code:

> Whoever invents or discovers any new and useful process, machine, manufacture, or composition of matter, or any new and useful improvement thereof, may obtain a patent therefor, subject to the conditions and requirements of this title.

A patent may be issued to two or more persons as joint inventors. Mere partnership in an enterprise or financial assistance does not make one a joint inventor. To be a joint inventor, one must collaborate with the other co-inventor(s) and contribute to the conception of an invention.

There are two statutory requirements for a patent to issue.

1. It must be an invention or discovery, e.g., the subject matter must be an invention or discovery not obvious to someone with skill in the particular field of invention or discovery.
2. The invention must be new and useful, e.g., the invention or discovery cannot be known or used by others (except for experiments), and there must be no public use; there must be no publication or sale or offer to sell more than one year prior to the patent application.

Invention or Discovery.   To obtain a patent, the subject matter must constitute an invention. Invention and discovery are often used synonymously, despite differences in meanings. **Discovery** refers to the recognition of something in existence that has never before been recognized, whereas **invention** refers to the production of something that did not exist before. For ease, this chapter follows the more common usage of referring to an invention and a discovery as the same thing. To be patentable, an invention must not only be new or previously unrecognized, it must also be useful, and something which is not obvious to or normally predictable by a person of ordinary skill in the field involved (35 USC §103).

There is no fixed yardstick of unobviousness in law. Each case is judged on its own merits in the light of similar cases and the present state of the art in the field of the subject matter of the proposed patent. That which might have been patentable at the turn of the 20th century is now commonplace to practitioners in the field involved. In the final analysis, it is the judgment of the court that determines whether true invention or discovery is present.

New and Useful.   A patentable invention cannot be something that is already known and used by others in the art involved (35 USC §102). Description of the invention or discovery in a printed publication either in the United States or abroad can be a bar to obtaining a patent. Certainly, if the subject of the invention was published by another prior to the "inventor's" conceiving it, it cannot be said to be new, or that person's invention. Publication by the inventor more than a year prior to his application also operates to bar him from obtaining patent rights. Publication consists of making information available to the general public. It is not held to occur if the information is given to a restricted group of persons with the express or implied condition that the information is not to become public knowledge. Examples of prior art publications include previously issued patents, which are publicly available.

For a publication to bar patenting an invention, it must contain more than a mere reference to a new idea. Sufficient details must be present to enable the reader to make practical use of the idea. Black, for instance, might safely refer in a published article to a new internal combustion engine in very general terms. As long as he does not reveal the invention, the year's limitation "clock" does not run.

Public use of the invention more than a year prior to filing the patent application also defeats patentability. If Black started manufacturing his engines for public sale, time would start to run against the year at least from the date of his first sale (or the date he first offered to sell one of the engines). Of course, there are many shades of gray between black and white. For instance, manufacture of a limited number of machines for experimental use by certain persons is not likely to be held public use. This is especially true where the inventor has told the users of the need for secrecy regarding the invention. Court cases have held that use can be experimental even though the inventor may have made some profit in the transactions. Experimental use is determined in part by the intent of the inventor and the nature of the invention. Some things must be tried publicly. Experimental use of a new roadsurfacing material, for example, would almost necessarily require public use in a street or highway if truly typical conditions are to be encountered.

The invention must also be useful. (Note that a design patent, however, must be ornamental, not useful. Most patents are "utility" patents; hence, the requirement of usefulness.) The word **useful** is given a rather broad interpretation; essentially, it is required that no harmful effect on society would result from the invention. The primary purpose of the entire patent system is to benefit the public. It follows, then, that the inventor of something that offers the public no benefit, or something that is harmful or immoral, should not be entitled to a patent.

An invention must be operable or capable of use if is it to satisfy the test of usefulness. If there is a question of the ability of the invention to perform as claimed, the Patent Office may require a model to be built for demonstration purposes. Thus, if the examiner had difficulty understanding Mr. Black's use of internal combustion, he might demand a model from Black. However, the requirement of a model is exceedingly rare. It is much more likely that the examiner would require an affidavit from someone who had actually seen the engine in operation. Of course, "perpetual motion" machines are viewed with suspicion. Models of such inventions may be required. (See Appendix E)

Patent Transferability

Patent rights are property rights. Hence, patent rights may be bought or sold, assigned, mortgaged, licensed, given, or willed to another nearly as easily as any other personal property. The two primary methods of patent transfer are discussed below.

Assignment.   Because a United States patent has the attributes of personal property, its ownership may be assigned to another person or business entity.[1] Assignment of the rights is accomplished using a written document, and may take place either before or after the inventor obtains his or her patent. If the Patent Office is not informed of an assignment (through specific procedures for recording assignment) within three months and the patent is subsequently assigned to another, the first assignee to record the transaction will be the new owner of the patent.

A patent may be issued directly to the assignee. When this is done, it is the patent application that is assigned by the inventor. The inventor must still execute the appropriate oath (discussed later), but any patent resulting from the application will be granted to the assignee.

It is, of course, possible to assign a portion of the rights acquired in a patent. But the interest assigned must be specified in the assignment. That is, assignment of a half interest in the royalties resulting from a patent could be determined and upheld in court.

Shop Rights and Contracts to Assign.   Generally, patent rights belong to the inventor. If the inventor used his employer's time and equipment in his creative activities, it seems only fair that the employer should have some benefit from the result.[2] The employer's right to benefit under these circumstances is well established in the law.[3] Some difficult questions arise in this connection, however, in regard to contracts made in anticipation of invention, and inventions created wholly or partly on the employee's own time.

---

[1] See 35 USC § 261 that says: "Subject to the provisions of this title, patents shall have the attributes of personal property. . . . Applications for patents, patents, or any interest therein, shall be assignable in law by an instrument in writing. The applicant, patentee, or his assigns or legal representatives may in like manner grant and convey an exclusive right under his application for patent, or patents, to the whole or any specified part of the United States."

[2] Absent a written agreement to the contrary, an employer has to show that the employee was hired to invent for a specific product or purpose or it had assigned the employee to do work that may become the basis for a patent.

[3] The seminal case for the boundaries of shop rights as well as the basis for the "employed-to-invent" rule, is *Standard Parts v. Peck,* 264 US 52, 44 S.Ct. 239 (1924) where the employee, Peck improved upon some machines on which he was hired to work. The company for which he worked (Axle Company) was acquired by Standard Parts. Under the acquisition agreement, all of the assets, including rights in machinery and devices were transferred. Peck claimed that his employment contract did not include an assignment and that his employer had no rights to the letters patent he obtained while working for them. The Supreme Court vehemently disagreed: "By the contract Peck engaged to devote his time to the development of a process and machinery and was to receive therefor a stated compensation. Whose property was the process and machinery to be when developed? The answer would seem to be inevitable and resistless— of him who engaged the services and paid for them, they being his inducement and compensation, they being not for temporary use but perpetual use, a provision for a business, a facility in it and an asset of it, therefore, contributing to it whether retained or sold—the vendee (in this case the Standard Company) paying for it and getting the rights the vendor had (in this case, the Axle Company)."

A **shop right** is the employer's nonexclusive, nonassignable license to use an employee's invention. It is limited to the particular employer involved and does not imply that compensation or royalties will be paid to the inventor. It arises from the employment relationship rather than from an express agreement to assign or license patent rights. Both the inventor and the employer have certain rights relating to the inventions; neither can totally exclude the other.

An employer's shop right to inventions created by his employees on company time and with the employer's facilities is an implied right derived from equitable principles. In some cases, the right may be based on the idea that it would be unfair to allow the employee to use the employer's resources and then try to refuse the employer any right to use the invention. However, the duty to assign patent rights may be made the subject matter of a contract, and as such are subject to state law. Often, an employee signs an agreement in which any patent rights are assigned to the employer. The duration of the agreement may be made to run for a period beyond termination of the employee's services; the scope of the agreement may be made to go well beyond things the employee might invent that would immediately benefit his employer. There is a limit, however, to the all-encompassing extent to which such agreements may be taken. An excessive length of time which may practically force a scientist or engineer to be tied to one employer is considered against public policy. It is common for such agreements to run for a year beyond termination of the employee's services. In essence, however, the question of an employer's rights to a patent obtained by a former employee is determined by when the creative work was done. If the work on which the patent was based was done for the former employer and if he had a right to the assignment of patents obtained by the employee, he may maintain an action for assignment of the patent.

It has become almost standard practice to require new engineers and scientists to agree to assign patent rights. Some companies even include an agreement to assign patent rights in their employment application form. In many instances this is merely an added precaution taken by the company. If an employee has been specially hired and retained to do research aimed toward obtaining patents, the patents so obtained must be assigned to the employer, even without an agreement to assign. On the other hand, the right of an employer to patents obtained by someone not hired to invent is a matter of shop rights, unless there exists a contract to assign. Because companies frequently move their technical personnel into and out of research as occasions demand, the precaution of an agreement as to patents seems well founded.

Consider Mr. Black and his engine. Assume him to be an engineer for the White Manufacturing Company. If Black developed an engine as a result of research and development endeavors for which the White Company paid him a monthly salary, Black's patent would have to be assigned to the company. If he was hired as a manufacturing engineer, but spent part of his time developing the engine with company facilities, the company would be entitled to at least shop rights in the engine. With shop rights the White Company could still make, use, and sell the engine even though the patent was granted to Black.

Mr. Black may have used his own facilities in developing the engine at home in the evenings. Would the White Company then have a right to his invention? Yes, if Black was hired to invent or if he had agreed to assign patents to the White Company as a condition of his employment. An oral contract to assign future patent rights may be enforced if the contract terms can be proved.

The Patent Application Procedure

The procedure involved in applying for and obtaining a patent ostensibly allows an individual inventor to obtain a patent on his invention without outside help. Why, then, should a prospective patentee spend money to obtain the services of a patent attorney or a patent agent? There are two main reasons for the expenditure. First, although the preparation and filing of a patent application (as outlined here) may appear fairly simple, technicalities and complications may arise which require knowledge that only a person with specialized knowledge and skill as to the applicable rules could possess. Second, unless the patent is properly prepared to begin with, the patentee is likely to find that the wording he used (which was so clear to him) actually offers little or no protection.

Usually, the patentee hopes to make money on the potential patent by selling or licensing it to another, by using it himself, or by manufacturing and selling the product himself. An application for a patent, then, is usually based on economic motives. Inventors try to get the greatest possible coverage for their inventions. The extent of protection for an invention is largely determined by the wording of the claims. An experienced patent attorney or agent always tries to word claims in such a manner that the broadest possible coverage is obtained.

There is another prominent motive for obtaining patents, one in which the profit resulting from the patent is not so apparent. It stems from the fact that realistic limitations must be imposed upon the claims in any patent. The coverage in the claims of any patent can extend only so far as justified by the nature of the invention and its novelty. The novelty of the invention, in turn, is primarily a function of the amount of prior art on point. Hence, an improved widget will likely have narrower claims if widgets have been around for a long time. To prevent others from entering the field in which the patent is useful, some companies attempt to patent things similar to what is already covered. A simple example might be a patent covering a DVD recorder, a programmable DVD recorder, a DVD recorder and player, and so on. One company might file a patent for each embodiment, thereby getting lots of patents and making it more difficult for others to enter the market. This practice, known as **blocking,** is commonly used by many large enterprises. By obtaining numerous patents that cover lots of various features, the owner of the patents makes it harder for someone to enter the market without infringing. Because the practice causes frustration to outsiders, many criticisms have been leveled against it. It is lawful, however, unless the result would tend to give the patent holder an unlawful monopoly over an entire industry.

Preliminary Search for Prior Art.   One cannot assume that, just because something is not being manufactured and sold commercially, it is not disclosed by a patent or other publication. Somewhere in the several million United States patents, or in the multitude of foreign patents, there is quite likely to be something similar to the subject matter under consideration. The purpose of a **prior art search** is to ferret out any patents or publications relating to the subject matter of an invention to find out if a patent can be obtained on the invention, and, if a patent can be obtained, the potential limits of its claims. Such a search is not a legal requirement; rather, it is a practical expedient. It is quite possible to apply for a patent without a preliminary search, but it generally is not advisable.

The Application.   The key of any patent application is the specification (which includes the claims and drawings). The proper fee must accompany the application. Additional fees may be required for various amendments, responses, and the like, and still more fees are due for the allowance of a patent. Once issued, still more fees are due to "maintain" the patent in force.

Specification.   The purpose of a **patent specification** is to clearly describe the invention. The clarity of description must be such that any skilled person in the field to which it pertains could, by using the specification, reproduce the invention and use it. One or more drawings (or possibly a model in unusual situations) may be required to make the invention clear.

Drawings are required in all cases in which they are meaningful. This includes almost every type of invention except compositions of matter and processes. In applications for processes, flow

---

For a very thorough report on this issue referred to in *eBay Inc. v. MercExchange,* see FTC, To Promote Innovation: The Proper Balance of Competition and Patent Law and Policy (Oct. 2003), www.ftc.gov/os/2003/10/innovationrpt.pdf (last viewed 9/25/07).

charts are often quite important to show the method clearly. There are special rules, which are quite detailed pertaining to patent drawings and, for this reason, most applicants hire specialists to make them. Currently, applicants may submit "informal" drawings which, although not meeting all the detailed requirements, nonetheless provide an appropriate description of the invention during the examination of the patent application. A substantial portion of the body of a specification is usually devoted to describing the various views shown in the drawings and the functions of the components. Prior to the explanation of the drawing details, however, there should be a background of the invention (discussing the area to which it pertains) and a brief summary of the substance and nature of the invention and, possibly, its purpose. The claims follow the detailed description of the invention. The detailed description must set out the best mode of the invention contemplated by the inventor for accomplishing the goals of the invention.

The claims essentially state what is considered to have been invented and, therefore, what is reserved to the patent holder. From the applicant's point of view, the broader the coverage in the claims, the greater the rights that the patent will provide. It would be to Black's great advantage, for instance, to claim and patent "an internal combustion engine." Of course, such a claim would not be allowed; if it were, Black could force manufacturers of any kind of internal combustion engines to cease manufacturing the engines or pay him royalties. The principles of internal combustion engines are well known, however, so Black's claim would have to be restricted to a particular kind of engine. His claim might read "An internal combustion engine comprising . . ." with the remainder restricting the claim to the elements of the engine that are truly inventive. Copies of issued patents are available to the public. If a particular patent appears highly profitable, more people are likely to try to approach the invention as closely as possible without infringement. It is here that the scope of a patent's claims are really tested. Prospective producers of patented things may go to great lengths to avoid the payment of royalties and may try to "design around" a patent's claims. Of course, if a way around the claims exists, there is no reason why it should not be used.

Because it is difficult to tell in advance just what portion of an invention will turn out to be most important in the ensuing 20 years, a considerable amount of imagination must be used in drafting the claims. Sometimes an apparently insignificant component of a device becomes more lucrative than the invention itself.

The original claims in the patent application should include some claims written as broadly as the preliminary investigation will allow. If they are too broad, such claims may be narrowed through amendment. Anything not claimed will be abandoned to the public when the patent issues.

Oath.   Another essential component of every patent application is an oath taken by the applicant that he believes himself to be the originator of the thing for which he requests a patent. If appropriate, a number of persons may execute oaths stating themselves to be joint inventors. There are also a number of specific requirements for such oaths.

Examination of the Application.   After the patent application is prepared, the applicant submits it to the Patent Office. There, the application will be sent to one of the examining divisions (specializing in particular types of subject matter) and will be handled by a patent examiner. Generally, examinations take place according to the filing dates on the applications. An application may wait for six months or even a year before the examiner gets to it and sends a first "Office Action."

The application is examined for compliance with the law, and the invention, as set forth in the specification and the claims, is closely scrutinized. If the form and content of the application comply with the law, the examiner turns next to the questions of novelty and obviousness. The patent examiner searches not only what already has been patented, but also trade publications, newspaper articles, and even mail order catalog descriptions and goods for sale in local stores. The result of the examiner's work is communicated to the applicant or his attorney in a letter known as an **Office Action.** If the examiner quarrels with the novelty or nonobviousness of the invention or claims, all of the claims may be rejected.

The Office Action may allow some or all of the claims. If all claims are allowed, the patent will be issued promptly. If some claims are allowed and others rejected, the applicant may obtain a patent including the approved claims, or he may reword or further restrict the rejected claims to resolve the conflict between them and the prior art cited in the Office Action by the examiner. In this case, the applicant may consider amending the application. As a last resort, the applicant may choose to abandon the application.

Amendment.   Amendment of an application that has been wholly or partly rejected as a result of its first examination is nearly always complex. Sometimes a visit to the examiner by the applicant's attorney can clear up misunderstandings or indicate rewording of certain claims to make them satisfactory. Much of the attorney's time will be spent in studying the claims in the application and comparing them with other patents and publications with which the examiner has noted a conflict. Amendments to the application are then made. A separate document must be prepared that answers all of the examiner's objections and makes the appropriate deletions or substitutions to cure the defects. The examiner then makes the corrections upon submission of the amendment. The original numbering of the claims is retained, although some of the claims may have been deleted. That is, if there were twelve claims in the original application and Number 11 were to be eliminated, the last claim would still be numbered "12."

Response.   A response to an Office Action (such as the submission of the amendment) must take place no later than six months from the mailing date of the Office Action. If no response is made within the time limit, the application is considered abandoned. Another waiting period of from six months to a year may take place before the examiner again responds with an Office Action. These rounds of Office Action followed by response may continue indefinitely. If a year per round is assumed, considerable time may elapse between the original application and the final outcome. If the applicant persists in his pursuit of a patent, the rounds of Office Action and response will end either with the examiner finally allowing the claims and issuing a patent, or with the applicant's receipt of a final rejection.

Appeal.   A Patent Office "final rejection" is not always final. The examiner's decision may be appealed to a higher tribunal. Relatively few applications are finally rejected by the examiner and then appealed by the applicant. Generally, the applicant's attorney and the examiner can find common ground before an appeal is necessary.

There exists within the Patent Office a Board of Appeals and Interferences to which a finally rejected application may be taken. Although the Board of Appeals is a part of the Patent Office, there is no definite tendency for it to support the examiner's position. Frequently, the Board will overrule the examiner and allow claims that were previously rejected.

If the Board of Appeals upholds the examiner's position, a further appeal is still possible. The applicant may take his case to the United States Court of Appeals for the Federal Circuit, or he may file a civil action against the Commissioner of Patents in a United States District Court. If either action is successful, the court will order the patent to be issued. However, just as with court actions involving administrative boards, the Patent Office must be shown to be wrong for the court to overrule the examiner's decision.

Interferences.   The United States patent system is unique in that it is a "first to invent" system— the first inventor is entitled to the patent. Most other countries are based on a "first to file" system, where the inventor who files a patent application first is entitled to the patent. A few (very few, actually) of the patent applications filed in the United States become involved in interference proceedings. An **interference** arises when two patent applications claim the same thing. It may also arise as a conflict between an application and a patent that has been issued for less than a year. An interference is designed to determine who first invented the subject of the patent. As the result of the interference, one adversary wins and may proceed to prosecute the application; the other has nothing but experience as a reward for the trouble.

In an interference, just as in any other court case, the winner is determined by the evidence presented to the reviewing body (the Board of Patent Appeals and Interferences). Records of the conception of and the efforts to reduce to practice the invention must have been made and kept if a contestant is to have a chance of winning. This is one reason why nearly all companies require research staff members to keep notebooks of activities (with pages serially numbered and entries dated), and why the periodic witnessing of the contents is also required. Anyone capable of understanding the contents, by the way, may act as a witness to research notes. Interference actions are taken first to the Patent Office Board of Patent Interferences. Later appeals may be taken to the Court of Appeals for the Federal Circuit or to a United States District Court.

In determining who was the first inventor, two dates become important: the date of the conception of the invention, and the date of its reduction to practice. If one party proves that it first conceived the invention and also first reduced it to practice, that person is entitled to the patent.

Proof by one party that he or she first conceived the invention is not enough. The purported inventor may also be called upon to prove that he or she pursued the idea with reasonable diligence. If the other party was not the first to conceive of the invention, but was the first to reduce it to practice, the party may obtain the desired patent by proving that his adversary temporarily abandoned the idea. Inventions are often developed in secrecy, but too much secrecy can preclude obtaining a patent. Proof of reduction to practice requires testimony of someone who actually saw the invention in operation. (Remember that the filing of a patent application constitutes a **constructive reduction to practice,** as opposed to actually building an operable model, which constitutes an **actual reduction to practice.**)

Assume that Mr. Black's claims in his application for his engine have met with an interference. That is, one Ms. White filed an application for a similar engine and includes claims covering the same engine. The case is tried before the Board of Interferences. White proves by records that she has the earlier date of conception of the invention. If White can now prove an earlier date of reduction to practice, the patent will issue to her. Suppose Black built and ran his engine before witnesses while White was still at the drawing board. White may still get the patent if she pursued her invention with reasonable diligence, since she first conceived the engine. Black's hope for the patent rests on his ability to prove that White temporarily abandoned her endeavor for a substantial period—even four or five weeks might be enough. Here, White's records (if she made and kept them) will protect her rights; without records she could lose what is rightfully hers.

Allowance and Issue.   Few patent applications become involved in appeals; fewer yet in an interference. Most patent issues between applicant and examiner are settled satisfactorily in the early stages, and either a patent results or the application is abandoned. If the patent is allowed, the applicant is sent a notice of allowance and the patent will issue in due course.

Infringement and Remedies

The grantee of a patent has a lawful monopoly. The right would be virtually worthless, however, if it could not be enforced. A patentee may prevent others from encroaching upon the invention that is the subject of the patent. Often, the owner of the patent may recover damages and seek an injunction to prevent additional infringements of the patent.

In **patent infringement** actions, the defendant is charged with having made, used, or sold the invention in violation of the monopoly granted by the patent. In such actions, the claims are compared with the allegedly infringing device or process. It is here that unnecessary restrictions in claims are likely to prove costly; if the claim includes one or more elements that are absent from an accused device, there is no infringement. What the patentee has given up in the Patent Office cannot be rewarded in court. For instance, where a patent claim requires the use of a microprocessor, the use of a hard-wired control circuit in the defendant's device may mean that there is no infringement. If the original claims covered both the use of a microprocessor and a hard-wired circuit, and the patent owner deleted (or canceled) the hard-wired circuit claims in the face of their rejection by the Examiner, the hard-wired device clearly should not infringe.

Defenses.   The defendant in an infringement action essentially has two areas of defense. A defendant usually attempts to prove either that the patent is invalid or that the defendant's device or process is not an infringement. All patents are presumed valid. The defendant often attacks the validity of the patent, however, by arguing that the invention was not novel or was obvious or that, during the prosecution of the Patent Office, some "inequitable conduct" occurred, such as a failure to inform the Examiner of relevant prior art. The defendant may also claim there is no infringement. For example, the defendant may admit the acts complained of, but show that the claims are not sufficiently broad to cover these acts. (See Appendix B, Case No. 01-4, Patents—Dispute over Rights to Specify.)

Remedies.   Once patent infringement is found, the patent holder is entitled to damages. Until the Supreme Court's decision in *eBay Inc. v. MercExchange,* there was a general rule that a permanent injunction would always issue once patent infringement and patent validity had been determined by a judge. In *eBay,* the Court held that the decision to grant or deny injunctive relief is within the equitable discretion of the courts, but that such discretion should be consistent with traditional principles of equity, e.g., the four factor test that the plaintiff must show:

1. irreparable injury;
2. no adequate remedy at law;
3. in balancing the hardships between the plaintiff and defendant, an equitable remedy is warranted; and
4. and that the public interest would not be disserved

The Court found support for this position within federal copyright cases applying equitable principles for granting injunctions. While U.S. Courts may continue to grant injunctions to prohibit patent infringement, they cannot do so automatically according to the Supreme Court. Hence, the U.S. Courts often grant injunctions to prohibit patent infringement.

Patent law allows the court to go beyond mere compensatory damages and assess up to triple damages against the defendant if there is "willful" infringement. The patent owner's attorney's fees can be covered if the case is "exceptional." Infringement may be considered "willful" if the defendant knew of the patent but did nothing to avoid its coverage or make sure that the defendant did not infringe.

The measure of compensatory damages in an infringement case is the plaintiff's losses. However, the statute requires that the patentee be compensated at least in the amount lost or that would have been lost in royalties because of the infringement. A question may arise as to the amount of royalties to be assessed if the plaintiff has not made previous royalty arrangements with others. The court then determines and awards a "reasonable royalty."

Patent Markings.   If the plaintiff in an infringement case is to get any damages for the injury, the plaintiff must have informed the defendant of the patent. Marking patented goods with the patent number serves this purpose. If the goods are not marked, the plaintiff must prove that the defendant was informed of the existence of the patent and that the infringement continued.

Goods sold are often marked "Patent Pending" or "Patent Applied For." These terms have no legal force so far as infringement is concerned. However, if the terms are used when, in fact, no application has been filed, it may be assumed that the purpose was to defraud the public. The patent statute prohibits such acts.

Common Criticisms

One common criticism is the time and money required to obtain a patent. At first, the criticism seems well founded. As to time, few patents are issued in much less than a year or two after the application is first filed; some have required twenty years or more. This sounds like a long time to wait for patent protection but, as a practical matter, the delay is seldom a hardship, and it may be beneficial to the applicant. The greater the delay in securing a patent, the greater is the period of practical patent protection. Theoretically, the applicant is not protected while his application is being processed. However, should the applicant be threatened with an infringement, the patent office, upon request, will usually expedite examination of the patent to permit legal action upon the infringement.

The monetary outlay required for a patent is extremely variable, depending upon the complications that may arise. It is possible for the patent to cost only a nominal amount. However, it is often likely to cost somewhere around $10,000 or more, just for the attorney's fees. A patent is somewhat like other business ventures; its probable cost in time and money should be weighed against the expected results. If economic analysis shows that the venture is likely to be profitable, it should be undertaken. If the likelihood of profit is remote, perhaps it is better kept confidential and used as a trade secret.

Occasionally, the objection is heard that a small inventor has no fair chance to compete with the research staffs of big business. It is true that research staffs are brought together to form a talented team, and many times the effort pays off. But even in a research team the best ideas often flow from one individual or a small nucleus of people. Once the idea occurs it is often pursued more rapidly by teamwork, but if an independent inventor has the first date of conception and has pursued it diligently, the speed of a corporate research team is no help.

Members of industrial research staffs sometimes complain that they are not adequately compensated for patents they obtain and assign to their employers. Perhaps the criticism is valid, for many receive little or no compensation for the assignment. However, the opposite argument is that the staff member is paid not only for the time during which he had produced something worthwhile, but also for the other time when he was not so productive. It is also argued that the research job, considering only the activities involved, is more enjoyable to many people than the alternatives. The employer also has an investment in research facilities, which must return a profit.

Corporations often are criticized because, with large funds available, they can indulge in long legal battles to win infringement or interference suits from less fortunate competitors. Sometimes the mere threat of a long legal battle is sufficient to intimidate a small inventor. While this criticism has merit, there is still an end to all legal battles. There is another aspect to this particular problem. Because of the large costs involved and the unpredictability of the outcome of such cases, large businesses frequently pay out-of-court settlements rather than undertake a long, drawn-out court battle. That is, if an inventor claims that a manufactured product infringes his patent and demands a small royalty, the manufacturer may purchase the patent or pay the royalty rather than go to court, even though the particular patent might be only remotely related to his product.

One criticism that seems to have greater validity than most is concerned with the public welfare. Suppose someone invents something of great public benefit—perhaps AIDS medication. Under the present law, it would be possible for the inventor to obtain a patent monopoly, make very limited amounts of the medication, and sell it at enormous prices. All but wealthy people would be deprived of the medicine—a rather appalling prospect. (See *GlaxoSmithKline v. AIDS Health Care Foundation,* U.S. District Court, Western District of California, case number CV-03-02792-TJH decision rendered July 25, 2003, for an interesting review of the public welfare issue.)

Other criticisms are directed at attempts to extend the scope of patent monopolies by pooling arrangements, tying clauses, marketing agreements, and the like. Some of these arrangements are lawful; others are not. The courts carefully scrutinize such arrangements under the antitrust laws. Despite shortcomings in the law, protection is afforded to the inventor, and the rate of technological advances testify to its success.

The U.S. Constitution succinctly explains why patent laws are so important—they promote the progress of science and engineering. In essence, the legal patent monopoly acts as an economic incentive to encourage innovation, as evidenced by better medicines, less expensive goods, and more efficient means of production.

TRADEMARKS

The good will attached to a trademark or a trade name is an example of intangible property. Trademarks are like other forms of property in that they can be bought, licensed, assigned, sold, and used by the owner. A trademark can be a word, design, or combination, that is used by the trademark owner to identify goods or services. A trademark differs from other property in that its value is in identifying the products of a particular manufacturer and distinguishing the goods of one manufacturer from similar goods of other manufacturers.

History

Trademarks are products of the Industrial Revolution. When goods were manufactured primarily for local sale, there was little need to identify the manufacturer. With specialization in manufacture and expanded marketing, successful advertising came to require that the manufacturer's products be identified in some way.

Who Can Be Protected by Trademark?

In the United States, the law regarding trademarks is based primarily on the common law. In addition, the federal Lanham Act of 1946 and state statutes provide trademark protection. Under the Lanham Act, a trademark is defined as including any word, name, symbol, or device or any combination thereof adopted and used by a manufacturer or merchant to identify his goods and distinguish them from the goods manufactured or sold by others. Similarly, a service mark may be any word, name, symbol, or device used to identify and distinguish services.

Trademark Rights

The common law established rights in the use of trademarks as well as penalties for abuses. Federal and state legislation (where statutes have been passed) have modified the established rules only slightly. The federal Lanham Act, which covers marks used in interstate commerce, provides for the registration of marks. Under the common law, trademark rights extended only to the geographic areas in which the mark was used. Trademark rights are established by the first user of the mark; such rights arise out of the use of the mark in connection with the sale of the goods.

A state registration expands those common law rights to the boundaries of the state. Similarly, a federal registration expands those common law rights to the boundaries of the United States. Since 1989, it has been possible to apply for federal trademark registrations of marks that had not been used as of filing of the application. To obtain the registration, however, the applicant must begin using the mark and submit proof of such use to the Trademark Office. Registration of the mark is valid for 10 years; if the mark is still in use after the 10-year period, it is renewable upon application by the party with authority to use it (15 USC § 1058).

A unique feature of the trademark right is that it can be established simply by using the mark in commerce. For example, if Company A designed a logo for a new product and sent out post-cards to potential buyers with the logo imprinted on them, this would be enough to establish the use of the trademark in commerce, as long as Company A had a "bona fide" intention to use the logo as a trademark and not merely to prevent others from using it.

Another unique feature of trademark rights is that they can continue indefinitely as long as the trademark is in use and not abandoned. If the trademark owner has filed with the USPTO, however, the owner must renew registration every 10 years.

Requirements.   In order for trademark rights to be secured, one has to use the mark, as indicated above. However, if a person wants to secure federal trademark protection, the applicant must check whether that word or phrase comes under one of the following four categories.

- Generic
- Descriptive
- Suggestive
- Arbitrary or fanciful

Generic.   Many marks and names are successfully used to identify various products, but the more common the mark or expression used, the more difficult it is to qualify for a trademark. It would, for instance, be difficult to obtain a trademark right for the word "pencil" or "automobile" because these words are generic of those types of goods. A word or phrase that is generic is incapable of qualifying for trademark protection.

Unfortunately, some words that once qualified for trademark protection have fallen into common usage and no longer have the strength they once had; therefore, rights in any trademark or trade name may be lost if the word, after use, becomes established as a term descriptive of many products. "Aspirin" was once a trademark of only one product; so was "Cellophane." For this reason, companies sometimes advertise in an attempt to persuade the public not to use the mark in ways that generically describe its goods or services.

Descriptive.   A **descriptive trademark** is a mark that describes the goods, their characteristics, functions, uses, or their ingredients. Whereas the word "restaurant" may be generic, the mark "Steve's Great Chinese Restaurant" is descriptive. The distinction between generic marks and descriptive marks is important, because descriptive marks can, in some situations, be legally protected; generic marks are never protected. The Lanham Act prohibits the registration of a mark that is "merely descriptive," that is, a mark that simply describes the good or service (15 USC §1052). However, a merely descriptive term may qualify if it has established **secondary mean-**

ing. Secondary meaning can be established if the word or phrase of the mark is capable of denoting to the consumer "a single thing coming from a single source.[1]

Just as descriptive terms would be difficult to establish as trademarks, geographical names and family names present problems. Descriptive terms, as well as geographical and family names, may become protected as trademarks if they are used and advertised sufficiently so as to identify a source of the goods or services. A new company manufacturing automobiles could not use the Ford name to identify its products, even though the owner might be named Ford. Some states, however, still take the view that everyone has an "absolute" right to use their name in their business. New competitors in a field have a right to identify themselves and the location of their businesses, but not in such a way that the trademarks or trade names of others are infringed.

Suggestive.   Because some marks fall between descriptive and arbitrary or fanciful, a separate category was created. Words or phrases may qualify for trademark protection if they are suggestive of the product or service provided. A useful way to determine whether a phrase or word qualifies for trademark protection because it is **suggestive** is to compare it with this standard:

"A term is suggestive if it requires imagination, thought, and perception to reach a conclusion as to the nature of the goods."[2]

Arbitrary or fanciful.   The strongest trademarks are newly coined terms that have nothing to do with the goods, that is, the terms are arbitrary or fanciful. Strong trademarks also often result from the use of words ordinarily not connected with a description of the specific product. The use of "Bluebird," for instance, would be an arbitrary or fanciful term for a television set. Many companies have made up their own trademarks or trade names. For example, "IVORY" is an arbitrary term for soap.

Licensing and Franchising.   One right that is frequently exercised is the right to license a mark. The term **license** means a type of permission to do something or, in the case of a trademark (or other type of intellectual property), the right to use the mark (or other property). A **franchise** usually refers to a specific type of license. The complexities of license arrangements range from a basic license to use a trademark, to a complex system in which the licensee receives the right to use the trademark, confidential know-how useful in the production of a good or service, and the benefit of widespread promotion. The latter arrangement probably would be considered a "franchise." Licensees (those who have a license from another) must submit to strict control of the quality of their product. Licenses and franchises are generally deemed beneficial to the public in that they tend to promote competition and assure a certain standard of quality. They provide an efficient means of combining centralized planning, direction, and standard-setting with a degree of local control and initiative.

The franchising coin has another side, however. If the use is such as to defraud the public, the trademark involved will not be protected. Consider the Brown Construction Company, a company so eminently successful in construction contracting in a particular city that it expands to new markets. Because its trade name and logo have become symbolic of high quality in construction, it expands by a licensing arrangement in which it reserves the right to dictate the company management and quality of construction, requiring a monetary return for the use of its well-promoted name. (Similar arrangements are common in the fast-food, automotive-repair, and numerous other

---

[1]*Zatarain's, Inc. v. Oak Grove Smokehouse, Inc.*, 698 F.2d 786 (5th Cir. 1983).

[2]*Stix Products, Inc. v. United Merchants and Manufacturers, Inc.*, 295 F.Supp. 479, 488 (SDNY 1968).

industries.) Suppose the services provided by the Brown Construction franchisees come to be far different from what the original reputation and the promotion imply. Under such circumstances, Brown risks not only legal actions against it for the poor performances of its franchisees, but also the loss of good will of the trademark. Thus, Brown has legal as well as practical incentives to be cautious in licensing its trademark, and in monitoring the quality of its licensees.

Trademark Transferability. Like patents, trademark rights are property rights, and can be sold, bought, assigned, etc. Most often when trademark rights are transferred, it is pursuant to a company sale where the purchaser also acquires any company "brands" which are frequently protected as trademarks.

The Trademark Application Procedure

The basic requirements for a federal trademark application include identifying the mark, selecting and specifying the mark's classification, listing the applicant's name and address, identifying the goods or services, and noting whether or not the mark was used in commerce at the time of the application. And, of course, paying the required fees.

There are two registration categories for classes in the USPTO: (1) U.S. Classification, and (2) International Classification. When you fill out the application, you can apply for coverage under both. The trademark applied for can be one of four types:

1. Brand name
2. Service mark
3. Certification mark
4. Collective mark

A **brand name** identifies a particular good, for example, Reebok® for athletic shoes. **Service marks** identify what the name implies, for example, Jiffy Lube® for oil change and car maintenance services. **Certification mark** identify those goods or services that meet certain qualifications. An example of a service mark is "UL," which stands for Underwriters Laboratories, Inc. A UL® mark indicates that a particular appliance meets the safety standards of Underwriters Laboratories. **Collective marks** identify goods, services, or members of a collective organization, such as the AFL-CIO.

Registration Refusal. As noted above, trademark registration may be refused if the term or phrase is generic or merely descriptive. The most common reason for registration refusal is if the mark is likely to cause confusion, that is, if the mark will confuse the public as to the source of the product or service provided.

Domain Name Registration. For the past several years, domain name owners can register a domain name as a trademark. The USPTO has published guidelines to address potential problems related to domain name registration.[3] The USPTO only considers the word to the left of the top level domain, (e.g., .com; .org; .gov; etc.) to determine whether the mark qualifies for registration, and then applies the same principles it uses for determining whether a con-

---

[3]Examination Guide No. 2-99, Marks Composed, In Whole or In Part of Domain Names (USPTO September 29, 1999) available at www.uspto.gov/web/offices/tac/notices/guide299.htm.

ventional mark would qualify. Specifically, whether the term or phrase is generic, descriptive, suggestive, or fanciful, as well as whether the mark is "confusingly similar" to a previously registered mark.

The Internet Corporation for Assigned Names and Numbers (ICANN) is tasked with the technical management of the domain name system. It approved the Uniform Domain Name Resolution Policy (UDRP) that governs alternative means to resolve domain name disputes.[4] There are four recognized alternative dispute settlement bodies that can resolve domain name complaints, including the World Intellectual Property Organization (WIPO), the National Arbitration Forum (NAF), eResolution, and CPR Institute for Dispute Resolution.

The UDRP only applies to disputes where "a domain name is identical or confusingly similar to a trademark or service mark in which the complainant has rights, the domain name registrant has no rights or legitimate interests in respect to the domain, and the domain name has been registered and is being used in bad faith" [UDRP §4.1.(i)]. Through the UDRP, there are two remedies:

1. Cancellation of the registration, or
2. Transfer of the domain name.

Infringement and Remedies

A trademark may be infringed if a person uses the same or a "confusingly similar" term or phrase to describe the same or a related product in the same geographical area (state-wide or nationwide). Section 32 of the Lanham Act allows a party to enter a cause of action against another who uses its trademark in commerce.

A mark is "confusingly similar" to another if there is a likelihood of confusion, mistake, or deception. The legal determination of whether a likelihood of confusion, mistake, or deception exists is a subjective determination by the trier-of-fact. The determination is made following the consideration of many facts. Elements considered in making such a determination (oftentimes called the "Polaroid Factors") include:

- The fame of the prior mark, including information about sales, advertising, and the length of time the mark has been used, i.e., the strength of the mark.
- The similarity of the names or marks in their appearance, sound, connotation, and commercial expression.
- The similarity in the nature of the goods or services of businesses in question.
- The likelihood that the defendant will "bridge the gap" between the goods or services it offers and those of the trademark holder.
- The nature and extent of any actual confusion.
- Whether or not the defendant adopted the mark in "good faith."
- The quality of the defendant's product or service.
- The conditions under which, and the buyers to whom, sales are made, i.e., the sophistication of the buyers.[5]

---

[4] www.icann.org/udrp/udrp-rules-24oct99.htm.

[5] *Polaroid Corp. v. Polarad Elects. Corp.*, 287 F.2d 492 (2d Cir.), *cert. Denied*, 368 U.S. 820 (1961). The elements have been elaborated upon in several of the circuits.

The relative "strength" or "weakness" of a mark also becomes important in determining whether a likelihood of confusion, mistake, or deception exists. If many different business entities exist that are using a particular word in connection with their goods and services so that the word has become very diluted or "weak," then a user of that word will have great difficulty convincing a court that some subsequent third-party use of that same word should be prohibited on the basis that it is likely to cause confusion. The courts generally conclude that, since so many people are already using the word, no harm can result from additional uses of the same word.

A person who subsequently uses an established mark or one that is confusingly similar may be enjoined from continuing the use of the mark. If the use of the mark was an intentional violation of another's rights, the court may award damages in excess of actual damages (including profits) if the acts were committed with knowledge that the person's acts would cause confusion, mistake, or deception, as well as the plaintiff's attorney's fees. If the trademark is registered with the USPTO, the plaintiff is entitled to treble damages if the infringement is willful (15 USC §1117).

COPYRIGHTS

Copyright laws protect expression manifested in a tangible medium. It does not protect an idea, but rather the manner in which the idea is expressed. The United States Constitution provides Congress with the right to enact copyright laws. Over the years, Congress has exercised that right and created numerous different copyright acts.

Perhaps the easiest way to grasp the concept of a copyright is to visualize the distinction between the original physical copy of the author's work and imagine, hovering over that work, the intangible rights of copyright provided by the federal copyright system. While someone might purchase the original physical copy of the author's work, such as an original painting, and have the right to look at and enjoy the painting, unless they also own the intangible rights of copyright in that work, they do not have the right to make copies of the work, to use the work to prepare derivative works, or to distribute copies of the work to the public. They do not even have the right to publicly display the work except under certain conditions.

History

The first known copyright statute was the Statute of Anne, instituted in England in 1710. It gave writers the rights to their own works, rights that had previously been held exclusively by stationers. The statute was in response to the perceived monopoly of the press by stationers and publishers. Authors were given protection of their work for 14 years from the date of first publication. It was designed in an effort to promote the public interest.

The adoption of similar copyright laws in the United States was based on two theories.

1. Copyrights should serve the interest of the public.
2. Copyright protection could serve as encouragement to authors and others engaged in artistic endeavors, because their rights in the creative works would be protected, and therefore, used to make money.

The first United Stated copyright act, passed in 1790, had features similar to the Statute of Anne. It was revised substantially in 1909 at the request of President Theodore Roosevelt. The next major overhaul, in 1976, settled some confusing areas of copyright law, including the duration of a copyright, whether and how a copyright notice was required, the subject matter of a copyright, and federal preemption of common law copyright.

What Is Copyrightable?

Under the copyright system, the copyright laws automatically provide copyright protection to all original works of authorship that are fixed in any tangible medium of expression. The requirement of "originality" is met by most works, so long as they embody some creative effort. Occasionally, however, a court will hold that a work fails to meet the originality requirements, especially if the work consists of factual or historical information. As soon as the work is fixed in some tangible medium of expression, the copyright laws come into play. If the work is kept only in the mind of the author and is not fixed in some tangible medium of expression, the federal Copyright Act does not apply. But once the work is reduced to writing, or recorded on a CD or in a photograph or the like, the federal Copyright Act automatically applies and the intangible rights of copyright exist.

Congress specifically provided that the federal laws apply to all "works of authorship" fixed in a tangible medium of expression. Specifically, the following may be eligible for protection under the copyright laws:

- Literary works
- Musical works, including any accompanying words
- Dramatic works, including any accompanying music
- Pantomimes and choreographic works
- Pictorial, graphic, and sculptural works
- Motion pictures and other audiovisual works
- Sound recordings
- Architectural works

The Idea/Expression Dichotomy.   Congress specified that copyright protection for an original work of authorship does not extend to any idea, procedure, process, system, method of operation, concept, principle, or discovery, regardless of the form in which it is described, explained, illustrated, or embodied in an original work of authorship. This statutory pronouncement by Congress maintained a longstanding distinction evolved by the courts in their interpretation of the federal copyright laws; namely, the federal copyright laws do not protect ideas. They only protect the form of expression of an idea.

> See the *Suntrust Bank v. Houghton-Mifflin* 268 F. 3d 1257; 2001 U.S. App. LEXIS 21690 (11th Cir. 2001), where a parody of Margaret Mitchell's *Gone With the Wind*, entitled *The Wind Done Gone*, by Alice Randall survived a copyright infringement claim in part by effectively using the idea/expression dichotomy.

Perhaps the best way to understand this distinction is by example. Suppose that one came up with the idea of writing a novel about a debonair gentleman living in the South at the time of the Civil War. This would be an idea. But taking that idea and fleshing it out into a novel about Rhett Butler constitutes the form of expression of that idea. By way of another example, suppose that one came up with the idea for a new system of accounting. The idea for the new system would not be covered by the federal copyright laws. But descriptions in a textbook of the system constitute a form of expression of that idea, and are protectable under the federal copyright laws.

Copyright Rights

Exclusive Rights.   The rights afforded a copyright owner are exclusionary in nature; those rights allow the copyright owner certain exclusive rights with respect to the work and the right to prohibit others from engaging in those actions. Under the federal copyright laws, the owner of a copyright obtains six intangible exclusive rights to:

1. Reproduce the copyrighted work in copies or phonorecords;
2. Prepare derivative works based upon the copyrighted work;
3. Distribute copies or phonorecords of the copyrighted work to the public by sale or other transfer of ownership, or by rental, lease, or lending;
4. Perform publicly in the case of literary, musical, dramatic, and choreographic works, pantomimes, and motion pictures and other audiovisual works;
5. Display in the case of literary, musical, dramatic, and choreographic works, pantomimes, and pictorial, graphic, or sculptural works, including the individual images of a motion picture or other audiovisual work; and
6. Perform the work by means of a digital audio transmission in the case of sound recordings (17 USC §106).

This bundle of rights can be broken up, and each right can be sold or licensed independently of the others.

Unlike the federal patent system, the rights afforded a copyright owner do not apply as against everyone in the world irrespective of how they may have happened to have created an identical work. For the copyright owner to recover against an alleged third-party infringer, the copyright owner must prove that the infringer actually copied the copyrighted work, or actually distributed copies of the copyrighted work, or actually used the copyrighted work to prepare derivative works. If the alleged infringer can show that he independently created an identical work, there is no copyright infringement.

Moreover, once the copyright owner has sold a copy of the work, unless specifically provided otherwise in an enforceable contract or provided by the copyright laws themselves, there is no limitation on certain of the buyer's rights to "use" the work. For example, when a copyrighted novel such as *Gone With the Wind* is sold to a buyer, the buyer, although he cannot copy the work or use the work to prepare derivative works based thereon, can "use" the work by reading it as many times as he wishes, or he can sell, lease, or rent that copy of the work to another, or he can mutilate or destroy the work. The same is true of copyrighted drawings, manuals, or the product literature a company sells to third parties without any specific limitations on how the third parties may use those works.

Architectural Works.   Congress passed the Architectural Works Copyright Protection Act of 1990. For many years, our copyright laws protected architectural plans, drawings, and blueprints. However, protection for the design features of a building itself were not protected. Hence, you could freely copy the overall shape, style, or design of a building, but you could not copy, for instance, the blueprints for the same building.

When the United States decided to join the Berne Convention, the United States was required to protect "architectural works" to meet its treaty obligations. Thus, the Architectural Works Copyright Protection Act was adopted to meet the requirements of the Berne Convention. An "architectural work" is defined by our copyright law as

The design of a building as embodied in any tangible medium of expression, including the building itself, architectural plans, or drawings. The work includes the overall form as well as the arrangement and composition of spaces and elements in the design, but does not include individual standard features.

Obviously, deciding what constitutes the "design" or "overall form" of a building sometimes is easy. In other cases, it will be difficult. Also, what are the "individual standard features" that should not be protected? Are these items things such as doors, windows, or stairs?

The owner of a copyright in an architectural work has the exclusive right to make and distribute copies and to prepare derivative works. However, these rights are subject to "fair use" by others. In addition, Congress adopted laws to specifically provide that making, distributing, or displaying pictures, paintings, or photographs of a building that has been built is not an infringement. However, this provision applies only when the building is located in or is ordinarily visible from a "public place." For similar practical reasons, Congress adopted a provision to allow the owner of a building to make or authorize alterations or the destruction of the building. (See *Moser Pilon Architects, LLC, et al. v. HNTB Corporation, et al.* at the end of the chapter.)

Copyright Term.   The duration of the intangible rights of copyright in a work created on or after January 1, 1978, is as follows:

- In General—Copyright in a work created on or after January 1, 1978, subsists from its creation and, except as provided by the following subsections, endures for a term consisting of the life of the author and 70 years after the author's death.
- Joint Works—In the case of a joint work prepared by two or more authors who did not work for hire, the copyright endures for a term consisting of the life of the last surviving author and 70 years after such last surviving author's death.
- Anonymous Works, Pseudonymous Works, and Works Made for Hire—In the case of an anonymous work, a pseudonymous work, or a work made for hire, the copyright endures for a term of 95 years from the year of its first publication, or a term of 120 years from the year of its creation, whichever expires first.

Once the term of the copyright expires, anyone who lawfully obtains possession of the work or any copy thereof can slavishly copy it. For example, if one could lawfully obtain access to an original Remington western painting hanging in a museum, because the term of the rights of copyright therein has expired one could photograph the painting (thereby creating a derivative work that carries its own independent right of copyright) and sell the prints. For this reason, many museums are very careful to whom they loan their paintings, and prohibit their patrons from photographing the works. This does not preclude a person who reproduces a visual work in its entirety from attributing the work to the original creator under the copyright laws. (See §106A.)

Exclusive Rights Exception—Fair Use.   One important limitation on the exclusive rights granted to the copyright owner is the doctrine of **fair use**. The idea that certain uses of copyrighted works were "fair" and therefore should not be an infringement of copyright in a work first developed in the courts. However, as part of the comprehensive redevelopment of the federal copyright law, Congress codified the fair use doctrine in what is now Section 107 of the Copyright Act of 1976.

Section 107 provides that:

[T]he fair use of a copyrighted work, including such use by reproduction in copies or phonorecords or by any other means specified by that section, for purposes such as criticism, comment, news reporting, teaching (including multiple copies for classroom use), scholarship, or research, is not an infringement of copyright. In determining whether the use made of a work in any particular case is a fair use the factors to be considered shall include:

1. The purpose and character of the use, including whether such use is of a commercial nature or is for nonprofit educational purposes;
2. The nature of the copyrighted work;
3. The amount and substantiality of the portion used in relation to the copyrighted work as a whole; and
4. The effect of the use upon the potential market for or value of the copyrighted work.

The fact that a work is unpublished shall not itself bar a finding of fair use if such finding is made upon consideration of all the above factors. Note that the four factors listed are only examples, not an exclusive list.

Of course, the determination of what constitutes fair use varies from case to case, thereby lessening the precedential value of any particular case. In addition, none of the four factors set forth in Section 107 is determinative, and the weight given to each of those factors varies from case to case. On the other hand, the effect of the use of the work on the commercial market for the work often is regarded as a most important factor. Because the purpose of copyright law is to provide an incentive for authors, uses that adversely impact the commercial market for an author's work are viewed as frustrating the intent of the federal copyright laws.[6] Consequently, the parameters of fair use are narrowed when a particular use of a copyrighted work is commercial in nature, and seems likely to eliminate or adversely affect the market for the copyright owner's work.

Copyright Transferability

The federal copyright laws do not govern the ownership and transfer of the original physical copy of the work. Normal state concepts of contract law govern those matters. However, the federal copyright laws do govern the ownership, transfer, and protection of the intangible rights of copyright that flow from the creation and fixation of the work in a tangible medium of expression.

Works Made For Hire.   The federal Copyright Act contains specific provisions governing the ownership and transfer of the rights of copyright. The act generally provides that the individual who creates the work is deemed to be the author of the work and, as such, is the initial owner of all rights of copyright. There are only two situations when some other entity is deemed to be the "author" of the work. Both of those exceptions are referred to as **works made for hire.**

The first exception occurs when the individual who created the work did so in the course and scope of his employment. In that situation, the employer is deemed to be the "author" of the work and is the owner of all rights of copyright. The various federal courts that have addressed this issue recently have reached somewhat different results as to when certain persons are to be considered an "employee" for purposes of the copyright laws. The second exception occurs when the individual who creates the work was commissioned to create the work (on an independent-contractor

---

[6]See *Sony Corp. of America v. Universal City Studios,* 464 U.S. 417 (1984) and *Campbell v. Acuff-Rose Music, Inc.,* 114 S.Ct. 1164 (1994).

basis) for one of nine specified purposes, and a written contract is signed by the independent contractor stating that the work is a "work made for hire." The nine specified purposes are

1. A contribution to a collective work
2. A part of a motion picture or other audio visual work
3. A translation
4. A supplementary work
5. A compilation
6. An instructional text
7. A test
8. Answer material for a test or
9. An atlas if the parties expressly agree in a written instrument signed by them that the work shall be considered a work made for hire (17 USC §101).

Unlike the prior version of the law, the Copyright Act provides that the sale or other transfer of the physical object (such as the manuscript, tape, etc.) in which the copyrighted work is initially embodied (or the sale or transfer of any copy thereof) does not in and of itself transfer any of the rights of copyright. The Copyright Act provides that the only way to effect a valid transfer (other than by operation of law, such as when the owner dies) is by a written instrument of conveyance signed by the owner of the rights conveyed, or his duly authorized agent.

Applying these principles, if an individual employed by a company creates a copyrightable work in the course and scope of his employment, then the company is deemed to be the "author" of that work, and is the owner of the work, as well as the exclusive rights of copyright associated with the expression used in the work. This means that the company has the right to prepare derivative works from the original work, to make copies of the work, and to distribute those copies by sale, license, or otherwise. The employee who created the work retains none of these rights.

However, if an employee of the same company creates a copyrightable work outside of the scope and course of his employment, then the employee is deemed to be the "author" of the work and the initial owner of all of the exclusive rights of copyright. The company may, depending on whether the individual used the company's time, facilities, and materials, have a right to use the ideas embodied in the work. However, the company will not have the right to reproduce, copy, or market the form of expression of the ideas covered by the rights of copyright.

The result is perhaps even more dramatic when a company commissions an independent contractor to create a work under his own control that is the subject matter of copyright. Assume that a corporation commissions a professional photographer to make photographs of the corporation's equipment for use in the corporation's company report. The photographer provides the corporation with the negatives. Assume that a written contract that included the magic words "work made for hire" was not utilized and that such photographs and the company report do not fall within any of the nine specific uses provided by the Copyright Act. In this case, because the photographer is not an employee of the corporation, the photographer is deemed to be the "author" of the works and the initial owner of all of the rights of copyright. While the corporation will have the right to use the photographs for the specific limited purpose authorized by the photographer, the corporation will not have the right in the future to reproduce the work, or to use the negatives to create additional works, or even to make copies of the work without the approval of the photographer. At an extreme, the photographer might be able to prevent the copying and distribution of the reports with his or her photographs.

Copyright Registration Procedure

The Copyright Act provides that at any time during the subsistence of copyright in any published or unpublished work, the owner of copyright in the work may obtain registration of the copyright claim by delivering to the copyright office the required deposit, together with the application and the required fees. The federal Copyright Act also provided that no action for infringement of the copyright in any work shall be instituted until registration of the copyright claim has been made. However, the Berne Act changed U.S. Laws to allow actions to be brought for the infringement of "Berne Convention works whose origin is" outside the United States, even if no registration has been obtained. Thus, although rights under the Copyright Act are maintained for new works even if a statutory copyright notice is omitted, another legal action that must be taken for a U.S. author to file suit to enforce rights of copyright is registration of the work if it is of U.S. origin.

One advantage in obtaining **early** registration of copyright is the ability to obtain **statutory damages.** The act provides that a copyright infringer is liable for either (1) the copyright owner's actual damages **and any** additional profits of the infringer, or (2) statutory damages. The act defines the **statutory** damages that the court may award to the copyright owner as a sum of not less **than** $750 or more than $30,000 as the court considers just for each infringing work. If the court finds that the infringement was "willful," the court may increase the award of statutory damages to the sum of not more than $150,000 for each infringing work.

The act also provides that the court may also award reasonable attorneys' fees to the prevailing party. No award of statutory damages or of attorneys' fees, however, may be made for any infringement of copyright commenced after first publication of the work and before the effective date of its registration, unless such registration is made within three months after first publication of the work. Consequently, if enforcing a copyright in the courts appears possible, registration is desirable, to avoid losing the opportunity to recover statutory damages and attorneys' fees.

Infringement and Remedies

Because it is often difficult for the copyright owner to prove that the infringer actually copied the work or used the work to prepare derivative works, the courts have developed the doctrine that copyright infringement may be established by circumstantial evidence. The only evidence required to support a finding of infringement is that (a) the alleged infringer had access to the copyrighted work, and (b) the alleged infringing work is "substantially similar" to the copyrighted work. Determining whether or not two songs, books, movies, or other works are "substantially similar" may involve detailed expert testimony, as well as the impression that the two works have on the fact finder.

Remedies include injunctions, impounding and disposition of infringing articles, damages (actual and statutory), profits, costs, and attorneys' fees. Due to the rampant copyright infringement occurring over the Internet, Congress needed additional enforcement measures to combat the problem.

The Digital Millennium Copyright Act (DMCA) went into effect October 28, 1998. This act addresses specifically those who thwart a lawful copyright owner's protection against infringement, including computer hardware and embedded code designed for such purposes. The DMCA, in addition to amending several sections of the Copyright Act and adding a new chapter 12, states in its section 1201 regarding the circumvention of copyright protection systems that:

> No person shall circumvent a technological measure that effectively controls access to a work protected under this title. . . .
> No person shall manufacture, import, offer to the public, provide, or otherwise traffic in any technology, product, service, device, component, or part thereof, that—
> (A) is primarily designed or produced for the purpose of circumventing a technological measure that effectively controls access to a work protected under this title;
> (B) has only limited commercially significant purpose or use other than to circumvent a technological measure that effectively controls access to a work protected under this title; or
> (C) is marketed by that person or another acting in concert with that person with that person's knowledge for use in circumventing a technological measure that effectively controls access to a work protected under this title (17 USC §1201).

Moreover, in section 1202, regarding the integrity of copyright management information:

> No person shall, without the authority of the copyright owner or the law—
> (1) intentionally remove or alter any copyright management information,
> (2) distribute or import for distribution copyright management information knowing that the copyright management information has been removed or altered without authority of the copyright owner or the law, or
> (3) distribute, import for distribution, or publicly perform works, copies of works, or phonorecords, knowing that copyright management information has been removed or altered without authority of the copyright owner or the law, knowing, or, with respect to civil remedies under section 1203, having reasonable grounds to know, that it will induce, enable, facilitate, or conceal an infringement of any right under this title (17 USC §1202).

The DMCA is similar to the rest of the Copyright Act in that there are penalties for infringement. Specifically, actual and statutory damages may be awarded civilly, and criminal penalties may be imposed. However, the criminal penalties for violating the DMCA are far more severe than under any other provisions in the Copyright Act. Specifically, any person who violates either section 1201 or 1202 willfully and for purposes of commercial advantage or private financial gain may be fined up to $500,000 or be imprisoned for up to 5 years or both and if the person violates the sections again, can be fined more than $1,000,000 or imprisoned for up to 10 years or both.

There are some limited liability provisions and "safe havens" for those who do not directly violate the DMCA, that is, internet service providers and those who maintain and repair computers, as well as libraries and nonprofit institutions.

TRADE SECRETS

Discussed above are three federally regulated intellectual property regimes with roots in the common law. One of the essential requirements for federal protection is disclosure, which allows the public access to the fruits of another's creative endeavor after the statutory monopoly period is expired. But what about intellectual property where the inventor or holder relies on secrecy to maintain a competitive edge?

This is the domain of **trade secrets.** The owner of a trade secret may find that trademark, patent, and copyright laws do not provide enough protection for their business' intellectual property. Often, the subject of a trade secret would not qualify for federal intellectual property protection. In order to keep the business secret under wraps, some companies require non-disclosure agreements to protect themselves. To accommodate such businesses, laws were created.

Trade secrets laws serve two broad policies:

1. They provide a vehicle for the maintenance of standards of commercial ethics, and
2. They encourage innovation by providing protection that is supplemental to that offered by the federal patent and copyright laws.

Trade secrets laws offer legal protection for an extremely wide variety of different types of information. Trade secret protection offers several advantages over other forms of legal protection. Unlike a patent or a copyright, a trade secret has no limited statutory life; trade secret protection may last forever. Moreover, trade secret protection extends to practically all types of information. Another advantage of trade secret protection over other forms of protection is that a trade secret is protected without any need to file any application for a registration or without any other formalities. Moreover, trade secret protection is secured without publicly disclosing the secret, in contrast to a patent, which discloses the information to the public.

In the United States, Section 757 of Restatement of Torts has been widely followed by the courts in trade secrets cases. More than 44 states have adopted the Uniform Trade Secrets Act (the "Uniform Act") since 1985. The Uniform Act was formulated by a private organization called the Commission on Uniform State Laws. Under both the Restatement and the Uniform Act, one who discloses or uses another's trade secret is generally liable if he or she discovered the secret by improper means or the disclosure or use constitutes a breach of confidence.

What Qualifies as a Trade Secret?

The Restatement defines a trade secret as follows:

> A trade secret may consist of any formula, pattern, device, or compilation of information that may be used in one's business, and that gives him an opportunity to obtain an advantage over competitors who do not know or use it. It may be a formula for a chemical compound; a process of manufacturing, treating, or preserving materials; a pattern for a machine or other device, or a list of customers. . . .

As one would expect, some element of secrecy must exist to obtain relief. The information claimed to be a trade secret must not be a matter of public knowledge or generally known in an industry. In fact, the court will look toward what efforts the company made to ensure secrecy, as well as how much money was spent to develop the trade secret. If the information is obvious upon the viewing of an object embodying the "secret" or if the information is merely a base of experiential knowledge that is readily available to others in the same field, it appears unlikely that a court would protect the information as a trade secret (but see, *Coca-Cola Bottling Company of Shreveport, Inc. v. The Coca-Cola Company,* at the end of the chapter, where The Coca-Cola Company was required to disclose its trade secret to comply with a court order). Generally, however, the availability of information through publicly available sources is irrelevant if the information is shown to have been acquired through a breach of a confidential relationship or through the use of improper means. Evidence of the unique and highly secret nature of the information at issue often increases the likelihood that the information will be held to constitute a trade secret.

What Constitutes Liability?

Section 757 lists what constitutes liability:

One who discloses or uses another's trade secret, without a privilege to do so, is liable to the other if
(a) he discovered the secret by improper means, or
(b) his disclosure or use constitutes a breach of confidence reposed in him by the other in disclosing the secret to him, or
(c) he learned the secret from a third person with notice of the facts that it was a secret and that the third person discovered it by improper means or that the third person's disclosure of it was otherwise a breach of his duty to the other, or
(d) he learned the secret with notice of the facts that it was a secret and that its disclosure was made to him by mistake.

The first two liability causes are discussed below.

Improper Means.   In a famous case involving the question of **improper means,** a competitor hired a photographer to conduct aerial reconnaissance of a chemical plant under construction. In holding that the complaint stated a cognizable claim, the court emphasized the value of the secret and the efforts expended to develop it. The court noted that reverse engineering and independent development were proper ways of acquiring information, but

one may not avoid these labors by taking the process from the discoverer without his permission at a time when he is taking reasonable precautions to maintain its secrecy. To obtain knowledge of a process without spending the time and money to discover it independently is improper unless the holder voluntarily discloses it or fails to take reasonable precautions to ensure its secrecy.

The court emphasized that the predicate of liability for such acts was the maintenance of standards of commercial morality.

Confidence.   The general rule of liability refers to a **breach of confidence.** A confidential relationship may arise not only from technical fiduciary relationships such as attorney-client, partner and partner, and so forth—which as a matter of law are relationships of trust and confidence—but may arise informally from moral, social, domestic, or purely personal relationships. Courts have found that a confidential relationship may exist when the parties' relationship is that of partners, joint adventurers, licensor and licensee, and employer and employee.

Generally, not all employment relationships are confidential. When an employee acquires an intimate knowledge of the employer's business, however, the relationship can be deemed confidential. The law thus implies as a part of the employment relationship an agreement not to use or disclose information that an employee receives incident to the employment when the employee knows that the employer desires such information be kept secret or if, under the circumstances, the employee should have realized that secrecy was desired. As noted above, the use or disclosure of a trade secret learned through a confidential relationship constitutes a misappropriation of the secret. Thus, the common law prohibits an employee's use or disclosure of the employer's trade secrets, whether the use or disclosure occurs during or after the employment relationship ends. The courts are usually careful to ensure that the information at issue is indeed a secret. The courts thus have developed the rule that a former employee is free to use the general knowledge, skills, and experience gained during prior employment. This is true even when the employee has increased his or her skills or has received complex training during the former employment. Such knowledge, skills, and experience may even be used to compete against a former employer.

Remedies

Although injunctive relief is probably the most common remedy sought by trade secret owners, the courts also allow the recovery of actual damages and punitive damages in trade secret cases. There are essentially three measures of actual damages in trade secrets cases:

1. The plaintiff's actual losses due to the defendant's use or disclosure of the secret;
2. The defendant's profits resulting from its use of the secret; and
3. A reasonable royalty, which is to constitute the value of what has been appropriated.

INTELLECTUAL PROPERTY AND CRIMINAL LAW

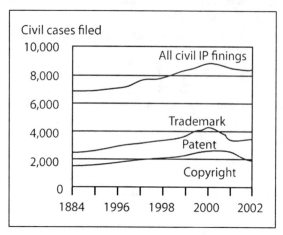

Figure 20.1 Bureau of Justice Statistics Special Report, October 2004

There is an ever increasing interest in intellectual property law for several reasons, not the least of which is the value intellectual property ownership brings to companies and the prestige reflected on the countries whose inventors file the most patents. Interest in intellectual property has also increased the diligence with which individuals and companies have pursued infringement actions. From 1994 to 2002, civil intellectual property suits increased 20 percent. (See Figure 20.1.) The median recovery award amount was $965,000 in 2002. Significantly, the success rates for complaints that actually went to trial are pretty high—plaintiffs won 59 percent of the time.

The United States Government not only allows intellectual property owners to avail themselves of the civil courts to seek remedies for intellectual property infringement, but imposes federal criminal penalties for copyright, trademark and trade secret theft.[1] (See Figure 20.2.) The conviction rate for criminal defendants with intellectual property offenses was 88 percent in 2002. The conviction rate is likely to increase over time, although finding and imposing criminal penalties on those who use sophisticated technological tools to avoid detection will continue to challenge law enforcement efforts.

---

[1]Federal civil penalties are assessed for patents, copyright, and trademark violations. Federal criminal penalties are *not* assessed for trade secret violations for two reasons: (1) trade secrets are not filed with the federal government (thereby protecting trade secret status because the secret is not disclosed) and (2) sufficient remedies are available through state unfair competition laws.

Selected Federal criminal intellectual property theft statutes and maximum penalties:

Criminal IP theft offenses in this report are defined according to copyright trademark, and trade secrets statutes:

Copyright

17 U.S.C. § 506 & 18 U.S.C. § 2319
Criminal infringement of a copyright. Statutory maximum penalty of 5 years in prison and $250,000 fine and 10 years in prison for repeat copyright offenders.

18 U.S.C. § 2318
Trafficking in Counterfeit Labels for phonograph records, copies of computer programs, and similar materials. Maximum penalty of 5 years in prison and $250,000 fine.

18 U.S.C. § 2319A
Unauthorized Fixation of and Trafficking in Sound Recordings and Music Videos of Live Musical Performances. Maximum penalty of 5 years in prison and $250,000 fine for first time offender and 10 years in prison for repeat "bootlegging" offender.

17 U.S.C. § 1201-1205
Circumvention of copyright protection systems. Maximum penalty of 5 years in prison and $500,000 fine for first time offender and 10 years in prison and $1,000,000 fine for repeat offender.

47 U.S.C. § 553
Unauthorized reception of cable services. Maximum penalty of 6 months in prison and $1,000 fine for individual use and 2 years in prison and $50,000 fine for commercial/financial gain with first time offender and 5 years in prison and $100,000 fine for repeat offender.

47 U.S.C. § 605
Unauthorized publication or use of communications. Maximum penalty of 6 months in prison and $2,000 fine for individual use and 2 years in prison and $50,000 fine for commercial/financial gain with first time offender and 5 years in prison and $100,000 fine for repeat offender.

Trademark

18 U.S.C. § 2320
Trafficking in Counterfeit Goods of Services. Maximum penalty of 5 years in prison and 10 years in prison and $5,000,000 fine for repeat offenders. Corporations subjects to fines up to $15,000,000.

Trade secrets

18 U.S.C. § 1831
Economic espionage. Maximum penalty of 15 years in prison and $500,000 fine for individual and $10,000,000 fine for corporate offender.

18 U.S.C. § 1832
Theft of trade secrets. Maximum penalty of 10 years in prison and $250,000 fine for individual and $5,000,000 fine for corporate offender.

Figure 20.2   Bureau of Justice Statistics Special Report, October 2004

EBAY INC., et al., Petitioners v. MERCEXCHANGE, L. L. C., 126 S, Ct, 1837; 164 L, Ed. 2d 641; 2006 U.S. LEXIS 3872 (2006)

Study terms: Patent infringement, injunctive relief, business method patent, equity

Justice Thomas delivered the opinion of the Court.

Ordinarily, a federal court considering whether to award permanent injunctive relief to a prevailing plaintiff applies the four-factor test historically employed by courts of equity. Petitioners eBay Inc. and Half.com, Inc., argue that this traditional test applies to disputes arising under the Patent Act. We agree and, accordingly, vacate the judgment of the Court of Appeals.

I

Petitioner eBay operates a popular Internet Web site that allows private sellers to list goods they wish to sell, either through an auction or at a fixed price. Petitioner Half.com, now a wholly owned subsidiary of eBay, operates a similar Web site. Respondent MercExchange, L. L. C., holds a number of patents, including a business method patent for an electronic market designed to facilitate the sale of goods between private individuals by establishing a central authority to promote trust among participants. See U.S. Patent No. 5,845,265. MercExchange sought to license its patent to eBay and Half.com, as it had previously done with other companies, but the parties failed to reach an agreement. MercExchange subsequently filed a patent infringement suit against eBay and Half.com in the United States District Court for the Eastern District of Virginia. A jury found that MercExchange's patent was valid, that eBay and Half.com had infringed that patent, and that an award of damages was appropriate.

Following the jury verdict, the District Court denied MercExchange's motion for permanent injunctive relief. 275 F. Supp. 2d 695 (2003). The Court of Appeals for the Federal Circuit reversed, applying its "general rule that courts will issue permanent injunctions against patent infringement absent exceptional circumstances." 401 F.3d 1323, 1339 (2005). We granted certiorari to determine the appropriateness of this general rule. 546 U.S. ____, 126 S. Ct. 733, 163 L. Ed. 2d 567 (2005).

II

According to well-established principles of equity, a plaintiff seeking a permanent injunction must satisfy a four-factor test before a court may grant such relief. A plaintiff must demonstrate: (1) that it has suffered an irreparable injury; (2) that remedies available at law, such as monetary damages, are inadequate to compensate for that injury; (3) that, considering the balance of hardships between the plaintiff and defendant, a remedy in equity is warranted; and (4) that the public interest would not be disserved by a permanent injunction. See, *e.g.*, *Weinberger v. Romero-Barcelo*, 456 U.S. 305, 311-313, 102 S. Ct. 1798, 72 L. Ed. 2d 91 (1982); *Amoco Production Co. v. Gambell*, 480 U.S. 531, 542, 107 S. Ct. 1396, 94 L. Ed. 2d 542 (1987). The decision to grant or deny permanent injunctive relief is an act of equitable discretion by the district court, reviewable on appeal for abuse of discretion.

These familiar principles apply with equal force to disputes arising under the Patent Act. As this Court has long recognized, "a major departure from the long tradition of equity practice should not be lightly implied." *Ibid.*; see also *Amoco, supra*, at 542, 107 S.Ct. 1396, 94 L. Ed. 2d 542. Nothing in the Patent Act indicates that Congress intended such a departure. To the contrary, the Patent Act expressly provides that injunctions "may" issue "in accordance with the principles of equity." 35 U.S.C. § 283.[2]

---

[2]Section 283 provides that "[t]he several courts having jurisdiction of cases under this title may grant injunctions in accordance with the principles of equity to prevent the violation of any right secured by patent, on such terms as the court deems reasonable."

To be sure, the Patent Act also declares that "patents shall have the attributes of personal property," § 261, including "the right to exclude others from making, using, offering for sale, or selling the invention," § 154(a)(1). According to the Court of Appeals, this statutory right to exclude alone justifies its general rule in favor of permanent injunctive relief. 401 F.3d, at 1338. But the creation of a right is distinct from the provision of remedies for violations of that right. Indeed, the Patent Act itself indicates that patents shall have the attributes of personal property "[s]ubject to the provisions of this title," 35 U.S.C. § 261, including, presumably, the provision that injunctive relief "may" issue only "in accordance with the principles of equity," § 283.

This approach is consistent with our treatment of injunctions under the Copyright Act. Like a patent owner, a copyright holder possesses "the right to exclude others from using his property." *Fox Film Corp. v. Doyal*, 286 U.S. 124, 127, 52 S. Ct. 546, 76 L. Ed. 1010 (1932); see also *id.*, at 127-128, 52 S. Ct. 546, 76 L. Ed. 1010 ("A copyright, like a patent, is at once the equivalent given by the public for benefits bestowed by the genius and meditations and skill of individuals, and the incentive to further efforts for the same important objects" (internal quotation marks omitted)). Like the Patent Act, the Copyright Act provides that courts "may" grant injunctive relief "on such terms as it may deem reasonable to prevent or restrain infringement of a copyright." 17 U.S.C. § 502(a). And as in our decision today, this Court has consistently rejected invitations to replace traditional equitable considerations with a rule that an injunction automatically follows a determination that a copyright has been infringed. See, *e.g.*, *New York Times Co. v. Tasini*, 533 U.S. 483, 505, 121 S. Ct. 2381, 150 L. Ed. 2d 500 (2001) (citing *Campbell v. Acuff-Rose Music, Inc.*, 510 U.S. 569, 578, n. 10, 114 S. Ct. 1164, 127 L. Ed. 2d 500 (1994)); *Dun v. Lumbermen's Credit Assn.*, 209 U.S. 20, 23-24, 28 S. Ct. 335, 52 L. Ed. 663 (1908).

Neither the District Court nor the Court of Appeals below fairly applied these traditional equitable principles in deciding respondent's motion for a permanent injunction. Although the District Court recited the traditional four-factor test, 275 F. Supp. 2d, at 711, it appeared to adopt certain expansive principles suggesting that injunctive relief could not issue in a broad swath of cases. Most notably, it concluded that a "plaintiff's willingness to license its patents" and "its lack of commercial activity in practicing the patents" would be sufficient to establish that the patent holder would not suffer irreparable harm if an injunction did not issue. *Id.*, at 712. But traditional equitable principles do not permit such broad classifications. For example, some patent holders, such as university researchers or self-made inventors, might reasonably prefer to license their patents, rather than undertake efforts to secure the financing necessary to bring their works to market themselves. Such patent holders may be able to satisfy the traditional four-factor test, and we see no basis for categorically denying them the opportunity to do so. To the extent that the District Court adopted such categorical rule, then, its analysis cannot be squared with the principles of equity adopted by Congress. The court's categorical rule is also in tension with *Continental Paper Bag Co. v. Eastern Paper Bag Co.*, 210 U.S. 405, 422-430, 28 S. Ct. 748, 52 L. Ed. 1122, 1908 Dec. Comm'r Pat. 594 (1908), rejected the contention that a court of equity has no jurisdiction to grant injunctive relief to a patent holder who has unreasonably declined to use the patent.

In reversing the District Court, the Court of Appeals departed in the opposite direction from the four-factor test. The court articulated a "general rule," unique to patent disputes, "that a permanent injunction will issue once infringement and validity have been adjudged." 401 F.3d, at 1338. The court further indicated that injunctions should be denied only in the "unusual" case, under "exceptional circumstances" and "'in rare instances . . . to protect the public interest.'" *Id.*, at 1338-1339. Just as the District Court erred in its categorical denial of injunctive relief, the court of Appeals erred in its categorical grant of such relief. Cf. *Roche Products v. Bolar Pharmaceutical Co.*, 733 F.2d 858, 865 (CAFed 1984) (recognizing the "considerable discretion" district courts have "in determining whether the facts of a situation require it to issue and injunction").

Because we conclude that neither court below correctly applied the traditional four-factor framework that governs the award of injunctive relief, we vacate the judgment of the Court of Appeals, so that the District Court may apply that framework in the first instance. In doing so, we take no position on whether permanent injunctive relief should or should not issue in this particular case, or indeed in any number of other disputes arising under the Patent Act. We hold only that the decision whether to grant or deny injunctive relief rests within the equitable discretion of the district courts, and that such discretion must be exercised consistent with traditional principles of equity, in patent disputes no less than in other cases governed by such standards.

Accordingly, we vacate the judgment of the Court of Appeals, and remand for further proceedings consistent with this opinion.

It is so ordered.

CONCUR BY: ROBERTS; KENNEDY

Concur

Chief Justice Roberts, with whom Justice Scalia and Justice Ginsburg join, concurring.

I agree with the Court's holding that "the decision whether to grant or deny injunctive relief rests within the equitable discretion of the district courts, and that such discretion must be exercised consistent with traditional principles of equity, in patent disputes no less than in other cases governed by such standards," *ante*, at ____, 164 L. Ed. 2d, at 647, and I join the opinion of the Court. That opinion rightly rests on the proposition that "a major departure from the long tradition of equity practice should not be lightly implied." *Weinberger v. Romero-Barcelo*, 456 U.S. 305, 320, 102 S. Ct. 1798, 72 L. Ed. 2d 91 (1982); see *ante*, at ____, 164 L. Ed. 2d, at 646.

From at least the early 19th century, courts have granted injunctive relief upon a finding of infringement in the vast majority of patent cases. This "long tradition of equity practice" is not surprising, given the difficulty of protecting a right to *exclude* through monetary remedies that allow an infringer to *use* an invention against the patentee's wishes—a difficulty that often implicates the first two factors of the traditional four-factor test. This historical practice, as the Court holds, does not *entitle* a patentee to a permanent injunction or justify a *general rule* that such injunctions should issue. The Federal Circuit itself so recognized in *Roche Products, Inc. v. Bolar Pharmaceutical Co.*, 733 F. 2d 858, 865-867 (1984). At the same time, there is a difference between exercising equitable discretion pursuant to the established four-factor test and writing on an entirely clean slate. "Discretion is not whim, and limiting discretion according to legal standards helps promote the basic principle of justice that like cases should be decided alike." *Martin v. Franklin Capital Corp.*, 546 U.S. ____, ____, 126 S. Ct. 704, 163 L. Ed. 2d 547 (2005). When it comes to discerning and applying those standards, in this area as others, "a page of history is worth a volume of logic." *New York Trust Co. v. Eisner*, 256 U.S. 345, 349, 41 S. Ct. 506, 65 L. Ed. 963, T.D. 3267 (1921) (opinion for the court by Holmes, J.).

Justice Kennedy, with whom Justice Stevens, Justice Souter, and Justice Breyer join, concurring.

The Court is correct, in my view, to hold that courts should apply the well-established, four-factor test—without resort to categorical rules—in deciding whether to grant injunctive relief in patent cases. The Chief Justice is also correct that history may be instructive in applying this test. *Ante*, at____,-____, 164 Ed. 2d, at 648 (concurring opinion). The traditional practice of issuing injunctions against patent infringers, however, does not seem to rest on "the difficulty of protecting a right to *exclude* through monetary remedies that allow an infringer to *use* an invention against the patentee's wishes." *Ante*, at ____, 164 L. Ed. 2d, at 648 (Roberts, C. J., concurring). Both the terms of the Patent Act and the traditional view of injunctive relief accept that the existence of a right to exclude does not dictate the remedy for a violation of that right. *Ante*, at ____-____, 164 L. Ed. 2d, at 646-647 (opinion of the Court). To the extent earlier cases establish a pattern of granting an injunction against patent infringers almost as a matter of course, this pattern simply illustrates the result of the four-factor test in the contexts then prevalent. The lesson of the historical practice, therefore, is most helpful and instructive when the circumstances of a case bear substantial parallels to litigation the courts have confronted before.

In cases now arising trial courts should bear in mend that in many instances the nature of the patent being enforced and the economic function of the patent holder present considerations quite unlike earlier cases. An industry has developed in which firms use patents not as a basis for producing and selling goods but, instead, primarily for obtaining licensing fees. For these firms, an injunction, and the potentially serious sanctions arising from its violation, can be employed as a bargaining tool to charge exorbitant fees to companies that seek to buy licenses to practice the patent. When the patented invention is but a small component of the product the companies seek to produce and the threat of an injunction is employed simply for undue leverage in negotiations, legal damages may well be sufficient to compensate for the infringement and an injunction may not serve the public interest. In addition injunctive relief may have different consequences for the burgeoning number of patents over business methods, which were not of much economic and legal significance in earlier times. The potential vagueness and suspect validity of some of these patents may affect the calculus under the four-factor test.

The equitable discretion over injunctions, granted by the Patent Act, is well suited to allow courts to adapt to the rapid technological and legal developments in the patent system. For these reasons it should be recognized that district courts must determine whether past practice fits the circumstances of the cases before them. With these observations, I join the opinion of the Court.

YOUGOTTAEAT, INC v. CHECKERS DRIVE-IN RESTAURANTS, INC. 2003 U.S. App. LEXIS 24251 (USCA 2d Cir. 2003)

Study terms: Trademark infringement, abandonment, *Polaroid* factors, consumer confusion

Generally, YGE is a small Internet business that provides discount coupons for restaurants through its website. Due to financial difficulties, its website was shut down during most or all of 2001 and 2002 and only resumed in late 2002 or early 2003. Checkers is the largest chain of double drive-thru restaurants in the United States, with approximately 800 restaurant locations operating under the brand Checkers or Rally's. In June 2000, Checkers, unaware of YGE's website, adopted a new slogan for its advertising campaign, "You Gotta Eat" (the "Slogan"). Before launching the new campaign, Checkers requested its outside intellectual property attorney to investigate the legal availability of the Slogan. Because the YGE business was not operational at that time, when the attorney went to YGE's web address he only received a generic error message.

On November 10, 2000, Checkers applied to the United States Patent and Trademark Office ("PTO") to register the Slogan. However, four months prior, YGE applied to the PTO to register its service mark "YOUGOTTAEAT." YGE's trademark application was granted on March 12, 2002. On June 27, 2002, the PTO informed Checkers that its application to trademark the Slogan was denied because it and YGE's marks were too similar. Checkers' attorney contacted YGE's attorney listed on its PTO application in an attempt to negotiate this issue but neither YGE nor its attorney responded. On August 26, 2002, Checkers initiated a PTO Cancellation Proceeding on the grounds that YGE had abandoned its mark. On February 13, 2003, YGE filed this lawsuit alleging trademark infringement and unfair competition under the Lanham Act and New York state law, and immediately moved for a preliminary injunction enjoining Checkers' use of the Slogan. The district court denied YGE's motion, finding that YGE failed to establish that it would prevail on the merits of its trademark claim.

This Court reviews a grant or denial of a preliminary injunction for abuse of discretion. *TCPIP Holding Co., Inc. v. Haar Communications, Inc.*, 244 F.3d 88, 92 (2d Cir. 2001). On appeal, this Court reviews the district court's factual findings in its order denying a motion for a preliminary injunction. . . . Here, the district court stated the following in denying YGE's preliminary injunction motion:

After applying the Polaroid factors to the facts in the present case, this Court determines that Plaintiff has failed to make a showing that [Checkers'] continued use of the phrase "you gotta eat" in its advertising campaign will infringe [YGE's] trademark by causing consumer confusion.

This order failed to sufficiently set forth the factual basis for its denial of YGE's motion. Nonetheless, this Court may still proceed with our review despite inadequate factual findings if we can discern facts from the record to enable us to render a decision. . . .

The district court's order denying YGE's motion for a preliminary injunction was proper because YGE did not establish that it was likely to prevail on its trademark claims. In cases involving claims of trademark infringement under the Lanham Act, "as in other types of cases, a party seeking a preliminary injunction must demonstrate (1) the likelihood of irreparable injury in the absence of such an injunction, and (2) either (a) likelihood of success on the merits or (b) sufficiently serious questions going to the merits to make them a fair ground for litigation plus a balance of hardships tipping decidedly toward the party requesting the preliminary relief." *Federal Express Corp. v. Federal Espresso, Inc.*, 201 F.3d 168, 173 (2d Cir. 2000).

To prevail on its trademark claim, Plaintiff must demonstrate that its registered trademark (1) "is entitled to protection" and (2) that "defendant's use of the mark is likely to cause consumers confusion as to the origin or sponsorship of the defendant's goods." *Virgin Enter. Ltd. v. Nawab*, 335 F.3d 141, 146 (2d Cir. 2003). In *Polaroid Corp. v. Polarad Electronics Corp.*, 287 F.2d 492 (2d Cir.), cert. denied, 368 U.S. 820, 7 L. Ed. 2d 25, 82 S. Ct. 36 (1961), this Court "outlined a series of nonexclusive factors" in analyzing the second prong. Each Polaroid factor is addressed and applied below.

1. Strength of the Mark: A mark's strength refers to two concepts: (1) inherent strength or "inherent distinctiveness;" and (2) fame or "acquired distinctiveness." *Virgin*, 335 F.3d at 147. It is debatable whether "YouGottaEat" is inherently distinctive, as opposed to merely "generic, descriptive or suggestive as to [certain]] goods." Id. The phrase, "You Gotta Eat" is a generic phrase that commonly appears in a variety of contexts, such as in article headings and in song lyrics. It is likewise debatable whether YGE's mark has acquired distinctiveness. Due to YGE's limited commercial success and short duration of operation, consumers are unlikely to associate YGE with its mark, "YouGottaEat." Accordingly, YGE possesses a weak mark.

2. The Marks are Identical: Although YGE's mark and the Slogan are highly similar, they are presented in different ways. The Slogan is used in conjunction with the brand name Checkers and Rally's, and with their corresponding logo design (checkered tablecloth). YGE's mark is presented with the logo of a man wearing a sandwich-board placard. Although in some circumstances graphic differences are irrelevant, see, e.g., *Virgin*, 335 F.3d at 149, in this case YGE operates a website, which necessarily relies on its visual displays. Thus, because the visual differences between YGE's and Checkers' marks are significant, this factor is neutral.

3. Proximity of the Products: This factor weighs decidedly in Checkers' favor. This factor considers the "proximity of the products being sold by plaintiff and defendant under identical (or similar) marks. . . . When the two users of a mark are operating in completely different areas of commerce, consumers are less likely to assume that their similarly branded product comes from the same source." Id. at 149-50. Here, the two services at issue are considerably different. Checkers is a large drive-thru fast food restaurant chain while YGE operates a website that offers discount dining certificates.

4. Likelihood of Bridging the Gap: This factor refers to the "senior user's interest in preserving avenues of expansion and entering into related fields." *Hormel Foods Corp. v. Jim Henson Prods., Inc.*, 73 F.3d 497, 504 (2d Cir. 1996) (citation omitted). The expansion must occur or be likely to occur in the "reasonably near future." *Virgin*, 335 F.3d at 150. Although YGE, financially strapped and unable to maintain even a website, asserts that it may consider franchising a restaurant chain under its name, this is not likely to occur in the reasonably near future.

5. No Actual Confusion: Actual confusion between the two marks at issue is highly unlikely. As discussed, Checkers and YGE are two very different establishments, one is a restaurant chain while the other operates a small internet website. Consumers would not likely associate, much less confuse, the two marks at issue.

6. Customer Base: The parties do not share the same customer base because Checkers maintains fast-food drive-thru restaurants and therefore appeals to consumers seeking to purchase food immediately and often on a whim. By contrast, YGE users are people who go online, research restaurants and print discount coupons to be used at a later time.

4 Quality of product or service
4 Sophistication of the buyers

7. Good Faith: YGE contends that Checkers' use of the Slogan was in bad faith because it had notice of YGE's website, and therefore was aware of its mark, "YOUGOTTAEAT," before it ran its first commercial on January 3, 2001. YGE, however, does not dispute Checkers' contention that when Checkers' attorney visited YGE's website, it was not accessible because YGE ceased operating the site at that time. Therefore, Checkers' reasonably ignored YGE's web address. In fact, Checkers exhibited good faith in this matter. It enlisted an outside attorney to conduct a trademark search and then attempted to negotiate its claims with YGE. Only when that proved unsuccessful, and upon independently determining that its use did not infringe on YGE's rights, did Checkers begin using the Slogan. Such measures indicate Checkers' good faith. See, *e.g.*, *Lang v. Retirement Living Publ'g Co.*, 949 F.2d 576, 583 (2d Cir. 1991) ("request for a trademark search and reliance on the advice of counsel" suggests a defendant's good faith).

The record below indicates that YGE was not likely to succeed on the merits of its trademark claim because it was unable to establish a substantial similarity between its mark and Checkers' Slogan. Accordingly, the Court hereby AFFIRMS the district court's order denying the preliminary injunction.

MOSER PILON NELSON ARCHITECTS, LLC, et al. v. HNTB CORPORATION, et al., 2006 U.S. Dist. LEXIS 58334; 80 U.S.P.Q.2D (BNA) 1085.

Study terms: Author, architectural works, Copyright Protection Act of 1990, preemption, unfair trade practices, conversion, unjust enrichment

In this action, Plaintiffs—losers of a publicly-bid competition to design a new parking garage for Central Connecticut State University (CCSU)—claim that Defendants—winners of that competition—modified their original design for the garage and constructed a garage along the lines of Plaintiffs' losing design, in violation of the Copyright Act, the Lanham Act, the Connecticut Unfair Trade Practices Act ("CUTPA"), and state common law prohibiting unjust enrichment and conversion.

Defendant HNTB Corporation (" HNTB") has filed three motions for summary judgment, each of which has been adopted by co-Defendant Downes Construction Co. ("Downes"). In these motions, Defendants argue: (1) that Plaintiffs' claims of conversion, unjust enrichment, and violation of CUTPA and the Lanham Act are preempted by the Copyright Act; (2) that Plaintiffs' Copyright Act claim fails because the copyright registration relied upon is invalid; and (3) that Plaintiffs other than Kenneth Pilon should be dismissed for lack of standing because they are not owners of the claimed copyright.

For the reasons that follow, the Court DENIES Defendants; Motion for Summary Judgment of Invalidity; GRANTS IN PART Defendants' Motion for Summary Judgment of Non-Ownership, and GRANTS IN PART Defendants' Motion for Summary Judgment of Preemption. Accordingly, only Kenneth Pilon and James Brockman may pursue Count I for copyright infringement, all Plaintiffs may pursue Count V for violation of CUTPA, and summary judgment is granted against all Plaintiffs on Counts II, III, and IV under the Lanham Act, state law of unjust enrichment, and state law of conversion, respectively.

I.

The summary judgment standard is a familiar one. Summary judgment is appropriate only when "the pleadings, depositions, answers to interrogatories, and admissions on file, together with the affidavits, if any, show that there is no genuine issue as to any material fact and that the moving party is entitled to judgment as a matter of law." Fed. R. Civ. P. 56(b). . . .

II.

The following forms the factual background to Plaintiffs' claims. As required on motion for summary judgment, the Court recites the facts in the light most favorable to the non-moving party—here Plaintiffs.

In November 2001, a team composed of Plaintiffs Macchi Engineers ("Macchi"), Moser Pilon Nelson Architects ("Moser Pilon" or "MPN"), and general contractor O&G Industries of Torrington, Connecticut, was shortlisted as a finalist in a competition to design and ultimately to construct a parking garage for Central Connecticut State University (the "CCSU garage"). Macchi was the design consultant to O&G, with responsibility for providing "the requisite engineering design," Plaintiffs' Local Rule 56(a)(2) Statement Supplement at 2, P 35, while Moser Pilon, which had agreed to work as the architectural sub-consultant to Macchi, was tasked with providing "the requisite architectural design input," id.

The collaborative effort of Defendants Downes and the HNTB was also shortlisted as a finalist in the CCSU garage competition. Although Plaintiffs' and Defendants' proposals shared the same basic rectangular design with a stair-tower at each corner of the garage, the layout and appearance of each team's proposed stair-tower differed substantially. The proposed costs for design and construction were also different: Defendants' proposal estimated a total cost of $16,969,810 or $16,160 per space for 1050 parking spaces. Plaintiffs' proposal estimated a lower total cost of $16,452,000, but a higher per space cost of $16,651 for either 988 or 997 spaces.

Plaintiffs' claim that the selection process was biased in favor or Downes. In support of this theory they point to the fact that Kenneth Russo of Downes dined socially every month with both the Deputy Commissioner and the Bureau Chief of the Connecticut Department of Public Works ("DPW")—the state agency that was running the competition. In addition, on November 20, 2001, a project manager for DPW unofficially told Kenneth Russo of Downes that his team's construction price was ranked second, and that Downes should "get rid of the architect," because their design submission was "not good."

On November 21, 2001, the President of CCSU sent an email to CCSU's representative on the selection panel for the competition stating that he did not like the Downes proposal, that in his view it did not comply with the design guidelines for the CCSU garage, and that he would convey his opinion to the Deputy Commissioner of the Department of Public Works, who was managing the competition. A few hours after the President's email, the three finalist teams were interviewed by a selection panel comprised of representatives of the DPW and CCSU. That same day, each panelist completed a score sheet for each proposal. In evaluating the proposed designs, the panelists assigned the following weights to the designated [sic] criteria: building aesthetics—15%; compliance with the "program for design"—15%; landscaping—10%; total project cost—20%; and construction schedule—25%.

Defendants' design had a lower per-parking-space cost and a better schedule than Plaintiffs' proposal. Accordingly, Defendants' design scored more highly on the criteria used by the selection panelists. Plaintiffs' proposal scored more highly than Defendants' on design aesthetics but fared less well on compliance with the "program for design." In line with these scores, the selection panel unanimously ranked Defendants' proposal first out of the three finalists' submissions; Plaintiffs' proposal ranked second. Accordingly, by letter dated November 29, 2001, Defendants were awarded the contract for the CCSU garage. The final contract was not signed until May 2002.

In December 2001, after the award of the CCSU garage contract to Defendants, but before execution of the of the final contract, CCSU staff showed Defendant Downes renderings of the CCSU garage that Downes understood to be authored by Moser Pilon. CCSU staff asked Downes to change its proposed design to adopt some of the features of the design shown on the Moser Pilon renderings. The record further reflects that Defendant HNTB was aware of CCSU's preference for the aesthetic aspects of the Moser Pilon proposal; and CCSU's representative on the selection panel testified that CCSU would have been unwilling to go forward with the Project unless Downes made the requested design changes.

Defendants agreed to revise their design but observed that the revised design of the stair-towers would increase the cost by $395,500. In order to complete the revised design for the Project at the price that Defendants had quoted in their original submission, Defendants requested and received permission to substitute thin brick tile finish in lieu of field laid brick on all spandrel panels. Defendants submitted renderings of their revised design in April 2002. The layout and appearance of the stair-towers in Defendants' re-design is similar to the design of the stair-towers in Plaintiffs' losing submission.

In February 2003, as final construction of the CCSU garage was underway, Plaintiffs learned for the first time that Defendants had altered their original proposal and adopted a final design for the stair-towers resembling that of the Plaintiffs' losing bid. Thereafter, in November 2003, Plaintiffs applied for copyright protection of their design for the CCSU garage.

Two features of Plaintiffs' application are critical to the legal analysis that follows. First, although Plaintiffs submitted technical drawings in support of their copyright application, they did not register the drawings themselves for protection as "pictorial, graphic, and sculptural works" under 17 U.S.C. 102(a)(2). Rather, Plaintiffs registered the design of the CCSU garage as an architectural work under a newly-minted provision of the Copyright Act, 17 U.S.C. 102(a)(8). That provision had significantly altered the traditional copyright framework—under which only architectural drawings were protected, and not the finished work—by extending protection to the shape of the three-dimensional structure and thus giving the owner of an architectural work copyright the exclusive right to build the structure (thereby bringing the United States into compliance with its obligations under the Berne Convention for the Protection of Literary and Artistic Works). Thus, the drawings themselves are not protected and this lawsuit is not about the copying of the drawings themselves; rather, the claim is one for infringement of the design of the CCSU garage as an architectural work.

Second, the application names only Kenneth Pilon as an author of the registered work. Although the application originally named as authors both James Brockman and Kenneth Pilon, Mr. Brockman's name was stricken by Mr. Pilon from the author field of the application after an official at the Copyright Office advised him that it was customary for the copyright registration of an architectural work to appear in the architect's name alone. Thus, although Plaintiffs assert that they had always understood themselves to be joint authors of the CCSU Project, and although Mr. Pilon claims that he understood the copyright would be jointly-owned, the copyright in the Macchi/MPN/O&G design for the CCSU Project was in fact registered in Mr. Pilon's name alone.

III.

■ ■ ■

Because a determination that the copyright is invalid affects the scope of analysis required on Defendants' other claims, the Court begins with the question of validity.

A. Validity of the Copyright

To succeed on their claim of copyright infringement, Plaintiffs must demonstrate the following: (1) ownership of a valid copyright; and (2) unauthorized copying of the copyrighted work. In their Motion for Summary Judgment of Invalidity, Defendants argue that Count I fails because Plaintiffs' copyright registration is invalid. Plaintiffs have presented a certificate of copyright registration, and that constitutes prima facie evidence of a valid copyright under 17 U.S.C. § 410(c). However, the presumption of validity created by a certificate of registration may be rebutted through presentation of evidence that the allegedly copyrighted work is not copyrightable. Here, Defendants argue that Plaintiffs' design for the CCSU garage is not a copyrightable item, on the grounds that the proposed CCSU garage is not a "building" within the meaning of the architectural works provision of the Copyright Act.

In relevant part, the Copyright Act provides as follows:

An "architectural work" is the design of a building as embodied in any tangible medium of expression, including a building, architectural plans, or drawings. The work includes the overall form as well as the arrangement and composition of spaces and elements in the design, but does not include individual standard features.

17 U.S.C. § 101. Because the Copyright Act does not define the key term "building," the Court must begin its interpretive task by consulting the term's ordinary meaning. . . .

The Eighth Edition of Black's Law Dictionary defines a building as "[a] structure with walls and a roof, esp. a permanent structure." Black's Law Dictionary (8th ed. 2004). Similarly, the Second Edition of Webster's Unabridged Dictionary defines building as "a relatively permanent enclosed construction over a plot of land, having a roof and usually windows and often more than one level, used for any of a wide variety of activities, as living, entertaining, or manufacturing." Webster's Unabridged Dictionary 274 (2d ed. 2001). Under either of these definitions, as well as in common parlance, the parking garage designed by Plaintiffs qualifies as a "building."

■ ■ ■

The Copyright Office defines the term "building" as follows: "humanly habitable structures that are intended to be both permanent and stationary, such as houses and office buildings, and other permanent and stationary structures designed for human occupancy, including but not limited to churches, museums, gazebos, and garden pavilions." 37 C.F.R. § 202.11(b)(2). It further states that "[s]tructures other than buildings, such as bridges, cloverleafs, dams, walkways, tents, recreational vehicles, mobile homes, and boats," "cannot be registered." 37 C.F.R. § 202.11(d)(1). Thus, the regulation divides structures into three broad categories: (1) structures that are "humanly habitable[,] . . . permanent and stationary, such as houses and office buildings," 37 C.F.R. § 202.11(b)(2) (emphasis added); (2) structures that are "permanent and stationary structures designed for human occupancy, including but not limited to churches, museums-gazebos, and garden pavilions," id. (emphasis added); and (3) structures that are not buildings at all, "such as bridges, cloverleafs, dams, walkways, tents, recreational vehicles, mobile homes, and boats," id.

■ ■ ■

Defendants also argue that the proposed CCSU garage is analogous to certain of the structures expressly excluded from the definition of a building by the Copyright Office's regulation—namely bridges, cloverleafs, dams, and walkways—and for that reason should be excluded from protection. Defendants' analogy is unconvincing. The CCSU garage may be distinguished from bridges, dams, walkways, and colverleafs on any number of grounds. . . . Most significantly in the Court's view, the CCSU garage is a building within the ordinary meaning of the term; bridges, cloverleafs, dams, and walkways self-evidently are not.

Finally, the Court notes that its understanding of the statutory term and of the Copyright Office's definition of that term is consistent with the legislative history of the architectural works provision. . . .

For the reasons explained above, the Court concludes that the design for the CCSU garage is a building within the meaning of the Architectural Works Copyright Protection Act of 1990, and that Plaintiffs' copyright registration may not be invalidated on the theory advanced by Defendants. Accordingly, Defendant's Motion for Summary Judgment of Invalidity is DENIED.

B. Ownership

This suit was brought by four Plaintiffs: Kenneth Pilon (the architect of the disputed design); Pilon's firm—Moser Pilon; James Brockman (an engineer); and Brockman's firm—Macchi. In their Motion for Summary Judgment of Non-ownership, Defendants assert that Plaintiffs Moser Pilon, James Brockman, and Macchi lack standing to pursue claims founded on alleged copying of the design for the parking garage because the copyright in that design is owned solely by Kenneth Pilon—Mr. Brockman's name having been stricken by Mr. Pilon from the author field of th application, and the corporate entities never having been listed on the application at all.

Although not named on the copyright registration, Plaintiffs Brockman, Macchi, and Moser Pilon are entitled to joint ownership of the copyright if they can demonstrate: (1) the mutual intent of the parties to be joint authors; and (2) that they made independently copyrightable contributions. The requirement of an independently copyrightable contribution requirement derives from the principle that ideas themselves may not be copyrighted, so that "the author is the party who actually creates the work, that is, the person who translates an idea into a fixed, tangible expression entitled to copyright protection." *Community for Creative Non-Violence v. Reid*, 490 U.S. 730, 737, 109 S. Ct. 2166, 104 L. Ed. 2d 811 (1989). Accordingly, mere collaboration is insufficient, and a putative joint author must show that he or she "independently created" a contribution demonstrating "some minimal degree of creativity." *Mattel, Inc. v. Goldberger Doll Mfg. Co.*, 365 F.3d 133, 135 (2d Cir. 2004).

1. Joint Authorship

As a preliminary matter, Plaintiffs' counsel conceded at oral argument that there was nothing in the record to support the notion that either of the Plaintiff corporate entities—Moser Pilon and Macchi—could satisfy the requirement of having contributed independently copyrightable work distinct fro the contributions of the individual Plaintiffs Kenneth Pilon and James Brockman. Thus, neither of these entities can claim authorship of the design for the parking garage.

The question remains whether Mr. Brockman made a contribution to the disputed design that would be independently copyrightable as an architectural work. In this regard, the Court must inquire not only whether he contributed original concepts to the design, but also whether he fixed them in a tangible medium of expression—in this case a building, architectural plans, or drawings. See 17 U.S.C. § 101 (defining an architectural work). The Plaintiffs' submissions are not a model of clarity with respect to exactly who did what in the development of the CCSU garage design proposal but, drawing all inferences in Mr. Brockman's favor as the Court must on a motion for summary judgment, the Court concludes that there are material issues of fact regarding Mr. Brockman's alleged contributions that preclude summary judgment on this issue. . . .

[I]n order for Mr. Brockman to be considered an author of any "arrangement and composition of spaces and elements," within the meaning of the Copyright Act, the jury must also be able to find that Mr. Brockman "translate[d] [his] idea into a fixed, tangible expression entitled to copyright protection." *Reid*, 490 U.S. at 737 (1989). In the context of architectural works, the protectable forms of expression are the building itself and any architectural plans and drawings. The record reflects that the Plaintiffs' copyright application contained not only color renderings of the final building design produced by Mr. Pilon, but also a variety of plans produced by James Brockman that speak to the arrangement and composition of spaces and elements within the stair-towers—something that is not at all apparent from the color renderings, which illustrate only the exterior of the proposed garage. Accordingly, the Court concludes that there are genuine and disputed issues of material fact regarding whether Mr. Brockman made an independently copyrightable contribution to the design of the CCSU garage.

2. Mutual Intent

As explained above, the Court agrees with Defendants that the claims of Plaintiffs Moser Pilon and Macchi Engineers fail because they did not make any independently copyrightable contributions. Accordingly, the Court need not, and does not, address the mutual intent prong of joint authorship with respect to the claims of these two Plaintiffs.

With respect to the mutual intent of Mr. Pilon and Mr. Brockman to be joint authors, the Court finds that the sworn affidavits of these two Plaintiffs to the effect that they always intended joint ownership of the copyright, and the evidence that Mr. Brockman was originally listed as a joint author on the certificate of registration, at least raise genuine issues of material fact sufficient to preclude summary judgment on this prong of the Childress test for joint authorship.

For the foregoing reasons, the Court concludes that neither Macchi nor Moser Pilon may be considered joint authors of the disputed design, but that there are dispute issues of material fact regarding whether Mr. James Brockman satisfies the requirements for joint authorship. Accordingly, Defendants' Motion for Summary Judgment of Non-Ownership must be DENIED IN PART, that is, denied with respect to Mr. Brockman and granted with respect to the entities Moser Pilon and Macchi.

The Court having determined that neither Moser Pilon nor Macchi is an author of the disputed design, a further issue arises as to their standing to pursue the remaining claims in this case. Since neither is the author of the disputed design, it follows that neither has standing to pursue claims that see redress for violations of an author's rights. Since, as discussed below, the Lanham Act, unjust enrichment and conversion claims, Counts II, III and IV of the Amended Complaint, are predicated on the same acts of alleged copying, and contingent on a possessory interest in the thing copied, the non-author entities Moser Pilon and Macchi lack standing to pursue those claims and Defendants are entitled to summary judgment against Moser Pilon and Macchi on those claims. However, the CUTPA claim does not depend upon a claimant having a possessory interest in the allegedly copied design. Rather, the CUTPA claim asserts that by agreeing to copy the design proposal that comprised part of Plaintiffs' overall submission, Defendants gained an unfair trade advantage over Plaintiffs' entire bid team, not just over the author of the design. The entities Moser Pilon and Macchi therefore have standing to pursue the CUTPA claim.

C. Preemption

Defendants' Motion for Summary Judgment of Preemption argues that Plaintiffs' claim under the Copyright Act preempts all of their other claims. The Copyright Act is a plaintiff's exclusive recourse whenever: (1) the work upon which the non-copyright claim is base is protected by the Copyright Act; (2) the non-copyright claim involves acts that violate federal copyright law, i.e, "acts of reproduction, adaptation, performance, distribution, or display"; and (3) the non-copyright claim does not contain any "extra elements that make it qualitatively different from a copyright infringement claim." *Briarpatch Ltd. v. Phoenix Pictures, Inc.*, 373 F. 3d 296, 305 (2d Cir. 2004). Applying this test, the Court concludes that Plaintiffs' federal claim under the Lanham Act and state law claims of conversion and unjust enrichment ar preempted by the Copyright Act, but Plaintiffs' CUTPA claim is not preempted.

1. Lanham Act

In Count II of the Amended Complaint, Plaintiffs assert a claim for false designation of origin and misleading description of fact under the Lanham Act, 15 U.S.C. § 1125(a)(1), which makes actionable false or misleading representations that are likely to cause confusion or deception as to the origin of goods or services. According to Plaintiffs, Defendants held out Plaintiffs' design for the CCSU garage as their own, and thereby caused confusion in the public and potential clients as to the origin of the design services actually provided by Plaintiffs. For the reasons explained below, the Court concludes that no reasonable juror could find for Plaintiffs on their Lanham Act claim against either Defendant.

Plaintiffs' claim of false designation of origin turns on "the ongoing publication on HNTB's website . . . of plaintiffs' design as representing a design produced by the defendants for the CCSU parking garage project." As a preliminary matter, the Court notes that this allegation in no way involves wrongful behavior by Defendant Downes. Accordingly, Downes is entitled to summary judgment on Plaintiffs' Lanham Act claim.

With respect to Defendant HNTB, Plaintiffs' claim turns on HNTB's alleged publication on its website of a copy of Plaintiffs' design. Plaintiffs do not claim that HNTB tried to pass off Plaintiffs' physical renderings as its own. The problem for Plaintiffs is that in *Dastar Corp. v. Twentieth Century Fox Film Corp.*, 539 U.S. 23, 123 S. Ct. 2041, 156 L. Ed. 2d 18 (2003), the Supreme Court held that copying of creative content like that alleged in this case is not protected by the "origin of work" provision of the Lanham Act, which covers only "the producer of the tangible good that are offered for sale, and not . . . the author of any idea, concept, or communication embodied in those goods." Id. at 37. . . . It appears to the Court that *Dastar* applies with equal force to architectural works and that Plaintiffs' claim under the Lanham Act therefore fails.

Plaintiffs assert that parallel claims under the Lanham Act and Copyright Act by architects have been upheld by the Sixth Circuit and that *Dastar* should not be read as overruling the Sixth Circuit on this point. But *Johnson v. Jones*, 149 F.3d 494 (6th Cir. 1998), upon which Plaintiffs rely, is simply inapposite: Defendant architect in Johnson not only held out as his own copies of some the predecessor architect's plans, but physically "took the [original] drawings and site plan that he had obtained from the city inspector, removed [plaintiff architect's] name and seal, and replaced them with his own name and seal." Id. at 499. No such physical misappropriation is alleged against Defendants in this case. Accordingly, the Court concludes that Plaintiffs' reliance on Johnson is misplaced, and Plaintiffs' claim under the Lanham Act is preempted by the Copyright Act in accordance with *Dastar.*

2. Unjust Enrichment

In Count III of the Amended Complaint, Plaintiffs claim that Defendants were unjustly enriched by wrongfully appropriating Plaintiffs' design. Defendants countered that Plaintiffs' unjust enrichment claim is identical to, and therefore preempted by, Plaintiffs' copyright claim. The court agrees. . . .

3. Conversion

Count IV of the Amended Complaint is a state law claim for conversion of Plaintiffs' design. Plaintiffs do not allege that Defendants took possession of the original renderings, but rather that Defendants copied Plaintiffs' work. Accordingly, the "conversion claim [is] preempted by the Copyright Act since it is based solely on copying, i.e., wrongful use, not wrongful possession." *A Slice of Pie,* 392 F. Supp 2d at 317.

As with respect to their enrichment claim, Plaintiffs argue that, even if their conversion claim may be preempted by a valid copyright claim, the conversion claim should survive dismissal as a claim in the alternative until the copyright claim is adjudicated. This argument fails for the reasons stated above in Subsection 2. Accordingly, the Court concludes that Plaintiffs' conversion claim is preempted by the Copyright Act.

4. CUPTA

Count Five of the Amended Complaint alleges that Defendants violated the Connecticut Unfair Trade Practices Act ("CUPTA"), Conn. Gen. Stat. § 42-110b (a). According to Plaintiffs, Defendants engaged in unfair competition by agreeing to abandon their proposed design and copy Plaintiffs' design in exchange for CCSU's agreement to execute the contract. Defendants argue that Plaintiffs' CUTPA claim is "grounded solely in the copying of a plaintiff's protected expression" and therefore preempted by the Copyright Act. See *Kregos v. Assoc.* Press, 3 F.3d 656, 666 (2d Cir. 1993). Plaintiffs counter that their CUTPA claim involves an extra element that qualitatively distinguishes it from a copyright claim—namely, that Defendants undermine the public bidding process for the contract by agreeing to copy Plaintiffs' design in return for being awarded the contract. The Court agrees and concludes that Plaintiffs CUTPA claim is not preempted by the Copyright Act. . . .

IV.

For the foregoing reasons, Defendants' Motion for Summary Judgment of Invalidity is DENIED, Defendants' Motion for Summary Judgment of Non-Ownership is GRANTED IN PART and DENIED IN PART, and Defendants' Motion for Summary Judgment of Preemption is GRANTED IN PART and DENIED IN PART.

Judgment shall enter for Defendants on Counts II, III, and IV as against all Plaintiffs. Judgment shall enter for Defendants on Count I as against Plaintiffs Moser Pilon Nelson Architects and Macchi Engineers only.

The following two claims remain in this case: (1) The claims of Mr. Pilon and Mr. Brockman against both Defendants under the Copyright Act; and (2) the claims of all Plaintiffs against both Defendants under CUTPA.

IT IS SO ORDERED.

COCA-COLA BOTTLING COMPANY OF SHREVEPORT, INC., ET AL. v. The COCA-COLA COMPANY, et al., 107 R.R.D. 288;1985 U.S. Dist. LEXIS 16644; 227 U.S.P.Q. (BNA) 18

Study terms: Trade secret, discovery, motion to compel, disclosure

The complete formula for Coca-Cola is one of the best-kept trade secrets in the world. Although most of the ingredients are public knowledge, see *Coca-Cola Bottling Co. of Shreveport, Inc. v. Coca-Cola Co.*, 563 F.Supp. 1122, 1132 (D Del. 1983), ingredient that gives Coca-Cola its distinctive taste is a secret combination of flavoring oils and ingredients known as "Merchandise 7X." The formula for Merchandise 7X has been tightly guarded since Coca-Cola was first invented and is known by only two persons within The Coca-Cola Company ("the Company"). The only written record of the secret formula is kept in a security vault at the Trust Company Bank in Atlanta, Georgia, which can only be opened upon a resolution from the Company's Board of Directors.

The impregnable barriers which the Company has erected to protect its valuable trade secret are now threatened by pretrial discovery requests in two connected cases before this Court. Plaintiffs in these lawsuits are bottlers of Coca-Cola products who seek declaratory, injunctive and monetary relief against the Company based upon allegations of breach of contract, violation of two 1921 Consent Decrees, trademark infringement, dilution of trademark value, and violation of federal antitrust laws, all of which allegedly occurred when the Company introduced diet Coke in 1982. Stripped to bare essentials, the plaintiff's contention is that the Company is obligated to sell them the syrup used in the bottling of diet Coke under the terms of their existing contracts covering the syrup used in the bottling of Coca-Cola. The primary issue arising from this contention is whether the contractual term "Coca-Cola Bottler's Syrup" includes the syrup used to make diet Coke. Plaintiffs contend that in order to prevail on this issue, they need to discover the complete formula, including the secret ingredients, for Coca-Cola, as well as the complete formulae, also secret, for diet Coke and other Coca-Cola soft drinks. Accordingly, plaintiffs have filed a motion to compel production of the complete formulae under Fed. R. Civ. P. 37(a). Defendant, which has resisted disclosure of its secret formulae at every turn, contests the relevance of the complete formulae to the instant litigation and avers that disclosure of the secret formulae would cause great damage to the Company.

The issue squarely presented by plaintiffs' motion to compel is whether plaintiffs' need for the secret formulae outweighs defendant's need for protection of its trade secrets. . . . I am also aware that an order compelling disclosure of the Company's secret formulae could be a bludgeon in the hands of plaintiffs to force a favorable settlement. On the other hand, unless defendant is required to respond to plaintiffs' discovery, plaintiffs will be unable to learn whether defendant has done them a wrong. Except for a few privileged matters, nothing is sacred in civil litigation; even the legendary barriers erected by The Coca-Cola Company to keep its formulae from the world must fall if the formulae are needed to allow plaintiffs and the Court to determine the truth in these disputes.

I. Factual Background

The history of The Coca-Cola Company and its bottlers has been set forth at length in earlier opinions in these cases, and that history does not bear repeating here. Instead, only a brief description of these two cases is warranted in order to establish the issues involved.

Since the turn of the century, Coca-Cola has been produced in a two-stage process: the Company manufactures "Coca-Cola Bottler's Syrup" ("Bottler's Syrup") and sells it to bottlers, who add carbonated water to the syrup and place the resulting product in bottles and cans. In 1921, following litigation between bottler groups and the Company concerning their contracts for Bottler's Syrup, the Company entered into Consent Decrees which established certain contractual terms between the Company and its bottlers. . . .

Beginning in 1978, due to inflationary pressures and declining sales, the Company sought price relief from the existing price formula in its contracts with bottlers. After negotiations, most of the bottlers agreed to an amendment ("the 1978 Amendment") to their contracts in exchange for a clause requiring the Company to pass on any cost savings if the Company decided to substitute a lower cost sweetener for granulated sugar. The 1978 Amendment established a new price formula for Bottler's Syrup which utilizes a "sugar element," a "base element," and the Consumer Price Index. The sugar element provides for adjustments based on the quoted market price of any sweetening ingredient used in Bottler's Syrup. The great majority of the bottlers, representing approximately 90 percent of domestic sales, have signed the 1978 Amendment. These bottlers are generally known as the "amended bottlers." The remaining bottlers, know as the "unamended bottlers," refuse to sign the amendment and continue to operate under Bottler's Contracts which basically conform to the contracts entered into after the 1921 Consent Decrees. In 1980, the amended bottlers began obtaining some benefit from the 1978 Amendment when the Company decided to substitute high fructose corn syrup ("HFCS-55"), a less expensive sweetener than granulated sugar, for approximately 50 percent of the granulated sugar in Bottler's Syrup.

On July 8, 1982, the Company introduced diet Coke to the market with great fanfare. The name was chosen carefully and focused on the descriptive nature of the word "diet" and the tremendous market recognition of "Coke." The advertising emphasized the taste of the new cola and its relationship to Coke. The public response to diet Coke has been phenomenal—in just three years, it has become the third largest selling soft drink in the United States and the best-selling diet soft drink in the world.

The introduction of diet Coke immediately gave rise to a dispute between Coke bottlers and the Company over what price bottlers must pay for diet Coke syrup. . . . This dispute led to the filing of these lawsuits in early 1983.

Since that time, there have been two significant and widely-publicized changes in Coca-Cola. First, in April, 1985, the Company announced that it would stop producing Coca-Cola under the existing formula ("old Coke") and immediately start producing "new" Coke, which, the Company proclaims, tastes even better than old Coke. According to the promotional materials that accompanied the announcement of new Coke, the formula for new Coke was derived from the research that led to the development of diet Coke. The secret ingredient in new Coke, called "7X-100," is different than the secret ingredient in old Coke, but it is still only known to a handful of individuals and is kept locked in a bank vault in Georgia.

The second significant change came when the Company, in response to consumer demand, announced in July, 1985, that it would bring back old Coke under the name "Coca-Cola Classic." The Company will now provide bottlers with two kinds of sugar-sweetened cola syrups—old Coke syrup, to be package as Coca-Cola Classic, and new Coke syrup. The Company has informed its bottlers that for the present, it will supply them with Coca-Cola Classic syrup under the terms of their contracts for Coca-Cola, but without prejudicing the Company's rights. As defendant's supplemental brief makes clear, the Company is ostensibly reserving the right to decide at a later time that the syrup for Coca-Cola Classic is not Coca-Cola Bottler's Syrup, even though the identical syrup was considered Coca-Cola Bottler's Syrup a few months ago. The merits of this remarkable position, however, are not before the Court at this time.

II. Plaintiffs' Motion to Compel

After extensive discovery, plaintiffs filed the instant motion that, in essence, seeks to compel the Company to produce the complete formulae, including secret ingredients, for Coca-Cola, diet Coke, caffeine free Coca-Cola, caffeine free diet Coke, TAB, and every experimental cola formula developed and tested by the Company for possible marketing under the Coca-Cola or Coke trademarks. Defendant's responses to the discovery requests at issue, which plaintiffs filed as an appendix to their motion, demonstrate that defendant has objected to plaintiffs' discovery wherever it approached matters related to the secret formulae. Thus, plaintiffs have been foreclosed both from learning the formulae themselves and from learning about other matters that relate to the formulae.

In support of their motion to compel, plaintiffs have contended that the secret formulae are relevant and necessary to prove their contentions and respond to defendant's argument that Coca-Cola and diet Coke are two different products. In response, the Company denies that the formulae are relevant and essential to resolve the central issues in these cases, and also contends that disclosure of these trade secrets is inappropriate at this stage of the litigation. . . . A stipulation that the secret ingredients in diet Coke and Coke are identical was used to avoid the disclosure of the secret formulae at the preliminary injunction stage, but plaintiff indicated at the hearing that the same stipulation was no longer acceptable, in part because it is not true. After the hearing, the parties conducted extensive negotiations on the formula issue but were unable to resolve it by stipulation. In addition, the parties were granted an opportunity to file supplemental submissions on the significance of the Company's decision to reintroduce old Coke as Coca-Cola Classic. Thus, the discoverability of the secret formulae is squarely presented and ripe for decision.

A. The Legal Standard Applicable to Discovery of Trade Secrets

It is well established that trade secrets are not absolutely privileged from discovery in litigation. In order to resist discovery of a trade secret, a party must first demonstrate by competent evidence that the information sought through discovery is a trade secret and that disclosure of the secret might be harmful. If this showing is made, "the burden shifts to the party seeking discovery to establish that the disclosure of trade secrets is relevant and necessary to the action." *Centurion Industries,* 665 F.2d at 325; see *Pennwalt Corp. v. Plough, Inc.,* 85 F.R.D. at 259. When disclosure of trade secrets is sought during discovery, the governing relevance standard that the movant must satisfy is the broad relevance standard applicable to pre-trial discovery, i.e., the movant must show that the material sought is relevant to the subject matter of the lawsuit. The level of necessity that must be shown is that the information must be necessary for the movant to prepare its case for trial, which includes proving its theories and rebutting its opponent's theories.

Once relevancy and need have been established, the Court must balance the need for the information against the injury that would ensue if disclosure is ordered. Because protective orders are available to limit the extent to which disclosure is made, the relevant injury to be weighed in the balance is not the injury that would be caused by public disclosure, but the injury that would result from disclosure under an appropriate protective order. In this regard, it is presumed that disclosure to a party who is not in competition with the holder of the trade secret will be less harmful than disclosure to a competitor.

The balance between the need for information and the need for protection against the injury caused by disclosure is tilted in favor of disclosure once relevance and necessity have been shown. . . . A survey of the relevant case law reveals that discovery is virtually always ordered once the movant has established that the secret information is relevant and necessary. . . . As Judge Learned Hand stated in one of the earliest trade secret cases.

> It is true that the result may be to compel the defendant to disclose [trade secrets], and that that may damage the defendant. . . . That is, however, an inevitable incident to any inquiry in such a case; unless the defendant may be made to answer, the plaintiff is deprived of its right to learn whether the defendant has done it a wrong.

Grasselli Chemical Corp. v. National Aniline & Chemical Co., 282 F. 379, 381 (S.D.N.Y. 1920).

B. The Coca-Cola Formulae Are Trade Secrets

To satisfy its burden of proving that the Coca-Cola formulae qualify for trade secret protection, defendant has submitted the affidavit of Robert A. Keller, Senior Vice President and General Counsel of the Company. According to the Keller affidavit, the Company has taken every precaution to prevent disclosure of the formula for "Merchandise 7X," the secret ingredient in old Coke. The written version of the secret formula is kept in a security vault at the Trust Company Bank in Atlanta, and that vault can only be opened by a resolution from the Company's Board of Directors. It is the Company's policy that only two persons in the Company shall know the formula at any one time. . . . The Company refuses to allow the identity of those persons to be disclosed or to allow those persons to fly on the same airplane at the same time. The same precautions are taken regarding the secret formulae of the Company's other cola drinks. . . .

The Keller affidavit further states that these secret formulae are highly valued assets of the Company and have never been disclosed to persons outside the Company. As an indication of the value the Company places on its secret formulae, Kellers avers that the Company elected to forego producing Coca-Cola in India, a potential market of 550 million persons, because the Indian government required the Company to disclose the secret formula for Coca-Cola as a condition of doing business there. The affidavit concludes by stating that because of intense competition in the soft drink industry, the disclosure of any information reflecting the formulae or the Company's research and development would be extremely damaging to the Company.

New Coke was introduced after the Keller affidavit was filed, but the materials supplied to the Court about new Coke provide information about the secret formula in new Coke. To develop new Coke, Merchandise 7X was "optimized," i.e., changed, and the new secret ingredient is called "7X-100," to commemorate the one hundredth year of Coca-Cola. This is the first change in the secret ingredient in Coca-Cola since the invention of Coke in 1886. . . .

Although some of the mystique surrounding the Coca-Cola formulae is the creation of marketing hype, it is beyond dispute that, behind the hype, the Company possesses trade secrets which have been carefully safeguarded and which are extremely valuable. It is also evident that any disclosure of those trade secrets would be harmful to the Company. Accordingly, I find that defendant's secret formulae are trade secrets and subject to the maximum protection that the law, as set forth above, allows.

C. Relevance and Necessity of the Formulae

Plaintiffs contend that discovery of these secret formulae is required because they are relevant and necessary to the presentation of plaintiffs' case. In order to determine whether these trade secrets are in fact relevant and necessary, a review of the issues in the two cases is warranted.

1. Unamended Bottlers

The unamended bottlers claim that defendant must furnish diet Coke syrup to them pursuant to the terms of their Bottler's Contracts and the 1921 Consent Decrees. The standard form contract for unamended bottlers states, in pertinent part: "COMPANY agrees to furnish to BOTTLER. . . . sufficient syrup for bottling purposes to meet the requirements of BOTTLER in the territory herein described. . . . COMPANY does hereby select BOTTLER as its sole and exclusive customer and licensee for the purpose of bottling the Bottlers' syrup, COCA-COLA, in the territory herein described." The contract further provides that "BOTTLER agrees. . . . to bottle COCA-COLA in the following manner: to have it thoroughly carbonated, put in bottles, using one ounce of Bottlers' Coca-Cola syrup in a standard bottle for Coca-Cola. . . . decorated with the name Coca-Cola in the characteristic script. . . ." The terms "bottle syrup" and Bottlers' Coca-Cola syrup" are not defined in the contract.

Plaintiffs contend that the terms "Bottlers' Coca-Cola syrup" and "bottle syrup" include any syrups manufactured by the Company for the purpose of providing any packaged soft drink sold under the names "Coca-Cola" or "Coke," including diet Coke. In addition, plaintiffs allege that Coca-Cola and diet Coke are just two versions of the same product, except that one is sweetened with caloric sweeteners and the other with non-caloric sweeteners. Defendants' response to these contentions has been that only syrup for sugar-sweetened Coca-Cola is covered by the unamended bottlers' contracts, and that diet Coke and Coca-Cola are two separate products.[4]

---

[4]On Plaintiff's motion for a preliminary injunction, this Court determined that the contractual phase "Bottler's Coca-Cola Syrup" is not likely to include diet Coke Syrup. Because that finding was made in the preliminary injunction context, it is not a final decision on the merits and does not preclude plaintiffs from proving at trial that diet Coke syrup is encompassed by the 1921 contracts.

2. Amended Bottlers

The amended bottlers rely on different contractual language to argue that the Company must furnish them diet Coke syrup on the same terms as Coca-Cola syrup. The 1978 Amendment, which all of the amended bottlers signed, replaced the pricing formula used for the unamended bottlers with one that was tied to the "Sugar Element," a term defined in the contract. The Amendment then provides: "In the event that the formula for Bottle Syrup is modified to replace sugar, in whole or in part, with another sweetening ingredient, the Company will modify the method for computing the Sugar Element in such a way as to give the Bottler the savings realized as a result of such modification through an appropriate objective quarterly measure of the market price of any such sweetening ingredient." The amended bottlers have contended that "another sweetening ingredient" includes saccharin or aspartame, the sweeteners that the Company has used in diet Coke.

The Company argues that this contractual language is inapplicable because diet Coke is a new and different product and is not modified Coca-Cola. Plaintiffs' response is that diet Coke is "simply a version of a product which has undergone evolutionary change but which retains its identity as Coke," and "that any differences between Coke and diet Coke Bottler's Syrup are either insignificant or reflect attempts to achieve taste identity."

On plaintiffs' motion for preliminary injunction, the Court conducted an extensive analysis of the product identity question, including a comparison of the publicly-disclosed ingredients of diet Coke and Coca-Cola, before concluding that "for at least some purposes diet Coke may be Coke."

3. Relevancy of the Secret Ingredients

A major issue common to both actions is whether diet Coke and Coca-Cola are the same product. The Company's primary defense has been that Coca-Cola and diet Coke are two separate and distinct products. Plaintiffs contend that the complete formulae for diet Coke and Coca-Cola would be relevant to rebut this defense by showing that the two colas share common attributes and that any differences between the two are insignificant and merely reflect attempts to achieve taste identity. With the introduction of new Coke, plaintiffs argue that because new Coke was derived in part from the secret formula for diet Coke, it may be true that new Coke is more like diet Coke than new Coke is like old Coke. In response, defendant argues that except for the difference in sweeteners, ingredient similarities and differences are not relevant to the determination of whether diet Coke and Coca-Cola are the same product. Instead, defendant relies upon the difference in taste, different essential characteristics of the beverages, different consumer markets for the beverages, and different consumer perception of the beverages.

Defendant's response is unavailing. When this Court previously addressed the merits of this litigation in the context of a motion for preliminary injunction, the first issue addressed was whether Coke and diet Coke are two versions of the same product. Although the parties' contentions have evolved in the intervening two years, the issue of product identity remains a part of these lawsuits. Although defendant has attempted to define the issues so that the only relevant ingredient is the sweetener, all the ingredients are relevant to determine whether the two colas are the same product. In fact, the secret ingredients may be the most relevant ones because the secret ingredients are what gives these drinks their distinctive tastes.

Plaintiffs could use the secret formulae to prove one of several product identity theories. An analysis of the secret ingredients in diet Coke and old Coke might show that diet Coke was designed to taste as much like old Coke as a low calorie cola could, and that any differences in secret ingredients reflect defendant's attempts to achieve taste identity. Alternatively, plaintiffs might use the secret formulae for diet Coke, old Coke, and new Coke in the following way: The syrup for old Coke and new Coke have both been sold as Coca-Cola Bottler's Syrup by the Company. It has been publicly disclosed, however, that the formula for new Coke was derived from the research used for diet Coke. If plaintiffs, armed with the complete formulae, can show that diet Coke is very similar to new Coke, and that diet Coke is more like new Coke than new Coke is like old Coke, that fact could tend to show that diet Coke is within the range of syrups that have been sold as Coca-Cola Bottler's Syrup. These examples, based only on speculation as to what plaintiffs might learn through discovery, illustrate that the complete formulae for diet Coke, old Coca-Cola, and new Coca-Cola are relevant to one of the primary issues in this litigation—product identity. The complete formulae, once known, will tend to make a disputed fact more (or less) likely: that, for purposes of this litigation, diet Coke syrup is Bottler's Syrup.

The complete formula for caffeine free Coca-Cola is not as directly relevant. Plaintiffs elicited deposition testimony from Dr. Anton Amon, the Company's Technical Director, that regular Coca-Cola and caffeine free Coca-Cola are the same product, despite the fact that Coca-Cola has, and caffeine free Coke does not have, kola nut extract (from which the term "cola" is derived), vanilla extract, and caffeine. See Tr. at 16. Plaintiffs contend that the complete caffeine free Coke formula is relevant to the product identity issue, because a comparison of the formulae for diet Coke, regular Coke, and caffeine free Coke might show that diet Coke is more similar to regular Coke than is caffeine free Coke. Plaintiffs could then link that finding to Dr. Amon's statement to argue that diet Coke is therefore the same product as Coke.

In response, defendant points out that plaintiffs and defendant have signed letter agreements which allow caffeine free Coke syrup to be priced and supplied on the same terms as regular Coke syrup, but which state that this agreement would not prejudice either side in litigation. This response is a non sequitur. The letter agreements do not foreclose plaintiffs from arguing caffeine free Coke and regular Coke are similar products; they only prevent plaintiffs from using the letter agreements themselves as an admission that they are the same product. Plaintiffs' ability to use Dr. Amon's testimony to their advantage is simply not affected by the letter agreements. Moreover, the reasons offered by plaintiffs demonstrate that the complete formula for caffeine free Coke is relevant to the product identity issue in conjunction with the testimony of Dr. Amon.

■ ■ ■

4. Necessity of Discovery of This Information

As in most disputes over the discoverability of trade secrets, the necessity of the discovery of the complete formulae follows logically from the determination that the formulae are relevant. Plaintiffs need the complete formulae in order to address the product identity issue comparing the ingredients of the various soft drinks involved. Plaintiffs cannot respond to the assertions of defendant's experts that diet Coke and Coca-Cola are two products unless plaintiffs' experts can analyze the complete formulae and explain why the products are the same.[8] Merely using the publicly-disclosed ingredients is obviously insufficient, because they would present an incomplete picture, and because the secret ingredients are the key to the taste of Coca-Cola. The differences in the public ingredients, including sweeteners, cannot be understood unless they are put in context through disclosure of the similarities and differences in the secret ingredients. Without the complete formulae, plaintiffs will be foreclosed from presenting all the relevant evidence in support of their position.

In addition, plaintiffs need the complete formulae in order to explore on cross-examination the bases for the opinions of Company witnesses that Coca-Cola and diet Coke are two separate products. As plaintiffs' counsel stated at oral argument, plaintiffs' cross-examination of defendant's witnesses had been foreclosed by defendant's objections that plaintiffs; questions relate to trade secrets. Plaintiffs cannot be expected to discover the truth without full cross-examination. Moreover, the formula information is not available from any other source, and no adequate substitute exists for this information. It follows that discovery of the complete formulae is necessary.

After the hearing, defendant attempted to remove the necessity for the information by offering to stipulate that the secret ingredients in old Coke, new Coke and diet Coke are identical. Defendant contends that this result is the most favorable set of facts that plaintiffs could hope to find through discovery. It is evident, however, that the actual formulae could be more favorable to plaintiffs than this stipulation. For example, if the secret ingredients in new Coke and diet Coke are very similar but different than the secret ingredients in old Coke, that would favor plaintiffs, as indicated above. On the other hand, discovery may show that the secret ingredients in old Coke were modified to make diet Coke, and that those modifications were intended to counterbalance the taste change caused

---

[8]Although the Company probably would not reveal its formulae to outside experts, some Company witnesses who know the formulae may testify. Given the Company's policy of not revealing the names of persons knowing the formulae, plaintiffs would not even be able to discover whether Company experts are basing their opinions on a comparison of the secret formulae.

by substituting artificial sweeteners for sugar. The effect of the secret ingredient change may have been to cancel out the changes in other ingredients and make diet Coke taste like Coke. Further, defendant's proposed stipulation does not reveal the number ingredients that are secret ingredients. If the secret ingredients in Coke and diet Coke are composed of the same 100 ingredients, so that the vast majority of all the ingredients in the two colas are identical, that fact would be more favorable to plaintiffs than if the secret ingredients were only a few in number. Finally, defendant's proposed stipulation does not solve the problem of plaintiffs being foreclosed from full cross-examination by defendant's assertion of trade secret privilege. In sum, defendant's proposed stipulation is not as favorable to plaintiffs as discovery might be and does not remove the necessity for disclosure.

5. Need Balanced Against Harm

The final part of the test for discoverability of trade secrets is to balance the need for disclosure against the harm that would ensue from disclosure. The potential harm that would come from public disclosure of the formulae for old Coke, new Coke, diet Coke, and caffeine free Coke is great, but virtually all of that harm can be eliminated with stringent protective orders and other safeguards. Because plaintiffs are Coca-Cola bottlers, they will have an incentive to keep the formulae secret. The likelihood of harm is less than if defendant's trade secrets were disclosed in litigation to competitors. See *United States v. United Fruit Co.*, 410 F.2d 553, 556 (5th Cir.), *cert. denied*, 396 U.S. 820, 24 L. Ed. 2d 71, 90 S. Ct. 59 (1969); 2 R. Milgrim, *Trade Secrets* § 7.06[1][b], at 7-88 (1984). The potential for harm from protected disclosure of the formulae for old Coke, new Coke, diet Coke, and caffeine free Coke is outweighed by the plaintiffs' need for the information. While plaintiffs; need for the experimental cola formulae is less strong, this lesser need is counterbalanced by the fact that the harm resulting from disclosure of these formulae would be less severe, because those colas have never been marketed and are less valuable trade secrets.

In sum, the product identity issue is important in these two cases, and analyses of the complete formulae will be a significant part of the proof on that issue. Plaintiffs' need for this information outweighs the harm that disclosure under protective order would cause. Disclosure will be ordered.

■ ■ ■

IV. Conclusion

It has been held that defendant must disclose its complete formulae, including secret ingredients, for diet Coke, old Coke, new Coke, caffeine free Coke, and certain experimental low calorie colas, but not the formulae for TAB and caffeine free diet Coke. . . . Given the proprietary nature of the formula information, however a more stringent protective order than the one currently in effect is warranted to prevent public disclosure of the formulae. For example, it may be advisable to limit the disclosure of the formulae to plaintiffs' trial counsel and independent experts. Because the parties have not addressed what additional protective measures would be satisfactory, the Court will not enter a new protective order at this time. Instead, the parties shall negotiate a protective order that both allows access to information and prevents disclosure of trade secrets. The parties shall submit that order within twenty days.

Review Questions

1. Why is the patent right the right to exclude others from making, using, or selling the invention rather than the right to make, use, or sell it?

2. White, an electrical engineer, goes to work for the ABC Company to design automation circuitry. No patent agreement is required of him. Using company time and facilities (in part), he develops a new product and makes application for a patent on it. If he obtains a patent, what rights, if any, will the ABC Company have in the product?

3. In question 2, assume that White had signed an agreement to assign patent rights to the ABC Company but had developed the product at home, using only his own time and facilities. What rights, if any, will the ABC Company have in the patent?

4. Outline the steps of the patenting procedure.

5. Why are records of the development of an invention necessary?

6. Consider trademarks and copyrights. Can you conceive of something that might be the subject of both a trademark registration or a copyright?

7. Which section in the Patent Act declares that "patents shall have the attributes of personal property"?

8. To what extent may an inventor publicly use or sell an invention prior to patent application without destroying its patentability?

9. "According to the court in *Moser Pilon Nelson Architects, LLC, et al. v. HNTB Corporation, et al.*, the plaintiffs were entitled to joint copyright ownership if they could demonstrate what?"

10. Gray, an engineer, learned a manufacturing process as a trade secret from White, his employer. Later, Gray quit and went to work for Black. Upon hearing that White had sold his plant to Brown, Gray quit Black's employ and went into business for himself, using the trade secret. Brown is suing for an injunction to prevent Gray from using the trade secret, claiming that he (Brown) bought the secret process with the rest of the business. Gray claims that he has respected the secret he learned from White because he did not use it until White sold the business. Would the court be likely to issue the injunction? Why or why not?

11. Why do you think You Gotta Eat (YGE) did not reply to Checker's attorneys' request to negotiate? Based on the court's application of the Polaroid factors, do you think YGE's claims will prevail? Why or why not? ~ Broke, abandoned ; wanted to show bad faith
~ No

12. "In *Coca-Cola Bottling of Shreveport, Inc. v. The Coca-Cola Company*, the Coca-Cola Company's senior vice-president and general counsel used what anecdote as an example of the lengths the company would go through to protect its trade secret formula of Coca-Cola?" Indra

Torts

Nearly all engineers are concerned with contracts and property; therefore, our time is well spent considering the rights and liabilities comprising the law in these areas. Just as important (perhaps even more important to some engineers), however, is the law concerning personal rights.

We all work with other people, so it follows that what we do may infringe upon their rights, and what they do may infringe upon ours. As is true for other professionals, an engineer's activities encounter **tort risks.** That is, what the engineer designs—a product, a process, or a system of some sort—may cause injury to someone's rights. For example, a design is based upon a mental image of the future. But designing a thing or a combination that has never before existed can be hazardous because one cannot possibly conceive of all ramifications of a new concept. Consequently, the price of progress in engineering is sometimes injury to someone—a tort.

One need only look at newspaper headlines to see how pervasive tort litigation has become. Automobile manufacturers have been sued for faulty design, resulting in multi-million-dollar verdicts for plaintiffs in class action suits. Medical malpractice claims have stretched hospital budgets as well as individual medical practitioners to the breaking point. Many state legislatures are looking to tort reform to put caps on the awards granted to plaintiffs in many tort-related cases.

A **tort** can be defined as a wrongful injury to another person or another's property rights, whereas a **crime** is an injury to society. A plaintiff may bring a civil action to obtain compensation for the injury suffered or to seek a decree that will prevent harm to personal or property rights, whereas the state brings a criminal action on behalf of the citizen. Just as the law undertakes to prevent acts against society by prosecuting those who have committed crimes, it also provides procedures to redress wrongs to an individual. Such wrongs fall into two general categories—breaches of contract and torts.

Breach of contract cases arise when there has been an agreement of some sort between two or more parties. Such cases were treated in earlier chapters. In contrast, tort actions arise from the duties existing as a matter of law between parties. The driver of a car, for instance has a duty to avoid hitting others while driving. If the driver is negligent and, as a result, injures another, a tort action is available to the victim to make the driver pay for the damage. In Chapter 21, we discuss a number of torts generally. The following chapter focuses on a particular type of tort action—products liability.

Common Torts

is common law

CHAPTER

21

HIGHLIGHTS

■ Torts are injuries (either physical or economic) to persons for which damages may be recoverable by law.

■ Negligence is the failure to fulfill obligations that give rise to responsibilities for damages as a result.

■ In order to establish negligence, the claimant needs to show:
1. The existence of a duty,
2. Breach of that duty,
3. An injury, and
4. Recoverable damages.

■ Common torts include:
1. Torts against a person,
2. Torts against one's reputation,
3. Torts against property rights and
4. Torts against economic rights

As with our other laws of contract and property, a great deal of our tort law can be traced to the common law developed in England over the centuries. As a result, different types of torts sometimes overlap. For example, negligence is itself a kind of tort, yet other kinds of torts (against persons or against property) may also involve negligence. Torts can also be intentional (as opposed to negligent), and intent can be an element in torts against persons, property, reputation, and business. This chapter addresses the torts that arise most often.

Most tort cases are based on negligence—someone did something negligently, or neglected to do something he or she should have done. Although different torts often overlap, it is often helpful to distinguish between torts that involve personal injuries or death and those that involve property damage. Obviously, however, this distinction becomes blurry or nonexistent in situations (such as a car accident) involving both personal injury and damage to property. In the following sections, we will first explore the tort of negligence and its elements. The discussion then turns to other common torts, which will be treated under the following four general headings:

1. Torts against a person,
2. Torts against one's reputation,
3. Torts against property rights, and
4. Torts against economic rights.

WHO CAN BE SUED UNDER TORT LAW

Any person who has committed a tort against another can be sued. Because a corporation is an artificial person, it cannot personally commit a tort, but its agents can. Through the laws of agency, then, a corporation may be held liable for tortious injury to another by one of its employees. The employee must, of course, be engaged in the corporation's business at the time the tort is committed. As a general rule, the corporation is still liable, even though the injury arose from an *ultra vires* act. As you learned in previous chapters, minors are granted some leniency with regard to contract law. The tort laws, however, are not as permissive toward minors because the injured party usually cannot avoid the tortuous act.

There is a popular misconception about parents' and guardians' responsibility for the torts committed by infants in their charge. The usual position taken by the law is that an infant is responsible for his or her own torts. Unless the infant is either acting under the parents' direction or should have been restrained by the parents, the parents are free of liability. The law seems reasonable in this stance.

An adult can choose not to contract with a minor and thereby avoid difficulty. However, it is often impossible to avoid tortious harm instigated by a minor (for instance, from an object propelled toward one's back). Knowledge that the tortfeasor was a minor rather than an adult doesn't help much after the injury occurs. (A **tortfeasor** is a person who commits a tort.) When the occurrence of tortious injury is inevitable, knowing the age of the tortfeasor is of little value to the victim.

NEGLIGENCE

The basic idea of negligence is that a failure to live up to one's duty is wrong and gives rise to liability for damages caused by that failure. *Black's Law Dictionary* defines negligence as "the omission to do something which a reasonable man, guided by those ordinary considerations which ordinarily regulate human affairs, would do." To establish the tort of negligence, the plaintiff needs to provide evidence supporting the following:

1. The existence of a duty,
2. A breach of that duty,

3. An injury caused by the breach, and

4. Recoverable damages.

In some cases when an injury results from negligence, the defendant must "take the plaintiff as he finds him," otherwise known as the "eggshell skull rule" where the defendant may be held liable for unforeseeable injuries to the plaintiff. (See *Benn v. Thomas* in this chapter).

Duty

Whether a legal duty exists under any particular set of facts and circumstances usually involves a question of law for the courts. There are two major standards under the duty's umbrella: reasonableness and foreseeability. The measure of **reasonableness** is whether a reasonable person would have behaved in a way similar to that of the defendant. The measure of **foreseeability** is whether the defendant could have anticipated the damage resulting from breach of a duty of care. In other words, under negligence theory, liability should not be imposed unless the event could have been foreseen by the defendant. Thus, the foreseeability of injury to a particular person or a group of persons is often considered an important basis for concluding that a legal duty does or does not exist. (See *Guin v. Brazos Higher Education Services* at the end of this chapter.)

Another factor sometimes considered by the courts is whether the conduct involved was "active" or "passive." Thus, a defendant who failed to prevent an injury by failing to act may be found to have no duty; on the other hand, a defendant who creates a dangerous condition by its conduct has a duty to do something to prevent any injury. However, the courts have eroded this particular distinction in more recent years. Some courts have adopted the view that "reasonable care" is required; the amount of care viewed as "reasonable" varies with the circumstances.

Another potential basis for the existence of a duty arises when one renders services recognized as necessary for the protection of persons or things. The service provider may have a legal duty if he or she fails to exercise care, and that failure increases the risk of harm to others, or if someone suffers harm because he or she relied on the service. A duty also can arise from the existence of contractual obligations. For example, contract duties may be considered tort duties in situations in which tort damages result from a breach of contractual duties, such as a breach of warranty.

Suppose Black Machinery Co. hires White to install a new lathe. White does so badly that he manages to tip the lathe over, destroying it and another piece of equipment. White's contract may provide the basis for imposing a duty to perform in a reasonably careful manner. Because White breached this duty by performing negligently, Black can sue for breach of contract and for negligence (in some states).

Breach of Standard of Care

Another element a plaintiff has to prove is that the defendant breached a duty owed to the injured party. Specifically, that the defendant owed some standard of care to the plaintiff and such care was not taken, which is a general standard for duty of care developed at common law over the centuries. However, many circumstances tend to modify the standard of care.

The general standard of care, as it has developed, is the degree of care that would be exercised by a reasonably prudent person in like circumstances. The average, or reasonably prudent, person is one possessing normal intelligence, memory, capacity, and skill. This person would have no handicaps, either physical or mental, that would set him or her aside as exceptional. It is, of course, easy to talk about an average person, but much harder to find one. Most people have something other than average intelligence or average physical structure. Reaction time (for a visual stimulus), for instance, averages somewhere around 0.19 seconds, but most people are either slower or faster. Nevertheless, a standard must be established, and then allowance must be made for the exceptions.

Modifications of the general standard of care exist in appropriate situations. The standard of care required of a surgeon in removing an appendix would be considerably more strict than the standard required of a person whose only claim to a knowledge of medicine came from a course in anatomy, but who, because of an emergency, had to attempt surgery. An engineer works under an exceptional standard of care when he or she designs or supervises the construction of a machine or structure. On the other end of the scale, people who have a physical or mental handicap cannot be held to the same standard as the average, or reasonably prudent, person.

Proximate Cause

The main cause of most tort injuries is usually quite apparent when the facts are established. On occasion, however, more than one act or omission may cause the injury. For example, a motorist driving along a highway at night at a lawful speed may be so blinded by oncoming headlights that he will not see an object he is approaching. If he strikes another car in the rear, is he liable, or would the person who failed to dim the headlights be liable? The question is one of proximate cause. The failure of the oncoming driver to dim his lights would probably be posed as a defense, but with a probable lack of success. The driver who couldn't see properly should have slowed down.

To rise to the level of a **proximate cause,** something must have been a substantial factor in bringing about the injury. A factor may be characterized as either cause in fact or legal causation. **Cause in fact** is met if the "but for" standard is satisfied. Would the injury have occurred but for the defendant's conduct? The issue of **legal causation,** however, depends on the connection between the cause and effect; the injury must be part of a natural and continuous sequence.

For example, Black and White are engaged in the electrical repair of an overhead crane. Before starting the work, Black turned off the electricity at the switch box. White is working as the ground man of the pair. Gray, requiring electricity for a job he is doing, throws the wrong switch at the box. Black, receiving a shock from the conductor, drops a wrench on White, thereby injuring him. The immediate cause of White's injury was the force of the blow from the wrench dropped by Black. The proximate cause was Gray's action in throwing the wrong switch. If the case went to court, the probable result would be a finding for White against Gray.

Now suppose that when White is struck, he falls over, knocking over a can of gasoline. The gasoline spreads, evaporates, and when the crane is activated, a fire starts. The fire spreads, and sparks move to a neighboring house. The sparks move from that house to a nearby garage, in which a Rolls Royce is kept. The garage roof collapses and destroys the Rolls.

What was the proximate cause of the loss of the Rolls? At this point, the fire would probably be considered a proximate cause, but it seems difficult to go all the way back to Gray. Proximate cause is the doctrine by which the courts view certain factual causes as just too far removed to be the basis for legal liability.

Res Ipsa Loquitur – alternative to proximate cause

In most tort cases, the plaintiff can point to specific negligent acts by the defendant. In some instances, however, it is difficult or impossible to show defendant's specific acts or omissions, but common sense tells us that someone was negligent. For example, a patient goes into an operating room and has surgery; a sponge is left in the patient's body. No one knows how it got there.

In such situations, the plaintiff may still make a case based on the doctrine known as ***res ipsa loquitur***. Essentially, this means "the thing speaks for itself." To use this doctrine, the plaintiff must show that

1. the injury would not have occurred unless someone had been negligent,
2. the defendant had control of the instrumentality causing the injury,
3. plaintiff was not responsible for his or her own injury and has negated any other explanations, and
4. the explanation is most likely in the defendant's possession than the plaintiff's.

The result is not direct proof, but circumstantial evidence that suggests to the jury that someone was negligent. The defendant, of course, has an opportunity for rebuttal, which may take one of several forms, but usually consists of showing that he or she did exercise the requisite care.

Gross Negligence

As stated earlier, doing something that should not be done, or neglecting to do something that should be done, thereby causing injury to another, amounts to negligence. When the act is done or neglected intentionally or with a reckless disregard for the consequences, it ceases to be the common variety of negligence and becomes **gross negligence.** The victim is much more likely to recover punitive damages if he or she can show gross negligence. For example, many jurisdictions will not allow a plaintiff to recover punitive damages based upon a claim of negligence. Such damages, however, often can be recovered if gross negligence is shown.

Negligence Defenses

As you may guess, there are many situations in which it is unclear just why or how something happened. In some situations, the injury was actually caused by the injured party. The discussion below explains the defenses that have developed in negligence cases.

Assumption of Risk.   People do not always do what is best for them. Occasionally they assume risks for the experience or thrill of the very danger involved. If one is injured or dies as a result of the risk assumed (e.g., death of heart attack during a roller coaster ride), there can be no recovery. Since the assumption-of-risk defense acts to relieve the defendant of liability for negligence, it has usually been applied in a narrow fashion. For instance, the courts limited the doctrine to situations in which the plaintiff knew and understood the risks involved. It's one thing to risk a scratch; it's quite another to literally risk a limb. In addition, the courts required the assumption of the risk to be free and voluntary.

The picture is a little more complicated where a person accepts employment in a risky occupation. For many years, the courts prevented employees from recovering damages for injuries because of the holding of assumption of risk. Some courts rejected this view, however, instead holding that economic realities often compel employees to engage in risky conduct. These courts thus viewed the risks as involuntary. Under the Worker's Compensation laws, however, the employer is deprived of this defense (see Chapter 25).

Assumed risks are only the risks normally and naturally involved with the undertaking. If, for example, the roller coaster suddenly became unsupported, with the resulting crash killing and injuring people, recovery would be quite possible.

A person often assumes a risk (of sorts) when he or she becomes a "good Samaritan" volunteer. If a person gives aid to another who is in distress, and this aid results in further injury to the distressed person, the volunteer may be liable for such injury.

[handwritten: →100% your fault]   *[handwritten: ft partially your own fault]*

Contributory and Comparative Negligence.   Assumption of risk ordinarily arises from a contractual situation of some sort, but **contributory negligence** comes from the negligent act of an injured party; that is, the victim failed to exercise ordinary care. At common law, if both the plaintiff and defendant were found to have acted negligently, the plaintiff could not recover. But this rule led to harsh results. Over time, the doctrine of contributory negligence gave way to today's rule of comparative negligence. Under **comparative negligence**, if both parties were negligent, and injury to the plaintiff would have occurred anyway, recovery may be allowed but diminished by an amount by which plaintiff's neglect contributed to the total damage.

TORTS AGAINST A PERSON

There are three traditional intentional torts—assault, battery, and false imprisonment—where the plaintiff can seek money damages. By the middle of the 20th century, however, with increasing awareness of the emotional aspects of human existence, the courts gave rise to a fourth cause of action: the intentional infliction of emotional distress. This cause of action was first elaborated upon by William L. Prosser in his seminal publication, *Intentional Infliction of Mental Suffering: A New Tort.*[1] False imprisonment and malicious prosecution are also included as torts against a person. These torts qualify as **intentional torts**.

Assault *[handwritten: - threat of immediate harm]*

The term **assault** is quite frequently used improperly. Its legal meaning refers to a threat of violence. It consists of one or more acts intended by the tortfeasor (the person who commits the tort) to create an apprehension of bodily harm in the victim. To rise to the level of assault, the tortfeasor must have an apparent, present means of inflicting the bodily harm. For instance, a knife or a gun (it would make no difference that the pistol was not loaded if the victim had reason to believe that it was) would scare most people. The tort of assault requires that the threat be concerned with immediate injury—not next week or in the future, such as the case where someone threatens injury if they ever see you again. Harm from the tort of assault frequently occurs when the victim has a weak heart or when a pregnant woman receives threats.

Battery

[handwritten: Sometimes consented ← 4 sports]

Assault ends and battery begins when the threat is carried out. **Battery** is the intentional and unlawful touching of another in an offensive manner. Battery is often incorrectly reported as assault; the two torts do often go together, but there is a distinction between them. A battery need not involve breaking a leg or hitting someone with a fist. An offensive contact amounting to battery could occur when someone gives another an unwelcome kiss or spits in another's face.

---

[1] 37 Mich. L. Rev. 874 (1939).

Intentional Infliction of Emotional Distress

Almost all of the states have laws governing this tort and use their own version of Restatement (Second) of Torts §46. In order to prevail under this cause of action, the plaintiff must show extreme and outrageous conduct that intentionally or recklessly caused severe emotional distress. This cause of action has been used in cases where a person has been a witness to a horrible accident[2] as well as when a person was in proximity to an electrically charged wire.[3] More recently, this cause of action is used in lawsuits involving sexual harassment in an employment context, as well as hostile work environment suits.

False Imprisonment

False imprisonment occurs when one is intentionally confined within limits set by tortfeasor. The victim must be aware that he or she is being confined, and the victim must not have consented to the confinement. The means of imprisonment is incidental as long as the victim's personal liberty is restricted. For example, confinement of a person in a car that is traveling too rapidly for exit to be made safely would be imprisonment. A particular means of confinement might be imprisonment to one, but not to another; that is, an athletic young man might escape through a window, whereas a person confined to a wheelchair could not.

Malicious Prosecution

The right to resort to the courts is sometimes abused. Occasionally, a plaintiff brings a suit against another merely as an annoyance or harassment. The tort involved in such an action is **malicious prosecution** (or abuse of process if based on a civil case). The tort hinges principally on the presence or absence of "probable cause." If there were reasonable grounds to believe that the facts warranted the action complained of, this is a defense against a suit for malicious prosecution. Usually, if a reputable attorney recommends an action at law after learning the facts, the plaintiff had "probable cause" to proceed.

Defenses to Intentional Torts

There are at least five defenses to an intentional tort to a person including:

- **Consent**—The defendant must show that the plaintiff consented to an intentional interference. The person giving consent must have the mental and legal capacity to do so. As discussed in earlier chapters, evidence fraud or duress eliminates consent as a defense.
- **Self-defense**—A person is justified in striking another in self-defense. In this situation, a person defending against the attack of another, for example, is justified in his defense up to the point where he becomes the aggressor. If he goes beyond a certain point, the person becomes answerable for the injury caused.
- **Defense of other**—An offshoot of self-defense, a person can claim that he or she was using force to protect a third party.

---

[2]*McMahon v. Bergeson*, 101 NW 2d 63 (Wis. 1960).

[3]*Orlo v. Connecticut Co.*, 21 A.2d 402 (Conn. 1941).

- **Arrest**—Police officers and private citizens are allowed to make an arrest in certain circumstances. In these situations, legal authority may be given to allow the commission of tortious acts—such as those necessary to make an arrest.
- **Justification**—The general catch-all provision "justification" can also be used as a defense if it does not fit into one of the other categories. For example, a person may claim that the tort was the result of an inevitable accident that caused the injury. Here, the defendant must show that he or she did everything reasonable under the circumstances to prevent injury. Accidents resulting from natural causes such as lightning, storms, and earthquakes, for example, are inevitable accidents.

TORTS AGAINST REPUTATION

A person has a right to whatever reputation he or she earns in day-to-day dealings with others. If false and malicious statements are expressed (orally or in writing), such statements may constitute defamation. **Defamation** occurs when false statements made about a person tend to expose him or her to public ridicule, contempt, or hatred. For example, defamation occurs when a statement falsely attributes a criminal act to a person. It also occurs in statements that tend to injure one in a job or profession. Defamation takes two forms. Oral defamation is known as **slander**. Printed or written defamation or defamation by pictures or signs is known as **libel**.

It is slander to falsely state that White is embezzling company funds; it is libel if the statement is written. Slander would not occur, however, if the statement were made only to White with no one else present. Someone other than White would have to hear the statement for slander to occur.

Libel results from printed matter. Even a radio or television broadcast of a speaker who reads a defamatory statement from a written article may constitute libel rather than slander. The damages recoverable for libel are usually greater than for slander, because of the lasting impression created.

Truth and privilege constitute complete defenses to defamation suits. Regardless of the malicious manner in which the statements are made, if they are true, there is no defamation. **Privilege** refers to the right of one person to defame another. A judge has this right, and so does a sworn witness on the stand. Privilege is usually found when the otherwise defamatory statements are made in carrying out a judicial, political, or social duty. It arises from the necessity of making a full and unrestricted communication. At common law, slander and libel imposed strict liability. That is, the speaker was held liable if the statement was false, no matter how strongly the speaker believed in the truth of the statement. However, the Supreme Court has held that the First Amendment prevents this rule in situations where the speaker is part of the media, or the subject is a "public figure." Hence, when a newspaper (falsely) reports that a politician is known to have alcohol problems, the statements are not actionable unless the newspaper acted with malice in printing the story.

TORTS AGAINST PROPERTY

When one owns property, what he or she really owns is a set of rights. The owner has the right to possess the property, to use it (as long as the use does not infringe upon the rights of others), and to dispose of it. One can normally exclude others from using one's property or from taking possession of it. Tort actions result from the invasion of these rights.

Trespass

The tort of **trespass to land** occurs whenever a person without license enters on the land of another. Even simply walking across a person's lawn is a tort. The law, however, does not concern itself with trifles, and a single instance of trespass such as the invasion of one's lawn probably would not be actionable. Even if it were, the result would likely be only nominal damages. An action for trespass is more likely when the trespass has been repeated numerous times, or when material damage can be shown. Such damage to real property can be shown, for instance, where the foundation of a structure encroaches upon the property of another.

Traditionally, there were three classes of persons who could enter upon someone's land:

1. Those who were invited, or "invitees";
2. Those who had a license or permission; and
3. Trespassers.

In this order also is the level of duty of care the land possessor show have in relation to these persons, that is, the invitee is owed a higher duty of care than that of a trespasser. Under the Restatement (Second) of Torts § 332(2) 1965, there are two subcategories of those who were invited:

1. A public invitee, that is, a customer in a department store; and
2. A "business visitor," that is, someone who was invited for purposes connected with the owner of the land.

These traditional categories are the subject of debate in several state jurisdictions. Nevertheless, they still form the basis of trespass law.

Recently, the courts have allowed plaintiffs to use the trespass laws in order to support a cause of action if the plaintiffs' computer systems were accessed without authority. In order for a plaintiff to prevail on this type of trespass claim, the plaintiff must show that the defendant interfered with the plaintiff's interest in the computer system without authority, and that, because of the unauthorized use, the plaintiff suffered damages. There continues to be some controversy about the damage level required, especially in a case where such computer trespass activity is minimal. (See *Oyster Software, Inc. v. Forms Processing, Inc.* in this chapter.)[4]

Consider this example. Black builds a pond near the edge of his property line (according to a survey). White, the owner of the adjacent property, later has another survey made. The later survey shows part of Black's pond to be on White's property. The court concludes that White's survey is correct. An equity court (where such a case would likely wind up) has the right to order the reconstruction or draining of the pond.

The person who is in possession of the land has the right to exclude others from trespassing. In other words, those who rent or lease property also have the right to exclude others from it (even the owner) as long as they lease the property. Still, personal rights take precedence over property rights. One does not ordinarily have the right to shoot trespassers. Usually the force used must be no more than sufficient to remove the person from the property.

Trespass Exceptions

At common law, a trespass involved no finding of intent or even negligence. If you entered or caused something to enter onto another's land, you were liable. Because this led to harsh results,

---

[4]See also *Intel Corporation v. Kourosh Kenneth Hamidi*, 71 P.3d 296 (CA 2003).

the courts developed exceptions, whereby certain actions (which might otherwise amount to a trespass) did not have the harsh results that the old common law rules sometimes caused.

Easement or License.   If numerous members of the public use a person's property as they desire, an **easement** may result. If Black, for instance, owns lakeside property and the public crosses his property to reach the lake, he may eventually be prevented from excluding the public. The period of time for such a public easement to occur often runs from 15 to 20 years, depending on the state. To create the right of easement, the public use must be continuous. It is for this reason that one occasionally sees a road blocked off for one day per year. See Chapter 19 for a more thorough discussion of easements.

The right to go upon another's land can be given by the person in possession. Such permission is known as a **license**. A caller at a home, for example, has the right to go as far as the door by a direct route. If a person must enter another's property to recover his or her own property, that person also has a right to do so.

Attractive Nuisance.   Ordinarily, one who trespasses upon another's property assumes whatever risk may be inherent in the trespass. If the trespasser is injured by some hidden danger, that person has little chance of recovery against the owner. But, just as is true with many other general rules of law, this one has its exceptions. Probably the most prominent exception is known as **attractive nuisance**, pertaining to children of tender age. This principle is of recent origin as legal doctrines go, and it has been rejected by some courts, but the number and size of recoveries prompt its consideration. The doctrine of attractive nuisance began in the United States with a case involving an injury to a child playing around a railway turntable. In the century or so since then, a multitude of property conditions and instrumentalities (including swimming pools) have come to be considered attractive nuisances for children.

In jurisdictions where the attractive-nuisance doctrine is applied, a property owner or occupant may be held liable for injuries sustained by children on his or her premises under the following conditions:

1. If the owner knew or should have known that the dangerous instrumentality or property condition would be attractive to children and failed to reasonably guard against injury to them, or
2. If the owner had reason to expect children to play there (that is, having seen them play in the vicinity) and did not warn them or take other suitable precautions.

The owner's (or occupant's) risk of attractive-nuisance liability is removed by taking reasonable precautions. For example, the owner is not expected to foresee very unlikely events—only those that might befall a normal, inquisitive child; and he or she would not be expected to guard something of danger obvious even to a child. The doctrine is aimed at conditions that would be inherently dangerous to a child but that the child could not be expected to foresee. Thus, an unguarded piece of machinery could easily be an attractive nuisance, whereas an open pit in a field would less likely be one. Generally, the attraction must be something unusual, uncommon, or artificial, as opposed to a natural hazard.

Attractive-nuisance cases could involve children of any age, but children between 5 and 10 years old seem particularly susceptible. The courts also consider such things as the child's intelligence, state of mental health, and other conditions as significant in such cases. The largest factor, however, is the presence or absence of proper precautions by the owner or occupant of the premises.

Defenses to Trespass to Property.   Defense of property, recapture of chattels, and necessity are defenses to an intentional trespass to property. For example, a person may use reasonable force to defend property. Notice that the key word is "reasonable." If a court finds that the person used excessive force to protect property, the defense is not allowed. In addition, a person may be justified in trespassing upon another's land if he or she must do so to regain possession of personal property (chattels). Finally, if the property of a third person is threatened, a person may be able to show that the trespass was necessary.

↳ out sailing when a storm hits & you dock at someone else's property to be safe

Conversion

John Deere skidder case

The tort counterpart to the crime of theft is conversion. **Conversion** usually constitutes the wrongful retention of another's personal property. It also includes the wrongful alteration of property and the wrongful use of property by persons other than the owner. Conversion may arise in instances of bailment, where something left with another is used or sold by the bailee. For example, Black leaves a television set with White (as bailee) to be repaired. White sells the television set to Gray. White's tortious act is conversion, for which Black may maintain a conversion action in court. A successful suit in conversion normally nets the true owner of the property the market value of the converted property, and vests title in the converter when the judgment amount has been paid. The owner, however, has two possible remedies available. He or she may sue on the tort of conversion or maintain an action in replevin to obtain the return of the property. In a **replevin action**, the owner sues to gain possession of the property (as opposed to suing for damages). If the owner wants the property back before the replevin action, he or she may usually obtain it by posting a bond to be forfeited in case the property is found to belong to the other party.

TORTS AGAINST ECONOMIC RIGHTS

As discussed in Chapter 20, intellectual property rights, while monopolistic in nature, are also economic rights, and infringement actions are considered tortious. In contrast, the United States government and the courts protect the right to compete with others in a business venture. Despite the likelihood that entrance into a particular field by an efficient newcomer may injure or even eliminate an established concern, such competition is favored. Usually, the result is healthy. The general tendency is to encourage efficiency, since the public benefits from it in lower consumer costs.

Competition, however, can lead to its own destruction. If a business drives all less efficient concerns from the field, a monopoly results. Since unregulated monopoly is usually associated with excessive prices, inefficient operation, and other undesirable effects, laws have been designed to preserve competition. In many situations, the courts have to balance the rights of those holding them lawfully with those who want to compete. It may be useful to consider the difference between a **product monopoly**, which may be lawful if the product is the result of a patent or trademark and a **market monopoly**, which may not be if no other product may compete in a particular market.

Since the field of law treating competitive practices is very large, no attempt will be made to cover the entire field here. Rather, a few of the most common torts against economic rights will be mentioned.

Fraud and Misrepresentation

The tort of **fraud** occurs all too often. As you may recall from chapter 8, in order to establish fraud, there must have been

1. a false representation,
2. of a material fact,
3. made with the intent that it be relied on, and
4. reliance on the misrepresentation by the injured party,
5. to his or her detriment.

The false representation must be about facts. Statements by a used car dealer, for example, that a car is "great" will probably not give rise to a claim for fraud. Courts tend to view such statements as opinion, and false statements of opinion generally are not actionable.

Other important concepts relating to fraud are its emphasis on the speaker's knowledge and intent, and on the recipient's reliance. If the speaker truly believes what he or she says, there's no intent to mislead. In short, the speaker simply may not know that the statement is false. Second, the recipient must rely on the false statement before fraud can occur. If there's no reliance, then the recipient can't claim that he or she was harmed by the statement.

As you read earlier in Chapter 8, the tort of negligent misrepresentation, like fraud, is intended to protect against damages due to reliance on another's false statements. As you know, only the negligence of the speaker is required. If the speaker made false statements negligently, then there may be liability. Recall that negligence requires that the plaintiff prove:

1. the existence of a duty,
2. its breach,
3. an injury and
4. damages.

The courts have been careful to limit this tort to situations where one party is clearly relying on the other in connection with some other transaction or relationship. In short, the courts are reluctant to impose blanket liability for any misstatement made negligently.

Suppose that Black, a civil engineer, is hired by White Mortgage to survey some land along an interstate highway. Black drinks too much at lunch one day and, as a result, his survey is flawed. Nonetheless, Black's survey is used by White Mortgage, which issues title insurance for the purchaser of the property. Because surveys are commonly relied on in connection with real estate transactions, Black probably could have foreseen that both White Mortgage and the buyer of the land would rely on the survey.

Consequently, surveyors have been held liable to purchasers of property. Note, however, that it would be practically impossible to show fraud by Black; Black's actions would instead be a negligent misrepresentation.

Inducing Breach of Contract

Although breach of contract is treated under the law of contracts, inducing another to breach a contract is a tort. This is usually referred to as **tortious interference with contract**. (See *Advanced Marine Enterprises, Inc. et al. v. PRC Inc.* in Chapter 14.) According to the ancient common law, inducing a breach of contract was not actionable unless it was accompanied by violence or fraud. This concept was changed by the case of *Lumley v. Gye*, in which an opera singer was induced to breach her contract and work for another [Ellis & Blackburn 216, 118 Eng. Rep. 749 (1853)]. Though no fraud or violence had occurred, the court stated that a right of action against the person inducing the breach of contract existed.

Since that time, the courts have developed a significant body of law about when someone may encourage another to breach a contract. For example, merely advising a prospective buyer of the merits or properties of a product is not inducing breach of contract. The end result may be a breach of contract, but the seller must have actively persuaded the customer to breach if the seller is to be justly accused of having a hand in it. (See *Monarch Industries v. Model Coverall Service*, in this chapter). As a further example, Black has a contract to buy parts from the White Screw Machine Products Company. Gray offers to sell Black better parts at a lower cost. Black breaches his contract, but Gray cannot be said to have induced the breach of contract unless he actively advocated Black's breach.

Courts have broadened this tort to cover conduct where a person interferes with another's prospective contracts (i.e., an interference with the expectation of a contract). As you might imagine, however, the courts tend to view normal competition as privileged. In other words, a company is free to submit a lower bid to obtain a contract. However, doing more to advocate a breach, such as by disparaging a competitor's services or resources, runs a risk of moving from competition to a tortious interference. In situations where a contract already exists, however, the courts do not always view competition as a sufficient basis for interfering with the contract.

False Advertising

The presence of false or misleading advertising is often apparent in our daily lives. Under common law, the only remedy afforded a person injured by such advertising was an action for fraud or deceit. However, federal and state statutes have modified the common law, and the courts have expanded the possible grounds for suits, as well as the remedies available. The Federal Trade Commission also seeks to prevent deceptive advertising, as do many state statutes. Enforcement, however, is often a major problem.

The federal Lanham Act governs federal trademark registration and trademark infringement actions. It also includes a provision that allows a company to sue a competitor if the competitor engages in false advertising. (For historical reasons, the Lanham Act was viewed as preventing **unfair competition**. One type of unfair competition is trademark infringement; another type is false advertising.) Not all false advertising is actionable, however. First, the statements must be false, misleading, or deceptive. Second, consumers must be confused or misled, or there must be a likelihood of confusion or deception. Third, the false statements must be material; they must be something that would influence the purchasing decisions of consumers. Under the Lanham Act, the court may enter an injunction to prevent further false advertising. In addition, a winning plaintiff may be able to recover damages and, in some situations, an accounting of the defendant's profits due to the false advertising.

Closely akin to false advertising is the disparagement of another's product. The tort resembles libel and is often called **trade libel**. Essentially, the law prevents a person from making false and misleading statements about a competitor's products. In addition to the preventive relief of injunction, damages for lost profits may be obtained if special damages (such as lost profits) can be shown. If any relief for disparagement is to be forthcoming, the plaintiff usually must prove the following about the statements:

1. They were untrue.
2. They were alleged to be fact (rather than opinion).
3. They were published.
4. There was no privilege.
5. They were made with malice.
6. The plaintiff suffered some type of special damages, such as lost profits.

Disparagement is a little more difficult to establish than libel. In a **disparagement** case, it is the plaintiff who must prove all the elements. If the plaintiff cannot prove each of the required elements, the plaintiff loses. In a libel case, however, the defendant must bear part of the proof burden. That is, in a case for disparagement, the plaintiff must prove that the statements made were untrue, whereas in a libel case the defendant would have the burden of proving statements true.

Just as in libel and slander, certain persons have the **privilege of disparagement.** If, for example, in the interest of preserving life and health, a doctor warns against the use of certain foods or drugs, the doctor does so with privilege. The same is true if a family member warns another member of the same family against using certain things. Furthermore, consumer organizations usually act without malice; they expect the public to benefit from their services. Each case involves a personal interest in benefiting others.

As you might imagine, some tension exists between the promotion of free speech under the First Amendment and the discouragement of speech that is misleading or deceptive. Hence, the courts are careful not to unduly prohibit advertising that is truthful.

NUISANCE

As you know, property owners have the right to use their property as they choose as long as they do not, in some way, injure the person or property of another. If one property owner does use his or her property in a way that injures the person or property of another, and the tort fits no other category, the tort of **nuisance** may cover it. A nuisance can be just about anything that interferes with the enjoyment of life and property. It may take the form of smoke or sulphur fumes, or pollution of a stream, or excessive noise, to mention only a few types of nuisances.

Nuisances are either public or private. A **public nuisance** is one that interferes with the rights of a substantial number of the persons in a community. A **private nuisance** produces special injuries to the private rights in real property of one or a very few people. Any citizen may successfully lodge a complaint about a public nuisance, but only the person injured can maintain a successful action based on a private nuisance.

Consider this example. Black owns a factory in which semi-trailers are manufactured. Since the manufacture requires the use of rivets, the process is quite noisy. When Black first built the factory, several years ago, the building site was a cornfield, and the nearest neighbor was some two miles distant. With the passage of time, Black's trailer business expanded. Moreover, adjoining land was sold to a land development company, and houses have been built and sold. Recently, orders for trailers have forced Black to put on a third shift at the factory, from 11 p.m. to 7 a.m. Some of the new homeowners have complained to Black about the noise; one (White) has even instituted a nuisance suit. In answer, Black contends that he was there first, that an injunction would force him to close down his plant and deprive workers of jobs, and that the noise just isn't great enough to injure anyone anyway.

A variety of judgments would be possible in such a situation. Black's claim that he was there first and, thus, acquired a right to maintain the "nuisance" is doubtful to succeed as a defense. Proof by White that the noise had increased in time and intensity would be likely to defeat that defense. Such a defense might only succeed if White had full knowledge of the noise problem (as to degree and time of day) and bought his house in spite of the noise. This is often known as **coming to the nuisance**.

As to Black's second claim—that an injunction would impose a hardship on him and his business—the response would differ from state to state. Certain state legislatures have adopted policies that encourage business migration into their states. In such places, the courts are reluctant to shut down an industry or company. The balance of hardships would probably also be considered here; for example, is it a greater hardship to the homeowners if operations continue, or is it a greater hardship to the company if it must eliminate the third shift (or perhaps move operations)? Hardship in terms of job loss and loss of income to the community would be weighed against the noise annoyance. The likely result of such a case in our example is a decree requiring Black to do all in his power to abate the noise problem. Many measures can be taken to attenuate such industrial noises.

TIME LIMITATIONS

An injured party should usually take prompt action against a tortfeasor. Most states have statutes of limitations for tort actions—the action must be instituted within so many years after the tortious act. If not timely filed, the action cannot be brought. Tort actions undertaken in a court having equity jurisdiction also run the risk of losing out to time. The equity term **laches** indicates a cause in which the plaintiff has "slept on his rights" too long. Stale causes are not popular, the feeling being that the court should not be more protective of the plaintiff's rights than the plaintiff was of his or her own.

DISCHARGE OF TORTS

A defendant may discharge the obligation to pay for damage from a tortious act in a couple of ways. First, not all causes of action find their way into court. Many are discharged by a simple agreement between the parties. Thus, the out-of-court settlement agreement (accord and satisfaction) is a common means of discharge. Rather than take the case to court, the tortfeasor agrees to pay the injured party for the damage done, thus avoiding court costs (in both time and money) and lawyer's fees.

Under common law, only the injured party was allowed to bring a tort action. If the injured party died, the cause of action ended. This has been changed by the almost complete adoption of survival statutes, which allow others to sue in the name of the deceased.

If the case does goes to a jury, a judgment results. If the judgment is for the defendant, you might say the court has discharged the potential tort liability. The amount of the judgment in a tort case is generally made up of two elements:

1. The out-of-pocket cost to the plaintiff, such as medical costs, lost wages, and the like; and
2. Compensation for pain and suffering (if any). Here, bankruptcy of the tortfeasor may act as a discharge of sorts for tort obligations. If a tort action has been instituted, a judgment rendered, or the obligation reduced to a contract before the bankruptcy proceedings are begun, the injured party shares in the bankrupt's estate as any other creditor. If no suit has been brought or contract made on the tort obligation, however, the tortfeasor's assets go to meet his or her obligations to creditors. The injured party's cause of action remains after bankruptcy, and he or she may elect to sue the tortfeasor for whatever remains.

CAROL A. BENN, As Executor of the Estate of LORAS J. BENN v. LELAND R. THOMAS, K-G, LTD. et al. 512 N.W.2d 537 (Iowa 1994)

Study terms: "Eggshell plaintiff" rule, jury instructions, special verdict, proximate cause

The main question here is whether the trial court erred in refusing to instruct the jury on the "eggshell plaintiff" rule in view of the fact that plaintiff's decedent, who had a history of coronary disease, died of a heart attack six days after suffering a bruised chest and fractured ankle in a motor vehicle accident caused by defendant's negligence. The court of appeals concluded that the trial court's refusal constituted reversible error. We agree with the court of appeals and reverse the judgment of the trial court and remand for a new trial.

I. Background facts and proceedings. On February 15, 1989, on an icy road in Missouri, a semi-tractor and trailer rear-ended a van in which Loras J. Benn was a passenger. In the accident, Loras suffered a bruised chest and a fractured ankle. Six days later he died of a heart attack.

Subsequently, Carol A. Benn, as executor of Loras's estate, filed suit against defendants Leland R. Thomas, the driver of the semi-tractor, K-G Ltd., the owner of the semi-tractor and trailer, and Heartland Express, the permanent lessee of the semi-tractor and trailer. The plaintiff estate sought damages for Loras's injuries and death. For the purposes of simplicity, we will refer to all defendants in the singular.

At trial, the estate's medical expert, Dr. James E. Davia, testified that Loras had a history of coronary disease and insulin-dependent diabetes. Loras had a heart attack in 1985 and was at risk of having another. Dr. Davia testified that he viewed "the accident that [Loras] was in and the attendant problems that it caused in the body as the straw that broke the camel's back" and the cause of Loras's death. Other medical evidence indicated the accident did not cause his death.

Based on Dr. Davia's testimony, the estate requested an instruction to the jury based on the "eggshell plaintiff" rule, which requires the defendant to take his plaintiff as he finds him, even if that means that the defendant must compensate the plaintiff for harm an ordinary person would not have suffered. The district court denied this request.

The jury returned a verdict for the estate in the amount of $17,000 for Loras's injuries but nothing for his death. In the special verdict, the jury determined the defendant's negligence in connection with the accident did not proximately cause Loras's death. The estate filed a motion for new trial claiming the court erred in refusing to instruct the jury on the "eggshell plaintiff" rule. The court denied the motion, concluding that the instructions given to the jury appropriately informed them of the applicable law.

The plaintiff estate appealed. The court of appeals reversed the trial court, concluding that the plaintiff's evidence required a specific instruction on the eggshell plaintiff rule. . . .

II. Jury instructions and the "eggshell plaintiff" rule. The estate claims that the court erred in failing to include, in addition to its proximate cause instruction to the jury, a requested instruction on the eggshell plaintiff rule. Such an instruction would advise the jury that it could find that the accident aggravated Loras's heart condition and caused his fatal heart attack. The trial court denied this request, submitting instead a general instruction on proximate cause. The court of appeals reversed, concluding that the trial court erred in refusing to specifically instruct on the eggshell plaintiff doctrine.

Under Iowa rule of civil procedure 244(h), an aggrieved party may, on motion, have an adverse verdict or decision vacated and a new trial granted for errors of law occurring in the proceedings only if the errors materially affected the party's substantial rights. When jury instructions contain a material misstatement of the law, the trial court has no discretion to deny a motion for a new trial. We find reversible error when the instructions given to the jury, viewed as a whole, fail to convey the applicable law.

A tortfeasor whose act, superimposed upon a prior latent condition, results in an injury may be liable in damages for the full disability. This rule deems the injury, and not the dormant condition, the proximate cause of the plaintiff's harm. This precept is often referred to as the "eggshell plaintiff" rule, which has its roots in cases such as *Dulieu v. White & Sons*, [1901] 2 K.B. 669, 679, where the court observed:

> If a man is negligently run over or otherwise negligently injured in his body, it is no answer to the sufferer's claim for damages that he would have suffered less injury, or no injury at all, if he had not had an unusually thin skull or an unusually weak heart.

The proposed instruction here stated:

If Loras Benn had a prior heart condition making him more susceptible to injury than a person in normal health, then the Defendant is responsible for all injuries and damages which are experienced by Loras Benn, proximately caused by the Defendant's actions, even though the injuries claimed produced a greater injury than those which might have been experienced by a normal person under the same circumstances.

Defendant contends that plaintiff's proposed instruction was inappropriate because it concerned damages, not proximate cause. Although the eggshell plaintiff rule has been incorporated into the Damages section of the Iowa Uniform Civil Jury Instructions, we believe it is equally a rule of proximate cause.

Defendant further claims that the instructions that the court gave sufficiently conveyed the applicable law.

The proximate cause instruction in this case provided:

The conduct of a party is a proximate cause of damage when it is a substantial factor in producing damage and when the damage would not have happened except for the conduct. "Substantial" means the party's conduct has such an effect in producing damage as to lead a reasonable person to regard it as a cause.

Special Verdict Number 4 asked the jury: "Was the negligence of Leland Thomas a proximate cause of Loras Benn's death?" The jury answered this question, "No."

We agree that the jury might have found the defendant liable for Loras's death as well as his injuries under the instructions as given. But the proximate cause instruction failed to adequately convey the existing law that the jury should have applied to this case. The eggshell plaintiff rule rejects the limit of foreseeability that courts ordinarily require in the determination of proximate cause. Once the plaintiff establishes that the defendant caused some injury to the plaintiff, the rule imposes liability for the full extent of those injuries, not merely those that were foreseeable to the defendant.

The instruction given by the court was appropriate as to the question of whether defendant caused Loras's initial personal injuries, namely, the fractured ankle and the bruised chest. This instruction alone, however, failed to adequately convey to the jury the eggshell plaintiff rule, which the jury reasonably could have applied to the cause of Loras's death.

Defendant maintains "the fact there was extensive heart disease and that Loras Benn was at risk any time is not sufficient" for an instruction on the eggshell plaintiff rule. Yet the plaintiff introduced substantial medical testimony that the stresses of the accident and subsequent treatment were responsible for his heart attack and death. Although the evidence was conflicting, we believe that it was sufficient for the jury to determine whether Loras's heart attack and death were the direct result of the injury fairly chargeable to defendant Thomas's negligence.

Defendant nevertheless maintains that an eggshell plaintiff instruction would draw undue emphasis and attention to Loras's prior infirm condition. We have, however, explicitly approved such an instruction in two prior cases. Moreover, the other jurisdictions that have addressed the issue have concluded that a court's refusal to instruct on the eggshell plaintiff rule constitutes a failure to convey the applicable law.

To deprive the plaintiff estate of the requested instruction under this record would fail to convey to the jury a central principle of tort liability. . . .

The record in this case warranted an instruction on the eggshell plaintiff rule. We therefore affirm the decision of the court of appeals. We reverse the judgment of the district court and remand the cause to the district court for a new trial consistent with this opinion.

STACY LAWTON GUIN v. BRAZOS HIGHER EDUCATION SERVICE CORPORATION, INC., 2006 U.S. Dist. LEXIS 4846 (Minnesota)

Study Terms: Identity theft, Federal Trade Commission, breach of fiduciary duty, negligence, Gramm-Leach-Bliley Act

Introduction

Plaintiff Stacy Guin alleges that Defendant Brazos Higher Education Service Corporation, Inc. ("Brazos") negligently allowed an employee to keep unencrypted nonpublic customer data on a lap-top computer that was stolen from the employee's home during a burglary on September 24, 2004. This matter comes before the Court on Brazos's Motion for Summary Judgment pursuant to *Federal Rule of Civil Procedure 56*. For the reasons set forth below, the Court will grant the Motion.

Background

Brazos, a non-profit corporation with headquarters located in Waco, Texas, originates and services student loans. Brazos has approximately 365 employees, including John Wright, who has worked as a financial analyst for the company since November 2003. Wright works from an office in his home in Silver Spring, Maryland. As a financial analyst for Brazos, Wright analyses loan portfolios for a number of transactions, including purchasing portfolios from other lending organizations and selling bonds financed by student loan interest payments. Prior to performing each new financial analysis, Wright receives an electronic database from Brazos's Finance Department in Texas. The type of information needed by Wright to perform his analysis depends on the type of transaction anticipated by Brazos. When Wright is performing asset-liability management for Brazos, he requires loan-level details, including customer personal information, to complete his work.

On September 24, 2004, Wright's home was burglarized and a number of items were stolen, including the laptop computer issued to Wright by Brazos. Wright reported the theft to the local police department, but the police were unable to apprehend the burglar or recover the laptop. After the police concluded their investigation, Brazos hired a private firm, Global Options, Inc., to further investigate the details the burglary. Global Options was unable to regain possession of the computer.

With the laptop missing, Brazos sought to determine what customer data might have been stored on the hard drive and whether the data was accessible to a third party. Based on internal records, Brazos determined that Wright had received databases containing borrowers' personal information on seven occasions prior to September 24, 2004. Upon receiving the databases, Wright typically saved the information to his hard drive, depending on the size of the database and the likelihood that he would need to review the information again in the future. However, Wright did not keep records of which databases were permanently saved on his hard drive and which databases were eventually deleted, so Brazos was not able to determine with any certainty which individual customers had personal information on Wright's laptop when it was stolen.

Without the ability to ascertain which specific borrowers might be at risk, Brazos considered whether it should give notice of the theft to all of its customers. In addition to contemplating guidelines recommended by the Federal

Trade Commission ("FTC")[1], Brazos learned that it was required by California law to give notice to its customers residing in that State. Brazos ultimately decided to send a notification letter (the "Letter") to all of its approximately 550,000 customers. The Letter advised borrowers that "some personal information associated with your student loan, including your name, address, social security number and loan balance, may have been inappropriately accessed by the third party." The Letter also urged borrowers to place "a free 90-day security alert" on their credit bureau files and review consumer assistance materials published by the FTC. In addition, Brazos established a call center to answer further questions from customers and track any reports of identity theft.

Plaintiff Stacy Guin, who acquired a student loan through Brazos in August 2002, received the Letter. Shortly thereafter, Guin contacted the Brazos call center to ask followup questions. Guin also ordered and reviewed copies of his credit reports from the three credit agencies listed in the Letter. Guin did not find any indication that a third party had accessed his personal information and, to this date, has not experienced any instance of identity theft or any other type of fraud involving his personal information. To Brazos's knowledge, none of its borrowers has experienced any type of fraud as a result of the theft of Wright's laptop.

On March 2, 2005, Guin commenced this action asserting three claims: (1) breach of contract, (2) breach of fiduciary duty, and (3) negligence. On September 12, 2005, Guin voluntarily dismissed his breach of contract and breach of fiduciary duty claims. Guin brings the remaining negligence claim under *Fed. R. Civ. P. 23*, on behalf of "all other Brazos customers whose confidential information was inappropriately accessed by a third party. . . ."

Standard of Review

Summary judgment is appropriate where there is no genuine issue of material fact, and the moving party is entitled to judgment as a matter of law. *Fed. R. Civ. P. 56(c)*. . . .

Analysis

In his negligence claim, Guin alleges that "[Brazos] owe[d] him a duty to secure [his] private personal information and not put it in peril of loss, theft, or tampering," and "[Brazos's] delegation or release of [Guin's] personal information to others over whom it lacked adequate control, supervision or authority was a result of [Brazos's] negligence. . . ." As a result of such conduct, Guin allegedly "suffered out-of-pocket loss, emotional distress, fear and anxiety, consequential and incidental damages."

In order to prevail on a claim for negligence, a plaintiff must prove four elements: (1) the existence of a duty of care, (2) a breach of that duty, (3) an injury, and (4) the breach of the duty was the proximate cause of the injury. In support of its instant Motion, Brazos advances three arguments: (1) Brazos did not breach any duty owed to Guin, (2) Guin did not sustain an injury, and (3) Guin cannot establish proximate cause. The Court will address each in turn.

1. Breach of Duty   *failed — unforseen act of theft*

In order to prove a claim for negligence, Guin must show that Brazos breached a legal duty owed to him under the circumstances alleged in this case. A legal duty is defined as an obligation under the law to conform to a particular standard of conduct towards another. The standard for ordinary negligence is "the traditional standard of the reasonable man of ordinary prudence." *Seim*, 306 N.W.2d at 810. In some negligence cases, however, a duty of care may be established by statute. In such cases, violation of a statutory-based duty may constitute negligence *per se*.

---

[1]The Federal Trade Commission guidelines recommend that when "deciding if notification [to customers of an identity theft threat] is warranted, [a company should] consider the nature of the compromise, the type of information taken, the likelihood of misuse, and the potential damage arising from misuse."

Guin argues that the Gramm-Leach-Bliley Act (the "GLB Act"), 15 U.S.C. § 6801, establishes a statutory-based duty for Brazos "to protect the security and confidentiality of customers' nonpublic personal information." For the purposes of this Motion only, Brazos concedes that the GLB Act applies to these circumstances and establishes a duty of care. The GLB Act was created "to protect against unauthorized access to or use of such records which could result in substantial harm or inconvenience to any customer [of a financial institution]." 15 U.S.C. § 6801(b)(3). Under the GLB Act, a financial institution must comply with several objectives, including:

> Develop, implement, and maintain a comprehensive written information security program that is written in one or more readily accessible parts and contains administrative, technical, and physical safeguards that are appropriate to your size and complexity, the nature and scope of your activities, and the sensitivity of any customer information at issue;

> Identify reasonably foreseeable internal and external risks to the security, confidentiality, and integrity of customer information that could result in the unauthorized disclosure, misuse, alteration, destruction or other compromise of such information, and assess the sufficiency of any safeguards in place to control these risks; and

> Design and implement information safeguards to control the risks you identify through risk assessment, and regularly test or otherwise monitor the effectiveness of the safeguards' key controls, systems, and procedures.

16 C.F.R. § 314.4(a)-(c).

Guin argues that Brazos breached the duty imposed by the GLB Act by (1) "providing Wright with [personal information] that he did not need for the task at hand," (2) "permitting Wright to continue keeping [personal information] in an unattended, insecure personal residence," and (3) "allowing Wright to keep [personal information] on his laptop unencrypted." Brazos counters that Guin does not have sufficient evidence to prove that it breached a duty by failing to comply with the GLB Act.

The Court concludes that Guin has not presented sufficient evidence from which a fact finder could determine that Brazos failed to comply with the GLB Act. In September 2004, when Wright's home was burglarized and the laptop was stolen, Brazos had written security policies, current risk assessment reports, and proper safeguards for its customers' personal information as required by the GLB Act. Brazos authorized Wright to have access to customers' personal information because Wright needed the information to analyze loan portfolios as part of Brazos's asset-liability management function for other lenders. Thus, his access to the personal information was within "the nature and scope of [Brazos's] activities." See 16 C.F.R. § 314.4(a). Furthermore, the GLB Act does not prohibit someone from working with sensitive data on a laptop computer in a home office. Despite Guin's persistent argument that any nonpublic personal information stored on a laptop computer should be encrypted, the GLB Act does not contain any such requirement.[2] Accordingly, Guin has not presented any evidence showing that Brazos violated the GLB Act requirements.

In addition, Guin argues that Brazos failed to comply with the self-imposed reasonable duty of care listed in Brazos's privacy policy—that Brazos will "restrict access to nonpublic personal information to authorized persons who need to know such information." Brazos concedes that under this policy, it owed Guin a duty of reasonable care, but argues that it acted with reasonable care in handling Guin's personal information. The Court agrees. Brazos

---

[2]While it appears that the FTC routinely cautions businesses to "provide for secure data transmission" when collecting customer information by encrypting such information "in transit," there is nothing in the GLB Act about this standard, and the FTC does not provide regulations regarding whether data should be encrypted when stored on the hard drive of a computer.

had policies in place to protect the personal information, trained Wright concerning those policies, and transmitted and used data in accordance with those policies. Wright lived in a relatively "safe" neighborhood and took necessary precautions to secure his house from intruders. His inability to foresee and deter the specific burglary in September 2004 was not a breach of Brazos's duty of reasonable care. Because Guin has failed to raise a genuine issue of material fact regarding whether Brazos breached its duty of care, summary judgment is appropriate.

Although Guin's failure to show that Brazos breached its duty of care provide sufficient grounds for granting Brazos's Motion for Summary Judgment, the Court will address Brazos's other two arguments.

2. Injury *— failed — no injury sustained*
— unknown if Guin's data was on stolen laptop

In order to prove a claim for negligence, Guin must show that he sustained an injury. A plaintiff must suffer some actual loss or damage in order to bring an action for negligence.

Guin argues that he has been injured by identity theft. Under both federal and Minnesota law, identity theft occurs whenever a person "transfers, possesses, or uses" another person's identity "with the intent to commit, aid, or abet any unlawful activity." 18 U.S.C. § 1028(a)(7); Minn. Stat. § 609.527(2). Guin argues that the circumstances of this case fulfill the definition of identity theft because "the burglars [in Wright's home in September 2004] had a criminal intention when they broke in and gained possession of [Guin's] identity information."

In response, Brazos contends that "any finding that a third party accessed [Guin's] personal information [is] sheer speculation." Brazos points out that the evidentiary record is completely devoid of any disputed facts indicating that Guin's personal information was actually on Wright's laptop at the time it was stolen, or that Guin's personal information is now in the possession of the burglar. Therefore, Brazos argues that Guin cannot show that he has been a victim of identity theft.

■ ■ ■

[I]n this case Guin has failed to present evidence that his personal data was targeted or accessed by the individuals who burglarized Wright's home in September 2004. The record shows that Brazos is uncertain whether Guin's personal information was even on the hard drive of Wright's laptop computer at the time it was stolen in September 2004. To this date, Guin has experienced no instance of identity theft or any other type of fraud involving his personal information. In fact, to Brazos's knowledge, none of its borrowers has been the subject of any type of fraud as a result of the theft of Wright's laptop computer. Furthermore, Guin has provided no evidence that his identity has been "transferred, possessed, or used" by a third party with "with the intent to commit, aid, or abet any unlawful activity." See 18 U.S.C. § 1028(a)(7); Minn. Stat. § 609.527(2). No genuine issue of material fact exists concerning whether Guin has suffered an injury. Accordingly, he cannot sustain a claim for negligence.

3. Causation

To prevail on his negligence claim, Guin must also show that Brazos's alleged breach of duty was the proximate cause of his alleged injury. Proximate cause is defined as "consequences which follow in unbroken sequence, without an intervening efficient cause, from the original negligent act." *Hilligoss v. Cross Cos.*, 304 Minn. 546, 228 N.W.2d 585, 586 (Minn. 1975). As a general rule, the criminal act of a third party is "an intervening efficient cause sufficient to break the chain of causation," provided that the criminal act was not foreseeable and there was no special relationship between the parties. "The question of foreseeability of an intervening act is normally one for the trial court and should be submitted to a jury only where there might be a reasonable difference of opinion." *Hilligoss*, 228 N.W.2d at 586.

Guin contends that the September 2004 theft of Brazos's laptop from Wright's home was reasonably foreseeable because "allowing confidential information to remain unencrypted on unsecured laptop computers increase[s] the risk of theft." Guin argues that "the test of foreseeability is whether the defendant was aware of facts indicating [that] the plaintiff was being exposed to [an] unreasonable risk of harm." Guin points to similar laptop thefts in the financial industry and the increasing problem of widespread identity theft. Based on this, Guin argues that the theft of Wright's laptop was reasonably foreseeable to Brazos because "a reasonable jury could conclude that the risk of information compromise is common knowledge in the financial industry."

The Court concludes that the September 2004 theft of Wright's laptop from his home was not reasonably foreseeable to Brazos. . . . Wright lived in a relatively "safe" neighborhood and took necessary precautions to secure his house from intruders. Wright was unaware of any previous burglaries on his block or in his immediate neighborhood. There is no indication that Wright or Brazos could have possibly foreseen the burglary which took place on September 24, 2004. A reasonable jury could not infer that the burglary caused Guin any alleged injury; such a conclusion would be the result of speculation and conjecture, not a reasonable inference. Guin cannot establish proximate cause in this case and therefore, his negligence claim fails.

Conclusion

Based on the foregoing, and all of the files, records and proceedings herein, it is ORDERED that Defendant's Motion for Summary Judgment is GRANTED, and the Complaint is DISMISSED WITH PREJUDICE.

LET JUDGMENT BE ENTERED ACCORDINGLY.

OYSTER SOFTWARE, INC. v. FORMS PROCESSING, INC., ET AL 2001 U.S. Dist. LEXIS 22520 (N.D. Ca. 2001)

Study Terms: Trespass, trademark, copyright, infringement, "metatags", vicarious liability, *respondeat superior*, intentional interference with the possession of personal property

I. Introduction

Plaintiff in this action, Oyster Software, Inc. ("Oyster"), alleges that Defendant Forms Processing, Inc. ("FPI") infringed upon its trademarks and copyright by copying metatags from Oyster's web site and using them in FPI's web site. Defendant FPI brings a motion for partial summary judgment on the ground that Plaintiff has failed to raise a genuine issue of material fact with respect to damages resulting from Defendant's alleged infringement.

I. Background

A. Facts

Oyster was founded in February of 1996. Oyster "develops software products for customers to process electronic or paper documents, using imaging system technologies to capture data contained in the documents . . . process the data, and store it in electronic form." Oyster offers its customers both custom-built and "off-the-shelf" systems. Between February and April of 1996, Oyster created and registered its web site, www.Oystersoftware.com.

Oyster was founded by Barry Bhangoo. Prior to forming Oyster, Bhangoo worked at Teknekron Customer Information Solutions ("Teknekron"), until that company stopped doing business, in January of 1996. While at Teknekron, Bhangoo was involved in the development of a software system called FormsPro. FormsPro was registered by Teknekron as a trademark on February 6, 1996, and transferred by assignment to Oyster on August 12, 1997. Oyster's first two customers, Fingerhut Corporation and Maritz Marketing Corporation, were acquired by Oyster between February and April of 1996 and were former customers of Teknekron.

Defendant FPI is a Florida corporation which offers document management services. In November of 1999, Oyster discovered that FPI was using metatags[5] copied from Oyster's web site on its own web site, www.formsprocessing.com. Bhangoo learned FPI was using its metatags after he received a call from someone using the search engine cnet.com who had found a description of Oyster but was unable to get to Oyster's web site.

On November 23, 1999, Barry Bhangoo sent an e-mail message to FPI informing it that its web site contained metatags that included Oyster's trademarked words, causing "people to find our company's description with a link to [FPI's] web site." FPI's president, Barry Matz, responded on November 24, 1998 in two e-mails to Barry Bhangoo. In the first, Matz stated that he had been "assured" by "people in the know around here" that the files containing Oyster's metatags were not FPI's. Later in the day, Matz sent another e-mail message to Bhangoo stating that he had learned that the files were created by a company called "D.J. Distributors in Roanoke Rapid, N.C. dba Top-Ten Promotions," a company that FPI had hired to increase its prominence with search engines. Matz stated that he was unaware of the pages containing Oyster's metatags until he received Bhangoo's e-mail message and that FPI had removed "all text and will continue to remove any reference that is not directly related to our specific business."

B. Complaint

In the [First Amended Complaint], Plaintiff alleges the following claims: . . .

Claim Six: Trespass (all defendants); . . .

C. FPI's Motion For Partial Summary Judgment

FPI brings a Motion for Partial Summary Judgment (the "Motion") on the following grounds:

■ ■ ■

5) Oyster's trespass claim should be dismissed because there is no evidence of obstruction of the basic function of Oyster's computer system by FPI; . . .

III. Analysis

A. Legal Standard on Summary Judgment

Rule 56 provides that summary judgment "shall be rendered forthwith if the pleadings, depositions, answers to interrogatories, and admissions on file, together with the affidavits, if any, show that there is no genuine issue as to any material fact and that the moving party is entitled to judgment as a matter of law." Fed. R. Civ. P. 56(c).

E. Trespass Claim

FPI asserts that Oyster has failed to raise a genuine issue of material fact with respect to its claim for trespass for two reasons: (1) it was Top-Ten and not FPI that sent robots to Oyster's web site and copied its metatags; and (2) regardless of who copied the metatags, there is no evidence that the basic function of Oyster's computer system was ever interfered with.

---

[5]Metatags are Hypertext Markup Language ("HTML") code which describe the contents of an Internet web site to a search engine. There are two types of metatags: 1) "Description" metatags, which "are intended to describe the web site;" and 2) "Keyword" metatags, which are "at least in theory . . . keywords relating to the contents of the web site." Id. "The more often a term appears in the metatags and in the text of the web page, the more likely it is that the web page will be 'hit' in a search for that keyword and the higher on the list of 'hits' the web page will appear" in search engine results.

The Court rejects FPI's first argument because FPI failed to address and presented no authority on the issue of whether FPI could be held vicariously liable for the alleged trespass by Top-Ten. Under the doctrine of respondeat superior, California courts have held that an employer is liable for the torts of his employees committed within the scope of their employment. On the other hand, "the general rule [in California] . . . is that a principal is not liable for torts committed by an independent contractor." *Yanez v. United States*, 63 F.3d 870, 872 (9th Cir. 1995). Thus, even if FPI knew nothing about Top-Ten's initial act of sending robots to Oyster's web site and copying its metatags, it may still be liable for Top-Ten's trespass if Oyster can persuade a jury that Top-Ten was an employee rather than a consultant. Regardless of whether or not Oyster will be able to meet this burden at trial, the Court cannot hold as a matter of law that Oyster will not be able to prevail on its trespass claim where FPI did not demonstrate in its motion that there is no material issue of fact with respect to the relationship between FPI and Top-Ten.

The Court also rejects FPI's second argument, that Oyster cannot prevail on its trespass claim because the interference by Top-Ten's robots was negligible. In order to prevail on a claim for trespass, a plaintiff must show that an intentional interference with the possession of personal property has proximately caused injury. *eBay Inc. v. Bidder's Edge, Inc.*, 100 F. Supp. 2d 1058, 1069 (N.D. Cal. 2000). Where a trespass claim is based on unauthorized access to a computer system, the plaintiff must demonstrate that 1) "defendant intentionally and without authorization interfered with plaintiff's possessory interest in the computer system; and 2) defendant's unauthorized use proximately resulted in damage to plaintiff." Id at 1069-1070. In Bidder's Edge, the plaintiff sought a preliminary injunction prohibiting defendant's use of "web crawlers" to search its web site. Id. at 1063. The defendant in that case accessed eBay's web site approximately 100,000 times a day for several months. Id. at 1071. In addressing whether or not the defendant's conduct constituted a sufficient interference to establish a likelihood of prevailing on the merits, the court rejected the argument that the plaintiff was required to present evidence of "substantial interference" with possession. Id. at 1070. Rather, the court held, it was sufficient to show that the defendant's conduct was at least "intermeddling with or use of another's personal property." Id. The court went on to note, "although the court admits some uncertainty as to the precise level of possessory interference required to constitute an intermeddling, there does not appear to be any dispute that eBay can show that BE's conduct amounts to use of eBay's computer system." Id.

In contrast, in *Ticketmaster Corp. v. Tickets.com, Inc.*, 2000 U.S. Dist. LEXIS 12987, 2000 WL 1887522 (C.D. Cal.), which also involved a motion for a preliminary injunction, the court reached the opposite conclusion. There, as in Bidder's Edge, the defendant used web crawlers to monitor the contents of the plaintiff's web site. The court found, however, that plaintiff was not entitled to a preliminary injunction in that case. Drawing on the reasoning of the court's decision in Bidder's Edge, the court explained its holding as follows:

> The computer is a piece of tangible property. It is operated by mysterious electronic impulses which did not exist when the law of trespass to chattels was developed, but the principles should not be too different. If the electronic impulses can do damage to the computer or to its function in a comparable way to taking a hammer to a piece of machinery, then it is no stretch to recognize that damage as trespass to chattels and provide legal remedy for it. Judge White [in Bidder's Edge] found the damage in occupation of a portion of the capacity of the computer to handle routine business and conjectured that approval of that use would bring many more parasitic like copies of the defendant feeding the computer to a clogged level upon the information expensively developed by eBay, the net result likely being severe damage to the function of the computer and thus the business of eBay.

Id. In contrast, the court found, the defendant's use of plaintiff's computer system in Ticketmaster was "very small" and there was "no showing that the use interfered to any extent with the regular business of" the plaintiff.

Here, Oyster has presented no evidence that the use of Top-Ten's robot interfered with the basic function of Oyster's computer system. Indeed, Oyster concedes that Top-Ten's robot's placed a "negligible" load on Oyster's computer system. Oyster asserts that Top-Ten's copying of its metatags is, nonetheless, sufficient to prevail on its trespass claim. The Court agrees. While the eBay decision could be read to require an interference that was more than negligible (as did the court in *Ticketmaster*), this Court concludes that eBay, in fact, imposes no such requirement. Ultimately, the court in that case concluded that the defendant's conduct was sufficient to establish a cause of action for trespass not because the interference was "substantial" but simply because the defendant's conduct amounted to "use" of Plaintiff's computer. 100 F. Supp. 2d at 1070; see also *Compuserve Inc. v. Cyber Promotions Inc.*, 962 F. Supp. 1015, 1022 (9th Cir. 1997) (relying on California law and holding that "[a] plaintiff can sustain an action for trespass to chattels, as opposed to an action for conversion, without showing a substantial interference with its right to possession of that chattel"). Therefore, the Court declines to dismiss Oyster's trespass claim on the grounds that Oyster has shown only a minimal interference because Oyster has presented evidence of "use" by Top-Ten.

IV. Conclusion

For the reasons stated above, FPI's Motion For Partial Summary Judgment is GRANTED in part and DENIED in part. . . .

MONARCH INDUSTRIES, ETC. v. MODEL COVERALL SERVICE
381 N.E.2d 1098 (Ind. Ct. App. 1978)

Study Terms: Tortious interference with contract, inducement to breach   *economic tort*

Plaintiff-appellant Monarch Industrial Towel and Uniform Rental, Inc. (Monarch), filed a complaint of tortious interference with a contractual relationship against Model Coverall Service, Inc. (Model). Monarch alleged that Model committed said tort by inducing Emmert Trailer Company (Emmert) to breach its uniform rental service contract with Monarch. (Emmert executed a similar contract with Model after terminating the contract with Monarch).

Following Monarch's presentation of evidence at trial, the trial court granted Model's motion for judgment on the evidence, pursuant to Ind. Rules of Procedure, Trial Rule 50. Monarch perfected this appeal arguing that the court erred in granting Model's motion for judgment on the evidence and in refusing to allow Monarch to reopen its case.

Indiana recognizes the tort of interference with contract relationships by inducing a breach of contract and has defined the essential elements which must be proved for recovery under such an action in Daly v. Nau . . ., as follows:

1. existence of a valid and enforceable contract;
2. defendant's knowledge of the existence of the contract;
3. defendant's intentional inducement of breach of the contract;
4. the absence of justification; and
5. damages resulting from defendant's wrongful inducement of the breach.

Model's T.R. 50 motion was grounded upon the lack of proper evidence as to damages. But it is well established that the judgment of the trial court will be affirmed on appeal if sustainable on any basis. . . . The evidence presented by Monarch failed to prove that Model intentionally induced Emmert to breach the contract with Monarch. Rather, the evidence shows that Emmert was dissatisfied with the service that Monarch was providing and had been looking at other rental services. The evidence is that Model approached Emmert to inquire as to whether Emmert was being supplied with a uniform rental service and that the Model sales representative was told that Emmert was not happy with the present supplier and intended to cancel the contract. And there was no evidence that Model offered any inducements to Emmert in the form of a better price or better services. Rather, the evidence is that there was no discussion of prices until Emmert had already given Monarch notice of termination of the contract and that the price Emmert was paying under its contract with Model was higher than it had paid under the contract with Monarch.

To sustain a judgment for defendant on the evidence, the evidence must be without conflict and susceptible of but one inference in favor of the moving party. If there is any evidence or legitimate inference therefrom tending to support at least one of plaintiff's allegations, a directed verdict should not be entered. . . . The evidence presented at trial clearly showed that Emmert elected to terminate the contract with Monarch and made the decision independent of Model's approach and subsequent contract with Emmert. There was no evidence from which an inference of intentional inducement to breach on the part of Model could be drawn. . . . .

Inasmuch as Monarch failed to prove inducement to breach, this Court need not discuss the sufficiency of the evidence as to the damages incurred by Monarch due to the breach of contract by Emmert or Monarch's argument that granting the T.R. 50 motion denied Monarch its rights to have the jury consider punitive damages.

Finally, whether the court erred in refusing to allow Monarch to reopen its case to present additional evidence as to the net profit it would have earned had the contract been completed is a question this Court does not decide. Error, if any, was harmless since that evidence could not have prevented the failure of the plaintiff's case on the essential element of intentional inducement to breach the contract. . . .

The judgment is affirmed.

Affirmed:

STATON, J., and CHIPMAN, P. J., participating by designation, concur.

Advanced Marine

Similarities   defendant approached + attracted others
    ↳ employees contractors leaving in favor of another's services their
    ↳ contracts involved → issue is tortuous interferer w/ contract
    ↳ both on appeal

Differences
    ↳ In Monarch, Emmert terminated contract w/o breach
       In Advanced, employees breached employment contract + violated its terms (e.g., 2 week notice)
    ↳ New employer induced breach in Advanced by weighing cost of lawsuit +
       bargaining terms of new contract while under old contract
       New vendor did intentionally not discuss prices until Emmert had already provided notice of termination.
    ↳ Conclusions differed.

Review Questions

1. Name a tort that is not a crime.

2. What generally must be proved in a tort action?

3. Black, an engineer, is injured while visiting the White Manufacturing Company. Black was splattered in the face with hot metal from a die-casting machine he was observing at the time. What complaint and reasoning might Black use in a tort claim? What reply would White be likely to use?

4. The Green Paper Company has responded to an invitation to set up a plant in a particular community. Several millions of dollars have been spent for buildings and equipment. However, in the first few months of operation, numerous complaints have been lodged and injunctions requested. It is claimed that the odors peculiar to the industry have lowered local property values and that the discharge of "black liquor" and dyes in the local stream has eliminated fishing. The plant employs approximately 1,000 people. How would the court be likely to treat the problem?

5. How does nuisance differ from attractive nuisance?

6. How does disparagement differ from defamation?

7. In the case of *Benn v. Thomas*, compare the proposed jury instruction and the actual jury instruction and Special Verdict Number 4 used to determine Benn's cause of death. Do you think the jury would have come to a different conclusion had they been instructed to use the proposed instruction?

8. In *Guin v. Brazos Higher Education Services*, what kind of duty or duties was the plaintiff trying to show? What the plaintiff successful?

9. Although not discussed directly in *Oyster Software, Inc. v. Forms Processing, Inc.*, what are the differences and similarities between a "virtual" and "physical" trespass? What potential damages could Oyster claim? Could there be an argument for Forms Processing under the license exception?

10. In *Monarch Industries v. Model Coverall Service*, what added evidence did Monarch need to win its case and recover from Model?

11. Compare and contrast *Advanced Marine* in Chapter 14, with *Monarch Industries*.

Products Liability

HIGHLIGHTS

■ There are three legal theories that may be used successfully in product liability cases:
1. Negligence theory
2. Warranty theory and
3. Strict liability theory

As we have seen, tort law consists of a body of rules, statutes, legal theories, and principles designed to assure recovery for private injury. The plaintiff makes a case by showing that the defendant neglected a duty owed to the plaintiff and that this was the proximate cause of the plaintiff's injury. But now, suppose the defendant is the producer of a product that injured the plaintiff. To what extent may the plaintiff recover? What defense could the producer use? Can the plaintiff obtain evidence held exclusively by the defendant? Must the plaintiff show negligence? The answers to these and many related questions comprise the expanding field of products liability.

A few decades ago one could easily have said that products liability and engineering had little in common. This changed in the early 1980s as the federal and state governments, as well as individuals, sought more ways to protect consumers against product defects. As Richard M. Morrow put it in his contribution to the book entitled *Product Liability and Innovation,* published by National Academy Press in 1994: "The message our product liability system conveys to engineers is that they must design and produce safe products. What it does not tell them is how to be safe or how safe to be."

How liability affects innovation in engineering fields is a topic separate from that discussed below, but as you read the following material, consider how your practice as an engineer may or may not be influenced by product liability laws.

HISTORY

The idea of recovering from the producer for a product-related injury is of quite recent origin. A century or so ago, such a case would have been virtually inconceivable. Even several decades ago, products liability cases and recoveries were fairly rare. But in the 1960s, revolutionary changes in the field occurred. To understand how and why these changes came about, we must review the common law.

Early History

Early English social and legal philosophy reflected the manufacturing nature of the economy. Producers of goods and services were respected, since their success meant success of the nation. The legal climate fostered their growth. Both logic and social philosophy supported the legal defense available if someone complained about a product—*caveat emptor* (the buyer beware). The logic was simple: People should examine what they are to receive before they buy it.

Allowing a buyer to recover for a bad product was viewed as supporting a buyer's negligence, and the law usually will not aid those who are negligent. But then, of course, the products produced in those days were somewhat more easily examined than what we buy today. One can easily see defects in a shovel or wheelbarrow, but automobiles, television sets, automatic washers, and the like have created a rather different product climate. It is a rare consumer who would understand the internal components and functions of a new car even if that consumer were to dismantle it before buying it. So, is the consumer being negligent when he or she buys in reliance on those who produced the vehicle?

Another traditional defense available to producers was the requirement of privity of contract—the idea that one who is not a party to a contract should have no rights arising from it. In other words, if a consumer were injured by a product, but the consumer had not bought it directly from the manufacturer, he or she could not act against that manufacturer to recover for the injury. The producer or manufacturer only needed to interpose a middleman—a wholesaler or retailer—as an insulator. Then, if the injured person could prove he or she bought the product, that person might sue the middleman but could not reach the "deep pocket." Although there is something to be said for the legal generality, the result does not seem quite right. After all, it was the producer, and not the middleman, who produced the faulty product. Nevertheless, such was the law in the past.

The 20th Century

Around the turn of the 20th century, the law of products liability began to change. Injured plaintiffs occasionally recovered from producers of faulty products. One decision stands out as being especially predictive of future events. This is the decision as written by Judge Cardozo in *McPherson v. Buick Motor Co.,* 111 N.E. 1050 (N.Y. 1916). In that case, a wheel on a new Buick collapsed, and the plaintiff was injured in the resulting accident. The plaintiff did not sue the car dealership, but rather the manufacturer, Buick, even though there was no privity of contract between plaintiff and defendant. The decision rendered by the court represented a sea-change in the law. In finding for the plaintiff in spite of the defenses of the lack of privity and *caveat emptor,* mentioned above, Cardozo's reasoning is a statement of products liability law as it evolved years later. However, during its development over the years, the law was not entirely predictable or "settled." Some courts continued to follow the McPherson decision. During this time, also, the courts developed three legal philosophies that came to be used successfully in products liability cases:

1. Negligence,
2. Warranty, and
3. Strict liability.

Negligence.   The idea in negligence here is the same as it is elsewhere in tort law. The plaintiff makes a case against the product's producer by showing:

1. That the producer owed the plaintiff a duty to carefully design and produce the product,
2. That this duty was neglected,
3. That this neglected duty was the proximate cause of plaintiff's injury, and
4. The plaintiff suffered damages.

The plaintiff's problems often arose in attempting to prove the producer's negligence in design or manufacture, and the producer who could prove contributory negligence by the plaintiff had a formidable defense. Even so, showing negligence did win cases for plaintiffs.

Warranty.   Plaintiffs won other cases based on a warranty theory. Specifically, the warranty is the promise that the thing bought will do the job for which it is intended. As we saw, the implied warranty of merchantability includes the promise that the product will not injure the product's purchaser while that job is being done. If injury occurs during the intended use of the warranted product, the injury gives the plaintiff a cause of action against the seller of the product; the implied warranty is breached, and this breach, in turn, amounts to a breach of contract.

But a problem occurs in taking a products liability action under warranty. Under the Uniform Sales Act, a predecessor to the Uniform Commercial Code, the only person protected by this warranty was the purchaser. Later, under the Uniform Commercial Code, this limitation was eased. As noted in Chapter 15, different states have adopted different versions of the UCC sections 2-314 and 2-315, which vary in just how broadly the warranty applies. But note also that the action was based on contract. The purchaser might act against the merchant who sold the product, but the purchaser still had no contract basis to act against the product's producer.

Thus, negligence was often difficult to prove, and warranty restricted the parties who might be plaintiff and defendant. Several influential judges and legal scholars believed that those injured by defective products deserved better treatment.

Strict Liability.   Surprisingly, it was a case decided by the House of Lords in England in 1868 that formed the foundation of our laws concerning strict liability, *Rylands v. Fletcher,* L.R. 3 H.L. 330 (1868). The defendants in this case owned a mill. Water from the mill escaped from the defendants' land and flooded the plaintiff's mine, causing considerable damage. The Law Lords decided that it was not necessary to show that the mill owners were at fault in order to be liable. In fact, they had acted "at their own peril." The fact that the water escaped and caused damage rendered the owner responsible regardless of the care he may have taken to prevent it. Sections 519 through 520 of the Restatement (Second) of Torts incorporated the lessons from this case, thereby building the foundation in the United States.

By 1963, two events changed the landscape of American law with respect to products liability. The first event was the American Law Institute's acceptance of a subsection A to §402 of the Restatement (Second) of Torts to allow special liability for sellers of products for personal use.

The second event was a decision, reported in *Greenman v. Yuba Power Products, Inc.,* by the California Supreme Court that recognized an independent cause of action for strict liability in tort, separate from that of warranty.[1] After the *Greenman* decision, the Restatement (Second) of Torts §402A was revised to incorporate features of the decision, allowing an independent cause of action for strict liability in tort. Most states generally follow the rule of products liability as it appears in section 402A of the Restatement (Second) of Torts. As with statutes that set out legal rules, the wording of section 402A is important:

> Interestingly, North Carolina is the only state that has rejected specifically strict liability in tort.[2]

1. One who sells any product in a defective condition unreasonably dangerous to the user or consumer or to his property is subject to liability for physical harm thereby caused to the ultimate user or consumer, or to his property, if (a) the seller is engaged in the business of selling such a product, and (b) it is expected to and does reach the user or consumer in the condition in which it is sold.
2. The rule stated in subsection (1) applies although (a) the seller has exercised all possible care in the preparation and sale of his product, and (b) the user or consumer has not bought the product from or entered into any contractual relation with the seller.

Although this **statement of strict liability,** as section 402A is often called, seems fairly clear, courts have interpreted and applied its provisions differently. Ultimately, strict liability means that no fault is required by the seller to establish liability. In addition, the products it covers is broad and includes food, prescription drugs, and automobiles. But this is how things currently stand in the products liability arena—the setting where opposing attorneys (and expert witnesses) prove the defendant liable or not for a product and where "how much" is determined. The situation is further confused when the UCC conflicts with strict liability. Where warranty rules and strict liability conflict, some states conclude that UCC provisions take precedence over a cause of action arising in tort that is subject to a contract. This appears to make sense, because state legislatures adopt the UCC, whereas strict liability rules are applied by the courts. The debate continues.

PRODUCT LIABILITY CLAIMS

To win against the producer/defendant, the plaintiff must prove that the product was defective when it left the producer's control. Evidence must then prove a causal connection between the product defect and the injury to the plaintiff. Usually, the case is made before a jury.

Defective Product

Proving that a product was defective when it left the producer's control may be difficult. At times, it is very simple and so obviously true that the plaintiff may rely on the doctrine of *res ipsa*

---

[1] *Greenman v. Yuba Power Products, Inc.,* 59 Cal. 2d 57, 377 P 2d 897 (1962).

[2] See *Bryant v. Adams,* 116 NC App. 448, 448 SE 2d 832 (1994), *rev. denied,* 339 NC 736, 454 SE 2d 647 (1995).

loquitur (the thing speaks for itself), discussed in Chapter 20. At other times, it may or may not be obvious, with expert witnesses contending both ways and with plaintiffs trying to convince the jury to believe their expert rather than the defendant's expert.

The case of *Barker v. Lull Engineering Company, Inc.,* is particularly thorough in its reasoning as to defects, and presents a logical approach to the definition.[3] The Barker standard requires that for the product to be defective, the plaintiff must:

1. Establish that the product failed to perform as safely as an ordinary consumer would expect when used in an intended or reasonably foreseeable manner, or
2. Show that the product proximately caused the injury; and following this, the defendant must fail to establish that the benefits outweigh the risk of danger.

Although the risk-benefit, or cost-benefit, approach has often been a tacit component of products liability cases, *Barker* appears to have been the first case to advocate the risk analysis as a formal criterion. Some courts devise their own tests derived from a list of strict liability factors created by Dean John Wade. This list includes considerations such as the product's usefulness, the likelihood that it will cause injury, whether or not the plaintiff could have avoided the injury, whether the user should have been aware of the dangers inherent in the product, and whether the manufacturer could have mitigated its damages by changing the price or getting insurance.[4] The presumption is that a manufacturer or seller that puts a defective product into the "stream of commerce" cannot later disregard any liability resulting from the use of that product. As described in the *McPherson* case above, and as opposed to contract law, the manufacturer is liable for these defects.

Defects can be classified according to three basic types:

1. Design defects,
2. Manufacturing defects, and
3. Failures to warn as defined by the Restatement of Torts.[5]

While there is some controversy about whether strict liability applies to design defects and failures to warn, most jurisdictions support strict liability claims as they apply to manufacturing defects. The following paragraphs discuss these types of defects.

Design Defects.   Of the three types of defects, often the most serious from the producer's point of view and the most difficult to define is the **design defect.** A design found to be defective usually involves not just one unit but an entire run or lot or, perhaps, many years' production. Furthermore, a defective design decision made against one product unit opens the door for suits against manufacturers of other product units. A great deal hinges, then, on determining when a design is defective. In some jurisdictions, the plaintiff must show that there was a safer alternative design that could have reduced the injury and was technologically and economically feasible at the time of manufacturer.

The controversy around design defects, and hence, the reluctance of some courts to apply strict liability to product design, centers on the fact that the manufacturer intended the design after considering design alternatives as well as price considerations and usefulness.

---

[3]*Barker v. Lull Engineering Co., Inc.,* 573 P.2d 443 (Cal. 1978).

[4]John Wade, Nature of Strict Torts Liability for Products, 44 Miss. L.J. 825 (1973).

[5]Restatement (Third) of Torts: Products Liability §2 (1998).

Manufacturing Defects.   In contrast to a design defect, a manufacturing defect may exist in a properly designed product. Manufacturing defects occur when "the product departs from its intended design even though all possible care was exercised in the preparation and marketing of the product." [Restatement (Third) of Torts: Products Liability §2(a)]. In other words, the manufacturer had no intention of allowing this defect into the product and would have removed it had the manufacturer known of it. For example, an airplane wing may be properly designed, but if flaws are introduced due to improper joining of components, the wing is defective. Such a flaw would be considered a manufacturing defect.

Failures to Warn.   Some products are inherently dangerous to use. In this circumstance, the manufacturer has a duty to adequately warn prospective users of the nature of the product. In some jurisdictions, the failure to warn is called a **marketing defect.** In one case of this nature, a man was spreading mastic prior to laying a parquet floor.[6] He had read the label and was aware of the message contained there, which cautioned that the mastic was flammable. The substance was actually more than merely flammable, however—it was explosive and should have been so labeled. The plaintiff was injured in the explosion and collected damages because of the inadequacy of the label.

Most products of more than a trivial nature are accompanied by user instructions of some sort. These may be installation instructions, maintenance procedures, repair details, or perhaps marketing material.[7] Obviously, following the instructions should not cause the product to injure someone. Suppose, however, that the instruction is meant to cure a known problem of the product. Suppose further that one who follows the instruction is, nevertheless, injured by the product.[8] The result can be devastating to the product's manufacturer, since the label or instructions probably refer to the dangerous nature of the product.

As with design defects, failure to warn cases are controversial. Some jurisdictions follow a line of cases supporting strict liability only if the manufacturer has been negligent. Others apply strict liability regardless of the manufacturer's negligence or even awareness of the product's hazards.[9]

As an example, suppose the manufacturer of a respirator provides instructions about its use for sandblasting. A court would no doubt conclude that the use of the respirator for sandblasting operations was entirely feasible. One can easily imagine that the instructions would emphasize the dangers if guidelines were not followed.

What should the producer do if a hazard is not discovered until a very large number of the products have been distributed? Very simply, the producer probably should do everything that can reasonably be done to remedy the situation. Automobile **recalls** are examples of the extent to which these remedies may be taken. Mailed warnings to all known users could be adequate if sufficient efforts were made to identify the likely victims. The producer's position is far from secure in such a circumstance, but the degree and sincerity of its efforts to warn people likely to be injured at least provides a reasonable defense.

Hidden Defects

The spirit of *caveat emptor* lives on. However, a manufacturer relies on this spirit's guidance to its peril. Use of an obviously defective product may set up a defense of assumption of the risk or

---

[6]*Murray v. Wilson Oak Flooring Co., Inc.,* 475 F.2d 129 (1973).

[7]See *McClanahan v. California Spray-Chemical Corp.,* 194 Va. 842 (1953).

[8]See *Byrd v. Hunt Tool Shipyards, Inc.,* 650 F.2d 44 (1981).

[9]See *Beshadu v. Johns-Manville Prods. Corp.,* 447 A.2d 539 (N.J. 1982).

product misuse. After liability has been established, a manufacturer may try to mitigate damages by introducing evidence of comparative negligence. However, manufacturers cannot use the contributory negligence defense in strict liability cases, and most jurisdictions do not distinguish between hidden and obvious defects in such cases.

Discovery

Very often, the evidence the plaintiff needs is in the defendant's hands. Products liability cases often involve evidence of quality control records, customer complaints, designs and design changes, research and product engineering records, and the like. In fact, documents like these are often vital to the plaintiff's case. As noted in an earlier chapter, document requests and other discovery procedures can and should be used to obtain such documents.

Damages

Products liability cases are usually intended to obtain money from the defendant to compensate for plaintiff's losses. Occasionally, because of extremely hazardous or careless behavior by a defendant, a court may go further and award an additional amount to the plaintiff as a punishment or penalty against the defendant. Damages are discussed in greater detail in Chapter 14.

Defenses

Sometimes the plaintiff's case is so overpowering that the defendant has no alternative but to settle the matter out of court. The defendant's expert witness may be extremely helpful in making the decision to settle. If a faulty design or manufacture is readily evident to the defendant's expert, a professional engineer and expert in the field, the fault would presumably be evident to the plaintiff's expert as well. In such a case, the best advice the expert can give a client is to settle out of court.

In contrast, when a defendant decides to go to trial, the defendant usually believes that it stands a high chance of winning or reducing the amount of damages to be paid. The decision as to the best course of action depends on what the evidence shows. The nature of the evidence may range from a cause for outright dismissal of the case to inconclusive opinions or speculation that a plaintiff could successfully counter.

Product Alteration.   The producer of a product should be held responsible only for those products he or she has produced. Close examination of a component may indicate that it was repaired after manufacture—perhaps, for example, replacement brake linings or a rebuilt wheel cylinder failed. If a rebuilt part was the cause of the injury, it may no longer be the automobile producer's responsibility. The question is basically this: Who produced the product that failed?

Suppose it was the original brake linings (or other component) that failed and caused plaintiff's injury. Suppose further that these linings or other components had been supplied to the automobile producer by another manufacturer. Shouldn't the component supplier be held responsible for the injury rather than the producer who assembled these components into a finished product? The simple answer is no. It is the producer of the final product as sold to the consumer who is responsible for its safe performance. Whether the producer made the parts or bought them is irrelevant as far as responsibility to the injured plaintiff is concerned. The producer may be able to pass part or all of the loss along to the supplier, but the producer of the final product has primary responsibility.

Proximate Cause.   If no causal connection exists between the plaintiff's injury and the producer's product, the producer should have no liability. One might be able to show a faulty design in the location of the fuel tank in a car, for example, but if this flaw is in no way connected to plaintiff's injury, the faulty fuel tank location has no bearing on the case. Just as a showing of product alteration usually serves as a complete defense, a successful showing of the lack of any proximate cause is generally an effective defense.

Assumption of the Risk.   In order for this defense to succeed, the plaintiff must have known there was a risk and voluntarily accepted it. The court will consider the plaintiff's age, intelligence, and experience, among other factors.

Misuse.   Misuse of a product may include one of the following:

1. Use of a product in a way not intended nor foreseeable by the manufacturer;
2. Use of a product by a person unforeseen or unintended by the manufacturer, e.g., a product intended for a professional's use that was used by an unprofessional, often in the pharmaceutical and beauty product fields; and
3. Using the product in a way contrary to the warnings provided.[10]

Producers must expect their products to be subjected to all sorts of abuses as well as the uses for which they were intended. Consider the uses to which you have put screwdrivers. Even television and newspaper advertisements show product misuse as persuasive proof of the rugged nature of products. Still, there is a logical limit to which anticipated abuse may be taken. For example, a plaintiff's injury resulting from the use of a computer keyboard as a hammer would probably not be tolerated as reasonable or foreseeable misuse. This sort of question is also often left for a jury to decide.

Unavoidably Unsafe Product and State of the Art.   In our present state of technological development, it is impossible to accomplish certain functions without some risk. A completely safe car would not move; a completely safe rotary lawn mower would not mow grass; a completely safe knife would not cut. But people need these pieces of equipment, so we cannot hold a manufacturer of such things to a standard of complete safety. Some danger is inherent in the operation of most products; perhaps the best we can expect is a condition we might consider reasonably dangerous. This is to say that because of the product's nature, it cannot be made safe. As you can imagine, this is not a very popular defense in the courts and is becoming increasingly less so when the product involves disease or contaminated blood. Thus some states have adopted **blood shield laws** which protect suppliers of blood and blood-related products from strict liability.[11] It has even been suggested by some courts that such a defense only be used when the technology has not been developed sufficiently to prevent danger.

The **state-of-the-art defense** is similar to that of the unavoidably safe product. This defense tends to hinge on the availability of a technological solution for the danger. One way to state this standard is as follows: The manufacturer should produce a product as safely as the state of the art will permit. In other words, the manufacturer should review the state of the art at the time he or she produces the product. If at that time, a survey of the literature and a comparison with similar

---

[10]See *Pavlides v. Galveston Yacht Basin, Inc.*, 727 F.2d 330 (5th Cir. 1984).

[11]*Weishorn v. Miles-Cutter*, 721 A.2d 811 (Pa. 1998).

products show no safer way of accomplishing the function, it seems unreasonable to expect more from the defendant.

Proof of compliance with the state of the art, however, may not necessarily be a complete defense for a producer. The decision could favor the plaintiff for the simple reason that the court or a jury believes the state of the art should have advanced more rapidly.

Standards.   Various federal agencies have become concerned with hazard reduction and safety issues, as have the military and various industrial organizations. Almost invariably, each of these organizations begins by formulating and publishing minimum standards to be met by whatever products are being considered. But suppose that Black Manufacturing Company, for example, has complied with all available standards in the manufacture of its product—say, a riding lawn mower. Green, in attempting to operate the lawn mower in very heavy grass, manages to injure himself. Green files a products liability action against Black Manufacturing. Is Black Manufacturing protected because it complied with the applicable standards?

The most common answer is no. The question of liability should probably still be answered according to the section 402A criteria. In fact, if the hazard were so improbable that those who wrote the standards completely missed it, this might suggest that what caused the injury was truly a hidden design defect. On the other hand, the defendant Black Manufacturing will argue to the jury that the standards show that problem was not foreseeable and the mower was not "unreasonably" dangerous.

If a producer fails to comply with published standards, and that failure results in injury to a plaintiff, the plaintiff certainly has strong evidence against the producer. Furthermore, such evidence might well be used as a basis for punitive damages. In addition, avoidance of standards published by federal agencies may lead to fines and/or other sanctions against the offending manufacturer.

Plaintiff's Negligence.   In products liability actions under strict liability, there is no longer an absolute defense of contributory negligence. This has been replaced by comparative fault or comparative negligence. The court does not consider whether the plaintiff contributed to his or her injury at the outset of the case. Rather, after the defendant's liability has been established, the court looks to the plaintiff's actions, offered as evidence by the defendant, to mitigate damages. In products liability cases under negligence, the jury may take into consideration the plaintiff's contributory negligence to the injuries suffered.

STATUTES

State Statutes

No one should suffer because of faulty products. Truly defective products should not have been produced or should have been removed by the producer before they could injure someone. A products liability problem does exist today, of course, but some would contend that it suffers from "overkill." For example, cases arise where products were not really faulty, but were abused, or perhaps were produced many decades ago. The cost of defending products liability suits and paying judgments and out-of-court settlements for such cases is added to production costs, and the public pays by way of higher prices. In recognition of such problems, some states have enacted products liability laws.

State products liability acts vary considerably in their content and their restrictions. However, such laws generally set out the following standards:

- That after a certain period of time the producer's responsibility for the product ends,
- That the "state of the art" can be used as a defense,
- That compliance with recognized standards (e.g., military standards) is a reasonable defense,
- That alteration and/or abuse by the plaintiff is a defense, and
- That a failure to warn is not a component of strict products liability.

Such restrictions as these may become more common in the future. In addition, advocates have suggested that there should be a federal products liability law. The primary advantage of a federal law would be its uniformity. Thus, a company making products in Illinois, for example, and selling them throughout the United States would not have to be concerned with the laws (and related risks) of every state. Opponents tend to argue that each state should be free to choose for itself what is appropriate for protecting people within its territory from injuries.

Federal Statutes

There are two dominant agencies that set product safety standards: The Federal Trade Commission and the Consumer Product Safety Commission.

The Federal Trade Commission.   The Federal Trade Commission (FTC) was granted authority to regulate warranty liability under the Magnuson-Moss Warranty–Federal Trade Improvement Act.[12] The Act was adopted in 1975.[13] Under this statute, the federal government can establish minimum standards for warranty protection. The standards promulgated under this Act supplement, and do not replace, state laws regarding warranty. If an individual wants to bring a claim under this act, he or she may file a case in a federal court or go to the FTC which has rules governing dispute settlement.

The Consumer Product Safety Commission.   The Consumer Product Safety Commission (CPSC) was granted authority under the Consumer Product Safety Act.[14] It was created in 1972.[15] The CPSC has authority to set standards for potentially dangerous products including, among others, hazardous substances, flammable fabrics and poisons. The CPSC can conduct tests, train, and investigate products to measure their safety. It can also establish standards for performance, as well as standards for adequate warnings and instructions.

These two agencies are not the only federal consumer protection agencies. The Food and Drug Administration, the National Highway Traffic Safety Administration as well as a host of other smaller agencies, also have authority to set standards and establish regulations and guidelines.

---

[12]15 USCA §§ 2301–2312 (1988).

[13]Pub. L. No. 93-637, §101 *et seq.*, 88 Stat. 2183 (1975).

[14]15 USCA §§ 2051–2083 (1988).

[15]Consumer Product Safety Act, Pub. L. No. 92-573 §§ 1–34, 86 Stat 1207-33 (1972), current version at 15 USCA §§ 2051–2083 (1988).

SUMMARY

⤷ Prevention

When all is said and done, how can engineer or engineering company avoid the sometimes draconian results of a strict liability case? Some prevention measures have been discussed in previous chapters, but among the things an engineer and his company can do include the following:[16]

- **Get Insurance!** Of course this is an obvious protective measure against financial ruin if a company is faced with a substantial damages at the conclusion of a strict liability case. It is also something that can be obtained in advance of any litigation. Insurance should cover not only potential product liability claims, but should also include professional insurance. In addition, the company should review contractual provisions with suppliers and vendors to protect themselves in the event of gaps in coverage, and perhaps adjust the contract to include insurance requirements and indemnity provisions.
- **Education.** This means not only providing educational materials and guidelines to employees about personal safety, but also providing education about how to report and resolve existing and potential problems.
- **Product Testing.** Not only is this a great preventative measure, the results obtained can be used as evidence in the event of litigation.
- **Record Retention.** Keep track of how products are designed an manufactured. Develop a record retention policy, stick with it, and enforce it.
- **Check the Warning Labels and Instructions.** This is especially important when the product has to comply with federal and state regulations. Confirm accuracy and whether the labels conflict with any known laws. This also includes reviewing advertising materials that may make assertions about product use and safety.

JAMES EDWIN HODGES; BEVERLY HODGES v. MACK TRUCKS, INC, *474 F.3d 188; 2006 U.S. App. LEXIS 31879*

Study terms: Product liability, "seatbelt evidence", diversity jurisdiction, *Daubert*, safer alternative design, judgment as a matter of law, subrogation

These three appeals arise out of a product-liability, diversity action for injuries sustained because of a secondary collision in Texas, involving a tractor-trailer manufactured by Mack Trucks, Inc. (Mack). Mack seeks judgment as a matter of law and, in the alternative, a new trial, claiming, *inter alia*, the district court improperly both admitted expert testimony and excluded evidence concerning the use, or nonuse, of his seatbelt by James Hodges (Hodges), the injured driver of the truck. Hodges received a multi-million dollar verdict. He and his wife, Beverly Hodges (the Hodges), contest her not also being awarded damages and seek a new trial on that issue. Finally, ABF Freight Systems, Inc. (ABF), Hodges' employer and workers'-compensation provider, challenges the district court's rulings on its subrogation claim. A new trial and ABF's claim's being reconsidered are required. **VACATED AND REMANDED.**

---

[16]These suggestions are derived in part from Paul Sherman's book *Products Liability for the General Practitioner,* McGraw-Hill, Inc. 1981 and May 2007 supplement.

I.

On 1 November 2002, a 16-year-old drove her vehicle into the path of an oncoming Mack truck, driven by Hodges, a 34-year veteran driver of large trucks. His cab was pulling two trailers, and the other vehicle hit the right front wheel of Hodges' truck, causing extensive damage. The truck swerved into the path of an oncoming car, breached a guard rail, and jack-knifed down an embankment. It came to rest with the nose of the tractor pointed up; the passenger-side door was damaged but the door frame and the cab were not deformed.

Hodges was ejected through the passenger side and sustained severe and permanent injuries, including paraplegia. (It is undisputed that, had he remained in the cab, his injuries would have been far less serious.) ABF, Hodges' employer, was self-insured and began paying Hodges workers' compensation.

ABF owned the truck. Its seatbelts were manufactured by Indiana Mills & Manufacturing (Indiana Mills). Its door latches, manufactured by KSR International, were installed by Mack.

In May 2003, the Hodges filed this action against Indiana Mills and Mack, claiming a design defect in the seatbelt caused Hodges to be ejected. (The Hodges had settled with the 16-year-old driver for $50,000.) In early 2004, the Hodges added a design-defect claim for the passenger-side door latch, asserting the defect caused the latch to fail after Hodges' truck was hit. That June, ABF intervened to protect its subrogation interests in workers' compensation paid to Hodges.

■ ■ ■

On 14 August, Indiana Mills settled with the Hodges on the seatbelt claim for $1.4 million. The settlement structure provided for James and Beverly Hodges to each receive half of the settlement amount. Accordingly, only the defective-door-latch issue remained for trial, with Mack as the sole defendant.

On the eve of trial, as a result of that settlement, the Hodges moved to exclude all evidence of Hodges' use, or nonuse, of his seatbelt, pursuant to § 545.413(g) of the Texas Transportation Code, claiming the statute proscribed introducing such evidence in civil trials (seatbelt evidence). The motion was granted without written reasons being given.

During trial, the Hodges introduced expert testimony by Steven Syson. He testified: the door latch failed; and there was a safer alternative design available that would have substantially reduced the likelihood of Hodges' injuries. Mack's pretrial motions to exclude this testimony had been denied.

On 26 August, following approximately two and one-half days of testimony, the jury returned its verdict, finding Mack and the 16-year-old driver 60% and 40% liable, respectively, for Hodges' injuries. It awarded $7.9 million in damages, but awarded the entire amount to Hodges. . . .

In October, Indiana Mills had interpled its $1.4 million in settlement funds into the court's registry. As noted, under the agreed settlement terms, James and Beverly Hodges were to each receive $700,000. ABF claimed it was entitled to the entire amount, not just the $700,000 Hodges was to receive, for workers' compensation it had paid, as well as would pay in the future. That December, the district court held an evidentiary hearing on the funds' disbursement. Among other rulings, it denied ABF's request for reapportionment of the settlement amount, holding, *inter alia*, the intent of the settlement scheme was not to deprive ABF of its rights to subrogation or future credit. The funds have been disbursed.

II.

For this diversity-jurisdiction action, arising out of an accident in Texas, its substantive law applies. At issue is whether the district court erred in: (1) admitting Syson's expert testimony; (2) denying Mack judgment as a matter of law (JML); (3) excluding the seatbelt evidence; (4) failing to grant a new trial on Beverly Hodges' damages; (5) approving the apportionment of the Indiana Mills settlement amounts between the Hodges; and (6) assessing attorney fees and litigation expenses out of ABF's subrogation recovery and calculating its right to future credit.

We hold, *inter alia*: JML was properly denied; the court reversibly erred, however, by excluding the seatbelt evidence; and, therefore, a new trial is required. Accordingly, we need *not* address Beverly Hodges' damages claim, nor fully address ABF's claims. ABF's claims are remanded to the district court for it, *inter alia*, to consider whether the effect of the settlement was to settle around ABF's subrogation lien.

A.

Mack maintains: Syson's testimony should have been excluded, pursuant to *Daubert v. Merrell Dow Pharms.*, 509 U.S. 579, 593-95, 113 S. Ct. 2786, 125 L. Ed. 2d 469 (1993); and, even if admissible, it failed, as a matter of law, to prove the requisite safer alternative design for the Mack door latch. Therefore, Mack contends judgment should be rendered in its favor. In the alternative, it seeks a new trial. . . .

An appellate court, in deciding whether JML should have been awarded, must first excise inadmissible evidence; such evidence "contributes nothing to a legally sufficient evidentiary basis". Weisgram v. Marley, 528 U.S. 440, 454, 120 S. Ct. 1011, 145 L. Ed. 2d 958 (2000). Therefore, we first address the contested admission of Syson's testimony. . . .

1.

The admission of expert testimony is reviewed for an abuse of discretion. . . .

Daubert interpreted *Federal Rule of Evidence 702* (admissibility of expert testimony) and assigned the trial court a gatekeeper role to ensure such testimony is both reliable and relevant. *Daubert*, 509 U.S. at 598. . . .

Rule 702 was amended in 2000, in response to the Supreme Court's decisions in Daubert and Kumho Tire[v. Carmichael, 526 U.S. 137 (1999)]. A party seeking to introduce expert testimony must show "(1) the testimony is based upon sufficient facts or data, (2) the testimony is the product of reliable principles and methods, and (3) the witness has applied the principles and methods reliably to the facts of the case". *FED. R. EVID. 702*.

In analyzing the Mack latch at issue, Syson: reviewed relevant Mack cab and door designs; examined numerous patents for latches and door designs in order to provide a safer alternative design; directed a third-party engineering firm to conduct force tests on the Mack latch; and analyzed the Federal Motor Vehicle Safety Standards (FMVSS) data published by the National Highway Traffic Safety Administration to determine the strength of the Mack latch as compared to an alternate design.

Mack challenges Syson's testimony as unreliable for a number of reasons, including: he is not a door-latch specialist; he was previously found to be an unreliable expert witness by a Texas court; he has not published any peer-reviewed articles purporting to show the weaknesses in the Mack latch; and he did not conduct his own tests or force calculations on the latches, but instead relied upon third-party testing.

Of course, whether a proposed expert should be permitted to testify is case, and fact, specific. . . . Syson, an engineer with many years experience working in, and testifying against, the automobile industry, presented very complex and technical testimony about the Mack latch and how it failed. He opined: Hodges was injured because Mack's passenger-side door latch failed (Mack does *not* dispute the latch failed at some point); and a safer alternative design existed which would not have broken and, thus, would have prevented Hodges' injuries. . . .

At trial, many of Mack's challenges to Syson's testimony were developed by its cross-examination of him; the judge and jury were able to determine his credibility. The trial (second) judge denied Mack's renewed request to exclude that testimony and denied Mack's two JML requests during trial based in part on that challenge. . . .

Based on our review of the record, and as reflected *infra*, it was *not* manifestly erroneous for the district court to find Syson's testimony relevant and reliable. Therefore, it did *not* err in admitting it pursuant to *Rule 702*.

2.

As noted, Mack next contends: even if Syson's testimony was properly admitted, Mack should be awarded JML because the testimony failed to prove the existence of a safer alternative design. . . .

To establish a design-defect claim under Texas law, the following must be proved by a preponderance of the evidence: (1) a safer alternative design existed; and (2) the design defect caused the injury. *TEX. CIV. PRAC. & REM. § 82.005*. A safer alternative design is

a product design other than the one actually used that in reasonable probability

(1) would have prevented or significantly reduced the risk of the claimant's personal injury. . . .without substantially impairing the product's utility; and

(2) was economically and technologically feasible at the time the product left the control of the manufacturer or seller by the application of existing or reasonably achievable scientific knowledge.

Id. *§ 82.005.*

A design is *not* a safer alternative if, "under other circumstances, [it would] impose an equal or greater risk of harm" than the design at issue. ***Uniroyal Goodrich Tire Co. v. Martinez,*** *977 S.W.2d 328, 337, 42 Tex. Sup. Ct. J. 43 (Tex. 1998), cert. denied, 526 U.S. 1040, 119 S. Ct. 1336, 143 L. Ed. 2d 500 (1999).* Similarly, the plaintiff must show "the safety benefits from [the] proposed design are foreseeably greater than the resulting costs, including any diminished usefulness or diminished safety". ***Uniroyal,*** *977 S.W.2d at 337*; see also ***Smith v. Louisville Ladder Co.,*** *237 F.3d 515, 520 (5th Cir. 2001).*

Mack relies upon ***Louisville Ladder,*** which concerned whether an extension ladder's cable-hook assembly mechanism was defective. As reflected above, our court held: the plaintiff's expert's testimony was insufficient to establish a safer alternative design; and, therefore, as a matter of Texas law, the plaintiff was unable to prove the ladder was defective. **Id.** *at 520.* The action at hand, however, differs. In ***Louisville Ladder,*** the expert testified that the proposed design "was a preliminary concept" not currently in use and "not ready to [be] recommend[ed] [] to a manufacturer". **Id.** *at 519.* Moreover, as noted supra, the expert never evaluated the risk associated with the proposed design and did *not* conduct a risk-benefit analysis. **Id.** Ultimately, he was unable to opine whether the proposed alternative would have prevented the injury in question. **Id.**

Unlike the expert's testimony in ***Louisville Ladder,*** Syson's was *not* mere speculation. Instead, he described in detail the latch at issue and how, and why, the proposed alternative latch would be safer. Syson examined several hundred door-latch patents on file with the Patent and Trademark Office to determine whether suitable alternative designs existed. When he found possible alternative designs, he examined how they performed compared to the Mack latch in the FMVSS-206 test, which examines the maximum longitudinal and transverse forces a door latch will maintain before it breaks. Based on that information and an analysis of the accident, Syson calculated the maximum amount of force required before deformation of the Mack latch would break it.

Syson concluded: the door latch used by Mack was defective; and another latch, the Eberhard latch, was a safer alternative and would have prevented Hodges' injuries. Among other things, Syson noted the Eberhard latch is 25% thicker at the stress point and provides 12,000 pounds of additional holding strength compared to the Mack latch, all factors that, in his opinion, would have prevented it from breaking in the accident.

Syson also testified that, based on his review of the above-discussed FMVSS-206 tests, Mack's latch was weaker than the latches used by 75 to 80% of similar vehicles. Based on his experience working with, and designing, parts for vehicles, Syson testified it would be easy, and inexpensive, for Mack to switch to the Eberhard latch. Along that line, he noted that, at the time of the accident, the Eberhard latch existed and was used in fire trucks.

Syson also conducted the requisite risk-utility analysis. He testified: a driver faces a significant risk if a door opens during an accident; engineers do not, and cannot, design for one particular accident; and the Eberhard latch would not impair the door's usefulness. In other words, part of a latch's utility is its ability to keep a door shut during a vehicle crash and using the Eberhard latch would *not* diminish the door's utility. Therefore, there was sufficient evidence for a jury to find Syson's testimony satisfied the requisite risk-utility test.

Syson provided the analysis required to allow the Hodges' to establish, by a preponderance of the evidence, that, under Texas law, a safer alternative design existed. Based upon his testimony, and drawing all reasonable inferences in the non-movant's favor, the evidence was sufficient to support the verdict. Accordingly, the district court did *not* err in denying JML to Mack.

B.

Concerning Mack's contesting the seatbelt-evidence exclusion, Texas began mandating seatbelt use in 1985. A person greater than 15 years of age is guilty of a traffic violation if he or she "is riding in the front seat of a passenger car while the vehicle is being operated . . . and . . . is not secured by a safety belt". *TEX. TRANSP. CODE § 545.413(a)*. The statute provides defenses for failure to wear a seatbelt, including, *inter alia*, a medical reason evidenced by a doctor's note. *Id.* at *§ 545.413(e)(1)*.

Pertinent to this issue, *subsection (g) of § 545.413* provided: "Use or nonuse of a safety belt is *not admissible evidence in a civil trial,* other than a proceeding under Subtitle A or B, Title 5, Family Code". *Id..* In 2003, however, the Texas legislature repealed *subsection (g)*. In doing so, the legislature specified: *subsection (g)* is not applicable to "action[s] filed on or after July 1, 2003. [But a]n action filed before July 1, 2003, is governed by the law in effect immediately before the change in law . . . and that law is continued in effect for that purpose."

As noted, this action was filed in May 2003, a few weeks before the 1 July 2003 effective date for the repeal of *subsection (g)*. In other words, its repeal is *not* applicable to this action. Accordingly, Texas statutory law proscribed the use of seatbelt evidence.

Nevertheless, Mack contends such evidence should be admitted because this action involves a secondary, not a primary, collision. (A primary collision concerns injuries sustained in the collision with another vehicle; a secondary collision concerns enhanced injuries caused by a collision with the interior of the vehicle or with an exterior object, if ejected.) This interpretation, Mack claims, is in line with a Texas Court of Appeals decision that seatbelt evidence is admissible in secondary-collision cases.

Mack notes *subsection (g)* was repealed only approximately one month after the Hodges filed this action and well before they added the defective-door-latch claim. At the time of trial on that claim, according to Mack, the intent of the Texas legislature was to allow seatbelt evidence, particularly in a crashworthiness action such as this. . . . According to Mack, without seatbelt evidence, the jury received a distorted view of the evidence, especially in the light of the Hodges' counsel's telling the jury: Hodges was ejected from the truck solely due to the defective door latch; and he did nothing to contribute to his injuries.

In addition, Mack also claims this circuit has affirmed the introduction of such evidence under other States' laws, despite statutory prohibition. Finally, Mack insists it is sound public policy to permit such evidence because federal law mandates truck drivers' wearing seatbelts.

Noting that, when they filed this action, *subsection (g)* was effective, and remained effective for all actions filed prior to 1 July 2003, the Hodges contend the district court properly excluded the seatbelt evidence because *subsection (g)* and Texas case law mandate its prohibition. They maintain: under Texas law, seatbelt evidence is admissible only under one rare exception—where the plaintiff makes a product-liability claim against a seatbelt manufacturer alleging a defective restraint system and must introduce evidence of his seatbelt use to prove causation. During trial, the district court stated it had based its eve-of-trial exclusion ruling on a similar understanding of Texas law. . . .

In addition, the Hodges claim: had the district court allowed seatbelt evidence, they would have offered "substantial evidence" that Hodges was belted at the time of the accident. In that regard, prior to settlement of the seatbelt claim, they contended the seatbelt was defective because it became *unlatched during* the accident.

■ ■ ■

Texas law mandates drivers wear a seat belt. The statute "was enacted to mandate the use of seat belts and to provide a criminal penalty for the failure to wear [one]". **Glyn-Jones**, *878 S.W.2d at 134*. The use, or nonuse, of a seatbelt's *not being* allowed in evidence in a civil trial was "to make clear that the sole legal sanction for the failure to wear a seatbelt [was] the criminal penalty provided by the statute and that the failure could not be used against the injured person in a civil trial". **Id.** As discussed *infra*, however, the Texas Supreme Court noted in **Glyn-Jones:** when viewed in the context of the entire statute, there is "ambiguity about the legislature's purpose"; this is because the seatbelt-evidence prohibition for civil trials falls within the criminal penalties of the Texas Transportation Code, see **id.** *at 133-34,* an unlikely place for a provision that has been read to have such an expansive scope.

■ ■ ■

The Texas Supreme Court's narrow holding in **Glyn-Jones** supports *subsection (g)'s* proscription *not* precluding the introduction of seatbelt evidence in the case at hand by Mack, the defendant. The Texas Supreme Court held the proscription did *not* bar all use of such evidence. On the other hand, contrary to the dissent's analysis, the Texas Supreme Court's opinion can *not* be read as holding—or even suggesting—such evidence cannot be introduced by a defendant, such as Mack.

In 2003, **Vasquez** reiterated the holding in **Glyn-Jones** that *subsection (g)* was intended to preserve the status quo concerning failure to wear a seatbelt not being contributory negligence. In **Vasquez,** the parents of a child killed by a deploying airbag in an automobile accident pursued a product-liability action against the manufacturer on a *crashworthiness* theory. See **Vasquez,** *119 S.W.3d at 850*. Although the Texas Court of Appeals, *en banc,* decided the case on other grounds, and therefore did not reach whether seatbelt evidence should be allowed in civil trials, it nonetheless noted: [the statute was never intended to exclude evidence of seatbelt use in "secondary collision" cases; as in **Vasquez,** where the functionality of the passenger's passive restraint system (which included the seatbelt) is at issue, seatbelt evidence is relevant to proving causation and the ultimate effectiveness of the restraint system; and the manufacturer's interest in offering seatbelt evidence was *not to mitigate* the "product defendant's liability for damages, but [was] offered . . . to support [its] defense that the air bag, in conjunction with seatbelt use, was *not* defective as designed". **Id.** *at 850, n.2.* (emphasis added).

■ ■ ■

In sum, . . . **Glyn-Jones,** together with **Vasquez,** are instructive. *Subsection (g)* prohibits the introduction of seatbelt evidence to show the plaintiff was *contributorily negligent*. On the other hand, in secondary-collision product-liability actions, such evidence may be admissible to show, or, as in this action, rebut, the essential element of causation. Seatbelt evidence was necessary for Mack to rebut the essential element of causation—whether its door latch was the proximate cause of Hodges' injuries—and, ultimately, to defeat a crashworthiness claim. Such evidence is *not* prohibited by *subsection (g)*. Arguably, this is also demonstrated by the repeal of *subsection (g),* even though that subsection applies here.

Therefore, the district court abused its discretion when it categorically excluded seatbelt evidence. Needless to say, this error was *not* harmless. Therefore, a new trial is required.

[The Court determined the following with regard to remanding the case back to the district court: 1) the district court is to reconsider the reasonability and fairness of Beverly Hodges apportionment; 2) the district court should consider both the pretrial settlement amount and ABF's pro-rata share of litigation expenses; 3) because the Court did not decide on any future-credit issues as they relate to ABF's subrogation lien, the issue is left for the district court to resolve.]

III.

For the foregoing reasons, the judgment as to Mack and the order as to ABF's subrogation amount are **VACATED** and this matter is **REMANDED** for a new trial and other proceedings, all consistent with this opinion.

VACATED AND REMANDED

> **HONDA OF AMERICA MANUFACTURING, INC. AND HONDA R&D CO., LTD., v. BRIAN NORMAN, INDIVIDUALLY AND AS SOLE ADMINISTRATOR OF THE ESTATE OF KAREN LESLIE VIVIENNE NORMAN, DECEASED, AND MARY NORMAN, INDIVIDUALLY 104 S.W.3d 600; 2003 Tex. App. LEXIS 1263(2003), *rev. denied*, 2003 Tex. LEXIS 368 (Sept. 11, 2003), *mot. for reh. for rev. denied*, 2003 Tex. LEXIS 542 (2003)**

Study terms: Products liability, compensatory damages, safer alternative design, economically and technologically feasible

This is a products liability suit. A jury awarded Brian Norman, individually and as sole administrator of the estate of Karen Leslie Vivienne Norman, deceased, and Mary Norman, individually, (the Normans) $65 million in compensatory damages in a suit the Normans brought against Honda of America Manufacturing, Inc. and Honda R&D Co., Ltd., after the Normans' daughter, Karen Norman, drowned in her Honda Civic automobile. The jury found that a design defect in the car's seatbelt was the producing cause of Karen's death.

In five issues, Honda argues (1) the evidence was legally and factually insufficient to prove causation; (2) the evidence was legally and factually insufficient to prove a safer alternative design; (3) the trial court abused its discretion in admitting unqualified and unreliable expert testimony; (4) the trial court abused its discretion in admitting evidence of other incidents; and (5) the evidence was legally and factually insufficient to support the jury's award of "grossly excessive damages, which were influenced by improper motivations." We reverse and render a take-nothing judgment.

Facts and Procedural Background

The Accident

At approximately 2:00 a.m. on December 2, 1992, Karen attempted to back her car up to turn around, and she accidentally backed down a boat ramp into the water in Galveston Bay. Her passenger, Josel Woods, was not wearing a seatbelt and was able to get out of the car by crawling out the passenger side window. After escaping, Woods reached back into the sinking car to get her purse. Woods testified that Karen was calm and did not appear scared. As Woods was swimming to the ramp, she heard Karen say, "Help me. I can't get my seatbelt undone." Woods testified that, after she reached the ramp, she heard Karen yell to her again that she could not get out of her seatbelt.

A dive team located Karen's car at 8:53 a.m. All of the windows were rolled up, including the one Woods testified she had escaped through, and all the doors were closed. Karen's body was found in the back seat. An autopsy revealed Karen's blood-alcohol level was .17.

The Car

At the time of the accident, Karen's four-door 1991 Honda Civic was equipped with a two-point passive restraint system—an automatic seatbelt that was mechanically drawn up over the shoulder when the door was closed—supplemented with a manual lap belt. The automatic seatbelt fastened itself with no action on behalf of the occupant. Robert Hellmuth, a former National Highway Traffic Safety Administration employee, testified that, in 1990, all cars were required to have either a passive belt system or an air bag. Hellmuth also testified that a two-point passive restraint system was the most expensive seatbelt system in use at the time Karen's car was manufactured.

The shoulder belt on both front seats was attached to a "mouse" that ran along a rail above the door. . . . When the door was opened or the ignition was turned off, the mouse moved forward, allowing the occupant to get out of the car. . . . Like most seatbelts, the shoulder belt was naturally taut across the body, but it was easy to spool out more belt to allow the occupant to lean forward and/or sideways. If the car experienced rapid deceleration (such as that encountered here when the car hit the water) or substantial tilting of the vehicle, however, the belt's emergency locking retractor would engage, preventing spooling of the belt and holding the occupant in her seat.

The Lawsuit

The Normans sued Honda, alleging that the seatbelt system in Karen's car was defectively designed and prevented her from getting out of the sinking car. The case was retried after the original trial resulted in a hung jury.

The Normans contend that the emergency locking retractor locked as Karen backed down the ramp and that she pulled on the door latch, causing the mouse to move and then stall and the seatbelt to pin her to her seat. . . . The Normans argue that the evidence showed the seatbelt system was defectively designed because (1) the mouse was able to move even when the retractor was locked, allowing the seatbelt to pin an occupant in the seat; (2) the seatbelt, when fully extended, could not be released easily and rapidly by pressing the emergency release button; and (3) the emergency release button was improperly located, in that Honda failed to provide an easy and rapid way to get out of the seatbelt under conditions it knew would occur.

The jury found that Karen was 25% contributorily negligent, awarded Karen's parents $60 million in actual damages, and awarded $5 million to Karen's estate. The trial court reduced the award to $20 million for Karen's mother and $18 million for Karen's father, and it denied Honda's motion for remittitur as to the estate.

Design Defect

The Civil Practice and Remedies Code prescribes two elements—a safer alternative design and producing cause—that must be proved, but are not alone sufficient, to establish liability for a defectively designed product. TEX. CIV. PRAC. & REM. CODE ANN. § 82.005 (Vernon 1997);[18] A claimant not only must meet the proof requirements of the statute but must show, under the common law, that the product was defectively designed so as to be unreasonably dangerous, taking into consideration the utility of the product and the risks involved in its use.

Evidence of Safer Alternative Design

In its second point of error, Honda argues that the judgment should be reversed because there was insufficient evidence of a safer alternative design to the seatbelt restraint system used in the Honda.

---

[18]The Civil Practice and Remedies Code provides, in relevant part: . . .
  (b) In this section, "safer alternative design" means a product design other than the one actually used that in reasonable probability:
    (1) would have prevented or significantly reduced the risk of the claimant's personal injury, property damage, or death without substantially impairing the product's utility; and
    (2) was economically and technologically feasible at the time the product left the control of the manufacturer or seller by the application of existing or reasonably achievable scientific knowledge.

Standard of Review

When reviewing the legal sufficiency of the evidence, we consider all of the evidence in the light most favorable to the prevailing party, indulging every reasonable inference in that party's favor.

We will sustain a factual sufficiency challenge only if, after viewing all the evidence, the evidence is so weak or the verdict so contrary to the overwhelming weight of the evidence as to be clearly wrong and unjust. This Court cannot substitute its opinion for that of the trier of fact and determine that it would have weighed the evidence differently or reached a different conclusion.

Safer Alternative Design

Honda argues that the Normans failed to meet their threshold statutory burden because they failed to prove there was a safer alternative design to the Honda's seatbelt restraint system. To prove a design defect, the Normans had to show, among other things, that (1) there was a safer alternative; (2) the safer alternative would have prevented or significantly reduced the risk of injury, without substantially impairing the product's utility; and (3) the safer alternative was both technologically and economically feasible when the product left the control of the manufacturer. The Normans had the burden of demonstrating by a preponderance of the evidence that a safer alternative design existed at the relevant time. In addition, a plaintiff complaining of a design defect is required to show that "the safety benefits from its proposed design are foreseeably greater than the resulting costs, including any diminished usefulness or diminished safety"—that is, that the alternative design not only would have reduced the risk of harm in the instant case, but also would not, "under other circumstances, impose an equal or greater risk of harm." *Uniroyal Goodrich Tire Co. v. Martinez*, 977 S.W.2d 328, 337, 42 Tex. Sup. Ct. J. 43 (Tex. 1998). Thus, the Normans had to prove that an economically and technologically feasible alternative seat belt and release system was available and would have prevented or significantly reduced the risk of Karen's death without substantially reducing the utility to the "intended users" of the product—namely, all automobile drivers. If no evidence is offered that a safer design existed, a product is not unreasonably dangerous as a matter of law.

The jury charge read, in part, as follows:

Was there a design defect in the seat belt restraint system at the time it left the possession of Honda that was a producing cause of the occurrence in question?

A "design defect" is a condition of the product that renders it unreasonably dangerous as designed, taking into consideration the utility of the product and the risk involved in its use. For a design defect to exist[,] there must have been a safer alternative design.

"Safer alternative design" means a product design[,] other than the one actually used[,] that in reasonable probability—

1. would have prevented or significantly reduced the risk of the occurrence in question without substantially impairing the product's utility and

2. was economically and technologically feasible at the time the product left the control of Honda acting by and though [sic] its agents and/or employees by the application of existing or reasonably achievable scientific knowledge.

The charge followed the language of section 82.005 and the Texas Pattern Jury Charge. The Normans elicited testimony from several experts, but only two of these experts commented on alternative designs for the seatbelt. Although they did not agree about the feasibility of the various alternative designs, two of the Normans' experts, Thomas Horton, a mechanical engineer, and Kenneth Ronald Laughery, a human factors expert, testified that there were three potential alternative seatbelt system designs: (1) the mouse could be on a timer; . . . and (3) there could be two release buttons—one near the hip and one over the shoulder.

(1) Mouse timer

Horton was the only witness who testified about a "mouse timer." Horton testified that his main criticism of the Honda seatbelt system was the tightening of the belt. . . . A timer could have been programmed so that, if the mouse did not travel its entire cycle within a certain period of time, it would reverse and return to its original position. Horton explained that it would have been "simple within the electronics" to have created such a system.

To prove that the mouse timer constituted a safer alternative design to the overhead manual seatbelt release in the Honda, the Normans had to show that (1) a mouse timer existed or that the scientific knowledge to produce it was reasonably achievable, and (2) a mouse timer was economically and technologically feasible at the time Karen's car left Honda's control. On cross-examination, however, Horton admitted that he had never drawn up schematics for a mouse timer system. The Normans argue that Horton testified that this design was feasible. He did not. Instead, he simply answered "yes" to the general question of whether, in his opinion, economically and technologically feasible designs were available at the time Honda manufactured Karen's car. Horton did not identify any such available design, nor did he discuss the economic or technological feasibility of such a hypothetical alternative design.

Nor did Horton give any testimony to support a finding that, considering all relevant risks, his alternative design would be safer than the one employed by Honda in that it would not "under other circumstances impose an equal or greater risk of harm." Horton further testified that he was not critical of Honda for the location of the emergency release button. In fact, he testified that he was "not going to express [an opinion] in this case" with respect to the location or functioning of Honda's emergency release button.

We conclude, based on the record, that the Normans failed to prove that a timer-controlled mouse was a safer alternative design to the seatbelt system in Karen's Honda. . . .

(3) Two release buttons

Finally, Laughery testified that having two release mechanisms, one over the shoulder and one near the hip, was another possibility. Laughery admitted, however, that he did not know whether his proposed design—which he conceded had never been used in any vehicle—was technologically feasible. When Laughery was asked how it would work, he replied, "That's an engineering question. I don't know the answer to that. I don't have engineering design opinions." Horton, the Norman's engineering expert, explained that having two release mechanisms was not technologically feasible.

We conclude, on the basis of the record, that the Normans failed to prove that a two-release-button design was technologically and economically feasible and thus a safer alternative design.

No Evidence

The Normans failed to show that the mouse timer . . . or the two-release-button system existed, were technologically and economically feasible, and were safer under relevant circumstances than the seat belt release system present in the Honda. There was, therefore, no evidence that a reasonably safer alternative design existed for Honda's passive restraint system when the car was manufactured. Having failed to present such evidence, the Normans failed to carry their threshold statutory burden of proving a safer alternative design. We hold that the evidence was legally insufficient to support the jury's finding that there was a design defect in Karen's Honda.

We sustain Honda's second point of error.

Having determined that the evidence was legally insufficient to support the jury's design defect finding, which is a threshold condition of a finding of liability, we need not address Honda's remaining issues on appeal.

We reverse and render a take-nothing judgment.

SAUPITTY v. YAZOO MFG. CO., INC. 726 F.2d 657 (1984)

Study terms: Compensatory damages, punitive damages, design defect, modification

Logan, Circuit Judge

Defendant Yazoo Manufacturing Company, Inc. appeals from a jury verdict awarding plaintiff, James Saupitty, $560,000 compensatory and $440,000 punitive damages. Plaintiff was injured while operating a lawnmower man-ufactured by defendant. Plaintiff brought suit on the theory of manufacturer's product liability, alleging that design defects rendered the mower unreasonably dangerous to the user. Oklahoma law applies in this diversity case.

Plaintiff was a civilian employee of the United States at Fort Sill, Oklahoma, in the Grounds Maintenance Depart-ment. One of his responsibilities was to cut the grass on the Fort Sill grounds. To do so he used a six-year-old Yazoo YR-60 riding lawnmower that defendant had manufactured. Plaintiff was riding the mower down a hill when the mower began bouncing and shaking. He attempted to slow or stop the mower by placing the machine into reverse gear. The gear shifting momentarily locked the mower's drive wheels, bucking plaintiff forward over the top of the machine. In his attempt to stop his fall his thumb and two fingers of his left hand were severed and his arm was injured.

Plaintiff alleged that the mower was defectively designed and unreasonably dangerous in a number of respects: It did not have a proper rear weight, a stabilizer bar, a dead man switch, or adequate brakes; the operator's seat was in an unsafe position; the controls allowed the operator to shift directly from forward to reverse; and the mower lacked sufficient warnings.

On appeal defendant asserts that the trial court should have directed a verdict in its favor because the mower's brakes and belt guard had been removed before the accident. Oklahoma cases have adopted the rule of *Re-statement (Second) of Torts* sect. 402A (1)(b) (1965), which imposes liability only when the product "is expected to and does reach the user or consumer without substantial changes in the condition in which it is sold." Thus, a manufacturer is not liable when an unforeseeable subsequent modification alone causes the plaintiff's injury. *Texas Metal Fabricating Co. v. Northern Gas Products Corp.* The manufacturer is liable, however, if the subsequent mod-ification was foreseeable . . . or if it was not a cause in fact of the injury. See *Blim v. Newbury Industries, Inc.* . . . In *Blim*, a plastic injector press was originally equipped with mechanical drop bars designed to prevent injuries to operators' hands. The plaintiff's employer testified that he removed the bars because the press did not operate properly with the drop bars in place. In a suit against the manufacturer the jury found for the plaintiff, who was in-jured when the press closed on her hand. The manufacturer, relying on *Texas Metal*, contended that the removal of the drop bars constituted a material alteration of the product and thus was an intervening and independent cause of the injury as a matter of law. This court disagreed, stating: *Problems w/ lawnmower were known*

"Appellant's reliance upon *Texas Metal* is misplaced. In that case there was no demonstrated relationship between the rattle (the alleged defect) and the ultimate explosion. . . . Here, the mechanical drop bars were safety fea-tures designed to prevent just such an injury as that sustained by appellee. Since evidence demonstrated that they were already ineffective, their removal could not even exacerbate the hazard; a *fortiori*, it could not, as a matter of law, constitute a superseding, intervening cause of the injury. . . ."

Similarly, in the instant case, plaintiff's coworkers and his expert witness testified that the scrubber brakes would not stop the mower. Even the manufacturer's brochure declared that the foot brake was ineffective while the mower was in gear. Thus, the jury could find that to slow or stop the mower plaintiff would have had to shift into reverse even if the brakes had not been removed. Furthermore, plaintiff testified that his hand was injured by the mower's cutting blades rather than by the belts exposed by the removal of the mower's belt guard. Plaintiff did testify that he disengaged the mower's blades shortly before the accident, but his expert testified that the model of mower involved in the accident had a history of failing to disengaged the blades when the operator placed the blade control in neutral. Thus, plaintiff presented evidence from which the jury could reasonably have concluded that the cutting blades continued to turn and cut plaintiff's hand, as he testified, and that the removal of the belt guard was not a cause of plaintiff's injury. Therefore, construing all of the evidence and the inferences therefrom in the light most favorable to the plaintiff, we cannot conclude that the removal of either the mower's brakes or its belt guard constituted a superseding, intervening cause of plaintiff's injury as a matter of law.

We see no merit in defendant's other arguments for reversal. The plaintiff produced evidence from which the jury could reasonably have concluded that the mower possessed the alleged design defects when defendant sold it to the government. Defendant's argument that the district court improperly instructed the jury is unmeritorious, particularly since its counsel did not object to the instructions at the time they were given. The Oklahoma Supreme Court has held that punitive damages are recoverable in products liability cases. . . . Plaintiff produced sufficient evidence of the manufacturer's reckless disregard for public safety to justify submitting the issue of punitive damages to the jury. . . . The trial court did not err in refusing to instruct the jury on the theory of comparative negligence, since the Oklahoma Supreme Court has said that comparative negligence statutes have "no application to manufacturers' products liability. . . . Finally, although the award of $560,000 compensatory and $440,000 punitive damages is high, the amounts do not shock our conscience.

Poor design led to modifications. Product was defective when sold

AFFIRMED.

Seth, Chief Judge, dissenting:

I must respectfully dissent. The majority concludes that the modifications to the six-year-old mower made by the government should be disregarded, and would apply tort standards and tort doctrines. Reading the record in the light most favorable to the plaintiff, I disagree. The plaintiff was awarded one million dollars for the loss of three fingers from his left hand and for punitive damages.

The government made two material alterations to the mower, either of which should constitute grounds for a directed verdict for the defendant on the strict liability claim under applicable doctrines. First, the mower's belt guards had been removed. At trial an issue was whether the mower's blades had cut Mr. Saupitty's fingers or whether his hand had become entangled in the swiftly moving belt and his fingers severed when they were pulled into contact with the belt's pulley. The majority accepts the plaintiff's assertion that his fingers were cut by the mower blades. The plaintiff testified that he had disengaged the mower's blades before starting down the hill. The majority assumes that the disengagement was unsuccessful because "his expert testified that the model of mower involved in the accident had a history of failing to disengage the blades when the operator placed the blade control in neutral." However, that theoretical possibility was directly refuted by Mr. Saupitty's own testimony that the blades were off and had stopped rotating. In these circumstances it is hard to understand why theory and not the actual facts were used. If the blades had indeed stopped turning, then the only other possibility was that Mr. Saupitty's fingers were severed when they were caught in the belt. Since the belt guard had been removed by the government after it bought the mower it must be concluded that a material alteration had occurred preempting strict liability.

The government also removed the mower's brakes sometime during its six years' use of the machine. That removal necessitated stopping the mower by pulling it into reverse. The machine could not even be otherwise slowed down because both the brakes and the throttle control had been removed. The degree of efficiency of the brakes before they were removed is not necessarily pertinent because although inefficient they nevertheless would have provided an alternative to the necessity of throwing the mower into reverse. In any event, the brakes originally on the mower met the standards for the industry. The plaintiff did not show that he had to stop the mower to avoid an accident. The mere slowing of the machine by use of the brakes could easily have prevented the accident. Their removal must constitute a material alteration. The proof of the plaintiff on both the brake and belt guard removals did not meet the requirements of *Stuckey v. Young Exploration Co.,* . . . to permit the trial judge to submit the case to the jury.

Even were strict liability to be imposed in this case, there is no evidence whatever in the record to support an award of punitive damages.

Since the defendant did not raise a contract specification defense, we cannot consider it on appeal.

I thus dissent.

Lots of variation in judgments!

Review Questions

1. Why was caveat emptor an appropriate defense 200 years ago but is not appropriate today?

2. What is the meaning of privity of contract? Under what circumstances does it make sense? Why isn't it an appropriate defense in a products liability case?

3. Contrast strict liability as stated in the Restatement (Second) of Torts, section 402A, with negligence and warranty as they would be applied to products liability.

4. The White Company produces a combination tool for home use that may be converted to a lathe, drill press, grinder, and many other power tools. Gray bought one of these tools and was injured while using it as a lathe. The injury occurred because the lathe chuck loosened, and the piece on which Gray was working flew out of the chuck, striking him in the head. Gray was out of work for several weeks and has lost partial sight of his right eye. What would you as plaintiff's expert witness look for to make a case for Gray to recover? What would you do as defendant's expert witness in the case?

5. Can you make the cost-benefit ideas of *Barker v. Lull Engineering* agree with the section 402A definition of strict liability? Does it agree better with negligence or warranty?

6. In *Hodges v. Mack Trucks, Inc.,* the Court differentiates the expert testimony in this case from that in *Smith v. Louisville Ladder Co.* What were the differences between the two experts' testimony?

7. Considering the damage award on the line in the *Honda v. Norman,* how much weight do you think the court gave the testimony of plaintiffs' witness Laughery with regard to the mouse timer and two-release button system design? Answer: The court may have considered his alternative seatbelt system designs insufficient to support evidence of a safer alternative design, considering he had not investigated, or may have not been asked to investigate, whether his proposed designs were technologically feasible. This evidence, along with the evidence of the Norton's engineering expert, failed to support the Norton's contention that there was a safer alternative design.

8. Both *Hodges v. Mack Trucks, Inc.* and *Honda v. Brian Norman* involve automobile design defects. How are the cases similar? Different?

9. After reading both the majority opinion and the dissent in *Saupitty v. Yazoo Mfg. Co.,* decide this million-dollar case for yourself and justify your decision.

---

8. Similar
- Texas, early 2000's, on appeal
- Product liability, considering safer alternative design
- Ruling in favor of producer
- Expert testimony figured prominently; appellant challenged court's admission of testimony

Different
- Hodges/Mack, expert testimony was supported by engineering analysis
  Norman/Hodges, " " " not " "
  ↳ theoretical w/o design, literature review, testing, "stress", "risk-utility test"
  ↳ unprepared
- Showed ~~state of the art~~ safer alternative design
  ↳ examined patents   ↳ another stronger latch was available at time
  ↳ used a force test   ↳ met risk-utility test
  ~~Both showed producing cause~~

Governmental Regulation

Political considerations often dictate engineering activity to a considerable degree. This is obviously true in projects run by the federal government (for example, engineering for NASA and the military) and by state and local governments (such as engineering for roads, bridges, and water supplies). Political intervention is an expected component of such projects. A major problem in the political control of engineering activities, however, is the lack of effective communication between politicians and engineers. On one hand, many politicians lack training in the sciences or else are not concerned with the technical aspects of a project. On the other hand, many, if not most, engineers find political involvement and motivations difficult to understand and reconcile with engineering requirements. The results are frequently less than desirable.

Administrative agencies exercise a great deal of control over engineering activities pursuant to administrative rules and laws. The basic idea is that an administrator with some knowledge of the engineering and business aspects involved in the project will act as the control mechanism according to established rules. Of course, most regulatory statutes are politically inspired (and the administrative rules are often made according to that inspiration). Unfortunately, too, like politicians, many regulatory personnel lack the background and the training that would allow them to do an effective job. Under these circumstances, it is not surprising that engineers often complain about the controls imposed on them.

Here, we will look at the law and activities of administrative agencies in general. Then we will discuss how administrative agencies affect three industrial functions: labor, worker's compensation, and safety.

Administrative Law

HIGHLIGHTS

■ Administrative agencies are governed by the administrative procedures they adopt. These procedures are usually based on the Administrative Procedure Act under which federal and state government agencies can conduct hearings and promulgate rules.

■ Agency hearings and rulemaking procedures have two advantages over the judicial system:
  1. Administrative law judges, as well as the agency's other administrative personnel, are likely to be experts in the relevant field; and
  2. The findings and final rules apply to the public at large, unlike judicial rulings that are unique to a set of facts and parties particular to a single dispute.

We have always had administrative agencies, bureaus, or commissions. For the first century and a half (or so) after our government was formed, agencies were few in number. Since then, however, their number has grown and grown. The federal government has established the NTSB, the FCC, the SEC, the NLRB, the IRS, and the FTC, to name only a few. The state government often has some sort of industrial commission, a governing board for the universities, a public utilities commission, a tax commission, numerous licensing boards, and many other agencies. Local agencies include zoning boards, school boards, welfare agencies, and others. All these entities have powers to investigate, make rules and rulings, and supervise activities in some limited geographical area and some sphere of activity. All have powers delegated to them by a legislative body or a chief executive (such as the president, governor, or mayor). These bodies have an ever-increasing influence on what we may do and how we may go about doing what we do. It seems only reasonable, then, that we should consider what or who controls these agencies, why they exist, and how one learns to live with them.

WHY AGENCIES?

Agencies and their associated rule-making authority were not anticipated by the separation of powers so carefully set forth in our Constitution. Rather, agency authority is a hybrid of powers delegated to them in order to govern a segment of the population or business world in the name of the public. Administrative agencies thus tend to have judicial, legislative, *and* executive powers.

Administrative agencies act like courts when they rule for or against an individual or a business. For example, Brown may be granted or denied a license to operate a business, or his license may be taken from him because he failed to follow an agency rule. Gray's tax return may be investigated by the Internal Revenue Service, and deductions allowed or denied. White Company may be required to cease and desist its pollution of a stream. Green Trucking Company may need agency approval of its rates before it can offer its services to customers. Whenever an agency renders a decision such as these, its actions resemble those of a court. Unlike courts, however, agencies do not have to wait until an injury has occurred before they act.

Agencies are also legislative in nature. They have the power to investigate situations that come to their attention and then make rules based on what they find. For example, the Internal Revenue Service may find it desirable to require people like Green to keep another kind of record. After following the appropriate rule-making procedure, the IRS's new rule requiring taxpayers (like Green) to keep such a record would appear in an issue of the *Federal Register*. Publication in the *Federal Register* is deemed by law to provide adequate notice to anyone affected by the proposed rule.

In many ways, agencies also act like an executive or supervisor. They generally have the right not only to investigate, but to act to prevent occurrences that they anticipate will harm someone. Specifically, many agencies have the power to prosecute violations of the rules adopted by the agency. As a practical matter, the threat of active supervision may be sufficient to obtain the objective sought by the agency. For example, companies will sometimes voluntarily withdraw a product from the market rather than try to meet the requirements of a new rule from the FDA.

The suggestion is sometimes made that courts could do better what agencies are supposed to do. But one need only consider the traditional nature of courts to see the magnitude of changes that would be required for courts to replace agencies.

The main function of a court is to decide specific disputes. Our traditional court procedures are designed to obtain all pertinent information regarding some past event. Then, armed with this information, the court makes a judgment. Only rarely, and then in the nature of an equity action, will courts concern themselves with things that have not yet occurred. In numerous cases, motions to dismiss are granted because the action is moot, meaning no real and present issue remains to be decided. Often a court's judgment refrains from going beyond the minimum reasoning required to answer a present question. Anything further is considered **dicta,** language that is not binding on other courts in future disputes.

On the other hand, promulgating rules is one of several functions of an agency. In addition, the rules should apply to the public at large. In fact, a claimant often does not have standing unless the public may benefit from the relief sought.

A point in favor of administrative agencies is that they usually include experts in a particular area of activity. For example, court supervision of the nuclear reactors throughout the United States seems less comforting than that provided by the Nuclear Regulatory Commission (NRC). The NRC includes experts, can investigate issues, and can make and enforce rules for future application. Thus, an agency's value tends to be its ability to expertly handle situations as they arise and, indeed, sometimes before they arise. The point is simply this: Judges are experts in law, and usually not expert in the technicalities of a given kind of activity.

Judge Brown, for example, probably was an attorney for a period of years before being elected or appointed to her present position. She may have specialized within the law, but even so, she might be classed as expert or semi-expert in almost any area of civil or criminal law. By brushing up on recent cases in a given area, she could make intelligent judgments on questions arising in those areas. But now, suppose we place her in a position where a knowledge of nuclear energy or electrical power distribution questions arise. Suppose further that we ask her to supervise activities by companies involved with these questions. By so doing, we are probably asking too much of the judge. A better choice might be Black (a nonlawyer), who has been involved with the field for a substantial period. By choosing Black, we presumably reduce the chances of horrible and costly errors, and increase the probability of sound decisions.

Administrative Abuses

Despite the advantages of having our government operate through agencies, newspaper reports, investigators' comments, legal cases, and comments from friends who have been victims often reveal problems with administrative agencies, frequently manifested in abuses of administrative power and unethical behavior. Of course, not all commissioners and administrative agency officials have questionable ethics. Fortunately, perceived ethical violations can be investigated and the abusers relieved of their duty.

ADMINISTRATIVE AGENCY ACTIVITIES

Administrative agencies are primarily responsible for:

- Holding hearings
- Making rules
- Supervising

In carrying out these functions, the agencies often have broad discretionary powers. If an agency steps beyond the boundaries of these powers as outlined by the statute that created the agency, it may be called upon to engage in a fourth activity—defending itself in court. Each administrative agency has its own version of the Administrative Procedure Act to guide its rule making activities. Title 5 is the section under which the APA was enacted in 1966.[1]

---

[1]Pub. L. 89-554, Sept. 6, 1966, 80 Stat. 378.

Hearings

In acting in their judicial role, agencies generally hold **hearings,** not trials. In doing so, they often follow their own rules. Proceedings before an agency usually involve less formalities than court proceedings. The result of a hearing is a **finding,** which the hearing officer submits to the agency for appropriate action. Most hearings are adjudicative, that is, they involve a plaintiff (often the agency itself) and a defendant. Some, however, are legislative, that is, they are simply a device for collecting evidence prior to making a general rule of some sort. The legislative hearing is appropriately a part of the rule-making procedure.

Adjudicative hearings are conducted (often before an administrative law judge) to determine what happened in some instance or set of instances in question (5 USC §554). Under the adjudicative section of the APA, the agency gives all interested parties an opportunity to submit and have considered facts, arguments, offers of settlement, or proposals of adjustment when time, the nature of the proceeding and the public interest permit. The employee who presides is responsible for making a decision based on these submissions. (See *Appeal of Advance Construction Services, Inc.* at the end of this chapter.)

For example, the Gray Company has been accused by the National Labor Association (NLA—a union) of an unfair labor practice. Specifically, the NLA says that just before the recent representation election, one or more members of management made statements amounting to threats if the NLA became the elected representative of the employees. (Obviously, the NLA lost the election or there would be no hearing.)

The unfair labor practice complaint was made to the National Labor Relations Board, which reacted to the complaint by appointing a hearing officer to investigate. A time and place for the hearing was set, the date came, and a hearing was held to elicit the facts of the case. Witnesses were called, sworn in, examined, and cross-examined. Hearsay may have been allowed, and sometimes, something less than the best evidence may have been used.

The presiding officer (again, this is usually an administrative law judge) sent the results of the hearing on to the NLRB, along with his or her recommendations for action (or inaction). The NLRB is not bound by the presiding officer's recommendation, that is, it may or may not order a new election, regardless of the recommendation it receives. If it does order an election, the Gray Company or the union that won the election may wish to appeal the order. If the NLRB rules against the NLA, it may wish to appeal. Depending on the nature of the question and other circumstances, appeal may be taken to a U.S. district court or a U.S. court of appeals.

The hearing procedure described above is similar to that required by most federal administrative agencies and some state agencies. In a local or state government agency, however, the hearing is likely to take place before the entire agency (that is, before all of the appointed members of the agency) rather than a hearing officer, and the appeal procedure will go through the state or a local court system rather than the federal system.

Rules and Rule Making

Rule making is a component of legislative procedure to the extent that it follows the APA. Administrative agencies are charged with making rules in the public interest to regulate the activities of the entities controlled. (See Appendix F). One problem of the procedure is that sometimes the public interest seems to get lost in the process and the objectives of special interests prevail. In some ways, this is not surprising: The daily operations of administrative agencies bring the agencies in close contact with the businesses they supervise. Thus, it is often the businesses themselves that propose the rules. In such circumstances, one should not expect the resulting regulations to bring excessive hardships to the businesses. A rule so made may very well benefit the business community supervised, more than it does the public.

Rules are usually made pursuant to directives in the statute creating the agency and delegating certain powers to the agency. If a rule is one called for by statute, it is known as a **legislative rule.** Such legislative rules, if adopted by the agency pursuant to the required procedures, have the same force as laws. The other general category of rules is referred to as **interpretive rules**— rules that interpret something. An interpretative rule may interpret a statute, a policy, another rule, a rate schedule, or something else. Interpretive rules generally do not have the full force of law, but they usually are treated with deference by the courts. (See *BP Exploration & Oil, Inc. v. U.S. Department of Transportation and U.S. Coast Guard,* in this chapter).

Procedural Fairness.   Rules could be made in such a way as to be arbitrary and completely unfair to one or more of those supervised or to the public in general. The APA was designed to prevent such unfairness. The APA sets forth general procedures to be followed by an agency in making rules. The primary means of assuring fairness exists in the requirement that agency actions be open to public scrutiny.

In keeping with procedural fairness, agencies usually do not make rules without first giving notice that they are considering such rules. An opportunity is given to those who would be adversely affected to attempt to counter the rule. Generally, an agency publishes a proposed rule in the Federal Register, along with an invitation to interested parties to make written comments.[2]

The agency's staff reviews the comments it receives and then may revise the proposed rules. The finalized rules are then published. (See Appendix F). In many situations, the finalized rules are published with a summary of the comments and, in some cases, a response by the agency's staff as to its views on the issues raised by the comments. Generally, such **finalized rules** become effective 30 days after their publication. Legislative hearings are sometimes scheduled, and those who would be affected are invited to attend. In addition, an agency may publish the proposed rule and make its effective date an extended period of time from publication. Such a publication acts as a **notice.** If no one objects, the rule may then become effective on the published date.

Rule making may have both beneficial and adverse effects. Consider Brown, who is licensed by the FCC to operate a television station reaching a given audience. Brown is very conscientious in his occupation. He has taken numerous surveys of his audience, and they all show a strong preference for old movies and game shows. Suppose that the FCC, in an effort to "clean up" television, passes a rule banning the types of programs Brown televises. Or suppose the FCC rule simply requires all network stations to televise only network programs during the prime viewing times. Such rules might help the television networks to sell greater amounts of advertising, since they would reach larger audiences. But Brown and his audience might well suffer because of the rule. Brown's audience would be depleted, and his ability to sell local advertising would be injured because of his smaller audience. Obviously, it is the smaller businesses that often cannot afford to hire lawyers to monitor proposed rules, provide comments, and lobby the agency.

Ex Post Facto Laws.   The United States Constitution prohibits Congress from passing ex post facto laws. Such laws make an action or event unlawful and set the law's effective date before the law was passed. Accordingly, to be "unlawful," an act must have violated a law that existed when the act occurred. The courts have interpreted this ex post facto provision to apply only to criminal laws, and our courts generally view administrative rules as civil in nature. So Gray, with his

---

[2] 5 USC §553 ("[g]eneral notice of proposed rule making shall be published in the Federal Register . . .").

income tax problem, may feel that his troubles are over when he pays what the IRS requires. But suppose the Internal Revenue Service passes a new regulation two years from now that would require Gray to pay a penalty in addition to the tax he has already paid. Gray may have to pay that additional amount or suffer whatever sanctions are specified in the law. "Final" statements by administrative agencies are not always final.

Supervision

Administrative agencies generally have the right to supervise those entities with which they are concerned. In other words, the agency may find fault with one of those entities and then discipline it for whatever fault was found. Or they may simply require those supervised to **cease and desist** a practice found faulty. The means that agencies use to accomplish their supervisory function vary from one agency to another. The administrative investigation is one commonly used device.

Investigation.   Investigating a person or a company may sound like a perfectly innocent and necessary activity, and it is in many situations. In some hands, however, the investigation turns into something far from innocent. Such an investigation can be very much akin to arresting and trying someone for a crime. The fact that a company or a person is under investigation can cause customers to drop away, lenders to balk, and stockholders to flee. Publicized investigations can become a means of coercion; the threat to "straighten up or we'll investigate you" often has teeth.

Of course, one who is subjected to such coercion may resort to the courts. A person or business may even be successful in getting the agency to drop the investigation. The chances of such success are usually low, however. The court is likely to uphold the agency's investigation if the agency can show a lawful purpose in conducting it. The court may even support the investigation if no reasonable purpose can be shown. After all, agencies exist to supervise activities; how can one effectively supervise without being aware of the facts in a given situation? To learn the relevant facts, the agency will argue, it must investigate. The agency may even have cause to investigate just to make sure everything is operating as it should.

Publicity about an investigation can be harmful, but so can some of the tactics used in the investigation. Consider the discovery procedures we mentioned earlier. An agency can serve a set of written **interrogatories,** or questions, on the subject of an investigation. When used, such interrogatories may ask a number of broad questions. Truthful answers often require the recipient to reveal detailed, private, and sometimes confidential information. Answering a properly phrased interrogatory can be both time-consuming and expensive. Failure to answer may subject the recipient to extreme sanctions. If one must answer many interrogatories phrased in rather broad fashion, he or she may find it necessary to hire a staff simply for this purpose, and of course, accept responsibility for paying the staff salaries. Similar problems arise when agencies subpoena numerous documents.

Discovery procedures by an agency may turn up unexpected (and sometimes trivial) violations that lead to further investigation. Gradually, the company's attention is diverted to the defense of the investigation rather than to the production of whatever goods or services it deals in. The imposition of such a burden by an agency may possibly be necessary and desirable—but it can also be abusive.

Supervisory Variations.   The threat of investigation is one way agencies control those supervised. The use of cease-and-desist orders is another. Such an order usually amounts to an order entered by an administrative law judge that requires the defendant to stop (that is, to cease and desist) actions specified in the order.

But other, more practical techniques exist as well. Certain agencies, the Environmental Protection Agency (EPA), for example, are empowered to levy fines against offenders. Suppose the law gives the EPA the right to assess a fine of $10,000 a day for a given violation. Suppose further that fines of this nature are frequently levied, but that the agency nearly always reduces them to a much lower figure—perhaps in the neighborhood of $1,000—when payment is made. We now have a handy device not only for supervision, but also to keep those supervised from complaining. If a company refuses to pay the assessed fine, a court action is the natural result, with the amount to be paid $10,000 a day, not $1,000.

Another practical means of supervising and avoiding court conflicts is known as the **consent decree.** On discovering a violation, an agency may threaten to take court action against the alleged offender and, at the same time, offer an alternative—a means to avoid the action. All the alleged offender needs to do is to, essentially, admit guilt and agree not to violate the rule again. Since court actions in which an agency is a party are often extended, costly, and easily lost (even if one is innocent), there is a pronounced tendency to negotiate and enter into a consent decree. In this way, the agency can avoid the time and expense of protracted court battles and still create an image of success in doing its supervisory job.

Of course, as noted earlier, with so much power available for supervisory use, there is a possibility of abuse. If the agency itself is not effectively controlled, it may choose arbitrarily which of its supervised entities to investigate and which to leave alone. An agency may well decide to overlook infractions by large companies with numerous governmental and political connections and concentrate its supervision on others.

Advisory Opinions.   Since an agency's sanctions may be extreme, sometimes it is better for a business to ask the agency's advice, in the form of a document called an **advisory opinion,** when a question arises. Suppose Green Trucking Company, for example, has been engaged in a certain practice for an extended period. Then, after discussing the practice with others, Green is led to believe that this practice may violate a particular rule. Would it be better to clear the doubt by asking the advice of the appropriate agency?

Some agencies are willing to give advisory opinions, but others are not. However, even if the agency is willing, at least two significant risks arise in asking for an advisory opinion. First, if Green's past practice is truly a transgression, the inquiry would attract attention to it and may lead to an investigation of the practice. Second, even if Green's inquiry results in a stamp of approval by someone in the agency, advisory opinions generally are not binding on the agency. Even if the agency were bound, it could still issue an interpretative rule holding the questionable practice to be a violation. So perhaps the less said about Green's customary behavior the better. Of course, this leaves Green in the unenviable position of perhaps continuing to violate an existing rule.

AGENCY OVERSIGHT

Courts exercise a limited degree of control over agencies, and legislative committees add a bit more. What an agency may or may not do is spelled out by legislation, and the legislation presumably responds to a public need. Most agencies are created by a legislative enactment; then, as a need becomes apparent, the behavior of agencies is modified by further legislation. To stem allegations related to agency abuses, Congress enacted the "Truth in Regulating Act of 2000"[3] This Act's purposes are to:

1. Increase the transparency of important regulatory decisions;
2. Promote effective congressional oversight to ensure that agency rules fulfill statutory requirements in an efficient, effective, and fair manner; and
3. Increase the accountability of Congress and the agencies to the people they serve.

APPEALING AN AGENCY DECISION

One who is injured usually has a right to take action to recover for the wrong that has been done. For example, an applicant who applies for a license and whose application is denied has an **interest** to be protected and can show injury to that interest. But how about members of the public who might have benefited if the license had been issued? Can they bring an action against the agency? Since actions by agencies affect large segments of the population (at least to a remote degree), the question of standing, or the degree of interest required to bring an action, often arises.

Standing

A few years ago, according to the courts, the people in our license-denial example would not have had sufficient interest to give them **standing** to act. In order to have standing, the issue must be of public interest. Recent cases, however, have revealed a tendency of the courts to lower this barrier. Still, courts are usually careful to avoid "opening the floodgates" of litigation by allowing too large a group to have standing. Simply having standing is not sufficient for the courts to review an agency decision. The issue must be "ripe" for such review.

Ripeness

Ordinarily the harvesting of an agricultural product must wait until that product is ripe. Roughly the same idea applies to court proceedings in which an administrative agency's actions are challenged. The action must be **ripe for review,** that is, there must be a concrete problem for the court to decide before it will evaluate the agency's action (or inaction).

The doctrine of ripeness was designed to "prevent the courts, through avoidance of premature adjudication, from entangling themselves in abstract disagreements over administrative policies, and also to protect the agencies from judicial interference until an administrative decision has been formalized and its effects felt in a concrete way by the challenging parties. The problem is best seen in a twofold aspect, requiring us to evaluate both the fitness of the issues for judicial decision and the hardship to the parties of withholding court consideration.[4] Generally, an agency's action is not ripe for a challenge unless it substantially threatens or injures someone.

---

[3]Pub. L. 106-312, Oct. 17, 2000, 114 Stat. 1248.

[4]*Abbott Laboratories v. Gardner,* 387 U.S. 136, 148-49, 87 S.Ct. 1507, 1515 (1967).

Once standing and "ripeness" thresholds have been crossed, the focus goes toward the scope of judicial review.

Scope of Review

An issue that often arises is whether an agency's action is subject to review by a court. The degree of **reviewability** of an agency action depends upon the laws under which the agency operates. As a general rule, if the legislature has entrusted a regulatory function to the agency, a court has no right to replace the action taken by the agency. Usually, judges are reluctant to decide such cases, anyway, in recognition of the agency's expertise in an area in which the judge is not likely to be an expert. Despite this reluctance, the Administrative Procedure Act permits judicial review to anyone who has suffered a legal wrong due to agency action, as well as anyone who is adversely affected or aggrieved by it within the meaning of the relevant statute (5 USC §702).

Boundaries and limits are either established by statutes or practice, and when an agency exceeds these limits, a call for review of the agency's action may be successful. Some statutes expressly state that certain agency decisions are not subject to review by any court. Challenges based on alleged procedural unfairness often have a better chance for success, assuming that the agency did not follow the required procedures.

One standard basis for review of agency action alleges that the agency action was not supported by substantial evidence. When a court examines this type of challenge to agency action, it is reviewing in its own area of expertise—the use of evidence and drawing inferences from it. The court is not substituting its judgment for the agency's in an area delegated by statute to the agency. Rather, the court is in a position to criticize the manner in which the agency did its job and require the agency to give a better performance.

The court's control of agency actions can only be limited, since a court cannot review those features of agency action delegated exclusively to the agency. Courts are thus left reviewing agency actions to see if they are arbitrary or capricious, or an abuse of discretion. The only remaining legal control of unwise agency action is by the legislative body that created the agency—either by legislative **oversight** committees or by legislative enactment.

In creating agencies we have created controlling devices necessary in our society, but devices that themselves must be controlled to limit or eliminate their abusive practices. The courts offer only a partial answer to agency control. Perhaps the most important control is requiring that agencies act via hearings open to the public and by maintaining a free press to bring injustice to the attention of the voters.

APPEAL OF ADVANCE CONSTRUCTION SERVICES, INC. ASBCA (Armed Services Board of Contract Appeals) No. 55232, Under Contract No. DACW38-03-C-0004

Study terms: Fixed price construction contract, convenience termination, excusable delay, cure period

The captioned contract was to perform construction and improvement to an existing embankment and levee along the Mississippi River in Louisiana. Appellant moves for summary judgment, arguing that the contracting officer (CO) gave appellant from 8 August to 22 September 2005 to show that it could complete performance by the contract completion date. There were nine days of excusable delay during that 45-day period, but the CO did not extend that period and terminated the contract for default on 23 September 2005. Therefore, movant concludes that the termination was procedurally defective and must be converted to a convenience termination. Respondent opposed the motion. Movant submitted a surreply thereto.

Statement of Facts (SOF) for the Purposes of the Motion

1. On 20 December 2002 the U.S. Army Corps of Engineers (COE), Vicksburg, awarded Contract No. DACW38-03-C-0004 (the contract) to Advance Construction Services, Inc. (Advance or appellant) for levee enlargement and berm construction on the west bank of the Mississippi River in Louisiana.

2. The contract included the FAR 52.249-10 DEFAULT (FIXED-PRICE CONSTRUCTION) (APR 1984) clause, which provided in pertinent part:

> (a) If the Contractor refuses or fails to prosecute the work or any separable part, with the diligence that will insure its completion within the time specified in this contract including any extension, or fails to complete the work within this time, the Government may, by written notice to the Contractor, terminate the right to proceed with the work (or the separable part of the work) that has been delayed. . . .

> (b) The Contractor's right to proceed shall not be terminated nor the Contractor charged with damages under this clause, if—

> (1) The delay in completing the work arises from unforeseeable causes beyond the control and without the fault or negligence of the Contractor. Examples of such causes include . . . (x) unusually severe weather . . .; and

> (2) The Contractor, within 10 days from the beginning of any delay (unless extended by the Contracting Officer), notifies the Contracting Officer in writing of the causes of delay. The Contracting Officer shall ascertain the facts and the extent of delay. If, in the judgment of the Contracting Officer, the findings of fact warrant such action, the time for completing the work shall be extended. . . .

This clause did not prescribe any cure or show cause notice.

3. The contract required Advance to complete performance within 450 calendar days after receipt of notice to proceed, but stated that the Exclusion Period "between 1 January and 31 May inclusive . . . has not been considered in computing the time allowed for completion".

4. Advance received notice to proceed on 19 March 2003. Thus, the contract required completion of performance by 22 June 2005.

5. The contract prescribed time extensions for unusually severe weather exceeding 24 calendar days of "anticipated adverse weathers delays" allocated among the months of June through December after the annual Exclusion Period.

6. The COE extended the contract completion date by 145 calendar days for 2003 and 2004 in Modifications A00002 (28 days), A00005 (51 days) and A00006 (59 days) and the CO's 2 December 2005 final decision (7 days).

7. The CO's 21 June 2005 letter to Advance stated that the contractor was approximately 10 percent behind schedule and was not diligently prosecuting the work, its continued lack of progress would prevent it from completing the contract within the required time, and gave Advance 14 days after receiving this letter to cure its delinquent performance.

8. The CO's 12 July 2005 letter to Advance stated that the contractor had not improved its progress and cured the conditions endangering performance and the COE was considering default termination of the contract. The CO gave Advance 10 days after receipt of this letter to present causes beyond its control and without its fault or negligence to excuse its delay.

9. Advance's 1 August 2005 letter to the CO disputed some of the CO's statements, stated that it was reassessing the number of days of excusable delay for unusually severe weather, high river stages and unnatural drainage from adjacent farmland into borrow area 2B that required considerable pumping, and proposed to meet with the COE on 2 August 2005 to review progress.

10. The CO's 4 August 2005 letter to Advance stated that at the parties' 2 August 2005 meeting, Advance had requested 45 days in which to demonstrate its commitment and capability to complete performance timely, and the CO decided to delay her final decision on contract termination and to grant those 45 days to take the steps Advance promised, namely:

> (a) bring a subcontractor on board, (b) augment your spread of production equipment, (c) operate two 10-hours shifts per day seven days a week, (d) satisfactorily place approximately 10,000 cubic yards of levee and berm embankment per day, and (e) either Mr. Bob Najor or Mr. Lew Najor will be on site at all times to personally oversee the construction operations as well as management of job site personnel.

The CO stated that she would closely monitor Advance's efforts, by granting this 45-day opportunity she did not establish a new completion date, and at the end of the 45-day period she intended "to promptly invoke appropriate sanctions provided under the terms of the contract in the event you do not perform in accordance with your plan."

11. The CO's 22 August 2005 letter to Advance stated that "based upon your [8 August] receipt date of our August 4, 2005 letter, the 45[-]day period in which you have been granted an opportunity to perform in accordance with your plan will expire on September 22, 2005."

12. Starting 8 August 2005 Advance brought more earth moving equipment on site and placed embankment. On 29 August 2005 Hurricane Katrina struck. Advance's 2 September 2005 letter to the COE stated that, due to Katrina, Advance had run out of diesel fuel and stopped site operations. The COE's 6 September 2005 letter to Advance stated that days disrupted by Katrina were excusable.

13. COE inspectors reported Advance's progress from 9 to 21 August 2005 was "satisfactory, but could be better with more hauling and support equipment"; start of night shift and equipment breakdowns from 22 to 28 August 2005; Katrina delay for lack of fuel from 2 to 5 September 2005; and reduced equipment in operation and production from 6 to 22 September 2005. . . .

14. On 23 September 2005 the CO issued to Advance: (a) unilateral Modification No. P00017 which extended the contract performance time by 41 calendar days due to unusually severe weather and fuel supply disruption caused by Hurricane Katrina during the period 1 June through 23 September 2005, of which 22-23 and 29-31 August and 2-5 September 2005 occurred during the 45-day period she had granted on 4 August 2005 (R4, tab E-51 at 7-8) and (b) a notice of termination of the contract for default, stating that she had granted Advance "an additional 45 day period in which to cure the condition that is endangering performance of the contract," but it had "failed to prosecute the work with such diligence that would ensure its completion within the time specified" in the contract, "failed to cure the condition that is endangering performance of the contract and . . . failed to show that your non-performance is due to excusable causes". . . .

15. The CO's Determination and Finding accompanying Modification No. P00017 determined that approximately 50% of the contract work had been completed by 23 September 2005 and considering only the 179 calendar days of excusable delay provided in Modification Nos. A00002, A00005, A00006 and P00017, the revised contract completion date was 18 December 2005.

16. The following colloquy occurred in the CO's December 2006 deposition:

> Q. Now, would you agree that between August 9th of 2005 and September 22nd of 2005 that Advance had at least nine more days of excusable delay?"
> A. Yes.
> Q. The cure period was not extended by those nine days, was it?
> A. That's correct.
> Q. It should have been, shouldn't it?
> A. Yes.

DECISION

Summary judgment is appropriate when there is no genuine issue as to any material fact and the moving party is entitled to judgment as a matter of law. FED. R. CIV. P. 56(c); *Anderson v. Liberty Lobby, Inc.*, 477 U.S. 242, 247 (1986).

Respondent contends that that 45-day period was not a "cure period" which it was required to extend for excusable delays. However, the undisputed purpose of the 45-day period was to enable appellant to demonstrate its commitment and capability to complete performance timely, and the CO's termination notice expressly described it as a period in which "to cure the condition that is endangering performance of the contract" and that Advance had "failed to cure the condition". In these circumstances, respondent's contention that the 45-day period was not a "cure" period is not a genuine dispute of material fact, but rather is a legal characterization that is wholly unpersuasive.

We agree with movant that the CO should have extended the cure period as well as the contract completion date for the days delayed by Hurricane Katrina. Indeed, the CO in 2006 testified that the cure period should have been extended for those nine days of excusable delay, and respondent's opposition does not contend to the contrary.

As a general rule, it is improper to terminate a contract for default before the cure period ends. *Fred Schwartz*, ASBCA No. 23183, 80-1 BCA ¶ 14,272 at 70,303. This rule applies when a CO gratuitously allows the contractor a cure period. The rationale for this rule is that the CO must consider a contractor's steps taken to cure delinquent performance during the entire cure period.

However, there are exceptions to this general rule. A default termination before the end of a mandatory or gratuitous cure period is not improper when it is clear that the default has not been cured and could not be cured before the end of the cure period.

The key issue in this motion is, whether the CO's failure to extend the 45-day period deprived Advance of the opportunity to establish a satisfactory rate of progress in placing embankment at the job site so as to ensure contract completion by the amended completion date? When a contractor is so delinquent that its performance in the time remaining in the cure period can make no difference, default termination is proper. But when a contractor has nearly completed performance and remaining work can be performed in a few hours or days, a premature default termination is improper.

The material facts pertinent to this motion lie on some point on the spectrum illustrated by the two foregoing examples. Movant points to a trend towards an improved rate of placement of embankment material from August to September 2005, particularly in the last few days before the termination. Respondent disputes that trend and movant's average placement data. We hold that the material facts with respect to this motion are not undisputed, and hence preclude summary judgment.

We deny the motion for summary judgment.
Dated: 30 May 2007

BP EXPLORATION & OIL, INC. v. U.S. DEPARTMENT OF TRANSPORTATION and U.S. COAST GUARD 44 F. Supp. 2d 34; 1999 U.S. Dist. LEXIS 3880; 48 ERC (BNA) 2014 (D.DC.1999)

Study terms: Abuse of descretion, jurisdiction, "mootness"

This matter is before the Court on the parties' cross-motions for summary judgment. BP Exploration & Oil, Inc. ("BP") challenges the Coast Guard's assessment of a $5,000 penalty against it under the Clean Water Act for the discharge of a harmful quantity of oil into Curtis Bay in Baltimore, Maryland. Because the Court concludes that the Coast Guard's determination that BP is liable for the discharge is supported by substantial evidence in the administrative record and was not an abuse of discretion, it upholds the Coast Guard's decision.

I. Background

BP operates an oil terminal facility on the banks of Curtis Bay in Baltimore, Maryland. BP's terminal includes an Oil Water Separator ("OWS"), a storm-sewer line, storage tanks, and a truck loading rack. The storm-sewer line carries storm water and entrained oil from nearby storage tanks and the truck loading rack to the OWS. The OWS, in turn, treats the water in three stages or compartments to ensure removal of oil and sediment before the water is discharged into Curtis Bay. BP possesses a valid permit under the National Pollution Discharge Elimination System ("NPDES") to discharge the treated storm water into the bay.

During the night of July 27, 1994, the Baltimore area received heavy rain, which flooded BP's storage tank areas. Early the next morning, BP began draining four storage tank areas through its OWS. At the same time, a BP customer spilled approximately 10 gallons of oil on the truck loading rack, which also empties into the OWS. While the storage tanks were draining, the storm water flowed through the OWS at a rate of approximately 1,865 gallons per minute ("gpm"). It is undisputed that the OWS functions most efficiently at a flow rate of 300 gpm or less and that, at that rate, oil will not collect in the third compartment of the OWS and will not discharge into the Bay. As the flow rate increases above 300 gpm, however, the water has less retention time in the OWS.

On July 28, 1994, BP was draining its storage tanks. About 1:40 p.m., the Coast Guard received a call from Star Enterprises, which is located just to the east of BP's facility, reporting that oil and debris had accumulated on the west-northwest side of Star's boom.[5] Coast Guard investigators arrived at Star Enterprises around 3:00 p.m. and found that a patch of oil had in fact accumulated against the western side of Star's boom. The investigators confirmed that the oil had not come from Star and proceeded to BP's facility, where for about thirty minutes BP employees had been using sorbent pads manipulated by long poles to remove oil from within OWS stage three. Coast Guard investigators inspected OWS discharge pipes and determined that intermittent patches of a light sheen were being discharged from stage three into Curtis Bay. The Coast Guard concluded that the sheen was oil that had been discharged.

---

[5]A "boom" is a temporary floating barrier used to contain oil spills.

On February 21, 1995, the Coast Guard notified BP that the agency was initiating civil proceedings against BP for discharging oil into Curtis Bay in violation of the Clean Water Act. On September 25, 1995, the Coast Guard held a hearing, at which BP argued that the discharge resulted from "operator error" when BP employees disturbed already-separated oil in the OWS during their attempt to clean it up with sorbent pads. Although the hearing officer found that BP's clean-up attempts did contribute to the problem, he also concluded that the increased flow rate caused the oil spill and therefore assessed a Class I administrative penalty of $5,000. BP appealed the hearing officer's decision to the Coast Guard Commandant, who found through his designee that "waste oil in a quantity that may be harmful was discharged from an oily [sic] water separator (OWS), causing a sheen on the water," in violation of the Clean Water Act, and that the statutory exemption on which BP relied did not apply because "the record [did] not indicate that BP's OWS at the facility was designed to, nor was it capable of, processing the spilled oil and rainwater." BP submitted a $5,000 check in payment of the penalty and filed this action seeking a refund.

II. Discussion

A. Jurisdiction and Mootness

The Coast Guard asserts that the Court lacks jurisdiction because the Clean Water Act has no refund provision and BP has failed to identify any waiver of sovereign immunity that would permit its refund suit. The Administrative Procedure Act, however, provides a mechanism for review of a final agency action. Both agency actions made reviewable by statute and final agency actions for which there is no other adequate remedy are subject to judicial review. See 5 U.S.C. § 704. While the APA may not allow an action at law for damages against the United States without a specific waiver of sovereign immunity, it does authorize "an equitable action for specific relief," which may in some cases include "the recovery of specific property or monies . . ." *Bowen v. Massachusetts*, 487 U.S. 879, 893, 101 L. Ed. 2d 749, 108 S. Ct. 2722 (1988).

In this case, the Commandant stated that his decision to impose a Class I penalty in the amount of $5,000 "constitutes final agency action," . . . and BP subsequently filed a notice of appeal in this Court as directed by the statute. BP therefore is entitled to a review of the Coast Guard's final determination under the APA. In addition, the relevant statute, the Clean Water Act, expressly permits "any person against whom a civil penalty is assessed" to obtain review of such assessment in the United States District Court for the District of Columbia or in any district in which the violation allegedly occurred. BP properly followed the requirements of the statute in order to obtain review of the Coast Guard's administrative decision, and this Court therefore has jurisdiction.

The Coast Guard also suggests that BP's action is moot because BP already has paid its $5,000 penalty. This is a specious argument. BP was under a statutory and regulatory duty to pay the fine immediately, see 33 U.S.C. § 1321(b)(6)(H); 33 C.F.R. § 1.07-85(b), but that does not preclude it from taking steps to seek judicial review and a refund. Indeed, as already noted, the Clean Water Act expressly provides for judicial review of the assessment of a Class I penalty. See 33 U.S.C. § 1321(b)(6)(G). Whether BP is entitled to a refund is a very real issue that directly affects BP's rights, as is the issue of the binding nature of the Marine Safety Manual. The case therefore is not moot.

B. Challenges to the Coast Guard's Decision

1. Marine Safety Manual

BP first argues that the Coast Guard's reliance on the Marine Safety Manual both in its administrative decision and in this Court are impermissible because the Manual is an invalid rule or regulation that was not promulgated in accordance with the notice and comment requirements of the Administrative Procedure Act. While Section 553 of the APA expressly requires agencies to afford notice of a proposed rule-making and an opportunity for public comment prior to promulgating a substantive rule, 5 U.S.C. § 553(b), (c), the issuance of interpretive rules and policy statements and guidance is not so constrained. Although the distinction between a substantive rule and an interpretive one is not always clear, an interpretive rule is usually non-binding, instructional and explanatory, whereas a substantive rule grants rights, imposes obligations or effects a change in existing law. "An agency pronouncement is not deemed a binding regulation merely because it may have 'some substantive impact,' as long as it 'leave[s] the administrator free to exercise his informed discretion.'" *Brock v. Cathedral Bluffs Shale Oil Co.,* 254 U.S. App. D.C. 242, 796 F.2d 533, 537 (D.C. Cir. 1986).

An agency must act consistently with its own pronouncements, procedures and policies only when "the agency intended to establish a 'substantive' rule, one which is not merely interpretative but which creates or modifies rights that can be enforced against the agency." *National Latino Media Coalition v. FCC,* 259 U.S. App. D.C. 481, 816 F.2d 785, 788 n.2 (D.C. Cir.1987). In the absence of an intent to be bound to a particular legal position, "the agency remains free in any particular case to diverge from whatever outcome the policy statement or interpretative rule might suggest."

There is no indication that the Marine Safety Manual is binding, or was intended to be binding, upon the Coast Guard. The Manual provides in the first chapter that while it is "the primary policy and procedure statement" for the marine safety programs of the Coast Guard, the Manual should only be used as "a guide for consistent and uniform administration of marine safety activities, without undue hampering of independent action and judgment by marine safety personnel." U.S. COAST GUARD MARINE SAFETY MANUAL ("Manual"). It further provides that in any case of apparent conflict between provisions of the Manual and any statute or regulation, "the legal requirements shall be observed." Neither the Coast Guard hearing officer nor the Commandant therefore was free to disregard the law or regulations and neither cited to the Manual as authority for his decision or suggested that he felt bound by it in deciding this case. Indeed, the hearing officer expressly mentioned the Manual only once and then stated that he was not bound by its "guidance" in determining the appropriate penalty. Because the Coast Guard has not demonstrated any intent to be bound by the Manual and quite clearly is not bound, the Court finds that the Coast guard was not required to subject the Manual to the notice and comment procedures of the APA before issuing it.

2. Decision of the Agency

BP argues that the Coast Guard's decision to penalize BP was not based on substantial evidence in the record and that it was an abuse of discretion because the hearing officer failed to consider certain evidence related to the flow rate. BP has properly framed the issue under both the Clean Water Act and the APA. Congress has provided that a court cannot overturn a decision imposing a civil penalty under the Clean Water Act "unless there is not substantial evidence in the record, taken as a whole, to support the finding of a violation or unless the Administrator's or the Secretary's assessment of the penalty constitutes an abuse of discretion." 33 U.S.C. § 1321(b)(6)(G). Both of these standards—"substantial evidence" and "abuse of discretion"—mirror those found in the Administrative Procedure Act, 5 U.S.C. § 706(A)(2), and should be interpreted and applied in the same way that they are interpreted and applied under the APA. . . .

The primary issue before the Coast Guard in this case was whether the discharge was allowable under the Clean Water Act because it was "caused by events occurring within the scope of relevant operating or treatment systems." 33 U.S.C. § 1321(a)(2). "Operator error" is one such excusable cause of an unpermitted discharge. According to BP, its OWS was capable of processing the water at flow rates of 1,865 gpm, the rate of flow on the day of the discharge, so there must have been a reason other than flow rate why the system failed and oil was discharged. BP cites the opinion of Mr. C.T. Cinko, a Coast Guard investigator, to suggest that oil will not discharge until the flow rate reaches 15,932 gpm and argues that the Coast Guard abused its discretion by not even considering this evidence. Primarily on the basis of this allegedly excluded evidence, BP maintains that the cause of the discharge must have been operator error in manipulating the poles and sorbent pads used to remove oil from stage three of the OWS, not from the design of the OWS or its ability to process the increased flow rate on the day in question.

Contrary to BP's argument, the Coast Guard considered the relevant evidence and simply rejected BP's position. Based on the statements of BP's own oil manager that the discharge occurred during the draining of the dikes, the Commandant found that the increased flow rate—at least six times the agreed upon normal flow rate of

300 gpm—caused the oil discharge into Curtis Bay. Even BP's in-house expert would not state with absolute certainty that the increased flow rate could not have caused the discharge.

As for Mr. Cinko, the Coast Guard official on whom BP relies, he opined that the 1,865 flow rate did not include the initial surge rate and that this initial surge was likely a major cause of the discharge. In addition, as the Coast Guard points out, Mr. Cinko was not addressing the rate of oil already entrained in the storm water but rather was discussing the possibility that already-separated oil that had settled to the bottom of the OWS would be taken back up and discharged. Specifically, Mr. Cinko concluded that "the sediment in the [third] compartment would have been flushed out with or without the disturbing of the pea gravel [on the bottom of the third compartment] due to the fact of the buildup of the sediment up to the discharge pipes. . . ."

In addition, the Coast Guard found that BP's alternative interpretation of the events of July 28, 1994 was inconsistent with the temporal evidence. Star Enterprises reported the patch of oil against its loading pier at 1:40 p.m. BP did not begin its attempt to clean the OWS until 2:30 p.m. The Coast Guard therefore concluded that the clean-up efforts did not cause the discharge of oil because the oil was observed *before* any clean-up began. Because the cumulative evidence in the record and considered by the Coast Guard more than rises to the level of "substantial evidence," and because there is no indication that the Coast Guard abused its discretion in evaluating it, the Court will not disturb the Coast Guard's finding of liability.

Nor was it an abuse of discretion for the Coast Guard to impose a $5,000 penalty in this case. The $5,000 penalty was half the amount allowable by statute for a Class I violation. The Coast Guard hearing officer determined that $5,000 was justified in light of a similar incident involving the BP facility at Curtis Bay that occurred less than five months before the July 28, 1994 incident, a permissible consideration under the statute. There was also evidence that BP's boom was placed on the shoreline and not in the water to contain the discharge, thereby heightening BP's culpability. Because the amount assessed is below the statutory maximum permitted and appears to be fair in light of the facts in the record, the penalty imposed by the Coast Guard was within its discretion.

An Order and Judgment consistent with this Opinion will issue this same day.

Review Questions

1. Why must we have agencies? Why, for example, must we have an agency to control air traffic? Is there a better way to regulate air traffic?

2. What is the *Federal Register?*

3. How are agency hearings similar to and different from court trials?

4. How does rule making differ from a legislative procedure?

5. Who controls agency activities? Is the control adequate?

6. FAR 52-249-1 states that the contractor's right to perform the work will not be terminated if a delay arises from unforeseeable causes beyond the control of the contractor, including severe weather conditions. Why do you think the CO in *Advance Construction Services, Inc.* terminated Advance's contract even though there were excusable delays, including Hurricane Katrina?

7. According to the court in *BP Exploration & Oil, Inc. v. U.S. Department of Transportation,* the U.S. Coast Guard Marine Safety Manual was not required to undergo notice and comment procedures for what reason?

Labor

HIGHLIGHTS

■ The right to peaceably assemble is a right guaranteed by the First Amendment to the Constitution.

■ The right to form unions is an extension of First Amendment rights, and, through those rights, historically unavailable privileges and protections have been granted to employees.

■ The National Labor Relations Act was created to encourage peaceful resolution of labor-management problems by formulating guidelines to that end.

A necessary component of everything we buy or sell is labor. According to one view, what we really pay for when we buy something is labor in one form or another. Capital, it is said, merely represents stored compensation for past labor; management, too, is just another form of labor. Thus, labor is a vital, and sometimes volatile, portion of all projects. For this reason, we will consider here some of the problems posed when people's services are bought and sold.

In this discussion, problems such as the cost, availability; and transportation of labor are not addressed; instead, this chapter addresses some of the legal problems involved in labor conflicts. The discussion will center around unions because without unions, or the threat of them, there is usually little conflict. Two powers must be fairly evenly matched for a conflict to occur. A single worker usually cannot successfully dispute a management edict, however wrong it may seem. By leaving, the employee can deprive a company of his or her services but, in most cases, others can be hired to do the job as well.

The laborer's answer to this rather unbalanced state of affairs is organization. A concerted refusal to work, in which all employees participate, can be a powerful weapon. To accomplish such organized activity, workers formed unions. Occasionally, the balance of power shifts very heavily to the union side. When this occurs, management cannot afford the conflict that might be caused by refusing a union request, especially a refusal likely to promote action that could destroy the company.

HISTORICAL BACKGROUND

"Congress shall make no law respecting an establishment of religion, or prohibiting the free exercise thereof; or abridging the freedom of speech, or of the press; or the right of the people peaceably to assemble, and to petition the Government for a redress of grievances." United States Constitution, First Amendment.

Working people now have the right to form unions. It can be argued that this right is constitutional. Through unions, they may petition their employer for improvements in wages or working conditions. Under certain circumstances, employees may resort to strikes, to enforce their demands. Although strikes are generally lawful today, such was not always the case.

Criminal Conspiracy

Under the English common law existing at the time of the American Revolution, the formation of an association of employees engaged in similar employment was unlawful. To a considerable extent, the judge made common law reflect the economic beliefs of the time—the law of supply and demand and the wage-fund theory, for example. Reasoning from basic notions of supply and demand, the courts held that it was unlawful to "artificially" regulate the price of labor. Such decisions viewed the only fair and lawful price as that established by competition for a particular type of labor. Moreover, the **wage-fund theory** essentially held that a worker could only improve his wages or working conditions by stealing from another worker. Since organizations of working people clearly intended to improve the lives of their members, such organizations were viewed as unlawful **criminal conspiracies.**

Over the years, the courts gradually came to recognize the right to organize. In the meantime, organizations of employees made some improvements in the working conditions and wages of their members. The demise of the criminal-conspiracy doctrine in its application to labor unions came about during the second half of the nineteenth century. This demise also brought with it a rise of union activities. New unions were formed, and attempts were made to organize on a national scale. Gradually, the use of economic force to bring about improvements in working conditions began to win recognition in the courts.

Consider this comment: "The right of voluntary association is, therefore, a natural right. It is an endowment of man's nature, not a privilege conferred by civil society. It arises out of his deepest needs, is an indispensable means to reasonable life and normal self-development . . . the State has no right to prohibit any individual action, be it ever so unnecessary which is, from the public point of view, harmless."[1]

---

[1]James A. Ryan, Right of Voluntary Association, www.newadvent.org.

Antitrust Laws

In 1890, the Sherman Antitrust Act was passed with the apparent purpose of preventing monopolistic practices of "big business."

However, the act's application to unions was unclear. In an early case involving a union boycott, the United States Supreme Court held that the boycott was a violation of the antitrust laws, and the union was fined triple damages. Subsequent use of the antitrust laws, and the threat of their further use, stifled unions.

Other Union Constraints

By this time, employers came to rely on two legal devices for minimizing union activity. The **yellow-dog** contract prevented employees from joining a union or from forming a union among themselves, and the **labor injunction** could be used to prevent anticipated damage from threatened union activities.

Yellow-Dog Contract.   The yellow-dog contract was a clause in an employment contract stating, in effect, that the employee agreed, as a condition of continued employment, not to engage in union activities. Under contract law, the courts generally enforced such agreements. If a union tried to persuade employees to become members, it was inducing the employees to breach their contracts, and was subject to either an injunction to prevent the activity or a claim for damages if it was successful. Moreover, the employees were terminated if they breached their contracts.

Labor Injunction.   Generally, anyone who is threatened with harm to his or her property has a right to request an injunction from a court having equity powers. This right, of course, applies to employers, too. In the case of employers, not only are their physical assets, such as the plants and machines, considered their property; the concept extends to an employer's good will in relations with customers and the public as well. Thus, for example, delayed filling of customers' orders could be such an injury. Therefore, an employer who was threatened with a strike or other economic action could ask a court for an injunction to prohibit the strikes or to stop the union from picketing. Courts responded to these requests quite freely with temporary injunctions or **labor injunctions,** as they came to be known. Some courts even went so far as to direct the injunctions against "whomsoever" might harm the employer.

Union-Supportive Legislation

Unions saw the courts as a friend of management. Hence, the unions began to lobby for legislation to curtail abusive legal tactics by management. The Clayton Act, passed in 1914, appeared to be the answer. Many advocates viewed the Clayton Act as the key to eliminating injunctions against unions under the antitrust laws. Its wording seemed clear. However, the Supreme Court interpreted the Clayton Act as applying only to strikes and picketing by employees against their employer. In other words, a court could still issue an injunction against non-employees who picketed. Thus, the courts were still able to issue injunctions against secondary economic pressure or boycotts (discussed later).

In 1932, Congress passed the Norris-LaGuardia "Anti-Injunction" Act. The Norris-LaGuardia Act eliminated nearly all labor injunctions by federal courts. Employee organizations could now strike and picket much more freely. Essentially, the balance of legal power had shifted toward the unions.

National Labor Relations Act

Congress went even further in 1935, with the passage of the National Labor Relations Act (NLRA) (29 USCA §151 et seq.), sometimes called the Wagner Act. The only unfair labor practices (i.e., unlawful practices) spelled out in the act were acts of employers. The Wagner Act set up the National Labor Relations Board (NLRB) to determine whether the federal law should be applied to a particular case, to hold elections to certify employee representatives, and to investigate and prevent unfair labor practices.

With the new protections afforded by the Norris-LaGuardia and Wagner Acts and the sympathetic attitude of government and the courts, the labor movement thrived. Old unions expanded, and new ones were born. Collective bargaining and enduring labor-management contracts became almost a way of industrial life.

Taft-Hartley Act

Unions became unpopular in the period right after World War II, in part because they had called strikes in important industries. In an effort to more evenly balance the parties' respective bargaining powers and balance the power back toward management, Congress passed the Taft-Hartley Act in 1947.

The Taft-Hartley Act prohibited certain union actions as unfair labor practices. Among other things, Taft-Hartley (and later NLRB rulings) made unlawful the creation of a closed shop arrangement with an employer—an arrangement in which a new employee had to be a union member before he or she could be employed. Moreover, the secondary boycott, jurisdictional strikes, and featherbedding were all made illegal. Furthermore, a 60-day notice of intent to strike was required in most cases before a union could lawfully strike. National emergency procedures were also set up in an attempt to avoid national disasters and still preserve free collective bargaining. A federal mediation and conciliation service was created to aid in reaching collective bargaining settlements. Finally, the act made unions responsible for their torts and allowed them to sue and be sued in court actions.

It is an understatement to note that unions objected to the new legislation; one of the more complimentary descriptions for it was the "slave labor act." Despite union objections to the act, unions continued to expand during the period following its passage. They grew and combined, becoming nationwide and even international in their affiliations. The American Federation of Labor (an association of trade unions) and the Congress of Industrial Organizations (built along industry lines) merged to present a united front.

Unions themselves became big business. With the increased size came certain abusive practices. For example, people who handle money belonging to others are often tempted to divert some of it to personal gain, disguising the diversion or obscuring it by simply not keeping records. Here, accumulating union dues, initiation fees, and assessments provided such a temptation. A senate investigating committee, headed by Senator McClellan, uncovered proof of such practices in some unions and strong suspicions of them in others. In at least one union, many officers were found to have long prison records. The work of the McClellan Committee was a main cause of passage of a second National Labor Relations Act amendment, the Labor-Management Reporting and Disclosure Act of 1959, known as the Landrum-Griffin Act.

Landrum-Griffin Act

While the main purpose of the Landrum-Griffin Act was to cause employers, labor organizations, and their officials to adhere to high standards of responsibility and ethical conduct, it also amended many sections of the Taft-Hartley Act. It imposed, for example, a $10,000 fine or imprisonment up to one year, or both, on those who did not comply. Enforcement of the Taft-Hartley Act, on the other hand, had been achieved mainly through cease-and-desist orders and contempt of court for noncompliance. In addition, the Landrum-Griffin Act restricted the ability to picket for purposes of recognition or organization.

CONTEMPORARY LABOR RELATIONS LAW

Today, the main source of law governing the relationships between labor and management is the NLRA. A secondary source is the Labor Management Relations Act (LMRA), formerly known as the Taft-Hartley Act. Of course, other federal acts such as the Fair Labor Standards Act, the Walsh-Healey Public Contracts Act, the Davis-Bacon Act, and the Contract Work Hours and Safety Standards Act address labor relations as well. Moreover, in the last forty years, employment discrimination based on race, sex, age, and disabilities has been made unlawful. Numerous state and local laws also have a bearing on the employment relationship. However, in the event of conflict between state law and the NLRA, the NLRA preempts such state regulatory laws that govern conduct pertaining to labor relations. The two leading cases outlining federal preemption as it relations to the NLRA are *San Diego Building Trades Council v. Garmon*[2] and *Machinists v. Wisconsin Employment Relations Commission.*[3] The central law of labor relations remains the National Labor Relations Act. (See *Baldwin, et al. v. Pirelli Armstrong Tire Corp.* in this chapter).

Over the years, decisions by the courts and the NLRB have refined and interpreted the act. Many of today's controversies are decided upon the basis of the "settled law" of that particular area of labor relations. But, then, with technological change and the seesaw bargaining of union-management contracts, new problems continue to arise. Thus, although the general rules are fairly well settled, the law continues to evolve.

Purpose, Policy, and Jurisdiction

The avowed purpose and policy of the National Labor Relations Act is to minimize industrial strife and thereby promote the full flow of commerce. The means chosen to accomplish this purpose was to define certain rights of employees, rights of employers, and orderly procedures for settling disputes between them.

Employees are guaranteed the right to organize and to bargain collectively with their employers. They are also guaranteed the right to refrain from such activities.

The agency charged with resolving conflicts involving such organizing and collective bargaining is the NLRB. It functions to prevent and remedy unfair labor practices and to conduct

---

[2]359 U.S. 236, 3 L. Ed. 2d 775, 79 S. Ct. 773 (1959).

[3]427 U.S. 132, 49 L. Ed. 2d 396, 96 S. Ct. 2548 (1976)

secret-ballot elections to select employees' bargaining representatives. In doing so, it engages in a wide variety of administrative procedures and practices. It must receive petitions and allegations, determine its own jurisdiction, investigate, hold hearings, issue orders, and make rules, and it often becomes a party to controversies in the federal courts. In short, it functions in much the same manner as do most federal administrative agencies. (See Chapter 23.)

The NLRB decides whether it has authority over a matter when it responds to petitions either claiming an unfair labor practice or requesting an election. In fact, its first order of business in response to such a petition is to decide whether it has jurisdiction in the case. According to its rules, the employer in the petition must be involved in interstate commerce for the NLRB to take the case, but involvement with interstate commerce does not necessarily mean that the NLRB must assert jurisdiction. Usually, the NLRB will not assert its jurisdiction unless the employer's volume of business exceeds set dollar limits for a particular industry. If the NLRB refuses to take jurisdiction, the petitioner may be able to turn to a state act for relief.

Employee Representation

In industries affecting interstate commerce, certification by the NLRB as the bargaining agent is largely a matter of winning an election. Section 9 of the Taft-Hartley Act (as amended by the Landrum-Griffin Act) describes the procedure. The following material gives an overview of that procedure.

Petition.   The process begins with a petition to the NLRB for an election to be held. The petition may come from an employee, a labor organization, or the employer.

Investigation.   On receipt of the petition, the NLRB must determine whether interstate commerce is affected and determine the appropriate bargaining unit. Various types of bargaining units are used to represent the employees involved. The unit might be an employer unit, that is, it would consist of all the employees of a particular employer or a substantial portion of them. The unit also could be composed of the employees of one plant or of one department within it. Alternatively, the choice might be a craft unit, if the employees involved in a particular trade or craft desire such representation. A group often chooses the unit that it believes will generate a majority vote. If the employer challenges the union's choice of a unit, review by a hearing officer, and possibly the courts, may occur. The bargaining-unit determination should be made in such a manner as to assure employees the fullest freedom in exercising their rights.

In determining bargaining units, special consideration is given to two groups of employees: professional employees and guards. Professional employees (such as engineers, accountants, buyers, and the like) are not to be included in a bargaining unit with other employees unless the majority of the professional group votes for inclusion. Guards employed for protection of company property cannot be included in the same bargaining unit with other employees of the company under any circumstances. Generally, the guards' labor organization may not even be affiliated with unions representing other employees. Combining representation could partially defeat the guards' functions.

Elections.   The NLRB regulates elections to determine who will be the employees' bargaining representative. If the only choice is between a particular union and no union at all, a simple majority of the votes determines the outcome. If the union fails to win more than half, the results of the vote will be certified as showing that the union was not the choice of a majority of the voting employees. If more than two choices exist and none receives more than half the votes cast, a runoff is held between the two choices receiving the most votes. Once a union has been selected and certified, that choice of a bargaining representative settles the issue for at least a year; after a year, certification can be removed by petition and another election.

Who May Vote.   Each employee in the appropriate bargaining unit is entitled to one vote to determine the bargaining representative. Shortly after passage of the Taft-Hartley Act, the question arose as to whether employees on strike could vote in an election. Rulings on this question proved difficult to follow in some situations, and the law was clarified in the Landrum-Griffin Act. Employees on a valid economic strike against their employer retain the right to vote in an election for a year after the beginning of the strike. Seasonal or temporary employees generally are not eligible to vote unless they have a legitimate expectation of being reemployed and an interest in the working conditions.

Campaigning.   In ensuring the free election of a bargaining representative, the NLRB is often concerned with preelection conduct by both the employer and the union. Neither the employer nor a union may use coercion or threats of reprisal to attempt to influence an employee's vote. The employer cannot promise benefits to be given if the union is unsuccessful. Furthermore, the employer must be particularly careful under NLRB rulings during the 24 hours directly preceding the election. In short, the employer cannot do anything to create an atmosphere that would discourage an election.

The limitations imposed on the employer, however, are not intended to inhibit the normal conduct of business. The employer can still hire new employees and discipline employees, even to the point of firing an employee for cause. The employer can speak to the employees on company time, or on their own time if attendance is voluntary. He or she can voice an opinion of unions or of the result to be expected from joining a union as long as threats or promises of benefit could not be implied from such statements. The employer can state the company's legal position, the dangers and costs of union membership, and what the union can and cannot do for employees.

In the context of a campaign, it is useful to consider just who is an employer. The term employer applies to the management of a company. This usually includes not only the president and general manager, but generally all salaried line or staff employees outside the bargaining units. Thus, what a foreman, an engineer, or a production control clerk says or does prior to an election may be interpreted as the words or deeds of the employer.

The Contract

The contract between the company and the bargaining representative of the company's employees is not an employment contract; no one holds a job by virtue of it. Rather, it is an agreement that sets forth the conditions under which those who are employed will work.

It has become almost standard practice to include certain provisions of importance in a **collective bargaining agreement,** or CBA. Besides addressing wages and items such as sick days, holidays, and vacations, for example, it usually contains a statement to the effect that the union agrees not to strike and the company agrees not to lock the union out during the life of the agreement. Most agreements set up a step-wise grievance procedure—an orderly process for settling disputes without resort to coercive tactics or to the courts.

If, for example, the paint gang at Black Manufacturing believes a newly issued incentive rate is unfair, its recourse is to petition for a change pursuant to the agreement, and follow the route of the grievance procedure. A strike by the paint gang as a result of the rate issuance would constitute an unlawful wildcat strike. Such a strike would probably allow Black Manufacturing to permanently replace any or all of the paint gang members. However, if the paint gang members are allowed to return to work, they come back with the same rights they would have had if they had not engaged in the wildcat strike.

In a similar vein, Black Manufacturing must bargain with its employees; it cannot lock them out. If, for example, Black's bargaining sessions with its employees' representative reveals what Black considers to be unreasonable demands, Black must still bargain. Black may not even be able to go out of business to avoid bargaining. For example, if Black is a multi-plant organization, it could not go out of business at, say, Dayton, where it has labor problems and move its Dayton operations to one of its other plants. To lawfully go out of business at Dayton to avoid the union, it would have to go completely out of business—at all its plants. Otherwise, its behavior constitutes **an unlawful lockout**—an unfair labor practice. Still, the duty to bargain is not a duty to agree to terms perceived as unreasonable.

UNFAIR LABOR PRACTICES

Section 8 of the NLRA lists a number of unfair labor practices. It is, for example, an unfair labor practice for either a union or a company to restrain or coerce employees as the employees exercise the right to form or join labor organizations or to choose not to join. It is also an unfair labor practice for the union and the employer to refuse to bargain with the other. In this regard, it is noteworthy that neither party is required to accept any proposal by the other or to agree to any concession. Bargaining simply means meeting at reasonable times and conferring in good faith on employment-related subjects. Other unfair labor practices are elucidated below.

Closed Shops

The **closed shop** is unlawful. This is an arrangement in which a new employee must be a union member before being employed; thus, in a closed shop, the union does the selection of new employees. In 1947, the Taft-Hartley amendment replaced the closed shop with the union shop, in which new employees are selected by the employer, but may be required to become union members after a probationary period of at least 30 days (seven days in construction). This means that employees must pay union initiation fees and dues; in fact, the only reason the union may require their termination is for nonpayment of these fees and dues.

Another lawful arrangement is the **agency shop,** in which employees are not required to be union members, but must pay a "service fee" (usually equivalent to union dues) to the union for union efforts in their behalf. Nonpayment of this fee is the only reason for the union to require their discharge.

A third arrangement is the open shop, in which dues-paying union members may work side by side with nonunion and nondues-paying employees. A manufacturing plant might lawfully operate with a union shop security agreement, an agency shop, or an open shop, but not a closed shop. The union might lawfully require an employee to be terminated for nonpayment of union fees in a union shop or agency shop, but for no other reason. On the other hand, coercive action by a manufacturing plant against one or more employees could be cause for union-employer controversy.

Secondary Boycotts

In this country and elsewhere in the world, people may choose freely between dealing with one competitor or another. If, for example, the Black Company chooses to buy a die-cast machine from Gray Enterprises rather than from Green, it is Black's choice. In fact, if Black wishes to deal exclusively with Gray rather than Green, that is Black's prerogative. The problem arises when someone else forces Black to make a particular decision. The situation is shown graphically in Figure 24.1. If Union A wishes to receive some concession from Management C, A may bring added coercive force against Management C by forcing Customer B to refuse to deal with Management C. This is known as a **secondary boycott,** and it takes many forms. Such conduct is secondary in the sense that it is directed not at management C (the primary target of the union), but at someone else.

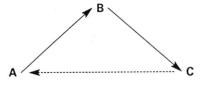

Figure 24.1

Under the NLRA, it is an unfair labor practice for a union to engage in or induce its members to strike where the objective is to force one company to cease dealing with another.[4] However, not all conduct that may influence a second company to curtail business with the employer amounts to a secondary boycott. For example, picketing directed at the target employer is probably acceptable.

It is an unfair labor practice for a union to force or threaten to force a company to assign work to members of one trade union rather than another (a **jurisdictional dispute**). For example, a contractor might be required to assign the job of hanging metal doors to metal workers rather than carpenters, or the reverse.

Hot Cargo. It is an unfair labor practice to force or threaten to force an employer to refuse to handle the products of some other company. A union-employer agreement of this nature is known as a **hot cargo agreement** and is unlawful for both parties. Such agreements were sometimes made as an attempt to get around the restrictions on secondary boycotts. Specifically, unions tried to negotiate agreements with the employers to the effect that the employees could not be required to handle or transport products from an "unfair" company.

Suppose, for example, that Black Company usually buys products from White Manufacturing. A strike against White occurs. If a hot cargo agreement exists, the union at Black may insist on enforcing Black's agreement to stop dealing with White to help force a favorable settlement with White. Here, the union could argue that its dispute with Black was "primary" (and therefore not a secondary boycott) because it was an attempt to enforce its contract with Black. Such agreements are unfair labor practices under the NLRA.

Allies. Sometimes, what appears to be a secondary boycott may not really be one. For example, a union may have a legitimate strike against Black, but it chooses to picket White, who is Black's main customer. This sounds like a secondary boycott against White. But suppose Black and White are owned and controlled by a common parent company. Or suppose they have formed a pact to the effect that they will assist each other in opposing union pressures. What if their products form an integrated link in a production network? In short, such situations or acts would make it apparent that White is no longer a "neutral." In such circumstances, the union usually has a right to act against White as Black's ally.

---

[4] 29 USCA §158(b)(4)(ii)(B).

Common Site.   Another problem involving secondary pressure may arise when a company hires a contractor to do work on its premises. Suppose this contractor has labor problems while the contracted work is being done. The union's primary and, presumably, legitimate target is the contractor, but the company for which the work is being done may well get splattered in the fray too. The common-site situation occurs frequently enough so that special rules exist to deal with it. The rules are intended to minimize injury to the second company without depriving the contractor's employees of their rights. If the contractor's employees have no way to exercise their rights without injuring the second company, that company may well expect to feel some pressure.

Subcontracting.   What about subcontractors? Accepting subcontract work is legal under most circumstances. However, if a subcontractor enters into an agreement with company on the verge of or embroiled in an employee strike, it may find lawful picketers at its site.

Lawful Union Coercion.   Of course, in certain situations, a union has a right to take **coercive action** against an employer, and in certain situations, what seems like a secondary boycott is really a lawful primary action. However, for a union's coercive action to be lawful, it must meet three tests:

1. The objective sought by the union must be lawful (the objective of forming a new closed shop, for example, would not be lawful).
2. The means used must be lawful (use of the secondary boycott, for example, would not stand this test).
3. The union must be the legitimate bargaining agent of the employees involved (for example, picketing by a union that had lost a representation election at the picketed plant would be unlawful).

Suppose the Black Company has a labor-management contract with the representative of its employees, the UWW (the United Wood Workers). Sixty days before the contract is due to end, Black receives a strike notice from the UWW. This does not necessarily mean that Black will have a strike on its hands in two months. Instead, it indicates that the union will enter negotiations for a new contract with a weapon it has a right to use when the current contract expires—the strike. The union also may picket Black's place of business and attempt to persuade employees of other employers to refuse to cross the picket line. In addition to the strike and such picketing, the UWW could use other means of informing the public of the strike against Black, and attempt to persuade members of the public not to do business with Black.

STRIKES AND PICKETING

The NLRA specifically preserves a union's right to strike as a "permissible economic weapon," but it limits this right. (See *Katz v. Westlawn Cemetery Association, Inc.* in this chapter). Generally, the limitations depend on the objective sought, the means used, and the existence of a labor-management contract with a no-strike provision.

If a no-strike provision in a contract is currently in force, the employees may still strike, but doing so may be a basis for discharge. Sometimes, the right to strike is contingent upon the presence of an immediate hazard to the employees. If a real danger threatens—for example, the assigning of a welding job near a paint booth—union members may not have time to resort to the grievance procedure. Under such a circumstance, it would be unreasonable to require the employees to continue to work.

Strikes

If the union has no contract provision barring it from a strike or if the contract time has run out, a strike may be quite lawful. In such a case, the strikers' rights depend to a considerable extent upon the objectives sought. An economic strike, for example, is one in which the objective sought is an increase in wages, improvement of working conditions, a change in overtime policy, or some other such concessions from the employer. In contrast, an unfair labor practice strike occurs in response to an unlawful labor practice by the employer.[5] Economic strikers are still employees, but they may be replaced by their employer. If the employer has hired permanent replacements by the time the economic strikers indicate a desire to return to work, the strikers are not entitled to reinstatement.

Strikers participating in an unfair labor practice strike are in a somewhat stronger position. If they are replaced by other employees, they retain their rights to reinstatement even though the replacement workers may have to be discharged.

Of course, strikers can act in ways that will bring their dismissal in any case. If strike activities go well beyond those allowed by the NLRA and other laws, the employees involved may lose their rights, whether they are economic strikers or unfair labor practice strikers, whether they are suffering from a mass "illness," refusing to work customary overtime, or simply engaging in a mass slowdown. Misconduct by the strikers—violence or threats of violence, for example—could be the cause of such a loss of rights.

The employer, just as anyone else, has the rights of dominion over and use of his or her property. The strike known as a sit-down strike—in which employees remain at the employer's premises but refuse to work—deprives the employer of these rights and is, therefore, unlawful.

Picketing

We all have the right to **picket.** Whether we are employees, union members, or just ordinary consumers, we may inform others of our grievances; whether they are real or fancied makes little difference. Removal of this right would require a drastic change in the First Amendment of the U.S. Constitution. Thus, picketing at an employer's place of business is generally seen as a constitutional right. However, the courts traditionally have not viewed picketing as speech that receives the full protection of the First Amendment. Of course, when people exercise their rights, they often run headlong into others' rights—this is what legal conflicts are made of.

Ordinarily the right of a union to picket an employer's place of business is a right reserved to the representative of the employer's employees. Particularly, organizational picketing by a noncertified union is barred as unlawful blackmail picketing. Sometimes, unions picket to inform the public that the employer does not employ union

> Picketing may also be prohibited if it takes place in a nonpublic forum and may interfere with traffic flow. See *Hawkins v. City and County of Denver*, 170 F.3d 1281, *cert. Denied*, 120 S.Ct. 172 (1999) where musicians picketed within a shopping mall interrupting the flow of pedestrians.

---

[5]29 USCA §151. See also *Clow Waters Systems Co., v. NLRB*, 92 F.3d 441 (CA 6 1996)

members or have a contract with a union. Such picketing is lawful, unless it interferes with the employer's business—preventing pick up and delivery of goods, for example.

METHODS, CHANGES, AND PROTECTED WORK

The management of a company selects the methods by which the company's product is produced. When a company is formed, it sets up whatever processes are needed to produce the goods and/or services it sells. It then arranges for labor to operate the equipment that performs the processes. From the management point of view, it seems right and reasonable that the company should also be able to change the processes when economic analysis indicates the desirability of such change. But unions sometimes attempt to block or delay such changes.

What happens if two processes are merged into one by a methods change? Or suppose it becomes more economical to subcontract a process or a set of processes. What if economics dictate closing a plant? Perhaps management wishes to buy or build a new plant in another state to house some of the present processes. Maybe a merger with another company seems desirable. Many people assume that all these decisions are a management prerogative. However, while management generally has the right to make such changes (unless it has bargained the right away), it usually must be willing to bargain over their effects on the employees.

For example, the merger of two companies into one is a management decision and doesn't require consultation with any related union. Some mergers would have little or no effect on employees. Even if employee status is seriously affected by the merger, however, management is not obligated to bargain with the union about the merger—only about its effects on the employees. In such situations, conflicts may arise about just which union has the right to bargain on behalf of the successor company.

Subcontracting work is usually a management decision as well, so long as it can be shown that the sole inspiration for farming the work out was economic. There are some exceptions to this rather broad generality, however. In one case, farming the work out to another precipitated a layoff, and the court supported the NLRB's contention that this was going too far.[6] In another case, the subcontracted work (maintenance) previously done by union members on the company premises was to be done by nonunion people. The NLRB decision, which was affirmed by the U.S. Supreme Court, indicated that this action was beyond reasonable limits of the management prerogative.[7]

In the construction industry, union-management contract provisions restricting subcontracting are permitted and are usually upheld. One version of such provisions has come to be known as a work preservation clause and has the full sanction of both the NLRB and the U.S. Supreme Court. Such clauses are designed to limit or prevent the subcontracting of work. Generally, a clause that limits subcontracting on grounds unrelated to the presence or absence of a union at the subcontractor is upheld. A clause that restricts subcontracting to other companies who have signed or contract with the union violates the NLRB. Such clauses are viewed as attempts to induce the subcontractors to unionize.

---

[6]*Weltronic Co. v. NLRB*, 419 F.2d 1120 (1969)

[7]*Fibreboard Paper Products Co. v. NLRB*, 379 U.S. 203 (1964)

Featherbedding

One form of union resistance to change is known as **featherbedding**, or the requiring of payment for work not done or not to be done. For an example, suppose the Black Company completely automates a process involving 10 people, and because of the change, the 10 jobs no longer exist. A reasonable move by management would be to retrain the people for other jobs and perhaps wait for normal attrition to take care of any excess personnel. The company might, of course, try to lay the people off or try to get them employment with another company. As long as Black Company treated the employees as fairly as it could under the circumstances, the union probably would have little to oppose. (Depending on the magnitude of the changes, the employer may need to bargain with the union about the changes.) The union could not require Black to retain and pay the employees as though they were still performing the eliminated jobs. Such a demand would be considered featherbedding, and is unlawful.

SUMMARY

While labor relations remains an extremely important area of the law, many other laws govern the employment relationship as well. A state's contract laws, for example, generally shape much of the relationship. As already noted, federal (and often state) statutes prohibit discrimination based on factors such as race, sex, age, handicaps or disabilities, etc. Other statutes govern pension funds, employee stock ownership plans, health insurance plans, and so on. Thus, the sum total of the laws relating to employment relationships is quite complex. Nonetheless, an understanding of these issues is important, because engineers often find themselves affected by collective bargaining and the laws of labor relations.

SCOTT BALDWIN, et al. v. PIRELLI ARMSTRONG TIRE CORPORATION, et al. 3 S.W.3d 1; 1999 Tenn. App. LEXIS 106; 160 L.R.R.M. 2541 (Tn. 1999)

Key Terms: Strike, collective bargaining agreement, retaliatory discharge, independent contract, preemption doctrines, tortious interference, pretext

When replacement workers were fired to make room for the returning union employees at the end of a strike, the replacement workers brought suit against the employer for breach of contract and retaliatory discharge, and against the local and international unions for intentional interference with their contract. The trial court granted the employers motion to dismiss the breach of contract count, because it believed the individual contracts had been subsumed into the collective bargaining agreement, but it overruled the motion to dismiss the retaliatory discharge count. The court granted the unions' motion to dismiss, because it believed the complaint did not state a cause of action for intentional interference with an employment at will contract. We reverse the judgment of the trial court.

I.

A. The Facts

Since this case was decided on a motion to dismiss we take the facts from a liberal construction of the complaint, and we assume the facts in the complaint are true.

The complaint alleges that Pirelli Armstrong Tire Corporation (Pirelli) operated a manufacturing plant in Madison, Tennessee where substantially all of the hourly workers were represented by the United Rubber, Cork, Linoleum and Plastic Workers Union. The Pirelli workers were members of Local 670. In July of 1994, the workers went out on strike. Pirelli hired some replacement workers, and after declaring that the parties had "bargained to an impasse," the company terminated the strikers and began to hire "permanent" replacement workers. All the replacement workers were hired with an express or implied promise that they would not be terminated solely to make room for the returning strikers.

In March of 1995, the union and the company entered into a new collective bargaining agreement (CBA). The company began to rehire the union members and, under pressure from the local and the national unions, to fire the replacement workers for pretextual reasons. Ultimately all of the replacement workers were fired.

B. The Procedural History.

The named plaintiffs brought an action against the company for a breach of contract and for a retaliatory discharge. The complaint sought to recover treble damages from the unions for an intentional interference with the plaintiffs' contract of employment with the company. . . .

II. Retaliatory Discharge

We will deal first with the company's contention that the trial judge erred in failing to dismiss the claim for retaliatory discharge. The elements of such a cause of action are fairly simple: "An employment at will relationship; a clear declaration of public policy which imposes duties on the employee or employer; and discharge of the employee for refusing to violate those duties." *Reynolds v. Ozark Motor Lines, Inc.*, 887 S.W.2d 822 at 825 (Tenn. 1994). . . .

In this case the complaint alleges the following facts:

> As of the filing of this complaint, on information and belief, none of the permanent replacement workers hired during the strike for hourly-rate positions remain employed by PIRELLI.

> On information and belief, when negotiating the new collective bargaining agreement ratified on March 27, 1995, representatives of PIRELLI, on the one hand, and the URW and URW LOCAL 670, on the other hand, discussed the issue of whether or not the jobs previously vacated by striking URW members and filled by the permanent replacement workers could be made available to URW members following the end of the strike.

> In the negotiations that resulted in the new collective bargaining agreement ratified on March 27, 1995, PIRELLI and the URW and URW LOCAL 670 agreed that all striking members of the URW would be rehired by PIRELLI, despite and with full knowledge of the fact that PIRELLI had already hired the permanent replacement workers to fill the positions previously occupied by striking members of the URW and URW LOCAL 670.

> On information and belief, after the end of the strike and through the termination of the permanent workers, the URW and URW LOCAL 670 did not permit any person they knew to have been a permanent replacement worker hired by PIRELLI during the strike to join the URW or URW LOCAL 670 in order to remain employed at PIRELLI following the end of the strike.

In addition, in count one the complaint makes the following allegations:

> Tennessee is a "right to work" state, in which it is contrary to public policy for an employer either to refuse to hire or to discharge an employee simply because the employee is not a member of a labor union.

> The "right to work" in Tennessee without being compelled to belong to a labor union is a clear public policy evidenced by the unambiguous statutory provision of T.C.A. § 50-1-201 (1991).

> PIRELLI's termination of the permanent replacement workers who are members of the Plaintiff Class violated the "right to work" public policy of Tennessee and was for this reason wrongful.

> The members of the Plaintiff Class have suffered damage as a result of their termination by PIRELLI.

> The damages suffered by the members of the Plaintiff Class include but are not limited to lost wages and benefits, consequential damages and emotional distress.

> In terminating the members of the Plaintiff Class, PIRELLI acted intentionally, fraudulently, maliciously and/or recklessly because PIRELLI knew that it was a violation of both Tennessee and federal law to terminate employees based on the fact the employees were not union members, but nevertheless did so alleging grounds for termination which were known to be false in an effort to cover up the real (but illegal) reason for termination. As a result, PIRELLI is liable for punitive damages.

As we read the complaint it does not state a claim for a retaliatory discharge. It does not allege that the replacement workers were fired for refusing to join the union. The complaint alleges that the union and the company agreed that the striking employees would be rehired. A necessary consequence of that agreement was that the replacement workers had to go, but the connection between that fact and union membership is not stated in the complaint.

We take no position on whether the right to work law, Tenn. Code Ann. §50-1-201, creates a private right of action for its violation. The appellants state emphatically that they are not asserting a cause of action for violating the statute, but they rely on the statute as a clear statement of public policy on which to base a claim of retaliatory discharge.

III. Breach of Contract

We are of the opinion that the complaint states a cause of action for breach of contract. A fair reading of the complaint reveals that the plaintiffs alleged (1) that they were hired with an express or implied promise that they would not be fired in order to make room for the returning strikers and (2) that the company breached that promise when it settled the strike with the union. The company's promises were more than a vague promise of "permanent" employment which creates no more than an employment at will.

The trial judge found, however, that the plaintiffs' contract with the company was subsumed into the collective bargaining agreement subsequently entered into by the company and the union. From a reading of the cases we find that the courts use that expression to indicate that the individual contracts were displaced by the subsequent collective bargaining agreement (CBA) or that the dispute was preempted by federal law.

A. The Displacement Issue

The only authority cited by either the company or the union that the individual contracts with the replacement workers were displaced by the CBA is *Beals v. Kiewit Pacific Co.,* 114 F.3d 892 (9th Cir. 1997). Beals is a pre-emption case in which the Ninth Circuit held that an employee could not maintain a state action seeking to enforce an employment contract entered into when the employer and the union were operating under an existing CBA. To the extent that the separate contract was inconsistent with the CBA, "the CBA controls and any claims seeking to enforce the terms of the [separate contract] are preempted. 114 F.3d at 894. . . .

That is the situation involved in this case. The appellants negotiated a contract at a time when neither they nor the company were under a CBA. Therefore, we think the contracts they negotiated were not subsumed into the subsequent CBA.

B. Preemption

For most of the same reasons appearing in the preceding section we hold that the appellants' independent contract claims are not preempted by the National Labor Relations Act (NLRA) nor by Section 301 of the Labor Management Relations Act (LMRA). Section 301 preempts state law claims that are based directly on rights created by a CBA or claims that are substantially dependent on an interpretation of a CBA. In *Belknap, Inc. v. Hale,* 463 U.S. 491, 498, 77 L. Ed. 2d 798, 103 S. Ct. 3172 (1983), the Supreme Court described two doctrines that determine whether state causes of action are preempted by the NLRA:

Under the first, set out in *San Diego Building Trades Council v Garmon,* 359 U.S. 236, 3 L. Ed. 2d 775, 79 S. Ct. 773 (1959), state regulations and causes of action are presumptively pre-empted if they concern conduct that is actually or arguably either prohibited or protected by the Act. The state regulation or cause or action may, however, be sustained if the behavior to be regulated is behavior that is of only peripheral concern to the federal law or touches interests deeply rooted in local feeling and responsibility. In such cases, the State's interest in controlling or remedying the effects of the conduct is balanced against both the interference with the National Labor Relations Board's ability to adjudicate controversies committed to it by the Act and the risk that the State will sanction conduct that the Act protects. The second pre-emption doctrine, set out in *Machinists v Wisconsin Employment Relations Comm'n,* 427 U.S. 132, 49 L. Ed. 2d 396, 96 S. Ct. 2548 (1976), proscribes state regulation and state-law causes of action concerning conduct that Congress intended to be unregulated, conduct that was to remain a part of the self-help remedies left to the combatants in labor disputes.

In Belknap the court decided that facts remarkably like the facts in this case did not bring the dispute within either of the NLRA preemption doctrines. . . .

Considering only the allegations in the amended complaint we see no allegations that the dispute involves rights created by a CBA, and without the CBA even being in this record we cannot see how the appellants' claims involve an interpretation of the CBA.

IV. Interference with Employment Contract

The trial judge dismissed the tortious interference count because it believed the appellants' contract had been subsumed within the CBA negotiated by the union. On appeal the only argument made by the local union is that one cannot interfere with a contract to which it is a party.

As we have pointed out, however, this action is not based on the CBA negotiated by Local 670. And the intentional interference with at-will employment by a third party without privilege or justification is actionable. The allegations in the complaint are sufficient to state a cause of action against the local union.

Although the national union argues on appeal that it is a separate entity from the local union, and that there are no specific allegations against it, the complaint does specifically refer to the "URW and URW Local 670". The complaint goes on to allege that both defendants created pressure on Pirelli to terminate the appellants and that in inducing Pirelli to terminate the appellants, both defendants acted with malice and fraudulent intent.

Whether the appellants can prove any of their allegations remains to be seen, but at this stage of the litigation, the complaint states a cause of action for intentional interference with contract against both union defendants.

The judgment of the court below is reversed as indicated herein and the cause is remanded to the Circuit Court of Davidson County for further proceedings. Tax the costs on appeal one-third to the appellants and two-thirds to the appellees.

PEAVEY CO. v. NLRB 648 F.2d 460 (7th Cir. 1981)

Study terms: Discharge, pretext, "dual motive"

Petitioner, Peavey Company, seeks review of an order of the National Labor Relations Board ("NLRB"). The NLRB cross-petitions for enforcement. We grant enforcement in part and deny it in part.

I

The Board found that Peavey violated section 8(a)(3) when it discharged Mellinda Snider. The Administrative Law Judge ("ALJ") found that Snider's discharge "was motivated at least in substantial part by her protected activities." While affirming the ALJ's conclusions of law, the Board labeled Peavey's reasons for the discharge a "pretext." The Board then found that "Snider was discharged solely because of her concerted and union activities." Peavey claims that legitimate business reasons justified Snider's discharge.

Since its decision in this case, the National Labor Relations Board issued its *Wright Line* decision. . . . *Wright Line* set forth definitive rules for resolving cases in which a "dual motive" discharge is alleged. The Board expressly rejected the "in part" test relied upon by the ALJ here. Instead, it applied the Supreme Court's test in *Mt. Healthy City School District Board of Education v. Doyle,* . . . to dual-motive discharge cases. Under the new test, the General Counsel must first make a prima facie showing that the employee's protected conduct was a motivating factor in the employer's decision to discharge the employee. Once this is established, the burden shifts to the employer to demonstrate that he would have discharged the employee even in the absence of the protected conduct. . . . The Board ruled that the *Mt. Healthy* test aimed to determine the causal relationship between the employee's protected activities and the employer's action. Once found, a causal relationship justifies liability under section 8(a)(3), without any quantitative label such as "in part" or "dominant motive." *Id.*

This is the first "dual-motive" case to reach us for decision since *Wright Line.* At least one other circuit has adopted the *Mt. Healthy* approach. . . . We have reviewed the decisions and have decided to follow the *Mt. Healthy/Wright Line* test in "dual motive" cases in this circuit.

Our review of the record as a whole convinces us that Peavey Company met its burden under the *Wright Line* decision. The evidence showed that, although a good typist, Snider was a sloppy worker and had a history of disputes with her supervisors. Peavey disciplined her in writing and advised her of possible discharge unless her work record and attitude improved. A few days before her discharge, she refused to insert notices of new pay scales into paycheck envelopes she was told to prepare. Snider's discharge was ultimately prompted by her refusal to retype some poorly typed letters.

The Board backed away from the ALJ determination that Snider's discharge was motivated "in part" by her union activity. In labeling Peavey's reasons as "pretextual," however, the Board relied on the same rationale as did the ALJ. It agreed that Peavey had "tolerated Snider's poor job performance for over eighteen months" until she began concerted activities. The Board also relied on Peavey's demonstrated "animus" towards the union and the timing of Snider's discharge.

Peavey's reasons here cannot be labeled pretextual. As *Wright Line* held, a pretext can be found to exist when "the purported rule or circumstances advanced by the employer did not exist, or was not, in fact, relied upon (sic)." . . . Here, however, it is undisputed that Snider had been disciplined, for cause, prior to her contact with the union. Snider testified, in fact, that she went to the union as a result of her discipline. Her prior discipline also undermines the Board's claim that Peavey had tolerated Snider's poor performance without complaint for eighteen months. Unlike in *St. Luke's Memorial Hospital, Inc. v. N.L.R.B.*, . . . there were independent acts of misconduct at the time of Snider's discharge which justified Peavey's action. Moreover, an employer's silence does not extinguish its right to discipline an employee whose conduct continues to worsen. . . .

Once Peavey's reasons for discharge are stripped of the label "pretext," it is apparent that Peavey met its burden under the *Wright Line* decision. Even though Snider engaged in some protected activity, Peavey showed that she would have been discharged even in the absence of the protected conduct. . . . The substantial evidence on the record as a whole does not support the Board's finding of an 8(a)(3) violation. We therefore deny enforcement of that portion of the Board's order calling for reinstatement of and back pay for Snider.

II

The Board also found five violations of section 8(a)(1) of the National Labor Relations Act. 29 U.S.C. Sec. 158(a)(1). The Board found that Peavey violated the Act when (1) its General Manager told Snider that she was a confidential employee and thus not entitled to participate in union activities; (2) Supervisor Sandy Noe followed employees and eavesdropped on their conversations; (3) it announced, after the union began organizing, that employees could see their personnel files; (4) its General Manager told an employee that persons who joined the union would be considered disloyal; (5) its General Manager promised the employees at a meeting that "he was going to see that things" got better.

After a review of the record as a whole, we conclude that substantial evidence supports the Board's finding of section 8(a)(1) violations. Peavey's contentions to the contrary are without merit. Accordingly, we enforce the rest of the Board's order.

Enforced in Part, Enforcement Denied in Part.

CAROLE J. KATZ, et al v. WESTLAWN CEMETERY ASSOCIATION, INC., et al. 285 Ill. App. 3d 695; 673 N.E.2d 1053 (1996)

Study terms: Collective bargaining agreement, tortious interference with property rights, civil conspiracy, intentional infliction of emotional distress, standing, preemption doctrines, Garmon doctrine

Plaintiff Carole J. Katz, individually, and on behalf of a class of similarly situated individuals, filed this class action against defendants, Westlawn Cemetery Association, Inc., on behalf of all others similarly situated. . . . In her fifth amended complaint, Katz alleged intentional infliction of emotional distress (count I), tortious interference with property rights (count II), breach of contract (count III) and civil conspiracy (count IV). On January 6, 1995, the trial court granted defendants' motion to completely dismiss the complaint under sections 2-615 and 2-619 of the Code of Civil Procedure; it is from this order that plaintiff appeals to this court. . . . For the reasons which follow, we affirm.

Factual Background

The genesis of this action is a labor strike by Chicago-area gravediggers, represented by their union, Local Union 106, Service Employees International Union, AFL-CIO, against the 27 defendant cemeteries and the subsequent lockout of strikers by defendants. On December 19, 1991, after contract renewal negotiations between defendants, represented by their collective bargaining agent, the Cemeteries Association of Greater Chicago, Inc. (hereinafter the "Association") and the union broke down, the union went on strike against four defendant cemeteries: Beverly Cemetery, Oak Hill Cemetery, Lincoln Cemetery and Memorial Park Cemetery. In response, defendants locked out their union employees. Consequently, burials in the ordinary course of business ceased.

On January 3, 1992, Mrs. Rose Michaels, plaintiff's mother and an Orthodox Jew, passed away. In 1973, Mrs. Michaels and her husband had purchased burial plots at Westlawn Cemetery. Orthodox Jewish religious tradition requires burial of the dead within 24 hours of death. During the lockout, plaintiff requested that her mother be buried at Westlawn Cemetery, which Westlawn allegedly refused to do due to the lockout. Consequently, plaintiff's family was unable to bury Mrs. Michaels in keeping with Orthodox Jewish tradition. The instant litigation followed.

Issues Presented for Review

On appeal, plaintiff posits that: (1) the trial court erred in holding that she did not have standing to sue the alleged co-conspirators for tortious acts that resulted in her alleged injury; (2) the trial court erred in holding that her allegations of civil conspiracy were not stated with sufficient specificity; (3) the trial court erred in holding that her claim for intentional infliction of emotional distress was time-barred; and (4) the trial court erred in holding that her claim for intentional infliction of emotional distress and tortious interference with property rights were preempted by the National Labor Relations Act.

Opinion . . .

The Supremacy Clause of the United States Constitution provides that the Federal Constitution, Federal statutes and treaties "shall be the supreme Law of the Land; and the Judges in every State shall be bound thereby, any Thing in the Constitution or Laws of any State to the Contrary notwithstanding." U.S. Const. art. VI, cl. 2. Thus, the Supremacy Clause forbids State encroachment on areas regulated by Congress, whether Congress mandate is "explicitly stated in the statute's language or implicitly contained in its structure and purpose." *Jones v. Rath Packing Co.*, 430 U.S. 519, 525, 51 L. Ed. 2d 604, 97 S. Ct. 1305 (1977). The NLRA, as an act of Congress, is the supreme law of the land and necessarily preempts any conflicting State law.

The NLRA reflects a congressional intent to create a national, uniform body of law regulating labor relations. The NLRA protects from State interference a labor union's right to strike an employer and withhold the labor of its members. The NLRA similarly protects an employer's use of both defensive and offensive lockouts as a proper economic weapon, whereby the employer may withhold employment either to resist its employees' demands or to gain concessions from them.

The Supreme Court has enunciated two preemption doctrines invalidating State laws that interfere with the NLRA's goals. First, under *San Diego Building Trades Council v. Garmon* (the Garmon doctrine) and its progeny state that State laws may not interfere with conduct regulated or even arguably regulated by NLRA, including collective bargaining. The second doctrine, created in Lodge 76, *International Association of Machinists & Aerospace Workers v. Wisconsin Employment Relations Commission* (the Machinists doctrine) holds that State laws may not interfere with the balance of power between unions and employers established by Federal Law, including those areas that Congress left under the control of the free play of economic forces. We agree with the circuit court that under either of these two doctrines, plaintiff's claims in counts I, II and IV were preempted by the NLRA.

In Garmon, . . . [t]he Supreme Court noted that State law is preempted whether a State has acted through "laws of broad general application" or "laws specifically directed towards the governance of industrial relations." Illinois courts have consistently recognized and applied the Garmon doctrine.

The Garmon court created two narrow exceptions to its holding, finding that the NLRA does not preempt State remedies: (1) where the activity regulated was "a merely peripheral concern" to the Federal labor laws, or (2) where the regulated conduct "touched interests so deeply rooted in local feeling and responsibility that, in the absence of compelling congressional direction, [it] could not [be] inferred that Congress had deprived the States of the power to act." *Garmon*, 359 U.S. at 243-44.

The Garmon court found that the first exception did not apply because picketing is specifically protected by section 7 of the NLRA and, therefore, is not an activity of "merely peripheral concern" to the NLRA. Similarly, the Supreme Court found that the second exception did not apply because the case did not involve a State interest deeply rooted in local feeling and responsibility such as the "maintenance of domestic peace" in the face of "conduct marked by violence and imminent threats to public order." *Garmon*, 359 U.S. at 247.

In case sub judice, plaintiff concedes that her claims, arising from a lockout protected by the NLRA, are "arguably subject to the jurisdiction of the" National Labor Relations Board, but contends that her claims are not preempted because they arise from "interests deeply rooted in local feeling and responsibility." Thus, plaintiff asserts that "it is hard to imagine anything more deeply rooted in local feeling and responsibility than burying the dead in accordance with religious and social customs." We disagree.

In *Cannon v. Edgar*, 33 F.3d 880 (7th Cir. 1994), the Seventh Circuit directly rejected plaintiff's contention. In Cannon, the Federal appeals court invalidated the Burial Rights Act, an act which was passed by the General Assembly in direct response to the labor dispute that is the subject of this litigation, which required cemeteries and gravediggers to negotiate for the establishment of a pool of workers designated to perform religiously required burials during labor disputes. The Seventh Circuit found the Burial Rights Act was preempted by the NLRA because it was "an invasion by a state into the collective bargaining process." *Cannon*, 33 F.3d at 884-86.

In so holding, the Cannon court found that the Burial Rights Act was not subject to either of the two *Garmon* exceptions:

"As to the first *Garmon* exception, the Burial Rights Act is not concerned with something peripheral to the NLRA. Indeed, performing burials, including burials during the course of a labor dispute, is the very subject that the cemeteries and the gravediggers negotiate about in the collective bargaining process. And as to the second Garmon exception, there is nothing deeply rooted in local feeling that would allow Illinois to encroach on the process of collective bargaining in the manner directed by the Burial Rights Act. The interment of deceased persons is something common to every community in the United States. The interment of the smaller category at issue here (particular religious groups) is not something deeply rooted in local feeling ***." 33 F.3d at 885.

Plaintiff submits that the Cannon court found the Garmon exceptions inapplicable because that case involved a civil statute rather than State criminal or tort law. We believe that argument misses the point. What is important here is the fact that plaintiff's allegations are inextricably linked to the labor dispute such that her claims would not exist but for that dispute.

In her fifth amended complaint, plaintiff alleges that because Westlawn and the other cemeteries did not perform interments as a result of the labor dispute that ipso facto they committed the various torts. This is not the law. In addition, plaintiff alleges defendants' joint lockout of the union members to be a civil conspiracy; however, this is an activity clearly protected by NLRA. Nowhere does plaintiff allege that any agent or employee of Westlawn Cemetery battered, insulted, defamed or committed any other tortious and/or criminal acts against her personally that traditionally would fall into the second Garmon exception. Accordingly, we find that none of plaintiff's tort claims fall within the exceptions to the Garmon doctrine. Therefore, the trial court did not err in dismissing plaintiff's complaint.

Alternatively, we find plaintiff's causes of action are preempted under the Machinists doctrine. In Machinists, the Supreme Court recognized that activities are protected by Federal labor law not only where they are specifically regulated by the NLRA, but also where the activity involves a "permissible economic weapon" that Congress intended to be unrestricted by any governmental power. Because the NLRA seeks to leave disputes between management and labor to the "free play of economic forces " without State interference, any State cause of action that affects the balance of economic weapons available to labor and management is preempted.

A union's right to strike and an employer's right to lockout are among the most significant of economic weapons. In *Golden State Transit Corp. v. City of Los Angeles,* the Supreme Court explained, "Congress left some forms of economic pressure unregulated, while it banned others. States are therefore prohibited from imposing additional restrictions on economic weapons of self-help, such as strikes or lockouts [citation] unless such restrictions presumably were contemplated by Congress." 475 U.S. at 614-16. The district court opinion in Cannon noted that the Machinists doctrine forbids the States from even "indirectly imposing additional burdens on economic weapons such as strikes and lockouts." *Cannon v. Edgar,* 825 F. Supp. 1349, 1360 (N.D. Ill. 1993), affirmed, 33 F.3d 880 (7th Cir. 1994). . . .

In the case sub judice, plaintiff's lawsuit is attempting to . . . use a State common law action to inject herself into a labor dispute and in the process deny both the union and the cemeteries of significant economic weapons, the power to strike and the power to lockout. This is impermissible. Thus, we hold no error was committed by the circuit court in dismissing plaintiff's cause under the Machinists doctrine.

In light of our disposition of plaintiff's appeal we need not reach the issues raised by defendants in their cross-appeal which were argued in the alternative had we reached a different disposition of plaintiff's appeal.

In light of the foregoing, the judgment of the circuit of Cook County is affirmed in toto.

Review Questions

1. Show graphically how Adam Smith's supply-and-demand theory might be used by a court to prove "unlawful" union action. Show by another graph how the wage-fund theory might be used for the same purpose.

2. What do the following terms mean? What is their present legal status?
   a. Yellow-dog contract
   b. Labor injunction
   c. Closed shop
   d. Picketing
   e. Strike
   f. Wildcat strike
   g. Featherbedding
   h. Jurisdictional dispute
   i. Agency shop
   j. Collective bargaining

3. What are the nature, purpose, function, and jurisdiction of the National Labor Relations Board?

4. Compare the effects on employers of the union shop and the closed shop.

5. What are the limits on the efforts that may be taken to persuade employees either to join or not to join a union prior to a union election?

6. Describe how a secondary boycott works. Give an example of a secondary boycott involving a union. Give an example of one not involving a union.

7. Distinguish between an economic strike and an unfair labor practice strike. What are the rights of employees involved in each?

8. Why didn't the NLRA apply in the *Scott Baldwin, et al. v. Pirelli Armstrong Tire Corporation, et al.* case?

9. According to the decision in *Peavey Co. v. NLRB*, what employee activities justified discharge? What specific acts by the company were censured?

10. What additional facts would have helped the plaintiffs in *Katz v. Woodlawn Cemetery?*

11. For quite some time the Amalgamated Embalmer's Union tried to win a representation election to represent the employees of the Explosion, Iowa division of the Black Mfg. Co. Last July, just before the election, "Weeping Jack" Black, the company president, came in from Stitch-in-time, Nebraska to speak to the employees. In his speech he mentioned that since the Explosion plant was operating so close to the break-even point, the added overhead caused by the union would make it necessary to reduce all wages by $.30 per hour if the union got in. Despite this, the union won by 122 votes to 121. Contracts were signed and the wage rate stayed the same. One day last March the eight employees of the paint shop walked off their jobs as a complaint about their piecework rates; a few minutes later they were joined by the other hourly employees. The group milled around the parking lot committing various injurious acts against management members' vehicles. An accountant had taken several pictures of the occurrence when he was accosted by group members and beaten, and his camera was smashed. The following morning all the workers re-

turned, but two of the paint shop employees were told they were fired. More work stoppages then occurred and interferences with production have continued to the present. The two employees who were fired have lodged a grievance which is pending final settlement. "Weeping Jack" Black has stated flatly that if these disturbances do not cease he will close the Explosion plant. The union has replied that if he does this it will strike the materials supplier that he uses at the home plant and force him out of business there, too. Identify the acts and proposals of the union and the company as either lawful or unlawful according to the interpretations of the federal labor law.

Worker's Compensation

CHAPTER

25

HIGHLIGHTS

- In the past, employer's had three defenses to a work related injury:
  1. Assumption of risk
  2. Contributory negligence
  3. Fellow Servant Rule

- The Worker's Compensation Laws changed the legal framework at the federal and state levels. Its primary purpose is to spread the risk of loss for injuries. It does this in two ways:
  1. Limiting an employee's ability to recover from the employer, and
  2. Requiring that employers make payments to insure their workers.

People who work for a living risk being injured on the job. Some of the machines and equipment they must use are inherently dangerous. Sometimes the building or the work situation itself is hazardous. Occasionally, it is dangerous to get to the job site or to return from it. Some degree of danger is associated with almost any human activity. Such being the case, who should bear the risk of injury on the job—the employee or the employer?

THE HISTORICAL PERSPECTIVE

The answer a century or so ago was that the employee bore all these risks. In other words, the law heavily favored employers. The injured worker might sue an employer for compensation for a work-related injury, but the chances of winning were virtually nonexistent. The employer had three strong defenses: assumption of risk, contributory negligence, and the fellow servant rule—basically those defenses available under negligence and tort law. Because of these three defenses, employees hardly ever won a suit against their employers. On the other hand, there was no monetary limit on what an employee could win. Thus, the occasional win by an employee usually involved a large amount.

Assumption of Risk

As you know, in tort law, one who voluntarily assumes a risk and is injured as a result cannot sue another and recover for that injury from the other person. Logically, then, one who accepts employment accepts the risks that are part of that employment. If Green, for example, applies for work at the Brown Company, and the Brown Company hires him as a machine operator, doesn't Green assume the risk of being injured? A century ago, the answer to such a question was clearly, yes. Therefore, if Green were injured during his employment in the operation of a Brown Company machine, Brown might express sympathy or even offer Green a payment as a gratuity, but Green probably could not win an award of compensation from Brown.

Contributory Negligence

If the assumption-of-risk defense should fail, an employer could use the same or a similar set of facts to establish the employee's contributory negligence. That is, if Green lost a finger or a hand in one of Brown Company's machines, Brown could simply argue that working around machines was obviously dangerous and that Green was negligent in not keeping his hand out of the machine. If Green was found at all negligent, the doctrine of contributory negligence would bar his claim against Brown.

Fellow Servant Rule

Assumption of risk and contributory negligence usually covered the employer in all worker-injury cases. But if they did not, the employer had a third line of defense. Returning to our example above, let's assume Green can establish that the machine that injured him was faulty, perhaps because of poor maintenance. Brown Company's answer could include as an added defense that a maintenance man or some other employee had been negligent, and this caused the mechanical fault. Brown's next step, then, would be to allege that the real defendant should be the negligent fellow employee—not the employer for whom both employees worked. Under the common law, any fellow employee's negligence was implied to be the cause of the injury. This was known as the **fellow servant rule.**

Employers' Liability Acts

Still, the specter of maimed and injured workers caused public concern. In various jurisdictions, legislatures changed the law in the late 1800s to what were known as employers' liability acts. Although these acts benefited the injured worker, he or she still had to sue the employer for compensation. This often required the worker to find an attorney willing to take a case on a contingency-fee basis. The contingency fee, in turn, would reduce the plaintiff's recovery by 25 to 50 percent of the amount (if any) recovered from the employer.

Presuming the injured employee could find an attorney willing to take the case, the employee usually had a better chance to win under employers' liability acts. Such acts often eliminated two of the employer's defenses and substantially changed the third one. Assumption of risk and the fellow servant rule were eliminated; contributory negligence was changed to comparative negligence. The resulting legal battles frequently left it up to a jury to decide the question of who was more negligent. Jury awards were often tempered to reflect a sharing of the negligence.

For example, if a jury found Brown and Green to be equally negligent, and the damages were $20,000, Green might be awarded $10,000. After paying an attorney, 40 percent of the award, Green would be left with $6,000 to compensate him for a $20,000 injury. Other expenses, such as expert witness fees, could reduce Green's award even further.

Under the acts, employers began to lose more cases and pay higher awards to injured employees. Employers developed ways to try to avoid liability. For example, suppose Green was severely injured while working. As he recovered in a hospital, he would receive a visit from a Brown Company representative, urging him to accept an amount (say, $500) from his employer to "help him in his time of need." Green would be required to sign a "receipt" (which turned out to be a release from liability) to obtain the cash.

The employers' liability acts were a significant improvement over the common law of torts, but they still left something to be desired. The injured worker still had to find an attorney and initiate action. The employee's award might be trimmed by comparative negligence and other expenses. Attempts by employers to obtain a release after paying only a paltry sum often were viewed by other workers and the public as reprehensible. In addition, the injured worker often had difficulty finding fellow workers who would risk their jobs by testifying for him. The system known as worker's compensation had been adopted in most European countries by 1900 or so. After a couple of false starts here, the federal government and a few states began to enact such laws. Between 1908 and 1948, all of our states adopted some version of the law.

WORKER'S COMPENSATION LAWS

It would certainly be surprising if the various federal and state versions of worker's compensation laws agreed on the details, considering that the laws were enacted over several decades. Even so, the coverage among the various laws is similar enough that we may discuss a general concept of worker's compensation, pointing out prominent exceptions along the way.

The laws attempt to reach the same general objective—to spread the risk of loss for injuries. Hence, these laws provide compensation to a worker for an injury "arising out of and in the course of employment." At the same time, these laws usually limit an employee's ability to recover from the employer. Specifically, they set up an insurance scheme through which employers pay "premiums." If an employee is injured in the course of employment, he or she will usually qualify for coverage and a payment.

Coverage

In almost all the laws, provision of coverage is either mandatory, with penalties for noncoverage, or is permissive, with a provision that those companies choosing not to cover workers are rendered virtually defenseless when sued. Such laws may provide that the employers who choose not to be covered may be sued in tort for an employee injury and cannot rely on the defenses of assumption of risk, contributory negligence, or the fellow servant rule. Some laws, however, require coverage, and state that each day of operation without coverage is a separate violation of the act, with a fine or imprisonment to result from each violation. At the same time, the statutes usually provide the employer with immunity to suits from employees, except for injuries caused by gross negligence or intentional misconduct. In either case, a company is under pressure to adopt worker's compensation for its employees.

The worker never has to pay for worker's compensation coverage. The employer bears the cost of the coverage in some way. Generally, an employer may contract with an insurance company for worker's compensation coverage. In many states, however, the employer may elect to self-insure. In a few states, the employer is required to buy this insurance from a state agency. In others, Uninsured Employer Funds have been established if an employer was not financially capable of making insurance payments. However, this does not relieve the employer from potential negligence liability or other claims the employee may have against it. In short, the employer must take the required steps to assure that he or she is capable of paying an injured worker's medical expenses and a weekly compensation for lost wages.

Each of the worker's compensation acts includes certain exceptions to general coverage. Many of the laws exempt agricultural employees, domestic servants, independent contractors, casual workers, and/or workers in religious or charitable organizations. Some require that the company have a minimum number of employees before the act applies. Sometimes owners are counted as employees. Sometimes an employer will voluntarily elect to obtain worker's compensation coverage for employees not automatically covered by the act. As time passes, the laws have tended to expand the coverage to more and more occupations.

Premiums

The cost to an employer for worker's compensation reflects the employer's insurance company's premium requirements. Generally, an employer pays the premium for coverage on a given job set forth in the insurance company's rate manual. Such premiums are usually based on $100 of payroll cost. If Green operates a rather hazardous machine for the Brown Company, the premium Brown has to pay for worker's compensation coverage may be as high as $10 for every $100 Green earns. On the other hand, if Green works in the office, where the only prominent hazard might be the possibility of tripping over a wastebasket, Brown's premium cost might be $.25 or $.50 per $100 of Green's earnings.

After the initial period of coverage, the premium cost is raised or lowered according to the employer's experience rating. Few severe injuries over the period may result in reduced premiums. Numerous and very severe injuries probably will increase the premiums.

Second Injury Funds

Suppose that Green has lost the sight of an eye. Would the Brown Company hire him? Suppose he lost the sight of the other eye while working for Brown. He would then be permanently, totally disabled. The cost of such a permanent, total disability is far greater than the cost of losing the sight of one eye, and Brown might well hesitate to risk such a heavy loss. For this reason, nearly all worker's compensation laws limit the second employer's responsibility to injuries suffered while the employee was with that second employer. If Green lost the sight of his other eye, Brown would be responsible only for the loss of sight in one eye. But Green, of course, would be totally blind, and this is where the second, or subsequent, injury fund comes in. Second Injury funds were created originally to encourage the hiring of partially disabled World War II veterans to make up the difference between what a company would have to pay and what an employee would receive if the employee became totally disabled. The most common way secondary funds are created is by legislative appropriation.

Amount of Compensation

The amount of compensation an injured worker is to receive for a work-connected injury depends upon the nature of the injury and the specifications of the worker's compensation law. Generally, worker's compensation laws require the payment of all medical expenses. Compensation for lost wages, however, usually depends upon the nature of the injury. Four categories of disability exist:

1. Temporary total disability,
2. Permanent partial disability,
3. Permanent total disability,
4. Death.

A **temporary total disability** usually requires the employer to pay compensation that is some percentage of the employee's wage. Upper and lower limits nearly always accompany this provision. That is, while an employee is waiting in the hospital for his broken legs to heal, he might receive two-thirds of his weekly wage providing this amount was no more than $300 per week and no less than $50 per week. Also, the law often has a holdback feature so that if an employee is out only a few days, he receives little or no compensation for lost wages. If he is out two or three weeks, the law may require an employer to compensate an employee for the first few days in addition to the continuing payments until an employee is ready to return to work.

A **permanent partial disability** usually involves the loss of function of some part of the worker's body. When an employee lost the sight of one eye, for example, that was a permanent partial disability. Loss of a hand, a finger, a foot, a leg, or the hearing in one ear would all constitute permanent partial disabilities. Worker's compensation laws generally have a schedule of the time period for which the worker will be paid a portion of the weekly wage for the disability. An employee's loss of an eye, for example, might require his employer to pay an employee two-thirds of his weekly wage (between the listed maximum and minimum) for perhaps 150 weeks. In addition, many of the worker's compensation laws provide for some form of rehabilitation for both permanent partial and permanent total disabilities.

When an employee lost the sight of both eyes he had a **permanent total disability.** Under many of the worker's compensation laws, all his related medical costs would be paid, and he would receive some compensation as lost wages for as long as he was disabled. If rehabilitation cannot be effective, this means wage compensation for life. Under other states' laws, some limitation (usually as to a maximum compensation total) is imposed. However, laws of this type tend to become more liberal over time.

If Green lost his life while working for the Brown Company, Brown would have to pay all an Green's medical expenses and then make payments to Green's wife and children, or to an employee's other beneficiaries. This wage compensation often continues until a statutory limit had been reached or until Green's widow has remarried. The amount of compensation might be modified when Green's children reach a certain age or are married. Brown usually must also pay for Green's burial up to a statutory maximum.

Determination of Compensability

As you have just seen, worker's compensation laws clear up the question of whether a worker injured while working will be compensated for the injury. They even go on to specify how much this compensation will be. Beyond that, the laws usually set up procedures for the filing and handling of claims. These schemes also set up procedures for appeals. The reason for appeals, of course, is that the laws still leave unanswered questions. A common question from which many, many cases arise is whether an injury actually was work-connected. In other words, did the injury arise out of and in the course of the worker's employment? (See *Neacosia v. New York Power Authority,* in this chapter). If the worker's injury occurred while he or she was at home mowing the lawn, painting the house, or playing tennis on a weekend, it is difficult to see why worker's compensation coverage should apply. If, instead, the worker is injured at work, there is usually little question as to the application of the law. But what if the worker is injured while going to work or while on lunch hour, for example? These are only two of numerous questions that the statutes left unclear.

Going and Coming.   Generally, an employee is not covered by worker's compensation on routine journeys to and from work. Still, there are enough exceptions to this generality that a rather imposing body of case law has developed on this topic alone. The general statement is still true, but numerous special circumstances have caused decisions for coverage.

For example, Green, traveling in his car at his customary time of travel over his customary route to or from work probably is not covered. However, if Green is a sales representative and is required to use his car in his work, he probably would be covered by worker's compensation in many jurisdictions. If Green is responding to a call to come in early or is returning home later than usual because of some unusual work requirement, he has a sound case for coverage. If he has taken an unusual route in response to a work requirement (even a mere suggestion by his boss), he also is very likely to be covered, as discussed in *Neacosia,* below. Furthermore, if his work has required him to be awake so long that he falls asleep at the wheel and is injured as a result, worker's compensation should cover him. In short, anything work-connected that causes him to alter his customary routine of travel to and from the job gives him a reasonable cause to claim worker's compensation. Such injuries can be viewed as resulting from a special mission given to the employee. In such situations, the employee's travel (and injury) may have arisen out of the employment and in the course of the employment.

Lunch Hour.   Suppose an employee leaves her place of business for lunch. Generally, the employee is not covered. However, there are plenty of exceptions. If the employee can show a connection between her job and her lunch hour activity, she has a case for worker's compensation coverage. Suppose that Green and Gray, Green's supervisor, go to a restaurant for a meeting with a vendor. Even if Gray suggested that Green ride with him, and Green is injured in an accident in Gray's car on the way, Green has a case for coverage. (For that matter, so does Gray.) The coverage is also likely where the lunch involves customers or co-workers on a business lunch, or where Green is asked to pick up materials or make a delivery in conjunction with his lunch period. In all these situations, Green has a reason to allege that injury arose out of and in the course of employment.

Intoxication.   If an employee comes to work so inebriated that she cannot properly control her actions and, as a result, becomes injured at her job, should this injury be compensated? The question has been raised many times, and compensation frequently has been denied. The logic for denial is that the injury was really caused by the employee's condition, into which he voluntarily placed himself. Thus, many worker's compensation acts provide that coverage does not extend to an employee who is intoxicated.

But let us now change Green's situation a little. Green is now a traveling salesman. His job requires him to entertain customers, and this often involves having a number of drinks. Brown Company actively encourages Green's drinking habit, or knows of it and tolerates it. Green is involved in an auto accident after having entertained customers and is injured. Should he be covered by worker's compensation or not? The answer to this is usually in the affirmative.[1] The entertainment of Brown's customers was part of Green's job, so Green's injury arose out of and in the course of his employment.

Injury by Co-Worker or Third Person.   In many situations, an error of another employee or an outsider causes the injury. If the injury arose out of and in the course of the employee's job, the employee has a right to worker's compensation. But this may not be the end of it. If strong evidence of a tort exists, the injured employee or the insurer (or both) may be justified in acting against the one who caused the injury. For example, Green is injured because a machine on which he was working malfunctioned. Green has a right to recover worker's compensation from Brown Company, his employer. He may also have a tort action, perhaps for products liability, against White Company, the manufacturer of the machine. Brown Company's insurer, Gray Insurance, also has an interest in the tort action. Under **subrogation** (i.e., if the contract provides that after making the payments, the insurance company stands in Brown's shoes), the insurer may seek to recover whatever has been paid to Green in worker's compensation payments. A worker's compensation case in which a set of payments of this nature occurred is *Moomey v. Massey Ferguson, Inc.*, 429 F.2d 1184 (10th Cir. 1970).

For worker's compensation coverage, injury by a co-worker or outsider must have arisen from the injured employee's work. That is, if an employee were injured in a strictly personal altercation with a non-employee, coverage would not be appropriate. If the altercation had nothing to do with the employee's work, the employee's only appropriate recourse is a tort action against the other party.

Other Situations.   It would be nearly impossible to describe all the cases that have occurred in the context of worker's compensation. A new line of cases involving work-related stress is winding its way through the courts. (See McCarron v. Workers' Compensation Appeal Board in this chapter). A few more general categories of cases might be mentioned, however, simply as examples of problems that arise. As in other types of circumstances, these cases tend to be decided on their own unique facts: The courts try to examine all relevant facts and decide whether the injury arose "out of and in the course of" the employment.

"Acts of God."   If an employee is injured at work by an "act of God," such as lightning or a tornado, the employee would be covered by worker's compensation, but only if his job exposed him to the hazard in a greater degree than members of the general public were exposed.

Tampering or Hazardous Operation.   If an employee disconnects a safety device on equipment with which he must work, the courts have gone both ways. If the company suggested or knew about the added hazard and did nothing about it, the chances for coverage are good. Otherwise, under most worker's compensation laws, coverage is likely to be denied.

Intentional Injury.   If an employee injures himself intentionally, and evidence supporting this allegation is found, worker's compensation coverage is likely to be denied.

---

[1]See, for example, *Boyd v. Francis Ford, Inc.,* 504 P.2d 1387 (Or. 1973).

Telecommuting or Working from Home.   Suppose an employee's employer permits him to work from home and the employee suffers an injury while working there. Whether the employee will be entitled to worker's compensation will depend on whether the there is sufficient evidence to show that the injury was sustained during and in the course of employment.

IN THE MATTER OF THE CLAIM OF MICHAEL NEACOSIA v. NEW YORK POWER AUTHORITY et al. WORKERS' COMPENSATION BOARD, APPELLANT 85 N.Y.2d 471; 649 N.E.2d 1188; 1995 N.Y. LEXIS 1041 (1995)

Study terms: Injuries arising out of and in the course of employment, "going and coming" rule, mixed purpose, special errand

Claimant Michael Neacosia was employed as a security officer by respondent New York State Power Authority at a nuclear power plant located near Oswego, New York. On May 17, 1991, Neacosia completed his shift at 1:00 P.M. and left the plant in his car. He stopped to deliver his uniforms to Karpinski's, a dry cleaner in Oswego. The cleaner was one of several recommended by respondent as part of an arrangement by which it provided its guards with uniforms, and required that the uniforms be kept clean and presentable. Although the uniform labels indicated that they could be machine washed, respondent assumed the expense of dry cleaning and maintained and paid directly accounts at the dry cleaning establishments. Alternatively, employees could use other establishments than those recommended and submit receipts to the employer for reimbursement.

After leaving some uniform shirts and trousers at Karpinski's, claimant headed home along his usual route. During the trip he was involved in an automobile accident and sustained severe injuries. The issue before the Court, broadly stated, is whether his injuries arose out of and in the course of his employment.

Neacosia submitted a claim for workers' compensation benefits contending that he was injured while "completing a work related trip." The employer denied liability claiming Neacosia was involved in an accident with his personal vehicle on a public highway outside of work hours. The Law Judge concluded upon stipulated facts that claimant's travel had a dual purpose which served to extend the scope of his employment, and accordingly awarded him workers' compensation benefits. Upon review by the Workers' Compensation Board the award was affirmed.

Respondent and its insurance carrier then appealed to the Appellate Division. A divided Appellate Division reversed, the majority concluding that in the absence of evidence that Neacosia was *required* to have his uniforms dry cleaned, the Board's finding was not supported by substantial evidence. The two dissenters believed the employer's policies with respect to claimant's uniforms and their maintenance established the requisite nexus between claimant's employment and his errand to support the claim. Neacosia and the Workers' Compensation Board appeal to this Court as of right.

I

An injury sustained by an employee is compensable under the Workers' Compensation Law if it "aris[es] out of and in the course of the employment" (see, Workers' Compensation Law § 10 [1]; § 2 [7]). The general rule is that injuries sustained during travel to and from the place of employment do not come within the statute. There are exceptions to this "going and coming" rule, however. For example, an outside employee, such as a travelling salesperson who does not have a fixed worksite, may be compensated for injuries sustained in the course of travel between home and appointments, an employee who has engaged in travel for dual purposes—both business and personal—may sustain compensable injuries during travel, and an employee whose home serves as an additional place of employment may also be compensated.

Claimant relies upon yet another, the "special errand" exception, which provides that when the employee's travel serves a purpose of the employer, injuries sustained during that travel may be. The question on this appeal is whether the special errand exception applies only if, as the Appellate Division held, the employer has expressly required or directed the employee to perform the errand.

A.

The test for determining whether specific activities are within the scope of employment or purely personal is whether the activities are both reasonable and sufficiently work related under the circumstances. Given the remedial nature of the Workers' Compensation Law, the courts have historically construed this requirement liberally to effectuate the statute's economic and humanitarian purposes.

Respondents urge us to apply the bright-line rule adopted by the majority at the Appellate Division to find compensable injury only when the employer expressly requires the employee to engage in the particular errand. However, the unpredictable and varied nature of work-related incidents render such claims unsuitable for any arbitrary rule. To attempt to apply a bright-line rule could produce results that are both inconsistent with our existing precedents and with the policy of liberally construing the statute.

Past decisions addressing exceptions to the "going and coming" rule have exhibited a common theme in which the courts have broadly applied considerations involving the nature of the employee's off-premises travel, whether it inured to the benefit of the employer, and whether the employer encouraged the employee's conduct. For example, in *Matter of Love v. N. Y. S. Craig School* a hospital attendant on a paid leave of absence was injured in an automobile accident while commuting to the nursing school she was attending. Although the claimant obtained an obvious personal benefit from the schooling, the court held that her attendance arose out of her employment because the employer induced her attendance by granting a fully paid leave of absence for education. Moreover, the employer stood to benefit from claimant's improved skills and knowledge because she had given a commitment to return to work following the schooling. Similarly, a high school teacher involved in a fatal auto accident on his way home from an off-premises night course he attended voluntarily was covered by compensation in *Matter of Bump v. Central School Dist. No. 3*. Although the teacher was not required to attend the course, his supervisors had approved his attendance, the employer contributed part of the cost, and both employer and employee stood to benefit from the schooling because the employee would acquire knowledge that he could apply to the development of a new curriculum for the school.

Other illustrations can be found in cases involving employee conduct with mixed purposes while off the employer's premises during the lunch hour, a time when an employee is normally outside the scope of his or her employment. In that case, the employee's injury was compensable because the employer facilitated the employee's errand—cashing her paycheck—by making arrangements with the bank to expand its hours and employ extra tellers during lunchtime on paydays, encouraged the errand by "looking the other way" if an employee returned from the bank a few minutes late, and benefitted from the errand because the employer no longer wished to use a cash payroll. In *Matter of Goldberg v. Gold Medal Farms*, it was established that the employer paid for and benefitted from employees taking lunch at a designated restaurant; thus injuries sustained while returning to work from that restaurant were compensable. In *Matter of Younger v. Motor Cab Transp. Co.*, compensation was awarded to a cabdriver, who divided his fares with his employer, when he parked his cab at an advantageous location and returned to the garage to pick up his paycheck and was fatally injured while returning to his cab. Notably, in the *Watson*, *Goldberg*, and *Younger* cases, none of the claimants were directed or required to perform the particular errand from which they were returning when injured, and each claimant had other options available to accomplish the task undertaken. In each, however, there was encouragement or inducement by the employer for conduct which benefitted the employer. . . .

The rule to be derived from these decisions is that an employee who engages in a work-related errand while travelling between work and home, and is injured during that trip, sustains injuries that arise out of and in the course of his or her employment if the employer both encouraged the errand and obtained a benefit from the employee's performance of the errand. Application of this two-part test assures the legitimacy of the claim that the employee was engaged in the service of his or her employer, as distinguished from having undertaken a personal errand that had some incidental work-related purpose. That the employer obtains some benefit from the employee's errand will not alone sustain the claim; there must be evidence of affirmative conduct by the employer soliciting or encouraging the errand.

B.

Because the determination whether an injury is compensable turns on the facts of a given case, the Workers' Compensation Board is afforded "wide latitude" in deciding whether the employee was engaged in a special errand. Here, the record supports the Board's conclusion that claimant was engaged in a special errand. He was required by his employer to wear a uniform and to keep it clean and was subject to discipline if he failed to do so. Although the employee could choose various means to clean his uniforms, the employer was not indifferent to how and where the cleaning was done. It encouraged guards to dry clean their uniforms by paying for dry cleaning services and by establishing accounts at selected establishments, such as Karpinski's. This not only induced employees to use the particular establishments but enabled them to receive free cleaning without the inconvenience of paying cash and submitting receipts for reimbursement. The benefit to the employer is equally apparent. Its security officers presented a neat and clean appearance and claimant's patronage at Karpinski's provided the employer with the extra benefit of consolidated accounting, because the employer was saved the inconvenience of reimbursing claimant for cash expenditures.

II

The question remains whether claimant's employment had terminated prior to the occurrence of the accident, thus foreclosing application of the special errand exception. Respondents contend that once Neacosia left Karpinski's and started home by his usual route he was no longer engaged in the service of his employer. However, the fact that the employee's injury occurs on the way home after having completed the work-related errand does not necessarily mean that the employee's travel is no longer within the scope of employment. Once an employee engages in the performance of a special errand, he or she is considered to be acting within the scope of employment during travel between the place of employment and home. The premise of the rule is that the errand for the employer has altered the usual geographical or temporal scheme of travel, thereby altering the risks to which the employee is usually exposed during normal travel.

Accordingly, the order of the Appellate Division should be reversed, with costs, and the decision of the Workers' Compensation Board reinstated.

FRANCIS McCARRON, SR. v. WORKERS' COMPENSATION APPEAL BOARD (DELAWARE COUNTY DISTRICT ATTORNEY'S OFFICE) 761 A.2d 668; 2000 Pa. Commw. LEXIS 597 (2000)

Study Terms: "Mental/mental" injury classification, "mental stimulus/mental" injury classification, "mental/physical" injury classification, abnormal working condition

Francis McCarron (Claimant) appeals from an order of the Workers' Compensation Appeal Board (Board) reversing the Workers' Compensation Judge's (WCJ) order granting Claimant's claim petition. For the reasons set forth herein, we affirm.

Claimant worked for the County of Delaware Office of the District Attorney (Employer) as a detective in its white collar crime unit. Claimant's job duties included white collar crime investigation, evidence collection and witness and defendant interviews. Claimant was also on call five weeks per year, which required that he work with other departments on drug interdiction efforts and homicide investigation.

Claimant filed a claim petition alleging that as of April 30, 1997, he suffered from work-induced elevated blood pressure, which precipitated work injuries in the form of hypertension, anxiety, nervousness, insomnia and loss of concentration. Claimant maintained that his work injury resulted from increased stress due to an increased work load and case complexity following the instatement of a new supervisor, Detective Joseph Ryan, eighteen months earlier. At hearings, Claimant presented his own testimony and the deposition testimony of his treating physician, Quintin Giorgio, D.O. Claimant testified that his increased work assignments and case complexity led to the development of his disabling high blood pressure, hypertension and other ailments. Claimant further stated that although his home life is stressful, his health problems did not develop until his job situation changed. Dr. Giorgio opined within a reasonable degree of medical certainty, that Claimant's job-related stress precipitated development of his disabling high blood pressure, hypertension, anxiety, nervousness, insomnia and loss of concentration.

In opposition to Claimant's claim petition, Employer presented the testimony of two of Claimant's fellow detectives and his supervisor, Detective Ryan. Each of these witnesses disputed that Claimant's workload and case complexity had increased. Employer also submitted various statistical reports disputing that Claimant's work load and case complexity had increased during Detective Ryan's tenure. Lastly, Employer presented the deposition testimony of Francis C. Kempf, M.D., who testified that although Claimant did require treatment for hypertension his condition was not causally related to his employment.

By decision and order dated June 25, 1999, the WCJ concluded that Claimant met his burden of proof by showing that he sustained physical injuries resulting from a psychological stimulus. The WCJ accepted as credible and persuasive the testimony of Claimant and Dr. Giorgio. The WCJ resolved that the testimony of Claimant's colleagues and supervisor did not adequately rebut Claimant's testimony. The WCJ rejected Employer's statistical reports as inaccurate and unpersuasive. The WCJ also rejected Dr. Kempf's testimony to the extent that it conflicted with Dr. Giorgio's causation opinion. The WCJ granted total disability benefits effective April 30, 1997 upon concluding that Claimant sustained a work injury as a result of increased work load and case complexity, which rendered him unable to return to his pre-injury job. Employer timely appealed to the Board.

By decision and order dated February 9, 2000, the Board reversed. The Board held that the WCJ had applied the wrong burden of proof. The Board reasoned that Claimant's maladies were of a psychiatric nature and were triggered by the added difficulty and volume of his workload. The Board concluded that Claimant's injuries arose under the mental/mental classification requiring that Claimant prove that his psychic injury is other than a subjective reaction to normal working conditions. The Board determined that Claimant did not carry his burden of proof based on the conclusion that increased volume and complexity of work does not qualify as an abnormal working condition necessary to support an award in a mental stimulus/mental injury case. Claimant now brings the instant appeal.

Claimant presents three issues for our consideration. First, Claimant asserts that the Board erroneously determined that his injury was caused by a mental stimulus. Claimant next argues that the Board erred by analyzing his claim petition under the burden of proof applicable to the mental/mental classification rather than the mental/physical classification designated by the WCJ. Alternatively, Claimant contends that even if the mental/mental classification is correct, he met his burden of proof because his increased workload and case complexity constitute an abnormal working condition.

Claimant first argues that the WCJ and the Board failed to recognize that his injury was caused by the physical demands of his job, and therefore, the Board erred by deciding the case under the abnormal working conditions standard. Initially, we note that Claimant has waived this issue pursuant to Pa. R.A.P. 1551 since he raises this argument for the first time on appeal. However, even if properly preserved, this argument would fail based on our review of the record. Although elevated, or high, blood pressure (as it is commonly called) may be caused by physical factors, such as, trauma, overexertion, excessive heat, morbid obesity, etc., all of the evidence presented by Claimant showed that his high blood pressure was caused by job stress rather than any physical factor.

We next turn to the question of whether the Board applied the correct burden of proof. While we disagree with the Board's determination that the facts of Claimant's case require that he satisfy the burden of proof applicable to a mental/mental rather than a mental/physical claim, we nevertheless affirm the Board's order since the Supreme Court decision in *Davis v. Workers' Compensation Appeal Board,* 561 Pa. 462, 751 A.2d 168 (2000), effectively nullified the mental/mental vs. mental/physical distinction. . . .

In our recent decision in *Daneker v. Workers' Compensation Appeal Board (White Haven Center),* 757 A.2d 429 (Pa. Cmwlth. 2000), we addressed a case with very similar facts. In Daneker, the claimant, Karen Daneker, worked at a nursing care facility as a clothing room aide. The employer eliminated Daneker's position and offered her alternate work as a residential service aide assisting with patient care. Daneker took the new position even though it resulted in a loss of seniority and the transfer to a less desirable shift. After approximately one year, Daneker filed a claim petition asserting that job stress produced a variety of physical ailments that ultimately rendered her disabled. The WCJ applied the abnormal working conditions standard and denied Daneker's claim petition. The Board affirmed and Daneker appealed to this Court asserting that the abnormal working condition standard did not apply to her case because she sustained physical injuries.

In Daneker, we summarized the historical assessment of the two mental stimulus injury classifications and the impact of the Davis decision as follows:

In analyzing a mental/physical claim, two common elements must be present: "(1) a psychological stimulus which causes a physical injury which continues after the psychological stimulus is removed; and (2) a disability, i.e., loss of earning power, caused by the physical condition. . . ." A mental/mental claim, however, explains a situation where a claimant is exposed to a psychological stimulus and subsequently develops a psychological injury.

Traditionally, the above mental/mental versus mental/physical analysis would have been germane to the result in this case. Recently, however, our Supreme Court in [Davis] held "that where a psychic injury is claimed, regardless of whether it is manifested through psychic symptoms alone or physical symptoms as well, the claimant must establish that the injury arose from abnormal working conditions in order to recover benefits." Thus, we need not decide whether [Daneker's] claim is mental/mental or mental/physical in nature, since, regardless of whether the psychological stimuli to which she was exposed resulted in physical and psychic, or merely psychic manifestations, the burden of proof, i.e., the requirement to prove abnormal working conditions, is the same in each instance. *Daneker,* 757 A.2d at 432-33.

Having resolved that Claimant is subject to the abnormal working condition standard, we next address Claimant's assertion that he satisfied this heightened burden of proof. Initially, we note that the question of whether working conditions are "abnormal" is a mixed question of law and fact and is fully reviewable by this Court.

The "abnormal working condition" approach is a method to distinguish psychological injuries that were compensable because the necessary causal relationship between employment and mental disability has been established from those psychological injuries that arise from an employee's subjective reaction to normal working conditions. *Martin v. Ketchum,* 523 Pa. 509, 568 A.2d 159 (1990). . . . Psychic injury cases are fact specific and are to be considered in the context of the specific employment.

Most workers' compensation claims involving abnormal working conditions are aptly described as arising from a series of work events occurring over time that cause stress, which then result in psychological harm. Unless a claimant proves that these events are so extraordinary and cannot be envisioned as part of the work, the psychiatric injury is not considered abnormal but rather normal stress flowing from the daily work events to which the claimant reacts in a subjective and abnormal fashion. Consequently, any ensuing psychological injury is non-compensable because it is recognized to be the result of employee's own subjective reaction to normal events rather than the work itself.

Claimant contends that he met this burden based on the WCJ's finding that his work injury resulted from increased stress due to an increase in work load and case complexity over an eighteen-month period following the instatement of his new supervisor. However, settled precedent establishes that stress induced injury resulting from an increased workload is not sufficient to satisfy the abnormal working condition standard.

On the facts here, Claimant worked as a detective in the Delaware County District Attorney's office. A change in supervisors brought on a restructuring of work assignments with Claimant being assigned to investigate a greater number of the white-collar crimes, which Claimant asserts are more complex cases from an investigation stand-point. We recognize the stress inherent in criminal investigation particularly when performed under time deadlines imposed by attorneys preparing for trial. However, in examining the facts as found by the WCJ, given the forgoing analysis we cannot conclude that Claimant experienced anything other than a subjective reaction to normal working conditions. Clearly, the District Attorney's office may structure the work assignments of its detectives as it deems necessary with each of the detectives equally subject to a change in volume or nature of their assignments. Additionally, Claimant's observation that white-collar crime investigation is more complex constitutes an individual perception making any commensurate response a subjective reaction to a normal working condition. Consequently, we cannot accept the conclusion that Claimant's working conditions were "abnormal."

Accordingly, the Board's order dated February 9, 2000 is hereby affirmed.

AE CLEVITE, INC. AND LIBERTY MUTUAL INSURANCE COMPANY v. LABOR COMMISSION AND CHARLES TJAS 2000 UT App 35; 996 P.2d 1072; 2000 Utah App. LEXIS 11 (2000)

Study Terms: Injuries arising out of and in the course of employment, "work at home"

Petitioners Ae Clevite, Inc., and its insurance carrier, Liberty Mutual Insurance Company, seek review from a final order of the Utah Labor Commission (Commission) entered on February 26, 1999, awarding Mr. Charles Tjas workers' compensation benefits from an injury occurring at his home. We affirm.

Background

Neither party disputes the facts of this case. In its ruling the Commission found that Mr. Tjas sustained a severe neck injury causing quadriplegia on January 13, 1997, while spreading salt on the driveway of his residence. When the injury occurred, Mr. Tjas was employed by Ae Clevite, an automotive supply company, as a district sales manager in Utah and several surrounding states. Because Ae Clevite did not have an office in Salt Lake City, it authorized Mr. Tjas to use his personal residence in Salt Lake City as a base of operations for his work. Ae Clevite provided Mr. Tjas with various office supplies, a car, and frequently delivered company correspondence and other materials to Mr. Tjas's home by U.S. mail or private courier. Part of Mr. Tjas's duties included making sales calls and performing office work at home.

The night before the accident, several inches of snow fell on Mr. Tjas's steep driveway. The next morning, Mr. Tjas drove to several local sales calls but did not clear the snow. Although Mr. Tjas's son removed the snow later that morning, the driveway remained icy. After returning home in the mid-afternoon, Mr. Tjas spent nearly an hour loading his car with material for an upcoming sales trip and waited for a large package to be delivered in connection with the business trip. When Mr. Tjas observed the mailman approaching, he decided to spread salt on the driveway so the postman could make his delivery more safely. In doing so, however, Mr. Tjas slipped on the ice and fell, suffering a severe neck injury.

Mr. Tjas subsequently filed a claim for workers' compensation benefits with the Utah Labor Commission for his injuries. The Commission's Administrative Law Judge (ALJ) concluded that Mr. Tjas's injury arose out of and in the course of his employment and awarded benefits. The Labor Commission subsequently affirmed the ALJ's decision awarding Mr. Tjas compensation pursuant to section 34A-2-401 of the Utah Code. Ae Clevite and its insurance carrier filed this petition for judicial review.

Issue and Standard of Review

This case involves the application of the Utah Workers' Compensation Act to a "work at home" situation. Specifically, we consider whether the Commission erred in determining that Mr. Tjas's injury "arose out of and in the course of" his employment with Ae Clevite, thus entitling him to workers' compensation benefits under Utah Code Ann. § 34A-2-401 (1997), the Utah Workers' Compensation Act.

The applicable standard of review for a formal adjudicative hearing is governed by the Utah Administrative Procedures Act. "When the Legislature has granted an agency discretion to determine an issue, we review the agency's action for reasonableness." *Caproz* 945 P.2d at 143; see *Cross v. Board of Review of Indus. Comm'n*, 824 P.2d 1202, 1204 (Utah Ct. App. 1992). Absent a grant of discretion, we use a correction-of-error standard "'in reviewing an agency's interpretation or application of a statutory term.'" *Cross*, 824 P.2d at 1204 (citation omitted).

In this case, the Legislature has granted the Commission discretion to determine the facts and apply the law to the facts in all cases coming before it. As such, we must uphold the Commission's determination that Mr. Tjas's injury "arose out of and in the course of" his employment, unless the determination exceeds the bounds of reasonableness and rationality so as to constitute an abuse of discretion under section 63-46b-16(h)(i) of the UAPA.

To qualify for workers' compensation benefits in Utah, a person must be an employee who suffers an injury caused by an accident. In addition, the employee must prove two essential elements under section 34A-2-401: (1) the accident occurred "in the course of" the employment, and (2) the accident "arose out of" the employment. Id. n3 An employee must prove both elements by a preponderance of the evidence. Petitioners do not dispute that Mr. Tjas sustained an accidental injury. Rather, petitioners argue that the injury does not satisfy either of the elements of section 34A-2-401 of the Utah Code.

First, petitioners argue that Mr. Tjas's injury did not arise "in the course of" his employment because Ae Clevite never requested, directed, encouraged, or reasonably expected Mr. Tjas to salt his driveway and because Mr. Tjas was not in an "employer controlled" area when the injury occurred. Utah courts, however, have recognized that an employee's injury arises in the course of employment even if these circumstances are not present. Indeed, "under Utah law, an accident occurs 'in the course of' employment when it 'occurs while the employee is rendering services to his employer which he was hired to do or doing something incidental thereto, at the time when and the place where he was authorized to render such service.'" *Buczynski*, 934 P.2d at 1172. An activity is "incidental to the employee's employment if it advances, directly or indirectly, his employer's interests." *Black*, 733 P.2d at 156.

In this case, the Commission concluded that Mr. Tjas's injury arose in the course of his employment because his efforts to make his driveway safe for the delivery of work-related materials was "reasonably incidental" to his work for Ae Clevite. Specifically, it ruled that the ability of Ae Clevite to have work-related materials delivered to Mr. Tjas's home by mail or courier service was an "integral part of the employment relationship," so that Mr. Tjas's activity was "reasonably incidental" to his business. We agree. Although Mr. Tjas was not performing a work-related duty or in an employer-controlled area when the injury occurred, he was removing an obstacle which could have impeded his work and was at the location of his regular place of work when the injury occurred. We recognize that Mr. Tjas may have decided to salt the driveway at some other time for his own non-job related purposes, yet the fact remains that when he did, it was in an attempt to remove a hurdle that could have prevented the delivery of the expected business package. In other words, Mr. Tjas's act of salting the driveway was motivated in-part by a purpose to benefit Ae Clevite and thus was reasonably incidental, rather than tangentially related, to his employment. As such, the Commission correctly concluded that Mr. Tjas's injuries arose "in the course of" his employment.

B. "Arising out of" Employment

Second, petitioners argue that Mr. Tjas's injury did not "arise out of" his employment with Ae Clevite. Specifically, petitioners contend that the injury arose from Mr. Tjas's duty as a homeowner to maintain his premises, a risk Mr. Tjas would have been equally exposed to apart from his employment.

In Buczynski we stated that in Utah,

> an accident arises out of employment when there is a causal relationship between the injury and the employment. Arising out of, however, does not mean that the accident must be caused by the employment; rather, the employment is thought of more as a condition out of which the event arises than as the force producing the event in affirmative fashion.

934 P.2d at 1172.

Under the facts of this case, we agree with the Commission that Mr. Tjas's injury arose from a risk associated with his work for Ae Clevite due to the parties' "work at home" arrangement. As such, we hold the Commission did not err in ruling that Mr. Tjas's injury arose from his employment with Ae Clevite.

Conclusion

As a general proposition, the Workers' Compensation Act, Utah Code Ann. § 34A-2-401 (1997), applies to "work at home" situations when a person sustains an injury by an accident "arising out of and in the course of" the employee's employment. Moreover, we hold that under these facts, Mr. Tjas's injury at his home falls within the category of compensability under section 34A-2-401 because it was an injury that arose out of and in the course of his employment.

Affirmed.

Review Questions

1. What three defenses did early employers have when sued by injured employees? How was this changed under the employers' liability acts? What further changes were made under worker's compensation laws?

2. Why should an employer be rendered virtually defenseless when an injured employee sues because of a work-connected injury, whereas the employer would have all the defenses normally available in a tort action by a nonemployee? Why should the employee be placed in such a preferred position?

3. Why were employers' liability acts unsatisfactory?

4. What basic proof is necessary for an injured employee to recover from an employer under worker's compensation laws?

5. What generally happens if an employer refuses to be covered by the state's worker's compensation system?

6. What effect does experience rating have on an employer's worker's compensation coverage purchased from an insurer?

7. What is a second (or subsequent) injury fund? Why is it necessary, and how does it work?

8. Give two examples for each: (a) temporary total disability, (b) permanent partia disability, and (c) permanent total disability.

9. Black, an engineer for the White Company, is injured while returning home from work. What circumstances might cause this injury to be covered by workers' compensation?

10. Of what importance is the method of cleaning in the *Neacosia v. New York Power Authority Case?* Did the employer benefit from the employee having his uniform dry cleaned as opposed to some other method?

11. Why do you think the employee in *Francis McCarron, Sr. v. Workers' Compensation Appeal Board* sought workers compensation as opposed to damages related to a hostile work environment, or intentional infliction of emotional distress (see Chapter 21)?

12. How can an employer protect itself in situations where an employee telecommutes or works at home, as in the case of *Ae Clevite v. Labor Commission and Charles Tjas?* Will a case like this discourage employers from allowing employees to work at home?

Safety

HIGHLIGHTS

- The Occupational Safety and Health Act is the basis for most of the standards and procedures related to worker injuries and illness incurred on the job.

- It is a supplement to state regulation and its jurisdiction is limited to companies that affect interstate commerce.

It seems only reasonable to expect that the environment in which people work should not be hazardous to them. Furthermore, since each employee often represents a significant investment in training (either formal or informal), an employer suffers a sizable economic loss when an employee is absent due to illness or injury. The actual monetary costs of work-related injuries go beyond those of the employer, however. The wealth of a society is measured by the goods and services it produces. Losses due to an employee's illness diminish that wealth. In addition, such injuries increase the public burden. Many people who rely on welfare and similar assistance can trace their plight to industrial accidents. Industrial injuries, then, are a source not only of personal misery, but also of social and economic problems.

In Chapter 25, we discussed worker's compensation and its historical perspective. Worker's compensation is a remedy—a device to partially compensate an injured employee for partial or total temporary or permanent loss. It is not and was not intended to be a total replacement for the loss suffered. For example, it includes no compensation for pain and suffering, despite the fact that the loss of a limb causes lots of pain. The fact that there is a loss should itself deter industrial injuries for both employee and employer. Still, the strength of the deterrent (or the incentive for safety) does not seem to have been sufficient.

In 2005 there were 5,703 work-related deaths, and 4,200,000 illnesses and injuries.[1] To increase the incentive for industrial safety, the United States Congress in December 1970, passed the Williams-Steiger Occupational Safety and Health Act (OSHA).[2] It took effect in April 1971. Congress's purpose in enacting OSHA was "to assure so far as possible every working man and woman in the Nation safe and healthful working conditions."

Of course, complete safety in a working environment is a virtual impossibility and probably would be undesirable even if it were achievable.[3] That is, few of us would be likely to accomplish much in a "padded cell" atmosphere. The work environment should be safe and healthful, yet not so safe as to represent a stultifying atmosphere. The real, practical goal, then, is to provide working conditions that are as safe and healthful as is technologically and economically feasible. (See Appendix B, Case No. 00-5, Public Welfare-Bridge Structure.) In addition, OSHA required a federal determination of an appropriate, minimum level of safety. The policy statement of the act acknowledges the practical limits of providing a safe working environment.

OSHA

The legal essence of OSHA is encapsulated in the following statement of employer duties:
Section 654, Duties:
(a) Each employer—
   (1) shall furnish to each of his employees employment and a place of employment which are free from recognized hazards that are causing or are likely to cause death or serious physical harm to his employees;
   (2) shall comply with occupational safety and health standards promulgated under this Act.
(b) Each employee shall comply with occupational safety and health standards and all rules, regulations, and orders issued pursuant to this Act which are applicable to his own actions and conduct.

Applicability

Congress passed OSHA in part because of the drain unsafe working conditions created on the national economy. As a federal law, OSHA is limited in its jurisdiction to companies that affect **interstate commerce.** Of course, involvement with interstate commerce is often a matter of judgment. Suppose White Company operates only in Oregon. If it uses or handles goods produced in Kentucky, for example, it may be said to operate in interstate commerce. Or Green Company may be a road builder building roads only in the vicinity of Helena, Montana; still, the argument could be made that the roads are intended for interstate travelers as well as those who live nearby. The courts tend to take a very expansive view of what constitutes interstate commerce.

---

[1]Bureau of Labor Statistics, 2005 www.bls.gov/news.release/pdf/osh.pdf, www.bls.gov/news.release/ pat/cfoi.pdf

[2]Pub. L. 91-596, December 29, 1970, 29 USC §651, *et seq.*

[3]In 2005 there were 5,703 work-related deaths, and 4,200,000 illnesses and injuries. Bureau of Labor Statistics, 2005 www.bls.gov/news.release/pdf/osh.pdf, www.bls.gov/news.release/pdf/cfoi.pdf

OSHA was not intended to replace coverage by other federal laws or the worker's compensation acts. OSHA therefore states that it does not apply to working conditions where other federal agencies exercise statutory authority affecting occupational safety and health. The meaning of this provision is not altogether clear, however, as may be noted from the cases.

Consider this example. Black Trucking is involved in interstate commerce. Its activities are, therefore, supervised by several federal agencies, including the Surface Transportation Board, Department of Transportation.[4]

Black might well think that it already has sufficient governmental supervision, and might resist what it sees as an attempt by OSHA to generate more paperwork and problems. A reviewing court, on the other hand, might conclude that the DOT is insufficiently involved with safety, and require Black to also comply with OSHA. If a serious safety problem is found or suspected, some device for exercising OSHA jurisdiction may also be found.

The OSHA Agency

When Congress passed OSHA, the Occupational Safety and Health Administration was created to enforce the safety standards. Congress delegated the power to set appropriate standards to the Department of Labor. The secretary of labor is the executive head of the OSHA. (Since the acronyms for the act and the agency are the same, we use the OSHA to mean the Occupational Safety and Health Administration and simply OSHA to refer to the statute.) A company's main contact with the OSHA usually comes from the local inspector (or compliance officer). If a company has a problem with the local officer, however, various appeals are possible. OSHA provides for a three-member review commission (the Occupational Safety and Health Review Commission, or OSHRC) to reassess contested citations for violations issued by the officer.

If White Company, for example, files an appropriate notice contesting an OSHA citation (discussed later), White will voice its complaint against the citation at an administrative hearing before an administrative law judge of the commission. If White Company is unhappy with the outcome there (or if the secretary of labor is unwilling to accept the result), a request may be made for a hearing before the full commission. The commission may, at its discretion, review any hearing. When all commission review possibilities have been exhausted, the next step is review of the case by a United States court of appeals. The final step available is to request review by the United States Supreme Court.

NIOSH

In 1978, OSHA created a new entity, the National Institute for Occupational Safety and Health (NIOSH). The director of NIOSH is immediately responsible to the secretary of the Department of Health and Human Services (HHS) and is part of its Centers for Disease Control and Prevention (CDC), but is required to work closely with the secretary of labor. NIOSH is charged with research and education functions for the prevention of work-related injury and illness, particularly as these functions are concerned with toxic substances. Of course, toxic substances also concern the Environmental Protection Agency (EPA), which suggests that cooperation between the two agencies is desirable.

---

[4]The Interstate Commerce Commission (ICC) was the prior agency from 1887 until it was abolished effective December 29, 1995, by 109 Stat. 932.

By itself, NIOSH cannot adopt or enforce standards. However, the NIOSH findings as to safety requirements (for example, maximum levels of toxic substances) are proposed to the secretary of labor as prospective industry standards. Approval and publication of the requirements in the *Federal Register* by the department of labor often lead to their adoption as rules. One example of a toxic substance with which both the EPA and OSHA are concerned is lead.

Section 669 of OSHA requires the HHS secretary to respond to requests by an employer or employee for information about the possible toxic nature of substances in the workplace. White Company or its employees may suspect a harmful concentration of lead in the air in the work environment, for example. In response to a request from either White Company or its employees, NIOSH would investigate the environment and report its results, assessing whether the lead would have potentially toxic effects in the concentrations found.

State Programs

Section 667 of OSHA provides for state occupational safety and health programs to replace those of the federal government. Essentially, if a state adopts a plan that is as effective as OSHA, the state will have jurisdiction over safety matters, not the OSHA. Of course, a proposed state program must meet certain requirements. Typically, the state prepares a plan and submits it to the OSHA. For three years after the approval of a state plan, the state program is considered developmental. During this time, the state must show it is capable of doing an effective job. If the state shows adequate legislative backing, standard-setting ability, standard-enforcing ability, and competent personnel, the OSHA may enter into an **operational status agreement** with the state. Final approval of the state plan may occur after an additional year or more of effective operation. One inducement for states to initiate their own programs is the provision that the OSHA will pay up to half the operating costs of state programs.

Provisions exist for decertification of state programs, and occasionally states voluntarily give up their programs.

On-Site Consultation

Much of the OSHA's efforts focus on work-environment standards and penalties for noncompliance, but this is not the complete story. Suppose, for example, that White Company is sincerely concerned about the safety of its employees and turns to the OSHA for advice. White Company's request will result in a visit by a consultant to identify hazardous conditions and recommend corrections to be made. The consultant sent by a state agency is likely to be a state employee, but if the OSHA is called upon, the consultant will be a private firm or individual. The consultation costs White Company nothing.

The procedures vary somewhat from state to state and in the federal agency, but the following provides an overview: The consultant arrives, has an opening conference, and then tours the workplace, identifying any hazards and possible violations. The consultant also may interview employees. After the tour, the consultant usually has a closing conference to review what he or she has seen and has been told. A written report follows. The report should identify observed hazards and instruct the employer about how to correct them. The consultant, however, cannot issue citations. For employers, the fact that the consultant's visit will not result in an investigation or enforcement proceeding by the OSHA is quite important. Thus, White Company could benefit from a free safety consultation without fear of being penalized by the consultant's findings.

Standards

The question of safety in a workplace is a relative one. As noted earlier, absolute safety is virtually impossible to achieve in any kind of practical circumstance. Thus, a decision must be made as to what is "safe enough." Congress directed that the standards should adequately assure "to the extent feasible, . . . functional capacity even if such employee has regular exposure to the hazard dealt with by such standard for the period of his working life."

When Congress created the OSHA, it gave the agency the first two years of its existence to examine and adopt (**promulgate**) so-called **consensus standards.** The result was wholesale adoption of existing industry standards, state standards, ANSI (American National Standards Institute) standards, and standards from various other sources. Most of these standards have been highly useful, some have had to be modified, and others have been fairly useless.

Some difficulty arises when the standard is ambiguous or its terms open to differing interpretations. In *Tierdael Construction Company v. Occupational Safety and Health Review Commission, et al.,* at the end of the chapter, the key issue was whether removing pipe fell under standards regulating the removal of asbestos from a "structure." Tierdael argued that the pipe was not a structure and therefore its removal did not qualify for regulation under the cited standard. The court disagreed and said that the "overall pipeline system is a structure." Needless to say, definitions are critical to standard drafting.

Standard Setting and Rule Making.   When the OSHA determines that a particular toxic material or harmful physical agent must be handled in a certain way for the sake of safety, it creates a standard. As with rules promulgated by other federal agencies, it does this by publishing the proposed standard or rule in the *Federal Register* along with a request for objections, data, or comments by interested parties. Those interested usually have a 30-day period in which to respond. If no one responds, the rule may be finalized by a second publication in the *Federal Register.* If objections or adverse comments do appear, there is often a hearing before the OSHA.

Variances.   Sometimes an employer may legally avoid compliance with the OSHA standard. Exceptions can be made under both temporary and permanent variances. Suppose, for example, that White Company (a) has a better (or equivalent) means of protecting its employees than the OSHA standard provides, or (b) cannot comply with the standard within the time allotted. As an example of the first situation, assume OSHA had passed the ground-fault electrical standard as first proposed and White Company already had an assured equipment grounding program. This would appear to be sound reason for White Company to request a **permanent variance,** which, if granted, would permit it to avoid the standard.

As an example of the second kind of variance (**a temporary variance**), White Company may find itself in the position of being required to comply with a standard in an impossibly short time. Reconstruction may be necessary, or personnel or materials to make the change may be hard to find. White Company's request for additional time to comply (i.e., temporary variance) may be honored by OSHA, but this is usually contingent upon White doing all it can to protect its employees from the hazard involved in the meantime.

Of course, other reasons exist for granting variances. For example, suppose an OSHA standard conflicts with an employee's religious views. For example, the OSHA requires carpenters to wear hard hats, but the Old Order Amish are required by their religion to wear wide-brimmed black felt hats. In this case, the OSHA issued a permanent variance to allow Old Order Amish carpenters to comply with their religion. Of course, if issuance of the variance might have endangered others, the problem would have been more difficult to resolve.

Standards Reforms.   One of the major complaints about the OSHA involves the issuing of citations about trivia. It is difficult for a company to understand why the design of a toilet seat, a 2″ variation in the location of a fire extinguisher, or a 1″ variation in the distance between the rungs of a wooden ladder (12″ for some ladders, 13″ for others) might be important to someone's safety. The explanation is that such regulations often came to be part of the OSHA during its wholesale adoption of consensus standards at its inception. Thus, many trivial regulations were adopted along with the truly important. Furthermore, Congress required inspectors from the OSHA to issue a citation for any violation they observed.

To avoid wasting resources on enforcing outmoded and trivial regulations, OSHA began a purge of such rules in 1978. The editing efforts have now resulted in the deletion of many unnecessary regulations, and OSHA's efforts to "clean up its act" are continuing.

Inspections

Congress apparently wanted the OSHA to make surprise inspections. OSHA allows a compliance officer to "enter without delay and at reasonable times any factory, plant, establishment, construction site, or other area, workplace, or environment where work is performed by an employee of an employer" and to "inspect and investigate during regular working hours, and at other reasonable times, and within reasonable limits and in a reasonable manner, any such place of employment and all pertinent conditions, structures, machines, apparatus, devices, equipment, and materials therein, and to question privately any such employer, owner, operator, agent, or employee" (29 USC §657).

These surprise inspections are probably the greatest single source of complaints against OSHA. White Company, for example, may have the best safety practices in its industry and, therefore, believe that any reasonable inspection would find no serious faults. But inspections have many disturbing aspects.

For example, many early investigators or compliance officers apparently had little or no industrial experience. They committed blunders of ineptness and engaged in investigations and arguments over trivia. Occasionally, a primary concern of such compliance officers was neatness—whether or not the floor was swept, for instance. Thus, the last thing White Company may desire is an unannounced interruption of its operations by a compliance officer who has no knowledge of the practical effects of inept inspection upon the production facilities.

The reasoning behind these rules was that surprise investigations are the only way to catch industries violating a rule. If the investigator must first ask, be refused, and return with further authority, the employer has an opportunity to change the conditions of the employment environment.

The Fourth Amendment to the United States Constitution indicates that warrantless searches are not allowed. The question thus arose as to whether a warrantless and unannounced inspection violated the Fourth Amendment. Such controversies as this generate work for the U.S. Supreme Court. In the case of *Marshall v. Barlow's, Inc.* (an electrical and plumbing installation business in Pocatello, Idaho), the issue was decided. Barlow's, it seems, had twice refused the OSHA inspector access to its premises. Some of the court's statements in finding the act unconstitutional in regard to warrantless searches are interesting: ". . . the businessman, like the occupant of a residence, has a constitutional right to go about his business free from unreasonable official entries upon his private, commercial property. The businessman, too, has that right placed in jeopardy if the decision to enter and inspect for violation of regulatory laws can be made and enforced by an inspector in the field without official authority evidenced by a warrant".[5]

---

[5]436 U.S. 307 (1978).

Thus, in nonconsensual instances (where the business refuses entry to the OSHA), the authority to inspect is not removed, but some of the intimidation from the threat of a warrantless search is gone. The OSHA officer can still get in to inspect, but he or she must first obtain a search warrant if the attempt to inspect is refused. The Court also held that OSHA provides statutory authority for the issuance of such warrants. But see *National Engineering & Contracting Company v. United States Of America,* Dept. Of Labor, Occupational Safety & Health Administration at the end of the chapter, where OSHA obtained permission from the U.S. Army Corps of Engineers.

Citations.   Just as an avid police officer could probably find something wrong with the way each of us lives, a zealous inspector from the OSHA can probably find rule violations present in almost any establishment. Just as traffic citations could result if you are observed speeding or failing to stop at a stop sign, citations could result from the OSHA inspections. OSHA and the OSHA's regulations require the compliance officer to issue citations for all observed violations (20 USC §658).

Of course, the employer is not required to meekly submit to the citation, posting it for the employees to see and paying the penalties involved. After receiving a citation, the employer has 15 working days to contest the citation, providing the employer has a disagreement with it and is willing to spend the necessary time and money to fight the citation. The employer's first step is to file a **notice of contest.** Generally, the employer does this by sending a written notice to the area director.

Employers appeal OSHA citations on many bases. One of the more common reasons is objection to the lack of specificity in the citation: Occasionally it is difficult to tell from the citation just what the inspector's complaint was. The penalties may also be challenged as excessive, or the time period allowed for abatement of the hazard may be inadequate. Even when the citation is specific, it may be that the employer's safety practice is superior to the agency's rule.

Penalties.   Citations may or may not include proposed penalties for the cited violation (29 USC §666). If the inspection reveals a serious violation, one involving a substantial probability of physical harm or death, there is a mandatory penalty assessment of up to $7,000. A willful violation runs the risk of a considerably greater penalty. In such cases where the employer knew of the hazard and did nothing to correct it or to protect against it, the employer risks substantial penalties. The maximum penalty for these willful violations runs up to $70,000 per violation and/or six months in jail. A second violation can double these maximum penalties.

Safety hazard citations usually state an abatement period, a time limit within which the employer is expected to remove the hazard. A failure to meet the time limit (or appeal the citation in this regard) may prompt the assessment of up to $7,000 per day of abatement-period violation.

Other penalties exist for falsifying records, failing to post notices of citations, and assaulting OSHA inspectors.

Record Keeping

Under OSHA, all but fairly trivial work-related injuries or illnesses must be recorded, and the records must be available for inspection by compliance officers. The purpose, of course, is to assemble a body of knowledge of work-related human problems, an apparently worthy and desirable motive. However, one wonders whether such record keeping is truly a cost-effective way of gathering such information.

With some exceptions, this requirement pertains to all employers having more than 10 employees. The Bureau of Labor Statistics, however, may ask even small businesses for a sampling of such information, so each such employer risks being required to keep those records. Besides small businesses, certain service industries are exempted, as are farmers with respect to records for members of their families.

When a company must keep records, those records are to include all occupational illnesses (assuming a determination can be made as to environmental cause). Also to be recorded are all occupational injuries resulting in

- Death
- Lost work-days
- Restriction of motion
- Loss of consciousness
- Transfer to another job
- Medical treatment other than first aid.

OSHA and the Political Environment

OSHA began in a healthy business climate. In such a climate people, including those who make laws, tend to be concerned with the health and well-being of workers. In an economic down-turn, people tend to be more concerned about the health and well-being of businesses and the unemployment rate. As the economic climate varies, the OSHA emphasis changes to reflect the change in public attitude. The OSHA's actions, thus, can be viewed as resulting not only from the law under which it was created, but also from the prevailing economic and political environment in which it operates.

TIERDAEL CONSTRUCTION COMPANY v. OCCUPATIONAL SAFETY AND HEALTH REVIEW COMMISSION, et al. 340 F.3d 1110 (10th Cir 2003)

Study terms: Citations, negative exposure assessment, due process

I. Introduction

Tierdael Construction Company ("Tierdael") was issued a citation alleging several serious violations of the Occupational Safety and Health Act of 1970, 29 U.S.C. §§ 651-678, after an inspection of their work site by a compliance officer from the Occupational Safety and Health Administration ("OSHA"). Tierdael appealed the citation to the Occupational Safety and Health Review Commission (the "Commission"). A four-day hearing was held before a Commission administrative law judge ("ALJ"). In his decision and order, the ALJ vacated four violations, affirmed four violations as other-than-serious, affirmed three violations as *de minimis,* and assessed penalties totaling $1,800.00. The decision of the ALJ became the final order of the Commission. Exercising jurisdiction pursuant to 29 U.S.C. § 660(a), this court denies Tierdael's petition for review, concluding that the plain and natural meaning of the regulation encompasses Tierdael's activity. Alternatively, OSHA's interpretation of 29 C.F.R. § 1926.1101 (the "OSHA Asbestos Standard") that Tierdael's activity falls within the definition of Class II asbestos work and that Tierdael was required to comply with the requirements of the OSHA Asbestos Standard is entitled to deference.

II. Background

On October 30, 2000, two Tierdael employees removed asbestos-containing cement pipe from an excavation trench located in the middle of a street intersection in a residential neighborhood in Littleton, Colorado. In order to remove the pipe, one of the employees used a two-pound hammer to break the pipe while the other employee sprayed the pipe. The pipe remained intact, in that it did not crumble or become pulverized. The pipe was then lifted out of the trench with a sling and backhoe.

In response to a complaint about Tierdael's removal of the asbestos-containing pipe, an OSHA compliance officer went to the construction site. After conducting an inspection, OSHA, acting under the authority of the Secretary of Labor (the "Secretary"), issued a citation alleging the following violations, which were classified as "serious": (1) failure to provide sufficient protection for the excavation in violation of 29 C.F.R. § 1926.651(j)(2); (2) failure to provide a regulated area for the Class II asbestos work in violation of 29 C.F.R. § 1926.1101(e)(1); (3) failure to conduct exposure monitoring in violation of 29 C.F.R. § 1926.1101(f)(1)(i); (4) failure to use required engineering controls and work practices, including wetting methods, a HEPA vacuum, and proper disposal methods, in violation of 29 C.F.R. § 1926.1101(g)(1); (5) failure to have a competent person supervise the Class II asbestos work in violation of 29 C.F.R. § 1926.1101(g)(7)(i); (6) failure to use work practices and controls for removal, including wetting, keeping the material intact during removal, and bagging and wrapping the material, in violation of 29 C.F.R. § 1926.1101(g)(8); (7) failure to provide respirators in violation of 29 C.F.R. § 1926.1101(h)(1); (8) failure to provide protective clothing in violation of 29 C.F.R. § 1926.1101(i)(1); (9) failure to establish an equipment room or area for decontamination of employees and their equipment in violation of 29 C.F.R. § 1926.1101(j)(2)(i); (10) failure to affix labels onto the water pipe or the bagged transite piping in violation of 29 C.F.R. § 1926.1101(k)(8)(i); and (11) failure to provide training to the employees exposed to the cement pipe in violation of 29 C.F.R. § 1926.1101(k)(9)(iv)(C).

Tierdael contested the citations and a four-day hearing was conducted before a Commission ALJ. At the hearing, Leary Jones, Tierdael's risk manager and controlling official at the work site on October 30, 2000, testified that the OSHA Asbestos Standard generally applied to Tierdael's work activities. He also testified that the pipe breakage and removal from the excavation constituted Class II asbestos work. He further stated that he believed the OSHA Asbestos Standard did not apply since Tierdael's construction activity did not meet or exceed the permissible exposure limit ("PEL") for asbestos. Jones testified that industry data revealed that the asbestos PEL is not exceeded when pipe is removed from an excavation. Moreover, Jones testified that Tierdael concluded that it had complied with the OSHA Asbestos Standard because an October 4, 1995 letter written by Western Environment and Ecology, Inc. ("Western"), which stated that the results of air monitoring conducted at a previous pipe removal construction project "were below current OSHA exposure standards," constituted objective data for a negative exposure assessment. Although the letter was the basis for Tierdael's "Asbestos Action Plan," Jones acknowledged that he did not have the 1995 letter with him at the work site and could not say whether the letter was reviewed prior to the work being performed. . . .

The ALJ ruled, consistent with OSHA's interpretation, that Tierdael removed the pipe from the Littleton water delivery system, which was a structure. The ALJ then concluded that under the plain meaning of the regulation, Tierdael's activity was Class II asbestos work.

The ALJ vacated the violations of failure to provide sufficient protection for the excavation, failure to require engineering controls and work practice of wetting methods and proper disposal, failure to use work practices and controls for removal, and failure to provide training. The ALJ affirmed the violations of failure to provide a regulated area, failure to use required engineering controls and work practices of a HEPA vacuum, failure to have a competent person supervise, and failure to affix labels but reduced the classification of the violations to "other-than-serious." The ALJ also affirmed violations of failure to conduct exposure monitoring, failure to provide respirators, failure to provide protective clothing, and failure to establish an equipment room, but reduced the classification of the violations from "serious" to "*de minimis.*" The ALJ assessed a penalty of $1,800.00.

Tierdael filed a petition for discretionary review with the Commission, appealing the ALJ's order. The Commission declined review. Thus, the ALJ's decision and order became the final order of the Commission.

Tierdael petitions this court for review from the Commission's order, claiming Tierdael's construction activity did not constitute Class II asbestos removal subject to the requirements of 29 C.F.R. § 1926.1101; there was no likely or actual exposure to asbestos because the pipe remained intact; and objective data established a negative exposure assessment, rendering the OSHA Asbestos Standard's requirements moot. Tierdael also asserts that the ALJ's decision and the Secretary's interpretation of the OSHA Asbestos Standard denies due process.

III. Standard of Review

This court reviews the Commission's findings of fact under a substantial evidence standard upon consideration of the record as a whole. The Commission's conclusions of law are reviewed "to determine if they are arbitrary, capricious, an abuse of discretion, or otherwise not in accordance with [the] law." *Universal Constr. Co. v. Occupational Safety & Health Review Comm'n*, 182 F.3d 726, 732 (10th Cir. 1999). . . .

IV. Discussion

A. Class II Asbestos Work

Tierdael first contends that the company did not engage in Class II asbestos removal work within the meaning of 29 C.F.R. § 1926.1101. To determine whether the OSHA Asbestos Standard is applicable, this court must first determine whether the regulation is clear or ambiguous. . . . If the regulation is ambiguous, this court must give substantial deference to the agency's interpretation of the regulation.

Tierdael argues that the OSHA Asbestos Standard is unambiguous and that the plain language of the regulation does not encompass its October 30, 2000 pipe removal activity. Section 1926.1101 regulates asbestos exposure in all construction, alteration, and/or repair work. The OSHA Asbestos Standard establishes four classifications for asbestos work, ranging from high risk Class I to low risk Class IV. 29 C.F.R. § 1926.1101(b). Class II asbestos work is defined as activity "involving the removal of [asbestos-containing material] which is not thermal system insulation or surfacing material." *Id.* Removal is further defined as "all operations where [asbestos-containing material] is taken out or stripped from structures or substrates, and includes demolition operations." *Id.*

Tierdael argues that its excavation work was not Class II asbestos work because the asbestos-containing pipe was not taken out or stripped from a structure or substrate since the removal took place from an excavated pipeline, not a building. On its face, the OSHA Asbestos Standard's definition of removal applies to Tierdael's activity on October 30, 2000. The plain and natural meaning of structure is "something made up of a number of parts that are held or put together in a particular way." The *American Heritage Dictionary of the English Language* (4th ed. 2000). Tierdael removed a section of pipe which was part of a larger water pipeline system. This pipeline system is made up of a number of smaller sections of pipe. Accordingly, the overall pipeline system is a structure. Because the plain and natural meaning of structure covers an underground pipeline, Tierdael's activity falls within the definition of Class II asbestos work. Therefore, the ALJ's decision which upheld OSHA's interpretation was not arbitrary, capricious, an abuse of discretion, or otherwise not in accordance with the law. . . .

Tierdael challenges OSHA's definition of removal as inconsistent with the OSHA Asbestos Standard and its preamble, the history of asbestos remediation in buildings and structures, and parallel Environmental Protection Agency ("EPA") standards. However, OSHA's interpretation is consistent with the purpose of the Occupational Safety and Health Act of 1970, which is "to assure so far as possible. . . . safe and healthful working conditions for every working man and woman in the Nation." *Id.* Tierdael's "Asbestos Action Plan" recognized the health risk associated with asbestos. This court simply cannot agree with Tierdael's suggestion, as presented by their expert, that the activity of breaking and removing the pipe even one inch outside of a building would not require compliance with the OSHA Asbestos Standard simply because the activity would not be Class II work, because the hazards of dealing with and handling asbestos-containing matter would still be present. OSHA's interpretation that removal of pipe from structures that are not buildings constitutes Class II work promotes safe and healthy working conditions for employees subject to potential asbestos exposure. Furthermore, OSHA's interpretation is consistent with the language of the regulation. Therefore, the Commission's conclusion was not unreasonable or against the purpose of the OSHA Asbestos Standard. By concluding that Tierdael's removal of the pipe constituted Class II asbestos work, neither the Commission nor the ALJ abused its discretion, acted arbitrarily or capriciously, or failed to act in accordance with the law.

B. Negative Exposure Assessment

Tierdael asserts that even if the activity constituted Class II asbestos work, it was not required to comply with the OSHA requirements because it conducted a negative exposure assessment. . . .

Tierdael claims that objective industry data, the preamble, and an OSHA compliance directive constitutes objective data supporting the negative exposure assessment. For those violations which can be complied with by producing a negative exposure assessment, the objective data must be comprised of sufficient information to conclude that the activity "will not result in fiber levels in excess of the PELs." Occupational Exposure to Asbestos, 60 Fed. Reg. 33,974, 33,975 (June 29, 1995). Specifically, the objective data should address whether concentrations above the PEL are unlikely under "foreseeable conditions." See Occupational Exposure to Asbestos, 59 Fed. Reg. 40,964, 40,983 (Aug. 10, 1994) (preamble).

After consideration of the entire record, this court upholds the ALJ's conclusion that Tierdael's negative exposure assessment was deficient under the OSHA Asbestos Standard. . . . Accordingly, the Commission did not abuse its discretion, act arbitrarily or capriciously, or fail to act in accordance with the law in concluding that the objective data was insufficient to support a negative exposure assessment in compliance with the OSHA Asbestos Standard.

C. Due Process Violation

Finally, Tierdael argues that its due process rights were violated because it did not receive adequate notice of a violation. As previously stated, OSHA's interpretation fits within the plain and ordinary language of the regulation. . . . Because the plain and ordinary meaning of structure would encompass a pipeline, there is no exclusion for work conducted outside of a building. Accordingly, Tierdael received adequate notice that its removal of the asbestos-containing pipe from the pipeline system on October 30, 2000 was Class II work.

Alternatively, even assuming the regulation could be considered ambiguous, OSHA's interpretation of the regulation remains entitled to deference even if the interpretation is advanced for the first time in an administrative adjudication. The adequacy of notice, however, must be considered when an interpretation is announced during a proceeding. After review of the record, this court concludes that even assuming the regulation was ambiguous, Tierdael had adequate notice that it was required to comply with the OSHA Asbestos Standard. . . . Accordingly, Tierdael was provided adequate notice that its construction activity constituted Class II asbestos work which required compliance with the OSHA Asbestos Standard.

Therefore, because Tierdael was adequately notified that its pipe removal activity required compliance with the OSHA Asbestos Standard, its due process rights were not violated.

IV. Conclusion

Based upon the foregoing reasons, the petition for review is denied.

NATIONAL ENGINEERING & CONTRACTING COMPANY v. UNITED STATES OF AMERICA, DEPT. OF LABOR, OCCUPATIONAL SAFETY & HEALTH ADMINISTRATION 687 F. Supp. 1219 (SD Ohio 1988)

Study terms: "High-hazard industry", inspection, third party consent, "plain sight" procedures

This matter is before the Court upon plaintiff's objections to the Order of the Magistrate Lifting the Stay Granted on June 3, 1988 and Motion for a Stay. . . .

Findings of Fact

1. National Engineering and Contracting Company is a construction company which is presently performing construction work involving encasing the Mill Creek in Cincinnati, Ohio, with concrete for purposes of flood control. This worksite is known as Section 4-A.

2. At the present time, National Engineering is engaged in a concrete pouring operation which involves emplacing concrete on the bottom of the Mill Creek. National Engineering utilizes a boom concrete pump truck to perform this operation. National Engineering's current worksite encompasses an area in length of approximately two city blocks along the Mill Creek, and with a width of approximately two hundred feet.

3. On October 21, 1986, National Engineering was engaged in a concrete pouring operation on this construction job, about two thousand feet upstream from the existing area of National Engineering's pouring operation. An accident occurred when the boom of the pump truck came into contact with energized overhead lines.

4. Whenever an accident occurs, which involves a "high-hazard industry", OSHA's policy and established procedures call for a comprehensive inspection. All construction operations are identified as "high-hazard industry" because of its high injury and illness rates.

5. OSHA began a mandatory inspection of that work area and that pouring operation on October 23, 1986. Because of the litigation initiated by the plaintiff, OSHA interrupted its own inspection. The inspection did provide OSHA with sufficient information to issue citations against National Engineering on April 20, 1987 alleging violations of OSHA standards applicable to this pouring operation. Subsequently, following a trial that occurred on January 5 and 6, 1988 before the Honorable James Burroughs of the Occupational Safety and Health Review Commission, these Citations were vacated by Judge Burroughs on May 3, 1988.

6. As of April 20, 1987, OSHA had not completed its comprehensive inspection of National Engineering's pouring operation because OSHA voluntarily stayed its inspection until the case filed in the Federal District Court had been fully litigated and decided. National Engineering's complaint was dismissed in the District Court on November 20, 1987. Plaintiff appealed, but voluntarily dismissed the appeal. The case was dismissed by the United States Court of Appeals for the Sixth Circuit on March 15, 1988.

7. On May 3, 1988, the Occupational Safety and Health Administration attempted to complete the comprehensive inspection of National Engineering's pouring operation which had been interrupted by the litigation of this case. The purpose for this inspection was to inspect National Engineering's concrete pouring operation as dictated by agency administrative procedures. OSHA did not seek nor obtain a new warrant for access to conduct such an inspection, but relied on the Court order. National Engineering refused OSHA access to the site.

8. OSHA obtained the consent of the U.S. Army Corps of Engineers to enter the work area to perform a "plain view" inspection. On June 2, 1988, OSHA began an inspection of National Engineering's work area, and specifically, its pouring operation.

9. Before entering National Engineering's work area on June 2, 1988, OSHA did not seek nor obtain a new warrant. OSHA's inspection of National Engineering's work area beginning on June 2, 1988 was pursuant to a general administrative plan. . . .

Conclusions of Law

1. Although no search warrant or other process is explicitly required by the Occupational Safety and Health Act of 1970 (29 U.S.C. § 651 *et seq.*), a search warrant or its equivalent is constitutionally necessary to conduct a *non-consensual* OSHA inspection. In this case, the equivalent of a warrant, a Court Order issued, after both parties had been afforded the opportunity to brief the matter.

2. A third party can consent to a search of jointly occupied property as long as the third party has "common authority" over the premises. The Corps has given valid third party consent for OSHA to conduct inspection of Section 4-A as specifically encouraged by the Supreme Court. This consensual search does not violate National's rights.

3. In *Marshall v. Barlow's*, the United States Supreme Court stated:

> Probable cause in the criminal law sense is not required. For purposes of an administrative search such as this, probable cause justifying the issuance of a warrant may be based not only on specific evidence of an existing violation but also on a showing that "reasonable legislative or administrative standards for conducting an. . . .inspection are satisfied with respect to a particular [establishment]." *Camara v. Municipal Court,* 387 U.S. 523 at 538, 87 S. Ct. 1727 at 1736, 18 L. Ed. 2d 930 (1967).

Further, the Occupational Safety and Health Act "authorizes two types of inspection: an inspection pursuant to a general administrative plan, 29 U.S.C. Section 657(a); and an inspection pursuant to an employee complaint, 29 U.S.C. Section 657 (f)". OSHA's inspection of National Engineering's work area was pursuant to a general administrative plan.

4. The accident which occurred on October 21, 1986 triggered the administrative policy of OSHA's requiring comprehensive inspection of National Engineering's present work area begun on June 2, 1988, pursuant to the Court Order.

5. The Magistrate's Order of March 11, 1987, which was subsequently affirmed by this Court, permitted OSHA to complete its inspection of National Engineering's work area within six months from the date of that order. The Court finds that the Magistrate has correctly and reasonably interpreted his own order as approved by this Court. The Magistrate has found that OSHA may inspect Section 4-A within six months of March 14, 1988, when National's appeal was dismissed before the Court of Appeals and the Order of March 11, 1987 became a final order.

6. OSHA had not completed its comprehensive investigation of National Engineering's pouring operation as of April 20, 1987 when it issued its Citation to National Engineering alleging violation of OSHA's standards. This Court will not penalize a party for voluntarily suspending its administrative procedure pending termination of litigation involving the validity of that procedure. But for the plaintiff's filing of this lawsuit, the inspection would have been completed within a short period after the accident which triggered it, pursuant to administrative procedures.

Conclusion

The Court finds that the inspection now being conducted is reasonable under the holding of *Marshall* because it was pursuant to an administrative standard procedure and the Court Order—the equivalent of a warrant. Further there was valid consent to the "plain view" inspection now being conducted, as found by the Magistrate and approved by this Court.

The plaintiff's Motion is hereby GRANTED to the extent that OSHA shall be allowed only to complete the standard administrative "plain sight" search interrupted by the filing of this lawsuit, in that the inspection shall not exceed five days in duration and shall proceed in accordance with normal "plain sight" procedures.

Review Questions

1. Why was the OSHA created?

2. In the business of the OSHA, what are the functions of the inspector, administrative law judge, the Occupational Safety and Health Review Commission, and a U.S. Court of Appeals?

3. How is NIOSH related to the OSHA?

4. In what ways could a state program for occupational safety and health be superior to the federal program? In what ways could the federal program be expected to be superior?

5. Consider a power press (punch press or brake press). List at least five ways to improve the safety of its human operator.

6. The U.S. Supreme Court decision in *Marshall v. Barlow's, Inc.* indicates that the OSHA must obtain a search warrant if an employer refuses to submit to inspection upon the inspector's first visit. What arguments can you propose for and against the warrant requirement?

7. In the *Tierdael* case, why did Tierdael object to the Commissioner's order, especially since the assessed penalties only totaled $1,800?

8. Do you think that National Engineering had sufficient notice that OSHA would continue its inspection? Why did OSHA obtain the consent to inspect from the U.S. Army Corps of Engineers?

Appendices

NSPE Code of Ethics for Engineers

PREAMBLE

Engineering is an important and learned profession. As members of this profession, engineers are expected to exhibit the highest standards of honesty and integrity. Engineering has a direct and vital impact on the quality of life for all people. Accordingly, the services provided by engineers require honesty, impartiality, fairness, and equity, and must be dedicated to the protection of the public health, safety, and welfare. Engineers must perform under a standard of professional behavior that requires adherence to the highest principles of ethical conduct.

I. FUNDAMENTAL CANONS

Engineers, in the fulfillment of their professional duties, shall:

1. Hold paramount the safety, health, and welfare of the public.
2. Perform services only in areas of their competence.
3. Issue public statements only in an objective and truthful manner.
4. Act for each employer or client as faithful agents or trustees.
5. Avoid deceptive acts.
6. Conduct themselves honorably, responsibly, ethically, and lawfully so as to enhance the honor, reputation, and usefulness of the profession.

Reprinted by permission of the National Society of Professional Engineers (NSPE) www.nspe.org

II. RULES OF PRACTICE

1. Engineers shall hold paramount the safety, health, and welfare of the public.
   a. If engineers' judgment is overruled under circumstances that endanger life or property, they shall notify their employer or client and such other authority as may be appropriate.
   b. Engineers shall approve only those engineering documents that are in conformity with applicable standards.
   c. Engineers shall not reveal facts, data, or information without the prior consent of the client or employer except as authorized or required by law or this Code.
   d. Engineers shall not permit the use of their name or associate in business ventures with any person or firm that they believe is engaged in fraudulent or dishonest enterprise.
   e. Engineers shall not aid or abet the unlawful practice of engineering by a person or firm.
   f. Engineers having knowledge of any alleged violation of this Code shall report thereon to appropriate professional bodies and, when relevant, also to public authorities, and cooperate with the proper authorities in furnishing such information or assistance as may be required.
2. Engineers shall perform services only in the areas of their competence.
   a. Engineers shall undertake assignments only when qualified by education or experience in the specific technical fields involved.
   b. Engineers shall not affix their signatures to any plans or documents dealing with subject matter in which they lack competence, nor to any plan or document not prepared under their direction and control.
   c. Engineers may accept assignments and assume responsibility for coordination of an entire project and sign and seal the engineering documents for the entire project, provided that each technical segment is signed and sealed only by the qualified engineers who prepared the segment.
3. Engineers shall issue public statements only in an objective and truthful manner.
   a. Engineers shall be objective and truthful in professional reports, statements, or testimony. They shall include all relevant and pertinent information in such reports, statements, or testimony, which should bear the date indicating when it was current.
   b. Engineers may express publicly technical opinions that are founded upon knowledge of the facts and competence in the subject matter.
   c. Engineers shall issue no statements, criticisms, or arguments on technical matters that are inspired or paid for by interested parties, unless they have prefaced their comments by explicitly identifying the interested parties on whose behalf they are speaking, and by revealing the existence of any interest the engineers may have in the matters.
4. Engineers shall act for each employer or client as faithful agents or trustees.
   a. Engineers shall disclose all known or potential conflicts of interest that could influence or appear to influence their judgment or the quality of their services.
   b. Engineers shall not accept compensation, financial or otherwise, from more than one party for services on the same project, or for services pertaining to the same project, unless the circumstances are fully disclosed and agreed to by all interested parties.
   c. Engineers shall not solicit or accept financial or other valuable consideration, directly or indirectly, from outside agents in connection with the work for which they are responsible.
   d. Engineers in public service as members, advisors, or employees of a governmental or quasi-governmental body or department shall not participate in decisions with respect to services solicited or provided by them or their organizations in private or public engineering practice.
   e. Engineers shall not solicit or accept a contract from a governmental body on which a principal or officer of their organization serves as a member.

5. Engineers shall avoid deceptive acts.
   a. Engineers shall not falsify their qualifications or permit misrepresentation of their or their associates' qualifications. They shall not misrepresent or exaggerate their responsibility in or for the subject matter of prior assignments. Brochures or other presentations incident to the solicitation of employment shall not misrepresent pertinent facts concerning employers, employees, associates, joint venturers, or past accomplishments.
   b. Engineers shall not offer, give, solicit, or receive, either directly or indirectly, any contribution to influence the award of a contract by public authority, or which may be reasonably construed by the public as having the effect or intent of influencing the awarding of a contract. They shall not offer any gift or other valuable consideration in order to secure work. They shall not pay a commission, percentage, or brokerage fee in order to secure work, except to a bona fide employee or bona fide established commercial or marketing agencies retained by them.

III. PROFESSIONAL OBLIGATIONS

1. Engineers shall be guided in all their relations by the highest standards of honesty and integrity.
   a. Engineers shall acknowledge their errors and shall not distort or alter the facts.
   b. Engineers shall advise their clients or employers when they believe a project will not be successful.
   c. Engineers shall not accept outside employment to the detriment of their regular work or interest. Before accepting any outside engineering employment, they will notify their employers.
   d. Engineers shall not attempt to attract an engineer from another employer by false or misleading pretenses.
   e. Engineers shall not promote their own interest at the expense of the dignity and integrity of the profession.
2. Engineers shall at all times strive to serve the public interest.
   a. Engineers are encouraged to participate in civic affairs; career guidance for youths; and work for the advancement of the safety, health, and well-being of their community.
   b. Engineers shall not complete, sign, or seal plans and/or specifications that are not in conformity with applicable engineering standards. If the client or employer insists on such unprofessional conduct, they shall notify the proper authorities and withdraw from further service on the project.
   c. Engineers are encouraged to extend public knowledge and appreciation of engineering and its achievements.
   d. Engineers are encouraged to adhere to the principles of sustainable development1 in order to protect the environment for future generations.
3. Engineers shall avoid all conduct or practice that deceives the public.
   a. Engineers shall avoid the use of statements containing a material misrepresentation of fact or omitting a material fact.
   b. Consistent with the foregoing, engineers may advertise for recruitment of personnel.
   c. Consistent with the foregoing, engineers may prepare articles for the lay or technical press, but such articles shall not imply credit to the author for work performed by others.

4. Engineers shall not disclose, without consent, confidential information concerning the business affairs or technical processes of any present or former client or employer, or public body on which they serve.

   a. Engineers shall not, without the consent of all interested parties, promote or arrange for new employment or practice in connection with a specific project for which the engineer has gained particular and specialized knowledge.

   b. Engineers shall not, without the consent of all interested parties, participate in or represent an adversary interest in connection with a specific project or proceeding in which the engineer has gained particular specialized knowledge on behalf of a former client or employer.

5. Engineers shall not be influenced in their professional duties by conflicting interests.

   a. Engineers shall not accept financial or other considerations, including free engineering designs, from material or equipment suppliers for specifying their product.

   b. Engineers shall not accept commissions or allowances, directly or indirectly, from contractors or other parties dealing with clients or employers of the engineer in connection with work for which the engineer is responsible.

6. Engineers shall not attempt to obtain employment or advancement or professional engagements by untruthfully criticizing other engineers, or by other improper or questionable methods.

   a. Engineers shall not request, propose, or accept a commission on a contingent basis under circumstances in which their judgment may be compromised.

   b. Engineers in salaried positions shall accept part-time engineering work only to the extent consistent with policies of the employer and in accordance with ethical considerations.

   c. Engineers shall not, without consent, use equipment, supplies, laboratory, or office facilities of an employer to carry on outside private practice.

7. Engineers shall not attempt to injure, maliciously or falsely, directly or indirectly, the professional reputation, prospects, practice, or employment of other engineers. Engineers who believe others are guilty of unethical or illegal practice shall present such information to the proper authority for action.

   a. Engineers in private practice shall not review the work of another engineer for the same client, except with the knowledge of such engineer, or unless the connection of such engineer with the work has been terminated.

   b. Engineers in governmental, industrial, or educational employ are entitled to review and evaluate the work of other engineers when so required by their employment duties.

   c. Engineers in sales or industrial employ are entitled to make engineering comparisons of represented products with products of other suppliers.

8. Engineers shall accept personal responsibility for their professional activities, provided, however, that engineers may seek indemnification for services arising out of their practice for other than gross negligence, where the engineer's interests cannot otherwise be protected.

   a. Engineers shall conform with state registration laws in the practice of engineering.

   b. Engineers shall not use association with a nonengineer, a corporation, or partnership as a "cloak" for unethical acts.

9. Engineers shall give credit for engineering work to those to whom credit is due, and will recognize the proprietary interests of others.

   a. Engineers shall, whenever possible, name the person or persons who may be individually responsible for designs, inventions, writings, or other accomplishments.

   b. Engineers using designs supplied by a client recognize that the designs remain the property of the client and may not be duplicated by the engineer for others without express permission.

c. Engineers, before undertaking work for others in connection with which the engineer may make improvements, plans, designs, inventions, or other records that may justify copyrights or patents, should enter into a positive agreement regarding ownership.

d. Engineers' designs, data, records, and notes referring exclusively to an employer's work are the employer's property. The employer should indemnify the engineer for use of the information for any purpose other than the original purpose.

e. Engineers shall continue their professional development throughout their careers and should keep current in their specialty fields by engaging in professional practice, participating in continuing education courses, reading in the technical literature, and attending professional meetings and seminars.

Footnote 1 "Sustainable development" is the challenge of meeting human needs for natural resources, industrial products, energy, food, transportation, shelter, and effective waste management while conserving and protecting environmental quality and the natural resource base essential for future development.

As Revised July 2007

"By order of the United States District Court for the District of Columbia, former Section 11(c) of the NSPE Code of Ethics prohibiting competitive bidding, and all policy statements, opinions, rulings or other guidelines interpreting its scope, have been rescinded as unlawfully interfering with the legal right of engineers, protected under the antitrust laws, to provide price information to prospective clients; accordingly, nothing contained in the NSPE Code of Ethics, policy statements, opinions, rulings or other guidelines prohibits the submission of price quotations or competitive bids for engineering services at any time or in any amount."

Statement by NSPE Executive Committee

In order to correct misunderstandings which have been indicated in some instances since the issuance of the Supreme Court decision and the entry of the Final Judgment, it is noted that in its decision of April 25, 1978, the Supreme Court of the United States declared: "The Sherman Act does not require competitive bidding."

It is further noted that as made clear in the Supreme Court decision:

1. Engineers and firms may individually refuse to bid for engineering services.
2. Clients are not required to seek bids for engineering services.
3. Federal, state, and local laws governing procedures to procure engineering services are not affected, and remain in full force and effect.
4. State societies and local chapters are free to actively and aggressively seek legislation for professional selection and negotiation procedures by public agencies.
5. State registration board rules of professional conduct, including rules prohibiting competitive bidding for engineering services, are not affected and remain in full force and effect. State registration boards with authority to adopt rules of professional conduct may adopt rules governing procedures to obtain engineering services.
6. As noted by the Supreme Court, "nothing in the judgment prevents NSPE and its members from attempting to influence governmental action . . ."

NOTE: In regard to the question of application of the Code to corporations vis-à-vis real persons, business form or type should not negate nor influence conformance of individuals to the Code. The Code deals with professional services, which services must be performed by real persons. Real persons in turn establish and implement policies within business structures. The Code is clearly written to apply to the Engineer, and it is incumbent on members of NSPE to endeavor to live up to its provisions. This applies to all pertinent sections of the Code.

NSPE Cases

APPENDIX

B

These opinions are for educational purposes only.

Employment—Offer of Employment by Vendor

Case No. 00-9

Facts

Upon graduation from an ABET/EAC-accredited civil engineering program, Engineer A is employed by U&I Construction Co., which is owned and operated by Engineer B and Engineer C, both licensed professional engineers. Engineer A is soon delegated the responsibility of preparing bills of materials for designs to be constructed, with appropriate allowance for waste, and negotiating the material procurement with suppliers. Engineer A negotiates quantity, schedule, specifications, and price, then submits a recommendation to his highly experienced, non-degreed supervisor to arrange for appropriate company approval authority for the procurement contract if the financial commitment to a supplier on a project exceeds $250. After two years, Engineer A expresses concern to his supervisor that his job seems repetitive and lacks the variety of experiences and challenges that draw on the breadth of his education. Engineer A is informed that he is providing an essential service to the company with exceptional proficiency, for which he seems very well paid, and that he will be considered for opportunities should they become available—if a replacement to cover his current activities can be found. Engineer A's financial authority is increased to $500 for any one supplier of the project.

Reprinted by permission of the National Society of Professional Engineers (NSPE) www. nspe.org

Another year passes and Engineer A is still performing the same level of assignments. Engineer A wants to remain in the locality, but there are limited alternative employment opportunities in engineering available. Engineer A has developed a highly respected reputation for knowledge, fairness, and integrity among the suppliers of U&I Construction Co. Engineer D, an employee of ACE Supplies, a frequent supplier to U&I, has developed a familiarity with Engineer A. When ACE has an opening for a civil engineer, Engineer D lets Engineer A know about it. Engineer A interviews for the position and after an evaluation period, Engineer A learns that he will receive an offer of employment with ACE.

The offer letter states that ACE is "looking forward to having Engineer A on its team commencing on a mutually agreed upon date . . . that Engineer A is not an employee of ACE until Engineer A physically reports to work at ACE's facilities, executes patent and proprietary information agreements at that time, and that the employer's physician confirms that Engineer A has no pre-existing health condition that would prevent Engineer A from performing the requirements of the position." In a subsequent discussion with Engineer A, Engineer D mentions to Engineer A that the position was one that the ACE Vice President of Engineering has the prerogative of filling, but on occasion the ACE CEO has eliminated the position opening even with outstanding offers pending until business conditions improved or when a major customer had expressed displeasure with the hiring of one of its employees.

Engineer A submits his resignation with a customary two weeks notice to U&I. Engineer A's supervisor Engineer E, is disturbed by the resignation and expresses a desire that Engineer A stay with U&I, saying that if Engineer E could prevent his leaving he would. Engineer A insists that his decision is firm. Engineer A is not asked and does not believe it is in his interests to mention that he will be employed by ACE. Engineer E requests that Engineer A should bring all of his work assignments to a point of completion that will facilitate his making an orderly transfer to other U&I employees and to conclude as many assignments as possible before departing. For the next two weeks before leaving U&I, Engineer A continues to negotiate and prepare recommendations on bids including those that had been submitted by ACE.

Questions

Question 1: Was it ethical for Engineer A not to volunteer to U&I the information that he would be employed by ACE within two weeks?

Question 2: Was it ethical for Engineer D to entice Engineer A to consider employment with ACE?

Question 3: Was it ethical for Engineer A to interview with a supplier of U&I without first advising U&I of his intent?

Question 4: Was it ethical for the ACE Vice President of Engineering to offer employment to Engineer A without first divulging the risk that the offer might be withdrawn by the ACE CEO?

Question 5: Would it have been ethical for U&I to have interfered in Engineer A's employment change had U&I become aware of the identity of the future employer and ACE's susceptibility to pressure from U&I?

Question 6: Would it have been ethical for ACE to have withheld an offer to Engineer A if ACE had become aware of U&I's displeasure?

Question 7: Was ACE's policy of withdrawing employment offers after they are made ethical?

References

Section I.4.—Code of Ethics: Act for each employer or client as faithful agents or trustees.

Section I.5.—Code of Ethics: Avoid deceptive acts.

Section I.6.—Code of Ethics: Conduct themselves honorably, responsibly, ethically and lawfully so as to enhance the honor, reputation and usefulness of the profession.

Section III.1.d.—Code of Ethics: Engineers shall not attempt to attract an engineer from another employer by false or misleading pretenses.

Section III.4.a.—Code of Ethics: Engineers shall not, without the consent of all interested parties, promote or arrange for new employment or practice in connection with a specific project for which the Engineer has gained particular and specialized knowledge.

Section III.7.—Code of Ethics: Engineers shall not attempt to injure, maliciously or falsely, directly or indirectly, the professional reputation, prospects, practice or employment of other engineers. Engineers who believe others are guilty of unethical or illegal practice shall present such information to the proper authority for action.

Discussion

Among some of the most challenging conflicts faced by professional engineers relates to situations involving changes or transitions in employment. Frequently, engineers who change employers are privy to information and involved in circumstances that raise ethical questions. A good illustration of this is the very recent BER Case No. 99-6. In that case Engineer A, a member of NSPE, was employed by the FGH Construction Company and worked closely with Engineer B who was an employee of LMN Supplies. LMN Supplies sold construction materials and supplies. Part of Engineer A's responsibilities were to negotiate and approve bids by LMN Supplies that were submitted by Engineer B. LMN Supplies offered, and Engineer A accepted, an employment position with LMN Supplies. Engineer A submitted his resignation and gave two weeks' notice to FGH Construction Company and was not asked and did not mention that he would be employed by LMN Supplies. For the next two weeks before leaving FGH Construction Company, Engineer A continued to negotiate and approve bids submitted by LMN Supplies. In deciding that it was not ethical for Engineer A to fail to mention to FGH Construction Company that he would be employed by its vendor, LMN Supplies, the Board noted that Engineer A's actions would most probably raise some doubt in the minds of the supervisors and, perhaps, owners of FGH Construction about whether Engineer A's continued negotiation and approval of bids submitted by LMN Supplies were somehow tainted and could have resulted in inflated costs to FGH Construction or other unearned competitive advantages for the benefit of Engineer A's new employer, LMN Supplies. The BER also noted that by failing to disclose the material conflict that exists concerning his new employment with LMN Supplies, Engineer A may have unwittingly planted "seeds of doubt" with FGH Construction and potentially damaged the goodwill that might have existed between FGH Construction and LMN Supplies. Based upon the facts as presented, FGH Construction might wrongly conclude that LMN Supplies somehow persuaded Engineer A not to disclose his new position with LMN Supplies during the two-week period in order to gain some advantages. The Board concluded, stating that Engineer A's failure to fully disclose his new position with LMN Supplies and to continue to negotiate and approve LMN Supplies' bids to his current employer, was not in accordance with the spirit or the intent of the NSPE Code.

An earlier BER case with a somewhat different perspective on this issue was BER Case No. 83-1. In that case, Engineer A worked as an employee for Engineer B. On November 15, 1982, Engineer B notified Engineer A that Engineer B was going to terminate Engineer A because of lack of work. Engineer A thereupon notified clients of Engineer B that Engineer A was planning to start another engineering firm and would appreciate being considered for future work. Meanwhile, Engineer A continued to work for Engineer B for several additional months after the November termination notice. During that period, Engineer B distributed a previously printed brochure listing Engineer A as one of Engineer B's key employees, and continued to use the previously printed brochure with Engineer A's name in it well after Engineer B did, in fact, terminate Engineer A. In ruling that Engineer B's actions were improper, the Board noted that an engineer is expected to act, at all times in professional matters for the client, as a faithful agent and trustee (See NSPE Code Section I.4.). That requires the engineer to recognize both a duty of loyalty and good faith. An essential aspect of those is the duty to disclose. Certainly it is not possible for an engineer to meet those obligations to the employer if the engineer is engaging in such promotional activity to the employer's detriment. The Board believes the salient points contained in BER Case Nos. 83-1 and 99-6 are instructive in the context of the present case.

Regarding Question 1, the Board is of the view that Engineer A had an obligation to inform his employer regarding the name and circumstances of his new position. Since Engineer A's new position was with a client of his present employer, there may be situations and circumstances that U&I Construction would need to take into account.

Regarding Question 2, the Board does not believe an engineer employed by a supplier of a client is unethical for seeking to recruit an employee of the client. We believe freedom of employment is the question at issue.

As for Questions 6 and 7, the Board believes that the engineering employment market is a dynamic one in which individuals make a variety of choices, determinations, and decisions concerning employment offers and acceptance. In the absence of any formal agreement, it is generally understood that the "employment at will" principle applies and no employee or employer is bound to another. Both employer and employee gain the advantages of this principle and also suffer from its shortcomings. In this context, companies and individuals will ultimately make certain decisions based at least partly upon self interest, budgetary considerations, salary expectations, employment benefits, corporate mission, and other factors. In this connection, the Board does not believe the issues raised by questions 6 and 7 involve ethical concerns and the Board therefore declines to address them. At the same time, it should be noted that as a general matter, the Board believes that as a good business practice, the offer of employment letter should have contained a full disclosure statement indicating that the offer of employment to Engineer A by ACE could be withdrawn at any time for any reason by ACE. Questions 6 and 7 relate to facts that are management decisions determined by ACE, which is not an engineering firm. Had it been an engineering firm involving the conduct of an individual engineer, the determination would have been different.

With respect to Question 3, as a practical matter, although it does happen under certain circumstances, it is generally not typical for an individual to inform his employer in advance of an interview for another position with another company. An employee may occasionally inform their employer in order to convey dissatisfaction with the employee's current position or in situations where the employee feels a special obligation to the employer due to length of service, loyalty, or other factors.

Regarding Question 5, the Board believes it would not be ethical for U&I to have interfered in Engineer A's employment change. The engineers of U&I were ethically bound by the Code of Ethics to avoid such conduct.

Regarding Question 4 of whether it was ethical for the ACE Vice President of Engineering to offer employment to Engineer A without first informing Engineer A that the ACE CEO could withdraw the offer, the Board believes it was improper. Clearly when sensitive questions of employment are concerned, those engineers in positions of authority to make decisions in this area must be mindful of the impact such decisions will have on clients as well as the individuals directly involved. The Board believes potential employers have an obligation to make all material facts that could impact a decision to proceed with employment known to the individual involved.

Conclusions

Question 1: It was not ethical for Engineer A not to volunteer the information that he would be employed by ACE within two weeks to U&I upon resignation.

Question 2: This did not involve an ethics issue.

Question 3: It was not unethical for Engineer A to interview with a supplier of U&I without first advising U&I of his intent.

Question 4: It was not ethical for the ACE Vice President of Engineering to offer employment to Engineer A without disclosing all material facts, i.e., the potential actions of the ACE CEO.

Question 5: It would not be ethical for U&I engineers to interfere with Engineer A's employment opportunity with ACE.

Question 6: This did not involve an ethics issue.

Question 7: This did not involve an ethics issue.

BOARD OF ETHICAL REVIEW

Lorry T. Bannes, P.E., NSPE
John W. Gregorits, P.E., F.NSPE
Louis L. Guy, Jr., P.E., F.NSPE
William J. Lhota, P.E., NSPE
Paul E. Pritzker, P.E., F.NSPE
E. Dave Dorchester, P.E., NSPE, Chair
(Harold E. Williamson, P.E., NSPE did not participate in this Opinion).

Each opinion is intended as guidance to individual practicing engineers, students and the public. In regard to the question of application of the NSPE Code to engineering organizations (e.g., corporations, partnerships, sole-proprietorships, government agencies, university engineering departments, etc.), the specific business form or type should not negate nor detract from the conformance of individuals to the NSPE Code. The NSPE Code deals with professional services—which services must be performed by real persons. Real persons in turn establish and implement policies within business structures.

This opinion is for educational purposes only. It may be reprinted without further permission, provided that this statement is included before or after the text of the case and that appropriate attribution is provided to the National Society of Professional Engineers' Board of Ethical Review.

Patents—Dispute Over Right to Specify

Case No. 01-4

Facts

Engineer A, a structural engineer, designs structural systems for large developers on hotel projects. Developer B would like to use a unique flooring system, but the system is patented by Inventor C, who is a professional engineer. Developer B contacts Attorney D, who tells Developer B that Inventor C has a legitimate patent and recommends that Developer B negotiate with Inventor C to obtain a license for Inventor C's patent. Developer B enters into negotiations with Inventor C, but the negotiations fail. Thereafter, Developer B hires Attorney E, who reviews the patent and indicates that he disagrees with Attorney D, and also indicates that, in his professional view, there is a genuine dispute as to the legitimacy of Inventor C's patent. Developer B tells Engineer A that he wants Engineer A to proceed with the project and have Engineer A specify the flooring system into the structural design of the project.

Question

Would it be ethical for Engineer A to proceed with the project and reference the flooring system of the project's structural design?

References

Section I.6.—Code of Ethics: Engineers, in the fulfillment of their professional duties, shall conduct themselves honorably, responsibly, ethically, and lawfully so as to enhance the honor, reputation, and usefulness of the profession.

Section III.1.b.—Code of Ethics: Engineers shall advise their clients or employers when they believe a project will not be successful.

Section III.9.—Code of Ethics: Engineers shall give credit for engineering work to those to whom credit is due, and will recognize the proprietary interests of others.

Section III.9.b—Code of Ethics: Engineers using designs supplied by a client recognize that the designs remain the property of the client and may not be duplicated by the Engineer for others without express permission.

Section III.9.c.—Code of Ethics: Engineers, before undertaking work for others in connection with which the Engineer may make improvements, plans, designs, inventions, or other records that may justify copyrights or patents, should enter into a positive agreement regarding ownership.

Discussion

Respect for the intellectual property rights of others, including patent and copyright, is fundamental to ethical professional practice. The NSPE Board of Ethical Review has long recognized the importance of intellectual property rights (see NSPE Code Sections III.9., III.9.a., III.9.b., and III.9.c.).

One of the earliest cases considered by the Board involving patent rights related to patent ownership. In BER Case No. 74-11, Engineer A, an expert in food processing machinery and systems, was retained by verbal agreement by a patent attorney who, in turn, had been retained by a food machinery manufacturing company in connection with a lawsuit in which the manufacturing company alleged that certain of its patented machinery was being infringed. Engineer A's assignment was to study the machines in question and reach a determination whether, in his opinion, there was an infringement. If so, he would testify to that effect at the trial. In the course of his study, Engineer A conceived an idea he believed to be patentable on the basis that it constituted an improvement in the state of the art of the particular food machinery involved in the pending litigation. Engineer A submitted the idea to the patent attorney and to the food machinery manufacturing company. After a three-month wait, Engineer A requested that the patent attorney and the manufacturer advise him, within a reasonable time, of the extent of the manufacturer's interest in his claimed improvement. This request was interpreted by the patent attorney and the manufacturer as being a "pressure move" for Engineer A to obtain additional fees. The patent attorney demanded that Engineer A assign all rights to his idea to the manufacturer without assurance of any additional compensation. Engineer A refused this demand and was released from further activity in connection with the patent infringement suit. In concluding that it was not ethically permissible for Engineer A to take the position indicated with regard to Engineer A's patent idea arising out of his original employment, the Board noted that the engineer's obligation to act in professional matters as a "faithful agent or trustee," runs to both clients and employers. Said the BER, "to be a 'faithful agent or trustee' means" . . . to act in a manner best calculated to serve his employer's interests." By withholding a patentable idea, Engineer A acted contrary to the interests of his employer."

The Board concluded that the criteria should be applied without regard to the relationship being one of consultant-client rather than employee-employer. The Board noted that on this issue, the NSPE Code may not be a model of clarity to indicate to the profession some helpful guidance in patent ownership matters arising out of a consultant-client relationship. On the surface it only requires that the parties spell out their respective rights. But the Board noted that it can be read to mean more in the context of the facts presented. The Board indicated that it is not without significance that the language calls upon the engineer to enter into a positive agreement regarding the ownership of such patent rights as might emerge from the relationship. In other words, it was incumbent upon the engineer to take the initiative during the negotiations for his services to reach an agreement on the patent rights of the parties. The Board recognized as a practical matter that under the facts, Engineer A did not likely think of the possibility that he might develop a related patentable idea from his studies for the client, but, even so, if he had the ethical duty to act in this regard and failed to do so he must bear the ethical burden when his position is challenged under these circumstances.

The Board concurs with much of the reasoning contained in BER Case No. 74-11 and believes that many of the issues discussed are applicable to the case at hand. Although the factual situations are somewhat different in the two cases, certain basic principles apply to the present case. For example, under the facts in the present case, Engineer A has an ethical obligation and must take the positive initiative in seeking to resolve the ethical issue. However, unlike BER Case No. 74-11, the ethical issue relating to the patent does not involve Engineer A's ownership of the patent in question or similar concerns. Instead, in the present case, the issue relates to a patent held by a third party (Inventor C) who has a potential adversarial position to Engineer A's client, Developer B. Under the facts in the present case, Engineer A does not claim any intellectual property rights. In addition, in view of the fact that Inventor C is also a professional engineer, Engineer A has an ethical obligation to respect the rights and give proper and due credit to Inventor C, and to respect the proprietary rights of Inventor C, however clouded.

Clearly, Engineer A has an obligation to consider and balance various ethical considerations. Under the facts, Engineer A is being placed in a particularly difficult position due to the conflicting opinions being offered by Attorneys D and E concerning the legitimacy of Inventor C's flooring system patent rights. As a professional engineer, Engineer A cannot be expected to make a competent professional judgment relating competing legal rights between Inventor C's patent rights and Developer B. Patent questions are highly technical legal issues and engineers are generally not competent in these areas.

Considering the issues at stake in the present case, the Board believes that there are at least two potential courses of action that Engineer A could take under the facts. First, Engineer A could explore with Developer B the possibility of using an alternative flooring system on the project in order to avoid the possibility of infringing upon Inventor C's patent rights. Although Developer B was particularly interested in the unique flooring system patent claimed by Inventor C, an experienced structural engineer should be resourceful enough to explore other possible comparable alternatives. A second option would be for Engineer A to communicate the importance of Developer B and Inventor C resolving the patent issue to permit Engineer A to proceed with the work without the ethical and legal clouds hanging over this project. Developer B would obviously need to determine how important the unique flooring system is to the success of the project and advise Engineer A.

The Board believes that these options appear to be the most practical approaches and place the responsibility for resolving the issues at hand before the parties in the best position to resolve those matters—Developer B and Inventor C.

The Board notes that this case presents two conflicting legal opinions, and before going forward, Engineer A must be convinced that whatever course of action he proceeds with is in accordance with the NSPE Code. The BER recognizes that this approach places engineers in a difficult position and requires the ethical high road, possibly at odds with the engineer's client.

Conclusion

It would be unethical for Engineer A to specify the flooring system into the project's structural design until the patent and proprietary rights of Inventor C are resolved.

BOARD OF ETHICAL REVIEW

E. Dave Dorchester, P.E., NSPE
John W. Gregorits, P.E., F.NSPE
Louis L. Guy, Jr., P.E., F.NSPE
William D. Lawson, P.E., NSPE
Roddy J. Rogers, P.E., F.NSPE
Harold E. Williamson, P.E., NSPE
William J. Lhota, P.E., NSPE, Chair

Conflict-of-Interest: Third Party Developer

Case No. 01-2

Facts

A developer, Mall Dev, has approached a town requesting approval to construct a development on a vacant site in Niceville. Based on the size of the development, Niceville is requesting that an environmental impact statement be prepared that will address traffic operations, as well as other issues.

Niceville requests an outside consultant, Engineer A, to assist the town in scoping out the necessary traffic analyses and to review and advise Niceville on possible traffic impacts of the proposed development. The development will be both retail and offices and will contain a supermarket.

The consultant, Engineer A, is also assisting other jurisdictions in review of proposals by Mall Dev. Engineer A has disclosed to the town all relationships, if any, with the proposed developer, Mall Dev with announced tenants, and with other customers that develop sites for retail development. Niceville is satisfied that there is no conflict of interest.

More specifically, Engineer A is not currently representing any other developers in the town, but in the past has prepared traffic impact studies for other developers on projects concerning other developments constructed in Niceville. Engineer A is currently providing traffic impact studies to other developers in other jurisdictions, as well as services to Mall Dev. These have all been disclosed to Niceville.

Mall Dev, however, has informed Niceville that it believes the use of the consultant Engineer A is a conflict of interest and breaches the code of professional ethics. Mall Dev bases its belief on the fact that Engineer A has worked in the past, and is currently working for, other developers who compete for the same tenants Mall Dev tries to attract to its developments.

Questions

1. Would Engineer A's work for the Niceville constitute a conflict of interest?

2. Was it appropriate for Mall Dev to raise an ethical issue relating to Engineer A's actions?

References

Section I.4.—Code of Ethics: Act for each employer or client as faithful agents or trustees.

Section II.4.a.—Code of Ethics: Engineers shall disclose all known or potential conflicts of interest which could influence or appear to influence their judgment or the quality of their services.

Section II.4.b.—Code of Ethics: Engineers shall not accept compensation, financial or otherwise, from more than one party for services on the same project, or for services pertaining to the same project, unless the circumstances are fully disclosed and agreed to by all interested parties.

Discussion

Fundamental to the practice of engineering is the duty of loyalty of the engineer to the client. This is an individual and personal obligation that all engineers owe uniquely to their clients. It includes the basic responsibility to perform professional services in a competent manner, considering a client's overall project requirements and needs. In performing their work, all engineers must seek to avoid situations or circumstances that call into question this basic duty. One of the most common examples in which an engineer's duty to the client is called into question is the area of conflict of interests.

A classic illustration of this was BER Case No. 88-1. In that case, Engineer A was retained by the county to perform a feasibility study and make recommendations concerning the location of a new power facility in the county. Two parcels of land located on a river had been identified by the county as the "candidates" for facility sites. The first parcel was undeveloped and owned by an individual who planned to build a recreational home for his family. The second parcel, owned by Engineer A, was a developed parcel of land. Engineer A disclosed that he was the owner of the second parcel of land and recommended that the county build the facility on the undeveloped parcel of land because (1) it was a better location for the power facility from an engineering standpoint, and (2) it would be less costly for the county to acquire. The county did not object to having Engineer A perform the feasibility study.

In determining that it was not ethical for Engineer A to perform a feasibility study and make recommendations concerning the location of a new power facility in the county, the Board noted that although Engineer A's professional opinion was supported by two important public policy considerations (e.g., that the undeveloped parcel was a better location for a power facility and that the county's cost of acquiring the developed property would be higher than the cost of acquiring the undeveloped tract of land), these reasons were not sufficient to justify Engineer A's decision to perform the feasibility study for the county. The Board noted that public perceptions play an important role in engineering ethics. The facts and circumstances of Engineer A's study may have appeared to suggest a benefit to the "common good" if his recommended course of action was followed but these same facts and circumstances allow for the appearance of impropriety, and this can easily damage public confidence in the engineering profession. Clearly there could have been public perception under the facts that Engineer A did not want to risk personal disruption of his developed property or possibly anticipated future appreciation of the value of the property. Engineer A should have followed the far simpler and more ethical approach recommended in the earlier BER Case No. 69-13 which stated, "(The Engineer) can avoid such a conflict under these facts either by disposing of his land holdings prior to undertaking the commission or by declining to perform the services if it is not feasible or desirable for him to dispose of his land at the particular time."

In the earlier cited BER Case No. 69-13, the Board reviewed a situation in which an engineer was an officer in an incorporated consulting engineering firm that was primarily engaged in civil engineering projects for clients. Early in the engineer's life, he had acquired a tract of land by inheritance, which was in an area being developed for residential and industrial use. The engineer's firm had been retained to study and recommend a water and sewer system in the general area of his land interest. The question faced by the Board under those facts was "May the engineer ethically design a water and sewer system in the general area of his land interest?"

The Board ruled that the engineer could not ethically design the system under those circumstances. The Board recognized that the issue was a difficult one to resolve, pointing to the fact that there was no conflict of interest when the engineer entered his practice. The conflict developed in the normal course of his practice, when it became apparent that his study and recommendation could lead to the location of a water and sewer system near his land. This could bring a considerable appreciation in the value of his land, depending upon the exact location of certain system elements in proximity to his land. The BER stated that while the engineer must make full disclosure of his personal interest to his client before proceeding with the project, such disclosure was not enough under the NSPE Code of Ethics. The Board concluded by saying, "This is a harsh result, but so long as men are in their motivations somewhat 'lower than angels,' it is a necessary conclusion to achieve compliance with both the letter and the spirit of the NSPE Code. The real test of ethical conduct is not when compliance with the NSPE Code comports with the interest of those it is intended to govern, but when compliance is adverse to personal interest."

In the more recent BER Case No. 85-6, the Board reviewed similar facts and circumstances and came to a different result. There, an engineer was retained by the state to perform certain feasibility studies relating to a possible highway spur. The state was considering the possibility of constructing the highway spur through an area adjacent to a residential community in which the engineer's residence was located. After learning of the proposed location of the spur, the engineer disclosed to the state the fact that his residential property might be affected and fully disclosed the potential conflict with the state. The state did not object to the engineer performing the work.

Engineer A proceeded with his feasibility study and ultimately recommended that the spur be constructed. In ruling that it was not unethical for the engineer to perform the feasibility study, despite the fact that his land might be affected thereby, the Board noted that the ethical obligations contained in NSPE Code Section II.4.a. do not require the engineer to "avoid" any and all situations that may or may not raise the specter of a conflict of interest. Such an interpretation of the NSPE Code, The Board said, would leave engineers without any real understanding of the ethical issues nor any guidance as to how to deal with the problem. The BER noted that the basic purpose of a code of ethics is to provide the engineering profession with a better awareness and understanding of the ethical issues that impact the public. The Board concluded that only through interacting with the public and clients will engineers be able to comprehend the true dimensions of ethical issues.

Turning to the facts in this case, while the circumstances described are somewhat different than the earlier cases considered, the Board believes some of the basic principles and issues considered are useful in understanding the present case. First, it is clear from the language in the NSPE Code and its application in the earlier cases that the obligation concerning conflicts of interest is owed to an "employer" or a "client." Therefore, under the facts presented, it appears that the duty would be owed solely to the "client," (e.g., the town Niceville) and would not extend to any third party (e.g., the developer Mall Dev). In other words, a conflict of interest cannot be asserted as a matter of ethical practice by a third party against an engineer. While Engineer A is performing work for Mall Dev in other jurisdictions and has obligations to Mall Dev, there is no factual assertion of a conflict of interest by Mall Dev other than a general, non-specific assertion of possible prejudice and bias. This, without more, is insufficient to raise a conflict. To conclude otherwise would result in subjecting an engineer's practice activities to a "veto" by any third party that might decide to allege some particular interest on a project.

The Board can easily imagine an endless list of speculative and baseless conflicts of interest alleged by third parties against engineers performing services for public agencies in order to improve the third party's business opportunities. We can also speculate that an ill-motivated client could assert in bad faith a conflict against an engineer for purely self-serving, and even malicious, motives. Under the NSPE Code, a third party, such as a developer, does not have a legitimate basis upon which to complain of an alleged conflict of interest, and once an "employer" or "client" is satisfied that no conflict of interest exists, the question of whether a conflict of interest exists should be resolved. Under the facts, it is clear that with full disclosure to all parties, Engineer A can pursue work with Niceville. As in all such cases, it is important that such situations be viewed in light of the total situation being contemplated.

Second, the NSPE Code language clearly recognizes that engineers frequently face conflicts of interest in their practice and are obligated to address them by disclosing all known or potential conflicts that could influence or appear to influence their judgment or the quality of their services. By doing so, the engineer fulfills his/her ethical obligation under the NSPE Code. Under the facts, this obligation appears to have been completely fulfilled with by Engineer A.

Conclusions

1. Engineer A's work for the town would not constitute a conflict of interest since there was full disclosure. Based upon the language in the NSPE Code, it is clear that no conflict of interest exists and the engineer has fulfilled his ethical obligation under the NSPE Code.

2. It was not appropriate for Mall Dev to raise an ethical issue relating to Engineer A's actions.

BOARD OF ETHICAL REVIEW

E. Dave Dorchester, P.E., NSPE
John W. Gregorits, P.E., F.NSPE
Louis L. Guy, Jr., P.E., F.NSPE

William D. Lawson, P.E., NSPE
Roddy J. Rogers, P.E., F.NSPE
Harold E. Williamson, P.E., NSPE
William J. Lhota, P.E., NSPE, Chair

NOTE: The NSPE Board of Ethical Review (BER) considers ethical cases involving either real or hypothetical matters submitted to it from NSPE members, other engineers, public officials and members of the public. The BER reviews each case in the context of the NSPE Code and earlier BER opinions. The facts contained in each case do not necessarily represent all of the pertinent facts submitted to or reviewed by the BER.

Public Welfare—Bridge Structure

Case No. 00-5

Facts

Engineer A was an engineer with a local government. Engineer A learned about a critical situation involving a bridge 280 feet long, 30 feet above the stream. This bridge was a concrete deck on wood piles built in the 1950's by the state. It was part of the secondary roadway system given to the counties many years ago.

In June 2000, Engineer A received a telephone call from the bridge inspector stating this bridge needed to be closed due to the large number of rotten piling. Engineer A had barricades and signs erected within the hour on a Friday afternoon. Residents in the area were required to take a 10-mile detour.

On the following Monday, the barricades were in the river and the "Bridge Closed" sign was in the trees by the roadway. More permanent barricades and signs were installed. The press published photos of some of the piles that did not reach the ground and the myriad of patch work over the years.

Within a few days, a detailed inspection report prepared by a consulting engineering firm, signed and sealed, indicated seven pilings required replacement. Within three weeks, Engineer A had obtained authorization for the bridge to be replaced. Several departments in the state and federal transportation departments needed to complete their reviews and tasks before the funds could be used.

A rally was held, and a petition with approximately 200 signatures asking that the bridge be reopened to limited traffic was presented to the County Commission. Engineer A explained the extent of the damages and the efforts under way to replace the bridge. The County Commission decided not to reopen the bridge.

Preliminary site investigation studies were begun. Environmental, geological, right-of-way, and other studies were also performed. A decision was made to use a design build contract to avoid a lengthy scour analysis for the pile design.

A non-engineer public works director decided to have a retired bridge inspector, who was not an engineer, examine the bridge, and a decision was made to install two crutch piles under the bridge and to open the bridge with a 5-ton limit. No follow-up inspection was undertaken.

Engineer A observes that traffic is flowing and the movement of the bridge is frightening. Log trucks and tankers cross it on a regular basis. School buses go around it.

Question

What is Engineer A's ethical obligation under these circumstances?

References

Section II.1.—Code of Ethics: Engineers shall hold paramount the safety, health and welfare of the public.

Section II.1.a.—Code of Ethics: If engineers' judgment is overruled under circumstances that endanger life or property, they shall notify their employer or client and such other authority as may be appropriate.

Section II.1.e.—Code of Ethics: Engineers having knowledge of any alleged violation of this Code shall report thereon to appropriate professional bodies and, when relevant, also to public authorities, and cooperate with the proper authorities in furnishing such information or assistance as may be required.

Section III.8.a.—Code of Ethics: Engineers shall conform with state registration laws in the practice of engineering.

Discussion

The obligation of a professional engineer to take action when faced with a situation involving a direct threat to the public health and safety has been addressed by this board on several other occasions. A review of the cases decided over the years by the NSPE Board of Ethical Review demonstrates a consistent approach regarding this fundamental obligation on the part of professional engineers.

For example, BER Case No. 92-6 involved Technician A serving as a field technician employed by a consulting environmental engineering firm. At the direction of his supervisor, Engineer B, Technician A sampled the contents of drums located on the property of a client. Based on Technician A's past experience, it was his opinion that analysis of the sample would most likely determine that the drum contents would be classified as hazardous waste. If the material was hazardous waste, Technician A knew that certain steps would legally have to be taken to transport and properly dispose of the drum, including notifying the proper federal and state authorities. Technician A asked his supervisor, Engineer B, what to do with the samples. Engineer B told Technician A only to document the existence of the samples. Technician A was then told by Engineer B that since the client did other business with the firm, Engineer B would tell the client where the drums were located but would do nothing else. Thereafter, Engineer B informed the client of the presence of drums containing "questionable material" and suggested that they be removed. The client contacted another firm and had the material removed.

In considering whether it was ethical for Engineer B merely to inform the client of the presence of the drums and suggest that they be removed, and whether Engineer B had an ethical obligation to take further action, the Board noted that the extent to which an engineer has an obligation to hold paramount the public health and welfare in the performance of professional duties (See NSPE Code Section I.1.) overlaps the duty of engineers not to disclose confidential information concerning the business affairs, etc. of clients (See NSPE Code Section III.4.). With regard to Case No. 92-6, the Board noted, that unlike the facts in the earlier cases, Engineer B made no oral or written promise to maintain the client's confidentiality. Instead, Engineer B consciously and affirmatively took actions that could cause serious environmental danger to workers and to the public, and were a violation of various environmental laws and regulations. Under the facts, it appeared that Engineer B's primary concern was not so much maintaining the client's confidentiality as it was in maintaining good business relations with a client. In addition, it appeared that, as in all cases that involve potential violations of the law, Engineer B's actions could have had the effect of seriously damaging the long-term interests and reputation of the client. In this regard, the Board noted that, under the facts, it appeared that the manner in which Engineer B communicated the presence of the drums on the property must have suggested to the client that there was a high likelihood that the drums contained hazardous materials. The Board noted that this subterfuge is wholly inconsistent with the spirit and intent of the NSPE Code of Ethics, because it makes the engineer an accomplice to what may amount to an unlawful action.

The Board noted that Engineer B's responsibility under the facts was to bring the matter of the drums possibly containing hazardous material to the attention of the client with a recommendation that the material be analyzed. To do less would be unethical. If analysis demonstrates that the material is indeed hazardous, the client would have the obligation of disposing of the material in accordance with applicable federal, state, and local laws.

In an earlier case, BER Case No. 89-7, an engineer was retained to investigate the structural integrity of a 60-year-old, occupied apartment building, which his client was planning to sell. Under the terms of the agreement with the client, the structural report written by the engineer was to remain confidential. In addition, the client made it clear to the engineer that the building was being sold "as is," and the client was not planning to take any remedial action to repair or renovate any system within the building. The engineer performed several structural tests on the building and determined that the building was structurally sound. However, during the course of providing services, the client confided in the engineer that the building contained deficiencies in the electrical and mechanical systems, which violated applicable codes and standards. While the engineer was not an electrical or mechanical engineer, he did realize that those deficiencies could cause injury to the occupants of the building and so informed the client. In his report, the engineer made a brief mention of his conversation with the client concerning the deficiencies; however, in view of the terms of the agreement, the engineer did not report the safety violations to any third parties. In determining that it was unethical for the engineer not to report the safety violations to appropriate public authorities, the Board, citing cases decided earlier, noted that the engineer "did not force the issue, but instead went along without dissent or comment. If the engineer's ethical concerns were real, the engineer should have insisted that the client take appropriate action or refuse to continue work on the project." The Board concluded that the engineer had an obligation to go further, particularly because the NSPE Code uses the term "paramount" to describe the engineer's obligation to protect the public safety, health, and welfare.

In BER Case No. 90-5, the Board reaffirmed the basic principle articulated in BER Case No. 89-7. There, tenants of an apartment building sued its owner to force him to repair many of the building's defects. The owner's attorney hired an engineer to inspect the building and give expert testimony in support of the owner. The engineer discovered serious structural defects in the building that he believed constituted an immediate threat to the safety of the tenants. The tenants' suit had not mentioned these safety-related defects. Upon reporting the findings to the attorney, the engineer was told he must maintain this information as confidential because it was part of the lawsuit. The engineer complied with the request. In deciding it was unethical for the engineer to conceal his knowledge of the safety-related defects, the Board discounted the attorney's statement that the engineer was legally bound to maintain confidentiality, noting that any such duty was superseded by the immediate and imminent danger to the building's tenants. While the Board recognized that there may be circumstances where the natural tension between the engineer's public welfare responsibility and the duty of nondisclosure may be resolved in a different manner, the Board concluded that this clearly was not the case under the facts.

The Board believes much of the same reasoning in the earlier cases applies to the case at hand. The facts and circumstances facing Engineer A involve basic and fundamental issues of public health and safety which are at the core of engineering ethics. For an engineer to bow to public pressure or employment situations when the engineer believes there are great dangers present would be an abrogation of the engineer's most fundamental responsibility and obligation. Engineer A should take immediate steps to contact the county governing authority and county prosecutors, state and/or federal transportation/highway officials, the state engineering licensure board, and other authorities. By failing to take this action, Engineer A would be ignoring his basic professional and ethical obligations.

Conclusion

Engineer A should take immediate steps to go to Engineer A's supervisor to press for strict enforcement of the five-ton limit, and if this is ineffective, contact state and/or federal transportation/highway officials, the state engineering licensure board the director of public works, county commissioners, state officials, and such other authorities as appropriate. Engineer A should also work with the consulting engineering firm to determine if the two crutch pile with five-ton limit design solution would be effective and report this information to his supervisor. In addition, Engineer A should determine whether a basis exists for reporting the activities of the retired bridge inspector to the state board as the unlicensed practice of engineering.

BOARD OF ETHICAL REVIEW

NOTE: The NSPE Board of Ethical Review (BER) considers ethical cases involving either real or hypothetical matters submitted to it from NSPE members, other engineers, public officials and members of the public. The BER reviews each case in the context of the NSPE Code and earlier BER opinions. The facts contained in each case do not necessarily represent all of the pertinent facts submitted to or reviewed by the BER.

This opinion is for educational purposes only. It may be reprinted without further permission, provided that this statement is included before or after the text of the case and that appropriate attribution is provided to the National Society of Professional Engineers' Board of Ethical Review.

CONFLICT OF INTEREST—UTILITY AUDITS FOR CITY Case No. 01-5

Facts

Engineer A receives a "Request for Qualifications (RFQ)" from City X for the review of unbilled and mis-billed water and wastewater service records. One paragraph of the RFQ reads as follows: "The consultant shall be entitled to receive X% of increased revenues generated. If the consultant fails to identify and document unbilled or mis-billed water and wastewater sewer service records, the City shall be under no obligation to compensate the consultant."

Question

Would it be ethical for Engineer A to enter into a contract under the circumstances described?

Reference

Section III.6.—Code of Ethics: Engineers shall not attempt to obtain employment or advancement or professional engagements by untruthfully criticizing other engineers, or by other improper or questionable methods.

Section III.6.a.—Code of Ethics: Engineers shall not request, propose, or accept a commission on a contingent basis under circumstances in which their judgment may be compromised.

Discussion

The manner in which engineers are compensated has long been a subject of the NSPE Code of Ethics. Over the years, the specific language contained in the NSPE Code, as well as the NSPE Board of Ethical Review Opinions, has significantly evolved. Many methods of compensation that have at one point in time been considered not ethical or improper are today deemed acceptable. This factor is a reflection of a variety of considerations, including changing ethical attitudes, competitive pressures, and legal requirements.

Over the years, the Board has considered ethical issues relating to contingent contracts based upon a variety of factors including speculative arrangements, savings to clients, identification of errors/omissions, lower construction costs, as well as other considerations. In each of these cases, among the key issues for the Board's review has been the question of whether the circumstances under which compensation was determined would have the effect of compromising the engineer's professional judgment.

Over the many years in which the Board has functioned, it has had several opportunities to review circumstances involving compensation arrangements with clients and others that could potentially call into question the independent judgment of the engineer. Such was the case in BER Case No. 73-4. There, an engineer, a specialist in utility systems, offered to industrial clients a service consisting of a technical evaluation of the client's use of utility services, including where appropriate, recommendations for changes in utility facilities and systems, methods of payment for such utilities, study of pertinent rating schedules, discussions with utility suppliers on rate charges, and renegotiation of rate schedules forming the basis of the charges to the client. The engineer was compensated by his client for these services solely on the basis of a percentage of money saved for utility costs. In finding that it was ethical for the engineer to be compensated solely on the basis of a percentage of savings to his client, the Board first noted that the NSPE Code does not rule out contingent contracts, pointing out that contingent contracts are improper only under circumstances in which the arrangement may compromise the professional judgment of an engineer. One example then cited by the Board as this type of restriction was found in BER Case No. 65-4, in which it was determined that it would not be ethical for an engineer to enter into a contingent contract under which his payment depends upon a favorable feasibility study for a public works project. The Board commented in BER Case No. 65-4 that "the import of the restriction . . . is that the engineer must render completely impartial and independent judgment on engineering matters without regard to the consequences of his future retention or interest in the project."

In another case, BER Case No. 66-11, an engineer expert was retained by an attorney to provide expert analysis and advice on the technical reasons for a failure which led to certain damage. Although the engineer had provided these services on a per diem basis in the past, it was proposed by the attorney that the engineer be compensated on the basis of being paid a percentage of the amount recovered by his client. If the judgment was in favor of the defendant, neither the engineer nor the attorney would be paid for their services. In concluding that it was not ethical for the engineer to provide technical advisory services or serve as an expert witness in a lawsuit on a contingent basis, the Board, agreeing with BER Case No. 65-4, noted that the "duty of the engineer as a technical advisor is to provide his client with all of the pertinent technical facts related to the case, favorable and unfavorable alike." The Board also stated that under the facts, the engineer "could not ethically serve on a contingent fee basis because his conclusions might be influenced by the fact that he stood to gain financially by having his conclusions coincide with his personal interest in his remuneration, which is dependent upon his client being successful in the litigation."

Comparing the facts to those in BER Case No. 65-4, the Board noted that the engineer must not be in a position whereby his form of compensation might tend to prevent him from being completely impartial, or from rendering a full and complete report containing both favorable and unfavorable facts or conclusions.

More recently in BER Case No. 91-2, Client, a non-engineer, retained Engineer A, a consulting engineer, to perform certain design services in connection with a waste-water treatment facility. Engineer A performed the design services and Client reviewed the documents prepared by Engineer A. Following the review, Client made the judgment that the documents prepared by Engineer A contained errors and omissions. Client terminated his relationship with Engineer A. Client then contacted Engineer B and proposed an arrangement whereby Engineer B would review the work prepared by Engineer A and identify errors/omissions contained in the documents in contemplation of a suit for breach of contract. Client proposed that Engineer B's fee would be dependent upon the ultimate court judgment or settlement made with Engineer A. Engineer B accepted the assignment under the terms proposed by Client. The Board concluded that it was apparent that Engineer B was being placed in a position of identifying errors/omissions in Engineer A's work in order to pressure Engineer A into a settlement which would result in a fee for Engineer B. By finding no errors/omissions in Engineer A's work there would be no fee.

The Board noted that "these circumstances appear to be just the very factors for which NSPE Code Section III.7.a. was intended to guard against." The Board determined that the circumstances in BER Case No. 91-2 were similar to those in BER Case No. 65-4 where the Board determined that a contingent arrangement based upon the results of a feasibility study was improper. Said the BER, "it would be difficult to imagine a clearer set of circumstances involving a contingent fee arrangement in which an engineer's professional judgment could risk becoming compromised." Importantly, the Board also noted that the circumstances in BER Case No. 73-4 involved the engineer's compensation being based on the money saved by clients for utility costs under circumstances which do not appear to involve a significant possibility of a compromise in judgment. However, in BER Case No. 91-2, the nature of the services and the related contingency arrangement suggested a strong possibility that the engineer's judgment could be compromised, or at the very least, create the appearance of being compromised.

Turning the facts in the present case, the Board believes that this case raises issues more analogous to BER Case No. 73-4. Under the facts, Engineer A would only be compensated if Engineer A satisfies her professional obligation to serve the client consistent with the public interest in making sure that utility fees are properly accounted for. It must be assumed that informed consent on the part of the public is appropriately obtained and this consent implies the existence of a public appeal process for an adverse decision made against a consumer. Within this context, the service is not subject to vague interpretation; it will in essence consist of properly documenting cases where the meter or measuring process is not functioning properly. The only apparent situation that could arise that might arguably compromise Engineer A's judgment could be circumstances where Engineer A might focus only on certain types of invoices (e.g., commercial vs. residential) that might yield the highest return, etc., as opposed to reviewing all bills equally. However, even in this situation, it would appear that such an approach would not adversely affect the client or the public, which would presumably support an approach that would obtain a higher revenue return. In the absence of some public policy concern that might impact on this issue differently, it is the Board's conclusion that no ethical concern exists. In view of the significant benefit to the client in this form of compensation, the Board can see the desirability of this approach.

Conclusion

It would be ethical for Engineer A to enter into a contract under the circumstances described.

BOARD OF ETHICAL REVIEW

E. Dave Dorchester, P.E., NSPE
John W. Gregorits, P.E., F.NSPE
Louis L. Guy, Jr., P.E., F.NSPE
William D. Lawson, P.E., NSPE
Roddy J. Rogers, P.E., F.NSPE
Harold E. Williamson, P.E., NSPE
William J. Lhota, P.E., NSPE, Chair

EMPLOYMENT PRACTICES—SOLICITING COMPETITOR'S EMPLOYEES
Case No. 99-5

Facts

Engineer A's firm is attempting to increase its staff capacity and after publishing a series of advertisements in local and national job classified publications, decides to send out recruitment postcards to engineers in the local and state engineering community. Using the state board registry of professional engineers, the firm sends the unsolicited postcards out to individual engineers at the address listed in the directory announcing Engineer A's firm's interest in recruiting new engineer employees. Such mailings are not prohibited by the state board. Many of the cards are sent to the individual engineers at their firm's address.

Question

Was it ethical for Engineer A's firm to send postcards out to individual engineers in the manner described?

References

Section III.3.—Code of Ethics: Engineers shall avoid all conduct or practice which deceives the public.

Section III.3.a.—Code of Ethics: Engineers shall avoid the use of statements containing a material misrepresentation of fact or omitting a material fact.

Section III.3.b.—Code of Ethics: Consistent with the foregoing, Engineers may advertise for recruitment of personnel.

Section III.7.—Code of Ethics: Engineers shall not attempt to injure, maliciously or falsely, directly or indirectly, the professional reputation, prospects, practice or employment of other engineers. Engineers who believe others are guilty of not ethical or illegal practice shall present such information to the proper authority for action.

Discussion

In today's employment environment, with employers of engineers scrambling to maintain a competent workforce, many employers are attempting more aggressive employment recruitment and retention approaches. As has been noted on numerous occasions, it appears that in times of heightened competition, whether for engineering services or for engineering employees, sometimes ethical considerations are minimized and even lost as firms attempt to do what is necessary to stay in business and prosper. However, failing to maintain reasonable standards of ethical conduct ultimately reflects poorly both on engineers who engage in such conduct, and also on the engineering profession as a whole.

The Board of Ethical Review has considered the issue of employment recruitment on numerous occasions. As early as BER Case No. 60-4, the Board noted that it was ethically proper for an engineer to discuss employment with another company, regardless of whether his company and the company are competitors, and to take such employment if in his professional interest. BER Case No. 60-4 involved a number of companies that required engineers for the design and development of their products and found it difficult in recent years to recruit a sufficient number of qualified engineers. As one means of contacting engineers interested in their type of work, these companies individually maintained temporary recruiting facilities in connection with various industrial exhibitions and meetings of professional and technical societies to interview those engineers in attendance who might be seeking employment or a change of employment.

Later in BER Case No. 68-4, Engineering Firm A sent to all engineers in Engineering Firm B a form letter reciting the history and policies of Firm A concluding with a statement, "We enclose for your consideration a summary of the aims and objectives of our firm, as well as the various advantages offered those who join us. We hope you will read it and perhaps refer to us those individuals whose professional philosophy matches our own." The enclosure referred to a 20-page booklet covering the history, aims, benefits, and rules of Firm A. In reviewing the NSPE Code and the facts, the Board concluded that the recruitment of engineering personnel through this method was ethical. In reaching its result, the Board suggested that this type of direct unsolicited contact with large numbers or employees of other firms who have not indicated any interest or desire to change employment is not in keeping with desirable professional standards and the proper relationship between firms within the profession.

Over the past decades, the culture of employment has changed from steady progression within a single company or firm to one where young engineers must plan on a substantial number of employment changes in the course of their careers. Employers, in an age of increasing competition, increasingly rely on relatively short term hiring of professionals to meet changing market demands. Free and open communication of available positions is critical in meeting today's market demands.

Reflecting this changing culture, since the 1970s, the Code has placed increasingly less emphasis on the style and methods of recruitment while steadfastly maintaining standards of integrity of recruitment statements. The facts in this case do not include any improprieties in the contents of the postcards. Instead, they center upon the method of distribution to prospective clients. Nevertheless the method used does raise at least two potential problems. By sending out a mass unsolicited mailing to an engineering licensure board list of professional engineers in the state, Engineer A would invariably be sending the solicitation in at least some cases to the business addresses of those professional engineers. The Board does not believe this type of solicitation crosses the line and employer should be expected to accept incoming correspondence from competing firms soliciting its employees for positions with that competing firm. Using company resources and time to process such material is a minor inconvenience which any employer can be expected to tolerate. This is much different than situations where employees use equipment, supplies, laboratory, or other office facilities of an employer to carry on an outside business. The Board does not believe the actions by Engineer A's firm rises to the level which allows the employer's place of business to be used as a "staging ground" for the "raiding" of its employees.

By sending unsolicited letters to firms of various sizes, Engineer A's actions might have the unintended effect of causing unknowing employers in firms receiving the letters to conclude, without more information, that their employees are soliciting information or actively seeking employment elsewhere and doing so on company time. This could have the effect of straining relations within the firm and cause misunderstanding and mistrust within the firm. However, the Board does not believe this issue rises to an ethical violation.

Conclusion

It was ethical for Engineer A's firm to send postcards out to individual engineers in the manner described.

BOARD OF ETHICAL REVIEW

Lorry T. Bannes, P.E., NSPE
E. Dave Dorchester, P.E., NSPE
John W. Gregorits, P.E., NSPE
Paul E. Pritzker, P.E., NSPE
Richard Simberg, P.E., NSPE
Harold E. Williamson, P.E., NSPE
(C. Allen Wortley, P.E., NSPE, Chair, did not participate in the consideration of this case)

The NSPE Board of Ethical Review (BER) considers ethical cases involving either real or hypothetical matters submitted to it from NSPE members, other engineers, public officials and members of the public. The BER reviews each case in the context of the NSPE Code and earlier BER opinions. The facts contained in each case do not necessarily represent all of the pertinent facts submitted to or reviewed by the BER.

World Legal Systems—CIA Factbook

APPENDIX

C

CIA Fact Book	
Country	**Legal system**
Afghanistan	based on mixed civil and Shari'a law; has not accepted compulsory ICJ jurisdiction
Akrotiri	the laws of the UK, where applicable, apply
Albania	has a civil law system; has not accepted compulsory ICJ jurisdiction; has accepted jurisdiction of the International Criminal Court for its citizens
Algeria	socialist, based on French and Islamic law; judicial review of legislative acts in ad hoc Constitutional Council composed of various public officials, including several Supreme Court justices; has not accepted compulsory ICJ jurisdiction
American Samoa	NA
Andorra	based on French and Spanish civil codes; no judicial review of legislative acts; has not accepted compulsory ICJ jurisdiction
Angola	based on Portuguese civil law system and customary law; recently modified to accommodate political pluralism and increased use of free markets
Anguilla	based on English common law
Antarctica	Antarctica is administered through meetings of the consultative member nations; decisions from these meetings are carried out by these member nations (with respect to their own nationals and operations) in accordance with their own national laws; US law, including certain criminal offenses by or against US nationals, such as murder, may apply extraterritorially; some US laws directly apply to Antarctica; for example, the Antarctic Conservation Act, 16 U.S.C. section 2401 et seq., provides civil and criminal penalties for the following activities, unless authorized by regulation of statute: the taking of native mammals or birds; the introduction of nonindigenous plants and animals; entry into specially protected areas; the discharge or disposal of pollutants; and the importation into the US of certain items from Antarctica; violation of the Antarctic Conservation Act carries penalties of up to $10,000 in fines and one year in prison; the National Science Foundation and Department of Justice share enforcement responsibilities; Public Law 95-541, the US Antarctic Conservation Act of 1978, as amended in 1996, requires expeditions from the US to Antarctica to notify, in advance, the Office of Oceans, Room 5805, Department of State, Washington, DC 20520, which reports such plans to other nations as required by the Antarctic Treaty; for more information, contact Permit Office, Office of Polar Programs, National Science Foundation, Arlington, Virginia 22230; telephone: (703) 292-8030, or visit their website at www.nsf.gov; more generally, access to the Antarctic Treaty area, that is to all areas between 60 and 90 degrees south latitude, is subject to a number of relevant legal instruments and authorization procedures adopted by the states party to the Antarctic Treaty
Antigua and Barbuda	based on English common law
Argentina	mixture of US and West European legal systems; has not accepted compulsory ICJ jurisdiction
Armenia	based on civil law system; has not accepted compulsory ICJ jurisdiction
Aruba	based on Dutch civil law system, with some English common law influence
Ashmore and Cartier Islands	the laws of the Commonwealth of Australia and the laws of the Northern Territory of Australia, where applicable, apply
Australia	based on English common law; accepts compulsory ICJ jurisdiction, with reservations

CIA Fact Book	
Country	**Legal system**
Austria	civil law system with Roman law origin; judicial review of legislative acts by the Constitutional Court; separate administrative and civil/penal supreme courts; accepts compulsory ICJ jurisdiction
Azerbaijan	based on civil law system; has not accepted compulsory ICJ jurisdiction
Bahamas, The	based on English common law
Bahrain	based on Islamic law and English common law; has not accepted compulsory ICJ jurisdiction
Baker Island	the laws of the US, where applicable, apply
Bangladesh	based on English common law; has not accepted compulsory ICJ jurisdiction
Barbados	English common law; no judicial review of legislative acts; accepts compulsory ICJ jurisdiction, with reservations
Belarus	based on civil law system; has not accepted compulsory ICJ jurisdiction
Belgium	based on civil law system influenced by English constitutional theory; judicial review of legislative acts; accepts compulsory ICJ jurisdiction, with reservations
Belize	English law
Benin	based on French civil law and customary law; has not accepted compulsory ICJ jurisdiction
Bermuda	English law
Bhutan	based on Indian law and English common law; has not accepted compulsory ICJ jurisdiction
Bolivia	based on Spanish law and Napoleonic Code; has not accepted compulsory ICJ jurisdiction
Bosnia and Herzegovina	based on civil law system; has not accepted compulsory ICJ jurisdiction
Botswana	based on Roman-Dutch law and local customary law; judicial review limited to matters of interpretation; accepts compulsory ICJ jurisdiction, with reservations
Bouvet Island	the laws of Norway, where applicable, apply
Brazil	based on Roman codes; has not accepted compulsory ICJ jurisdiction
British Indian Ocean Territory	the laws of the UK, where applicable, apply
British Virgin Islands	English law
Brunei	based on English common law; for Muslims, Islamic Shari'a law supersedes civil law in a number of areas; has not accepted compulsory ICJ jurisdiction
Bulgaria	civil law and criminal law based on Roman law; accepts compulsory ICJ jurisdiction with reservations
Burkina Faso	based on French civil law system and customary law; has not accepted compulsory ICJ jurisdiction
Burma	based on English common law; has not accepted compulsory ICJ jurisdiction

CIA Fact Book	
Country	**Legal system**
Burundi	based on German and Belgian civil codes and customary law; has not accepted compulsory ICJ jurisdiction
Cambodia	primarily a civil law mixture of French-influenced codes from the United Nations Transitional Authority in Cambodia (UNTAC) period, royal decrees, and acts of the legislature, with influences of customary law and remnants of communist legal theory; increasing influence of common law; accepts compulsory ICJ jurisdiction with reservations
Cameroon	based on French civil law system, with common law influence; accepts compulsory ICJ jurisdiction
Canada	based on English common law, except in Quebec, where civil law system based on French law prevails; accepts compulsory ICJ jurisdiction, with reservations
Cape Verde	based on the legal system of Portugal; has not accepted compulsory ICJ jurisdiction
Cayman Islands	British common law and local statutes
Central African Republic	based on French law
Chad	based on French civil law system and Chadian customary law; has not accepted compulsory ICJ jurisdiction
Chile	based on Code of 1857 derived from Spanish law and subsequent codes influenced by French and Austrian law; judicial review of legislative acts in the Supreme Court; has not accepted compulsory ICJ jurisdiction; note - in June 2005, Chile completed overhaul of its criminal justice system to a new, US-style adversarial system
China	based on civil law system; derived from Soviet and continental civil code legal principles; legislature retains power to interpret statutes; constitution ambiguous on judicial review of legislation; has not accepted compulsory ICJ jurisdiction
Christmas Island	under the authority of the governor general of Australia and Australian law
Clipperton Island	the laws of France, where applicable, apply
Cocos (Keeling) Islands	based upon the laws of Australia and local laws
Colombia	based on Spanish law; a new criminal code modeled after US procedures was enacted into law in 2004 and is gradually being implemented; judicial review of executive and legislative acts
Comoros	French and Islamic law in a new consolidated code
Congo, Democratic Republic of the	a new constitution was adopted by referendum 18 December 2005; accepts compulsory ICJ jurisdiction, with reservations
Congo, Republic of the	based on French civil law system and customary law
Cook Islands	based on New Zealand law and English common law
Coral Sea Islands	the laws of Australia, where applicable, apply
Costa Rica	based on Spanish civil law system; judicial review of legislative acts in the Supreme Court; has accepted compulsory ICJ jurisdiction
Cote d'Ivoire	based on French civil law system and customary law; judicial review in the Constitutional Chamber of the Supreme Court; accepts compulsory ICJ jurisdiction, with reservations

CIA Fact Book	
Country	**Legal system**
Croatia	based on Austro-Hungarian law system with Communist law influences; has not accepted compulsory ICJ jurisdiction
Cuba	based on Spanish civil law and influenced by American legal concepts, with large elements of Communist legal theory; has not accepted compulsory ICJ jurisdiction
Cyprus	based on English common law, with civil law modifications; accepts compulsory ICJ jurisdiction, with reservations
Czech Republic	civil law system based on Austro-Hungarian codes; has not accepted compulsory ICJ jurisdiction; legal code modified to bring it in line with Organization on Security and Cooperation in Europe (OSCE) obligations and to expunge Marxist-Leninist legal theory
Denmark	civil law system; judicial review of legislative acts; accepts compulsory ICJ jurisdiction, with reservations
Dhekelia	the laws of the UK, where applicable, apply
Djibouti	based on French civil law system, traditional practices, and Islamic law; accepts ICJ jurisdiction, with reservations
Dominica	based on English common law
Dominican Republic	based on French civil codes; Criminal Procedures Code modified in 2004 to include important elements of an accusatory system; accepts compulsory ICJ jurisdiction
Ecuador	based on civil law system; has not accepted compulsory ICJ jurisdiction
Egypt	based on Islamic and civil law (particularly Napoleonic codes); judicial review by Supreme Court and Council of State (oversees validity of administrative decisions); accepts compulsory ICJ jurisdiction with reservations
El Salvador	based on civil and Roman law with traces of common law; judicial review of legislative acts in the Supreme Court
Equatorial Guinea	partly based on Spanish civil law and tribal custom
Eritrea	primary basis is the Ethiopian legal code of 1957, with revisions; new civil, commercial, and penal codes have not yet been promulgated; government also issues unilateral proclamations setting laws and policies; also relies on customary and post-independence-enacted laws and, for civil cases involving Muslims, Islamic law; does not accept compulsory ICJ jurisdiction
Estonia	based on civil law system; accepts compulsory ICJ jurisdiction with reservations
Ethiopia	based on civil law; currently transitional mix of national and regional courts; has not accepted compulsory ICJ jurisdiction
European Union	comparable to the legal systems of member states; first supranational law system
Falkland Islands (Islas Malvinas)	English common law
Faroe Islands	the laws of Denmark, where applicable, apply
Fiji	based on British system
Finland	civil law system based on Swedish law; the president may request the Supreme Court to review laws; accepts compulsory ICJ jurisdiction with reservations

CIA Fact Book	
Country	**Legal system**
France	civil law system with indigenous concepts; review of administrative but not legislative acts; has not accepted compulsory ICJ jurisdiction
French Polynesia	the laws of France, where applicable, apply
French Southern and Antarctic Lands	the laws of France, where applicable, apply
Gabon	based on French civil law system and customary law; judicial review of legislative acts in Constitutional Chamber of the Supreme Court; has not accepted compulsory ICJ jurisdiction
Gambia, The	based on a composite of English common law, Islamic law, and customary law; accepts compulsory ICJ jurisdiction with reservations
Georgia	based on civil law system; accepts compulsory ICJ jurisdiction
Germany	civil law system with indigenous concepts; judicial review of legislative acts in the Federal Constitutional Court; has not accepted compulsory ICJ jurisdiction
Ghana	based on English common law and customary law; has not accepted compulsory ICJ jurisdiction
Gibraltar	the laws of the UK, where applicable, apply
Greece	based on codified Roman law; judiciary divided into civil, criminal, and administrative courts; accepts compulsory ICJ jurisdiction with reservations
Greenland	the laws of Denmark, where applicable, apply
Grenada	based on English common law
Guam	modeled on US; US federal laws apply
Guatemala	civil law system; judicial review of legislative acts; has not accepted compulsory ICJ jurisdiction
Guernsey	the laws of the UK, where applicable, apply; justice is administered by the Royal Court
Guinea	based on French civil law system, customary law, and decree; accepts compulsory ICJ jurisdiction with reservations
Guinea-Bissau	based on French civil law; accepts compulsory ICJ jurisdiction
Guyana	based on English common law with certain admixtures of Roman-Dutch law; has not accepted compulsory ICJ jurisdiction
Haiti	based on Roman civil law system; accepts compulsory ICJ jurisdiction
Heard Island and McDonald Islands	the laws of Australia, where applicable, apply
Holy See (Vatican City)	based on Code of Canon Law and revisions to it
Honduras	rooted in Roman and Spanish civil law with increasing influence of English common law; recent judicial reforms include abandoning Napoleonic legal codes in favor of the oral adversarial system; accepts ICJ jurisdiction with reservations
Hong Kong	based on English common law

CIA Fact Book	
Country	**Legal system**
Howland Island	the laws of the US, where applicable, apply
Hungary	based German-Austrian legal system; accepts compulsory ICJ jurisdiction with reservations
Iceland	civil law system based on Danish law; has not accepted compulsory ICJ jurisdiction
India	based on English common law; judicial review of legislative acts; accepts compulsory ICJ jurisdiction with reservations; separate personal law codes apply to Muslims, Christians, and Hindus
Indonesia	based on Roman-Dutch law, substantially modified by indigenous concepts and by new criminal procedures and election codes; has not accepted compulsory ICJ jurisdiction
Iran	based on Shari'a law system; has not accepted compulsory ICJ jurisdiction
Iraq	based on European civil and Islamic law under the framework outlined in the Iraqi Constitution; has not accepted compulsory ICJ jurisdiction
Ireland	based on English common law, substantially modified by indigenous concepts; judicial review of legislative acts in Supreme Court; has not accepted compulsory ICJ jurisdiction
Isle of Man	the laws of the UK, where applicable, apply and Manx statutes
Israel	mixture of English common law, British Mandate regulations, and, in personal matters, Jewish, Christian, and Muslim legal systems; in December 1985, Israel informed the UN Secretariat that it would no longer accept compulsory ICJ jurisdiction
Italy	based on civil law system; appeals treated as new trials; judicial review under certain conditions in Constitutional Court; has not accepted compulsory ICJ jurisdiction
Jamaica	based on English common law; has not accepted compulsory ICJ jurisdiction
Jan Mayen	the laws of Norway, where applicable, apply
Japan	modeled after German civil law system with English-American influence; judicial review of legislative acts in the Supreme Court; accepts compulsory ICJ jurisdiction with reservations
Jarvis Island	the laws of the US, where applicable, apply
Jersey	the laws of the UK, where applicable, apply and local statutes; justice is administered by the Royal Court
Johnston Atoll	the laws of the US, where applicable, apply
Jordan	based on Islamic law and French codes; judicial review of legislative acts in a specially provided High Tribunal; has not accepted compulsory ICJ jurisdiction
Kazakhstan	based on Islamic law and Roman law; has not accepted compulsory ICJ jurisdiction
Kenya	based on Kenyan statutory law, Kenyan and English common law, tribal law, and Islamic law; judicial review in High Court; accepts compulsory ICJ jurisdiction with reservations; constitutional amendment of 1982 making Kenya a de jure one-party state repealed in 1991
Kingman Reef	the laws of the US, where applicable, apply

CIA Fact Book	
Country	**Legal system**
Kiribati	NA
Korea, North	based on Prussian civil law system with Japanese influences and Communist legal theory; no judicial review of legislative acts; has not accepted compulsory ICJ jurisdiction
Korea, South	combines elements of continental European civil law systems, Anglo-American law, and Chinese classical thought; has not accepted compulsory ICJ jurisdiction
Kuwait	civil law system with Islamic law significant in personal matters; has not accepted compulsory ICJ jurisdiction
Kyrgyzstan	based on French and Russian laws; has not accepted compulsory ICJ jurisdiction
Laos	based on traditional customs, French legal norms and procedures, and socialist practice; has not accepted compulsory ICJ jurisdiction
Latvia	based on civil law system with traces of Socialist legal traditions and practices; has not accepted compulsory ICJ jurisdiction
Lebanon	mixture of Ottoman law, canon law, Napoleonic code, and civil law; no judicial review of legislative acts; has not accepted compulsory ICJ jurisdiction
Lesotho	based on English common law and Roman-Dutch law; judicial review of legislative acts in High Court and Court of Appeal; accepts compulsory ICJ jurisdiction with reservations
Liberia	dual system of statutory law based on Anglo-American common law for the modern sector and customary law based on unwritten tribal practices for indigenous sector; accepts compulsory ICJ jurisdiction with reservations
Libya	based on Italian and French civil law systems and Islamic law; separate religious courts; no constitutional provision for judicial review of legislative acts; has not accepted compulsory ICJ jurisdiction
Liechtenstein	local civil and penal codes based on civil law system; accepts compulsory ICJ jurisdiction with reservations
Lithuania	based on civil law system; legislative acts can be appealed to the constitutional court; has not accepted compulsory ICJ jurisdiction
Luxembourg	based on civil law system; accepts compulsory ICJ jurisdiction
Macau	based on Portuguese civil law system
Macedonia	based on civil law system; judicial review of legislative acts; has not accepted compulsory ICJ jurisdiction
Madagascar	based on French civil law system and traditional Malagasy law; accepts compulsory ICJ jurisdiction with reservations
Malawi	based on English common law and customary law; judicial review of legislative acts in the Supreme Court of Appeal; accepts compulsory ICJ jurisdiction with reservations
Malaysia	based on English common law; judicial review of legislative acts in the Supreme Court at request of supreme head of the federation; Islamic law is applied to Muslims in matters of family law and religion; has not accepted compulsory ICJ jurisdiction
Maldives	based on Islamic law with admixtures of English common law primarily in commercial matters; has not accepted compulsory ICJ jurisdiction

CIA Fact Book	
Country	**Legal system**
Mali	based on French civil law system and customary law; judicial review of legislative acts in Constitutional Court; has not accepted compulsory ICJ jurisdiction
Malta	based on English common law and Roman civil law; accepts compulsory ICJ jurisdiction with reservations
Marshall Islands	based on adapted Trust Territory laws, acts of the legislature, municipal, common, and customary laws
Mauritania	a combination of Islamic law and French civil law; has not accepted compulsory ICJ jurisdiction
Mauritius	based on French civil law system with elements of English common law in certain areas; accepts compulsory ICJ jurisdiction with reservations
Mayotte	the laws of France, where applicable, apply
Mexico	mixture of US constitutional theory and civil law system; judicial review of legislative acts; accepts compulsory ICJ jurisdiction with reservations
Micronesia, Federated States of	based on adapted Trust Territory laws, acts of the legislature, municipal, common, and customary laws
Midway Islands	the laws of the US, where applicable, apply
Moldova	based on civil law system; Constitutional Court reviews legality of legislative acts and governmental decisions of resolution; accepts many UN and Organization for Security and Cooperation in Europe (OSCE) documents; has not accepted compulsory ICJ jurisdiction
Monaco	based on French law; has not accepted compulsory ICJ jurisdiction
Mongolia	blend of Soviet, German, and US systems that combine "continental" or "civil" code and case-precedent; constitution ambiguous on judicial review of legislative acts; has not accepted compulsory ICJ jurisdiction
Montenegro	based on civil law system; has not accepted compulsory ICJ jurisdiction
Montserrat	English common law and statutory law
Morocco	based on Islamic law and French and Spanish civil law systems; judicial review of legislative acts in Constitutional Chamber of Supreme Court; has not accepted compulsory ICJ jurisdiction
Mozambique	based on Portuguese civil law system and customary law
Namibia	based on Roman-Dutch law and 1990 constitution
Nauru	acts of the Nauru Parliament and British common law; accepts compulsory ICJ jurisdiction with reservations
Navassa Island	the laws of the US, where applicable, apply
Nepal	based on Hindu legal concepts and English common law; has not accepted compulsory ICJ jurisdiction
Netherlands	based on civil law system incorporating French penal theory; constitution does not permit judicial review of acts of the States General; accepts compulsory ICJ jurisdiction with reservations
Netherlands Antilles	based on Dutch civil law system with some English common law influence

CIA Fact Book	
Country	**Legal system**
New Caledonia	based on French civil law; the 1988 Matignon Accords grant substantial autonomy to the islands
New Zealand	based on English law, with special land legislation and land courts for the Maori; accepts compulsory ICJ jurisdiction with reservations
Nicaragua	civil law system; Supreme Court may review administrative acts; accepts compulsory ICJ jurisdiction
Niger	based on French civil law system and customary law; has not accepted compulsory ICJ jurisdiction
Nigeria	based on English common law, Islamic law (in 12 northern states), and traditional law; accepts compulsory ICJ jurisdiction with reservations
Niue	English common law; note - Niue is self-governing, with the power to make its own laws
Norfolk Island	based on the laws of Australia, local ordinances and acts; English common law applies in matters not covered by either Australian or Norfolk Island law
Northern Mariana Islands	based on US system, except for customs, wages, immigration laws, and taxation
Norway	mixture of customary law, civil law system, and common law traditions; Supreme Court renders advisory opinions to legislature when asked; accepts compulsory ICJ jurisdiction with reservations
Oman	based on English common law and Islamic law; ultimate appeal to the monarch; has not accepted compulsory ICJ jurisdiction
Pakistan	based on English common law with provisions to accommodate Pakistan's status as an Islamic state; accepts compulsory ICJ jurisdiction with reservations
Palau	based on Trust Territory laws, acts of the legislature, municipal, common, and customary laws
Palmyra Atoll	the laws of the US, where applicable, apply
Panama	based on civil law system; judicial review of legislative acts in the Supreme Court of Justice; accepts compulsory ICJ jurisdiction with reservations
Papua New Guinea	based on English common law; has not accepted compulsory ICJ jurisdiction
Paraguay	based on Argentine codes, Roman law, and French codes; judicial review of legislative acts in Supreme Court of Justice; accepts compulsory ICJ jurisdiction
Peru	based on civil law system; accepts compulsory ICJ jurisdiction with reservations
Philippines	based on Spanish and Anglo-American law; accepts compulsory ICJ jurisdiction with reservations
Pitcairn Islands	local island by-laws
Poland	based on a mixture of Continental (Napoleonic) civil law and holdover Communist legal theory; changes being gradually introduced as part of broader democratization process; limited judicial review of legislative acts, but rulings of the Constitutional Tribunal are final; court decisions can be appealed to the European Court of Justice in Strasbourg; accepts compulsory ICJ jurisdiction with reservations

CIA Fact Book	
Country	**Legal system**
Portugal	based on civil law system; the Constitutional Tribunal reviews the constitutionality of legislation; accepts compulsory ICJ jurisdiction with reservations
Puerto Rico	based on Spanish civil code and within the US Federal system of justice
Qatar	based on Islamic and civil law codes; discretionary system of law controlled by the amir, although civil codes are being implemented; Islamic law dominates family and personal matters; has not accepted compulsory ICJ jurisdiction
Romania	based on civil law system; has not accepted compulsory ICJ jurisdiction
Russia	based on civil law system; judicial review of legislative acts; has not accepted compulsory ICJ jurisdiction
Rwanda	based on German and Belgian civil law systems and customary law; judicial review of legislative acts in the Supreme Court; has not accepted compulsory ICJ jurisdiction
Saint Barthelemy	the laws of France, where applicable, apply
Saint Helena	English common law and statutes, supplemented by local statutes
Saint Kitts and Nevis	based on English common law
Saint Lucia	based on English common law
Saint Martin	the laws of France, where applicable, apply
Saint Pierre and Miquelon	the laws of France, where applicable, apply
Saint Vincent and the Grenadines	based on English common law
Samoa	based on English common law and local customs; judicial review of legislative acts with respect to fundamental rights of the citizen; has not accepted compulsory ICJ jurisdiction
San Marino	based on civil law system with Italian law influences; has not accepted compulsory ICJ jurisdiction
Sao Tome and Principe	based on Portuguese legal system and customary law; has not accepted compulsory ICJ jurisdiction
Saudi Arabia	based on Shari'a law, several secular codes have been introduced; commercial disputes handled by special committees; has not accepted compulsory ICJ jurisdiction
Senegal	based on French civil law system; judicial review of legislative acts in Constitutional Court; the Council of State audits the government's accounting office; accepts compulsory ICJ jurisdiction with reservations
Serbia	based on civil law system; has not accepted compulsory ICJ jurisdiction
Seychelles	based on English common law, French civil law, and customary law
Sierra Leone	based on English law and customary laws indigenous to local tribes; has not accepted compulsory ICJ jurisdiction
Singapore	based on English common law; has not accepted compulsory ICJ jurisdiction

CIA Fact Book	
Country	**Legal system**
Slovakia	civil law system based on Austro-Hungarian codes; accepts compulsory ICJ jurisdiction with reservations; legal code modified to comply with the obligations of Organization on Security and Cooperation in Europe (OSCE) and to expunge Marxist-Leninist legal theory
Slovenia	based on civil law system; has not accepted compulsory ICJ jurisdiction
Solomon Islands	English common law, which is widely disregarded
Somalia	no national system; a mixture of English common law, Italian law, Islamic Shari'a, and Somali customary law; accepts compulsory ICJ jurisdiction with reservations
South Africa	based on Roman-Dutch law and English common law
South Georgia and the South Sandwich Islands	the laws of the UK, where applicable, apply; the senior magistrate from the Falkland Islands presides over the Magistrates Court
Spain	civil law system, with regional applications; accepts compulsory ICJ jurisdiction with reservations
Sri Lanka	a highly complex mixture of English common law, Roman-Dutch, Islamic, Sinhalese, and customary law; has not accepted compulsory ICJ jurisdiction
Sudan	based on English common law and Islamic law; as of 20 January 1991, the now defunct Revolutionary Command Council imposed Islamic law in the northern states; Islamic law applies to all residents of the northern states regardless of their religion; however, the CPA establishes some protections for non-Muslims in Khartoum; some separate religious courts; accepts compulsory ICJ jurisdiction with reservations; the southern legal system is still developing under the CPA following the civil war; Islamic law will not apply to the southern states
Suriname	based on Dutch legal system incorporating French penal theory; accepts compulsory ICJ jurisdiction with reservations
Svalbard	the laws of Norway, where applicable, apply
Swaziland	based on South African Roman-Dutch law in statutory courts and Swazi traditional law and custom in traditional courts; accepts compulsory ICJ jurisdiction with reservations
Sweden	civil law system influenced by customary law; accepts compulsory ICJ jurisdiction with reservations
Switzerland	civil law system influenced by customary law; judicial review of legislative acts, except with respect to federal decrees of general obligatory character; accepts compulsory ICJ jurisdiction with reservations
Syria	based on a combination of French and Ottoman civil law; Islamic law is used in the family court system; has not accepted compulsory ICJ jurisdiction
Taiwan	based on civil law system; has not accepted compulsory ICJ jurisdiction
Tajikistan	based on civil law system; no judicial review of legislative acts; has not accepted compulsory ICJ jurisdiction
Tanzania	based on English common law; judicial review of legislative acts limited to matters of interpretation; has not accepted compulsory ICJ jurisdiction
Thailand	based on civil law system, with influences of common law; has not accepted compulsory ICJ jurisdiction

CIA Fact Book	
Country	**Legal system**
Timor-Leste	UN-drafted legal system based on Indonesian law remains in place but are to be replaced by civil and penal codes based on Portuguese law; these have passed but have not been promulgated; has not accepted compulsory ICJ jurisdiction
Togo	French-based court system; accepts compulsory ICJ jurisdiction, with reservations
Tokelau	New Zealand and local statutes
Tonga	based on English common law
Trinidad and Tobago	based on English common law; judicial review of legislative acts in the Supreme Court; has not accepted compulsory ICJ jurisdiction
Tunisia	based on French civil law system and Islamic law; some judicial review of legislative acts in the Supreme Court in joint session; has not accepted compulsory ICJ jurisdiction
Turkey	civil law system derived from various European continental legal systems; note - member of the European Court of Human Rights (ECHR), although Turkey claims limited derogations on the ratified European Convention on Human Rights; has not accepted compulsory ICJ jurisdiction
Turkmenistan	based on civil law system and Islamic law; has not accepted compulsory ICJ jurisdiction
Turks and Caicos Islands	based on laws of England and Wales, with a few adopted from Jamaica and The Bahamas
Tuvalu	NA
Uganda	in 1995, the government restored the legal system to one based on English common law and customary law; accepts compulsory ICJ jurisdiction, with reservations
Ukraine	based on civil law system; judicial review of legislative acts; has not accepted compulsory ICJ jurisdiction
United Arab Emirates	based on a dual system of Shari'a and civil courts; has not accepted compulsory ICJ jurisdiction
United Kingdom	based on common law tradition with early Roman and modern continental influences; has nonbinding judicial review of Acts of Parliament under the Human Rights Act of 1998; accepts compulsory ICJ jurisdiction, with reservations
United States	federal court system based on English common law; each state has its own unique legal system, of which all but one (Louisiana, which is still influenced by the Napoleonic Code) is based on English common law; judicial review of legislative acts; has not accepted compulsory ICJ jurisdiction
United States Pacific Island Wildlife Refuges	the laws of the US, where applicable, apply
Uruguay	based on Spanish civil law system; accepts compulsory ICJ jurisdiction
Uzbekistan	based on civil law system; has not accepted compulsory ICJ jurisdiction
Vanuatu	unified system being created from former dual French and British systems
Venezuela	open, adversarial court system
Vietnam	based on communist legal theory and French civil law system has not accepted compulsory ICJ jurisdiction

CIA Fact Book	
Country	**Legal system**
Virgin Islands	based on US laws
Wake Island	the laws of the US, where applicable, apply
Wallis and Futuna	the laws of France, where applicable, apply
World	all members of the UN are parties to the statute that established the International Court of Justice (ICJ) or World Court
Yemen	based on Islamic law, Turkish law, English common law, and local tribal customary law; has not accepted compulsory ICJ jurisdiction
Zambia	based on English common law and customary law; judicial review of legislative acts in an ad hoc constitutional council; has not accepted compulsory ICJ jurisdiction
Zimbabwe	mixture of Roman-Dutch and English common law
This page was last updated on 19 July, 2007	
https://www.cia.gov/library/publications/the-world-factbook/fields/2100.html	

Specification Instructions— Outdoor Fireplaces

APPENDIX

D

Division 2
Site Construction
READ AND REMOVE THIS PAGE BEFORE PLACING
THE SPECIFICATION SECTION IN THE CONTRACT

■ ■ ■ ■ ■ ■ ■ ■

DIVISION 2: SITE CONSTRUCTION

SECTION 02877—OUTDOOR FIREPLACES

USE: Furnishing and installing outdoor fireplaces. Unique job requirements may require specific coverage in the following specification sections:

NOTES TO SPECIFIER:

A. Options within paragraphs are identified by enclosure in brackets; blank spaces enclosed by brackets [_____] provide for information to be inserted, as appropriate, to individual projects.

B. TO AVOID CONFLICTING REQUIREMENTS AND COSTLY MISTAKES, the specifier must edit (delete, substitute or add to the text). The remaining paragraphs must be renumbered.

C. DELETE FOR FINAL COPY: Notes to the Specifier, located between lines of asterisks; all [brackets], underlining, boldface, and italics, within the paragraphs; and this page. (Open Reveal Codes [alt F3] to do this accurately.)

D. To Edit "(Footer B)": (Corel 8). ENTIRE FOOTER SHOULD BE BOLDED. Click in the footer and delete the "Project Name (ftr B)" text and replace it with the Name of the Project (upper and lower case). Replace date with current MM/YY (WITHOUT BRACKETS).

E. The Guide Specifications are intended to be continuously reviewed and revised. User's technical comments are appreciated. Also, other related or similar specification that are written "from scratch" that could benefit others, or other questions can be sent to NSTC A/E Group, ST 110, or call (303) 987-6868

DRAWING DATA REQUIRED:

REVISIONS OF THIS SECTION:
CSI format, editorial changes 12/01
Metric: Not Applicable
Technical Update: 08/00

SECTION 02877

OUTDOOR FIREPLACES

PART 1: GENERAL

1.1 SUMMARY

A. Description: [Furnishing and installing] [Installing with Government-furnished property] [Installing with Contractor-furnished property] outdoor [flip-top camp] [pedestal] [grills] [and] [fire] [circles] [holes].

1.2 SUBMITTALS

A. General: Submittals shall be according to [section 01009—General Information and Requirements.] [Section 01300—Submittals.]

B. Manufacturer's Literature: Submit [_____] copies of the manufacturer's descriptive data for pedestal grills.

C. Installation Instructions: Submit [_____] copies of the manufacturer's installation instructions for pedestal grills.

D. Shop Drawings: Submit [_____] copies of shop drawings for [flip-top camp] [pedestal] [grills] [and] [fire] [cirles] [holes].

1.3 QUALITY ASSURANCE

A. Failure Criteria: Not limited to the following:
1. Poorly fitting or warped parts.
2. Un-level or unstable installation.
3. Cracking of welds.

1.4 WARRANTY:

A. Requirements: Finish a written warranty from the manufacturer covering the [flip-top camp] [pedestal] grills according to [Section 01009—General Information and Requirements] [Section 01700—Contract Closeout]. The warranty period shall extend for a minimum of 1 year.

PART 2: PRODUCTS

2.1 MATERIALS

A. Flip-Top Camp Grill: Shall be as shown on the drawing.

B. Pedestal Grill: The materials shall meet the following minimums:
1. Body: 3/16-inch hot-rolled steel plate, riveted or all-welded construction.
2. Fire Grate or Ash Pan Assembly: Fabricated from 10-ga steel.
3. Cooking Grate: 1/2-inch-diameter steel rods. Cooking area 270 in2.

4. Handles: 5/8-inch-diameter steel rods with 1/8-inch-diameter, spiral wound grips.

5. Pedestal (Support Post): 1-1/2-inch-diameter galvanized steel pipe according to ASTM A 53.

6. Finish: Heat-resistant black enamel or aluminum paint.

7. Hardware and Accessories: As shown on the drawings or in manufacturer's literature; galvanized.

C. Fire Circle: Shall be as shown on the drawings.

D. Fire Fole: Install according to the drawings.

E. Concrete: Shall be according to Section 03306—Minor Concrete.

PART 3: EXECUTION

3.1 PREPARATION

A. [Prepare base according to Section 02213—Site Grading.] [Prepare base by _____ to conform with [flip-top grills] [grill pedestals] [and] [fire] [circles] [holes].

3.2 INSTALLATION

A. Flip-Top Camp Grill: Install according to the drawings, Section 03306—Minor Concrete, and applicable manufacturer's instructions.

B. Pedestal Grill: Set pedestal in concrete base according to the drawings, Section 03306—Minor Concrete, and applicable manufacturer's recommendations. Insure that the pedestal is level and fully supported until concrete has set. Movement of the pedestal within the concrete after it has set, shall be cause for rejection. Set body so that cooking surface is adjustable within the range of 30 inches to 36 inches above the ground at the base. The grill shall be attached to the pedestal in such a way as to prevent removal without tools, but will allow the grill to rotate 360 degrees on the pedestal.

C. Fire Circle: Construct according to the drawings.

D. Fire Hole: Construct according to the drawings.

PART 4: MEASUREMENT AND PAYMENT

4.1 METHOD OF MEASUREMENT

A. Units: The work described in this section will [not be measured for payment.] [be measured and paid for] [on lump sum basis] [by each [grill] [and] [fire] [circle] [hole] installed.]

4.2 BASIS OF PAYMENT

A. Payment: [No direct payment for the work described under this section will be made. The Contractor shall include consideration for this item in the bid price for other items of the Contract.] [Prices and payment will be full compensation for the work described in this section. Payment will made under:]

Pay Item Pay Unit

02877(01) Pedestal Grill Lump Sum
02877(02) Pedestal Grill Each
02877(10) Grill and Fire Circle Lump Sum
02877(11) Grill and Fire Circle Each
02877(21) Fire Circle Lump Sum
02877(22) Fire Circle Each
02877(31) Fire Hole Lump Sum
02877(32) Fire Hole Each

Drawings—Outdoor Fireplaces

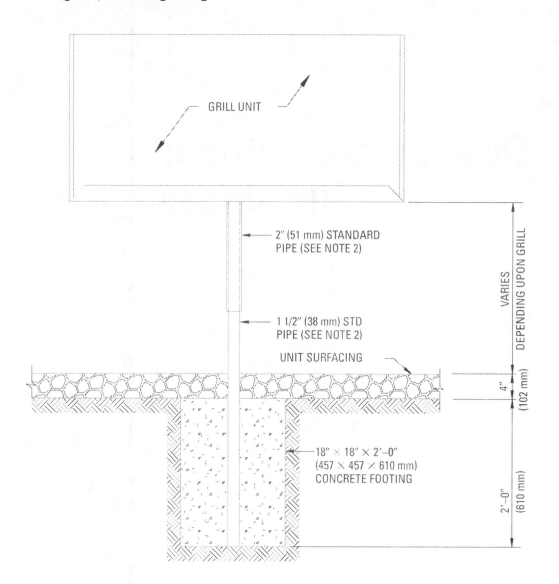

GRILL UNIT

2" (51 mm) STANDARD
PIPE (SEE NOTE 2)

1 1/2" (38 mm) STD
PIPE (SEE NOTE 2)

UNIT SURFACING

18" × 18" × 2'-0"
(457 × 457 × 610 mm)
CONCRETE FOOTING

VARIES DEPENDING UPON GRILL

4" (102 mm)

2'-0" (610 mm)

MOUNTING DETAIL

NOTES:

1. SEE SPECIFICATIONS FOR TYPE OF GRILL & UNIT SURFACING
   TO BE USED.

2. FOR SOME MODELS PIPE DIAMETER MAY VARY AND
   ONLY ONE PIPE BE USED TO CONNECT GRILL TO THE
   FOOTING.

3. THE METRIC CONVERSIONS ARE PROVIDED IN PARENTHESIS
   FOLLOWING THE ENGLISH UNITS.

SOIL INSIDE FIRE CIRCLE TO BE CLEARED OF ALL VEGETATIVE & BURNABLE MATERIAL

ROCK SIZE – 12"0 (305 mm) MIN.

2'–6"R (762 mm)

NATIVE ROCK

PLAN

8"–10" (203–254 mm)

36" (762 mm)

PLAN

EXISTING GROUND LINE

INSTALL LEVEL

6" (152 mm)

9" (229 mm)

3" (76 mm)

SOIL TO BE FIRMLY COMPACTED AROUND ROCKS

SOIL CLEARED 3" (76 mm) BELOW EXISTING GRADE

SECTION

FIRE CIRCLE

STANDARD 14– OR 16– GAUGE GALVANIZED CMP

4" (102 mm)

8" MIN. (203 mm)

SECTION

GRAVEL FILL WITHIN CMP

ROLL & WELD PLATE

3'–0" (914 mm)

PLAN

FIRE HOLE MADE OF CMP–RIVETED PIPE, 14– OR 18– GAUGE. LENGTHS OF PIPE OR COLLARS MAY BE USED.

3'–0" (914 mm)

PLAN

SURFACING

PERSPECTIVE

FIRE CIRCLE W/CORRUGATED METAL PIPE (CMP)

SURFACING (FLAT STONE OR AGGREGATE)

FIRE HOLE MADE OF 1/8" (3 mm) STEEL SHEET

GROUND LEVEL

GROUND LEVEL

2" (51 mm)

1'–0" ± (305 mm)

1'–0" ± (305 mm)

4" (102 mm)

SECTION

SECTION

FIRE HOLE

NOTE:

1. THE METRIC CONVERSIONS ARE PROVIDED IN PARENTHESIS FOLLOWING THE ENGLISH UNITS.

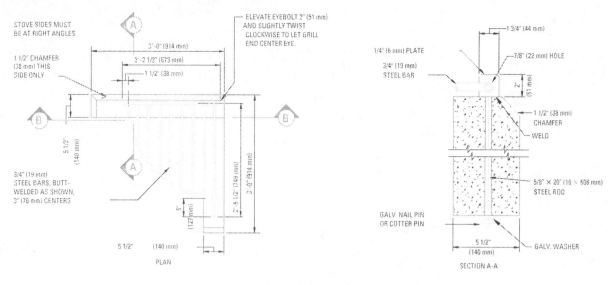

CONCRETE CAMP GRILL

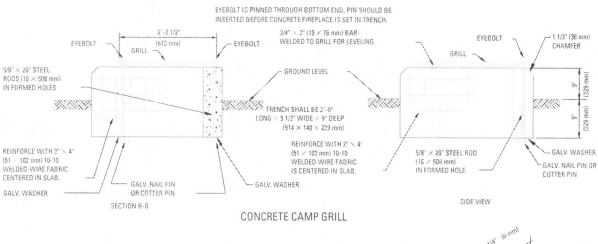

STEEL CAMP GRILL

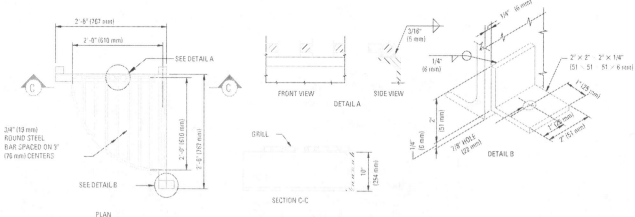

NOTES:

1. MILD-STEEL BARS SHALL BE S.A.E. 1020 GRADE.

2. STEEL HOLD-DOWN STAKES ARE PROVIDED WITH THE STEEL COMMERCIAL MODEL,
   HOWEVER, A METHOD OF SECURING THE GRILL TO PREVENT THEFT SHALL BE
   PROVIDED, SUCH AS MOUNTING ON A CONCRETE BASE.

3. THE METRIC CONVERSIONS ARE PROVIDED IN PARENTHESIS FOLLOWING THE
   ENGLISH UNITS

Patent No. 764957

A. T. PRATHER.
PENDULUM POWER.
APPLICATION FILED DEC. 3, 1903.

NO MODEL.

Fig.1

Fig.2

WITNESSES:

INVENTOR
Andrew T. Prather
BY
ATTORNEYS

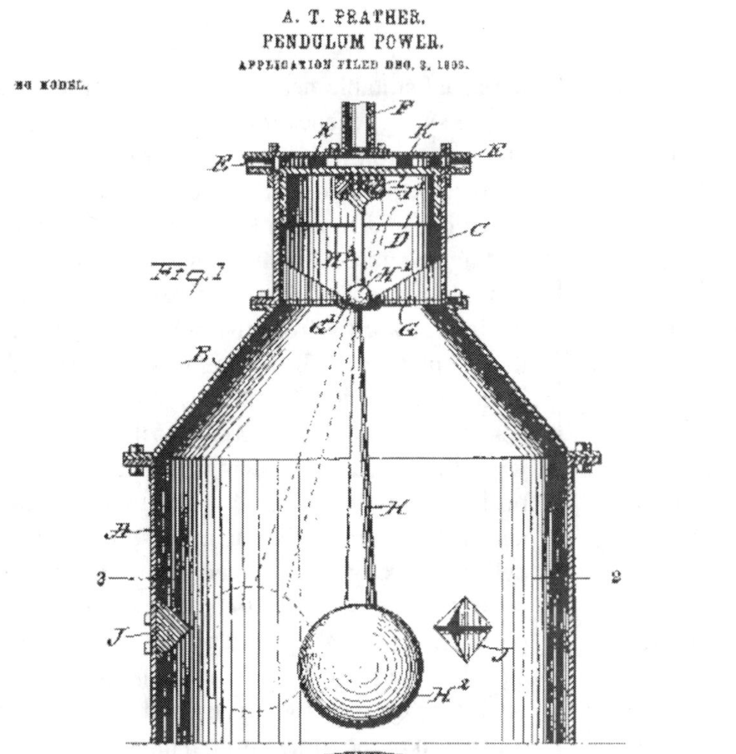

No. 764,957.   Patented July 12, 1904.

UNITED STATES PATENT OFFICE.

ANDREW THOMAS PRATHER, OF DOUGLAS, ARIZONA TERRITORY.

PENDULUM-POWER.

SPECIFICATION forming part of Letters Patent No. 764,957, dated July 12, 1904.
Application filed December 3, 1903. Serial No. 183,671. (No model.)

To all whom it may concern:

Be it known that I, ANDREW THOMAS PRATHER, a citizen of the United States, and a resident of Douglas, in the county of Cochise and Territory of Arizona, have invented a new and Improved Pendulum-Power, of which the following is a full, clear, and exact description.

The object of the invention is to provide a new and improved pendulum-power more especially designed for use on marine vessels, land-vehicles, and the like, and arranged to utilize the swaying motion of the vehicle for actuating an air-pump or like motor.

The invention consists of novel features and parts and combinations of the same, as will be more fully described hereinafter and then pointed out in the claims.

A practical embodiment of the invention is represented in the accompanying drawings, forming a part of this specification, in which similar characters of reference indicate corresponding parts in both views.

Figure 1 is a sectional side elevation of the improvement, and Fig. 2 is a sectional plan view of the same on the line 2 2 of Fig. 1.

The pendulum-power is mounted on a suitably-constructed support A, secured to a marine vessel, land-vehicle, or the like, so as to sway with the same. The dome-shaped top B of the said support A carries a vertically-disposed cylinder C, in which is mounted to reciprocate a piston D, and the said cylinder is provided with valved inlets E and a valved outlet F, as plainly illustrated in Fig. 1.

In the bottom of the cylinder C is secured a spider G, having a central socket G′, in which is hung the ball H′ of a pendulum H, provided at its lower end with

the usual suitable heavy ball H^2 to cause the pendulum to swing whenever a swaying motion is given by the vehicle to the support A. By the arrangement described the pendulum *h* is free to swing in any desired direction, and in order to use the pendulum for imparting a reciprocating motion to the piston D and said pendulum is provided with an extension-arm H^3, extending upwardly from the ball H′ and connected at its terminal by a ball-bearing with a socket I′, formed in the under side of a plate I, having a ball-bearing connection with the inner face of the piston D, so that when the pendulum H swings from its normal vertical position then the piston D is free to move downward, and thereby draws the air into the upper end of the cylinder through the valved inlets E, and when the pendulum swings back to a central position then the piston D is pushed upward by the action of the extension H^3 and plate I, and consequently the air previously drawn into the upper end of the cylinder is forced out of the same through the valved outlet F into a suitable reservoir or the like, from which the compressed air may be utilized for driving other machinery or for other purposes.

In order to prevent the pendulum from swinging around in the cylindrical support A, a plurality of projections J are arranged in a circle on the inner .face of the support A in alinement with the ball H^2, so that the latter in striking one of the projections J is caused to rebound to insure a constant up-and-down sliding movement of the piston D to pump air, as previously explained. The projections J are preferably pyramidal in shape, as indicated in the drawings, to insure a proper rebounding of the ball H^2. The piston D falls by its own weight; but, if desired, springs K may be used and interposed between the upper head of the cylinder C and the piston to insure a ready downward movement of the piston.

Having thus described my invention, I claim as new and desire to secure by Letters Patent—

1. A power comprising a pendulum mounted on a swaying vessel or vehicle, a pump in axial alinement with the vertical center line of the pendulum, and a sliding connection between the pendulum and inner face of the piston of the pump, as set forth.

2. A power comprising a support, a pendulum hung in the said support and having an extension beyond its fulcrum-point, a cylinder mounted on the said support and into which the extension of the pendulum projects, a piston reciprocating in the said cylinder, and a connection between the inner face of the piston and the extension of the pendulum, as set forth.

3. A power comprising a support, a pendulum hung on the said support and having an extension beyond its fulcrum, a cylinder mounted on the said support and having inlet and outlet valves, a piston reciprocating in the said cylinder, and a plate mounted to slide and engaging the said piston, and connected with the said extension, as set forth.

4. A power comprising a support, a pendulum hung in the said support and having an extension be-

yond its fulcrum-point, a cylinder mounted on the said support, a piston reciprocating in the said cylinder, a connection between the said piston and the extension of the pendulum, and projections in the said support, in alinement with the weighted end of the said pendulum, as set forth.

5. A power, comprising a support, a cylinder mounted on the support, a piston in the cylinder, and a pendulum hung on the support and having an extension beyond its fulcrum, said extension projecting into the cylinder and having a sliding connection with the inner face of the piston, as set forth.

6. A power, comprising a support, a cylinder of the support, a piston in the cylinder, a pendulum hung on the support and having an extension beyond its fulcrum, said extension terminating in a ball, and a plate in sliding engagement with the inner face of the piston and provided on its under face with a socket to receive the ball on the end of the extension of the pendulum, as set forth.

In testimony whereof I have signed my name to this specification in the presence of two subscribing witnesses.

ANDREW THOMAS PRATHER.

Witnesses:

S. F. FORBES,

E. A. VON ARNIM.

Federal Register

APPENDIX F

FEDERAL REGISTER

Vol. 68, No. 213

Proposed Rules

DEPARTMENT OF TRANSPORTATION (DOT)

National Highway Traffic Safety Administration (NHTSA)

49 CFR Part 571

[Docket No. NHTSA-1999-6550; Notice 3]

RIN 2127-AI63

Federal Motor Vehicle Safety Standards; Hydraulic and Electric Brake Systems

68 FR 62417

DATE: Tuesday, **November** 4, 2003

ACTION: Notice of proposed rulemaking (NPRM).

SUMMARY: In this document, NHTSA proposes to amend the Federal motor vehicle safety standard on hydraulic and electric brake systems to include an option for the use of a roll bar structure during specified testing of brake systems in single unit trucks and buses. This option is already available during similar testing of air braked trucks and buses. We tentatively conclude that permitting the use of a roll bar structure would help protect drivers and technicians in the event of a rollover during testing of hydraulically-braked trucks and buses. The safety of drivers and technicians is a primary concern during vehicle testing. The use of a roll bar structure would offer protection to the drivers and technicians performing brake tests conducted at lightly loaded vehicle weight.

DATES: You should submit comments early enough to ensure that Docket Management receives them not later than January 5, 2004.

ADDRESSES: You may submit comments [identified by DOT DMS Docket Number NHTSA-1999-6550] by any of the following methods:

. Web site: *http://dms.dot.gov*. Follow the instructions for submitting comments on the DOT electronic docket site.

. Fax: 1-202-493-2251.

. Mail: Docket Management Facility; U.S. Department of Transportation, 400 Seventh Street, SW., Nassif Building, Room PL-401, Washington, DC 20590-001.

. Hand Delivery: Room PL-401 on the plaza level of the Nassif Building, 400 Seventh Street, SW., Washington, DC, between 9 a.m. and 5 p.m., Monday through Friday, except Federal holidays.

. Federal eRulemaking Portal: Go to *http://www.regulations.gov*. Follow the online instructions for submitting comments.

Instructions: All submissions must include the agency name and docket number or Regulatory Identification Number (RIN) for this rulemaking. For detailed instructions on submitting comments and additional information on the rulemaking process, see the Submission of Comments heading of the Supplementary Information section of this document. Note that all comments received will be posted without change to *http://dms.dot.gov*, including any personal information provided. Please see the Privacy Act heading under Regulatory Notices.

Docket: For access to the docket to read background documents or comments received, go to *http://dms.dot.gov* at any time or to Room PL-401 on the plaza level of the Nassif Building, 400 Seventh Street, SW., Washington, DC, between 9 a.m. and 5 p.m., Monday through Friday, except Federal Holidays.

FOR FURTHER INFORMATION CONTACT: For non-legal issues, you may call Samuel Daniel Jr., Safety Standards Engineer, Office of Crash Avoidance Standards, Vehicle Dynamics Division, at (202) 366-4921, and fax him at (202) 493-2739.

For legal issues, you may call Christopher Calamita of the NHTSA Office of Chief Counsel, at (202) 366-2992, and fax him at (202) 366-3820.

You may send mail to both of these officials at the National Highway Traffic Safety Administration, 400 Seventh St., SW., Washington, DC, 20590.

SUPPLEMENTARY INFORMATION:

I. Background

NHTSA has two brake standards for medium and heavy vehicles. Federal Motor Vehicle Safety Standard (FMVSS) No. 105, *Hydraulic and electric brake systems*, applies to vehicles with hydraulic brakes. FMVSS No. 121, *Air brake systems*, applies to vehicles with air brakes.

FMVSS No. 105 and 121 have similar brake performance requirements, but the two standards differ with respect to their specifications concerning the use of a roll bar during these tests. Roll bars are sometimes added to vehicles for brake testing if there are concerns about a possible vehicle rollover.

Air braked vehicles-roll bar use in braking-in-a-curve test. On March 10, 1995, NHTSA published a final rule amending FMVSS No. 121 requiring all air braked vehicles to be equipped with antilock brake systems (ABS) (60 FR 13216). The amendments to FMVSS No. 121 included a braking-in-a-curve performance test for truck tractors. Due to concern of potential vehicle rollover, the agency also included a manufacturer's option for using a roll bar structure during performance of that test at lightly loaded vehicle weight (LLVW). Loading of a vehicle to test at the gross vehicle weight rating (GVWR) already afforded manufacturers the opportunity to use a roll bar structure.

Air braked vehicles-roll bar use in straight line stop and parking brake grade holding tests. In response to a petition from the Truck Manufacturers Association, we published a final rule correcting and clarifying the air brake standard (66 FR 64154; December 12, 2001). The December 2001 final rule permitted the use of a roll bar structure for vehicles tested at lightly loaded vehicle weight in certain FMVSS No. 121 tests, including the 60 mph straight-line stop and the parking brake grade holding tests. In extending the option [*62418] for using a roll bar structure to these tests, we determined that the roll bar option is equally appropriate for tractors as well as single-unit vehicles.

Hydraulic braked vehicles-roll bar use in braking-in-a-curve test. On August 11, 2003, NHTSA published a final rule for braking-in-a-curve test requirements for ABS equipped single-unit trucks and buses with a GVWR greater than 10,000 pounds (68 FR 47485). Again, the concerns regarding possible rollover led NHTSA to grant manufacturers the option to use a roll bar structure for single-unit trucks and buses undergoing the braking-in-a-curve test under FMVSS No. 105.

II. Proposal To Permit Use of Roll Bar in Additional Brake Performance Tests of Hydraulically-Braked Trucks and Buses

In this document, we are proposing to amend FMVSS No. 105 to give manufacturers the option of using a roll bar structure for medium and heavy vehicles during additional brake testing at lightly loaded vehicle weight. Performance testing of brake systems at LLVW on vehicles with a GVWR greater than 10,000 pounds may result in vehicle rollover because of the configuration of these vehicles. Trucks and buses with a GVWR greater than 10,000 pounds often have a high center of gravity resulting in a low rollover threshold. Rollover threshold is the lateral acceleration at which a vehicle will roll over and for trucks and buses with a GVWR greater than 10,000 pounds it is usually 0.5 g or less. In contrast, a typical light vehicle has a rollover threshold between 0.8 g and 1.2 g. For tests performed at GVWR, manufacturers can already include roll bar structure weight in the vehicle weight to provide test drivers and technicians additional safety. This proposal would permit, at manufacturer's option, the use of a roll bar structure on these vehicles undergoing testing at LLVW.

Hydraulically-braked vehicles with a GVWR greater than 10,000 pounds must meet the requirements of FMVSS No. 105, including 60 mph straight-line stopping distance requirements and, for heavy school buses, parking brake requirements. During straight line stop testing, an equipment malfunction or a problem with the ABS can create the potential for these trucks and buses to yaw. Because of the low rollover threshold, these vehicles may roll over if they experience yaw at test speeds. During the parking brake test, while the vehicle is in the forward direction on a 20 percent grade, a failure of the brake system on one side of the vehicle can also cause the vehicle to yaw and perhaps roll over.

Currently, heavy school buses are the only vehicles with a GVWR greater than 10,000 pounds required by FMVSS No. 105 to meet the parking brake requirements. However, the agency has requested comments on a proposal that would require all hydraulically braked vehicles with a GVWR greater than 10,000 pounds to have parking brakes that meet these same requirements (67 FR 66098).

The agency also notes that single-unit trucks with a GVWR greater than 10,000 pounds may undergo brake system testing either as completed trucks or as chassis-cabs without bodies or equipment that would normally be installed by a final-stage manufacturer. A completed vehicle is likely to have more structure to protect a test driver than an incomplete vehicle. If a completed truck were to roll over, the impact force would be distributed across the body and cab of the truck. In the absence of a body or additional equipment during testing of a chassis-cab, the vehicle cab would receive a greater impact force during a rollover, increasing the potential of harm to the driver. Permitting the use of a roll bar would allow manufacturers to provide additional protection for the test driver in the event of a rollover.

The same concerns for vehicle rollover present in testing for FMVSS No. 121 are present in testing for FMVSS No. 105. Under FMVSS No. 121, NHTSA gives manufacturers the option of using a roll bar structure on trucks and buses tested at LLVW to improve safety for test drivers and technicians. This proposed amendment would permit the use of a roll bar structure on any vehicle with a GVWR greater than 10,000 pounds during FMVSS No. 105 compliance testing of the parking brake system at LLVW, the service brake system at LLVW, and the service brake system in partial failure mode at LLVW.

III. Compliance Date

The amendments proposed here do not impose any new requirements. Instead, the agency proposal would simply allow manufacturers the option of a roll bar as an added safety measure during the specified compliance tests. Since these proposed amendments would relieve a restriction and promote safety for test drivers, NHTSA proposes that they become effective 30 days after publication of the final rule in the **Federal Register.**

IV. Rulemaking Analyses and Notices

A. Executive Order 12866 and DOT Regulatory Policies and Procedures

Executive Order 12866, "Regulatory Planning and Review" (58 FR 51735, October 4, 1993), provides for making determinations whether a regulatory action is "significant" and therefore subject to Office of Management and Budget (OMB) review and to the requirements of the Executive Order. The Order defines a "significant regulatory action" as one that is likely to result in a rule that may:

(1) Have an annual effect on the economy of $100 million or more or adversely affect in a material way the economy, a sector of the economy, productivity, competition, jobs, the environment, public health or safety, or State, local, or Tribal governments or communities;

(2) Create a serious inconsistency or otherwise interfere with an action taken or planned by another agency;

(3) Materially alter the budget impact of entitlements, grants, user fees, or loan programs or the rights and obligations of recipients thereof; or

(4) Raise novel legal or policy issues arising out of legal mandates, the President's priorities, or the principles set forth in the Executive Order.

This rulemaking document was not reviewed by the Office of Management and Budget under E.O. 12866. It is also not considered to be significant under the Department's Regulatory Policies and Procedures (44 FR 11034; February 26, 1979).

This document proposes to amend 49 CFR 571.105 by including a manufacturer's option for the use of a roll bar structure during the performance testing of hydraulic brake systems. The proposed amendment would allow at manufacturer's option the use of a roll bar structure when testing hydraulic braked vehicles with a GVWR greater than 10,000 pounds at lightly loaded vehicle weight. Because of the configuration of these vehicles they are susceptible to roll over during testing. We tentatively conclude that permitting the use of a roll bar structure would help protect drivers and technicians in the event of a rollover during these tests. As noted above, the amendments proposed here do not impose any new requirements. Instead, the agency proposal would simply allow manufacturers the option of a roll bar as an added safety measure during the specified compliance tests. The proposal's impacts are so small that a full regulatory evaluation was not prepared.

B. Regulatory Flexibility Act

In compliance with the Regulatory Flexibility Act, 5 U.S.C. 601 *et seq.*, NHTSA has evaluated the effects of this proposed action on small entities. I hereby certify that this notice of proposed rulemaking would not have a significant impact on a substantial number of small entities.

The following is the agency's statement providing the factual basis for the certification (5 U.S.C. 605(b)). The amendments proposed herein would primarily affect manufacturers of medium and heavy weight trucks. The Small Business Administration (SBA) regulation at 13 CFR part 121 organizes size standards according to the Standard Industrial Classification (SIC) codes. SIC code number 3711, *Motor Vehicles and Passenger Car Bodies*, prescribes a small business size standard of 1,000 or fewer employees. SIC codes No. 3714, *Motor Vehicle Part and Accessories*, prescribes a small business size standard of 750 or fewer employees.

Most of the intermediate and final stage manufacturers of vehicles built in two or more stages have 1,000 or fewer employees. However, the agency expects testing for FMVSS No. 105 to be conducted by the original equipment manufacturers, most, if not all, of which do not qualify as a small business under SBA guidelines. Further, if adopted, the proposed amendments would not require use of the roll bar structure and therefore would not require any increased costs or other burdens on truck manufacturers. The proposed amendments to FMVSS No. 105 would permit the use of a roll bar structure at the manufacturer's option, on test vehicles undergoing brake testing. Accordingly, there would be no significant impact on small businesses, small organizations, or small governmental units by these amendments. For these reasons, the agency has not prepared a preliminary regulatory flexibility analysis.

C. Executive Order No. 13132

NHTSA has analyzed this proposed rule in accordance with the principles and criteria set forth in Executive Order 13132, Federalism and has determined that this proposal does not have sufficient Federal implications to warrant consultation with State and local officials or the preparation of a Federalism summary impact statement. The proposal would not have any substantial impact on the States, or on the current Federal-State relationship, or on the current distribution of power and responsibilities among the various local officials.

D. National Environmental Policy Act

NHTSA has analyzed this proposal for the purposes of the National Environmental Policy Act. The agency has determined that implementation of this action would not have any significant impact on the quality of the human environment.

E. Paperwork Reduction Act

This proposed rule does not contain any collection of information requirements requiring review under the Paperwork Reduction Act of 1995 (Pub. L. 104-13).

F. National Technology Transfer and Advancement Act

Under the National Technology Transfer and Advancement Act of 1995 (NTTAA) (Pub. L. 104-113), "all Federal agencies and departments shall use technical standards that are developed or adopted by voluntary consensus standards bodies, using such technical standards as a means to carry out policy objectives or activities determined by the agencies and departments." Society of Automotive Engineers (SAE) Recommended Practice J1626 APR96, *Braking, Stability, and Control Performance Test Procedures for Air-Brake-Equipped Truck Tractors*, includes an option for using a roll bar structure for testing at LLVW. While the SAE practice applies to air braked trucks, the SAE tests performed at LLVW are similar to tests performed at LLVW under FMVSS No. 105. The proposed amendment would permit the use of a roll bar structure in a similar manner as the SAE recommended practice.

G. Civil Justice Reform

This proposal would not have any retroactive effect. Under 49 U.S.C. 21403, whenever a Federal motor vehicle safety standard is in effect, a State may not adopt or maintain a safety standard applicable to the same aspect of performance which is not identical to the Federal standard, except to the extent that the state requirement imposes a higher level of performance and applies only to vehicles procured for the State's use. 49 U.S.C. 21461 sets forth a procedure for judicial review of final rules establishing, amending or revoking Federal motor vehicle safety standards. That section does not require submission of a petition for reconsideration or other administrative proceedings before parties may file suit in court.

H. Unfunded Mandates Reform Act

The Unfunded Mandates Reform Act of 1995 requires agencies to prepare a written assessment of the costs, benefits and other effects of proposed or final rules that include a Federal mandate likely to result in the expenditure by State, local or tribal governments, in the aggregate, or by the private sector, of more than $100 million annually (adjusted for inflation with base year of 1995). This rulemaking would not result in expenditures by State, local or tribal governments, in the aggregate, or by the private sector in excess of $100 million annually.

I. Regulation Identifier Number (RIN)

The Department of Transportation assigns a regulation identifier number (RIN) to each regulatory action listed in the Unified Agenda of Federal Regulations. The Regulatory Information Service Center publishes the Unified Agenda in April and October of each year. You may use the RIN contained in the heading at the beginning of this document to find this action in the Unified Agenda.

J. Executive Order 13045

Executive Order 13045 (62 FR 19885, April 23, 1997) applies to any rule that: (1) Is determined to be "economically significant" as defined under E.O. 12866, and (2) concerns an environmental, health, or safety risk that NHTSA has reason to believe may have a disproportionate effect on children. If the regulatory action meets both criteria, we must evaluate the environmental health or safety effects of the planned rule on children, and explain why the planned regulation is preferable to other potentially effective and reasonably feasible alternatives considered by us.

This proposed rule is not subject to the Executive Order because it is not economically significant as defined in E.O. 12866 and does not involve decisions based on environmental, health, or safety risks that disproportionately affect children. The proposed rule, if made final, would permit manufacturers to use a roll bar structure when testing medium and heavy hydraulic braked trucks and buses at LLVW.

K. Executive Order 13211

Executive order 13211 (66 FR 28355, May 18, 2001) applies to any rule that: (1) Is determined to be economically significant as defined under E.O. 12866, and is likely to have a significant adverse effect on the supply of, distribution, or use of energy; or (2) that is designated by the Administrator of the Office of Information and Regulatory Affairs as a significant energy action. If made final, this rulemaking would permit the voluntary and limited use of a roll bar structure during brake testing. Therefore this proposal was not analyzed under E.O. 13211.

L. Plain Language

Executive Order 12866 and the President's memorandum of June 1, 1998, require each agency to write all rules in plain language. Application of the principles of plain language includes consideration of the following questions:

. Have we organized the material to suit the public's needs?

. Are the requirements in the rule clearly stated?

. Does the rule contain technical language or jargon that isn't clear?

. Would a different format (grouping and order of sections, use of headings, paragraphing) make the rule easier to understand?

. Would more (but shorter) sections be better?

. Could we improve clarity by adding tables, lists, or diagrams?

. What else could we do to make the rule easier to understand?

If you have any responses to these questions, please include them in your comments on this proposal.

M. Privacy Act

Anyone is able to search the electronic form of all comments received into any of our dockets by the name of the individual submitting the comment (or signing the comment, if submitted on behalf of an association, business, labor union, etc.). You may review DOT's complete Privacy Act Statement in the Federal Register published on April 11, 2000 (Volume 65, Number 70; Pages 19477-78) or you may visit http://dms.dot.gov.

V. Submission of Comments

How Do I Prepare and Submit Comments?

Your comments must be written and in English. To ensure that your comments are correctly filed in the Docket, please include the docket number of this document in your comments.

Your comments must not be more than 15 pages long (49 CFR 553.21). We established this limit to encourage you to write your primary comments in a concise fashion. However, you may attach necessary additional documents to your comments. There is no limit on the length of the attachments.

Please submit two copies of your comments, including the attachments, to Docket Management at the address given above under **ADDRESSES.** Comments may also be submitted to the docket electronically by logging onto the Dockets Management System Web site at *http://dms.dot.gov.* Click on "Help & Information" or "Help/Info" to obtain instructions for filing the document electronically. Please note, if you are submitting comments electronically as a PDF (Adobe) file, we ask that the documents submitted be scanned using Optical Character Recognition (OCR) process, thus allowing the agency to search and copy certain portions of your submissions.[1]

How Can I Be Sure That My Comments Were Received?

If you wish Docket Management to notify you upon its receipt of your comments, enclose a self-addressed, stamped postcard in the envelope containing your comments. Upon receiving your comments, Docket Management will return the postcard by mail.

---

[1] Optical character recognition (OCR) is the process of converting an image of text, such as a scanned paper document or electronic fax file, into computer-editable text

How Do I Submit Confidential Business Information?

If you wish to submit any information under a claim of confidentiality, you should submit three copies of your complete submission, including the information you claim to be confidential business information, to the Chief Counsel, NHTSA, at the address given above under **FOR FURTHER INFORMATION CONTACT.** In addition, you should submit two copies, from which you have deleted the claimed confidential business information, to Docket Management at the address given above under **ADDRESSES.** When you send a comment containing information claimed to be confidential business information, you should include a cover letter setting forth the information specified in our confidential business information regulation (49 CFR part 512).

Will the Agency Consider Late Comments?

We will consider all comments that Docket Management receives before the close of business on the comment closing date indicated above under DATES. To the extent possible, we will also consider comments that Docket Management receives after that date. If Docket Management receives a comment too late for us to consider it in developing a final rule (assuming that one is issued), we will consider that comment as an informal suggestion for future rulemaking action.

How Can I Read the Comments Submitted by Other People?

You may read the comments received by Docket Management at the address given above under **ADDRESSES.** The hours of the Docket are indicated above in the same location. You may also see the comments on the Internet. To read the comments on the Internet, take the following steps:

(1) Go to the Docket Management System (DMS) Web page of the Department of Transportation *(http://dms.dot.gov/).*

(2) On that page, click on "search."

(3) On the next page *(http://dms.dot.gov/search/),* type in the four-digit docket number shown at the beginning of this document. Example: If the docket number were "NHTSA-1998-1234," you would type "1234." After typing the docket number, click on "search."

(4) On the next page, which contains docket summary information for the docket you selected, click on the desired comments. You may download the comments. However, since the comments are imaged documents, instead of word processing documents, the downloaded comments are not word searchable.

Please note that even after the comment closing date, we will continue to file relevant information in the Docket as it becomes available. Further, some people may submit late comments. Accordingly, we recommend that you periodically check the Docket for new material.

Anyone is able to search the electronic form of all comments received into any of our dockets by the name of the individual submitting the comment (or signing the comment, if submitted on behalf of an association, business, labor union, etc.). You may review DOT's complete Privacy Act Statement in the **Federal Register** published on April 11, 2000 (Volume 65, Number 70; Pages 19477-78) or you may visit *http://dms.dot.gov.*

List of Subjects in 49 CFR Part 571

Imports, Motor vehicle safety, Motor vehicles, Rubber and rubber products, and Tires.

In consideration of the foregoing, NHTSA proposes to amend 49 CFR part 571 as set forth below.

PART 571—FEDERAL MOTOR VEHICLE SAFETY STANDARDS

1. The authority citation for Part 571 would continue to read as follows:

Authority: 49 U.S.C. 322, 30111, 30115, 30117 and 30166; delegation of authority at 49 CFR 1.50.

2. Section 571.105 would be amended by revising S6.1.2, S7.7.3, S7.8, and S7.9.1 to read as follows:

§ 571.105—Standard No. 105; Hydraulic and electric braking systems.

■ ■ ■

S6.1.2 For applicable tests specified in S7.5(a), S7.7, S7.8, and S7.9, vehicle weight is lightly loaded vehicle weight, with the added weight, except for the roll bar structure allowed for trucks and buses with a GVWR greater than 10,000 pounds, distributed in the front passenger seat area in passenger cars, multipurpose passenger vehicles, and trucks, and in the area adjacent to the driver's seat in buses.

■ ■ ■

S7.7.3 *Lightly loaded vehicle.* Repeat S7.7.1 or S7.7.2 as applicable except with the vehicle at lightly loaded vehicle weight or at manufacturer's option, for a vehicle with GVWR greater than 10,000 pounds, at lightly loaded vehicle weight plus not more than an additional 1,000 pounds for a roll bar structure on the vehicle.

■ ■ ■

S7.8 *Service brake system test-lightly loaded vehicle (third effectiveness) test.* Make six stops from 60 mph with vehicle at lightly loaded vehicle weight, or at the manufacturer's option for a vehicle with GVWR greater than 10,000 pounds, at lightly loaded vehicle weight plus not more than an additional 1,000 pounds for a roll bar structure on the vehicle. (This test is not applicable to a vehicle which has a GVWR of not less than 7,716 pounds and not greater than 10,000 pounds and is not a school bus.)

S7.9 *Service brake system test-partial failure.*

S7.9.1 With the vehicle at lightly loaded vehicle weight or at the manufacturer's option for a vehicle with a GVWR greater than 10,000 pounds, at lightly loaded vehicle weight plus not more than an additional 1,000 pounds for a roll bar structure on the vehicle, alter the service brake system to produce any one rupture or leakage type of failure, other than a structural failure of a housing that is common to two or more subsystems. Determine the control force, pressure level, or fluid level (as appropriate for the indicator being tested) necessary to activate the brake system indicator lamp. Make four stops if the vehicle is equipped with a split service brake system, or 10 stops if the vehicle is not so equipped, each from 60 mph, by a continuous application of the service brake control. Restore the service brake system to normal at completion of this test.

■ ■ ■

Issued on: October 29, 2003.

Stephen R. Kratzke,

Associate Administrator for Rulemaking.

[FR Doc. 03-27657 Filed 11-3-03; 8:45 am]

BILLING CODE 4910-59-P

FEDERAL REGISTER

Vol. 69, No. 11

Rules and Regulations

DEPARTMENT OF COMMERCE (DOC)

Bureau of Industry and Security

15 CFR Part 711

[Docket No. 0312113311-3311-01]

RIN 0694-AC97

Chemical Weapons Convention Regulations: Electronic Submission of Declarations and Reports Through the Web-Data Entry System for Industry (Web-DESI)

69 FR 2501

DATE: Friday, January 16, 2004

ACTION: Interim final rule.

SUMMARY: The Bureau of Industry and Security (BIS) published an interim rule, on December 30, 1999, that established the Chemical Weapons Convention Regulations (CWCR) to implement the provisions of the Chemical Weapons Convention (CWC) affecting U.S. industry and other U.S. persons. The CWCR include requirements to report certain activities, involving Scheduled chemicals and Unscheduled Discrete Organic Chemicals, and to provide access for on-site verification by international inspectors of certain facilities and locations in the United States. This interim final rule amends the CWCR by adding instructions on how to obtain authorization from BIS to make electronic submissions of declarations and reports through the Web-Data Entry System for Industry (Web-DESI), which can be accessed on the CWC Web site at http://www.cwc.gov. The rule also establishes procedures for the assignment and use of passwords for facilities, plant sites and trading companies (USC password) and procedures for the assignment and use of Web-DESI user accounts.

DATES: This rule is effective January 16, 2004.

FOR FURTHER INFORMATION CONTACT: For questions of a general or regulatory nature, contact the Regulatory Policy Division, telephone: (202) 482-2440. For program information on declarations and reports, contact the Treaty Compliance Division, Office of Nonproliferation Controls and Treaty Compliance, telephone: (703) 605-4400.

SUPPLEMENTARY INFORMATION:

Background

On April 25, 1997, the United States ratified the Convention on the Development, Production, Stockpiling and Use of Chemical Weapons and on Their Destruction, also known as the Chemical Weapons Convention (CWC or Convention). The CWC, which entered into force on April 29, 1997, is an arms control treaty with significant nonproliferation aspects. As such, the CWC bans the development, production, stockpiling or use of chemical weapons and prohibits States Parties to the CWC from assisting or encouraging anyone to engage in a prohibited activity. The CWC provides for declaration and inspection of all States Parties' chemical weapons and chemical weapon production facilities, and oversees the destruction of such weapons and facilities. To fulfill its arms control and non-proliferation objectives, the CWC also establishes a comprehensive verification scheme and requires the declaration and inspection of facilities that produce, process or consume certain "scheduled" chemicals and unscheduled discrete organic chemicals, many of which have significant commercial applications. The CWC also requires States Parties to report exports and imports and to impose export and import restrictions on certain chemicals. These requirements apply to all entities under the jurisdiction and control of States Parties, including commercial entities and individuals. States Parties to the CWC, including the United States, have agreed to this verification scheme in order to provide transparency and to ensure that no State Party to the CWC is engaging in prohibited activities.

The Chemical Weapons Convention Implementation Act of 1998 ("Act") (22 U.S.C. 6701 et seq.), enacted on October 21, 1998, authorizes the United States to require the U.S. chemical industry and other private entities to submit declarations, notifications and other reports and also to provide access for on-site inspections conducted by inspectors sent by the Organization for the Prohibition of Chemical Weapons (OPCW). Executive Order (E.O.) 13128 delegates authority to the Department of Commerce to promulgate regulations, obtain and execute warrants, provide assistance to certain facilities, and carry out appropriate functions to implement the CWC, consistent with the Act.

On December 30, 1999, the Bureau of Industry and Security (BIS), U.S. Department of Commerce, published an interim rule that established the Chemical Weapons Convention Regulations (CWCR) (15 CFR parts 710-722). The CWCR implemented the provisions of the CWC, affecting U.S. industry and U.S. persons, in accordance with the provisions of the Act. This interim final rule amends the CWCR by adding instructions on how to obtain authorization from BIS to make electronic submissions of declarations and reports through the Web-Data Entry System for Industry (Web-DESI), which can be accessed on the CWC Web site at *http://www.cwc.gov*. The rule also establishes procedures for the assignment and use of passwords for facilities, plant sites and trading companies (USC password) and procedures for the assignment and use of Web-DESI user accounts (user name and password).

Rulemaking Requirements

1. This interim final rule has been determined to be not significant for purposes of E.O. 12866.

2. Notwithstanding any other provision of law, no person is required to respond to, nor shall a person be subject to a penalty for failure to comply with, a collection of information subject to the requirements of the Paperwork Reduction Act (PRA), unless that collection of information displays a current, valid OMB control number. This rule amends an existing collection of information authority approved under OMB Control No. 0694-0091. The public reporting burdens for the collection of information are estimated to average 10.6 hours for Schedule 1 Chemicals, 11.9 hours for Schedule 2 chemicals, 2.5 hours for Schedule 3 chemicals, 5.3 for Unscheduled Discrete Organic Chemicals (UDOCs), and 0.17 hours for Schedule 1 notifications. The burden hours associated with completing a particular type of declaration or report package (e.g., Schedule 2 annual declaration on past activities) will change depending on the number of forms required to comply with the specific declaration or report requirement. Supplement 2 to parts 712, 713, 714, and 715 of the CWCR identifies the specific forms that must be included in each type of declaration or report package. The CWC Declaration and Report Handbook includes a "Guide to Submission of Forms" which also identifies the specific forms that must be included in a declaration or report package.

BIS will use the information contained in declarations and reports submitted by U.S. persons to compile the U.S. National Industrial Declaration in order to meet our obligations under the Chemicals Weapons Convention (CWC). BIS will submit the U.S. National Industrial Declaration to the United States National Authority who will forward the Declaration to the Organization for the Prohibition of Chemical Weapons (OPCW) as required by the Convention.

3. This rule does not contain policies with Federalism implications as this term is defined in Executive Order 13132.

4. Pursuant to 5 U.S.C. 553(b)(B), the provisions of the Administrative Procedure Act requiring a prior notice and an opportunity for public comment are waived for good cause, because it is unnecessary to provide public notice and opportunity for comment. This regulation does not impose any new regulatory requirements or effect a sub-

stantive change to any existing regulatory requirement. Submission of documents through the Web-DESI system is voluntary and provided for the convenience of submitters. No other law requires that a notice of final rulemaking and an opportunity for public comment be given for this rule. Because a notice of final rulemaking and an opportunity for public comment are not required to be given for this rule under the Administrative Procedure Act or by any other law, the analytical requirements of the Regulatory Flexibility Act (5 U.S.C. 601 et seq.) are not applicable.

List of Subjects in 15 CFR Part 711

Chemicals, Confidential business information, Reporting and recordkeeping requirements.

Accordingly, part 711 of the Chemical Weapons Convention Regulations is amended as follows:

PART 711—[AMENDED]

1. The authority citation for 15 CFR part 711 continues to read as follows:

Authority: 22 U.S.C. 6701 *et seq.;* E.O. 13128, 64 FR 34703.

2. Section 711.7 is added to read as follows:

§ 711.7—How to request authorization from BIS to make electronic submissions of declarations or reports.

(a) Scope. This section provides an optional method of submitting declarations or reports. Specifically, this section applies to the electronic submission of declarations and reports required under the CWCR. If you choose to submit declarations and reports by electronic means, all such electronic submissions must be made through the Web-Data Entry System for Industry (Web-DESI), which can be accessed on the CWC Web site at *http://www.cwc.gov.*

(b) *Authorization.* If you or your company has a facility, plant site, or trading company that has been assigned a U.S. Code Number (U.S.C. Number), you may submit declarations and reports electronically, once you have received authorization from BIS to do so. An authorization to submit declarations and reports electronically may be limited or withdrawn by BIS at any time. There are no prerequisites for obtaining permission to submit electronically, nor are there any limitations with regard to the types of declarations or reports that are eligible for electronic submission. However, BIS may direct, for any reason, that any electronic declaration or report be resubmitted in writing, either in whole or in part.

(1) *Requesting approval to submit declarations and reports electronically.* To submit declarations and reports electronically, you or your company must submit a written request to BIS at the address identified in § 711.6 of the CWCR. Both the envelope and letter must be marked "Attn: Electronic Declaration or Report Request." Your request should be on company letterhead and must contain your name or the company's name, your mailing address at the company, the name of the facility, plant site or trading company and its U.S. Code Number, the address of the facility, plant site or trading company (this address may be different from the mailing address), the list of individuals who are authorized to view, edit, or edit and submit declarations and reports on behalf of your company, and the telephone number and name and title of the official responsible for certifying that each individual listed in the request is authorized to view, edit, or edit and submit declarations and reports on behalf of you or your company. Additional information required for submitting electronic declarations and reports may be found on BIS's Web site at http://www.cwc.gov. Once you have completed and submitted the necessary certifications, you may be authorized by BIS to view, edit, or edit and submit declarations and reports electronically.

Note to § 711.7(b)(1): You must submit a separate request for each facility, plant site or trading company owned by your company (e.g., each site that is assigned a unique U.S. Code Number).

(2) Assignment and use of passwords for facilities, plant sites and trading companies (U.S.C. password) and Web-DESI user accounts (user name and password).

(i) Each person, facility, plant site or trading company authorized to submit declarations and reports electronically will be assigned a password (U.S.C. password) that must be used in conjunction with the U.S.C. Number. Each individual authorized by BIS to view, edit, or edit and submit declarations and reports electronically for a facility, plant site or trading company will be assigned a Web-DESI user account (user name and password) telephonically by BIS. A Web-DESI user account will be assigned to you only if your company has certified to BIS that you are authorized to act for it in viewing, editing, or editing and submitting electronic declarations and reports under the CWCR.

Note to § 711.7(b)(2)(i): When individuals must have access to multiple Web-DESI accounts, their companies must identify such individuals on the approval request for each of these Web-DESI accounts. BIS will coordinate with such individuals to ensure that the assigned user name and password is the same for each account.

(ii) Your company may reveal the facility, plant site or trading company password (U.S.C. password) only to Web-DESI users with valid passwords, their supervisors, and employees or agents of the company with a commercial justification for knowing the password.

(iii) If you are an authorized Web-DESI account user, you may not:

(A) Disclose your user name or password to anyone;

(B) Record your user name or password, either in writing or electronically;

(C) Authorize another person to use your user name or password; or [*2503]

(D) Use your user name or password following termination, either by BIS or by your company, of your authorization or approval for Web-DESI use.

(iv) To prevent misuse of the Web-DESI account:

(A) If Web-DESI user account information (i.e., user name and password) is lost, stolen or otherwise compromised, the company and the user must report the loss, theft or compromise of the user account information, immediately, by calling BIS at (703) 235-1335. Within two business days of making the report, the company and the user must submit written confirmation to BIS at the address provided in § 711.6 of the CWCR.

(B) Your company is responsible for immediately notifying BIS whenever a Web-DESI user leaves the employ of the company or otherwise ceases to be authorized by the company to submit declarations and reports electronically on its behalf.

(v) No person may use, copy, appropriate or otherwise compromise a Web-DESI account user name or password assigned to another person. No person, except a person authorized access by the company, may use or copy the facility, plant site or trading company password (U.S.C password), nor may any person steal or otherwise compromise this password.

(c) *Electronic submission of declarations and reports.* (1) *General instructions.* Upon submission of the required certifications and approval of the company's request to use electronic submission, BIS will provide instructions on both the method for transmitting declarations and reports electronically and the process for submitting required supporting documents, if any. These instructions may be modified by BIS from time to time.

(2) *Declarations and reports.* The electronic submission of a declaration or report will constitute an official document as required under parts 712 through 715 of the CWCR. Such submissions must provide the same information as written declarations and reports and are subject to the recordkeeping provisions of part 720 of the CWCR. The company and Web-DESI user submitting the declaration or report will be deemed to have made all representations and certifications as if the submission were made in writing by the company and signed by the certifying official. Electronic submission of a declaration or report will be considered complete upon transmittal to BIS.

(d) *Updating.* A company approved for electronic submission of declarations or reports under Web-DESI must promptly notify BIS of any change in its name, ownership or address. If your company wishes to have an individual added as a Web-DESI user, your company must inform BIS and follow the instructions provided by BIS. Your company should conduct periodic reviews to ensure that the company's designated certifying official and Web-DESI users are individuals whose current responsibilities make it necessary and appropriate that they act for the company in either capacity.

Dated: January 12, 2004.

Peter Lichtenbaum,

Assistant Secretary, for Export Administration.

[FR Doc. 04-938 Filed 1-15-04; 8:45 am]

BILLING CODE 3510-33-P

Index